Mathematical Physics

V. Balakrishnan

Mathematical Physics

Applications and Problems

V. Balakrishnan
Department of Physics
Indian Institute of Technology Madras
Chennai, India

ISBN 978-3-030-39682-4 ISBN 978-3-030-39680-0 (eBook)
https://doi.org/10.1007/978-3-030-39680-0

Jointly published with ANE Books India
The print edition is not for sale in South Asia (India, Pakistan, Sri Lanka, Bangladesh, Nepal and Bhutan) and Africa. Customers from South Asia and Africa can please order the print book from: ANE Books India.
ISBN of the Co-Publisher's edition: 9789386761118

This Springer imprint is published by the registered company Springer Nature Switzerland AG
The registered company address is: Gewerbestrasse 11, 6330 Cham, Switzerland

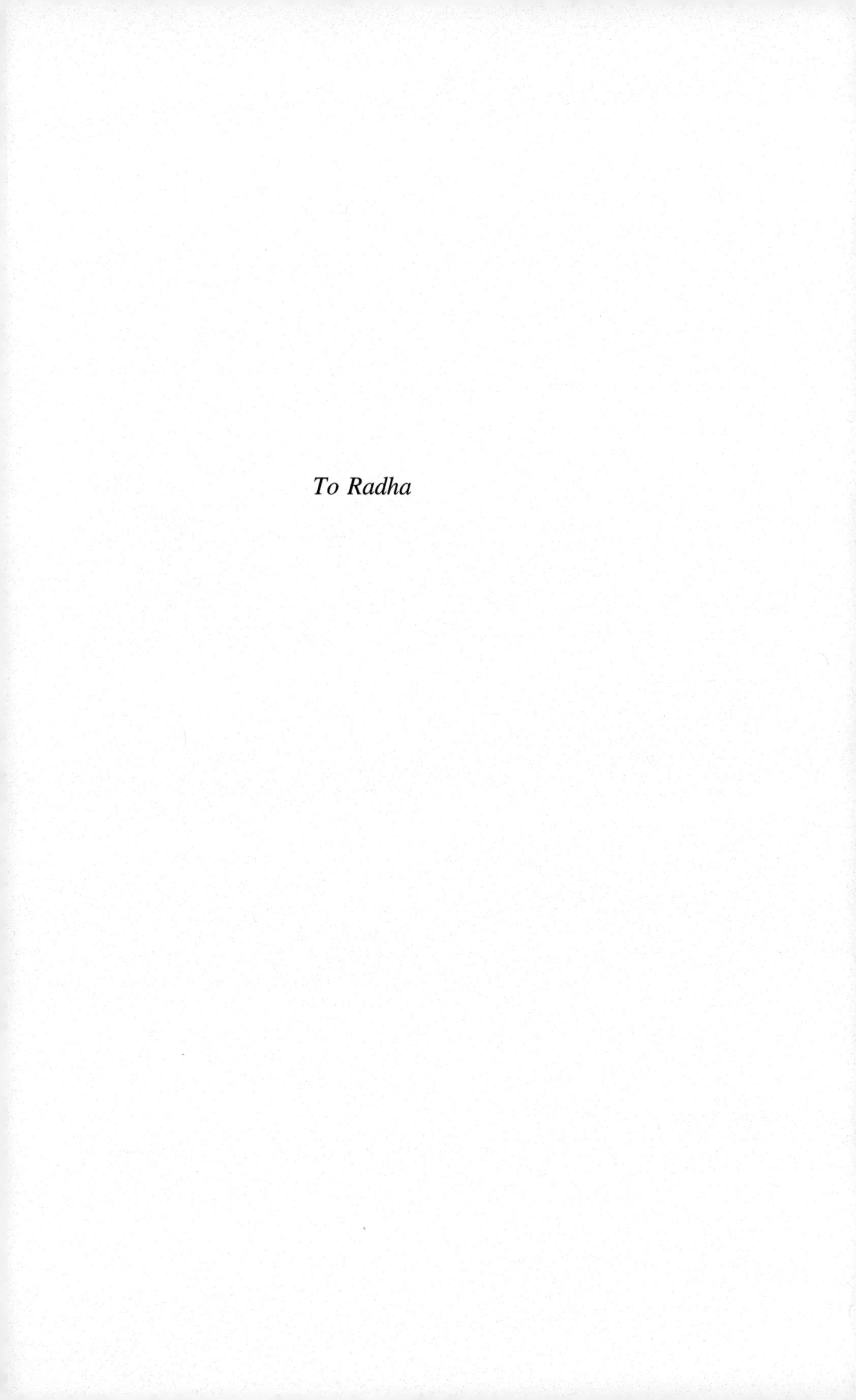

To Radha

Preface

"The book of nature is written in the language of mathematics." This is the essence of the profound observation made by Galileo Galilei as far back as 1623. Evidence in support of this insight has accumulated ever since. Much has been written about what has been termed "the unreasonable effectiveness of mathematics" in the description of physical phenomena. Is mathematics inherent in nature itself, or is it a construct of the human mind? This deep question has also been debated intensely among mathematicians, physicists and philosophers of science. Whatever be the answer, it is undeniably true that mathematical structures seem to be embedded deeply in the physical universe.

After approximately four hundred years of continuous development, physics is undoubtedly the most 'mathematized' of the sciences. Physics attempts to describe nature in precise and logical terms, and it requires a language that has logic built into it. As Richard Feynman put it, "Mathematics is language plus logic". A certain degree of facility in mathematics is therefore not only helpful, but also absolutely necessary, in order to really understand physics and to appreciate its concepts and laws even at an elementary level.

But what kind of mathematics does physics entail? Both physics and mathematics are very vast domains. The natural question that arises in the mind of a student beginning the study of physics is, "Exactly how much mathematics do I need to study and to understand physics?" There can be no definite or complete answer to this question, because it depends on the level at which one wishes to understand the laws of physics and the structure of the physical universe. The problems that physics addresses require the application of mathematics at all levels. These range from elementary algebra right up to some of the most advanced state-of-the-art developments in mathematical research.

Along the way, there are remarkable instances of an almost uncanny match between a physical context and the specific kind of mathematics needed in that context. The first such pairing was between the dynamics of motion and calculus. Indeed, calculus was developed for that very purpose. Subsequent instances are: electromagnetism and vector calculus; general relativity and Riemannian geometry; quantum mechanics and linear vector spaces; symmetries in (condensed matter, atomic, nuclear, and subnuclear) physics and group theory; Hamiltonian dynamics and symplectic geometry; and so on. In every one of these cases, the mathematical structure involved seems to have been tailor-made for the physical problem concerned.

What is mathematical physics? The fact that physics requires mathematics at all levels makes the very definition of mathematical physics as a subject in the university physics curriculum rather fuzzy. Over the years, however, there has emerged a set of mathematical topics and techniques that are the most useful and widely applicable ones in various parts of physics. It is this *repertoire* or collection that constitutes 'mathematical physics' as the term is generally understood in its pedagogical sense. This book has chapters devoted to most of the topics of this core set, and considerably more, besides.

Further, I have taken the phrase 'mathematical physics' literally. As a consequence, this book is not an applied mathematics text in the conventional sense. As a glance at the table of contents will show, it digresses into physics whenever the opportunity presents itself. Although numerous mathematical results are introduced and discussed, hardly any formal, rigorous proofs of theorems are presented. Instead, I have used specific examples and physical applications to illustrate and elaborate upon these results. The aim is to demonstrate how mathematics intertwines with physics in numerous instances. In my opinion, this is the fundamental justification for the very inclusion of mathematical physics as a subject in the physics curriculum.

To whom is this book addressed? It is my belief and hope that appropriate parts of the book will serve a wide spectrum of students, ranging from the undergraduate right up to the doctoral level. More than one route map can be drawn to navigate through the chapters to form courses at different levels and of different durations. I have not done so because I believe this choice is best left to the user. Likewise, fairly self-contained sets of chapters can be selected to provide short courses of study on specific topics in mathematical physics. Here are some examples of the possibilities in this regard:

Vector calculus and applications, Chaps. 5–9.
Linear vector spaces, matrices and operators, Chaps. 10–15.
Probability, statistics and random processes, Chaps. 19–21.
Complex analysis, Chaps. 22–27.
Special functions, Chaps. 16, 25, 26.
Basic partial differential equations of physics, Chaps. 29–32.

The chapters listed in each case do not stand in complete isolation from the rest of the book, of course. In framing short specific-topic courses, it would naturally be helpful to include appropriate sections from other chapters, as needed.

Exercises and problems comprise an indispensable component of any book on mathematical methods, and this book is no exception. There are 370 of these in this book, many of them with several parts and subparts. Most (but not all) of them are problems, rather than exercises of the drillwork type. They form an integral part of the text. In many cases, they require the reader to complete, or verify, or work out, or extend the details described in the text, in order to acquire a better understanding of the subject matter. In other cases, they explore sidelights and interconnections between different aspects. For this reason, I have made the problems contiguous with the text, indicating the beginning of each one with the symbol ★.

Solutions: Student readers are best served if a book containing problems also provides solutions. At the end of each chapter, solutions to the problems therein are given either in outline or in detail, except in those cases in which the solutions are obviously straightforward extensions of the text. The end of each solution is indicated by the symbol ►. I reiterate that working out the problems in each chapter is of paramount importance. Only after an honest attempt has been made should the reader consult the solutions provided.

The table of contents lists not only the chapters and sections of each chapter, as is customary, but also the subsections, which is somewhat less common. This has been done in order to provide the reader with a conveniently detailed list of all the topics discussed in the book, avoiding the need to hunt for these in the body of the text. Together with the Index, the table of contents should make it easy for you to navigate from place to place, back and forth, through this rather lengthy book.

The index at the end of the book runs to thirteen pages. It has intentionally been made rather extensive, because I believe that the reader should be able to refer quickly to any topic or theme, and the different contexts in which it occurs in the book, based on just a keyword or phrase.

Cross-Referencing: As one might expect, numerous topics, themes and equations appear more than once in the book. I have tried as far possible to cross-reference these with chapter, section and equation number, so that recall becomes easier. In a few necessary instances, equations have been repeated for ready reference.

Above all, it has been my intention *to make this book as comprehensive, self-contained and amenable to self-study as possible*. Naturally, there are many omissions in the book. Several important topics that I should like to have touched upon have had to be left out. Examples that stand out include the calculus of variations, functional integration, the elements of differential geometry, and a more systematic account of group theory in physics. But the need to keep the length of the book within a manageable limit necessitated these omissions.

Finally, I must point out that all the mathematics involved in this book is 'classical'. By and large, more modern and/or abstract parts of mathematics have not as yet become part of the standard repertoire referred to earlier, although some areas such as topology and differential geometry increasingly find application in various parts of physics such as quantum field theory, general relativity and condensed matter physics.

This book has been several years in the writing, having grown out of various courses and sets of lectures given by me over a considerable number of years. I owe a debt of gratitude to all the students who attended the lectures, asked questions that set me thinking, and enabled me get a better understanding of the subject matter.

I thank Suresh Govindarajan for his generous help and valuable assistance, and Ashok Velayutham for drawing all the figures in the book. As always, my wife Radha has been a pillar of support and encouragement, and her assistance has been invaluable. I dedicate this book to her with affection.

Chennai, India V. Balakrishnan

Contents

About the Author

Prof. V. Balakrishnan is currently an Adjunct Professor of Physics at the Indian Institute of Technology (IIT) Madras. He joined the Indian Institute of Technology Madras as a Professor in 1980, and retired as Professor Emeritus in December 2013. He was educated at the University of Delhi and received his PhD in theoretical high energy physics from Brandeis University, USA in 1970. He joined IIT Madras after working for a decade at TIFR (Mumbai) and RRC, DAE (Kalpakkam). His research interests have spanned many areas including particle physics, many-body theory, condensed matter physics, stochastic processes, quantum dynamics, nonlinear dynamics and chaos. He has published numerous research papers in these areas and a number of popular pedagogic articles. He has co-authored a monograph titled Beyond the Crystalline State and authored a monograph titled Elements of Non-equilibrium Statistical Mechanics. Over nearly four decades, he has taught a wide variety of extremely popular courses at the undergraduate, postgraduate and doctoral levels, comprising all the standard subjects in physics as well as some new and non-traditional ones. He has received high accolades for his teaching at IIT Madras, and his lectures at numerous courses sponsored by the UGC, DST (SERC) and ICTP (Trieste, Italy) have been well-received by their attendees. His video lecture courses under the auspices of the NPTEL programme of the HRD Ministry (available on YouTube) have received very high acclaim. He has held visiting positions at several institutions abroad. He is a Fellow of the Indian Academy of Sciences.

Chapter 1
Warming Up: Functions of a Real Variable

1.1 Sketching Functions

1.1.1 Features of Interest in a Function

Sketching functions *schematically* is one of the most basic skills that is required of any student of science or technology. The visual impact of a plot or graph is considerable. It enables us to understand, at a glance, a good deal about the essential structure of any given functional form. Of course, we can always write a program to plot a given function, or use readily available packages for this purpose. But this is not always as instructive as sketching the functions involved in the formulas. Doing so provides insight into the nature of results derived by more formal considerations.

Most functions are nonlinear, i.e., they have shapes that are far more interesting than a mere straight line. Given any formula or functional form that occurs in an application, you should be able to sketch the function *qualitatively*, such that its main features and interesting aspects are brought out. In particular, this includes its

- zeroes;
- maxima and minima;
- inflection points or points where the curvature of the function changes sign;
- asymptotes;
- discontinuities;
- infinities or divergences;
- its behavior near $x = 0$, i.e., for small values of the argument;
- its behavior as $x \to \pm\infty$, i.e., for very large values of the argument;

and so on.

It is also important to learn to generalize, that is, to draw conclusions that can take one from particular cases to general results. This implies that you must develop the ability to distinguish between specific details and general features. One way to do this is to try to regard, wherever possible, the specific problem at hand as a member of a

V. Balakrishnan, *Mathematical Physics*,
https://doi.org/10.1007/978-3-030-39680-0_1

family of similar problems: that is, you must attempt to *imbed* the problem in a general framework. This is very helpful in putting the problem in a proper perspective. Let us begin with three very simple examples of this approach.

1.1.2 Powers of x

We know that the graph of $y = x$ is a straight line, while x^2 is a quadratic function of x, x^3 is a cubic function, and so on. It is interesting to consider these as members of the *family* of functions x^α, and to draw them on the same plot. Figure 1.1 shows the cases when α is equal to the positive integers $1, 2$, and 3 in the region $x \geq 0$. Note how the function gets flatter at the origin, while it increases more steeply for $x > 1$, with increasing α. The functions x^α for $\alpha = \frac{1}{2}$ is also drawn on the same plot. Note how, as α decreases below the value 1, the function now gets steeper at the origin, while increasing less rapidly for $x > 1$. All the curves above pass through the point $(1, 1)$ in the plot, enabling an instructive comparison to be made. Clearly, the features just described are applicable to the function x^α for *any* positive exponent α, not necessarily integral or rational, in the region $x > 0$. The straight line graph corresponding to the value $\alpha = 1$ separates two *qualitatively* different kinds of behavior.

1.1.3 A Family of Ovals

As you know, the equation $x^2/a^2 + y^2/b^2 = 1$ where a and b are positive constants represents an ellipse centered at the origin, with its principal axes along the x and y axes. The semi-major axis is a, and the semi-minor axis is b. Now consider the *family* of functions given by the equation

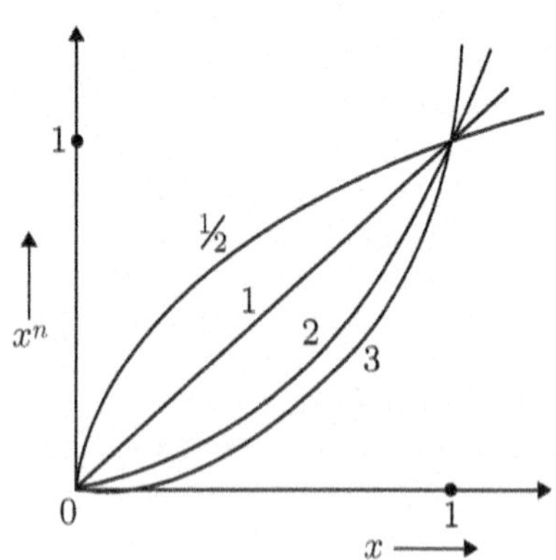

Fig. 1.1 Plots of x^α for $\alpha = \frac{1}{2}, 1, 2$ and 3. The value of α is indicated in each case

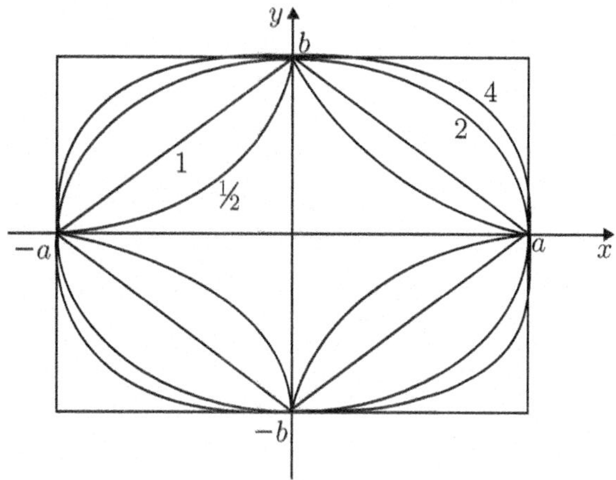

Fig. 1.2 Plots of the family of closed curves in Eq. (1.1) with $a > b > 0$. Values of the parameter α are indicated

$$\left|\frac{x}{a}\right|^{\alpha} + \left|\frac{y}{b}\right|^{\alpha} = 1, \tag{1.1}$$

where α is a positive number. It is easy to see that the graph is a simple closed curve that is symmetric with respect to reflections about both the coordinate axes. It is also confined to the rectangle formed by the straight lines $x = \pm a$ and $y = \pm b$. Figure 1.2 depicts the family for the parameter values $\alpha = 1, 2$, and 4, and also for $\alpha = \frac{1}{2}$. As α increases, the "oval" bulges out toward the rectangle. The rectangle may be regarded as the limiting curve in the limit $\alpha \to \infty$. Similarly, as α decreases, the curve shrinks toward the axes, which may be regarded as the limiting curves in the limit $\alpha \to 0$. Once again, the case $\alpha = 1$ separates two qualitatively different kinds of behavior.

1.1.4 A Family of Spirals

Plane polar coordinates (ϱ, φ) in the xy-plane are defined via the familiar relations $x = \varrho \cos \varphi$, $y = \varrho \sin \varphi$. It is easy to see that the family of curves $\varrho =$ a positive constant is a set of concentric circles centered at the origin. Similarly, the family of curves $\varphi =$ a positive constant is a set of half-lines starting at the origin and tending to infinity in different radial directions. The two families are *orthogonal* to each other, i.e., they intersect at right angles. Is there any way of "going continuously" from one family to the other by tuning some parameter?

Consider a curve such that, at each point on the curve, the tangent and radius vector are at the same angle ψ with respect to each other. Clearly, $\psi = 0$ for the radial lines themselves, and $\psi = \frac{1}{2}\pi$ for the circles. For values of ψ in between these extremes, we get a family of so-called **equiangular spirals** whose equation is

$$\varrho = \varrho_0 \, e^{\alpha(\varphi - \varphi_0)} \quad \text{where} \quad \alpha = \cot \psi, \tag{1.2}$$

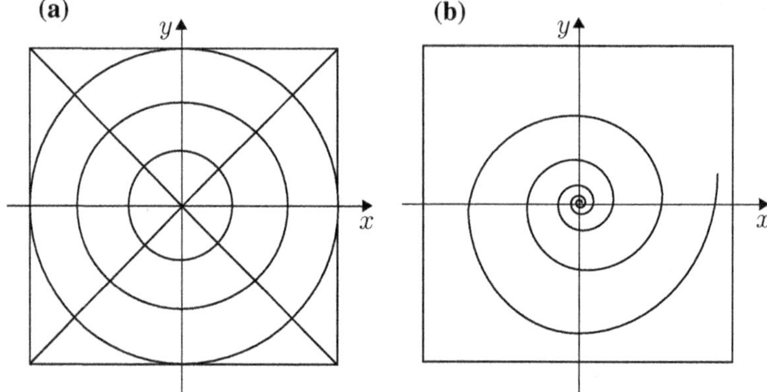

Fig. 1.3 **a** Orthogonal families of circles and radial lines. **b** An equiangular spiral

and the curve passes through any given point $(\varrho_0\,,\ \varphi_0)$. Figure 1.3a shows the orthogonal families of circles ($\psi = \frac{1}{2}\pi$ or $\alpha = 0$) and radial lines ($\psi = 0$ or $\alpha \to \infty$). Figure 1.3b shows one of the spirals that interpolates between these two families.

★ **1.** Schematically sketch each of the functions given below, paying attention to the features listed in Sect. 1.1.1. Unless otherwise specified, $-\infty < x < +\infty$. Wherever the function becomes indeterminate, take the value of the function to be its limiting value. (For example, if $f(x) = (\sin x)/x$, take $f(0) = 1$.)

(1) $x\,|x|$ (2) $|x|^{1/3}$ (3) $1/(x^2+1)$
(4) $(x-1)(x-2)/x$ (5) $\text{sech}\, x$ (6) $x\,\exp(-x^2)$
(7) $e^{-x}\cos x$ (8) $e^{-x}\sin x$ (9) $x^3/(x^2+1)$
(10) $e^{-|x|}\cos x$ (11) $e^{-|x|}\sin x$ (12) $\tanh x$
(13) $\coth x$ (14) $\tanh(x^2)$ (15) $\coth x - x^{-1}$
(16) $x - 1 + e^{-x}$ (17) $\ln x \ (x > 0)$ (18) $\ln\ln x \ (x > 1)$
(19) $(\ln x)/x \ (x > 0)$ (20) $x^x \ (x > 0)$ (21) $|x|^{|x|}$
(22) $x/(e^x - 1)$ (23) $1/(e^x + 1)$ (24) $\sin(x^2)$
(25) $(\sin^2 x)/x^2$ (26) $(x^2 - 1)\,e^{-x^2}$ (27) $\ln(e^x - 1) \ (x > 0)$
(28) $x^{1/x} \ (x > 0)$ (29) $(x\cos x - \sin x)/x^3$ (30) $(\sin 3x)/\sin x$
(31) $(3^x - 2^x)/x$ (32) $x^{-12} - x^{-6}$ (33) $|x|^{1/2}/(e^x + 1)$
(34) $4x\,(1 - x)$ (35) $\cosh^{-1} x$ (35) $\coth^{-1} x$
(37) $|x|^{1/2}/(1 + |x|^{1/2})$ (38) $(x^2 - 1)^2$ (39) $(1 - \cos x)/x^2$
(40) $x^3/(e^x + 1)$ (41) $x - \sin x$ (42) $\exp(1/x)$
(43) $x^{1/2}/(e^x + 1)$ (44) $x^{-1/2}\,e^{-1/x} \ (x > 0)$ (45) $\sin(\sin x)$
(46) $\exp(-1/x^2)$ (47) $\ln\dfrac{1-x}{1+x} \ (|x| < 1)$ (48) $\dfrac{1}{x}\ln\dfrac{1-x}{1+x} \ (|x| < 1)$.

Are there any periodic functions of x in this list?

1.2 Maps of the Unit Interval

In most problems in the physical sciences, the primary question is the variation of observable quantities as functions of time. An important fact, especially in the context of what is known as **deterministic chaos**, is the following: The time variation can be extremely complicated even when the underlying equations of motion are themselves quite simple in appearance. Useful models of dynamical systems that display such complex behavior are provided by relatively simple-looking functions. When the time evolution is assumed to occur in *discrete* time steps, the dynamics is specified by simply *iterating* such functions. In this context the functions are referred to as **maps**. In other words, if x_n is the value of the variable at the nth time step, its value after the next time step is given by $x_{n+1} = f(x_n)$, where $f(x)$ is a specified function of x. One of the simplest maps used in this context is the following:

$$\text{The symmetric tent map,}\quad f(x) = \big|1 - |2x-1|\big| = \begin{cases} 2x, & 0 \le x \le \frac{1}{2} \\ 2(1-x), & \frac{1}{2} \le x \le 1. \end{cases} \tag{1.3}$$

The map is depicted in Fig. 1.4. Here are some other one-dimensional maps that are used most frequently in the study of low-dimensional dynamical systems. In each case, the "phase space" is the unit interval $0 \le x \le 1$.

$$\text{The Bernoulli map}: \quad f(x) = 2x \bmod 1 = \begin{cases} 2x, & 0 \le x \le \frac{1}{2} \\ 2x-1, & \frac{1}{2} < x \le 1. \end{cases} \tag{1.4}$$

$$\text{The logistic map}: \quad f(x) = \big|1 - |2x-1|^2\big|. \tag{1.5}$$

$$\text{The cusp map}: \quad f(x) = \big|1 - |2x-1|^{1/2}\big|. \tag{1.6}$$

$$\text{The Gauss map}: \quad f(x) = x^{-1} - [x^{-1}]. \tag{1.7}$$

In Eq. (1.7), $[x^{-1}]$ stands for the largest integer $\le 1/x$.

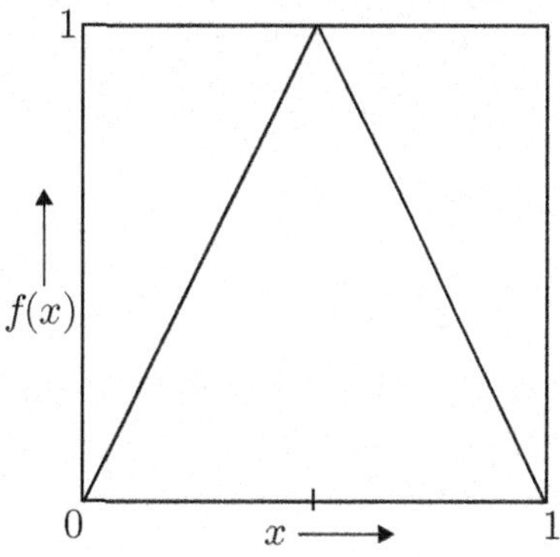

Fig. 1.4 The symmetric tent map of the unit interval, Eq. (1.3). The map is onto and noninvertible

★ **2.** Schematically sketch each of the functions in Eqs. (1.4)–(1.7) for $x \in [0, 1]$.

All the functions in Eqs. (1.3)–(1.7) are *onto* maps of the unit interval: that is, as x ranges over the unit interval, the value of $f(x)$ also extends over the whole of the unit interval. Observe that all the examples listed above are **noninvertible maps**, i.e., for a given $f(x)$ there is more than one value of x. There are two such values in the tent, Bernoulli, logistic, and cusp maps, and an *infinite* number of such values in the Gauss map.

The Bernoulli map is also known as the **Bernoulli shift**, for the following reason. The action of the map is best seen by writing x in the form of a binary "decimal",

$$x = 0.\, a_0\, a_1\, a_2 \cdots , \tag{1.8}$$

where each of the digits a_i is either 0 or 1. This means, of course, that the numerical *value* of x is

$$x = \frac{a_0}{2^1} + \frac{a_1}{2^2} + \frac{a_2}{2^3} + \cdots . \tag{1.9}$$

Hence the numerical value of $f(x)$ is given by

$$f(x) = 2x \bmod 1 = \frac{a_1}{2^1} + \frac{a_2}{2^2} + \frac{a_3}{2^3} + \cdots . \tag{1.10}$$

As a binary "decimal", this means that $f(x)$ is written as

$$f(x) = 2x \bmod 1 = 0.\, a_1\, a_2\, a_3 \cdots , \tag{1.11}$$

regardless of whether $a_0 = 0$ or 1. Thus, the effect of the map is simply to *shift* the "decimal point" by one digit to the right, and to delete the digit to the left of the decimal point. If the map is iterated, i.e., applied repeatedly to any initial value x, more and more digits of the initial value get "lost" in this manner. This fact is of significance in the behavior of the map as a discrete dynamical system that exhibits chaos.

The Gauss map, shown in Fig. 1.5, is also called the **continued fraction map**. The action of the map is best seen by writing $x \in [0, 1]$ as a continued fraction in standard form, with all the numerators equal to unity and successive denominators given by the positive integers $(a_0\, ,\ a_1\, ,\ a_2\, , \ldots)$. That is,

$$x = \cfrac{1}{a_0 + \cfrac{1}{a_1 + \cfrac{1}{a_2 + \cdots}}} \tag{1.12}$$

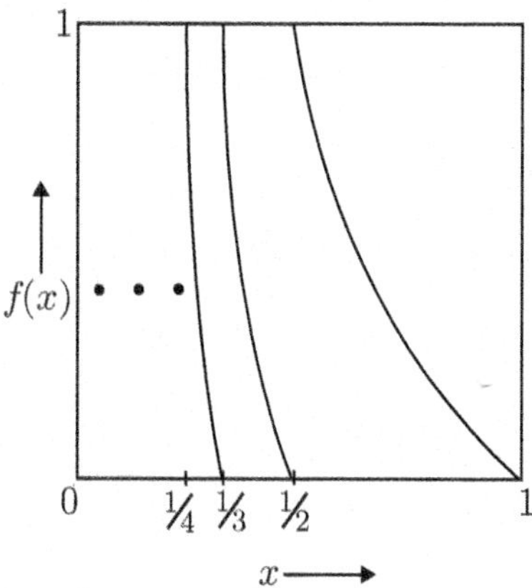

Fig. 1.5 The Gauss map of the unit interval, Eq. (1.7). The map has an infinite number of branches

Then $f(x)$ is the continued fraction with denominators given by the integers $(a_1, a_2, a_3, \ldots)$, i.e.,

$$f(x) = \frac{1}{x} - \left[\frac{1}{x}\right] = \frac{1}{x} - a_0 = \cfrac{1}{a_1 + \cfrac{1}{a_2 + \cfrac{1}{a_3 + \cdots}}} \tag{1.13}$$

Once again, information about the first integer a_0 is lost under the action of the map, i.e., in going from x to $f(x)$. As the map is iterated, more and more information about the initial value is lost, as the integers $a_1, a_2, \ldots$ are successively removed. This loss of information is a reflection of the noninvertibility of the map. In turn, the noninvertibility of all the onto maps listed above is closely related to their role as simple models of chaotic dynamical behavior.

Chapter 2
Gaussian Integrals, Stirling's Formula, and Some Integrals

2.1 Gaussian Integrals

2.1.1 The Basic Gaussian Integral

The function e^{-x^2}, called a Gaussian, appears everywhere in the mathematical sciences. It plays a fundamental role in probability and statistics. We will discuss this aspect further in Chap. 20, Sect. 20.2. The fundamental **Gaussian integral** in its simplest form is

$$I = \int_0^\infty dx\, e^{-x^2} = \tfrac{1}{2}\int_{-\infty}^\infty dx\, e^{-x^2} = \tfrac{1}{2}\sqrt{\pi}\,. \tag{2.1}$$

The integral cannot be evaluated by the usual method of integration by parts. Its value is determined as follows. Consider the *square* of the integral and change to plane polar coordinates (ϱ, φ), where $\varrho = (x^2 + y^2)^{1/2}$ and $\varphi = \tan^{-1}(y/x)$. (This trick is attributed to Poisson.) The region of integration is the first quadrant in the xy-plane. The area element in plane polar coordinates is $\varrho\, d\varrho\, d\varphi$. Thus

$$I^2 = \int_0^\infty dx \int_0^\infty dy\, e^{-x^2-y^2} = \int_0^\infty d\varrho\, \varrho\, e^{-\varrho^2} \int_0^{\pi/2} d\varphi = \tfrac{1}{4}\pi. \tag{2.2}$$

Hence $I = \frac{1}{2}\sqrt{\pi}$. It follows that

$$\int_0^\infty dx\, e^{-ax^2} = \tfrac{1}{2}\int_{-\infty}^\infty dx\, e^{-ax^2} = \tfrac{1}{2}\sqrt{(\pi/a)}\,, \tag{2.3}$$

where a is any positive constant. In fact, this remains true even if a is a *complex* number, as long as $\mathrm{Re}\, a > 0$.

V. Balakrishnan, *Mathematical Physics*,
https://doi.org/10.1007/978-3-030-39680-0_2

The simple result above has many interesting extensions that are useful in a remarkably large number of physical problems. I shall consider here just a few of these corollaries.

★ **1.** Find the value of the integral $\int_{-\infty}^{\infty} dx\, e^{-ax^2+bx}$ where $a > 0$ and b is a real constant.

Hint: The result just derived is actually valid for arbitrary finite *complex* values of a and b, provided only that $\text{Re}\, a > 0$. This leads to the very useful formula for a *shifted* Gaussian integral:

$$\boxed{\int_{-\infty}^{\infty} dx\, e^{-ax^2+bx} = \sqrt{(\pi/a)}\; e^{b^2/4a} \quad (\text{Re}\; a > 0).} \tag{2.4}$$

Here b any complex number. It then follows that, if $a > 0$ and k is a real constant,

$$\int_0^{\infty} dx\, e^{-ax^2} \cos kx = \tfrac{1}{2}\sqrt{(\pi/a)}\; e^{-k^2/4a}. \tag{2.5}$$

★ **2.** Establish Eq. (2.5). Satisfy yourself that the integral $\int_0^{\infty} dx\, e^{-ax^2} \sin(kx)$ *cannot* be evaluated by this procedure.

Some other integrals related to the Gaussian integral will be encountered in Chap. 3, Sect. 3.1.4.

2.1.2 A Couple of Higher Dimensional Examples

The generalization of the basic Gaussian integral to the multidimensional case is very important, as it occurs in a large variety of contexts. We will consider this extension in Chap. 11, Sect. 11.5, after matrices have been introduced. But it is instructive to work out, here, a couple of special cases of the general result that will be given in Eq. (11.56).

A two-dimensional Gaussian integral: The first of these is a two-dimensional integral. Let μ be a constant such that $-1 < \mu < 1$. Then

$$\int_{-\infty}^{\infty} dx \int_{-\infty}^{\infty} dy\; e^{-(x^2+2\mu x\, y + y^2)} = \frac{\pi}{\sqrt{1-\mu^2}}. \tag{2.6}$$

★ **3.** Establish Eq. (2.6).

Note that the right-hand side of Eq. (2.6), which gives the value of the integral, becomes infinite when $\mu = \pm 1$.

- This divergence is a signal that the left-hand side is a convergent integral only in the range $-1 < \mu < 1$ of the parameter μ.

You must understand the reason for the divergence of the integral at the values $\mu = \pm 1$. Setting $\mu = 1$ in the left-hand side of Eq. (2.6), the integrand becomes $\exp[-(x+y)^2]$. Change variables of integration from (x, y) to $u = (x+y)$ and $v = (x-y)$. We have

$$dx\, dy = \left| \frac{\partial(x, y)}{\partial(u, v)} \right| du\, dv = \tfrac{1}{2}\, du\, dv, \tag{2.7}$$

where

$$\left| \frac{\partial(x, y)}{\partial(u, v)} \right| \equiv \begin{vmatrix} \partial x/\partial u & \partial x/\partial v \\ \partial y/\partial u & \partial y/\partial v \end{vmatrix} \tag{2.8}$$

is the determinant of the Jacobian matrix of the transformation. As the original integral is over the whole of the xy-plane, the new range of integration is *also* over the whole uv-plane. The integral then becomes

$$\tfrac{1}{2} \int_{-\infty}^{\infty} du\, e^{-u^2} \int_{-\infty}^{\infty} dv,$$

which is obviously infinite.

★ **4.** Exactly the same sort of argument can be given in the case $\mu = -1$. Do so.

A three-dimensional Gaussian integral: The second example is a three-dimensional integral. Let ν be a constant such that $-\frac{1}{2} < \nu < 1$. Then

$$\int_{-\infty}^{\infty} dx \int_{-\infty}^{\infty} dy \int_{-\infty}^{\infty} dz\, e^{-(x^2+y^2+z^2+2\nu\, x\, y+2\nu\, y\, z+2\nu\, z\, x)} = \frac{\pi^{3/2}}{(1-\nu)\sqrt{1+2\nu}}. \tag{2.9}$$

★ **5.** Establish Eq. (2.9).

★ **6.** As before, the right-hand side of Eq. (2.9) suggests that the integral on the left-hand side diverges when $\nu = 1$ and also when $\nu = -\frac{1}{2}$.

(a) Give a reason why this should be so, on the lines of the argument I have given above in the case of Eq. (2.6) at the parameter values $\mu = \pm 1$.
(b) The divergence at $\nu = -\frac{1}{2}$ is of the inverse square root type, arising from the factor $(1+2\nu)^{-1/2}$. In contrast, the divergence at $\nu = 1$ is a "stronger" one, since it arises from the factor $(1-\nu)^{-1}$. Why should this be so? (The precise

reason for this difference will become clear after we discuss the general formula for a Gaussian integral in Chap. 11, Sect. 11.5.)

2.2 Stirling's Formula

The Gaussian integral is useful, for instance, in deriving the famous result known as **Stirling's formula** for the factorial (or the logarithm of the factorial) of a large positive integer n.

It is obvious that $n!$ is a very rapidly increasing function of n. It takes only a few seconds to establish that 69 is the largest integer whose factorial can be found with a standard pocket calculator, since $69! \approx 1 \cdot 7 \times 10^{98}$ and $70!$ exceeds 10^{99} (the display capacity of an ordinary calculator). The crudest estimate for $n!$ would be n^n. But this is obviously an *over*estimate, because $n! = n(n-1)(n-2)\cdots 1$. An approximate expression for n is readily obtained by an elementary argument, by considering the logarithm of $n!$ rather than $n!$ itself. We have $\ln n! = \sum_{k=1}^{n} \ln k$. For sufficiently large n, this can be approximated by the *integral* $\int_0^n dt\, \ln t = n \ln(n/e)$. Therefore $n! \approx n^n\, e^{-n}$.

To make this reasoning more systematic and accurate, a convenient starting point is the integral

$$\int_0^\infty dt\, t^n\, e^{-t} = n!\,, \tag{2.10}$$

where n is a positive integer. (Incidentally, setting $n = 0$ on the left-hand side yields 1 as the value of the integral. This result enables us to *define* $0!$ as 1.) Now, t^n is a rapidly *increasing* function of t for large n, while e^{-t} is a rapidly *decreasing* function of t. The product of the two factors is a function that starts at 0 when $t = 0$ and peaks to a maximum value, and then drops to zero as $t \to \infty$. (Draw a schematic sketch of the function.) The peak gets more and more pronounced as n increases. The integrand essentially looks like a bell-shaped curve, a Gaussian. The contribution to the definite integral then comes mainly from the immediate neighborhood of the peak. The trick, then, is to (i) approximate the integrand by a Gaussian centered about the maximum of the integrand, and (ii) extend the range of integration from $-\infty$ to ∞. The outcome is

$$\boxed{n! \simeq n^n\, e^{-n} \sqrt{2\pi n}.} \tag{2.11}$$

This is Stirling's approximation or Stirling's formula. The corrections to it will become clear in Eq. (2.13) below.

★ **7.** Derive the formula (2.11).

Stirling's formula is quite remarkable, because it is valid to an astonishing degree of accuracy even for relatively *small* values of n. In fact, even if n is as small as 1 itself, the formula is accurate to about 1 part in 12: the values of e and $\sqrt{2\pi}$ only differ by about 8%, as you can check easily! When n is 10, the accuracy is already about 99.2%, and for $n = 100$, this becomes 99.92%, and so on. Stirling's formula for the *logarithm* of the factorial of a large integer,

$$\boxed{\ln n! \simeq n \ln n - n + \tfrac{1}{2} \ln (2\pi n),} \tag{2.12}$$

is undoubtedly familiar to you from its widespread use in statistical mechanics. The accuracy of this formula for values of n as large as Avogadro's number is, of course, stupendous: it is correct to 1 part in 10^{24}. This estimate follows from the exact form of **Stirling's series** for $n!$, which reads

$$n! = n^n \, e^{-n} \, (2\pi n)^{1/2} \left\{ 1 + \frac{1}{12n} + \frac{1}{288n^2} + \mathcal{O}(n^{-3}) \right\}. \tag{2.13}$$

The curly brackets is an infinite series in powers of $1/n$. It is *not* a convergent power series, but rather, an **asymptotic series**. More precisely, Stirling's series is an asymptotic series for the **gamma function**, which we shall consider in Chap. 3, Sect. 3.1.4, and again in Chaps. 25 and 26. The numerical coefficients in the series are explicitly known in terms of the **Bernoulli numbers**. We will have occasion to define these numbers in Chap. 26, Sect. 26.2.4. An important property of an asymptotic series is the following: the error introduced by truncating such a series at any stage is as small as the next term in the series. In a series such as that in Eq. (2.13), for instance, for each given value of n there is an integer $k(n)$, such that the best approximation to $n!$ is provided by truncating the series at the kth term. Retaining terms beyond that actually worsens the approximation.

The Gaussian approximation used above for evaluating the integral is a special case of a general technique called Laplace's method that is very useful in finding asymptotic expansions of functions defined by integrals. Laplace's method is itself a reduced form of a more general technique called the method of steepest descent or the **saddle-point method**, involving integration in the complex plane.

2.3 The Dirichlet Integral and Its Descendants

Here is a very important integral, and a whole set of integrals that descend from it, all of which can be evaluated starting from very elementary considerations.

As you know, the integral $\int_0^\infty dx/x$ does not exist, because it diverges *logarithmically* at both the lower and upper limits of integration: That is, if the limits had been ϵ and L, respectively, the integral would have been equal to $\ln (L/\epsilon)$. This becomes infinite when either $L \to \infty$, or $\epsilon \to 0$, or both.

What about the integral $\int_0^\infty dx\,(\sin bx)/x$, where b is a nonzero, real constant? The singularity of the integrand at $x = 0$ is no longer present, since the integrand tends to the finite limit b as $x \to 0$. You might still expect the integral to diverge logarithmically at the upper limit of integration, however, because $|\sin bx|$ is always ≤ 1, and the magnitude of the integrand still falls off only as slowly as $1/x$ for large x. The integral $\int_0^\infty dx\,|\sin bx|/x$ is indeed divergent. Remarkably enough, however, the fact that $\sin(bx)$ oscillates between positive and negative values saves the situation, and makes the integral $\int_0^\infty dx\,(\sin bx)/x$ convergent. This integral is of great importance, as it and its relatives occur in numerous contexts. Here is one way to compute the value of this definite integral.

We know that

$$\int_0^\infty dx\, e^{-ax} = \frac{1}{a}, \quad \text{for any positive constant } a. \tag{2.14}$$

But this result remains valid even if a is replaced by a complex number $a + ib$, as long as its *real* part a is positive: It is only the real part that controls the convergence of the integral, because the factor e^{-ibx} is just the sum of a cosine and a sine, and these are oscillatory functions of magnitude ≤ 1. Thus

$$\int_0^\infty dx\, e^{-(a+ib)x} = \frac{1}{a+ib} \quad (a > 0, \quad b \text{ real}). \tag{2.15}$$

Equating the imaginary parts of the two sides, we get the well-known result

$$\int_0^\infty dx\, e^{-ax}\, \sin bx = \frac{b}{a^2+b^2}\,. \tag{2.16}$$

Now integrate both sides of this equation from 0 to ∞ *with respect to the parameter* a, and interchange the order of integration on the left-hand side. (This can be shown to be a legitimate operation.) Use the fact that

$$\int \frac{da}{a^2+b^2} = \frac{1}{|b|}\tan^{-1}\frac{a}{|b|}\,. \tag{2.17}$$

The result is

$$\boxed{\int_0^\infty dx\, \frac{\sin bx}{x} = \tfrac{1}{2}\pi\,\varepsilon(b).} \tag{2.18}$$

Here $\varepsilon(b)$ stands for the discontinuous **signum function** defined as

$$\varepsilon(b) \stackrel{\text{def.}}{=} \begin{cases} +1 & \text{for } b > 0 \\ -1 & \text{for } b < 0. \end{cases} \tag{2.19}$$

It is obvious that $\varepsilon(x)$ is not really a function in the conventional, elementary sense. It is an example of a **generalized function** or **distribution**. A few of the most common generalized functions will be discussed in Chap. 4.

The formula (2.18) is a rather important and useful one. The integral on the left-hand side is called the **Dirichlet integral**. We will encounter it several times in the rest of this book, and it will be evaluated by more than one method. The name "Dirichlet integral" is also given to several other integrals, including some of the related results to be deduced below. But I will use the name exclusively for the integral in Eq. (2.18). You may also recall that the integrand $(\sin bx)/x$ is often called the **sinc function**, especially in engineering applications.

Equation (2.18) leads at once to further results. Treat b as a variable and integrate both sides of this equation over b from 0 to any (real) number c, to get

$$\int_0^\infty dx\, \frac{(1-\cos cx)}{x^2} = \tfrac{1}{2}\,\pi\,|c| = \tfrac{1}{2}\,\pi\, c\,\varepsilon(c). \tag{2.20}$$

It follows easily that

$$\int_0^\infty dx\, \left(\frac{\sin cx}{x}\right)^2 = \tfrac{1}{2}\,\pi\,|c|. \tag{2.21}$$

Note that the integrand in each of the two integrals above has an apparent quadratic singularity at $x = 0$ due to the factor $1/x^2$. But the numerator also vanishes like x^2 in each case, so that the integrand is well-behaved and has a finite limit as $x \to 0$.

★ **8.** Work through the steps described above to derive Eqs. (2.20) and (2.21). Treat the cases $c > 0$ and $c < 0$ separately, so as to obtain the factor $|c|$ correctly.

By repeated use of the procedure above (integrating the integral at each stage with respect to the parameter from 0 up to some real number, and relabeling the latter as c), we can go further. The next two stages give

$$\int_0^\infty \frac{dx}{x^3}\,(\sin cx - cx) = -\frac{\pi\, c^2}{2\,(2!)}\,\varepsilon(c). \tag{2.22}$$

and

$$\int_0^\infty \frac{dx}{x^4}\left(1 - \frac{c^2x^2}{2!} - \cos cx\right) = -\frac{\pi\, c^3}{2\,(3!)}\,\varepsilon(c). \tag{2.23}$$

Once again, note how the apparent singularities of the integrand at the origin are actually canceled out. In each case, the integrand tends to a finite limit as $x \to 0$.

★ **9.** Derive Eqs. (2.22) and (2.23).

You should be able to see a pattern here. The integrands in the successive integrals represent the power series of the sine and cosine functions with the first n terms subtracted out, where n runs over the positive integers. The general formulas thus obtained are

$$\int_0^\infty \frac{dx}{x^{2n+1}} \left\{ \sin cx - \sum_{k=0}^{n-1} \frac{(-1)^k (cx)^{2k+1}}{(2k+1)!} \right\} = (-1)^n \frac{\pi\, c^{2n}}{2\,(2n)!}\, \varepsilon(c) \tag{2.24}$$

and

$$\int_0^\infty \frac{dx}{x^{2n}} \left\{ \sum_{k=0}^{n-1} \frac{(-1)^k (cx)^{2k}}{(2k)!} - \cos cx \right\} = (-1)^{n-1} \frac{\pi\, c^{2n-1}}{2\,(2n-1)!}\, \varepsilon(c), \tag{2.25}$$

where n is any positive integer. Again, the integrands in the two integrals above have finite nonzero limits as $x \to 0$. Clearly, as $n \to \infty$, the finite sums inside the curly brackets on the left-hand side tend to $\sin(cx)$ and $\cos(cx)$, respectively, and the integrands vanish. It is obvious that the right-hand sides of Eqs. (2.24) and (2.25) also tend to zero as $n \to \infty$, for any finite value of c.

★ **10.** Derive Eqs. (2.24) and (2.25). The method of induction suggests itself!

Interestingly, the sequence of integrals evaluated above finds an application in the problem of **random flights** in probability theory, as you will see in Chap. 20, Sect. 20.4.2.

2.4 Solutions

1. Complete the square in the exponent and shift the variable of integration by a suitable constant amount to obtain the standard Gaussian integral. ▶

2. Put $b = ik$ in Eq. (2.4); use the fact that $\cos kx$ is an even function of x. ▶

3. Consider the integral over y first. Complete the square in the quadratic exponent, and shift the variable of integration from y to $y' = y + \mu x$. The result is a Gaussian integral whose value can be written down. What is left is another Gaussian integral over x, whose value can also be written down. ▶

5. Once again, consider first the integral over z. Complete the square in the exponent, shift the variable of integration, and write down the value of the resulting Gaussian integral. Repeat the procedure for the integrals over y and x, in succession. ▶

7. Write the integrand in Eq. (2.10) in the form $e^{-f(t)}$, where $f(t) = t - n \ln t$. It is easy to see that $f(t)$ has a maximum at $t = n$. Expand $f(t)$ in a Taylor series about $t = n$ and retain terms up to second order in $(t - n)$. Thus $f(t) \simeq n - n \ln n + (t - n)^2/(2n)$. Use this and approximate the integral as described in the text, to obtain (2.11). ▶

Chapter 3
Some More Functions

3.1 Functions Represented by Integrals

The functions we come across in physical problems may not always occur in a completely explicit form. Quite frequently, a function is specified by means of an integral. This is termed an **integral representation** of the function concerned. A formula such as

$$f(x) = \int_a^b dy\ \phi(x, y), \tag{3.1}$$

where ϕ is a given function and a, b are constants, must be regarded as an integral representation of the function $f(x)$. In general, it is not possible to actually carry out the integration over y in closed form, so that the integral representation is about as close as one can get to an explicit form for the function.

The dependence on x in an integral representation may occur both in the integrand as well as in one or both limits of integration. The general form of an integral representation of this sort is thus

$$f(x) = \int_{a(x)}^{b(x)} dy\ \phi(x, y). \tag{3.2}$$

There are numerous important examples of functions specified by integral representations. All the so-called **special functions** of mathematical physics have integral representations that are very useful in practice. I will consider some of these in Chap. 26, after discussing analytic functions of a complex variable, because integral representations are generally contour integrals in the complex plane. In this chapter we shall restrict ourselves to a few simple cases involving real variables.

V. Balakrishnan, *Mathematical Physics*,
https://doi.org/10.1007/978-3-030-39680-0_3

3.1.1 Differentiation Under the Integral Sign

It is important to be able to write down the derivative of a function defined by a definite integral. Under suitable conditions of differentiability, it is evident that the derivative of $f(x)$ in Eq. (3.1) is again an integral, namely, $f'(x) = \int_a^b (\partial\phi/\partial x)\, dy$. In the more general case of Eq. (3.2), extra terms appear in $f'(x)$ owing to the dependence of the limits of integration upon x. The result is a very useful formula:

$$\boxed{\frac{df(x)}{dx} = \int_{a(x)}^{b(x)} dy\, \frac{\partial\phi(x, y)}{\partial x} + \frac{db(x)}{dx}\, \phi\big(x, b(x)\big) - \frac{da(x)}{dx}\, \phi\big(x, a(x)\big)\,.} \tag{3.3}$$

3.1.2 The Error Function

As we have seen in Chap. 2, the Gaussian integral $\int_0^\infty dt\, e^{-t^2} = \frac{1}{2}\sqrt{\pi}$. The *incomplete* Gaussian integral, namely, the integral of the Gaussian function e^{-t^2} from 0 up to any value x, defines the so-called **error function**. We have

$$\boxed{\text{erf}\,(x) \stackrel{\text{def.}}{=} \frac{2}{\sqrt{\pi}} \int_0^x dt\, e^{-t^2}.} \tag{3.4}$$

It is easy to show that

$$\text{erf}\,(0) = 0\,, \;\; \text{erf}\,(\infty) = 1\,, \;\; \text{erf}\,(-\infty) = -1\,. \tag{3.5}$$

The error function is an odd function of x, i.e.,

$$\text{erf}\,(-x) = -\,\text{erf}\,(x)\,. \tag{3.6}$$

Using Eq. (3.3), we have

$$\frac{d}{dx}\,\text{erf}\,(x) = \frac{2}{\sqrt{\pi}}\, e^{-x^2}\,, \tag{3.7}$$

which is positive for all finite x. Hence erf (x) is a monotonically increasing function of x. It increases monotonically from -1 to $+1$ as x increases from $-\infty$ to $+\infty$. The function

$$\text{erfc}\,(x) \stackrel{\text{def.}}{=} \frac{2}{\sqrt{\pi}} \int_x^\infty dt\, e^{-t^2} = 1 - \text{erf}\,(x) \tag{3.8}$$

is called the **complementary error function**.

The error function appears frequently in statistics. As you will see in Chap. 20, Sect. 20.2.1, the cumulative probability distribution function of a so-called **normal** or **Gaussian distribution** is essentially an error function.

★ **1.** Sketch erf (x), and erfc (x) versus x.

3.1.3 *Fresnel Integrals*

If the exponent in the integrand in Eq. (3.4) is pure imaginary, we get the **Fresnel integrals**. Their precise definitions are

$$\boxed{X(s) \stackrel{\text{def.}}{=} \int_0^s dt\, \cos\left(\tfrac{1}{2}\pi t^2\right) \quad \text{and} \quad Y(s) \stackrel{\text{def.}}{=} \int_0^s dt\, \sin\left(\tfrac{1}{2}\pi t^2\right), \quad (s \geq 0).} \tag{3.9}$$

You may recall that these functions occur, for instance, in the theory of Fresnel diffraction. The Cartesian coordinates of **Cornu's spiral** are given by $(X\,,\ Y)$, parametrized by the arc length s of a point on the spiral as measured from the origin of coordinates.

It is obvious from the definitions above that $X(0) = Y(0) = 0$. But it is not immediately clear whether $X(s)$ and $Y(s)$ remain finite as $s \to \infty$. Since cos $\left(\frac{1}{2}\pi t^2\right)$ and sin $\left(\frac{1}{2}\pi t^2\right)$ oscillate extremely rapidly between the values 1 and -1 when t becomes very large, there is some reason to hope that the Fresnel integrals might actually converge to finite values in the limit $s \to \infty$. This turns out to be so. It can be shown that

$$\lim_{s\to\infty} X(s) = \lim_{s\to\infty} Y(s) = \tfrac{1}{2}\,. \tag{3.10}$$

A non-rigorous, heuristic way to see this is as follows. $X(\infty)$ and $Y(\infty)$ are the real and imaginary parts, respectively, of the integral $\int_0^\infty dt\, \exp\left(\frac{1}{2}i\pi t^2\right)$. If we now assume that Eq. (2.3) for the value of the Gaussian integral is valid as it stands even when the parameter a is pure imaginary, then

$$\int_0^\infty dt\, e^{i\pi t^2/2} = \sqrt{(i/2)} = \tfrac{1}{2}(1+i). \tag{3.11}$$

Hence $X(\infty) = Y(\infty) = \frac{1}{2}$.

★ **2.** Sketch $X(s)$, $Y(s)$ and Cornu's spiral (Y versus X, eliminating s). You may want to write a suitable program to calculate $X(s)$ and $Y(s)$ and to plot the graphs.

3.1.4 *The Gamma Function*

Like the error function, the gamma function is a very basic function that is defined by an integral. It is a generalization of the factorial of a nonnegative integer. Recall that

$$\int_0^\infty dt\, t^n\, e^{-t} = n! \quad \text{for } n = 0,\ 1, \ldots . \tag{3.12}$$

Replacing n in the factor t^n on the left-hand side by a real variable $(x - 1)$ defines a function of x called the **gamma function**:

$$\boxed{\Gamma(x) \stackrel{\text{def.}}{=} \int_0^\infty dt\, t^{x-1}\, e^{-t}, \quad x > 0.} \tag{3.13}$$

The gamma function is also called the **Euler integral of the second kind.**[1] The condition $x > 0$ in Eq. (3.13) is necessary to ensure the convergence of the integral, because the factor t^{x-1} is integrable at the lower limit of integration (namely, 0) only if $x > 0$. Convergence at the upper limit of integration, ∞, does not pose any such problem, because of the presence of the decaying exponential factor e^{-t}.

When x equals any positive integer n, the value of the integral reduces to $(n - 1)!$. Hence $\Gamma(n) = (n - 1)!$ for $n = 1, 2, \ldots$. Thus,

- the gamma function $\Gamma(x)$ *interpolates* between positive integral values of its argument to provide a generalization of the factorial of a nonnegative integer.

It must be stated right away that such interpolations, in general, are never unique. For instance, a term like $\sin \pi x$ could have been present on the right-hand side of Eq. (3.13), without affecting Eq. (3.12). Extrapolations from the integers to the continuum can become unique only when further conditions are imposed and met. In the present case, it turns out that

- $\Gamma(x)$ is the unique interpolation of $n!$ such that its logarithm, the function $\ln \Gamma(x)$, has a positive curvature for all $x > 0$. In technical terms, $\ln \Gamma(x)$ is a **convex function** for $x > 0$.

Figure 3.1 shows $\Gamma(x)$ for $x > 0$. The function has a simple minimum at $x = 1 + e^{-1}$. The gamma function can be extended to negative values of x as well. In Fig. 25.5 of Chap. 25, Sect. 25.2.1, you will see what $\Gamma(x)$ looks like for all real x.

A very important property of the gamma function is the **functional equation** that it satisfies, namely,

$$\Gamma(x + 1) = x\, \Gamma(x). \tag{3.14}$$

---

[1]The Euler integral of the first kind is the **beta function**, which will be discussed in Chap. 25, Sect. 25.2.6.

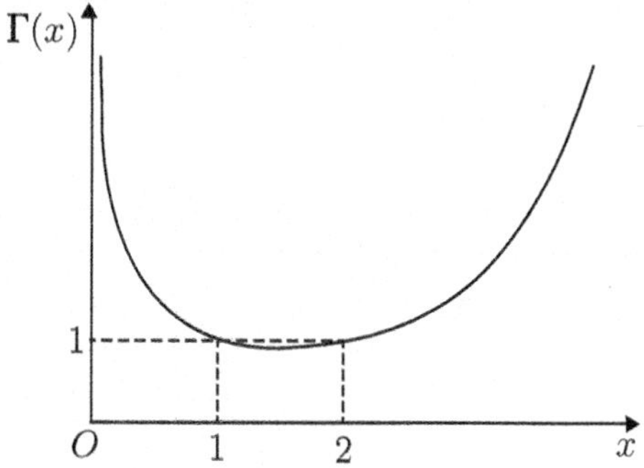

Fig. 3.1 $\Gamma(x)$ for real positive values of the argument x

★ **3.** Using integration by parts, establish Eq. (3.14) from Eq. (3.13). Note how it is *essential* to have $x > 0$ during this process, in order to be able to derive the result sought.

This functional equation is a fundamental property of the gamma function. It can be used to give meaning to the gamma function in the region in which the original defining integral, Eq. (3.13), is no longer convergent: namely, the region $x \leq 0$.

We shall study the gamma function $\Gamma(z)$ in greater detail in Chap. 25. As you will see in Sect. 25.2.1, replacing x by a complex variable z in Eq. (3.13) yields the definition of the gamma function $\Gamma(z)$ in the whole of the right half-plane, $\text{Re}\, z > 0$. The functional equation (3.14), too, remains valid when x is replaced by z. It then enables us to extend (or "analytically continue") the gamma function to the whole of the complex plane. A contour integral representation of $\Gamma(z)$ that is valid in the whole of the complex z-plane will be derived in Chap. 26, Sect. 26.2.1.

3.1.5 *Connection to Gaussian Integrals*

Gaussian integrals, that we considered briefly in Chap. 2, Sect. 2.1, can be expressed in terms of gamma functions. We know from Eq. (2.1) that $\int_0^\infty dx\, e^{-x^2} = \frac{1}{2}\sqrt{\pi}$. Setting $x^2 = u$ gives

$$\int_0^\infty du\, u^{-1/2}\, e^{-u} = \Gamma\left(\tfrac{1}{2}\right) = \sqrt{\pi}. \tag{3.15}$$

This is an important result. More generally, we have

$$\boxed{\int_0^\infty dx\, x^n\, e^{-ax^2} = \tfrac{1}{2}\Gamma\left(\tfrac{1}{2}(n+1)\right)\, a^{-(n+1)/2}, \quad a > 0,\ n > -1.} \tag{3.16}$$

★ **4.** Establish Eq. (3.16).

Note, in particular, that n need not be an integer in the formula of Eq. (3.16). In fact, as you might guess, the formula continues to be valid even for *complex* values of the parameters n and a, provided $\mathrm{Re}\, n > -1$ and $\mathrm{Re}\, a > 0$.

★ **5.** Let $\alpha > 0$. Show that $\displaystyle\int_0^\infty dx\, e^{-x^\alpha} = \Gamma(1+\alpha^{-1})$.

The gamma function of a half-odd-integer: From the recursion relation (3.14) and the fact that $\Gamma\left(\frac{1}{2}\right) = \sqrt{\pi}$, it follows at once that the value of the gamma function of a half-odd-integer can be written down explicitly. We have

$$\left.\begin{aligned} \Gamma\left(n+\tfrac{1}{2}\right) &= \frac{\sqrt{\pi}\,(2n)!}{2^{2n}\,n!}, \\ \Gamma\left(-n+\tfrac{1}{2}\right) &= \frac{\sqrt{\pi}\,(-1)^n 2^{2n}\,n!}{(2n)!} \end{aligned}\right\}\quad (n = 0, 1, 2, \ldots). \tag{3.17}$$

★ **6.** Derive Eqs. (3.17).

You will observe that the product of the two gamma functions above is particularly simple:

$$\Gamma\left(n+\tfrac{1}{2}\right)\Gamma\left(-n+\tfrac{1}{2}\right) = (-1)^n \pi. \tag{3.18}$$

This is not by chance. It is the special case, for $z = n + \frac{1}{2}$, of the "reflection formula" (Eq. (25.48)) that will be derived in Chap. 25, Sect. 3.19. The formula is

$$\Gamma(z)\,\Gamma(1-z) = \pi \operatorname{cosec} z. \tag{3.19}$$

Some useful trigonometric integrals: The gamma function for half-odd-integer values of the argument enables us to express the values of a number of very useful trigonometric integrals. I record these here, as they will be used in the succeeding chapters.

Let n and m be positive integers. Then

$$\int_0^{\pi/2} d\theta\,(\cos\theta)^{n-1}\,(\sin\theta)^{m-1} = \frac{\Gamma\left(\frac{1}{2}n\right)\Gamma\left(\frac{1}{2}m\right)}{2\Gamma\left(\frac{1}{2}(n+m)\right)}. \tag{3.20}$$

This formula is a special case of a more general result (Eq. (25.43)) that will be established after we discuss the beta function in Chap. 25, Sect. 25.2.6. It follows at once that

$$\int_0^{\pi} d\theta\,(\sin\theta)^l = 2\int_0^{\pi/2} d\theta\,(\sin\theta)^l = \frac{\sqrt{\pi}\,\Gamma\left(\frac{1}{2}(l+1)\right)}{\Gamma\left(1+\frac{1}{2}l\right)} \quad (l = 0,\,1,\,2,\,\ldots). \tag{3.21}$$

Further, Eq. (3.20) and the first of Eqs. (3.17) give, on simplification,

$$\int_0^{\pi/2} d\theta\,(\sin\theta)^{2l} = \int_0^{\pi/2} d\theta\,(\cos\theta)^{2l} = \frac{\pi(2l)!}{2^{2l+1}(l!)^2} \tag{3.22}$$

and

$$\int_0^{\pi/2} d\theta\,(\sin\theta)^{2l+1} = \int_0^{\pi/2} d\theta\,(\cos\theta)^{2l+1} = \frac{2^{2l}\,(l!)^2}{(2l+1)!}, \tag{3.23}$$

where $l = 0,\,1,\,2,\,\ldots$.

★ **7.** Starting from Eq. (3.20), verify Eqs. (3.21)–(3.23).

3.2 Interchange of the Order of Integration

Quite frequently, one encounters double or multiple integrals in which the limits of integration over a variable depend on the variables yet to be integrated over. Further, it may be necessary to interchange the order of integration to simplify or evaluate the multiple integral. Some care must be exercised in determining the correct limits of integration when such an interchange is done.

One of the most commonly occurring examples is the following double integral. Let a be a positive number. Then

$$\int_0^a dx \int_0^x dy\;\phi(x, y) = \int_0^a dy \int_y^a dx\;\phi(x, y). \tag{3.24}$$

Figure 3.2 shows how you can write down this identity by inspection. All one has to do is to note that the triangular region of integration can be "scanned" from left to right, or from bottom to top, as depicted by the lines in the shaded triangle.

If the integrand $\phi(x, y)$ is a *symmetric* function of its arguments, i.e., if $\phi(x, y) = \phi(y, x)$, then a further relation holds good:

$$\int_0^a dx \int_0^x dy\;\phi(x, y) = \frac{1}{2}\int_0^a dx \int_0^a dy\;\phi(x, y). \tag{3.25}$$

This equation merely says that the integral over the full square is twice the integral over each triangle, because the integrand is symmetric under a reflection about the diagonal line $y = x$.

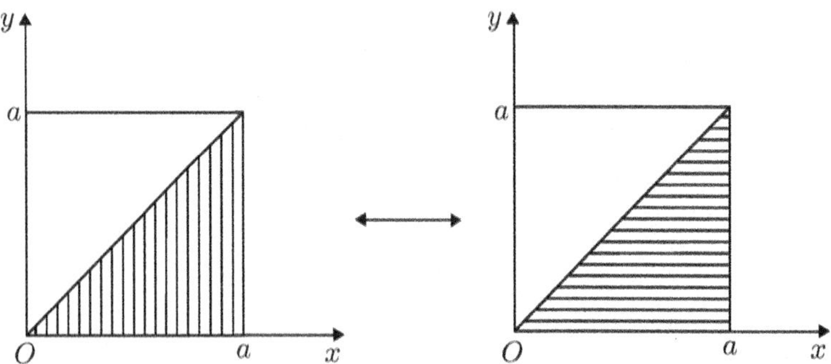

Fig. 3.2 Region of integration in Eq. (3.24) (shaded)

The relation (3.25) is a special case of a more general one in an arbitrary number of variables. Let $\phi(x_1, x_2, \ldots, x_n)$ be a totally symmetric function of its arguments, i.e., it remains unchanged in value under the interchange of any pair of its arguments. The region given by

$$0 \le x_1 \le a, \quad 0 \le x_2 \le a, \quad \cdots, \quad 0 \le x_n \le a, \tag{3.26}$$

where a is a positive number, is obviously an n-dimensional hypercube of side length a. The region specified by the conditions

$$0 \le x_1 \le x_2 \le \cdots \le x_n \le a \tag{3.27}$$

is contained in this hypercube. A convex region of this sort is called a **simplex**. It is the generalization, to n dimensions, of a triangle in two dimensions, a tetrahedron in three dimensions, and so on. There are obviously $n!$ such simplexes in the hypercube, comprising all permutations of the coordinates $x_1, x_2, \ldots, x_n$ in (3.27). A simple identity connects the integral of $\phi(x_1, x_2, \ldots, x_n)$ over any one of the simplexes to its integral over the hypercube. We have, for instance,

$$\begin{aligned}\int_0^a dx_n \int_0^{x_n} dx_{n-1} \cdots \int_0^{x_2} dx_1\, \phi(x_1, x_2, \ldots, x_n) \\ = \frac{1}{n!}\int_0^a dx_n \int_0^a dx_{n-1} \cdots \int_0^a dx_1\, \phi(x_1, x_2, \ldots, x_n).\end{aligned} \tag{3.28}$$

The multiple integral on the right-hand side has the advantage that the limits of integration in each integral are constants rather than variables. The identity (3.28) is very useful in numerous applications. One encounters it, for instance, in quantum mechanics, in the context of the **Dyson series** for the time-development operator in the case of a time-dependent Hamiltonian. It also appears in the derivation of **Wick's**

Theorem in quantum field theory.

★ **8.** Consider the double integral $I(t) = \int_0^t dt_1 \int_0^t dt_2\ \phi(|t_1 - t_2|)$, where $t > 0$ and ϕ is a given function of its argument.

(a) Show that the integral can be reduced to

$$I(t) = 2\int_0^t dt'\,(t - t')\,\phi(t').$$

(b) Assuming that the integral $\int_0^\infty dt'\ \phi(t')$ is finite, show that

$$\lim_{t\to\infty} \frac{I(t)}{2t} = D \quad \text{where} \quad D = \int_0^\infty dt'\ \phi(t').$$

Such an integral arises in the physical context of the simplest model for the random motion of a particle in a fluid. $I(t)$ then represents the mean squared displacement of the particle along any given direction in a time interval t. The function ϕ is called its velocity autocorrelation function, and D is the **diffusion coefficient**. The relation equating it to the time integral of the velocity autocorrelation function is called the **Kubo–Green formula** for the diffusion coefficient. We shall consider several aspects of diffusion in Chaps. 21 and 30.

★ **9.** Show that, if $\phi(x)$ is some given integrable function,

$$\int_0^x dx_n \int_0^{x_n} dx_{n-1} \cdots \int_0^{x_2} dx_1\ \phi(x_1) = \frac{1}{(n-1)!}\int_0^x dx_1\,(x - x_1)^{n-1}\,\phi(x_1).$$

★ **10.** Let $x > 0$. Define the function $f(x)$ as the multiple integral

$$f(x) = \iiiint_{\mathcal{R}} dx_1\, dx_2\, dx_3\, dx_4\,,$$

where $\mathcal{R}$ is the region given by

$$\mathcal{R}: \begin{cases} x_1 \ge 0,\ x_2 \ge 0,\ x_3 \ge 0,\ x_4 \ge 0; \\ x_1 + x_2 + x_3 + x_4 \le x\,. \end{cases}$$

Evaluate the integral, and sketch $f(x)$ as a function of x.

3.3 Solutions

5. Change the variable of integration to $t = x^\alpha$ and use the definition (3.13).

Remark With increasing α, the integrand $\exp(-x^\alpha)$ essentially remains close to unity for $x < 1$, but falls off to zero more and more rapidly for $x > 1$. The integral has a minimum at $\alpha = e$, and tends toward its limiting value of unity as $\alpha \to \infty$. ▶

9. Use the formula (3.24) repeatedly, starting with the integrations over x_1 and x_2. ▶

10. The integral can be written as

$$\int_0^x dx_1 \int_0^{x-x_1} dx_2 \int_0^{x-x_1-x_2} dx_3 \int_0^{x-x_1-x_2-x_3} dx_4\,,$$

which is easily evaluated to give $f(x) = x^4/4!$. Observe that this is just the "volume" of one of the 24 simplexes comprising the four-dimensional hypercube of side length x. ▶

Chapter 4
Generalized Functions

Functions that have finite or infinite discontinuities (or jumps) occur very frequently in mathematical models of physical systems. **Potential barriers** represent a commonly encountered example. **Impulse functions** in signal analysis represent another. Such discontinuous functions are not functions in the conventional sense of the term. There is, however, a rigorous mathematical theory of such **generalized functions** or **distributions**. The most common among these distributions are the step function and the Dirac delta function. This chapter is devoted to a discussion of these generalized functions at an elementary level.

4.1 The Step Function

The unit **step function**, also called the **Heaviside function** in earlier times, is the discontinuous function defined as

$$\boxed{\theta(x) \stackrel{\text{def.}}{=} \begin{cases} +1 & \text{for } x > 0 \\ 0 & \text{for } x < 0. \end{cases}} \tag{4.1}$$

It is obvious that $\theta(x)$ has a finite jump at $x = 0$. It is sometimes convenient to define $\theta(0)$ to be the average value $\frac{1}{2}$, but this is not always necessary. The sum $\theta(x) + \theta(-x)$ is evidently equal to 1. The *difference* $\theta(x) - \theta(-x)$ is

$$\theta(x) - \theta(-x) = \varepsilon(x), \tag{4.2}$$

the signum function already defined in Eq. (2.19) of Chap. 2, Sect. 2.3. The function $\varepsilon(x)$ looks like the limit of a tanh (or hyperbolic tangent) function as the "kink" in the function becomes more and more steep, i.e., as the slope at the origin tends to

V. Balakrishnan, *Mathematical Physics*,
https://doi.org/10.1007/978-3-030-39680-0_4

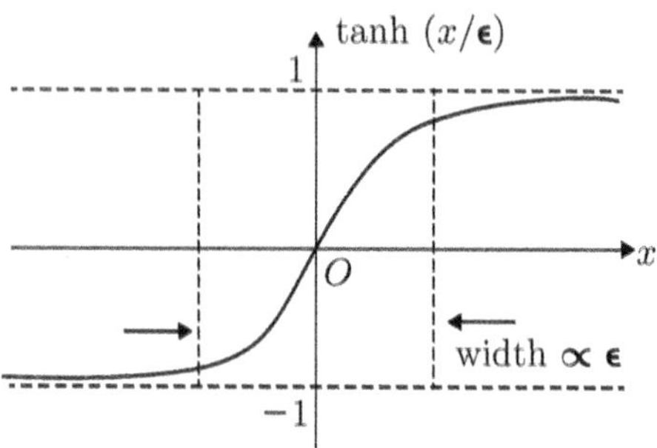

Fig. 4.1 The function $\tanh(x/\epsilon)$ tends to the distribution $\varepsilon(x)$ as $\epsilon \to 0$

infinity, as shown in Fig. 4.1. In fact, we could define $\varepsilon(x)$ as the limit of a continuous sequence of functions $\tanh(x/\epsilon)$ as the positive parameter $\epsilon \to 0$. That is,

$$\lim_{\epsilon \downarrow 0} \tanh(x/\epsilon) = \varepsilon(x). \tag{4.3}$$

As you will see shortly, this representation will help us relate the step function and the signum function to the Dirac delta function.

★ **1.** Sketch the following generalized functions:

(a) $\theta(1 - x^2)$ (b) $\theta(\tanh x)$ (c) $x\,\theta(\sin x)$

(d) $\theta(x+1) - \theta(x-1)$ (e) $\theta(1+x)\,\theta(1-x)$ (f) $(2 - |x|)\,\theta(2 - |x|)$.

A Dirichlet-type integral for a rectangular pulse is provided by the integral

$$\frac{2}{\pi}\int_0^\infty dt\,\frac{\sin t\,\cos xt}{t} = \begin{cases} 1 \text{ for } |x| < 1 \\ \frac{1}{2} \text{ for } |x| = 1 \\ 0 \text{ for } |x| > 1. \end{cases} \tag{4.4}$$

★ **2.** Derive Eq. (4.4) from the Dirichlet integral $\int_0^\infty dt\,(\sin xt)/t = \frac{1}{2}\pi\varepsilon(x)$.

★ **3.** Show that $\displaystyle\int_{-\infty}^{\infty} dx\,[\theta(x+2) - \theta(x-2)]\,\text{sech}\,x = 4\tan^{-1}(e^2) - \pi$.

4.2 The Dirac Delta Function

4.2.1 Defining Relations

Let $f(x)$ be a function that is well-defined and finite for all values of the real variable x. Can we construct some sort of filter or "selector" that, when operating on this function, *singles out* the value of the function at any prescribed point x_0 ?

A hint is provided by the discrete analog of this question. Suppose we have a sequence $(a_1, a_2, \dots) = \{a_j \mid j = 1, 2, \dots\}$. How do we select a particular member a_i from the sequence? We do so by summing over all members of the sequence with a selector called the **Kronecker delta**, denoted by δ_{ij} and defined as

$$\boxed{\delta_{ij} \stackrel{\text{def.}}{=} \begin{cases} 1 \text{ if } i = j \\ 0 \text{ if } i \neq j. \end{cases}} \tag{4.5}$$

It follows immediately that

$$\sum_j \delta_{ij}\, a_j = a_i\,. \tag{4.6}$$

Further, we have the normalization $\sum_j \delta_{ij} = 1$ for each value of i, and also the symmetry property $\delta_{ij} = \delta_{ji}$. We will consider the Kronecker delta more formally in Chap. 5, Sect. 5.1.3.

Reverting to the continuous case, we must replace the summation over j by an integration over x. The role of the specified index i is played by the specified point x_0. The analog of the Kronecker delta is written like a function, retaining the same symbol δ for it. (Presumably, this was Dirac's reason for choosing this notation for the delta function.) So we seek a "function" $\delta(x - x_0)$ such that

$$\boxed{\int_{-\infty}^{\infty} dx\, \delta(x - x_0)\, f(x) = f(x_0).} \tag{4.7}$$

Exactly as in the discrete case of the Kronecker delta, we impose the normalization and symmetry properties

$$\boxed{\int_{-\infty}^{\infty} dx\, \delta(x - x_0) = 1 \quad \text{and} \quad \delta(x - x_0) = \delta(x_0 - x).} \tag{4.8}$$

Equations (4.7) and (4.8) may be taken to define the **Dirac delta function**. The form of Eq. (4.7) suggests that $\delta(x - x_0)$ is more like *the kernel of an integral operator* than a conventional function. And indeed it is—the kernel of the *unit operator*. (We will discuss integral operators in Chap. 32.) Incidentally, it follows from the

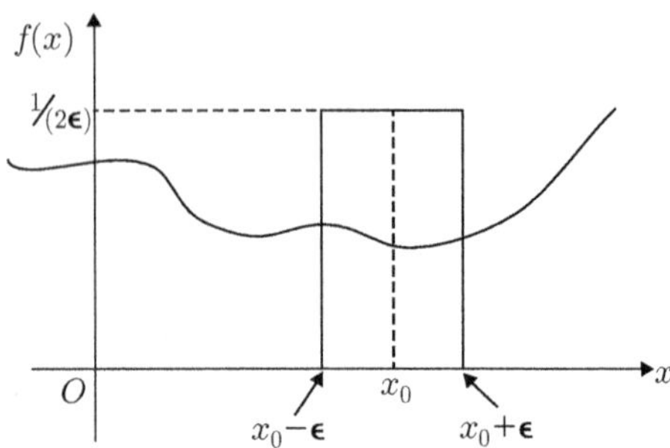

Fig. 4.2 As $\epsilon \to 0$, the rectangular window tends to $\delta(x - x_0)$

normalization that, if the variable x has some physical dimensions, then $\delta(x)$ has the reciprocal of those dimensions.

What can $\delta(x - x_0)$ possibly look like? A naive way of answering this question is as follows. Take a rectangular window of width 2ϵ and height $1/(2\epsilon)$, so that the area of the window is unity. Place it with its bottom edge on the x-axis and slide it along this axis. When the window is centered at the chosen point x_0, as shown in Fig. 4.2, the integral of $f(x)$ multiplied by this window function is simply

$$(1/2\epsilon) \int_{x_0-\epsilon}^{x_0+\epsilon} dx\, f(x).$$

This does not quite select $f(x_0)$ alone, of course. But it will do so if we take the limit $\epsilon \to 0$. In this limit, the width of the window becomes vanishingly small. Simultaneously, its height becomes arbitrarily large, so as to capture in full the ordinate in the graph of $f(x)$, no matter how large the value of $f(x_0)$ is. A possible explicit form for the Dirac delta function $\delta(x - x_0)$ is therefore given by

$$\delta(x - x_0) = \begin{cases} \lim\limits_{\epsilon \to 0} 1/(2\epsilon)\,, & \text{for } x_0 - \epsilon < x < x_0 + \epsilon \\ 0\,, & \text{for all other } x\,. \end{cases} \tag{4.9}$$

This cannot be a stand-alone definition. It cannot be taken literally. If we did so, then, formally, $\delta(x - x_0)$ must be zero for all $x \neq x_0$, while it must be infinite for $x = x_0$.

- An explicit form for the delta function is always to be understood as something that makes sense only when it occurs in an integral like

$$\int_{-\infty}^{\infty} dx\, \delta(x - x_0)\, f(x),$$

i.e., when it acts on ordinary functions like $f(x)$ and an integration over x is carried out.

- We know that the value of the integral above is $f(x_0)$. It is easy to see that we also have

$$\int_{-\infty}^{\infty} dx\, \delta(x - x_0)\, f(x) = \int_a^b dx\, \delta(x - x_0)\, f(x) = f(x_0), \tag{4.10}$$

  for any a and b such that $a < x_0 < b$. In other words, the delta function must "fire" within the range of integration. Otherwise the integral is equal to zero.

It is immediately clear that the so-called Dirac delta "function" cannot be a function in the conventional sense. In particular, $\delta(x - x_0)$ must be *singular* (formally infinite) at $x = x_0$, that is, at the point where its argument is zero. The Dirac delta function is a generalized function or distribution.

4.2.2 Sequences of Functions Tending to the δ-Function

Mathematically, an explicit form for the Dirac delta function is properly given in terms of a *sequence* or family of conventional functions. This is preferable to the "window" representation (4.9), which involves discontinuous functions. It can then be arranged that, in a suitable limit, the sequence of functions approaches a quantity that has all the properties desired of the delta function. An infinite number of such sequences may be constructed. For instance, take any family of continuous functions $\phi_\epsilon(x - x_0)$ parametrized by a positive constant ϵ, and with the following properties: Each member of the family

(i) has a peak at x_0;
(ii) is symmetric about the point x_0; and
(iii) is integrable, such that $\int_{-\infty}^{\infty} dx\, \phi_\epsilon(x) = 1$.

Matters are arranged such that, as the parameter ϵ is made smaller and smaller, the height of the peak in $\phi_\epsilon(x)$ increases while its width simultaneously decreases, keeping the total area under the curve equal to unity. Then

$$\lim_{\epsilon \to 0} \phi_\epsilon(x - x_0) = \delta(x - x_0). \tag{4.11}$$

Some of the simplest choices for such sequences are given below. For ease of writing, let us set $x_0 = 0$. One of the simplest possibilities is the family of **Lorentzians**, given by

$$\phi_\epsilon(x) = \frac{\epsilon}{\pi(x^2 + \epsilon^2)}. \tag{4.12}$$

Then $\lim_{\epsilon \to 0} \phi_\epsilon(x)$ is a representation of the Dirac delta function $\delta(x)$.

★ **4.** Here are some other functions $\phi_\epsilon(x)$ that tend to $\delta(x)$ as $\epsilon \to 0$ from above:

$$\text{(i)}\ \frac{1}{2\epsilon}\, e^{-|x|/\epsilon} \quad \text{(ii)}\ \frac{1}{2\sqrt{\pi\epsilon}}\, e^{-x^2/4\epsilon} \quad \text{(iii)}\ \frac{\operatorname{sech}^2(x/\epsilon)}{2\epsilon} \quad \text{(iv)}\ \frac{\sin(x/\epsilon)}{\pi x}.$$

(a) Sketch each these functions schematically, and check out what happens as smaller and smaller values of ϵ are chosen.
(b) As an amusing exercise, think up at least one other sequence of continuous functions that leads to the delta function as a limiting case.

4.2.3 *Relation Between $\delta(x)$ and $\theta(x)$*

In pictorial terms, differentiating a kink-shaped function produces a bell-shaped function. As the kink gets steeper, the bell curve gets narrower. More precisely,

$$\frac{d}{dx} \tanh\left(\frac{x}{\epsilon}\right) = \frac{1}{\epsilon} \operatorname{sech}^2\left(\frac{x}{\epsilon}\right). \tag{4.13}$$

In the limit $\epsilon \to 0$, this yields the formal relation

$$\boxed{\frac{d\varepsilon(x)}{dx} = 2\delta(x),} \tag{4.14}$$

where $\varepsilon(x)$ is the signum function. But we also have

$$\theta(x) = \tfrac{1}{2}\,[1 + \varepsilon(x)]. \tag{4.15}$$

We may therefore conclude that

$$\boxed{\frac{d\theta(x)}{dx} = \delta(x).} \tag{4.16}$$

4.2.4 *Fourier Representation of the δ-Function*

The fact that

$$\lim_{\epsilon \to 0} \frac{\sin(x/\epsilon)}{\pi x} = \delta(x) \tag{4.17}$$

leads to a most useful way of representing the delta function. If we put $\epsilon = 1/K$, we get

$$\delta(x) = \lim_{K\to\infty} \frac{\sin(Kx)}{\pi x} = \lim_{K\to\infty} \frac{e^{iKx} - e^{-iKx}}{2i\pi x}. \tag{4.18}$$

But this can be written as

$$\boxed{\delta(x) = \frac{1}{2\pi} \lim_{K\to\infty} \int_{-K}^{K} dk\, e^{ikx} = \frac{1}{2\pi} \int_{-\infty}^{\infty} dk\, e^{ikx}.} \tag{4.19}$$

This is a very important result. Since $|e^{ikx}| = 1$, it is obvious that the final integral above is not absolutely convergent. Nor is the integral well-defined in the ordinary sense, because $\sin kx$ and $\cos kx$ do not have definite limits as $kx \to \pm\infty$. These are just further reminders of the fact that the delta function is not a conventional function, as I have already emphasized. We shall study Fourier series and Fourier transforms in Chaps. 17 and 18, respectively. But if you are already familiar with Fourier transforms, you will recognize that the last equation above seems to suggest that

- the Fourier transform of the Dirac delta function is just unity.

This is indeed so, as you will see in Eq. (18.6) of Chap. 18, Sect. 18.1.3. It suggests, too, that one way of *defining* "singular" functions (or distributions) like the delta function might be *via* their Fourier transforms. For example, we could *define* $\delta(x)$ as the inverse Fourier transform of a constant—in this case, the constant is just unity.

As the delta function is a symmetric function of its argument, we have

$$\boxed{\delta(x-a) = \delta(a-x) = \frac{1}{2\pi} \int_{-\infty}^{\infty} dk\, e^{\pm ik(x-a)}.} \tag{4.20}$$

for any real number a. Note the $\pm$ sign of the exponent in the integrand: the result is valid for both signs.

4.2.5 Properties of the δ-Function

To reiterate what has already been said:

- All formulas involving $\delta(x)$ are to be understood as valid when both sides of the formula are multiplied by any suitable smooth function of x, and an integration over x is performed.

In the general context of distributions, these smooth functions are called **test functions**. In order to be mathematically precise we must also specify, for each distribution, the corresponding family or space of test functions.

Consider the delta function $\delta(ax)$ where $a\ (\neq 0)$ is a real constant. Using the properties of $\delta(x)$ already listed, it is easily shown that

$$\boxed{\delta(ax) = \frac{\delta(x)}{|a|}.} \tag{4.21}$$

In practice, we often encounter the δ-function in the form $\delta\big(g(x)\big)$, i.e., the argument of the δ-function is itself some function of x. This quantity can be given a meaning when the *real* zeroes of $g(x)$ are all *simple* zeroes, i.e, the equation $g(x) = 0$ does not have any repeated real roots. Let $x_i\ \ (i = 1,\ 2,\ \ldots)$ be the real roots of the equation $g(x) = 0$. Then, in the neighborhood of any root x_i, the function $g(x)$ has a Taylor series expansion

$$g(x) = (x - x_i)\, g\,'(x_i) + \mathcal{O}\big((x - x_i)^2\big), \tag{4.22}$$

provided $g\,'(x_i) \neq 0$, that is, x_i is a simple zero of $g(x)$. Here $g\,'(x)$ stands for the derivative of $g(x)$. Since $g\,'(x_i)$ is a constant, it follows from Eq. (4.21) that

$$\boxed{\delta\big(g(x)\big) = \sum_i \frac{\delta(x - x_i)}{|g\,'(x_i)|},} \tag{4.23}$$

where the sum runs over all the real zeroes of $g(x)$. Further:

- If $g(x)$ has no real zeroes, the integral $\int_{-\infty}^{\infty} dx\ f(x)\,\delta\big(g(x)\big) = 0$.
- If $g(x)$ has any multiple (or repeated) real zero, $\delta\big(g(x)\big)$ does not have any meaning.

A special case of the general formula (4.23) that occurs frequently is the following. If $a\ (\neq 0)$ is any real number, then

$$\boxed{\delta(x^2 - a^2) = \frac{\delta(x + a) + \delta(x - a)}{2|a|}.} \tag{4.24}$$

Note that $\delta(x^2)$ is meaningless, as the equation $x^2 = 0$ has only a double root at $x = 0$.

The derivatives of the δ-function may also be defined, as distributions. A rough argument to help you understand the nature of these distributions is as follows. Consider a suitable (bell-shaped) continuous differentiable function $\phi_\epsilon(x)$ that tends to the δ-function in the limit $\epsilon \to 0$. With increasing n, the function $d^n\phi_\epsilon(x)/dx^n$ oscillates more and more wildly between large positive values and large negative values. These oscillations are increasingly compressed into the neighborhood of $x = 0$ as $\epsilon \to 0$. The successive derivatives of the δ-function are therefore even more singular at $x = 0$ than $\delta(x)$ itself. They must be interpreted by using integration by parts a sufficient number of times, so that the derivative operator acts on the test function rather than the δ-function. Formally, $d^n\delta(x)/dx^n$ is an even or odd function of x according as n is even or odd. In particular,

$$\delta'(-x) = -\delta'(x). \tag{4.25}$$

Another useful formula is

$$x\,\delta'(x) = -\delta(x), \quad \text{or} \quad \boxed{\delta'(x) = -\frac{\delta(x)}{x}}, \tag{4.26}$$

which shows you that $\delta'(x)$ is an odd function of x, and also precisely how it is "more singular" at the origin than $\delta(x)$.

Some objects involving the δ-function may be too singular to be defined even as distributions. For instance:

- The square of the Dirac delta function, $\delta^2(x)$, does not exist as a distribution.

Thus $\delta^2(x)$ is meaningless, as far as we are concerned.

★ **5.** Show that $\int_{-\infty}^{\infty} dx\, e^{-|x|}\,\delta(\sin x) = \coth \frac{1}{2}\pi$.

★ **6.** Write $\delta\,(\sin x - \cos x)$ as a sum of δ-functions in x, i.e., find the points x_n, the coefficients c_n, and the range of values of the summation index n in the expansion

$$\delta\,(\sin x - \cos x) = \sum_n c_n\,\delta\,(x - x_n).$$

Another example of the use of the δ-function is provided by the Gauss map of the unit interval $[0\,,\,1]$, defined in Eq. (1.7) of Chap. 1, Sect. 1.2. Recall that the map is

$$f(x) = x^{-1} - [x^{-1}], \quad x \in [0, 1] \tag{4.27}$$

where $[x^{-1}]$ denotes the largest integer $\leq 1/x$. In the context of dynamical systems, it is of interest to solve the homogeneous integral equation

$$\rho(\xi) = \int_0^1 dx\,\delta\big(\xi - f(x)\big)\,\rho(x) \tag{4.28}$$

for the function $\rho(\xi)$, where the variable ξ is also in the range $[0, 1]$. (Equation (4.28) is called the **Frobenius–Perron equation.**) The solution $\rho(\xi)$ is the **invariant density** of the Gauss map. With the help of the formula (4.23), Eq. (4.28) is converted to the functional equation

$$\rho(\xi) = \sum_{n=1}^{\infty} \frac{1}{(\xi+n)^2}\, \rho\Big(\frac{1}{\xi+n}\Big). \tag{4.29}$$

This certainly does not appear to be an easy equation to solve. But its form prompts the guess that some rational function of ξ might be a possible solution. The actual solution is due to Gauss himself! It is given by

$$\rho(\xi) = \frac{1}{(1+\xi)\,\ln 2}\,. \tag{4.30}$$

The constant factor ln 2 is included in order to satisfy the normalization condition $\int_0^1 d\xi\, \rho(\xi) = 1$. It can be shown that Eq. (4.30) represents the unique nonnegative, integrable solution to the integral equation (4.28).

★ **7.** Show that, when $f(x)$ is given by the Gauss map (4.27), the integral equation (4.28) reduces to the functional equation (4.29). *Verify* that (4.30) is a solution of this equation.

★ **8.** Here are a couple of examples of multiple integrals involving products of δ-functions.

(a) Show that

$$\int_0^1 dx_1 \cdots \int_0^1 dx_n\, \delta\,(x_n - \sqrt{x_{n-1}})\,\delta\,(x_{n-1} - \sqrt{x_{n-2}}) \cdots \delta\,(x_2 - \sqrt{x_1}) = 1.$$

(b) Show that

$$\int_0^1 dx_1 \cdots \int_0^1 dx_n\, \delta\,(x_n - 2\sqrt{x_{n-1}})\,\delta\,(x_{n-1} - 2\sqrt{x_{n-2}}) \cdots \delta\,(x_2 - 2\sqrt{x_1}) = 2^{2-2^n}.$$

4.2.6 *The Occurrence of the δ-Function in Physical Problems*

Why does the δ-function appears so naturally in physical problems? Here is a familiar instance. Consider the basic problem of electrostatics: given a static charge density $\rho(\mathbf{r})$ in free space, what is the corresponding electrostatic potential $\phi(\mathbf{r})$ at any arbitrary point $\mathbf{r} = (x, y, z)$? From Maxwell's equations, we know that ϕ satisfies **Poisson's equation**, namely,

$$\nabla^2\phi(\mathbf{r}) = -\rho(\mathbf{r})/\epsilon_0\,, \tag{4.31}$$

where ϵ_0 is the permittivity of the vacuum. What does one do in the case of a *point* charge q located at some point $\mathbf{r}_0 = (x_0\,, y_0\,, z_0)$? A point charge is an idealization in which a *finite* amount of charge q is supposed to be packed into *zero* volume. The charge density must therefore be infinite at the point $\mathbf{r}_0$, and zero elsewhere. The delta function comes to our aid. We may write, in this case,

$$\rho(\mathbf{r}) = q\,\delta(x - x_0)\,\delta(y - y_0)\,\delta(z - z_0) \equiv q\,\delta^{(3)}(\mathbf{r} - \mathbf{r}_0)\,, \tag{4.32}$$

where the *three-dimensional* delta function $\delta^{(3)}$ is shorthand for the product of the three delta functions in the equation above. It is easy to verify that this expression for $\rho(\mathbf{r})$ has all the properties required of a point charge at the point $\mathbf{r}_0$.

This example illustrates how (and why) the delta function frequently appears as the right-hand side of fundamental equations of mathematical physics.

- The Dirac delta function models the density of a point source. (This density could be the charge density, or the mass density, or any other density.)
- A δ-function of the time variable t also models a so-called "unit impulse function", and hence occurs naturally in signal analysis and response theory.
- More generally, the δ-function represents the unit operator in function space, and therefore appears automatically as the right-hand side in equations satisfied by the **Green functions** of differential operators.[1]
- It turns out that the δ-function also appears as the singular part of fundamental *solutions* to basic equations such as the wave equation (as you will see in Chap. 31).

Representations of multidimensional δ-functions like $\delta^{(3)}$ are easily written down in Cartesian coordinates. For instance, the three-dimensional counterpart of the Fourier representation of Eq. (4.19) is just

$$\delta^{(3)}(\mathbf{r}) = \frac{1}{(2\pi)^3}\int_{-\infty}^{\infty} dk_1 \int_{-\infty}^{\infty} dk_2 \int_{-\infty}^{\infty} dk_3\, e^{i\,(k_1 x + k_2 y + k_3 z)}\,. \tag{4.33}$$

As in Eq. (4.20) for the one-dimensional case, the sign of each of the exponents in Eq. (4.33) can be either $+$ or $-$. In more compact notation, we therefore have

$$\boxed{\delta^{(3)}(\mathbf{r}) = \frac{1}{(2\pi)^3}\int d^3k\, e^{\pm i\,\mathbf{k}\cdot\mathbf{r}}.} \tag{4.34}$$

The notation d^3k is self-explanatory: it is the volume element in $\mathbf{k}$-space. I have mentioned already that the physical dimensions of a δ-function are those of the reciprocal of its argument. Hence $\delta^{(3)}(\mathbf{r})$ has the physical dimensions of the reciprocal of a volume, namely, $[\text{length}]^{-3}$.

---

[1]This will become clear when we consider function spaces in Chap. 13, Sect. 13.2.2. The Green functions of the most common partial differential operators of mathematical physics will be derived in Chaps. 29–32.

4.2.7 The δ-Function in Polar Coordinates

We often encounter higher dimensional δ-functions in non-Cartesian coordinates. These entail coordinate-dependent factors that multiply the individual δ-functions. The most commonly occurring among these are the two- and three-dimensional δ-functions in polar coordinates.

(a) Let (ϱ, φ) denote plane polar coordinates in the (x, y) plane. Then

$$\boxed{\delta^{(2)}(\mathbf{r} - \mathbf{r}') = \frac{1}{\varrho}\,\delta(\varrho - \varrho')\,\delta(\varphi - \varphi').} \tag{4.35}$$

(b) Let (ϱ, φ, z) denote cylindrical polar coordinates in three-dimensional space. Then

$$\boxed{\delta^{(3)}(\mathbf{r} - \mathbf{r}') = \frac{1}{\varrho}\,\delta(\varrho - \varrho')\,\delta(\varphi - \varphi')\,\delta(z - z').} \tag{4.36}$$

(c) Let (r, θ, φ) denote spherical polar coordinates in three-dimensional space. Then

$$\boxed{\delta^{(3)}(\mathbf{r} - \mathbf{r}') = \frac{1}{r^2 \sin\theta}\,\delta(r - r')\,\delta(\theta - \theta')\,\delta(\varphi - \varphi').} \tag{4.37}$$

The factor $\sin\theta$ in the denominator of Eq. (4.37) obviously comes from the simplification of $\delta(\cos\theta - \cos\theta')$. Note that the equation $\cos\theta - \cos\theta' = 0$ has a unique root given by $\theta = \theta'$ in the range $[0, \pi]$ of the polar angle. Further, $\sin\theta$ is nonnegative in this range, so that $|\sin\theta| = \sin\theta$.

★ **9.** Establish Eqs. (4.35)–(4.37).

4.3 Solutions

1. (a), (d), and (e) are different ways of representing a rectangular pulse of unit height ranging from $x = -1$ to $x = 1$. The function in (f) is a triangular pulse. ▶

3. Recall, from elementary calculus, that the indefinite integral $\int dx\, \mathrm{sech}\, x$ is easily done by changing variables to $u = e^x$. ▶

5. Pick up the contributions to the integral from all the points $x = n\pi$, where $n \in \mathbb{Z}$. The sum over n is just a geometric series. ▶

6. The argument of the delta function vanishes whenever $\tan x = 1$, or $x = (n + \frac{1}{4})\pi$, where n is any integer. Evaluating the derivative of $(\sin x - \cos x)$ at these points, we get

$$\delta\,(\sin x - \cos x) = \sum_{n=-\infty}^{\infty} \sqrt{2}\,\delta\left(x - n\pi - \tfrac{1}{4}\pi\right).$$

Each of the coefficients c_n is equal to $\sqrt{2}$ in this case. ▶

8. (a) The factor $\delta\,(x_n - \sqrt{x_{n-1}})$ can be used to perform the integration over x_n right away, because for each value of $x_{n-1} \in [0\,,\,1]$, there exists a value of $x_n \in [0\,,\,1]$. In this manner, all the integrals from $\int dx_n$ up to $\int dx_2$ can be carried out at once. This leaves just $\int_0^1 dx_1 = 1$ as the final result.
(b) You will find it helpful to sketch the graph of $x_n = 2\sqrt{x_{n-1}}$ for $x_{n-1} \in [0, 1]$. The factor $\delta\,(x_n - 2\sqrt{x_{n-1}})$ gives a nonzero contribution only as long as x_{n-1} lies in the range $[0\,,\,\frac{1}{4}]$. Hence, once the integration over x_n is carried out, the integral over x_{n-1} is confined to the range $[0\,,\,\frac{1}{4}]$. The factor $\delta\,(x_{n-1} - 2\sqrt{x_{n-2}})$ then restricts x_{n-2} to the range $[0\,,\,\frac{1}{64}]$. In this manner, the successive ranges of integration get more and more restricted, leading to the final answer quoted. ▶

Chapter 5
Vectors and Tensors

5.1 Cartesian Tensors

5.1.1 What Are Scalars and Vectors?

At school, we learn that a vector is a quantity with a magnitude and a direction—in contrast to a scalar, with which no direction is associated. We then proceed to physical examples of vectors such as velocity and force, which help us understand "intuitively" how to handle vectors.

But here is the question that should be asked immediately when one is told that a vector is a quantity "with both a magnitude and a direction." Direction with respect to what? With respect to a given, fixed set of coordinate axes prescribed once and for all? If so, why is it that no such set is ever prescribed at the start of any text on mechanics, for instance? The short answer is that relationships between vectors are valid in *every* frame of reference; and there is *no need* to specify any specific set of axes, precisely because *the way vectors change from one set of axes to another is encoded in the very definition of a vector*. I hasten to add that this short answer requires further elaboration, of course.

The fact is that the school-level "definition" quoted at the beginning of this section is seriously flawed. It does not give the true defining property of scalars and vectors; and it does not convey the fundamental need for introducing such quantities. In a nutshell:

- It turns out that the laws of physical science are unchanged in form under various choices of coordinate systems, frames of reference, etc. That is, they are **form-invariant** under various groups of transformations.
- In order to make this property manifest, these laws must be relationships between *quantities whose transformation properties are well-defined*. That is, they must be expressed in terms of **covariant quantities**.
- Scalars, vectors, tensors, etc., are precisely objects of this kind.

V. Balakrishnan, *Mathematical Physics*,
https://doi.org/10.1007/978-3-030-39680-0_5

The next question that arises is: what kinds of transformations? Several kinds of **transformations** are relevant to physics: those that involve the space–time coordinates, such as **rotations** of the spatial coordinate axes, **translations**, or shifts in the origin of the space–time coordinates, **boosts**, or transformations to moving frames of reference, etc. In addition, there are other "internal" transformations that are not induced by coordinate transformations, such as *gauge transformations*. There is a systematic way of introducing quantities with precise transformation properties under all these transformations.

- At the most elementary level, what we call scalars, vectors, and tensors are (sets of) quantities with specific transformation properties under *rotations* of the spatial coordinate axes.

Let us, therefore, restrict our attention to **Cartesian tensors** in Euclidean space of 3 dimensions (and more generally, of d dimensions). Here are the proper definitions of a scalar and a vector in this case:

(i) A **scalar** (or a tensor of rank 0) is a quantity that is unchanged under a rotation of the coordinate axes about the origin of coordinates.
(ii) A **vector** (or a tensor of rank 1) is a set of quantities (called its components) that transforms in exactly the same way as the coordinates themselves transform, under a rotation of the coordinate axes.
(iii) A **tensor** of rank ≥ 2 is a set of quantities whose transformation properties under a rotation of the coordinate axes generalize in a straightforward manner that of a vector, as we shall see shortly.

I reiterate that the precise definition of vectors, tensors, and other such objects is much more general. The space concerned need not be Euclidean, and many other groups of transformations may be considered. Of these, the most commonly occurring one is the Lorentz group of transformations in four-dimensional space–time arising from Special Relativity. More will be said about this in Chap. 9, Sect. 9.2.

5.1.2 Rotations and the Index Notation

Latin indices $i, j, k, l, \dots$ will be used to denote Cartesian components of vectors and tensors of higher rank. These indices will run over the values $1, 2, \dots, d$, where d is the dimensionality of the space concerned. We use the (Einstein) **summation convention** for repeated indices. It is an astonishingly useful and powerful notational device.

(i) If an index appears once on the left-hand side of any equation (it is then called a **free index**), it must appear exactly once on the right-hand side as well.
(ii) If an index appears twice on one side of any equation, it is called a **contracted index** or **dummy index**, and it is to be summed over the values $1, 2, \dots, d$. Therefore this index cannot appear explicitly in the result of the summation.

This is a trivial but very useful fact. In calculations, the exchange of labels or dummy indices is often helpful.

(iii) If it appears three times or more on the same side in any equation, there is a mistake somewhere!

At this level, much of vector and tensor algebra reduces to the simple task of manipulating indices correctly. In particular, you must remember to *use a fresh symbol for each distinct dummy index.*

To simplify matters, I shall loosely call 1-index objects like a_i a vector, 2-index objects like T_{ij} a tensor of rank 2, 3-index objects like S_{ijk} a tensor of rank 3, and so on—although these are, strictly speaking, just the *components* of a vector, a second-rank tensor, and a third-rank tensor, respectively.

A rotation of the coordinate axes takes a general point $\mathbf{r}$ to another point $\mathbf{r}'$, such that: (i) the origin of coordinates remains unchanged, (ii) the distance between any two points remains unchanged. Therefore:

- Every rotation of the coordinate axes about the origin in d-dimensional space (where $d \geq 2$) is a linear, homogeneous transformation of the (Cartesian) coordinates.
- Such a transformation is specified by a $(d \times d)$ **orthogonal matrix** R. In Sect. 5.1.4, you will see why any rotation matrix must be an orthogonal matrix.

That is, R satisfies the condition

$$\boxed{R R^{\mathrm{T}} = R^{\mathrm{T}} R = I,} \tag{5.1}$$

where the superscript T denotes the transpose, and I is the **unit matrix**.

Let us consider three-dimensional space, in order to be specific. (But the discussion that follows is easily generalized to any Euclidean space of d dimensions, where $d \geq 2$.) Here is one of the simplest examples of a rotation matrix in three-dimensional Euclidean space. The matrix corresponding to a rotation of the coordinate axes about the origin, in the xy-plane, and through an angle α, is given by

$$R(\alpha) = \begin{pmatrix} \cos\alpha & \sin\alpha & 0 \\ -\sin\alpha & \cos\alpha & 0 \\ 0 & 0 & 1 \end{pmatrix}. \tag{5.2}$$

It is easy to check that $R(\alpha)$ is an orthogonal matrix. A remark is in order here. We could also have called $R(\alpha)$ "a rotation about the z-axis through an angle α," because it is a rotation in three-dimensional space. The correct way to specify rotations in any number of dimensions is to specify the *plane* in which the rotation takes place, rather than the *axis* about which it occurs. This is because no such axis may exist in general, although in $d = 3$ it so happens that it always does. Basically, this is because the number of mutually orthogonal axes ($= d$) becomes equal to the number of mutually orthogonal planes ($= \frac{1}{2}d(d-1)$) only for $d = 3$. I will return to this point in Chap. 12, Sect. 12.2.2, in connection with the eigenvalues of a rotation matrix in d dimensions.

The general form of a rotation matrix R in three dimensions, corresponding to a rotation in an arbitrary plane through an arbitrary angle, will be derived in in Chap. 11, Sect. 11.3.2. (Equation (11.32) gives the matrix elements of R explicitly.) Under a general rotation, the components x_i of a point $\mathbf{r}$ change to the components x'_i of the vector $\mathbf{r}'$, given by

$$x'_i = R_{ij}\, x_j \;\; (\equiv R_{i1}\, x_1 + R_{i2}\, x_2 + R_{i3}\, x_3). \tag{5.3}$$

Here R_{ij} is the (ij)th element of the (3×3) matrix R. A summation over the values 1, 2, and 3 of the repeated index j is implied in the expression $R_{ij}\, x_j$. The definition of *any* vector $\mathbf{a}$ now follows:

- The triplet $\mathbf{a} = (a_1,\, a_2,\, a_3)$ is a vector if, under a rotation R of the coordinate axes, the new components are given by

$$a'_i = R_{ij}\, a_j\,. \tag{5.4}$$

Tensors of rank 2, 3, ... are sets of quantities that have transformation properties generalizing that for a vector. For example:

- Tensors of rank 2 and 3 transform, respectively, like

$$T'_{ij} = R_{ik}\, R_{jl}\, T_{kl} \tag{5.5}$$

and

$$S'_{ijk} = R_{il}\, R_{jm}\, R_{kn}\, S_{lmn}\,. \tag{5.6}$$

It is evident that a Cartesian tensor of rank ℓ has 3^ℓ components in three-dimensional space. In d-dimensional space, this becomes d^ℓ.

- A scalar is a single-component object, corresponding to $\ell = 0$. By definition, it remains unchanged under a rotation of the coordinate axes.

The transformation rule for a tensor of rank 2 is of special interest. Equation (5.5) can be written as

$$T'_{ij} = R_{ik}\, T_{kl}\, (R^{\mathrm{T}})_{lj} = R_{ik}\, T_{kl}\, (R^{-1})_{lj} = (RTR^{-1})_{ij}\,, \tag{5.7}$$

because the orthogonality condition on R implies that $R^{\mathrm{T}} = R^{-1}$. But we may also regard components T_{ij} of a tensor of rank 2 in d dimensions as the elements of a $(d \times d)$ matrix T. Thus the transformation rule in Eq. (5.7) can be written in the compact form

$$\boxed{T' = R\, T\, R^{-1}.} \tag{5.8}$$

In other words, T' is obtained from T by a **similarity transformation** involving R.

Rotations of the coordinate axes in Euclidean space of every dimension $d \geq 2$ form the **rotation group** in that space, as they have the following properties:

- Two rotations in succession are equivalent to a single "resultant" rotation.
- No rotation at all corresponds to the identity element of the group.
- For every rotation there is an inverse rotation that takes us back to the original orientation of the axes.

Equivalently, the *matrices* representing these rotations form a group that is isomorphic to the corresponding rotation group. (Two groups are isomorphic if there is a one-to-one correspondence between their respective elements.) The composition of two rotations corresponds to multiplying the respective matrices representing the two rotations. More will be said about rotation matrices in Chap. 11, Sect. 11.3. We will return to rotation transformations more than once in this book.

A remark is in order at this point. Representing a rotation in three-dimensional space by a (3×3) matrix *presumes* that we write the position vector of any point as a (3×1) column vector, with elements x_1, x_2, and x_3. This might appear to be obvious, but I must mention right away that there are other ways of representing the position of a point in space. In Chap. 15, Sect. 15.3.2, we will consider another important way of doing so, namely, as a (2×2) matrix. We will then need to use another way of representing rotations—as it turns out, in terms of (2×2) *unitary* matrices.

5.1.3 Isotropic Tensors

An isotropic tensor is one whose components remain unchanged in numerical value under rotations of the coordinate axes. There are only two independent isotropic Cartesian tensors in three-dimensional Euclidean space.

The Kronecker delta δ_{ij} is the first of these. Repeating Eq. (4.5) of Chap. 4, Sect. 4.2.1 for ready reference,

$$\delta_{ij} = \begin{cases} 1, & \text{if } i = j \\ 0, & \text{if } i \neq j. \end{cases} \tag{5.9}$$

Clearly, $\delta_{ij} = \delta_{ji}$, i.e., the Kronecker delta is a symmetric tensor. δ_{ij} is just the (ij)th matrix element of the (3×3) unit matrix I. It follows at once that it is an isotropic tensor, because $I' = R I R^{-1} = I$. Hence δ'_{ij} remains equal to δ_{ij} under any arbitrary rotation R of the coordinate axes. It is trivially seen that

$$\delta_{ij}\,\delta_{ik} = \delta_{jk}, \text{ and hence } \delta_{ii}\,(= \delta_{11} + \delta_{22} + \delta_{33}) = 3\,. \tag{5.10}$$

The Levi-Civita symbol (or "totally antisymmetric symbol") ϵ_{ijk} is the other isotropic tensor in three-dimensional Euclidean space. Also called the **permutation tensor**, it is defined as

$$\boxed{\epsilon_{ijk} \stackrel{\text{def.}}{=} \begin{cases} 1, & \text{if } ijk \text{ is an even permutation of } 123 \\ -1, & \text{if } ijk \text{ is an odd permutation of } 123 \\ 0, & \text{in all other cases.} \end{cases}} \tag{5.11}$$

By an "even permutation" or "odd permutation" of the natural order 123, we mean the following. Any permutation of an ordered set of objects can be decomposed into a succession of *transpositions*, in which a *pair* of objects is exchanged while the rest are left in their original positions. For example, the alphabetical order $ABCD$ can be recovered from the permutation $ACDB$ by first interchanging D and B, to get $ACBD$. Next, we interchange C and B to get $ABCD$. The number of transpositions in this case is 2, which is an even number. We therefore call $ACDB$ an even permutation of $ABCD$.

- The *number* of transpositions into which a permutation can be decomposed is not unique, but the *evenness* or *oddness* of this number is unique for every permutation.

The values of the components of ϵ_{ijk} are sometimes defined as follows. Instead of saying, "ijk is an even or odd permutation of the natural order 123," one says, "ijk is in cyclic or anticyclic order." This makes no difference *only* as long as we restrict ourselves to three dimensions. But the Levi-Civita symbol can be generalized to an arbitrary number of dimensions $d \geq 2$, as we shall see shortly. The correct definition of the Levi-Civita symbol, applicable in a space of an arbitrary number of dimensions d, is in terms of even and odd permutations of the natural order $12 \cdots d$. The definition is given in Eq. (5.29) below. Here is a trivial example to show that the definition based on "cyclic or anticyclic order" is not consistent: In 2 dimensions, $\epsilon_{12} = 1$ and $\epsilon_{21} = -1$, although 12 and 21 are both in cyclic order.[1]

In contrast to the Kronecker delta, ϵ_{ijk} changes sign if *any* two of its indices are interchanged, i.e., it is a *totally antisymmetric* tensor.

$$\epsilon_{ijk} = -\epsilon_{ikj} = \epsilon_{kij} = -\epsilon_{kji} = \epsilon_{jki} = -\epsilon_{jik} \,. \tag{5.12}$$

This rank-3 tensor has 27 components, of which only 6 are nonzero. These are $\epsilon_{123} = \epsilon_{231} = \epsilon_{312} = 1, \ \epsilon_{132} = \epsilon_{321} = \epsilon_{213} = -1$. All components in which any two indices are the same vanish identically.

★ **1.** Let S_{ijk} and A_{ijk} denote, respectively, tensors of rank three in three-dimensional Euclidean space that are totally symmetric and totally antisymmetric under the exchange of any pair of indices.

[1] I make a special mention of this point because a random sampling of standard texts shows that a sizable number of them use this definition, but (regrettably) without mentioning the fact that it does not extend to dimensions other than 3.

(a) How many *independent* components does S_{ijk} have, in general?
(b) Show that, in general, A_{ijk} has only *one* independent component.

The last result leads to an important conclusion:

- *Every* totally antisymmetric tensor A_{ijk} of rank 3 in three dimensions must necessarily be a multiple of the Levi-Civita symbol!

That is, A_{ijk} *must* be of the form $a\,\epsilon_{ijk}$. That is why we may speak of the Levi-Civita symbol as *the* totally antisymmetric tensor of rank 3, in three dimensions. We will get to the proof of the fact that ϵ_{ijk} is an isotropic tensor in a short while.

A fundamental relation between the Levi-Civita symbol and the Kronecker delta arises when one of the indices of the former is contracted. This relation is given by

$$\boxed{\epsilon_{ijk}\,\epsilon_{lmk} = \epsilon_{ikj}\,\epsilon_{lkm} = \epsilon_{kij}\,\epsilon_{klm} = \delta_{il}\,\delta_{jm} - \delta_{im}\,\delta_{jl}\,.} \tag{5.13}$$

It follows from this relation that

$$\epsilon_{ijk}\,\epsilon_{ljk} = 2\,\delta_{il}\,, \quad \text{and hence} \quad \epsilon_{ijk}\,\epsilon_{ijk} = 3! = 6. \tag{5.14}$$

★ **2.** Verify Eq. (5.13) by explicit enumeration of the components, and hence Eq. (5.14) as well.

★ **3.** Show that (i) $\epsilon_{ijk}\,\epsilon_{klm}\,\epsilon_{mni} = -\epsilon_{jln}$ (ii) $\delta_{ij}\,\delta_{jk}\,\delta_{kl}\,\delta_{li} = 3$.

Contracting a symmetric tensor with an antisymmetric one: Since δ_{ij} is a symmetric tensor and ϵ_{ijk} is antisymmetric, it follows that $\delta_{ij}\,\epsilon_{ijk} \equiv 0$. To prove this formally, first note that $\delta_{ij}\,\epsilon_{ijk} = \delta_{ji}\,\epsilon_{jik}$ because both i and j are dummy indices, and we can replace them with *any* other index symbol without changing the result; in particular, we can *exchange* the two symbols. But, having done this, we use the fact that $\delta_{ji} = \delta_{ij}$ because δ_{ij} is a symmetric tensor; while $\epsilon_{jik} = -\epsilon_{ijk}$ because ϵ_{ijk} is an antisymmetric tensor under the interchange of *any* pair of its three indices. Therefore $\delta_{ij}\,\epsilon_{ijk} = -\delta_{ij}\,\epsilon_{ijk}$. But this is only possible if this quantity vanishes identically.

The same argument remains valid for the contraction of any symmetric tensor of any rank with an antisymmetric tensor of any rank. If $S_{ij} = S_{ji}$ denotes a symmetric tensor of rank 2 and $A_{ij} = -A_{ji}$ an antisymmetric tensor of rank 2, then $S_{ij}\,A_{ij} = S_{ji}\,A_{ji} = -S_{ij}\,A_{ij} = 0$. More generally:

- Let $S_{..ij..}$ be a tensor (of any rank ≥ 2) that is symmetric in the indices i and j, and let $A_{..ij..}$ be a tensor (of any rank ≥ 2) that is antisymmetric in these two indices. Then the contraction $S_{..ij..}\,A_{..ij..}$ vanishes identically.

This property is an extremely useful one in practice.

5.1.4 Dot and Cross Products in Three Dimensions

The scalar (or dot) product and the vector (or cross) product of any two vectors **a** and **b** are very conveniently expressed using the index notation and the isotropic tensors. We have

$$\boxed{\mathbf{a}\cdot\mathbf{b} = \delta_{ij}\, a_i\, b_j = a_i\, b_i\,.} \tag{5.15}$$

For the cross product, we know that

$$\mathbf{a}\times\mathbf{b} = \mathbf{c} \quad \Rightarrow c_1 = a_2\, b_3 - a_3\, b_2,\;\; c_2 = a_3\, b_1 - a_1\, b_3,\;\; c_3 = a_1\, b_2 - a_2\, b_1\,. \tag{5.16}$$

This is compactly written, using the index notation, as[2]

$$\boxed{c_i = \epsilon_{ijk}\, a_j\, b_k\,.} \tag{5.17}$$

Let **a**, **b**, **c** be three non-coplanar vectors. As you know, the volume of the parallelepiped formed by **a**, **b**, and **c** is given, *up to an overall sign*, by the **scalar triple product** $\mathbf{a}\cdot(\mathbf{b}\times\mathbf{c})$. (See Fig. 5.1.) Moreover, this quantity remains unchanged under a cyclic permutation of the three vectors, i.e.,

$$\mathbf{a}\cdot(\mathbf{b}\times\mathbf{c}) = \mathbf{c}\cdot(\mathbf{a}\times\mathbf{b}) = \mathbf{b}\cdot(\mathbf{c}\times\mathbf{a})\,. \tag{5.18}$$

This triple product is written compactly in index notation as $\epsilon_{ijk}\, a_i\, b_j\, c_k$. In this form, its cyclic invariance follows trivially, because any pair of dummy index labels can be interchanged. Note that it can also be written in the form of a determinant, according to

$$\mathbf{a}\cdot(\mathbf{b}\times\mathbf{c}) = \epsilon_{ijk}\, a_i\, b_j\, c_k = \begin{vmatrix} a_1 & b_1 & c_1 \\ a_2 & b_2 & c_2 \\ a_3 & b_3 & c_3 \end{vmatrix} = \begin{vmatrix} a_1 & a_2 & a_3 \\ b_1 & b_2 & b_3 \\ c_1 & c_2 & c_3 \end{vmatrix}\,. \tag{5.19}$$

The **vector triple product** of the three vectors satisfies another well-known identity, namely,

$$\mathbf{a}\times(\mathbf{b}\times\mathbf{c}) + \mathbf{b}\times(\mathbf{c}\times\mathbf{a}) + \mathbf{c}\times(\mathbf{a}\times\mathbf{b}) = 0. \tag{5.20}$$

Equation (5.20) is actually a particular example of an important relationship called the **Jacobi identity** between the elements of a certain mathematical structure called a **Lie algebra**. We will encounter Lie algebras more than once in the sequel.[3]

---

[2]You should now abandon that rather misleading mnemonic for the cross product that is taught at school, involving a "determinant" in which the first row has unit vectors while the other two rows are the Cartesian components of the two vectors. It is obvious that such a hybrid cannot be a genuine determinant! It is even less so when the second row comprises differential operators, as in the case of the curl of a vector field.

[3]For instance, in Chap. 11, Sect. 11.3.1; Chap. 12, Sect. 12.4.2; and Chap. 15, Sect. 15.1.1.

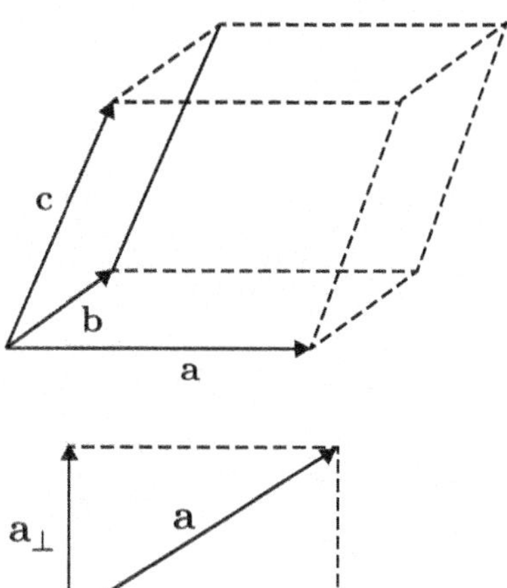

Fig. 5.1 The volume of the parallelepiped formed by three non-coplanar vectors **a**, **b**, and **c** is given by (the magnitude of) their scalar triple product

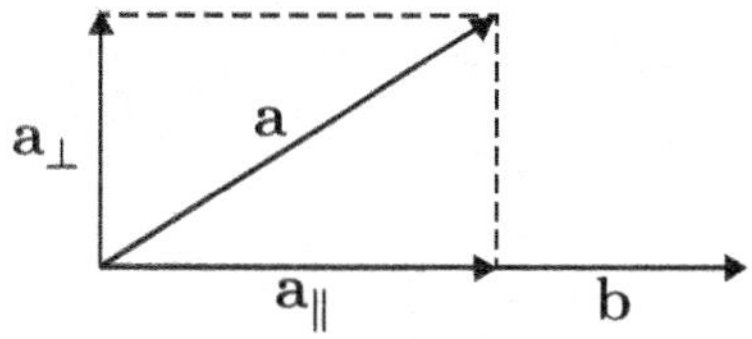

Fig. 5.2 The resolution of a vector **a** into a component along another vector **b** and a component normal to **b**

★ **4.** You can derive the standard identities of elementary vector algebra very easily with the help of the index notation. Using this notation and the summation convention, establish Eq. (5.20) as well as the following identities:

(a) $\mathbf{a} \times (\mathbf{b} \times \mathbf{c}) = (\mathbf{a} \cdot \mathbf{c})\, \mathbf{b} - (\mathbf{a} \cdot \mathbf{b})\, \mathbf{c}$
(b) $(\mathbf{a} \times \mathbf{b}) \cdot (\mathbf{c} \times \mathbf{d}) = (\mathbf{a} \cdot \mathbf{c})(\mathbf{b} \cdot \mathbf{d}) - (\mathbf{a} \cdot \mathbf{d})(\mathbf{b} \cdot \mathbf{c})$
(c) $(\mathbf{a} \times \mathbf{b}) \times (\mathbf{c} \times \mathbf{d}) = [\mathbf{a} \cdot (\mathbf{c} \times \mathbf{d})]\mathbf{b} - [\mathbf{b} \cdot (\mathbf{c} \times \mathbf{d})]\mathbf{a}$

$$= [\mathbf{d} \cdot (\mathbf{a} \times \mathbf{b})]\mathbf{c} - [\mathbf{c} \cdot (\mathbf{a} \times \mathbf{b})]\mathbf{d}$$

You will need to make frequent use of the identity in Eq. (5.13).

Resolution of a vector along and normal to another vector: Here is a very simple but useful result. Consider any two vectors **a** and **b**, as in Fig. 5.2. It is often required to resolve one of them, say **a**, into a part $\mathbf{a}_\parallel$ that is directed *along* the other vector **b** (the *longitudinal* part), and a part $\mathbf{a}_\perp$ that is *normal* to **b** (the *transverse* part). If $\mathbf{e}_b$ denotes the unit vector in the direction of **b**, this resolution is obviously given by

$$\mathbf{a} = \mathbf{a}_\parallel + \mathbf{a}_\perp = (\mathbf{e}_b \cdot \mathbf{a})\, \mathbf{e}_b + [\mathbf{a} - (\mathbf{e}_b \cdot \mathbf{a})\, \mathbf{e}_b]. \tag{5.21}$$

Using the fact that $\mathbf{e}_b = \mathbf{b}/b$ where b is the magnitude of b, this can be written in the convenient form

$$\boxed{\mathbf{a} = \frac{(\mathbf{b} \cdot \mathbf{a})\, \mathbf{b}}{b^2} + \frac{(\mathbf{b} \times \mathbf{a}) \times \mathbf{b}}{b^2}.} \tag{5.22}$$

★ **5.** Establish Eq. (5.22).

Orthogonality of rotation matrices: It is now very easy to see why an arbitrary rotation of the coordinate axes about the origin is specified by an orthogonal matrix.

Under such a rotation, the distance from the origin to any point remains unchanged. Therefore $r'^2 = r^2$, or

$$x'_i x'_i = x_j x_j = \delta_{jk} x_j x_k . \tag{5.23}$$

But we also have, as in Eq. (5.3), $x'_i = R_{ij} x_j$. Therefore

$$x'_i x'_i = R_{ij} x_j R_{ik} x_k = (R^{\mathrm{T}})_{ji} R_{ik} x_j x_k = (R^{\mathrm{T}} R)_{jk} x_j x_k . \tag{5.24}$$

The two expressions for r'^2 must be equal to each other for every point in space. Therefore we must have

$$(R^{\mathrm{T}} R)_{jk} = \delta_{jk} , \quad \text{or} \quad R^{\mathrm{T}} R = I. \tag{5.25}$$

For finite-dimensional square matrices, the left and right inverses are the same. Hence $R^{\mathrm{T}} R = I \Rightarrow R R^{\mathrm{T}} = I$.

Proof that ϵ_{ijk} is an isotropic tensor: Under a rotation of the coordinate axes that is described by an orthogonal matrix R, the Levi-Civita tensor transforms according to

$$\epsilon'_{ijk} = R_{il} R_{jm} R_{kn} \epsilon_{lmn} . \tag{5.26}$$

By direct verification it follows that, for every set of values of the indices i, j and k, we have

$$\epsilon'_{ijk} = (\det R)\, \epsilon_{ijk} . \tag{5.27}$$

But the orthogonality condition $R R^{\mathrm{T}} = I$ implies that $(\det R)^2 = 1$, or $\det R = \pm 1$. Therefore, as long as we restrict ourselves to the class of rotations corresponding to $\det R = 1$ (called *proper* rotations), the components of the Levi-Civita tensor remain unchanged under rotations. More will be said in Sects. 5.2.1 and 5.2.2 on the classification of rotations based on the sign of det R.

★ **6.** Verify that Eq. (5.26) leads to Eq. (5.27).

5.1.5 The Gram Determinant

If **a**, **b**, and **c** are three non-coplanar vectors, and no two of them are parallel, the volume of the parallelepiped formed by them cannot be zero. The converse of this statement is also true. This means that the three vectors are *linearly independent* of each other: that is, none of them can be written as a linear combination of the other two.

There is a very simple relationship between the determinant formed by the mutual scalar products of any three vectors **a**, **b**, **c** and the determinant representing the scalar triple product $\mathbf{a} \cdot (\mathbf{b} \times \mathbf{c})$ of these vectors (recall Eq. (5.19)). It is

$$G(\mathbf{a}, \mathbf{b}, \mathbf{c}) \stackrel{\text{def.}}{=} \begin{vmatrix} \mathbf{a} \cdot \mathbf{a} & \mathbf{a} \cdot \mathbf{b} & \mathbf{a} \cdot \mathbf{c} \\ \mathbf{b} \cdot \mathbf{a} & \mathbf{b} \cdot \mathbf{b} & \mathbf{b} \cdot \mathbf{c} \\ \mathbf{c} \cdot \mathbf{a} & \mathbf{c} \cdot \mathbf{b} & \mathbf{c} \cdot \mathbf{c} \end{vmatrix} = \begin{vmatrix} a_1 & b_1 & c_1 \\ a_2 & b_2 & c_2 \\ a_3 & b_3 & c_3 \end{vmatrix}^2 . \tag{5.28}$$

★ **7.** Verify Eq. (5.28).

$G(\mathbf{a}, \mathbf{b}, \mathbf{c})$ is called the **Gram determinant** of the three vectors. The last term in (5.28) is of course the square of the volume of the parallelepiped formed by the three vectors. It follows that the Gram determinant cannot be negative.

- If the three vectors **a**, **b**, and **c** are linearly independent, their Gram determinant must be strictly positive.

If any of the vectors is a linear combination of the other two, then it lies in the plane formed by these two vectors. The volume of the parallelepiped formed by the three vectors, therefore, collapses to zero. Hence so does the Gram determinant.

- The Gram determinant condition for the linear independence of a set of vectors is a general statement valid in any **linear vector space**.

We shall return to the general form of this condition in Chap. 10, Sect. 10.3.3. I mention here that the well-known Cauchy–Schwarz inequality (which will be discussed in Chap. 10) is a special case of the non-negativity of a Gram determinant.

5.1.6 Levi-Civita Symbol in d Dimensions

The definition of the Kronecker delta in Eq. (5.9) is valid as it stands in d-dimensional space for any $d \geq 2$. It remains a tensor of rank 2 in all dimensions. On the other hand, the Levi-Civita symbol in d-dimensional space is a tensor of rank d. It is defined as

$$\boxed{\epsilon_{i_1 i_2 \cdots i_d} \stackrel{\text{def.}}{=} \begin{cases} 1, & \text{if } i_1 i_2 \cdots i_d \text{ is an even permutation of } 12 \cdots d \\ -1, & \text{if } i_1 i_2 \cdots i_d \text{ is an odd permutation of } 12 \cdots d \\ 0, & \text{in all other cases} . \end{cases}} \tag{5.29}$$

$\epsilon_{i_1 i_2 \cdots i_d}$ is a completely antisymmetric tensor: it changes sign when *any* two of its indices are interchanged. The tensor has d^d components, of which all but $d!$ components are equal to zero. Of these, $\frac{1}{2}(d!)$ components are equal to 1, and $\frac{1}{2}(d!)$ components are equal to -1. In two dimensions, we have $\epsilon_{11} = \epsilon_{22} = 0$, while $\epsilon_{12} = 1$ and $\epsilon_{21} = -1$.

- The only *independent* isotropic tensors in d-dimensional Euclidean space are the Kronecker delta and the Levi-Civita symbol.
- As in the case of three-dimensional space, the Levi-Civita symbol is essentially the *only* totally antisymmetric tensor of rank d in d-dimensional Euclidean space. Every totally antisymmetric tensor of rank d in this space must be a scalar multiple of $\epsilon_{i_1 i_2 \cdots i_d}$.

★ **8.** Let ϵ_{ijkl} denote the Levi-Civita symbol in four-dimensional Euclidean space. (Hence the indices run over the values 1, 2, 3 and 4.)

(a) Show that the four-dimensional analog of Eq. (5.13) for the once-contracted product $\epsilon_{ijkl}\,\epsilon_{mnpl}$ is given by

$$\begin{aligned}\epsilon_{ijkl}\,\epsilon_{mnpl} &= \delta_{im}\,\delta_{jn}\,\delta_{kp} - \delta_{im}\,\delta_{jp}\,\delta_{kn} + \delta_{in}\,\delta_{jp}\,\delta_{km} \\ &\quad - \delta_{in}\,\delta_{jm}\,\delta_{kp} + \delta_{ip}\,\delta_{jm}\,\delta_{kn} - \delta_{ip}\,\delta_{jn}\,\delta_{km}\,.\end{aligned}$$

(b) Hence show that

$$\begin{aligned}\epsilon_{ijkl}\,\epsilon_{mnkl} &= 2!\,(\delta_{im}\,\delta_{jn} - \delta_{in}\,\delta_{jm}) \\ \epsilon_{ijkl}\,\epsilon_{mjkl} &= 3!\,\delta_{im} \\ \epsilon_{ijkl}\,\epsilon_{ijkl} &= 4!\end{aligned}$$

under successive contractions of the tensor with itself.

5.2 Rotations in Three Dimensions

Rotations in three-dimensional space are of great importance in physics, for many reasons. They occur everywhere, and on all scales, from subnuclear physics to cosmology. The algebraic and group theoretical aspects of rotations will be a recurring theme to which we will return in several places: Chap. 11, Sects. 11.1.2 and 11.3.1; Chap. 12, Sect. 12.4.2; and Chap. 15, Sects. 15.3.1 and 15.3.3. For the present, I turn to another aspect of rotations in three-dimensional space.

5.2.1 Proper and Improper Rotations

We have seen that a rotation of the coordinate axes about the origin is specified (in three-dimensional Euclidean space) by a (3×3) orthogonal matrix with real elements. These matrices form the orthogonal group, denoted by $O(3)$. Let R be a rotation matrix, i.e., a matrix whose elements tell you what linear combinations of the old coordinates yield the new coordinates. The orthogonality condition $R\,R^{\mathrm{T}} = I$ on the matrix R implies that $(\det R)^2 = 1$. Therefore $\det R = \pm 1$.

Rotations for which det $R = +1$ are called continuous or **proper rotations**. They are obtainable "continuously from the identity transformation"—that is, they can be built up by a succession of infinitesimal rotations, starting from the identity transformation (or no rotation at all). For this reason, the set of proper rotations is called the **connected component** of the rotation group. Proper rotations constitute a group of their own, the special orthogonal group $SO(3)$. (S stands for *special*, which means "unimodular" or "with unit determinant", in this context.) $SO(3)$ is a subgroup of $O(3)$. Proper rotations preserve the orientation or *handedness* of the coordinate system. That is, a right-handed coordinate system remains right-handed after a proper rotation; similarly, a left-handed one remains left-handed after a proper rotation.

In contrast, transformations with det $R = -1$ are called discontinuous or **improper rotations**. They cannot be built up continuously from the identity transformation: in general, they involve proper rotations *together with* reflections, such that a right-handed coordinate system transforms to a left-handed one or vice versa. Figure 5.3 shows what happens to the coordinate axes under a proper and improper rotation, respectively. Examples of such orientation-reversing transformations in three dimensions are the following:

(i) **Reflection** about any plane in space. For example, a reflection about the yz-plane corresponds to a transformation under which $x \mapsto -x,\ y \mapsto y,\ z \mapsto z$. In general, the plane need not be one of the three planes normal to the three Cartesian axes.
(ii) The **parity transformation** $\mathbf{r} \mapsto -\mathbf{r}$, i.e., $x \mapsto -x,\ y \mapsto -y,\ z \mapsto -z$.

Note that reversing the signs of any *two* of the three coordinates is actually a proper transformation: The determinant of the corresponding matrix remains equal to $+1$.

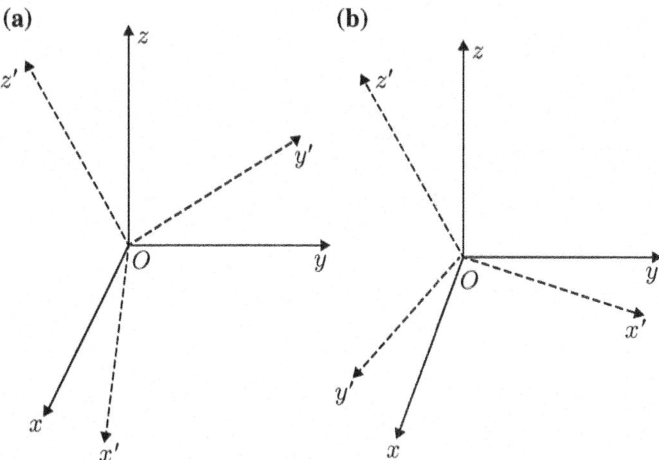

Fig. 5.3 **a** Proper and **b** improper rotation of the coordinate axes about the origin in three-dimensional space

For example, the transformation $x \mapsto -x$, $y \mapsto -y$, $z \mapsto z$ is just a rotation by an angle π in the xy-plane. *Improper rotations cannot form a subgroup of their own*, because they do not include the identity transformation. Moreover, the product of two matrices, each with with determinant equal to -1, is a matrix with determinant equal to $+1$.

The most important aspect of rotations in three dimensions is the following:

- Successive rotations do not *commute* with each other, unless they are rotations in the same plane.

In other words, the net result of two successive rotations depends on the order in which the two are carried out. This *non-commutativity* is crucial to the understanding of rotations. It has truly profound consequences for the way the physical universe is. In technical terms, the rotation group in three dimensions (in fact, in all dimensions $d \geq 3$) is a noncommutative or **non-abelian group**.

5.2.2 Scalars and Pseudoscalars; Polar and Axial Vectors

We can now make a finer distinction among scalars, depending on their transformation properties under proper and improper rotations, respectively. A true scalar is a quantity that remains unchanged under both proper and improper rotations; a **pseudoscalar**, on the other hand, remains unchanged under a proper rotation, but changes sign under an improper rotation. Similarly, the components of a vector transform like the coordinate themselves under both proper and improper rotations; a **pseudovector** behaves just like a vector under proper rotations, but has an extra change of sign under improper rotations. The same remark applies to tensors and **pseudotensors** of higher rank.

In the usual three-dimensional Euclidean space, one often uses the terms **polar vectors** and **axial vectors** for vectors and pseudovectors, respectively.

- The dot product of two polar vectors is a scalar. So is the dot product of two axial vectors.
- The dot product of a polar vector with an axial vector is a pseudoscalar.
- The cross product of two polar vectors, or that of two axial vectors, is an axial vector.
- The cross product of a polar vector and an axial vector is a polar vector.

The last two statements above follow from the fact that

- the Levi-Civita symbol ϵ_{ijk} is itself not a tensor, but rather, a pseudotensor.

This follows from Eq. (5.27), namely, $\epsilon'_{ijk} = (\det R)\, \epsilon_{ijk}$. Hence ϵ_{ijk} transforms like a tensor under proper rotations of the coordinate axes, but changes sign under an improper transformation. Examples of polar vectors include the position vector $\mathbf{r}$ of a point (naturally, since we have used this to *define* a polar vector), the velocity $\mathbf{v}$, the linear momentum **p**, the electric field **E**, etc. Common examples of axial vectors are the orbital angular momentum $\mathbf{L} = \mathbf{r} \times \mathbf{p}$ and the magnetic field **B**.

Here is an exclusive property of three-dimensional space:

- It is only in three dimensions that the "cross product" of two tensors of rank 1 leads to a (pseudo)tensor that is again of rank 1. The expression $\epsilon_{ijk}\, a_j\, b_k$ has only one free index, namely, i.
- Hence the "vector product" of two vectors, *that again yields a vector*, exists only in three dimensions!

The physical consequences of this circumstance are profound.

The generalization of the cross product: Although the *vector* product of two vectors is meaningful only in three-dimensional space, there does exist a generalization of the cross product of any two vectors **a** and **b** in a space of any dimensionality $d \geq 2$. It is defined as follows:

$$\text{The "cross product" of two vectors}\mathbf{a} \text{ and } \mathbf{b} \stackrel{\text{def.}}{=} \epsilon_{ij\cdots rs}\, a_r\, b_s\,. \tag{5.30}$$

Since two of the d indices of the Levi-Civita symbol get contracted, it is evident that what remains is a *totally antisymmetric pseudotensor* of rank $(d-2)$. As expected, this reduces to a pseudovector in the case $d = 3$. In two dimensions, the "cross product" of two vectors $\mathbf{a} = (a_1, a_2)$ and $\mathbf{b} = (b_1, b_2)$ is defined as $\epsilon_{ij}\, a_i\, b_j = a_1\, b_2 - a_2\, b_1$. This is a *pseudoscalar*, rather than a vector. The orbital angular momentum $L = x\, p_y - y\, p_x$ of a particle moving in two dimensions (i.e., in the xy-plane) is a physical example of such a quantity.

5.2.3 Transformation Properties of Physical Quantities

It is important and interesting to identify the transformation properties of all physical quantities under both proper and improper rotations. How can this be done?

- The physical input needed for this purpose is the *invariance*, under the transformations concerned, of certain relationships between these quantities.
- Ultimately, such invariance must be deduced from experimental observation.

Here is a simple example of how this argument typically works. Consider a non-relativistic particle of mass m moving under the influence of a force **F**. Newton's equation of motion (Newton's II Law) for the particle reads

$$m\frac{d\mathbf{v}}{dt} = \mathbf{F}. \tag{5.31}$$

We know that **r** is a polar vector, by definition. Therefore $\mathbf{v} \equiv d\mathbf{r}/dt$ is also a polar vector, and so is $d\mathbf{v}/dt$ as well. (Space and time are distinct and do not get "mixed up" in nonrelativistic physics.) If we now assume that *the equation of motion is invariant under both proper and improper rotations of the coordinate system* (a conclusion that is ultimately based on experimental observation), then, since m is a scalar constant, $\mathbf{F}$ must also be a polar vector. The assumption that m is a scalar is again an assumption, of course, and must be checked out for consistency. Let us take it that this has been done.

Recall the question asked at the beginning of Sect. 5.1.1, and the subsequent comments made there. It should now be clear that the invariance of Newton's equation of motion under coordinate rotations is made *manifest* by writing this equation as a relationship between vectors! If I find that $m\,d\mathbf{v}/dt = \mathbf{F}$ in my coordinate frame, and you find that the force and acceleration are $\mathbf{F}'$ and $d\mathbf{v}'/dt$, respectively, in your coordinate frame (which is tilted with respect to my set of axes), then it is *guaranteed* that $m\,d\mathbf{v}'/dt = \mathbf{F}'$. In fact, if we know precisely *how* your frame is oriented with respect to mine, we can *calculate* both $\mathbf{F}'$ and $d\mathbf{v}'/dt$ from a knowledge of $\mathbf{F}$ and $d\mathbf{v}/dt$, because these quantities are vectors.

Note, in passing, that t is the same in both frames of reference. A spatial rotation does not affect the time, of course. Further, t would continue to remain the same in both frames even if the frames were moving with a uniform velocity with respect to each other, *in Newtonian mechanics*—but not so when special relativity is brought in. This is because Newtonian mechanics corresponds to the limit in which the fundamental speed $c \to \infty$.

Returning to the deduction that $\mathbf{F}$ is a polar vector, we can use this fact to draw further conclusions. It is a manifest (but profound) fact that Newton's equation of motion remains valid for all *kinds* of forces—mechanical, electromagnetic, and so on. In particular, suppose the particle has a charge e (once again, assumed to be a scalar constant), and moves in an applied electric field $\mathbf{E}$ and magnetic field $\mathbf{B}$. The Lorentz force on it is given by the familiar expression

$$\mathbf{F} = e[\mathbf{E} + (\mathbf{v} \times \mathbf{B})]. \tag{5.32}$$

It follows that both $\mathbf{E}$ and $(\mathbf{v} \times \mathbf{B})$ must be polar vectors. But a polar vector changes its sign under the parity transformation $\mathbf{r} \mapsto -\mathbf{r}$. Since $\mathbf{v}$ itself changes sign under a parity transformation, $\mathbf{B}$ cannot do so. Hence the magnetic field $\mathbf{B}$ must be an *axial* vector, in contrast to the electric field $\mathbf{E}$. It now follows that, while $\mathbf{E}^2$ and $\mathbf{B}^2$ are scalars, $\mathbf{E} \cdot \mathbf{B}$ is a pseudoscalar; $(\mathbf{E} \times \mathbf{B})$ is again a polar vector, and so on. We can now go on to connect $\mathbf{E}$ and $\mathbf{B}$ to other physical quantities such as charge and current densities, scalar and vector potentials, etc. If these relationships are also invariant under both proper and improper rotations, we can deduce the transformation properties of those other quantities as well. I will return to electromagnetism in Chap. 9, after a discussion of vector calculus in Chaps. 6 and 8.

Another interesting improper (or discontinuous) transformation is **time reversal**, $t \mapsto -t$. The spatial independent variable $\mathbf{r}$ is unaffected by time reversal. Hence $\mathbf{v} = d\mathbf{r}/dt$ changes sign under this transformation, while the acceleration $d\mathbf{v}/dt$ does not. In the absence of dissipative forces like friction, we have reason to believe that Eq. (5.31) is time-reversal invariant. Therefore a conservative $\mathbf{F}$ does not change sign under time reversal. In turn, this means that **E** remains unchanged, while **B** changes sign, under time reversal.

A final point:

- It is important to realize that that the transformation properties of any given, well-defined physical quantity are fixed once and for all, independent of the specific circumstances in which it occurs. If it were not so, such properties would not be of much use.

For example, the electric field **E** is a polar vector that does not change sign under time reversal, regardless of whether **E** is produced by static charges (i.e., it is an electrostatic field) or by a time-varying magnetic flux (i.e., by electromagnetic induction).

★ **9.** Let $\mathbf{r}$, $\mathbf{p}$ and $\mathbf{L} = \mathbf{r} \times \mathbf{p}$ be the position, linear momentum, and angular momentum, respectively, of a particle moving in space. Consider the following quantities: (i) $\mathbf{r} \cdot \mathbf{p}$ (ii) $\mathbf{r} \times \mathbf{L}$ (iii) $\mathbf{p} \times \mathbf{L}$ (iv) $(\mathbf{r} \times \mathbf{L}) \cdot (\mathbf{p} \times \mathbf{L})$ (v) $(\mathbf{r} \times \mathbf{L}) \times (\mathbf{p} \times \mathbf{L})$.

(a) Identify the scalars, pseudoscalars, polar vectors, and axial vectors among the above.
(b) Find the behavior of each quantity under time reversal.

5.3 Invariant Decomposition of a 2nd Rank Tensor

5.3.1 Spherical or Irreducible Tensors

I conclude this chapter with a discussion of a feature of Cartesian tensors of rank 2 that is of considerable importance in applications. Numerous physical quantities are second-rank tensors. Examples include

- mechanical stress and strain;
- dielectric permittivity and magnetic permeability, and the associated electric and magnetic susceptibilities;
- the moment of inertia of a mass distribution;
- the quadrupole moment of a charge distribution;
- the Maxwell stress tensor of an electromagnetic field;
- the "order parameter" in various types of liquid crystals;

and so on. In fact, whenever two vector fields **u** and **v** are connected to each other by a linear relationship of the form $u_i = c_{ij}\, v_j$, a second-rank tensor appears naturally as the set of coefficients c_{ij}.

It turns out that second- and higher rank Cartesian tensors are *reducible*, in the following sense. Under a rotation of the coordinate axes, each component of the new tensor is a linear combination of all the components of the original tensor, in general.

However, certain *linear combinations* of components transform among themselves—that is, the primed version of each such combination involves the same components as in the original combination. These special linear combinations constitute what are called **spherical tensors**. They correspond to **irreducible representations** (or "irreps", for short) of the rotation group. A general Cartesian tensor of rank ℓ in three dimensions has 3^ℓ components. In contrast, a spherical tensor of rank ℓ only has $(2\ell + 1)$ components. For $\ell = 0$ and $\ell = 1$, the number of components is, respectively, 1 and 3 for both Cartesian and spherical tensors. But each Cartesian tensor of rank $\ell \geq 2$ is actually "made up" of *irreducible* spherical tensors of different ranks.

In the case $\ell = 2$, the break-up into irreps is as follows. Given an arbitrary Cartesian tensor T_{ij} (which can be written as a (3×3) matrix T), we can write it as the sum of a symmetric tensor and an antisymmetric tensor, according to

$$\boxed{T_{ij} = S_{ij} + A_{ij}\,,} \tag{5.33}$$

where

$$S_{ij} = \tfrac{1}{2}(T_{ij} + T_{ji}) = S_{ji} \text{ and } A_{ij} = \tfrac{1}{2}(T_{ij} - T_{ji}) = -A_{ji}\,. \tag{5.34}$$

The symmetric tensor S_{ij} has 6 independent components. The antisymmetric tensor A_{ij} has only 3 independent components, as its diagonal elements vanish identically.

- The break-up in Eqs. (5.33)–(5.34) is rotationally invariant: under an arbitrary rotation of the coordinates axes, the transformed tensors S'_{ij} and A'_{ij} remain symmetric and antisymmetric, respectively.
- The three independent components of the antisymmetric part can be identified with the components of a vector, according to

$$\mathbf{b} \equiv (A_{23},\, A_{31},\, A_{12}) \quad \text{or} \quad b_i = \epsilon_{ijk}\, A_{jk} = \epsilon_{ijk}\, T_{jk}\,. \tag{5.35}$$

  A vector is a spherical tensor of rank $\ell = 1$.
- The trace of the tensor,

$$\text{Tr } T = T_{11} + T_{22} + T_{33} = S_{11} + S_{22} + S_{33}\,, \tag{5.36}$$

  is a scalar, i.e., a spherical tensor of rank $\ell = 0$.
- When the trace part is subtracted out of the symmetric part of the tensor, we are left with a traceless symmetric tensor with 5 independent components. This is a spherical tensor of rank $\ell = 2$, given by

$$\widetilde{S}_{ij} = S_{ij} - \tfrac{1}{3}(\text{Tr } T)\,\delta_{ij} = \tfrac{1}{2}(T_{ij} + T_{ji}) - \tfrac{1}{3}(\text{Tr } T)\,\delta_{ij}\,. \tag{5.37}$$

To summarize:

- A general Cartesian tensor of rank 2 in three dimensions, T_{ij}, can be decomposed into a scalar Tr T, a vector $\epsilon_{ijk} T_{jk}$, and a traceless symmetric second-rank tensor (a spherical tensor of rank 2) $\widetilde{S}_{ij}$.

As a count of the number of components, this amounts to saying that

$$3 \times 3 = 1 + 3 + 5.$$

I will not digress here into the group-theoretic interpretation of this seemingly trivial fact.

★ **10.** The foregoing statements are easily established by using the fact that the transformation law for a second-rank tensor is given by Eq. (5.5) or (5.8).

(a) Show that the break-up of a second-rank tensor into a symmetric part and an antisymmetric part is rotationally invariant. That is, the symmetric part of the transformed tensor T'_{ij} is the transform of S_{ij}, and its antisymmetric part is the transform of A_{ij}.
(b) Show that the trace of the tensor is rotationally invariant.
(c) The fact that δ_{ij} is an isotropic tensor implies that

$$\widetilde{S}'_{ij} = S'_{ij} - \tfrac{1}{3}(\text{Tr } T')\, \delta_{ij}.$$

Check that $\widetilde{S}'_{ij}$ is also symmetric and traceless.
(d) Show that $\mathbf{b} \equiv (A_{23}, A_{31}, A_{12})$ transforms like a vector under rotations.

5.3.2 *Stress, Strain, and Stiffness Tensors*

A physical application of Cartesian tensors is provided by the theory of elasticity. The **stress tensor** $\boldsymbol{\sigma}$, with components σ_{ij}, is a symmetric tensor of rank 2. Consider a cube of the medium with its principal axes aligned along the coordinate axes. (See Fig. 5.4.) The diagonal components σ_{11}, σ_{22} and σ_{33} represent uniaxial **tension** (or compression) along each of the coordinate axes. This is, of course, the kind of stress one applies in an experiment to measure Young's modulus for a material. The off-diagonal terms σ_{12}, σ_{23} and σ_{31} represent the **shear** stresses between the respective opposite pairs of faces of the cube. The symmetry of the tensor implies that the shear components σ_{ij} and σ_{ji} are equal.

The tensor

$$\sigma^{\text{dil}}_{ij} = \tfrac{1}{3}(\sigma_{11} + \sigma_{22} + \sigma_{33})\, \delta_{ij} = \tfrac{1}{3}(\text{Tr}\, \boldsymbol{\sigma})\, \delta_{ij} \tag{5.38}$$

is called the dilatory or **hydrostatic stress**. The hydrostatic **pressure** is given by $P = -\frac{1}{3}\,\text{Tr}\, \boldsymbol{\sigma}$. Being proportional to the trace of a tensor, the pressure is a scalar, and

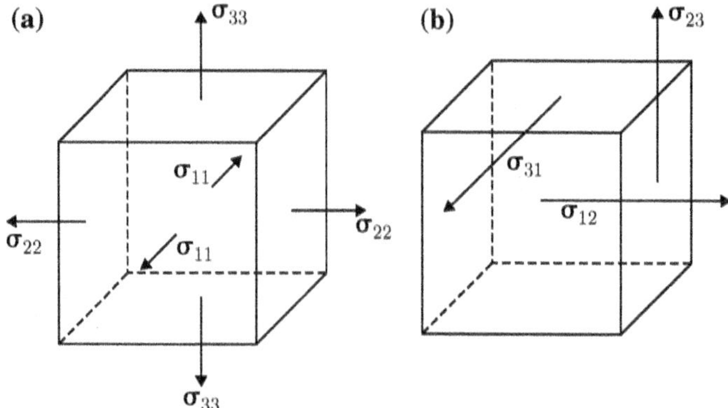

Fig. 5.4 **a** The tensile (or compressive) components of the stress tensor; **b** the shear components σ_{12}, σ_{23} and σ_{31}. For clarity, the other three shear components are not shown. In each shear, it is assumed that the face opposite the one on which the stress is applied is held fixed

hence rotation-invariant—alternatively, it is the same in all directions at any point in a fluid at rest.[4]

- This fact, in turn, leads to **Pascal's Law** and its numerous applications in hydraulics.

Subtracting out the hydrostatic stress from the stress tensor, we get a traceless tensor called the **deviatoric stress**:

$$\sigma_{ij}^{\text{dev}} = \sigma_{ij} - \sigma_{ij}^{\text{dil}} = \sigma_{ij} - \tfrac{1}{3}(\text{Tr}\,\boldsymbol{\sigma})\,\delta_{ij} = \sigma_{ij} + P\,\delta_{ij}\,. \tag{5.39}$$

Exactly the same sort of decomposition applies to the symmetric strain tensor $\boldsymbol{\varepsilon}$ (with components ε_{ij}) as well. The hydrostatic or **volumetric strain** is given by

$$\varepsilon_{ij}^{\text{dil}} = \tfrac{1}{3}\,(\text{Tr}\,\boldsymbol{\varepsilon})\,\delta_{ij}\,. \tag{5.40}$$

The **deviatoric strain** is the traceless tensor given by

$$\varepsilon_{ij}^{\text{dev}} = \varepsilon_{ij} - \varepsilon_{ij}^{\text{dil}} = \varepsilon_{ij} - \tfrac{1}{3}(\text{Tr}\,\boldsymbol{\varepsilon})\,\delta_{ij}\,. \tag{5.41}$$

[4]Recall the definition of pressure that you learnt at school: the pressure at a point in a fluid is the force per unit area on an infinitesimal area element immersed in the fluid. But forces and area elements are vectors, and it is obviously meaningless to divide by a vector. So how is one to understand the definition? The answer is that the area element may be oriented in *any* direction, and the force is along its normal. We can now understand the reason why this is so: the rotational invariance of the trace.

Hooke's Law is applicable in the regime of low strains, or linear elasticity. For a linearly elastic solid, the general form of Hooke's Law is given by

$$\sigma_{ij} = c_{ijkl}\, \varepsilon_{kl}\,, \tag{5.42}$$

where the fourth-rank tensor c_{ijkl} is called the (elastic) **stiffness tensor**. Its components represent the different elastic moduli of the medium concerned. A formal inversion of Eq. (5.42) leads to a relation of the form

$$\varepsilon_{ij} = s_{ijkl}\, \sigma_{kl}\,, \tag{5.43}$$

where s_{ijkl} is called the **compliance tensor**. At first sight, it appears that the stiffness tensor c_{ijkl} has $3^4 = 81$ independent components. But several symmetries exist, that help reduce this number considerably. Since $\sigma_{ij} = \sigma_{ji}$, and $\varepsilon_{kl} = \varepsilon_{lk}$, it follows from Eq. (5.42) that

$$c_{ijkl} = c_{jikl} = c_{ijlk} = c_{jilk}\,. \tag{5.44}$$

This reduces the number of independent components of c_{ijkl} to $6^2 = 36$.

Now, the stress is related to the force on a volume element of the medium. Recall that a conservative force field can be obtained as the gradient of a scalar potential energy. (We will define and discuss the gradient of a scalar field in Chap. 6, Sect. 6.2.1.) The gradient involves partial derivatives with respect to the displacement. The strain, in turn, is related to the displacement. It is not surprising, therefore, that the stress is also expressible as the derivative of some kind of "potential" with respect to the strain: $\sigma_{ij} = \partial\Phi/\partial\varepsilon_{ij}$, where Φ is a certain scalar function. It then follows from Hooke's Law that the compliance tensor is the second derivative of the potential, according to

$$c_{ijkl} = \frac{\partial^2\Phi}{\partial\varepsilon_{kl}\,\partial\varepsilon_{ij}}\,. \tag{5.45}$$

Since the partial derivatives can be taken in either order, we have the additional symmetry property

$$c_{ijkl} = c_{klij}\,. \tag{5.46}$$

★ **11.** Show that the symmetries implied by Eq. (5.44) reduce the number of independent components of the stiffness tensor from 81 to 36, and that those implied by Eq. (5.46) further reduce this number to $\frac{1}{2}(6 \times 7) = 21$.

Depending on the symmetries present in the medium, this number gets even further reduced. The elastic properties of crystals belonging to the various crystallographic classes have been classified. For crystals belonging to the lowest symmetry class, triclinic, the number of independent elastic moduli is nine. As the degree of symmetry increases from triclinic symmetry to cubic symmetry, the number of independent elastic moduli decreases to three. Finally,

- for a fully *isotropic* medium, the number of independent elastic moduli is just 2.

As you know, the most commonly used moduli in the case of an isotropic medium are **Young's modulus** Y, the **bulk modulus** K, the **shear modulus** G, and **Poisson's ratio**, ν. There are two independent relations between these four quantities. Recall the well-known relation between the first three of these,

$$\frac{3}{Y} = \frac{1}{G} + \frac{1}{3K}. \tag{5.47}$$

(You would have encountered this relation in an elementary course customarily called "Properties of Matter".) In terms of Poisson's ratio, one also has the relations

$$Y = 3K(1 - 2\nu) = 2G(1 + \nu). \tag{5.48}$$

Incidentally, these relations show why Poisson's ratio must always lie between $\frac{1}{2}$ (the upper bound) and -1 (the lower bound).[5] Bear in mind that Eqs. (5.47) and (5.48) are only valid for an isotropic medium.

Hooke's Law for an isotropic medium reduces to a particularly simple form:

$$\sigma_{ij} = \sigma_{ij}^{\text{dil}} + \sigma_{ij}^{\text{dev}} = 3K\, \varepsilon_{ij}^{\text{dil}} + 2G\, \varepsilon_{ij}^{\text{dev}}. \tag{5.49}$$

The volumetric strain thus represents the compression of the medium, while the deviatoric strain represents its shear. But the decomposition of the stress and strain into their dilatory and deviatoric parts is rotationally invariant, being a decomposition into irreps of the rotation group. We can therefore equate the respective dilatory and deviatoric parts in Eq. (5.49). It follows that, for an isotropic medium,

$$\boxed{P = -K\, \mathrm{Tr}\, \boldsymbol{\varepsilon} \quad \text{and} \quad \sigma_{ij}^{\text{dev}} = 2G\, \varepsilon_{ij}^{\text{dev}}.} \tag{5.50}$$

5.3.3 *Moment of Inertia*

Symmetric second-rank tensors also occur naturally in the description of mass and charge distributions.

Consider a mass distribution of volume V given by the density function $\rho(\mathbf{r})$. The moment of inertia tensor (about the origin of coordinates) is given by

$$\boxed{I_{ij} \stackrel{\text{def.}}{=} \int_V dV\, (r^2\, \delta_{ij} - x_i\, x_j)\, \rho(\mathbf{r}),} \tag{5.51}$$

[5] Yes, there do exist media with *negative* values of Poisson's ratio! Stretching such a medium in one direction causes it to bulge out in transverse directions as well. To get an idea of how this can happen, pull out a ball of loosely crumpled paper slightly by holding it at two diametrically opposite points. This is a (very) rough analogy, but it suggests how certain media comprising "loose" networks of bonds could have negative values of ν.

where the x_i are the position coordinates of the volume element dV, located at a distance r from the origin. Note that I_{ij} is a symmetric tensor (but not a traceless one). We may write the components of the tensor in the form of a symmetric matrix with real elements. These properties imply that the matrix can be diagonalized by a similarity transformation implemented by an orthogonal matrix. The physical meaning of this statement is as follows: we are guaranteed that there exists a rotation of the coordinate axes such that, in the new coordinate system, the moment of inertia tensor has only diagonal terms, with all the off-diagonal terms identically equal to zero. The new axes are called the **principal axes of inertia**, and the diagonal elements, which are denoted by I_1, I_2 and I_3, are called the **principal moments of inertia**. The rotational invariance of the trace implies that

$$I_{ii} = I_1 + I_2 + I_3 = 2 \int_V dV\, r^2\, \rho(\mathbf{r}). \tag{5.52}$$

Ellipsoid of inertia: Now, the equation

$$\frac{x^2}{a^2} + \frac{y^2}{b^2} + \frac{z^2}{c^2} = 1, \tag{5.53}$$

where a, b and c are positive constants, describes the surface of a solid figure called an *ellipsoid*. The ellipsoid is centered at the origin, with its principal axes along the coordinate axes, as shown in Fig. 5.5. Suppose the ellipsoid has a uniform mass density and a total mass M. Then, in this coordinate system, its moment of inertia tensor has only diagonal elements, given by

$$I_1 = \tfrac{1}{5}M\,(b^2 + c^2), \quad I_2 = \tfrac{1}{5}M\,(c^2 + a^2), \quad I_3 = \tfrac{1}{5}M\,(a^2 + b^2). \tag{5.54}$$

- Since the moment of inertia tensor of *any* mass distribution can be diagonalized, in the principal axes system the mass distribution is *effectively* that of an ellipsoid. The latter is called the **ellipsoid of inertia** of the mass distribution.

The equality of two (or all three) of the principal moments of inertia is an indication of additional special symmetries in the mass distribution. For instance, if $a = b > c$,

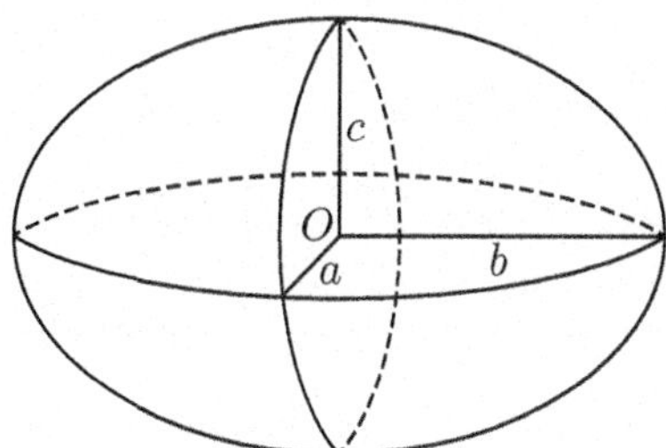

Fig. 5.5 The ellipsoid of inertia for a mass distribution

the ellipsoid becomes an **oblate spheroid** (which looks like a sphere flattened at the poles). This figure may be obtained by rotating the ellipse

$$\frac{x^2}{a^2} + \frac{z^2}{c^2} = 1 \tag{5.55}$$

about its minor axis (the z-axis). In this case $I_1 = I_2 = \frac{1}{5}M\,(a^2 + c^2)$, while $I_3 = \frac{2}{5}Ma^2$. Hence $I_1 = I_2 < I_3$. On the other hand, if $a > b = c$, we get a **prolate spheroid** (shaped like a rugby football) that is obtained by rotating the same ellipse about its major axis (the x-axis). In this case $I_1 = \frac{2}{5}Mc^2$ and $I_2 = I_3 = \frac{1}{5}M\,(a^2 + c^2)$. Hence $I_1 < I_2 = I_3$.

A transformation to the principal axes frame is very advantageous when we consider the rotational dynamics of a rigid body. Here is a short digression.

5.3.4 *The Euler Top*

The equations of motion of a rigid body rotating freely about a fixed point (the so-called **Euler top**) take on a particularly simple form in the principal axes frame. Let $\boldsymbol{\omega}$ denote the instantaneous angular velocity of the body. Then its components satisfy the equations of motion (called Euler's equations)

$$I_1\,\dot\omega_1 = (I_2 - I_3)\,\omega_2\omega_3,\;\; I_2\,\dot\omega_2 = (I_3 - I_1)\,\omega_3\omega_1,\;\; I_3\,\dot\omega_3 = (I_1 - I_2)\,\omega_1\omega_2, \tag{5.56}$$

where an overhead dot denotes the time derivative. These coupled nonlinear equations constitute an **integrable dynamical system** with interesting solutions. Observe that, in the general case in which I_1, I_2 and I_3 are all unequal, it is not possible for all three right-hand sides in Eqs. (5.56) to have the same sign. This fact has implications for the stability of rotational motion about the different principal axes—specifically, the so-called **tennis racquet theorem**. (Check it out in a text on mechanics.)

★ **12.** For the dynamical system specified by the equations of motion (5.56), show that there are two independent quadratic functions of the ω_i, call them $F_1(\omega_1, \omega_2, \omega_3)$ and $F_2(\omega_1, \omega_2, \omega_3)$, that are constants of the motion, i.e., they remain at their initial values as time elapses.

Equation (5.56) are three coupled first-order differential equations for the dynamical variables ω_1, ω_2 and ω_3. The phase space of this dynamical system is therefore three-dimensional. For any given set of initial values of the ω_i, the intersection of the (two-dimensional) surfaces $F_1 =$ constant and $F_2 =$ constant is one-dimensional. This intersection must therefore correspond to the phase trajectory of the system. Can there be a third, independent, time-independent constant of the motion in this case?

No, because the intersection of three surfaces is generically just a *point* (or discrete set of points), and that would imply that no motion occurs.

5.3.5 Multipole Expansion; Quadrupole Moment

The multipole expansion of the potential due to a general charge distribution in electrostatics provides another example of the use of spherical tensors.

Let $\phi(\mathbf{r})$ be the electrostatic potential at the point $\mathbf{r}$ due to a static charge distribution specified by a charge density $\rho(\mathbf{r}')$ in space. Now, it turns out that the basic partial differential equation satisfied by the potential is Poisson's equation, as you will see in Eq. (9.16) of Chap. 9, Sect. 9.1.5. This equation must be augmented with appropriate boundary conditions to deduce any specific solution of the equation. For our present purposes, let us assume the natural boundary condition $\phi(\mathbf{r}) \to 0$ as $r \to \infty$ along any direction in space. I will return to the formal solution of Poisson's equation in Chap. 29, Sect. 29.3.1 to Sect. 29.3.3. Here, I shall merely write down the solution based on Coulomb's Law plus the superposition principle. The latter is applicable because Poisson's equation is a linear equation for ϕ. The solution is

$$\boxed{\phi(\mathbf{r}) = \frac{1}{4\pi\epsilon_0}\int d^3r'\,\frac{\rho(\mathbf{r}')}{|\mathbf{r}-\mathbf{r}'|}}\,. \tag{5.57}$$

Observe that the boundary condition $\phi \to 0$ as $r \to \infty$ is implicit in this expression.

For simplicity, and so that we need not worry about questions of convergence, let us assume that the distribution is compact: that is, all charges are restricted to some finite volume V about the origin of coordinates (see Fig. 5.6). Now consider

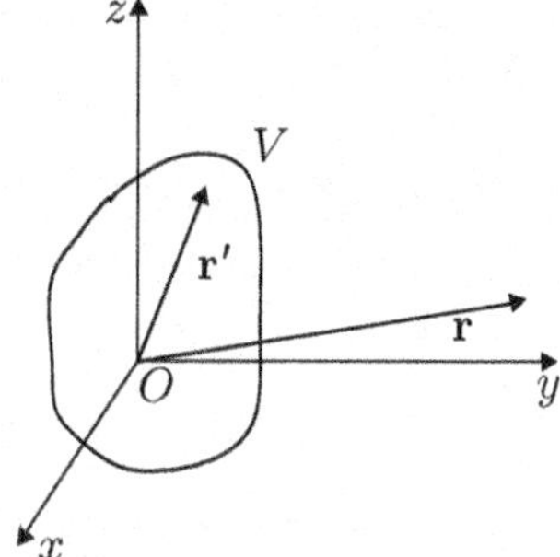

Fig. 5.6 Potential at **r** due to a charge distribution in V

the case when $r > r'$ for all $\mathbf{r}' \in V$. The **multipole expansion** of the potential is an expansion of $\phi(\mathbf{r})$ in inverse powers of r. It reads as follows[6]:

$$\boxed{\phi(\mathbf{r}) = \frac{1}{4\pi\epsilon_0}\left\{\frac{Q}{r} + \frac{P_i\, x_i}{r^3} + \frac{Q_{ij}\, x_i\, x_j}{r^5} + \frac{Q_{ijk}\, x_i\, x_j\, x_k}{r^7} + \ldots\right\},} \tag{5.58}$$

where the coefficients $Q, P_i, Q_{ij}, Q_{ijk}, \ldots$ of the successive terms are the **multipole moments** corresponding to the charge distribution.

- The multipole moments of a charge distribution are spherical tensors of rank $\ell = 0, 1, 2, \ldots$ that involve, respectively, the zeroth, first, second, ... moments of the charge density.

The first of these is the **monopole moment** or total charge. This is a scalar:

$$\boxed{Q = \int_V d^3r'\, \rho(\mathbf{r}').} \tag{5.59}$$

The second coefficient is the **dipole moment** of the distribution. This is a vector with components

$$\boxed{P_i = \int_V d^3r'\, x'_i\, \rho(\mathbf{r}').} \tag{5.60}$$

The third coefficient is the **quadrupole moment**. This is a symmetric traceless tensor of rank 2 with components

$$\boxed{Q_{ij} = \tfrac{1}{2}\int_V d^3r'\, \left(3x'_i x'_j - r'^2\, \delta_{ij}\right)\, \rho(\mathbf{r}').} \tag{5.61}$$

Very far away from the charge distribution, the potential looks like that of a point charge Q located at the origin, to leading order.

- The dipole, quadrupole and all higher moments of a spherically symmetric charge distribution vanish identically.
- The successive terms in the expansion of Eq. (5.58) represent the effects of the *departure from spherical symmetry* of the charge distribution.

★ **13.** The multipole moments defined above are the moments about the origin of coordinates. Except for the total charge, the others are dependent, in general, on the specific choice of the origin. It turns out, however, that a given moment is independent of the location of the origin if all the lower moments vanish identically.

[6]In Chap. 16, Sect. 16.4.8, I will discuss a related aspect: the expansion of the so-called **Coulomb kernel** $|\mathbf{r} - \mathbf{r}'|^{-1}$ in spherical harmonics, i.e., its "factorization" in terms of functions of the spherical polar coordinates of $\mathbf{r}$ and $\mathbf{r}'$, respectively. See Eq. (16.139).

(a) Show that the dipole moment **P** is unchanged under a shift of the origin of coordinates if the total charge $Q = 0$.
(b) Show that the quadrupole moment Q_{ij} is unchanged under a shift of the origin of coordinates if $Q = 0$ and $\mathbf{P} = 0$.

5.3.6 *The Octupole Moment*

Textbook discussions of electrostatics usually do not go beyond the quadrupole term in the multipole expansion of the potential. Let us therefore go a step further and consider the next term (the **octupole** term) in the expansion of the potential in Eq. (5.58).

What can the third-rank tensor Q_{ijk} possibly be? It must essentially involve the third moment of the charge density, given by

$$\int_V d^3r'\, x'_i x'_j x'_k\, \rho(\mathbf{r}'). \tag{5.62}$$

The third-rank tensor in (5.62) has $3^3 = 27$ components. On the other hand, the octupole moment must be an irreducible spherical tensor of rank $\ell = 3$. It must therefore have only $2\ell + 1 = 7$ *independent* components. The 27 components of the Cartesian tensor get reduced to the 7 independent components of the spherical tensor as described below.

The factor $x_i\, x_j\, x_k$ with which Q_{ijk} is contracted in Eq. (5.58) is totally symmetric in all three of its indices, i.e., under the exchange of any pair of indices. Hence Q_{ijk} must also be a totally symmetric tensor of rank 3. This symmetry is already manifest in the expression in (5.62). But you have already seen (in Sect. 5.1.3) that such a tensor has only 10 independent components. The final reduction from 10 to 7 independent components arises as follows. We can contract any two of the three indices in (5.62), to produce a vector. If i and j are contracted, for instance, we have $x'_i x'_i x'_k = r'^2 x'_k$, with one free index—that is, a vector. (The factor r'^2 is a scalar.) This vector part, which has 3 components, must be subtracted appropriately from the tensor, to get a 7-component spherical tensor of rank $\ell = 3$. Clearly, one must allow for the contraction of all three pairs of indices (i and j, j and k, k and i) in a symmetric fashion, so as to preserve the totally symmetric nature of the tensor. Moreover, the subtraction must be such that the contraction of any two of the three indices in the resultant tensor makes the tensor vanish. The irreducible tensor we seek is then given by

$$\boxed{Q_{ijk} = \tfrac{1}{2}\int_V d^3r' \left[5x'_i x'_j x'_k - r'^2\left(x'_i\,\delta_{jk} + x'_j\,\delta_{ki} + x'_k\,\delta_{ij}\right)\right]\rho(\mathbf{r}').} \tag{5.63}$$

It should be clear that the *odd*-rank tensor Q_{ijk} cannot contain any scalar part, which corresponds to the even value $\ell = 0$.

★ **14.** Following the argument outlined above, derive Eq. (5.63).

5.4 Solutions

1. (a) The independent (i.e., independently specifiable) elements of the totally symmetric tensor S_{ijk} can be classified as follows: those with (i) all three indices the same, (ii) two indices the same, and (iii) all three indices different. Since the tensor is totally symmetric, any component is unchanged in value under a permutation of the indices. Hence there are $3 + (3 \times 2) + 1 = 10$ *independent* elements. One possible set of these is given by

$$S_{111}, S_{222}, S_{333}\,;\; S_{112}, S_{113}, S_{221}, S_{223}, S_{331}, S_{332}\,;\; S_{123}\,.$$

Specifying S_{123}, for instance, also specifies the elements S_{132}, S_{213}, S_{231}, S_{312} and S_{321}, since they are all equal to each other by the symmetry of the tensor.

(b) In the case of the totally antisymmetric tensor A_{ijk}, it is clear that all elements in which any two indices are the same vanish identically. This leaves six nonzero elements. But once we specify any one of these, say A_{123}, the other five are also specified: we have

$$A_{123} = A_{312} = A_{231} = -A_{132} = -A_{213} = -A_{321}\,.$$

Hence the number of *independent* components of A_{ijk} is just 1. ▶

8. (a) It is evident that the first term must be $\delta_{im}\,\delta_{jn}\,\delta_{kp}$. Consider the array $\left(\begin{smallmatrix} i & j & k \\ m & n & p \end{smallmatrix}\right)$. Permute the symbols in the lower row among themselves, changing sign with every transposition. ▶

9. (a) The quantities in (i) and (iv) are scalars. Those in (ii) and (iii) are polar vectors, while that in (v) is an axial vector.

(b) Under time reversal, the quantity in (iii) does not change sign, while all the others do. ▶

10. (a) and (c) involve straightforward verification.

(b) is trivially demonstrated, since $T\,'$ is obtained from T by a similarity transformation. Use the cyclic property of the trace of a product of (finite-dimensional) matrices.

(d) It is evident that the components of $\mathbf{b}'$ are obtained from the components of $\mathbf{b}$ by a linear homogeneous transformation. Now start with

$$b'_i = \epsilon_{ijk}\, A'_{jk} = \epsilon_{ijk}\, R_{jl}\, R_{km}\, A_{km}$$

(since ϵ_{ijk} is an isotropic tensor). Use the orthogonality property of the rotation matrix R to show that $b'_i\, b'_i = b_i\, b_i$. Hence the components b_i transform like the coordinates themselves, which means that $\mathbf{b}$ is a vector. ▶

12. It follows from Eqs. (5.56) by inspection that the time derivatives of the following quantities vanish identically:

$$F_1 = I_1\,\omega_1^2 + I_2\,\omega_2^2 + I_3\,\omega_3^2 \quad \text{and} \quad F_2 = I_1^2\,\omega_1^2 + I_2^2\,\omega_2^2 + I_3^2\,\omega_3^2\,.$$

Observe that F_2 is the square of the total angular momentum of the body about the origin. ▶

13. Let the origin of coordinates be shifted to the point $\mathbf{a}$, where $\mathbf{a}$ is a constant vector. A point located at $\mathbf{r}'$ in the original coordinate system is now located at $\mathbf{r}'' = \mathbf{r}' - \mathbf{a}$. Set $\mathbf{r}' = \mathbf{r}'' + \mathbf{a}$ in the formulas (5.60) and (5.61), and calculate the new moments. ▶

14. In order to arrive at the overall factor of $\frac{1}{2}$ on the right-hand of (5.63), use the simple trick of looking at a special case! Instead of a charge distribution, take a unit point charge located at any point $\mathbf{r}'$ in space, other than the z-axis. Let $\mathbf{r}$ be a point on the positive z-axis, so that $\mathbf{r} = (0, 0, r)$, such that $r > r'$. Now expand the "Coulomb kernel"

$$|\mathbf{r} - \mathbf{r}'|^{-1} = \left(r^2 - 2\mathbf{r}\cdot\mathbf{r}' + r'^2\right)^{-1/2} = \left(r^2 - 2rr'\,\cos\,\theta' + r'^2\right)^{-1/2}$$

in powers of (r'/r) using the binomial theorem, and pick out the coefficient of the term proportional to (r'^3/r^4). In Chap. 16, Sect. 16.4.1, you will recognize that the quantity $\frac{1}{2}(5\cos^3\theta' - 3\cos\theta')$ is just the Legendre polynomial $P_3(\cos\theta')$. ▶

Chapter 6
Vector Calculus

In this chapter, we recapitulate the basic ideas of vector calculus in the familiar Euclidean space of three dimensions. As you are likely to be familiar with a good portion of this material, the discussion will be rather brief for the most part. Attention will be restricted, as in Chap. 5, to ordinary vectors under rotations of (3-dimensional) Euclidean space. I shall not consider curved spaces, covariant and contravariant vectors, etc., or discuss the differential geometric approach.

6.1 Orthogonal Curvilinear Coordinates

Orthogonal coordinate systems are those in which the basis vectors are normal to each other. The basic example is the Cartesian coordinate system, in which $(\mathbf{e}_x, \mathbf{e}_y, \mathbf{e}_z)$ comprise an orthogonal basis. Curvilinear coordinate systems are those in which the basis vectors are position-dependent, i.e., their directions do not remain the same at all points in space, unlike the Cartesian basis. However, at each point, the basis vectors form a mutually orthogonal triad.

6.1.1 Cylindrical and Spherical Polar Coordinates

Let us recall some elementary facts about the most common curvilinear coordinate systems. I will use standard notation:

$\mathbf{r}$ for the position vector of an arbitrary point P in 3-dimensional Euclidean space;
(x, y, z) for its Cartesian coordinates;
(ϱ, φ) for plane polar coordinates in the (x, y) plane;
(ϱ, φ, z) for its cylindrical polar coordinates;
(r, θ, φ) for its spherical polar coordinates, with the z-axis as the polar axis.

Referring to Fig. 6.1, we have the familiar relations

V. Balakrishnan, *Mathematical Physics*,
https://doi.org/10.1007/978-3-030-39680-0_6

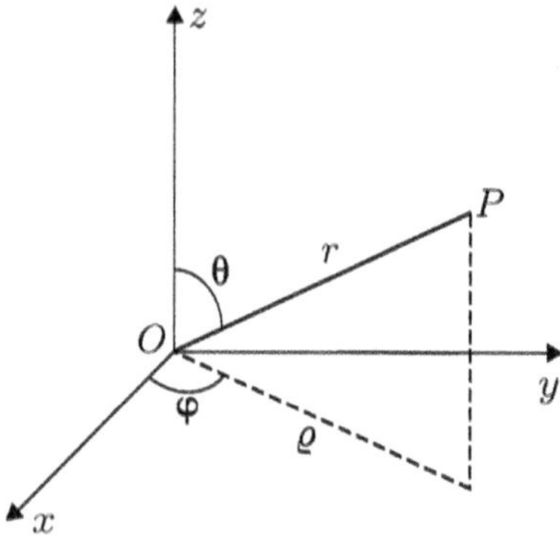

Fig. 6.1 Plane, cylindrical, and spherical polar coordinates

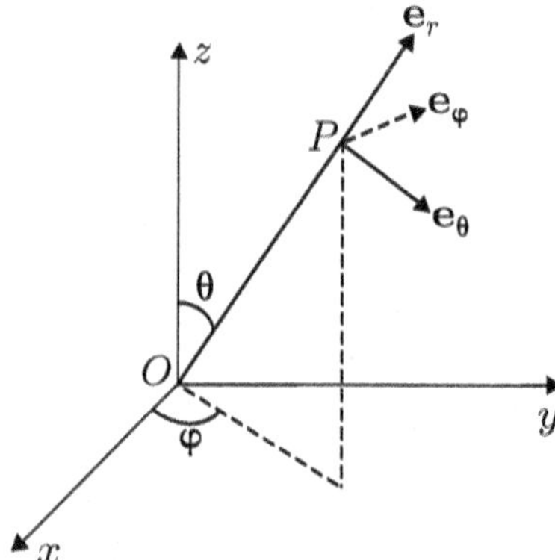

Fig. 6.2 Unit vectors in spherical polar coordinates

$$x = \varrho \cos\varphi = r \sin\theta \cos\varphi, \quad y = \varrho \sin\varphi = r \sin\theta \sin\varphi, \quad z = r \cos\theta. \tag{6.1}$$

The ranges of the non-Cartesian coordinates are

$$0 \le \varrho < \infty\,,\; 0 \le \varphi < 2\pi\,,\; 0 \le r < \infty\,,\; 0 \le \theta \le \pi. \tag{6.2}$$

The right-handed unit vector triads in the Cartesian, cylindrical polar, and spherical polar systems are $(\mathbf{e}_x, \mathbf{e}_y, \mathbf{e}_z)$, $(\mathbf{e}_\varrho, \mathbf{e}_\varphi, \mathbf{e}_z)$ and $(\mathbf{e}_r, \mathbf{e}_\theta, \mathbf{e}_\varphi)$, respectively. This implies, for instance, that

$$\mathbf{e}_r \times \mathbf{e}_\theta = \mathbf{e}_\varphi\,, \quad \mathbf{e}_\theta \times \mathbf{e}_\varphi = \mathbf{e}_r\,, \quad \mathbf{e}_\varphi \times \mathbf{e}_r = \mathbf{e}_\theta\,. \tag{6.3}$$

Figure 6.2 shows the unit vectors at an arbitrary point in spherical polar coordinates. The position vector of any point is given by

$$\boxed{\mathbf{r} = x\,\mathbf{e}_x + y\,\mathbf{e}_y + z\,\mathbf{e}_z = \varrho\,\mathbf{e}_\varrho + z\,\mathbf{e}_z = r\,\mathbf{e}_r\,.} \tag{6.4}$$

It is important to recognize that the unit vectors $(\mathbf{e}_x, \mathbf{e}_y, \mathbf{e}_z)$ in Cartesian coordinates are *constant* vectors, i.e., they are not functions of the coordinates themselves. In marked contrast,

- the unit vectors in curvilinear coordinate systems are functions of the coordinates, as the *directions* of these vectors differ from point to point.

★ **1.** Find all the partial derivatives listed below.

(a) $\dfrac{\partial \mathbf{e}_\varrho}{\partial \varrho}, \dfrac{\partial \mathbf{e}_\varrho}{\partial \varphi}, \dfrac{\partial \mathbf{e}_\varrho}{\partial z}, \dfrac{\partial \mathbf{e}_\varphi}{\partial \varrho}, \dfrac{\partial \mathbf{e}_\varphi}{\partial \varphi}, \dfrac{\partial \mathbf{e}_\varphi}{\partial z}.$

(b) $\dfrac{\partial \mathbf{e}_r}{\partial r}, \dfrac{\partial \mathbf{e}_r}{\partial \theta}, \dfrac{\partial \mathbf{e}_r}{\partial \varphi}, \dfrac{\partial \mathbf{e}_\theta}{\partial r}, \dfrac{\partial \mathbf{e}_\theta}{\partial \theta}, \dfrac{\partial \mathbf{e}_\theta}{\partial \varphi}, \dfrac{\partial \mathbf{e}_\varphi}{\partial r}, \dfrac{\partial \mathbf{e}_\varphi}{\partial \theta}, \dfrac{\partial \mathbf{e}_\varphi}{\partial \varphi}.$

(c) Express the unit vector $\mathbf{e}_\varrho$ in spherical polar coordinates.

The **line element** $d\mathbf{r}$ (sometimes also written as $d\boldsymbol{\ell}$) is given by

$$\left.\begin{aligned} d\mathbf{r} &= (dx)\,\mathbf{e}_x + (dy)\,\mathbf{e}_y + (dz)\,\mathbf{e}_z \\ &= (d\varrho)\,\mathbf{e}_\varrho + (\varrho\, d\varphi)\,\mathbf{e}_\varphi + (dz)\,\mathbf{e}_z \\ &= (dr)\,\mathbf{e}_r + (r\, d\theta)\,\mathbf{e}_\theta + (r\, \sin\theta\, d\varphi)\,\mathbf{e}_\varphi\,. \end{aligned}\right\} \tag{6.5}$$

The square of the distance between the points $\mathbf{r}$ and $\mathbf{r} + d\mathbf{r}$ is

$$\left.\begin{aligned} (ds)^2 &= (dx)^2 + (dy)^2 + (dz)^2 \\ &= (d\varrho)^2 + \varrho^2\,(d\varphi)^2 + (dz)^2 \\ &= (dr)^2 + r^2\,(d\theta)^2 + r^2\,\sin^2\theta\,(d\varphi)^2\,. \end{aligned}\right\} \tag{6.6}$$

The **gradient operator** ∇, sometimes referred to as the del operator, is a *vector* differential operator, given by

$$\left.\begin{aligned} \nabla &= \mathbf{e}_x \frac{\partial}{\partial x} + \mathbf{e}_y \frac{\partial}{\partial y} + \mathbf{e}_z \frac{\partial}{\partial z} \\ &= \mathbf{e}_\varrho \frac{\partial}{\partial \varrho} + \frac{\mathbf{e}_\varphi}{\varrho} \frac{\partial}{\partial \varphi} + \mathbf{e}_z \frac{\partial}{\partial z} \\ &= \mathbf{e}_r \frac{\partial}{\partial r} + \frac{\mathbf{e}_\theta}{r} \frac{\partial}{\partial \theta} + \frac{\mathbf{e}_\varphi}{r\,\sin\theta} \frac{\partial}{\partial \varphi}\,. \end{aligned}\right\} \tag{6.7}$$

It is very convenient to write the gradient operator in Cartesian coordinates as $\nabla = (\partial/\partial x_1\,, \partial/\partial x_2\,, \partial/\partial x_3) \equiv (\partial_1\,, \partial_2\,, \partial_3)$, so that

- the del operator is very compactly represented in index notation by ∂_i.

6.1.2 Elliptic and Parabolic Coordinates

While plane polar, cylindrical polar and spherical polar coordinates are the most frequently used curvilinear orthogonal coordinate systems, there exist numerous other systems of orthogonal coordinates. These are useful in exploiting specific symmetries (other than cylindrical or spherical symmetry) that may be present in some cases. A few of the general results are quoted below, for completeness.

The starting point is the expression for $(ds)^2$, the square of the distance between the points $\mathbf{r}$ and $\mathbf{r} + d\mathbf{r}$. Let (ξ_1, ξ_2, ξ_3) denote the orthogonal coordinates. Then $(ds)^2$ can be written in the form

$$\boxed{(ds)^2 = h_1^2\,(d\xi_1)^2 + h_2^2\,(d\xi_2)^2 + h_3^2\,(d\xi_3)^2\,,} \tag{6.8}$$

where the quantities h_i are functions of the coordinates, and are called **scale factors**. The gradient operator is given by

$$\boxed{\nabla = \frac{\mathbf{e}_\xi^{(1)}}{h_1}\frac{\partial}{\partial \xi_1} + \frac{\mathbf{e}_\xi^{(2)}}{h_2}\frac{\partial}{\partial \xi_2} + \frac{\mathbf{e}_\xi^{(3)}}{h_3}\frac{\partial}{\partial \xi_3}\,,} \tag{6.9}$$

where $(\mathbf{e}_\xi^{(1)}, \mathbf{e}_\xi^{(2)}, \mathbf{e}_\xi^{(3)})$ is the unit vector basis in the (ξ_1, ξ_2, ξ_3) coordinate system. In the Cartesian coordinate system, the scale factors (h_1, h_2, h_3) are obviously $(1, 1, 1)$. In cylindrical and spherical polar coordinates they are $(1, \varrho, 1)$ and $(1, r, r\,\sin\,\theta)$, respectively.

The reduction of such formulas to the case of two dimensions is straightforward. Here are a couple of examples of curvilinear coordinate systems in the plane.

Elliptic coordinates (u, v) in the xy-plane are given in terms of the Cartesian coordinates by

$$x = a\,(\cosh\,u)\,(\cos\,v) \quad \text{and} \quad y = a\,(\sinh\,u)\,(\sin\,v), \tag{6.10}$$

where $0 \leq u < \infty\,,\; 0 \leq v < 2\pi$, and a is a positive constant.

★ **2.** The curves $u =$ constant are ellipses, while the curves $v =$ constant are hyperbolas. They comprise two orthogonal families of curves (i.e., they intersect each other at right angles).

(a) Verify the statement made above. Observe that the pair of values $(v, 2\pi - v)$ together comprise a branch of a hyperbola. The values $v = 0, \frac{1}{2}\pi, \pi, \frac{3}{2}\pi$, and 2π are obviously limiting values when the hyperbola degenerates to straight line segments.

(b) Sketch the curves

(i) $u = 0$ (ii) $u = 1$ (iii) $v = 0, 2\pi$

(iv) $v = \frac{1}{4}\pi, \frac{3}{4}\pi$ (v) $v = \frac{1}{2}\pi, \frac{3}{2}\pi$ (vi) $v = \pi$.

(c) Show that the scale factors are $h_1 = h_2 = a(\sinh^2\,u + \sin^2\,v)^{1/2}$.

Parabolic coordinates (ξ, η) in the xy-plane are given in terms of the Cartesian coordinates by

$$x = \tfrac{1}{2}(\xi^2 - \eta^2) \quad \text{and} \quad y = \xi\eta, \tag{6.11}$$

where $-\infty < \xi < \infty$ and $0 \le \eta < \infty$.

★ **3.** The curves $\xi =$ constant and the curves $\eta =$ constant are two mutually orthogonal families of parabolas, with the origin as the common focus and the x-axis as the common axis of all the parabolas.

(a) Verify the statement made above.
(b) Sketch the curves

$$\text{(i) } \eta = 0 \quad \text{(ii) } \xi = 0 \quad \text{(iii) } \xi = \eta \quad \text{(iv) } \xi = -\eta$$

$$\text{(v) } \xi = 1 \quad \text{(vi) } \xi = -1 \quad \text{(vii) } \eta = 1.$$

(c) Show that the scale factors are $h_1 = h_2 = (\xi^2 + \eta^2)^{1/2}$.
(d) This system of coordinates can be extended to three-dimensional parabolic coordinates (ξ, η, φ) by rotating the parabolas about the x-axis (their common axis). We now have

$$x = \tfrac{1}{2}(\xi^2 - \eta^2), \quad y = \xi\eta \cos\varphi, \quad z = \xi\eta \sin\varphi.$$

Show that the scale factors are now $h_1 = h_2 = (\xi^2 + \eta^2)^{1/2}, \; h_3 = \xi\eta$.

6.1.3 Polar Coordinates in d Dimensions

The extension of "spherical" polar coordinates to an arbitrary number of dimensions $d > 3$ is useful in many applications. I, therefore, digress briefly to discuss this generalization.

Let $(x_1, x_2, \ldots, x_d)$ be the Cartesian coordinates of a point in d-dimensional Euclidean space. The "ultraspherical" polar coordinates comprise (i) the radial variable r, (ii) a set of $(d-2)$ "polar" angles $\theta_1, \ldots, \theta_{d-2}$, and (iii) an azimuthal angle φ. The ranges of these variables are

$$\boxed{0 \le r < \infty, \quad 0 \le \theta_j \le \pi \;\; (1 \le j \le d-2), \quad \text{and} \quad 0 \le \varphi < 2\pi.} \tag{6.12}$$

The Cartesian coordinates are related to the corresponding polar coordinates by

$$\left.\begin{aligned}
x_1 &= r \cos\theta_1 \\
x_2 &= r \sin\theta_1 \cos\theta_2 \\
x_3 &= r \sin\theta_1 \sin\theta_2 \cos\theta_3 \\
\cdots &= \cdots\cdots\cdots\cdots\cdots\cdots \\
x_{d-1} &= r \sin\theta_1 \sin\theta_2 \ldots \sin\theta_{d-2} \cos\varphi \\
x_d &= r \sin\theta_1 \sin\theta_2 \ldots \sin\theta_{d-2} \sin\varphi.
\end{aligned}\right\} \tag{6.13}$$

The inverse transformations are

$$\left.\begin{aligned} r &= (x_1^2 + \cdots + x_d^2)^{1/2} \\ \theta_j &= \tan^{-1}\left[(x_{j+1}^2 + \cdots + x_d^2)^{1/2}/x_j\right] \quad (1 \le j \le d-2) \\ \varphi &= \tan^{-1}\left(x_d/x_{d-1}\right). \end{aligned}\right\} \tag{6.14}$$

In the case $d = 3$, this corresponds to the identifications $x_1 = z,\ x_2 = x,\ x_3 = y$, with the z-axis chosen to be the polar axis.

The volume element is, of course, $dV = dx_1 \ldots dx_d$ in Cartesian coordinates. In polar coordinates, it is given by

$$\boxed{dV = r^{d-1}\,(\sin\theta_1)^{d-2}\,(\sin\theta_2)^{d-3}\ \ldots\ (\sin\theta_{d-2})\,dr\,d\theta_1\ \ldots\ d\theta_{d-2}\,d\varphi.} \tag{6.15}$$

A pair of important and useful formulas may now be deduced.

***Volume of a hypersphere*:** A **ball** or **hypersphere** of radius R in d-dimensional space, centered at the origin, is the set of points satisfying the condition

$$(x_1^2 + \cdots + x_d^2)^{1/2} \le R. \tag{6.16}$$

The "volume" of this ball is given by

$$V_d(R) = \int_0^R dr\,r^{d-1}\int_0^\pi d\theta_1\,(\sin\theta_1)^{d-2}\cdots\int_0^\pi d\theta_{d-2}\,\sin\theta_{d-2}\int_0^{2\pi} d\varphi. \tag{6.17}$$

We now need the value of the definite integral $\int_0^\pi d\theta\,(\sin\theta)^l$. This is given by the formula (3.21) of Chap. 3, Sect. 3.1.5. Equation (6.17) then yields, after simplification,

$$\boxed{V_d(R) = \frac{\pi^{d/2}R^d}{\Gamma\left(1+\frac{1}{2}d\right)}.} \tag{6.18}$$

***Surface of a hypersphere*:** The surface of the hypersphere above is given by the equation

$$(x_1^2 + \cdots + x_d^2)^{1/2} = R. \tag{6.19}$$

The "area" of this hypersurface is given by

$$S_d(R) = R^{d-1}\int_0^\pi d\theta_1\,(\sin\theta_1)^{d-2}\cdots\int_0^\pi d\theta_{d-2}\,\sin\theta_{d-2}\int_0^{2\pi} d\varphi, \tag{6.20}$$

which simplifies to

$$\boxed{S_d(R) = \frac{2\pi^{d/2}R^{d-1}}{\Gamma\left(\frac{1}{2}d\right)}.} \tag{6.21}$$

Note the obvious relations

$$V_d(R) = \int_0^R dr\, S_d(r) \quad \text{and} \quad S_d(R) = \frac{dV_d(R)}{dR}\,. \tag{6.22}$$

★ **4.** Carry out the simplification required to obtain Eqs. (6.18) and (6.21).
In $d = 2$ and 3, respectively, Eq. (6.21) reduces to $2\pi R$ and $4\pi R^2$ for the circumference of a circle of radius R and the surface area of a sphere of radius R. Setting $d = 1$ in Eqs. (6.18) and (6.21), respectively, we get $2R$ (the length of a line extending to a distance R on either side of the origin) and 2 (the "surface" of this line, namely, the two end points)! Note also that the "surface-to-volume" ratio of a hypersphere increases linearly with the dimensionality: we have

$$\frac{S_d(R)}{V_d(R)} = \frac{d}{R}\,. \tag{6.23}$$

- The hypersurface given by Eq. (6.19) is a smooth manifold whose topological dimensionality is $(d-1)$. In mathematics it is called the $(d-1)$-sphere, and is denoted by S^{d-1}.

6.2 Scalar and Vector Fields and Their Derivatives

We shall denote scalar fields by symbols such as $\phi(\mathbf{r}), \psi(\mathbf{r}), \ldots$, and vector fields by symbols such as $\mathbf{u}(\mathbf{r}), \mathbf{v}(\mathbf{r}), \ldots$. Unless otherwise specified, $\mathbf{a}, \mathbf{b}, \mathbf{k}, \ldots$ will denote constant vectors, i.e., they do not depend on $\mathbf{r}$.

6.2.1 The Gradient of a Scalar Field

When the del operator is applied to a scalar field $\phi(\mathbf{r})$, it yields a vector field called the **gradient** of the scalar field, written as grad $\phi(\mathbf{r})$ or $\nabla\phi(\mathbf{r})$. In geometrical terms, the direction of grad $\phi(\mathbf{r})$ is normal to the level surface $\phi(\mathbf{r}) =$ constant on which the point $\mathbf{r}$ lies, and is along the direction of *increasing* ϕ. See Fig. 6.3a. This implies that

- the gradient of a scalar field at any point is along the direction in which the rate of increase of the function is the highest.

Let $\mathbf{e}_\xi$ denote the unit vector in an *arbitrary* direction at the point $\mathbf{r}$, and ξ the coordinate along this direction, as in Fig. 6.3b. What is the change in ϕ for an infinitesimal displacement $d\boldsymbol{\xi} \equiv (d\xi)\,\mathbf{e}_\xi$ along this direction? On the one hand, it is given by $d\phi \equiv (\partial\phi/\partial\xi)\, d\xi$. On the other hand, it is also given by

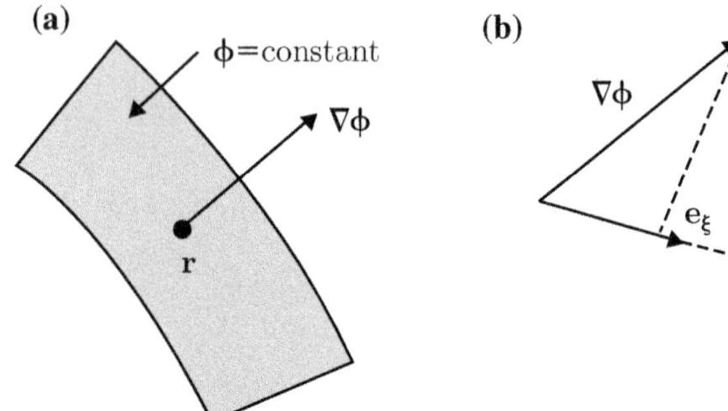

Fig. 6.3 **a** $\nabla\phi$ is normal to the surface ϕ = constant passing through the point $\mathbf{r}$. **b** The component of $\nabla\phi$ along any direction $\mathbf{e}_\xi$ gives the rate of change of ϕ in that direction

$$d\phi = \nabla\phi \cdot d\boldsymbol{\xi} = (\mathbf{e}_\xi \cdot \nabla\phi)\, d\xi. \tag{6.24}$$

Therefore we must have

$$\boxed{\frac{\partial\phi}{\partial\xi} = \mathbf{e}_\xi \cdot \nabla\phi,} \tag{6.25}$$

for any arbitrary direction $\mathbf{e}_\xi$. In other words:

- The rate of change of a scalar field along *any* direction is the component of the gradient of the field along that direction.
- This is the reason why the gradient is also called the **directional derivative**.

In index notation, the gradient of a scalar field ϕ is of course represented by the vector $\partial_i\,\phi$.

Using the expression for the gradient operator in spherical polar coordinates, it is clear that the gradient of any function $\phi(r)$ of the radial coordinate alone is given by $\nabla\phi(r) = \phi'(r)\,\mathbf{e}_r$, where ϕ' is the derivative of ϕ with respect to its argument. It is also useful to note that, if $\mathbf{c}$ is a constant vector, then

$$\nabla\,(\mathbf{c}\cdot\mathbf{r}) = \mathbf{c} \quad \text{and} \quad \nabla\,\phi(\mathbf{c}\cdot\mathbf{r}) = \phi'(\mathbf{c}\cdot\mathbf{r})\,\mathbf{c}. \tag{6.26}$$

★ **5.** Using Eqs. (6.7), find the gradient of each of the following scalar fields:

$$\text{(a) } \varrho^3 \cos 3\varphi \quad \text{(b) } r^2 \cos\theta \quad \text{(c) } \mathbf{a}\cdot(\mathbf{b}\times\mathbf{r}) \quad \text{(d) } (\mathbf{a}\times\mathbf{r})\cdot(\mathbf{b}\times\mathbf{r})$$

***A property of plane waves*:** The complex representation of a plane wave with wave vector $\mathbf{k}$ is given by the function $e^{i\mathbf{k}\cdot\mathbf{r}}$. The gradient of this scalar function is frequently required. It is straightforward to show that

$$\boxed{\nabla\, e^{i\mathbf{k}\cdot\mathbf{r}} = i\mathbf{k}\, e^{i\mathbf{k}\cdot\mathbf{r}}.} \tag{6.27}$$

Having seen how the del operator produces a vector field from a scalar field, let us see how it can produce a scalar field from a vector field.

6.2.2 The Flux and Divergence of a Vector Field

On any smooth surface, even if it is curved, an *infinitesimal* element of area located at any point $\mathbf{r}$ can be regarded as a *planar* area element. Hence a direction can be associated with it, namely, the direction normal to its plane. We write the area element as a vector $\delta\mathbf{S} \equiv \mathbf{n}\,\delta S$, where $\mathbf{n}$ is the unit normal to the element and δS is the magnitude of the area of the element. Then the **flux** of the vector field $\mathbf{u(r)}$ through the area element $\delta\mathbf{S}$ is a scalar quantity that is defined as follows:

$$\text{Flux of } \mathbf{u(r)} \text{ through } \delta\mathbf{S} \stackrel{\text{def.}}{=} \mathbf{u}\cdot\delta\mathbf{S} = (\mathbf{u}\cdot\mathbf{n})\,\delta S. \tag{6.28}$$

This flux is a measure of the way the field lines of the vector field pass through the area element. Its value changes from point to point because both $\mathbf{u}$ and $\mathbf{n}$ are functions of $\mathbf{r}$ in general, for a given magnitude δS of the area element.

You will observe at once that there is an ambiguity in the definition of the normal to an area element, because both $\mathbf{n}$ and $-\mathbf{n}$ can serve as normals. This is resolved by the following convention. Let δC be the *oriented* simple closed curve that represents the boundary of the area element. (That is, there is an arrow on the curve to tell us the sense in which we must move along the curve.) Then, as δC is traversed, the direction in which a right-handed screw moves forward is taken as the direction of the normal $\mathbf{n}$. See Fig. 6.4. For a *closed* surface, the convention is to take the *outward* normal at each point as the direction of the normal. It is worth remembering that

- a finite area is *not* a vector; but a vector can be associated with an *infinitesimal* area element.

It is a simple matter to see that

- the flux of a vector field is an additive quantity.

Given a surface S, we can break it up into infinitesimal area elements. The total flux Φ of the vector field through S is then given by the sum of the fluxes through the individual area elements:

$$\text{Total flux of } \mathbf{u} \text{ through } S,\ \Phi = \int_S \mathbf{u}\cdot d\mathbf{S} = \int_S (\mathbf{u}\cdot\mathbf{n})\,dS. \tag{6.29}$$

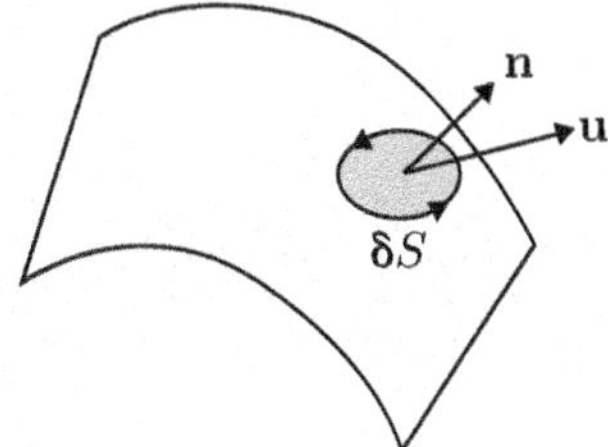

Fig. 6.4 The flux of the vector field $\mathbf{u}$ through the area element $\delta\mathbf{S} = \mathbf{n}\,\delta S$ is $(\mathbf{u}\cdot\mathbf{n})\,\delta S$

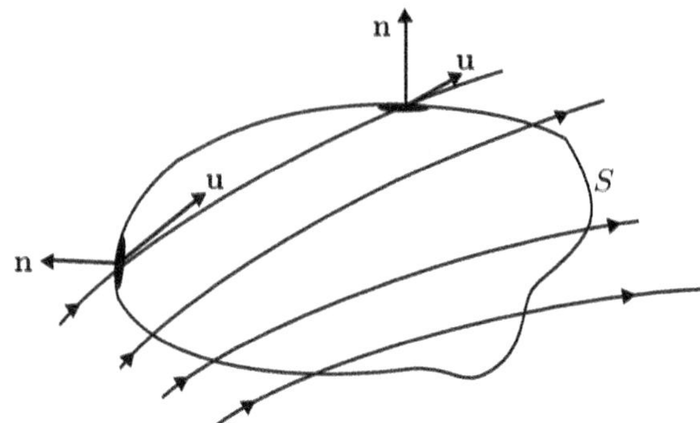

Fig. 6.5 Flux of a vector field through a closed surface S, with the field lines of a vector field **u** entering and leaving the surface

The flux of a vector field through a *closed* surface S is of particular interest (see Fig. 6.5). As the normal is always directed out of the surface for each surface element in this case, field lines that *leave* the surface contribute positive amounts to the flux (the angle between **u** and **n** is acute, so that its cosine is positive). On the other hand, field lines that *enter* the surface contribute negative amounts to the flux (the angle between **u** and **n** is now obtuse, so that its cosine is negative). The net flux is a measure of the difference between the outward flux and the inward one.

Now consider an infinitesimal volume element δV at any point **r**, bounded by the (closed) surface δS. The flux per unit volume of the vector field **u** at the point **r** is defined as the **divergence** of **u** at that point, and written as $\text{div}\,\mathbf{u}(\mathbf{r})$. Thus

$$\boxed{\text{div}\ \mathbf{u}(\mathbf{r}) \stackrel{\text{def.}}{=} \lim_{\delta V \to 0} \frac{\int_{\delta S} \mathbf{u}(\mathbf{r}) \cdot d\mathbf{S}}{\delta V}\,.} \tag{6.30}$$

The interesting fact is that the limit above is a *local* quantity, i.e., it is a function of **r** alone, independent of the shape or orientation of the volume element. This property enables us to *calculate* $\text{div}\,\mathbf{u}$ using any convenient shape for the volume element. This calculation is a standard exercise, and I shall not repeat it here. The outcome is a *formula* or algorithm for the computation of the divergence that reads, formally, as

$$\boxed{\text{div}\,\mathbf{u}(\mathbf{r}) = \nabla \cdot \mathbf{u}(\mathbf{r}).} \tag{6.31}$$

You must understand clearly that Eq. (6.30) is the *definition* of the divergence of a vector field, while the right-hand side of Eq. (6.31) is a formula or algorithm for *computing* it.

Let the components of $\mathbf{u}(\mathbf{r})$ be given by $(u_x\,,\ u_y\,,\ u_z)$ in Cartesian coordinates, $(u_\varrho\,,\ u_\varphi\,,\ u_z)$ in cylindrical polars, and $(u_r\,,\ u_\theta\,,\ u_\varphi)$ in spherical polars. Remember that *each* component of the vector field is, in general, a function of *all* the coordinates. Then, in explicit form,

$$\left.\begin{aligned}\nabla\cdot\mathbf{u}(\mathbf{r}) &= \frac{\partial u_x}{\partial x}+\frac{\partial u_y}{\partial y}+\frac{\partial u_z}{\partial z}\\ &= \frac{1}{\varrho}\frac{\partial(\varrho\, u_\varrho)}{\partial\varrho}+\frac{1}{\varrho}\frac{\partial u_\varphi}{\partial\varphi}+\frac{\partial u_z}{\partial z}\\ &= \frac{1}{r^2}\frac{\partial(r^2\,u_r)}{\partial r}+\frac{1}{r\,\sin\,\theta}\frac{\partial(\sin\,\theta\, u_\theta)}{\partial\theta}+\frac{1}{r\,\sin\,\theta}\frac{\partial u_\varphi}{\partial\varphi}\,.\end{aligned}\right\} \tag{6.32}$$

Once again, in index notation, we have the compact expression

$$\boxed{\operatorname{div}\mathbf{u}=\nabla\cdot\mathbf{u}=\partial_i\,u_i\,.} \tag{6.33}$$

It follows trivially that the divergence of the vector field $\mathbf{u}(\mathbf{r})=\mathbf{r}$ itself is

$$\boxed{\operatorname{div}\mathbf{r}=\partial_i\,x_i=3.} \tag{6.34}$$

This is just the number of dimensions of space.

★ **6.** Find the divergence of each of the following vector fields:

(a) $\mathbf{a}\times(\mathbf{b}\times\mathbf{r})$ (b) $(\mathbf{a}\times\mathbf{r})\times(\mathbf{b}\times\mathbf{r})$ (c) $\phi(\varrho)\,\mathbf{e}_\varrho$ (d) $\phi(r)\,\mathbf{e}_r$

***Another property of plane waves*:** A plane wave vector field (e.g., an electric or magnetic field) with a constant amplitude $\mathbf{a}$ and wave vector $\mathbf{k}$ is represented by the function $\mathbf{a}\,e^{i\mathbf{k}\cdot\mathbf{r}}$. The divergence of this vector field is of physical interest. It is a simple exercise to show that

$$\boxed{\nabla\cdot(\mathbf{a}\,e^{i\mathbf{k}\cdot\mathbf{r}})=(i\mathbf{k}\cdot\mathbf{a})\,e^{i\mathbf{k}\cdot\mathbf{r}}.} \tag{6.35}$$

Note that $\mathbf{e}_\varrho$ is not uniquely defined on the z-axis ($\varrho=0$), and that $\mathbf{e}_r$ is not uniquely defined at the origin $r=0$. In general, axially symmetric fields are *singular* on the axis of symmetry, and central fields are *singular* at the origin. It is easy to see that

- a central field whose magnitude is proportional to r^n has a vanishing divergence for all $r>0$ if and only if $n=-2$, so that $u_r\propto 1/r^2$, i.e., if it is an *inverse square* field.

A vector field whose divergence vanishes identically is said to be **solenoidal**.

6.2.3 *The Circulation and Curl of a Vector Field*

The del operator can also produce another vector field from a given vector field. For this purpose, we need the concept of the circulation of a vector field.

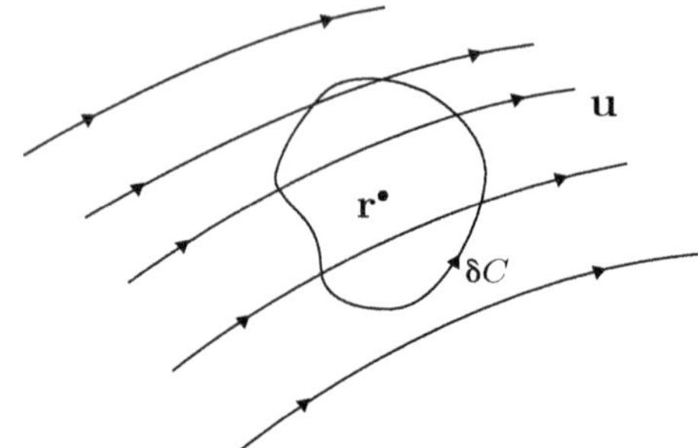

Fig. 6.6 The circulation of **u** over the infinitesimal closed contour δC bounding the area element δS is the line integral of **u** over δC. In the case illustrated, the normal **n** to the area element is directed out of the plane of the paper

The **line integral** of a vector field $\mathbf{u}(\mathbf{r})$ from a point P with position vector $\mathbf{r}_1$ to another point Q with position vector $\mathbf{r}_2$ along a directed contour C is defined as $\int_C \mathbf{u}(\mathbf{r}) \cdot d\boldsymbol{\ell}$. In general, the value of the integral depends on the particular path taken from P to Q. If the contour is an oriented *closed* contour C, starting and ending at the same point, then the line integral is called the **circulation** of the vector field over the contour C:

$$\text{Circulation of } \mathbf{u} \text{ over } C \stackrel{\text{def.}}{=} \oint_C \mathbf{u}(\mathbf{r}) \cdot d\boldsymbol{\ell}, \tag{6.36}$$

where the symbol $\oint$ denotes an integral over a *closed* path or contour.

In order to obtain a *local* quantity at any point $\mathbf{r}$, consider an infinitesimal area element $\delta\mathbf{S} = \mathbf{n}\,\delta S$ centered at $\mathbf{r}$. Let δC be the oriented closed contour that is the boundary of the area element (see Fig. 6.6). Then the circulation per unit area of the vector field $\mathbf{u}$ at $\mathbf{r}$ yields the magnitude of a certain vector field associated with $\mathbf{u}$. This field is called the **curl** of $\mathbf{u}$, and is written as curl $\mathbf{u}$. Its direction is given by $\mathbf{n}$. Thus

$$\boxed{\text{curl}\,\mathbf{u} \stackrel{\text{def.}}{=} \lim_{\delta S \to 0} \left(\frac{\oint_{\delta C} \mathbf{u}(\mathbf{r}) \cdot d\boldsymbol{\ell}}{\delta S}\, \mathbf{n} \right).} \tag{6.37}$$

Note that the sense or direction in which the contour δC is traversed is related to the direction of $\mathbf{n}$ by the right-hand screw rule.

Once again, it is a remarkable fact that the limit in Eq. (6.37) is actually independent of the shape and orientation of the area element $\delta\mathbf{S}$. This property permits us to use any conveniently shaped and oriented contour to evaluate the integral and pass to the limit concerned. The result is a formula or algorithm for the curl of a vector field that reduces, formally, to

$$\boxed{\text{curl}\,\mathbf{u}(\mathbf{r}) = \nabla \times \mathbf{u}(\mathbf{r}).} \tag{6.38}$$

As in the case of the divergence, you must appreciate the fact that Eq. (6.37) is the *definition* of the curl of $\mathbf{u}$, while Eq. (6.38) is a formula that enables us to calculate this quantity. In the Cartesian, cylindrical polar and spherical polar coordinate systems, respectively, we have

$$
\begin{aligned}
\nabla \times \mathbf{u}(\mathbf{r}) &= \left(\frac{\partial u_z}{\partial y} - \frac{\partial u_y}{\partial z}\right)\mathbf{e}_x + \left(\frac{\partial u_x}{\partial z} - \frac{\partial u_z}{\partial x}\right)\mathbf{e}_y + \left(\frac{\partial u_y}{\partial x} - \frac{\partial u_x}{\partial y}\right)\mathbf{e}_z \\
&= \left(\frac{1}{\varrho}\frac{\partial u_z}{\partial \varphi} - \frac{\partial u_\varphi}{\partial z}\right)\mathbf{e}_\varrho + \left(\frac{\partial u_\varrho}{\partial z} - \frac{\partial u_z}{\partial \varrho}\right)\mathbf{e}_\varphi + \frac{1}{\varrho}\left(\frac{\partial(\varrho\, u_\varphi)}{\partial \varrho} - \frac{\partial u_\varrho}{\partial \varphi}\right)\mathbf{e}_z \\
&= \frac{1}{r\,\sin\theta}\left(\frac{\partial(\sin\theta\, u_\varphi)}{\partial\theta} - \frac{\partial u_\theta}{\partial\varphi}\right)\mathbf{e}_r + \\
&\quad + \frac{1}{r}\left(\frac{1}{\sin\theta}\frac{\partial u_r}{\partial\varphi} - \frac{\partial(r\, u_\varphi)}{\partial r}\right)\mathbf{e}_\theta + \frac{1}{r}\left(\frac{\partial(r\, u_\theta)}{\partial r} - \frac{\partial u_r}{\partial\theta}\right)\mathbf{e}_\varphi\,. \qquad (6.39)
\end{aligned}
$$

The curl of a vector is very conveniently represented in index notation:

$$\boxed{\mathbf{v} = \text{curl}\,\mathbf{u} = \nabla \times \mathbf{u} \quad \Rightarrow \quad v_i = \epsilon_{ijk}\,\partial_j\,u_k\,.} \qquad (6.40)$$

The curl of the vector field $\mathbf{u}(\mathbf{r}) = \mathbf{r}$ itself vanishes identically:

$$\big(\text{curl}\,\mathbf{r}\big)_i = \epsilon_{ijk}\,\partial_j\,x_k = \epsilon_{ijk}\,\delta_{jk} = 0. \qquad (6.41)$$

A vector field whose curl vanishes is said to be **irrotational** or curl-free.

Here is a simple result that is quite useful in applications. If $\mathbf{b}$ is a constant vector, then

$$\boxed{\text{curl}\,\tfrac{1}{2}(\mathbf{b} \times \mathbf{r}) = \mathbf{b}.} \qquad (6.42)$$

It is a simple exercise to establish this formula using the index notation (see the exercise below). In magnetostatics (to be discussed in Chap. 9, Sect. 9.1.6), the expression

$$\mathbf{A} = \tfrac{1}{2}(\mathbf{B} \times \mathbf{r}) \qquad (6.43)$$

is often used for the **vector potential** corresponding to a constant magnetic field $\mathbf{B}$. It leads at once to the relation curl $\mathbf{A} = \mathbf{B}$. Another instance arises in the rotational dynamics of rigid bodies. The linear and angular velocities of a mass element on the body are related in this case by

$$\boxed{\mathbf{v} = \boldsymbol{\omega} \times \mathbf{r} \quad \Rightarrow \quad 2\boldsymbol{\omega} = \text{curl}\,\mathbf{v}.} \qquad (6.44)$$

Such a relation also occurs in fluid dynamics, as we shall see in Chap. 7, Sect. 7.3.1.

★ **7.** Use the index notation in the exercises that follow.

(a) Establish Eq. (6.42).
(b) Find the curl of each of the following vector fields:

$$\text{(i) } (\mathbf{a} \times \mathbf{r}) \times (\mathbf{b} \times \mathbf{r}) \quad \text{(ii) } \phi(\varrho)\,\mathbf{e}_\varrho \quad \text{(iii) } \phi(r)\,\mathbf{e}_r.$$

***Yet another property of plane waves*:** The curl of the vector field representing a plane wave is also of physical interest. It is easily shown that

$$\boxed{\nabla \times (\mathbf{a}\, e^{i\mathbf{k}\cdot\mathbf{r}}) = (i\mathbf{k} \times \mathbf{a})\, e^{i\mathbf{k}\cdot\mathbf{r}}.} \tag{6.45}$$

Equation (6.27) shows that the action of the operator ∇ on the function $e^{i\mathbf{k}\cdot\mathbf{r}}$ is simply to replace ∇ by $i\mathbf{k}$. Further, Eqs. (6.35) and (6.45) show that this remains true for operations $\nabla\cdot$ and $\nabla\times$ as well. These properties are of fundamental importance in the Fourier expansion of vector fields. They are very useful, for instance, in the analysis of Maxwell's field equations for electromagnetic fields, as you will see in Chap. 9.

6.2.4 Some Physical Aspects of the Curl of a Vector Field

The flow of fluids provides a number of insights into the behaviour of vector fields. We will discuss fluid dynamics in greater detail in Chap. 7, but it is helpful to make a couple of points here. These should help dispel some misconceptions regarding the notion of the curl of a vector field.

(a) Consider the velocity field on the surface of the water flowing in straight streamlines in a long straight canal with its banks parallel to the x-axis, given by the lines $y = -a$ and $y = a$, respectively. The central axis of the canal is along the x-axis. See Fig. 6.7. The velocity of the water at any point (x, y) has only an x-component, and is of the form

$$\mathbf{v}(x, y) = v_x(y)\,\mathbf{e}_x. \tag{6.46}$$

Here, $v_x(y)$ is a symmetric function of y that is maximum at $y = 0$ (the center of the canal) and drops to zero at $y = \pm a$. (In the simplest case,

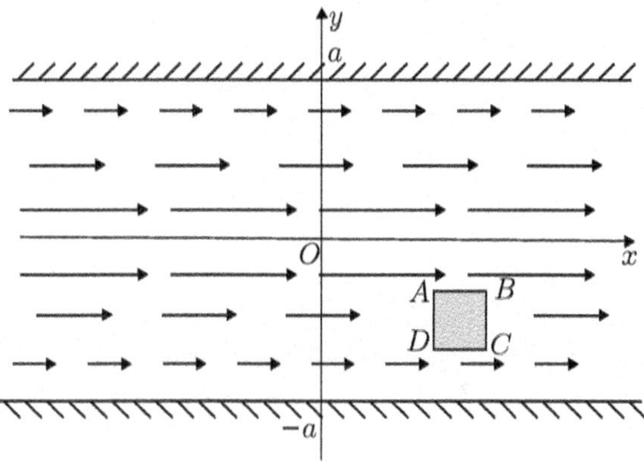

Fig. 6.7 Streamlines in fluid flow on the surface of a straight canal. The length of each arrow is meant to indicate the magnitude of the fluid velocity at that point. ABCD represents a small raft on the water. It undergoes a rotation about its center (clockwise, looking down from above) as it floats down the canal

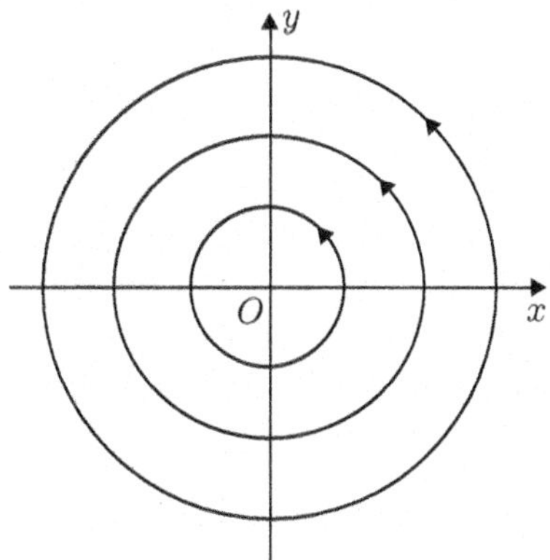

Fig. 6.8 Streamlines of fluid flow around a vortex on the z-axis (which is directed normal to the plane of the paper)

$$v_x(y) \propto (a^2 - y^2). \tag{6.47}$$

This is called **Poiseuille flow**, and is a consequence of the viscosity of the liquid.) Since the only nonzero component of the velocity is v_x, the curl of the velocity field is given by

$$\text{curl}\,\mathbf{v}(x, y) = -\frac{\partial v_x}{\partial y}\,\mathbf{e}_z \not\equiv 0. \tag{6.48}$$

Thus, the velocity field is *not* irrotational, even though the field lines of $\mathbf{v}$ are straight lines parallel to the x-axis.

- Field lines that are parallel straight lines do not necessarily imply an irrotational vector field!

(b) On the other hand, consider the velocity field

$$\mathbf{v} = K\,\nabla\varphi \tag{6.49}$$

where K is a constant and φ denotes the azimuthal angle in cylindrical polar coordinates, as usual. This is a simple model of the velocity field around a **vortex** located on the z-axis, as shown in Fig. 6.8. The field lines are concentric circles centered about the z-axis. The vector field is singular *at* the vortex itself, i.e., on the z-axis. Using the formula for the gradient operator in cylindrical polar coordinates (the second equation in (6.7)), we get

$$\mathbf{v} = K\,\nabla\varphi = (K/\varrho)\,\mathbf{e}_\varphi\,. \tag{6.50}$$

(The singularity of the field at $\varrho = 0$, i.e., on the z-axis, is now manifest.) Thus $\mathbf{v}$ has only an azimuthal component, $v_\varphi = K/\varrho$. Using the second formula in (6.39), we see that curl $\mathbf{v}$ vanishes identically in this case.

- Curved field lines do not necessarily imply a vector field with a nonvanishing curl!

An infinitesimal fluid element may be regarded as a "rigid" body to a first approximation. The relation $2\boldsymbol{\omega} = \text{curl}\,\mathbf{v}$ (the second equation in (6.44)) then implies that the curl of the velocity field of a fluid at any point is equal to twice the *local* angular velocity of a fluid element about its center. I will return to this point in Chap. 7, Sect. 7.3.1, when discussing vorticity. In example **(a)** above, a small raft floating down the canal (see Fig. 6.7) will *rotate* about the raft's center as it drifts downstream, because the side AB of the raft nearer the central axis of the canal will tend to move faster than the side CD that is nearer the bank. The curl of the velocity field, being directly proportional to the transverse velocity gradient $\partial v_x/\partial y$, is a measure of the angular velocity of this rotation. It is also easily seen that a small raft drifting around the vortex in example **(b)** actually undergoes no such "intrinsic" rotation about its own center—although it is rotated by exactly 2π as it completes every full revolution around the vortex, i.e., around the singularity of the vector field concerned. This is consistent with the vanishing of curl $\mathbf{v}$ everywhere (except on the singularity) in this case. Incidentally, this simple example illustrates a crucial point in vector calculus (and, ultimately, in differential geometry). I will return to it in Chap. 8, Sects. 8.1.3 and 8.3.

6.2.5 *Any Vector Field is the Sum of a Curl and a Gradient*

Vector fields satisfy two very basic and useful identities.
(i) *The divergence of the curl of a vector field vanishes identically.*
This identity is trivially established using the index notation. All that is needed is the fact that the contraction of an antisymmetric tensor with a symmetric one vanishes identically. If $\mathbf{u}(\mathbf{r})$ is a vector field, then

$$\boxed{\text{div curl}\,\mathbf{u}(\mathbf{r}) = \nabla\cdot(\nabla\times\mathbf{u}) = \epsilon_{ijk}\,\partial_i\,\partial_j\,u_k \;\equiv 0,} \tag{6.51}$$

because ϵ_{ijk} is antisymmetric under the interchange of the indices i and j, while $\partial_i\,\partial_j$ is symmetric.
(ii) *The curl of the gradient of a scalar field vanishes identically.*
If $\phi(\mathbf{r})$ is a scalar field, the ith component of the curl of grad ϕ is given by

$$\boxed{[\text{curl grad}\,\phi(\mathbf{r})]_i = (\nabla\times\nabla\phi)_i = \epsilon_{ijk}\,\partial_j\,\partial_k\,\phi \equiv 0,} \tag{6.52}$$

again because ϵ_{ijk} is antisymmetric in the indices j and k, while $\partial_j\,\partial_k$ is symmetric in these indices.

These are extremely useful identities. The first of them implies that every solenoidal vector field $\mathbf{v}(\mathbf{r})$ can be always be written as the curl of some other vector field, say $\mathbf{s}(\mathbf{r})$. That is,

$$\boxed{\text{div}\,\mathbf{v} = 0 \quad\Rightarrow\quad \mathbf{v} = \text{curl}\,\mathbf{s}.} \tag{6.53}$$

The second identity implies that every irrotational vector field $\mathbf{w}(\mathbf{r})$ can always be written as the gradient of a scalar field, say $\psi(\mathbf{r})$. That is,

$$\boxed{\text{curl}\,\mathbf{w} = 0 \quad \Rightarrow \quad \mathbf{w} = \text{grad}\,\psi.} \tag{6.54}$$

You might wonder whether there are any more identities of this kind. The answer is "no". I mention here that the most natural way of understanding this assertion is in the language of differential geometry, although the latter subject is beyond the scope of this book.

***Conservative vector field*:** Any vector field that is the gradient of a scalar field enjoys another useful property: its line integral from any point P with position vector $\mathbf{r}_1$ to another point Q with position vector $\mathbf{r}_2$ is *independent* of the actual path taken from P to Q. We have in this case

$$\int_{\mathbf{r}_1}^{\mathbf{r}_2} \mathbf{w}\cdot d\boldsymbol{\ell} = \int_{\mathbf{r}_1}^{\mathbf{r}_2} \nabla\psi\cdot d\boldsymbol{\ell} = \psi(\mathbf{r}_2) - \psi(\mathbf{r}_1). \tag{6.55}$$

It follows that

- the line integral of a gradient field around any closed path C vanishes:

$$\oint_C \mathbf{w}\cdot d\boldsymbol{\ell} = \oint_C \nabla\psi\cdot d\boldsymbol{\ell} = 0. \tag{6.56}$$

Such a vector field is also called a **conservative vector field**. Foremost among physical examples is a force field "derived" from a scalar potential $V(\mathbf{r})$ according to the familiar relation

$$\boxed{\mathbf{F}(\mathbf{r}) = -\nabla V(\mathbf{r}).} \tag{6.57}$$

Helmholtz's Theorem is of fundamental importance. It asserts that every vector field $\mathbf{u}(\mathbf{r})$ can be written as the sum of a solenoidal vector field $\mathbf{v}$ and an irrotational vector field $\mathbf{w}$:

$$\mathbf{u}(\mathbf{r}) = \mathbf{v}(\mathbf{r}) + \mathbf{w}(\mathbf{r}) \quad \text{where} \quad \text{div}\,\mathbf{v} = 0 \quad \text{and} \quad \text{curl}\,\mathbf{w} = 0. \tag{6.58}$$

Hence any vector field $\mathbf{u}(\mathbf{r})$ can be represented in the form

$$\boxed{\mathbf{u}(\mathbf{r}) = \text{curl}\,\mathbf{s}(\mathbf{r}) + \text{grad}\,\psi(\mathbf{r}).} \tag{6.59}$$

Now, three "pieces of information" are required to specify the original vector field $\mathbf{u}(\mathbf{r})$, namely, its three components. It might appear that this number has increased to four when we write $\mathbf{u}(\mathbf{r})$ as in Eq. (6.59): the three components of $\mathbf{s}$, as well as the scalar function ψ. But this is not so. Remember that the divergence of curl $\mathbf{s}$ is identically zero. This implies a relationship between the three components of curl $\mathbf{s}$.

Hence there are, in effect, just two independent pieces of information encoded in this term. Together with the scalar ψ, that makes a total of three, the same as the original number required to specify **u**.

6.2.6 The Laplacian Operator

The operator $\nabla \cdot \nabla$ formed by taking the scalar product of the del operator with itself is called the **Laplacian**, and is written[1] as ∇^2. It is important to note that ∇^2 is a *scalar* differential operator, in contrast to ∇, which is a vector operator. From the mathematical point of view, the Laplacian and its generalizations are of deep and profound importance. In Chap. 8, Sect. 8.2.3, I will describe in brief why this is so. In physics, too, the Laplacian appears everywhere, in an enormous variety of contexts. It is present in almost all of the fundamental equations of mathematical physics, such as *Laplace's equation*, *Poisson's equation*, the *Helmholtz equation*, the *diffusion equation*, the *wave equation*, the *Schrödinger equation*, the *Navier–Stokes equation*, etc. We will encounter these equations in Chaps. 7, 9, and 29–32.

When it acts on a scalar field, the Laplacian is the same as the divergence of the gradient of the field. That is,

$$\nabla^2 \phi = (\nabla \cdot \nabla)\, \phi = \nabla \cdot (\nabla\, \phi) = \text{div grad}\, \phi. \tag{6.60}$$

In index notation, we have

$$\nabla^2 \phi = \partial_i\, \partial_i\, \phi. \tag{6.61}$$

In general orthogonal coordinates $(\xi_1\,,\, \xi_2\,,\, \xi_3)$, the Laplacian of a scalar field ϕ is given by

$$\begin{aligned}\nabla^2 \phi = \frac{1}{h_1\, h_2\, h_3} \left\{ \frac{\partial}{\partial \xi_1} \left(\frac{h_2\, h_3}{h_1} \frac{\partial \phi}{\partial \xi_1} \right) + \frac{\partial}{\partial \xi_2} \left(\frac{h_3\, h_1}{h_2} \frac{\partial \phi}{\partial \xi_2} \right) \right. \\ \left. + \frac{\partial}{\partial \xi_3} \left(\frac{h_1\, h_2}{h_3} \frac{\partial \phi}{\partial \xi_3} \right) \right\},\end{aligned} \tag{6.62}$$

where $(h_1\,,\, h_2\,,\, h_3)$ are the scale factors introduced in Eqs. (6.8) and (6.9). The respective explicit expressions in Cartesian, cylindrical polar, and spherical polar coordinates, and in the three-dimensional parabolic coordinates (ξ, η, φ) defined in Sect. 6.1.2, are as follows:

---

[1] Texts on mathematics often use the notation $-\Delta$ for the Laplacian, or a generalization of the Laplacian to curved spaces. I shall stick to ∇^2, in order to avoid possible confusion.

$$\nabla^2\phi = \begin{cases} \dfrac{\partial^2\phi}{\partial x^2} + \dfrac{\partial^2\phi}{\partial y^2} + \dfrac{\partial^2\phi}{\partial z^2} \\[2ex] \dfrac{1}{\varrho}\dfrac{\partial}{\partial\varrho}\left(\varrho\dfrac{\partial\phi}{\partial\varrho}\right) + \dfrac{1}{\varrho^2}\dfrac{\partial^2\phi}{\partial\varphi^2} + \dfrac{\partial^2\phi}{\partial z^2} \\[2ex] \dfrac{1}{r^2}\dfrac{\partial}{\partial r}\left(r^2\dfrac{\partial\phi}{\partial r}\right) + \dfrac{1}{r^2\sin\theta}\dfrac{\partial}{\partial\theta}\left(\sin\theta\dfrac{\partial\phi}{\partial\theta}\right) + \dfrac{1}{r^2\sin^2\theta}\dfrac{\partial^2\phi}{\partial\varphi^2} \\[2ex] \dfrac{1}{(\xi^2+\eta^2)}\left\{\dfrac{1}{\xi}\dfrac{\partial}{\partial\xi}\left(\xi\dfrac{\partial\phi}{\partial\xi}\right) + \dfrac{1}{\eta}\dfrac{\partial}{\partial\eta}\left(\eta\dfrac{\partial\phi}{\partial\eta}\right)\right\} + \dfrac{1}{\xi^2\eta^2}\dfrac{\partial^2\phi}{\partial\varphi^2}. \end{cases} \tag{6.63}$$

★ **8.** Find the scalar fields that result by applying the Laplacian operator to the following scalar fields:

(a) $\varrho^3 \cos 3\varphi$ (b) $r^2 \cos\theta$ (c) $\mathbf{a}\cdot(\mathbf{b}\times\mathbf{r})$ (d) $(\mathbf{a}\times\mathbf{r})\cdot(\mathbf{b}\times\mathbf{r})$

∇^2 ***acting on a plane wave*:** You have already seen that, when the del operator acts on a plane wave $e^{i\mathbf{k}\cdot\mathbf{r}}$, ∇ is essentially replaced by $i\mathbf{k}$ (Eqs. (6.27), (6.35), and (6.45)). It follows at once from the first two of these equations that

$$\boxed{\nabla^2 e^{i\mathbf{k}\cdot\mathbf{r}} = -k^2 e^{i\mathbf{k}\cdot\mathbf{r}}.} \tag{6.64}$$

Again, this a very useful result.

The Laplacian operator may also act on a vector field, and on higher rank tensor fields. In particular, the quantity $\nabla^2\mathbf{u}$ is *defined* as the vector field with Cartesian components $\nabla^2 u_i$ or, in index notation, $\partial_j\,\partial_j\,u_i$. When working in curvilinear coordinates, you must remember that the unit vectors themselves are functions of the coordinates. We have, for instance,

$$\nabla^2\,\mathbf{e}_r = -2\mathbf{e}_r/r^2, \quad \text{or} \quad \nabla^2\,(\mathbf{r}/r) = -2\mathbf{r}/r^3. \tag{6.65}$$

★ **9.** Verify Eq. (6.65).

★ **10.** Find the vector fields that result by applying the Laplacian operator to the following vector fields. **a** and **b** are constant vectors, while $\phi(\varrho)$ and $\psi(r)$ are arbitrary twice-differentiable functions of their arguments.

(a) $\frac{1}{2}(\mathbf{b}\times\mathbf{r})$ (b) $(\mathbf{a}\times\mathbf{r})\times(\mathbf{b}\times\mathbf{r})$ (c) $\mathbf{a}\,e^{i\mathbf{k}\cdot\mathbf{r}}$ (d) $\phi(\varrho)\,\mathbf{e}_\varrho$ (e) $\psi(r)\,\mathbf{e}_r$.

Remember that the unit vectors $\mathbf{e}_\varrho$ and $\mathbf{e}_r$ are functions of the coordinates.

6.2.7 *Why Do div, curl, and del-Squared Occur so Frequently?*

You would have noticed that physical laws involving scalar and vector fields almost always involve the divergence, curl, and Laplacian of these fields. Why should this be so?

In Chap. 5, Sect. 5.1, I have explained why the invariance of physical laws under rotations of the coordinate axes requires that they be expressed in terms of quantities such as scalars, vectors, and tensors. These quantities have definite transformation laws under such rotations.

Physical laws generally involve differential equations. The broad reason is that the underlying "mechanisms" are *local* in space and time. The viewpoint that has emerged over the past 350 years or so, in our progression from nonrelativistic Newtonian mechanics to relativistic quantum field theory, is as follows:

- There is no "action at a distance". Events at a given space–time point are affected by what goes on in the immediate neighborhood of that point.
- This is implemented mathematically via differential operators, leading to the expression of physical laws in the form of differential equations.
- The form-invariance of such equations then requires that the differential operators be such that, when they act on the fields, the resulting quantities also have definite transformation properties.

The Laplacian ∇^2 is a scalar operator. Acting on a scalar field, it produces a scalar field. Acting on a vector field, it produces a vector field. I have already mentioned its importance and ubiquity (see also Chap. 8, Sect. 8.2.3). Equations such as Poisson's equation for the electrostatic potential, $\nabla^2\phi = -\rho/\epsilon_0$, are guaranteed to be rotationally invariant. Roughly speaking, the appearance of ∇^2 in an equation of mathematical physics is a reflection of the rotational invariance of the physical situation described by that equation.

The reason for the natural appearance of the divergence and curl of vector fields is a little more involved. Let $\mathbf{u}(\mathbf{r})$ be some physical vector field in three-dimensional space. There are nine possible first derivatives, comprising (the components of) the tensor $\partial_i\, u_j$. In Chap. 5, Sect. 5.3.1, we have seen that Cartesian tensors of rank $\ell \geq 2$ carry within them more than one irreducible part, i.e., that they are made up of several spherical tensors. In particular, a Cartesian tensor of rank 2 has a scalar part (the sum of its diagonal elements) and a vector part (comprising the 3 components of its antisymmetric part). In the case of the tensor $\partial_i\, u_j$, the trace is just the divergence of $\mathbf{u}$, which is a scalar:

$$\operatorname{div}\mathbf{u} = \partial_i\, u_i = \partial_1\, u_1 + \partial_2\, u_2 + \partial_3\, u_3 = \frac{\partial u_x}{\partial x} + \frac{\partial u_y}{\partial y} + \frac{\partial u_z}{\partial z}\,. \tag{6.66}$$

Moreover, the antisymmetric part of the tensor is $\frac{1}{2}(\partial_i\, u_j - \partial_j\, u_i)$. This is just half the curl of $\mathbf{u}$, which is a vector:

$$\begin{aligned}\text{curl}\,\mathbf{u} &= (\partial_2\, u_3 - \partial_3\, u_2\,,\ \partial_3\, u_1 - \partial_1\, u_3\,,\ \partial_1\, u_2 - \partial_2\, u_1)\\ &= \left(\frac{\partial u_z}{\partial y} - \frac{\partial u_y}{\partial z}\,,\ \frac{\partial u_x}{\partial z} - \frac{\partial u_z}{\partial x}\,,\ \frac{\partial u_y}{\partial x} - \frac{\partial u_x}{\partial y}\right).\end{aligned} \tag{6.67}$$

The divergence and curl of $\mathbf{u}$ are spherical tensors of rank 0 and 1, respectively. They have simple and definite transformation properties under rotations of the coordinate axes. It is therefore very appropriate, and not surprising, that these quantities appear in physical applications.

But there is more to it. It turns out that the divergence and curl of a vector field actually suffice to specify the vector field completely, in a manner that is *independent* of the choice of any particular frame of reference:

- If the divergence and curl of a vector field are specified at all points of a (three-dimensional) region, and the normal component of the curl is specified at all points on the boundary of the region, the vector field is essentially determined.

The way this works will become clear after we discuss Fourier transforms in Chap. 18. I shall, however, describe the general idea here. Any (integrable) function $\mathbf{u}(\mathbf{r})$ of the coordinates can be written as a superposition (or sum) of exponential functions $e^{i\mathbf{k}\cdot\mathbf{r}}$, summed over all possible values of the constant vector $\mathbf{k}$. Since the components of the vector $\mathbf{k}$ are continuous, the superposition is an integral rather than a sum. It reads

$$\mathbf{u}(\mathbf{r}) = \frac{1}{(2\pi)^3}\int d^3k\ \widetilde{\mathbf{u}}(\mathbf{k})\, e^{i\mathbf{k}\cdot\mathbf{r}}. \tag{6.68}$$

(As you will see in Chap. 18, Eq. (6.68) is the representation of $\mathbf{u}(\mathbf{r})$ as a Fourier integral.) Since $\mathbf{u}$ is a vector, the coefficient $\widetilde{\mathbf{u}}(\mathbf{k})$ of each $e^{i\mathbf{k}\cdot\mathbf{r}}$ is also a vector. It is labelled by $\mathbf{k}$ (and is independent of $\mathbf{r}$, of course). Now, setting $\mathbf{a} = \widetilde{\mathbf{u}}(\mathbf{k})$ in Eqs. (6.35) and (6.45), we have

$$\left.\begin{aligned}\nabla\cdot\big(\widetilde{\mathbf{u}}(\mathbf{k})\, e^{i\mathbf{k}\cdot\mathbf{r}}\big) &= \big(i\mathbf{k}\cdot\widetilde{\mathbf{u}}(\mathbf{k})\big)\, e^{i\mathbf{k}\cdot\mathbf{r}},\\ \nabla\times\big(\widetilde{\mathbf{u}}(\mathbf{k})\, e^{i\mathbf{k}\cdot\mathbf{r}}\big) &= \big(i\mathbf{k}\times\widetilde{\mathbf{u}}(\mathbf{k})\big)\, e^{i\mathbf{k}\cdot\mathbf{r}}.\end{aligned}\right\} \tag{6.69}$$

Therefore:

- For any given $\mathbf{k}$, the divergence of the vector field gives the component of the vector field in the direction of $\mathbf{k}$, i.e., its *longitudinal* component.
- The curl of the vector field gives its component in the plane normal to $\mathbf{k}$, i.e., its *transverse* component.
- This is true for *every* $\mathbf{k}$. Moreover, these statements are independent of any particular choice of the coordinate axes.
- Hence, a knowledge of the divergence and curl of a vector field amounts to a complete specification of the field *in a manner that is independent of the choice of the coordinate axes*.

That is why the fundamental laws of physics (e.g., Maxwell's equations of electromagnetism) involve the derivatives of vector fields in combinations that correspond to the divergence and curl of these fields.

6.2.8 *The Standard Identities of Vector Calculus*

I conclude the first part of our discussion of vector calculus with a list of standard identities. They are very conveniently established with the help of the index notation.

★ **11.** Using the index notation, show that

(a) $\nabla\,(\phi+\psi) = \nabla\phi + \nabla\psi$
(b) $\nabla\,(\phi\,\psi) = \phi\,(\nabla\psi) + (\nabla\phi)\,\psi$
(c) $\nabla\cdot(\phi\,\mathbf{u}) = \phi\,(\nabla\cdot\mathbf{u}) + (\nabla\phi)\cdot\mathbf{u}$
(d) $\nabla\times(\phi\,\mathbf{u}) = \phi\,(\nabla\times\mathbf{u}) + (\nabla\phi)\times\mathbf{u}$
(e) $\nabla\cdot(\mathbf{u}\times\mathbf{v}) = \mathbf{v}\cdot(\nabla\times\mathbf{u}) - \mathbf{u}\cdot(\nabla\times\mathbf{v})$
(f) $\nabla\times(\mathbf{u}\times\mathbf{v}) = \mathbf{u}\,(\nabla\cdot\mathbf{v}) - \mathbf{v}\,(\nabla\cdot\mathbf{u}) + (\mathbf{v}\cdot\nabla)\mathbf{u} - (\mathbf{u}\cdot\nabla)\mathbf{v}$
(g) $\nabla\,(\mathbf{u}\cdot\mathbf{v}) = (\mathbf{u}\cdot\nabla)\mathbf{v} + (\mathbf{v}\cdot\nabla)\mathbf{u} + \mathbf{u}\times(\nabla\times\mathbf{v}) + \mathbf{v}\times(\nabla\times\mathbf{u})$
(h) $\nabla^2\,(\phi\,\psi) = \phi\,\nabla^2\,\psi + \psi\,\nabla^2\,\phi + 2\,(\nabla\phi\cdot\nabla\psi)$
(i) $\nabla\times(\nabla\times\mathbf{u}) = \nabla(\nabla\cdot\mathbf{u}) - \nabla^2\mathbf{u}$
(j) $\mathbf{v}\times(\nabla\times\mathbf{v}) = \nabla\,(\tfrac{1}{2}v^2) - (\mathbf{v}\cdot\nabla)\,\mathbf{v}$

The symbol $(\mathbf{u}\cdot\nabla)$ in the above denotes the scalar differential operator

$$u_i\,\partial_i \equiv u_x\,\frac{\partial}{\partial x} + u_y\,\frac{\partial}{\partial y} + u_z\,\frac{\partial}{\partial z}\,.$$

It is the directional derivative along the direction of the vector $\mathbf{u}$. Among other applications, the identity in (j) is used frequently in fluid dynamics, as you will see in Chap. 7.

In Chap. 8, we shall resume our discussion of vector calculus, and consider the various integral theorems satisfied by vector fields.

6.3 Solutions

1. You can work this out geometrically, but there is also a straightforward way of doing so with a bit of algebra. Expand each of the unit vectors in cylindrical and spherical polar coordinates in terms of $\mathbf{e}_x$, $\mathbf{e}_y$, and $\mathbf{e}_z$. Now exploit the fact that the latter are constant in both magnitude and direction.
(a) $\mathbf{e}_\varrho$ and $\mathbf{e}_\varphi$ have the following expansions in terms of the unit vectors in Cartesian coordinates:

$$\mathbf{e}_\varrho = \mathbf{e}_x \cos\varphi + \mathbf{e}_y \sin\varphi,$$
$$\mathbf{e}_\varphi = -\mathbf{e}_x \sin\varphi + \mathbf{e}_y \cos\varphi.$$

It follows that

$$\frac{\partial \mathbf{e}_\varrho}{\partial \varrho} = 0, \quad \frac{\partial \mathbf{e}_\varrho}{\partial \varphi} = \mathbf{e}_\varphi, \quad \frac{\partial \mathbf{e}_\varrho}{\partial z} = 0, \quad \frac{\partial \mathbf{e}_\varphi}{\partial \varrho} = 0, \quad \frac{\partial \mathbf{e}_\varphi}{\partial \varphi} = -\mathbf{e}_\varrho, \quad \frac{\partial \mathbf{e}_\varphi}{\partial z} = 0.$$

(b) Similarly, $(\mathbf{e}_r, \mathbf{e}_\theta, \mathbf{e}_\varphi)$ have the following expansions in terms of $(\mathbf{e}_x, \mathbf{e}_y, \mathbf{e}_z)$:

$$\mathbf{e}_r = \mathbf{e}_x \sin\theta \cos\varphi + \mathbf{e}_y \sin\theta \sin\varphi + \mathbf{e}_z \cos\theta,$$
$$\mathbf{e}_\theta = \mathbf{e}_x \cos\theta \cos\varphi + \mathbf{e}_y \cos\theta \sin\varphi - \mathbf{e}_z \sin\theta,$$
$$\mathbf{e}_\varphi = -\mathbf{e}_x \sin\varphi + \mathbf{e}_y \cos\varphi.$$

It then follows that the nine partial derivatives sought are

$$\frac{\partial \mathbf{e}_r}{\partial r} = 0, \quad \frac{\partial \mathbf{e}_r}{\partial \theta} = \mathbf{e}_\theta, \quad \frac{\partial \mathbf{e}_r}{\partial \varphi} = \mathbf{e}_\varphi \sin\theta$$

$$\frac{\partial \mathbf{e}_\theta}{\partial r} = 0, \quad \frac{\partial \mathbf{e}_\theta}{\partial \theta} = -\mathbf{e}_r, \quad \frac{\partial \mathbf{e}_\theta}{\partial \varphi} = \mathbf{e}_\varphi \cos\theta$$

$$\frac{\partial \mathbf{e}_\varphi}{\partial r} = 0, \quad \frac{\partial \mathbf{e}_\varphi}{\partial \theta} = 0, \quad \frac{\partial \mathbf{e}_\varphi}{\partial \varphi} = -\mathbf{e}_r \sin\theta - \mathbf{e}_\theta \cos\theta.$$

(c) Since $\mathbf{e}_\varrho = -\partial \mathbf{e}_\varphi/\partial\varphi$, it follows that $\mathbf{e}_\varrho = \mathbf{e}_r \sin\theta + \mathbf{e}_\theta \cos\theta$. ▶

5. (a) $3\varrho^2(\mathbf{e}_\varrho \cos 3\varphi - \mathbf{e}_\varphi \sin 3\varphi)$. (b) $2\mathbf{e}_r\, r \cos\theta - \mathbf{e}_\theta\, r \sin\theta$.
(c) Writing ϕ as $(\mathbf{a} \times \mathbf{b}) \cdot \mathbf{r}$, it follows that $\nabla\phi = (\mathbf{a} \times \mathbf{b})$.
(d) Rewrite the function as $(\mathbf{a} \cdot \mathbf{b})\, r^2 - (\mathbf{a} \cdot \mathbf{r})(\mathbf{b} \cdot \mathbf{r})$. The chain rule of differentiation applies to the gradient operator as well! The final result can be written as

$$\nabla\big[(\mathbf{a} \times \mathbf{r}) \cdot (\mathbf{b} \times \mathbf{r})\big] = \mathbf{a} \times (\mathbf{r} \times \mathbf{b}) + (\mathbf{a} \times \mathbf{r}) \times \mathbf{b}.$$

Since the function given is symmetric under the interchange of the vectors **a** and **b**, its gradient must also have this symmetry. Check this out. ▶

6. (a) $-2\,(\mathbf{a} \cdot \mathbf{b})$ (b) $4\,\mathbf{r} \cdot (\mathbf{a} \times \mathbf{b})$ (c) $\dfrac{\phi(\varrho)}{\varrho} + \phi'(\varrho)$ (d) $\dfrac{2\phi(r)}{r} + \phi'(r)$

Remark The fields in (c) and (d) are, respectively, *axially symmetric* and and *spherically symmetric* fields. The latter is also called a **central field**. ▶

7. (a) The ith component of the vector curl $\frac{1}{2}(\mathbf{b} \times \mathbf{r})$ is

$$\tfrac{1}{2}\, \epsilon_{ijk}\, \partial_j\, \epsilon_{klm}\, b_l\, x_m = \tfrac{1}{2}\, \epsilon_{ijk}\, \epsilon_{klm}\, b_l\, \delta_{jm} = \tfrac{1}{2}\, \epsilon_{jki}\, \epsilon_{jkl}\, b_l = \delta_{il}\, b_l = b_i\,.$$

Hence the vector curl $\frac{1}{2}(\mathbf{b} \times \mathbf{r}) \equiv \mathbf{b}$.
(b) (i) $(\mathbf{a} \times \mathbf{b}) \times \mathbf{r}$ (ii) 0 (iii) 0

Remark The field in (ii) is axially symmetric, i.e., it has cylindrical symmetry. The field in (iii) is spherically symmetric (it is a central field). Observe that both these fields are irrotational. ▶

8. (a) 0 (b) $4\cos\theta$ (c) 0 (d) $4\,(\mathbf{a}\cdot\mathbf{b})$ ▶

9. Write $\mathbf{e}_r = \mathbf{r}/r = (\sin\theta\cos\varphi)\,\mathbf{e}_x + (\sin\theta\sin\varphi)\,\mathbf{e}_y + (\cos\theta)\,\mathbf{e}_z$. Note that the Cartesian unit vectors are independent of the coordinates. Use the formula in (6.63) for $\nabla^2\phi$ in spherical polar coordinates. ▶

10. (a) 0 (b) $2\,(\mathbf{a}\times\mathbf{b})$ (c) $-\mathbf{a}\,k^2\,e^{i\mathbf{k}\cdot\mathbf{r}}$

(d) $\left[\phi''(\varrho) + \dfrac{\phi'(\varrho)}{\varrho} - \dfrac{\phi(\varrho)}{\varrho^2}\right]\mathbf{e}_\varrho$

(e) $\left[\phi''(r) + \dfrac{2\phi'(r)}{r} - \dfrac{2\phi(r)}{r^2}\right]\mathbf{e}_r$.

Remark The results in cases (d) and (e) are the expressions obtained by the action of the Laplacian on an axially symmetric vector field and a spherically symmetric vector field, respectively. ▶

Chapter 7
A Bit of Fluid Dynamics

Fluid dynamics provides a beautiful physical illustration of numerous aspects of vector calculus. This chapter is devoted to a quick recapitulation of some of these aspects.

7.1 Equation of Motion of a Fluid Element

There are two broad approaches to fluid dynamics. In the first, attention is focused on a particular *particle* of the fluid, and its motion is tracked as a function of time. This must be done consistently for all the particles of the fluid. This is called the **Lagrangian approach**. In the second, called the **Eulerian approach**, we focus our attention on a particular *point* in the region occupied by the fluid, and keep track of the velocity of the fluid and its density *at that point* as a function of time. This must be done consistently for all the points of the region occupied by the fluid. Of course, *different* particles of the fluid arrive at any given point and depart from it as time elapses, but this does not matter. Each of the two approaches has its advantages and disadvantages, but they are, ultimately, equivalent to each other. I will use the Eulerian approach here, as it is simpler to describe quantitatively.

7.1.1 Hydrodynamic Variables

The Eulerian description is based on a set of local **hydrodynamic variables**. The basic ones among these (the ones that are relevant in the simplest situations) are the pressure $P(\mathbf{r}, t)$, the velocity $\mathbf{v}(\mathbf{r}, t)$, and the density $\rho(\mathbf{r}, t)$. Here $\mathbf{r}$ is the position vector of a general *point* in the region in which the fluid is flowing, and *not* the position vector of any particular physical particle of the fluid. The state of the entire

V. Balakrishnan, *Mathematical Physics*,
https://doi.org/10.1007/978-3-030-39680-0_7

fluid at any instant of time is described by a pair of fields, namely, the scalar field $\rho(\mathbf{r}, t)$ and the vector field $\mathbf{v}(\mathbf{r}, t)$. The basic problem is the following:

- Given the pressure field $P(\mathbf{r}, t)$ and the externally applied forces (if any), find the fields $\mathbf{v}(\mathbf{r}, t)$ and $\rho(\mathbf{r}, t)$.

We shall be concerned only with **streamlined flow**. Recall that

- a streamline is a field line of the velocity field $\mathbf{v}$. That is, the velocity vector $\mathbf{v}(\mathbf{r}, t)$ is tangential to the streamline passing through the point $\mathbf{r}$ at time t.

When the velocity field $\mathbf{v}(\mathbf{r}, t)$ is well-defined almost everywhere, and is sufficiently smooth, we have streamlined flow—as opposed to **turbulent flow**. When turbulence occurs, the variation of the velocity field in space and time has a degree of randomness or stochasticity. In spite of a great deal of progress over the past 300 years or so, the precise nature of this chaotic behavior is not fully understood, and the problem of fluid turbulence is far from fully solved.

It is important not to confuse streamlined flow with **steady flow**. In general, the streamlined flow may be unsteady—the streamline pattern of the fluid may change with time. If the flow is steady, this time dependence disappears, and we have a fixed streamline pattern throughout the fluid. The **equation of continuity**, connecting the density and velocity fields, expresses the conservation of matter. In a region without sources or sinks, we must have

$$\boxed{\frac{\partial \rho}{\partial t} + \nabla \cdot \mathbf{j} = 0\,, \quad \text{where the fluid current density } \mathbf{j} = \rho\,\mathbf{v}\,.} \tag{7.1}$$

For an *incompressible, homogeneous* fluid, $\rho(\mathbf{r}, t)$ is equal to a constant both in space and in time. Therefore the equation of continuity reduces to

$$\boxed{\nabla \cdot \mathbf{v} = 0 \;\; \text{(incompressible fluid).}} \tag{7.2}$$

Hence the velocity field of an incompressible fluid is solenoidal (in a region free of sources and sinks). This fact leads to considerable simplification. In practical terms, it represents one of the main differences between the flow of a gas and that of a liquid.

7.1.2 Equation of Motion

In order to obtain the equation of motion of the fluid, we apply Newton's second law to an infinitesimal volume element δV of the fluid that is located at the point $\mathbf{r}$ at time t. The mass of this element is $\rho\,\delta V$, and its acceleration is $d\mathbf{v}/dt$. (Remember that $\rho = \rho(\mathbf{r}, t)$ in general.) The fluid flows under the action of three types of forces:

(i) Any external or applied force; let $\mathbf{F}_{\text{ext}}$ denote this force, per unit volume of the fluid.
(ii) The force arising from any "pressure head" (i.e., pressure difference) in the fluid; this is given by $-\nabla P$ per unit volume of the fluid, where P is the pressure at the point concerned.
(iii) In addition to these forces there is always a dissipative force, due to **viscosity** or internal friction, whenever the fluid moves; let $\mathbf{F}_{\text{visc}}$ be the viscous drag force per unit volume of the liquid.

Then, canceling out the factor δV on both sides of Newton's equation, we have

$$\rho \frac{d\mathbf{v}}{dt} = \mathbf{F}_{\text{ext}} - \nabla P + \mathbf{F}_{\text{visc}} . \tag{7.3}$$

If the external force is a conservative one, it can be related to the gradient of a potential: we have $\mathbf{F}_{\text{ext}}(\mathbf{r}) = -\rho \nabla \Phi$, where Φ is a scalar potential. (The most common example is, of course, the force of gravity. Taking the z-direction to be the vertical one, we have $\Phi = gz$ in this case.) The equation of motion is then

$$\frac{d\mathbf{v}}{dt} = -\frac{\nabla P}{\rho} - \nabla \Phi + \frac{\mathbf{F}_{\text{visc}}}{\rho} . \tag{7.4}$$

It is important to note that $\mathbf{F}_{\text{visc}}$ *cannot* be written as a gradient of some scalar potential. In Sect. 7.4.1, you will see that it can be expressed in terms of the velocity field itself, in the framework of a simple model of the stresses in the fluid.

The acceleration $d\mathbf{v}/dt$ on the left-hand side of Eq. (7.4) is the *total* time derivative of the velocity $\mathbf{v}(\mathbf{r}, t)$. This total time rate of change has two contributions. First, $\mathbf{v}$ can have an *explicit* dependence on t: at a given fixed location in space, $\mathbf{v}$ may vary with time *if the flow is not steady*. Second, $\mathbf{v}$ has an *implicit* dependence on t through its dependence on $\mathbf{r}$. This arises because the particular fluid element that happens to be at $\mathbf{r}$ at time t has an instantaneous velocity $d\mathbf{r}/dt$. Hence

$$\frac{d\mathbf{v}(\mathbf{r}, t)}{dt} = \frac{\partial \mathbf{v}}{\partial t} + \frac{\partial \mathbf{v}}{\partial x}\frac{dx}{dt} + \frac{\partial \mathbf{v}}{\partial y}\frac{dy}{dt} + \frac{\partial \mathbf{v}}{\partial z}\frac{dz}{dt} = \frac{\partial \mathbf{v}}{\partial t} + (\mathbf{v} \cdot \nabla)\, \mathbf{v}. \tag{7.5}$$

Here the symbol $(\mathbf{v} \cdot \nabla)$ stands for the scalar operator that is given in Cartesian coordinates by

$$\mathbf{v} \cdot \nabla = v_x \frac{\partial}{\partial x} + v_y \frac{\partial}{\partial y} + v_z \frac{\partial}{\partial z} . \tag{7.6}$$

Thus, in a moving medium, the total time derivative of *any* function of $\mathbf{r}$ and t is made up of two parts, according to the formula

$$\boxed{\frac{d\,\cdot}{dt} = \frac{\partial}{\partial t} + \mathbf{v} \cdot \nabla} \tag{7.7}$$

The second term on the right-hand side is a direct effect of the motion of the fluid, and is called the **convective derivative**.

Returning to the equation of motion, (7.4) may now be written as

$$\frac{\partial \mathbf{v}}{\partial t} + (\mathbf{v} \cdot \nabla)\, \mathbf{v} = -\frac{\nabla P}{\rho} - \nabla \Phi + \frac{\mathbf{F}_{\text{visc}}}{\rho}\,. \tag{7.8}$$

Next, recall the last of the identities listed at the end of Chap. 6, Sect. 6.2.8, namely,

$$(\mathbf{v} \cdot \nabla)\, \mathbf{v} = \nabla \left(\tfrac{1}{2} v^2\right) - \mathbf{v} \times (\nabla \times \mathbf{v}). \tag{7.9}$$

Substituting this in the equation of motion, we finally obtain

$$\boxed{\frac{\partial \mathbf{v}}{\partial t} - \mathbf{v} \times (\nabla \times \mathbf{v}) = -\frac{\nabla P}{\rho} - \nabla \left(\Phi + \tfrac{1}{2} v^2\right) + \frac{\mathbf{F}_{\text{visc}}}{\rho}.} \tag{7.10}$$

This is the general form of the equation of motion of a fluid. Together with the equation of continuity

$$\boxed{\frac{\partial \rho}{\partial t} + \nabla \cdot (\rho\, \mathbf{v}) = 0,} \tag{7.11}$$

we thus have, in principle, a sufficient number of equations to solve for the fields $\mathbf{v}(\mathbf{r}, t)$ and $\rho(\mathbf{r}, t)$—provided the other quantities such as Φ, P and $\mathbf{F}_{\text{visc}}$ are specified, and appropriate initial and boundary conditions are given.

It is very important to note the presence of the quadratic nonlinearity in the velocity in the equation of motion (7.10). This nonlinearity is the primary reason why fluid motion, in general, exhibits such incredibly complex kinds of dynamical behavior—including, in particular, the phenomenon of turbulence.

7.2 Flow When Viscosity Is Neglected

7.2.1 Euler's Equation

In many physical situations, it is reasonable to assume that the viscous drag $\mathbf{F}_{\text{visc}}$ is negligible. We then have a "dry" or **inviscid fluid**. Clearly, this assumption cannot be rigorously true in general. But it is a good approximation that leads to a simplified theory for the understanding of phenomena in fluid dynamics in which viscosity plays no direct role. For example, if a cylindrical solid is rotated about its axis in a fluid, it sets up a circular motion of the fluid about the axis because the fluid has a viscosity and "sticks" to the surface of the solid. Having set up a circulation, we may be justified, in some situations, in considering the motion of the fluid over short time scales under the assumption that the viscosity is negligible. I shall assume henceforth that the viscous drag is negligible, and return to the case of a viscous fluid in Sect. 7.4. For an inviscid fluid, the equation of motion reduces to

$$\boxed{\frac{\partial \mathbf{v}}{\partial t} - \mathbf{v} \times (\nabla \times \mathbf{v}) = -\frac{\nabla P}{\rho} - \nabla \left(\Phi + \tfrac{1}{2}v^2\right).} \tag{7.12}$$

This is called **Euler's equation**.

7.2.2 *Barotropic Flow*

Even in the absence of $\mathbf{F}_{\text{visc}}$, the right-hand side in the equation of motion (7.12) is not a pure gradient term. In general, the quantity $(\nabla P)/\rho$ is not the gradient of a scalar function. However, there are two cases in which it is indeed a gradient. The first of these is a special case of the second.

(i) If the fluid is *incompressible* (and homogeneous), then of course $\rho =$ constant. It then follows trivially that $(\nabla P)/\rho = \nabla(P/\rho)$.

(ii) In the case of *gases*, we do not expect $\rho =$ constant to be a good approximation. Gases are quite compressible, and we must certainly take the variation of ρ into account.[1] In general, the equation of state of a fluid gives its pressure as a function of its density and temperature. At any fixed temperature, therefore, P is a function of ρ. In principle, this relationship may then be inverted to express ρ as a function of P alone. In such cases it *is* possible to write $(\nabla P)/\rho$ as a gradient of a scalar function, as follows. Since $\rho = \rho(P)$, the indefinite integral

$$\int \frac{dP}{\rho(P)} = \text{some function } f(P), \text{ say.} \tag{7.13}$$

Now, for any function $f(P)$, we have

$$\nabla f(P) = \frac{df(P)}{dP}\, \nabla P. \tag{7.14}$$

But, by the very definition of $f(P)$,

$$\frac{df(P)}{dP} = \frac{1}{\rho(P)}. \tag{7.15}$$

[1] Surprisingly, however, it turns out that in many problems in gas dynamics, the assumption of incompressibility does not lead to much error, as long as the speed of the gas is less than about a third of the speed of sound in the gas.

Hence

$$\frac{\nabla P}{\rho(P)} = \nabla f(P) = \nabla \left(\int \frac{dP}{\rho(P)} \right). \tag{7.16}$$

A flow in which $\rho = \rho(P)$ is called a **barotropic flow**. It is immediately obvious that incompressible flow is a simple special case of a barotropic flow.

Here onward, we shall consider the barotropic flow of an inviscid liquid, unless otherwise specified. Euler's equation for such a flow reads

$$\boxed{\frac{\partial \mathbf{v}}{\partial t} - \mathbf{v} \times (\nabla \times \mathbf{v}) = -\nabla \left(\int \frac{dP}{\rho(P)} + \Phi + \frac{v^2}{2} \right) \quad \text{(barotropic flow).}} \tag{7.17}$$

This equation will be our starting point in what follows. Note that the right-hand side in Euler's equation has finally been reduced to a pure gradient term. This fact has significant implications, as you will see shortly.

★ **1.** Write down Euler's equation in the case when $P = K\rho^n$ where K and n are positive constants.

7.2.3 Bernoulli's Principle in Steady Flow

In *steady* flow, the velocity field $\mathbf{v}$ has no explicit time dependence, i.e., $\mathbf{v} = \mathbf{v}(\mathbf{r})$, so that $\partial \mathbf{v}/\partial t = 0$. Equation (7.17) then reduces to

$$\mathbf{v} \times (\nabla \times \mathbf{v}) = \nabla \left(\int \frac{dP}{\rho(P)} + \Phi + \frac{v^2}{2} \right). \tag{7.18}$$

Equation (7.18) leads at once to a very useful result: namely,

$$\boxed{\int \frac{dP}{\rho(P)} + \Phi + \frac{v^2}{2} = \text{a constant along each streamline.}} \tag{7.19}$$

This is the famous relation known as **Bernoulli's Principle**. You must note that the value of the constant can be (and in general, is) different along different streamlines.

If the flow is *steady as well as irrotational*, then

$$\int \frac{dP}{\rho(P)} + \Phi + \frac{v^2}{2} = \text{a constant } \textit{throughout} \text{ the fluid.} \tag{7.20}$$

In other words, the constant is the *same* for all streamlines in this case.

★ **2.** Establish (a) Eq. (7.19) for steady flow, and (b) Eq. (7.20) for steady irrotational flow.

7.2.4 Irrotational Flow and the Velocity Potential

The equations governing the velocity field become particularly simple in the special case of the steady, irrotational flow of an inviscid, incompressible fluid. The continuity equation reduces to the statement that the divergence of **v** is zero (recall Eq. (7.2)); and "irrotational" means that its curl is zero. In this very special case, therefore. we have

$$\left.\begin{array}{l}\nabla\cdot\mathbf{v}=0,\\ \nabla\times\mathbf{v}=0\end{array}\right\}\quad\text{(steady, irrotational, inviscid, incompressible flow).}\tag{7.21}$$

Since $\mathbf{v}(\mathbf{r})$ is now an irrotational vector field, it can be written as the gradient of a scalar field, $\mathbf{v}(\mathbf{r}) = \nabla\,\phi(\mathbf{r})$. It is natural to call the scalar field $\phi(\mathbf{r})$ the **velocity potential**. It follows from $\nabla\cdot\mathbf{v}=0$ that ϕ satisfies the equation

$$\boxed{\nabla^2\,\phi = 0.}\tag{7.22}$$

This is, of course, **Laplace's equation**, which appears in many other contexts as well. The solutions of Laplace's equation are called **harmonic functions**. More will be said about this equation and its solutions in Chap. 8, Sect. 8.2 (see also Chap. 22, Sect. 22.3.2).

Two-dimensional flow: The velocity potential is particularly useful in the case of two-dimensional flow (in the xy-plane, say). Since $\mathbf{v}=\nabla\,\phi$, the streamlines are normal to the family of curves $\phi(x, y) = \text{constant}$, i.e., to the equipotentials.

- Streamlines and equipotentials thus constitute two mutually orthogonal families of curves.

Just as the equipotentials are given by $\phi(x, y) = \text{constant}$, the streamlines are given by an equation of the form $\psi(x, y) = \text{constant}$. The function $\psi(x, y)$ is called the **stream function**. Clearly, $\nabla\,\phi\cdot\nabla\,\psi = 0$, as illustrated in Fig. 7.1. It turns out that the stream function also satisfies Laplace's equation, i.e., it is also a solution of

$$\nabla^2\,\psi(x, y) = 0.\tag{7.23}$$

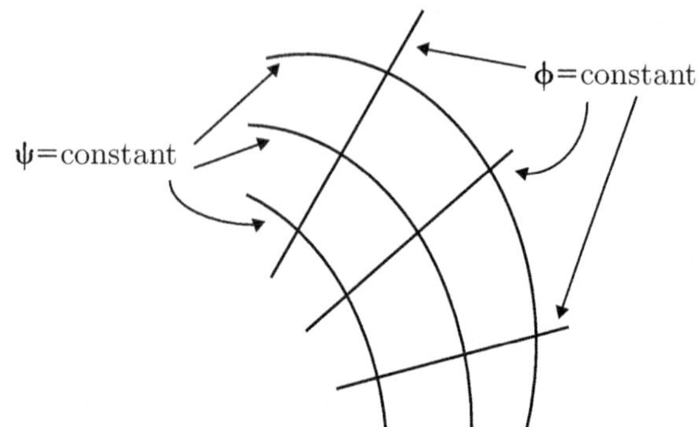

Fig. 7.1 Orthogonal families of equipotentials and streamlines in two-dimensional flow

★ **3.** Consider the steady, streamlined flow of an incompressible, inviscid fluid in the xy-plane.

(a) If the streamlines are given by the family of curves $x^2 - y^2 =$ constant, find the velocity potential $\phi(x, y)$. Schematically sketch the families of streamlines and equipotentials.
(b) If the velocity potential is given by $x^2 + y^2$, find the equation to the family of streamlines. Schematically sketch the families of streamlines and equipotentials.

The relationship between equipotentials and stream functions is actually even closer. It turns out that they constitute the real and imaginary parts of an analytic function of the complex variable $z = x + iy$. We will discuss analytic functions in some detail Chap. 22 and in subsequent chapters. I merely mention here that ϕ and ψ satisfy the **Cauchy–Riemann conditions** that make $\phi + i\psi$ an analytic function of z. These conditions are

$$\frac{\partial\phi}{\partial x} = \frac{\partial\psi}{\partial y}, \quad \frac{\partial\phi}{\partial y} = -\frac{\partial\psi}{\partial x}. \tag{7.24}$$

Observe that Eqs. (7.24) imply that both ϕ and ψ satisfy Laplace's equation, i.e., both of them are harmonic functions.

7.3 Vorticity

7.3.1 *Vortex Lines*

Closely associated with the velocity vector field we have another vector field, namely, its curl. This is called the **vorticity**, and is defined as

$$\boxed{\chi \stackrel{\text{def.}}{=} \operatorname{curl} \mathbf{v}.} \tag{7.25}$$

Physically, a direct manifestation of vorticity is the occurrence of vortexes or eddies (or "whirlpools") in the flow. As you have already seen in Eq. (6.44) of Chap. 6, Sect. 6.2.3, the vorticity at any point in the fluid is a direct measure of the *local* angular velocity $\boldsymbol{\omega}$ of a fluid element at that point. Repeating that equation for ready reference,

$$\boxed{\boldsymbol{\omega} = \tfrac{1}{2}\chi.} \tag{7.26}$$

The field lines of the vorticity field are called **vortex lines**. The direction of the vorticity vector at any point is tangential to the vortex line through that point. As in the case of streamlines, different vortex lines cannot intersect, because the velocity field and its curl must be unique at every point at any instant of time. If the velocity field is irrotational, there are no vortex lines at all. In that case, $\chi = 0$ everywhere.

You must not confuse a *vortex line*, which is just the field line of the vector field curl $\mathbf{v}$, with a *vortex*. The latter term is used for a *singularity* of the velocity field around which the circulation of $\mathbf{v}$ is nonzero. We have already encountered an example of a vortex in Chap. 6, Sect. 6.2.3. Equation (6.50), $\mathbf{v} = (K/\varrho)\, \mathbf{e}_\varphi$, describes a velocity field that has no vorticity, but has a vortex lying along the z-axis. Figure 7.2 schematically depicts a vortex in a fluid, a line singularity around which the circulation of the velocity is nonzero.

Going back to Eq. (7.18) for steady, barotropic flow, it is easy to derive a result that is analogous to Bernoulli's Principle, Eq. (7.19). We have

$$\boxed{\int \frac{dP}{\rho(P)} + \Phi + \frac{v^2}{2} = \text{constant along each vortex line.}} \tag{7.27}$$

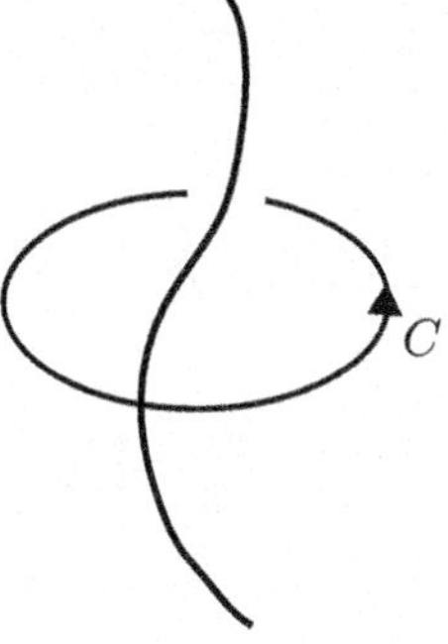

Fig. 7.2 A vortex or a line singularity of the velocity field. The circulation $\oint_C \mathbf{v} \cdot d\boldsymbol{\ell}$ around the singularity is a measure of the strength of the vortex

★ **4.** Derive Eq. (7.27).

★ **5.** If $\mathbf{v}(\mathbf{r}) = K\,xyz\,\mathbf{r}$ where K is a constant, show that χ is everywhere normal to $\mathbf{v}$ for this flow.

Beltrami flow: In a general flow, the curl of the velocity, $\nabla \times \mathbf{v}$, need not be normal to the velocity vector $\mathbf{v}$. Remember that ∇ is not an ordinary vector, but a vector differential operator! In fact, $\nabla \times \mathbf{v}$ can even be *parallel* to $\mathbf{v}(\mathbf{r})$. Such a flow is called a **Beltrami flow**.

A special case of Beltrami flow arises when $\chi(\mathbf{r}) = k\,\mathbf{v}(\mathbf{r})$ where k is a constant, and the fluid is incompressible. The velocity field then satisfies the **Helmholtz equation**

$$\boxed{(\nabla^2 + k^2)\,\mathbf{v} = 0.} \tag{7.28}$$

★ **6.** Establish Eq. (7.28), using the vector identity for $\nabla \times (\nabla \times \mathbf{v})$ and the continuity equation (7.2) for an incompressible fluid.

7.3.2 *Equations in Terms of* v *Alone*

Let us go back to Eq. (7.17) for barotropic flow. If the fluid is incompressible, we can use the vorticity vector to obtain simpler equations that completely determine the velocity field.

Taking the curl of both sides of Eq. (7.17) gives

$$\frac{\partial \chi}{\partial t} + (\mathbf{v}\cdot\nabla)\,\chi - (\chi\cdot\nabla)\,\mathbf{v} = 0 \tag{7.29}$$

where $\chi = \nabla \times \mathbf{v}$, as already defined. This equation, together with the corresponding equation of continuity $\nabla \cdot \mathbf{v} = 0$, completely specifies the velocity field of an inviscid, incompressible fluid.

★ **7.** Establish Eq. (7.29), using the vector identity for $\nabla \times (\mathbf{v} \times \chi)$.

Using Eq. (7.7) for the total time derivative, we can rewrite Eq. (7.29) as

$$\frac{d\chi}{dt} - (\chi\cdot\nabla)\,\mathbf{v} = 0. \tag{7.30}$$

This is a *first*-order differential equation in time for χ. Therefore:

- if the vorticity is zero everywhere at some instant of time t_0 in a barotropic flow, then it remains zero for all time.

An even stronger result can be derived. It can be shown that there is a kind of "conservation of vorticity", in the sense that *the total circulation in the fluid remains unchanged in time*, although vortex filaments can move about within the fluid. I do not go into the details here.

Note, however, that the foregoing results are based on the approximation that viscous damping can be neglected. In a real fluid, one of the effects of the viscosity of the fluid is to damp out vortices or eddies. Another effect is to cause transfer of energy between eddies of different strengths and sizes. Clearly, the vorticity cannot remain constant in time under such circumstances.

7.4 Flow of a Viscous Fluid

7.4.1 The Viscous Force in a Fluid

Viscosity is another name for the internal friction in a fluid. It is quantified by means of a coefficient of viscosity, η, defined as follows. Consider fluid flow in the x-direction between two layers of area δA separated by a distance δy (see Fig. 7.3). Newton's empirical relation or "law of viscosity" connects the frictional force in this volume element to the change in the velocity between the two layers, according to

$$\delta F = \eta\, \delta A\, \frac{\delta v_x}{\delta y}\,. \tag{7.31}$$

Here $\delta v_x/\delta y$ is the transverse velocity gradient. The constant of proportionality η is called the coefficient of viscosity of the fluid. It has the physical dimensions $[\eta] = ML^{-1}T^{-1}$.

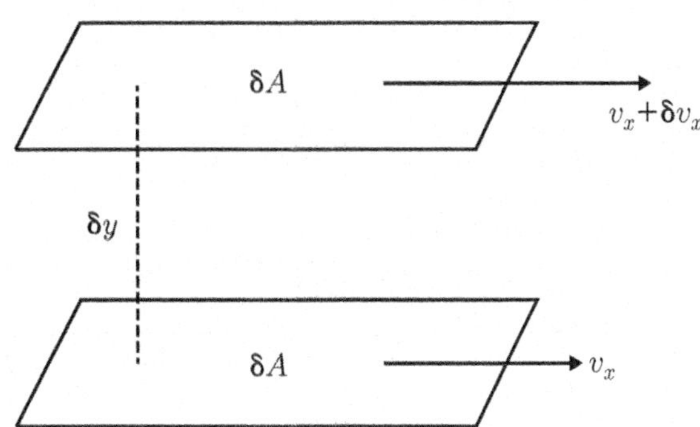

Fig. 7.3 Illustrating Newton's empirical formula for viscous drag in a fluid

Newton's law of viscosity is generalized as follows, to define the class of **Newtonian fluids**. The force per unit area, $\delta F/\delta A$, is the shear stress in the x-direction, acting on faces normal to the y-direction. It is equal to $\eta\,(\delta v_x/\delta y)$, which tends to $\eta\,\partial v_x/\partial y$ in the limit. This is generalized to the case of the shear components of the stress tensor $\boldsymbol{\sigma}$ according to

$$\sigma_{xy} = \eta\Big(\frac{\partial v_x}{\partial y} + \frac{\partial v_y}{\partial x}\Big), \quad \sigma_{yz} = \eta\Big(\frac{\partial v_y}{\partial z} + \frac{\partial v_z}{\partial y}\Big), \quad \sigma_{zx} = \eta\Big(\frac{\partial v_z}{\partial x} + \frac{\partial v_x}{\partial z}\Big). \tag{7.32}$$

The symmetrization in the indices is required because the stress tensor is a symmetric tensor (i.e., $\sigma_{ij} = \sigma_{ji}$), on general grounds. In physical terms, σ_{yx} is the shear stress in the x-direction on faces normal to the y-direction. σ_{zy} and σ_{xz} have analogous interpretations. (I have already mentioned this in Chap. 5, Sect. 5.3.2.) The (x-component of the) viscous force per unit volume is obtained, in the present case, by dividing $\delta F/\delta A$ by δy, because $\delta A\,\delta y$ is the volume of the element of fluid considered. This gives $(F_{\text{visc}})_x = \partial\sigma_{yx}/\partial y$. More generally, it is given by

$$(F_{\text{visc}})_x = \frac{\partial \sigma_{xx}}{\partial x} + \frac{\partial \sigma_{yx}}{\partial y} + \frac{\partial \sigma_{zx}}{\partial z}. \tag{7.33}$$

By direct analogy, the other two components of the viscous force per unit volume are

$$(F_{\text{visc}})_y = \frac{\partial \sigma_{xy}}{\partial x} + \frac{\partial \sigma_{yy}}{\partial y} + \frac{\partial \sigma_{zy}}{\partial z}, \quad (F_{\text{visc}})_z = \frac{\partial \sigma_{xz}}{\partial x} + \frac{\partial \sigma_{yz}}{\partial y} + \frac{\partial \sigma_{zz}}{\partial z}. \tag{7.34}$$

It only remains to substitute the expressions for the components of the stress tensor, Eqs. (7.32), in Eqs. (7.33) and (7.34). Doing so yields

$$\boxed{\mathbf{F}_{\text{visc}} = \eta\,\big[\nabla^2\mathbf{v} + \nabla\,(\nabla\cdot\mathbf{v})\big]. \quad \text{(Newtonian fluid)}} \tag{7.35}$$

Remember that $\mathbf{F}_{\text{visc}}$ is the viscous force per unit volume of the fluid.

★ **8.** Derive Eq. (7.35) from Eqs. (7.32)–(7.34).

7.4.2 The Navier–Stokes Equation

We are finally in a position to write down the fundamental equation of motion of a viscous fluid. Going back to Eq. (7.10) and using Eq. (7.35) for $\mathbf{F}_{\text{visc}}$, we get

$$\boxed{\frac{\partial \mathbf{v}}{\partial t} - \mathbf{v}\times(\nabla\times\mathbf{v}) = -\frac{\nabla P}{\rho} - \nabla\big(\Phi + \tfrac{1}{2}v^2\big) + \frac{\eta}{\rho}\big[\nabla^2\mathbf{v} + \nabla\,(\nabla\cdot\mathbf{v})\big].} \tag{7.36}$$

This is (one form of) the famous **Navier–Stokes equation**. Note that the equation has now become a *second*-order partial differential equation. From the mathematical point of view, this is a major complication, over and above the nonlinearity of the equation in $\mathbf{v}$. For these reasons, the Navier–Stokes equation presents very formidable mathematical challenges that are far from resolved, to this day.

If the fluid is incompressible, $\nabla \cdot \mathbf{v} = 0$. Equation (7.36) then simplifies to

$$\boxed{\frac{\partial \mathbf{v}}{\partial t} - \mathbf{v} \times (\nabla \times \mathbf{v}) = -\nabla \left(\frac{P}{\rho} + \Phi + \frac{v^2}{2} \right) + \frac{\eta}{\rho} \nabla^2 \mathbf{v}.} \tag{7.37}$$

Even in this simplified form, the mathematical complexity of the Navier–Stokes equation persists. Take the curl of both sides of Eq. (7.37) and write $\chi = \text{curl}\, \mathbf{v}$ as usual. Since $\nabla \times \nabla^2 \mathbf{v} = \nabla^2 (\nabla \times \mathbf{v})$, we obtain

$$\frac{\partial \chi}{\partial t} = \nabla \times (\mathbf{v} \times \chi) + \frac{\eta}{\rho} \nabla^2 \chi, \tag{7.38}$$

The quantity η/ρ is called the **kinematic viscosity**.

- It is the ratio η/ρ, rather than η or ρ separately, that controls the behavior of the velocity field.

Using the same vector identities as before, it is easy to see that

$$\boxed{\frac{d\chi}{dt} = (\chi \cdot \nabla)\, \mathbf{v} + \frac{\eta}{\rho} \nabla^2 \chi.} \tag{7.39}$$

Together with $\nabla \cdot \mathbf{v} = 0$, this equation governs the motion of a viscous incompressible fluid. Once again, its relatively simple appearance is deceptive. Its set of possible solutions is incredibly complex. Turbulent flows lurk among these possibilities.

Here is another indication of the difficulties involved in solving Eq. (7.38) or (7.39). Suppose we know the solution to the equation in the complete absence of viscous damping, i.e., for η set equal to zero, in some given situation (that is, for specified initial and boundary conditions). Now suppose the viscosity is nonzero, but very small. One might imagine that the solution in this case could be obtained by perturbation theory, with corrections of successively higher order in the small parameter η added to the zero-viscosity solution. But this would be completely wrong, in general, because η appears in the coefficient of the *highest* derivative (second derivative) in the partial differential equation. Setting it equal to zero changes the very *order* of the differential equation. Perturbation theory in powers of η is, therefore, not possible, in general. One needs more sophisticated techniques from what is known as **singular perturbation theory**. Fractional powers of η make their appearance in the solutions. Among other effects, diverse **boundary layer phenomena** show up as physical manifestations of these complications. It is necessary to reiterate that fluid dynamics presents problems of truly formidable mathematical complexity and difficulty.

7.5 Solutions

1. When $n \neq 1$, we have

$$\frac{\partial \mathbf{v}}{\partial t} - \mathbf{v} \times (\nabla \times \mathbf{v}) = -\nabla\Big(\frac{nK^{1/n}P^{(n-1)/n}}{n-1} + \Phi + \frac{v^2}{2}\Big).$$

When $n = 1$, we have

$$\frac{\partial \mathbf{v}}{\partial t} - \mathbf{v} \times (\nabla \times \mathbf{v}) = -\nabla\Big(K \ln P + \Phi + \frac{v^2}{2}\Big).$$

Remark The power-law dependence $P = K\rho^n$ is widely used as a model in various problems in fluid dynamics, including stellar dynamics in astrophysics. ▶

2. (a) Take the scalar product of each side of Eq. (7.18) with $\mathbf{v}$: the left-hand side vanishes identically. Recall that the gradient is the directional derivative: if the component of the gradient of a scalar quantity along some direction vanishes, it means that the quantity does not change as you move along that direction. Since the velocity field at any point is tangential to the streamline through that point, Eq. (7.19) follows.

(b) $\nabla \times \mathbf{v} = 0$ for irrotational flow. If the gradient of a scalar quantity vanishes identically in some region, all its components vanish. Hence the quantity does not change from point to point in *any* direction, i.e., it must be constant throughout the region. ▶

3. (a) On the streamline $x^2 - y^2 =$ constant, we have $x\,dx - y\,dy = 0$, or $dy/dx = x/y$. Recall (from elementary coordinate geometry!) that the normal to a straight line with slope m is a straight line with slope $-1/m$. The normal to the streamline at any point, therefore, has a slope given by $dy/dx = -y/x$. Integrating this equation, we get $xy =$ constant. The velocity potential is, therefore, given by $\phi(x, y) = xy$ (apart from irrelevant constant factors). The equipotentials are given by the family of curves $xy =$ constant. These are rectangular hyperbolas with the x and y axes as the asymptotes. They are orthogonal to the family of rectangular hyperbolas $x^2 - y^2 =$ constant, which have the lines $y = \pm x$ as their asymptotes.

(b) The curves $x^2 + y^2 = C$, where the constant C takes on positive values, are concentric circles centered at the origin. The slope at any point is given by $x\,dx + y\,dy=0$, or $dy/dx = -x/y$. The normal at any point, therefore, has a slope $dy/dx =$

y/x. Integrating this, we get $y = mx$. The streamlines are, therefore, radial lines from the origin to infinity in all directions in the plane. ▶

4. Take the dot product of both sides of Eq. (7.18) with χ, and use an argument similar to the one used in the case of a streamline. ▶

5. Write $\mathbf{r} = x\,\mathbf{e}_x + y\,\mathbf{e}_y + z\,\mathbf{e}_z$ and work in Cartesian coordinates. ▶

8. Here is yet another instance in which the index notation simplifies matters enormously. Equations (7.32) are equivalent to $\sigma_{ji} = \eta\,(\partial_j\, v_i + \partial_i\, v_j)$, while Eqs. (7.33) and (7.34) tell us that $(F_{\text{visc}})_i = \partial_j\, \sigma_{ji}$. Therefore

$$(F_{\text{visc}})_i = \eta\,(\partial_j\, \partial_j\, v_i + \partial_i\, \partial_j\, v_j) = \eta\,\big[\nabla^2\, v_i + \partial_i\,(\nabla \cdot \mathbf{v})\big],$$

which is just Eq. (7.35). ▶

Chapter 8
Some More Vector Calculus

8.1 Integral Theorems of Vector Calculus

From the point of view of physical applications, the most important among the theorems of vector calculus are **Gauss's Theorem** and **Stokes' Theorem**. You are undoubtedly already familiar with the elementary application of Gauss's Theorem in electrostatics, to find the electric field due to a charge distribution with spherical or cylindrical symmetry; and the application of Stokes' Theorem in magnetostatics, to find the magnetic field due to a steady current in a long straight wire. As you know, Gauss' Theorem relates the integral of the divergence of a vector field over a volume to the integral of the field over the closed surface enclosing the volume; while Stokes' Theorem relates the integral of the curl of a vector field over an open surface to its integral over the contour bounding the surface. Further, there is **Green's Theorem** in the plane, which is the planar counterpart of Stokes' Theorem.

8.1.1 The Fundamental Theorem of Calculus

At a basic level, *there is really just one theorem*! All the results mentioned above are special cases of this single theorem. In order to see how this is so, however, one needs some familiarity with calculus on manifolds in differential geometry, which I do not go into here. But it is worth understanding the origin of this theorem in simple, very informal terms. No claim to rigor is made here. I have left out details such as the precise continuity requirements on the functions concerned, the sense in which the integrals are to be defined (Riemann or Lebesgue), and so on.

Let x denote a real variable. We know that if $\phi(x)$ is the primitive of $f(x)$, i.e., if the indefinite integral $\int f(x)\,dx = \phi(x)$, then $\int_a^b f(x)\,dx = \phi(b) - \phi(a)$, where a and b are real numbers. This is the fundamental theorem of the calculus, in its most elementary form. $f(x)$ is the derivative of $\phi(x)$, so that $\int_a^b \phi'(x)\,dx = \phi(b) - \phi(a)$. Note that the boundary of the open interval (a, b) comprises the two points a and

V. Balakrishnan, *Mathematical Physics*,
https://doi.org/10.1007/978-3-030-39680-0_8

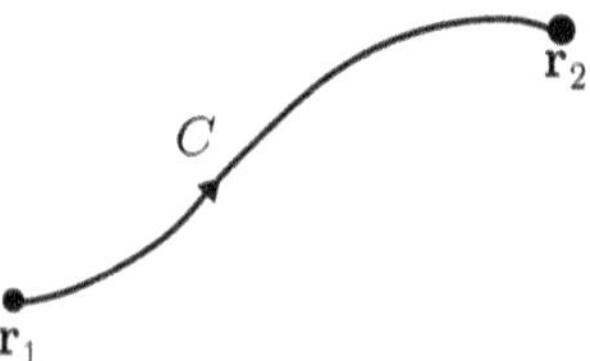

Fig. 8.1 Line integral of a gradient

b. This is a special case of a more general result. If $\phi(\mathbf{r})$ is a scalar field, and C is a contour running from the point $\mathbf{r}_1$ to the point $\mathbf{r}_2$, then

$$\boxed{\int_C \nabla\phi \cdot d\boldsymbol{\ell} = \phi(\mathbf{r}_2) - \phi(\mathbf{r}_1).} \tag{8.1}$$

The left-hand side is the integral of a certain derivative of the field over a *one*-dimensional region, the contour C. On the right-hand side, we have an "integral" (or algebraic sum) of the field over the boundary of the region—in this instance, just the *zero*-dimensional pair of end points of C, as illustrated in Fig. 8.1.

What emerges from calculus on manifolds is a profound extension of this relationship, the fundamental theorem of calculus: The integral of a certain quantity over a d-dimensional region is equal to the integral of a related quantity over the $(d-1)$-dimensional boundary of that region; and the first quantity is, in a certain sense, the "derivative" of the second. I have quoted the precise result in Eq. (8.19) below, purely for the sake of completeness.

8.1.2 *Stokes' Theorem*

Stokes' Theorem takes the simple relationship given by Eq. (8.1) to the next stage: it relates the *two*-dimensional or surface integral of a certain derivative of a vector field to a *one*-dimensional integral—namely, to the circulation of the vector field over the oriented closed curve representing the boundary of the surface. Let $\mathbf{u}(\mathbf{r})$ be a vector field that is well-defined and nonsingular at all points on an open surface S and on its boundary, namely, the closed contour C. Then

$$\boxed{\int_S (\nabla \times \mathbf{u}) \cdot d\mathbf{S} = \oint_C \mathbf{u} \cdot d\boldsymbol{\ell},} \tag{8.2}$$

where the direction in which C is traversed is such that the surface S which it bounds always lies to one's left. (Recall that $\oint_C$ denotes a line integral over the oriented closed contour C.) This ensures consistency with the right-hand rule already mentioned: the unit normal to an area element $d\mathbf{S}$ is along the direction in which a right-handed screw progresses as the boundary of the element is traversed. Note that S need not be a

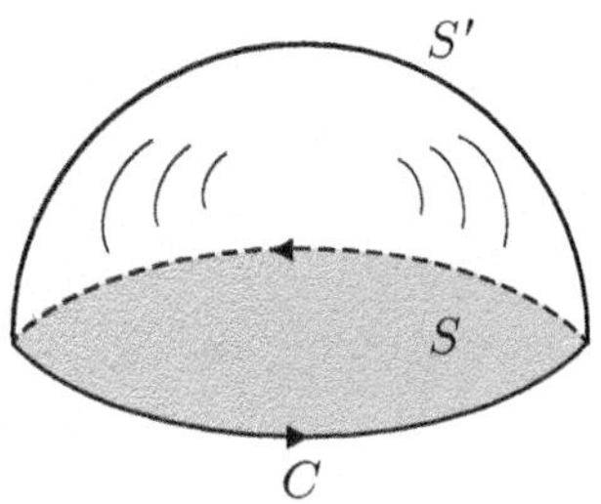

Fig. 8.2 Two open surfaces S and S' that are bounded by the contour C

planar surface. Moreover, given a closed contour C, the surface S that it bounds is certainly not unique. Equation (8.2) is applicable, then, to *any* surface bounded by C, as long as the vector field is well-defined at all points on it. Figure 8.2 shows two such surfaces, S and S', that share a common boundary C.

★ **1.** Several special cases of Stokes' Theorem follow readily. For instance, if $\phi(\mathbf{r})$ is a scalar field, show that

$$\int_S d\mathbf{S} \times \nabla\phi = \oint_C \phi\, d\boldsymbol{\ell}\,.$$

8.1.3 Green's Theorem

Green's Theorem in the plane is a special case of Stokes' Theorem. Let

$$\mathbf{u}(\mathbf{r}) = u_x(x, y)\, \mathbf{e}_x + u_y(x, y)\, \mathbf{e}_y \tag{8.3}$$

be a vector field in the xy-plane. If S is a region of the plane bounded by the contour C, then

$$\boxed{\int_S \left(\frac{\partial u_y}{\partial x} - \frac{\partial u_x}{\partial y}\right) dS = \oint_C \mathbf{u} \cdot d\boldsymbol{\ell} = \oint_C (u_x\, dx + u_y\, dy).} \tag{8.4}$$

The final equation above follows from the fact that $d\boldsymbol{\ell} = \mathbf{e}_x\, dx + \mathbf{e}_y\, dy$. Equation (8.4) is Green's Theorem in the plane.

Special case: When $\mathbf{u}$ is the gradient of a scalar field, i.e., $\mathbf{u} = \nabla\phi$, we have $u_x = \partial\phi/\partial x$ and $u_y = \partial\phi/\partial y$. The right-hand side in Green's Theorem is then given by

$$\oint_C \mathbf{u} \cdot d\boldsymbol{\ell} = \oint_C \nabla\phi \cdot d\boldsymbol{\ell} = \oint_C d\phi = 0, \tag{8.5}$$

because the scalar field $\phi(\mathbf{r})$ is supposed to be single-valued. The left-hand side of Eq. (8.4) also vanishes, because of the "integrability condition"

$$\frac{\partial^2 \phi}{\partial x\, \partial y} = \frac{\partial^2 \phi}{\partial y\, \partial x}. \tag{8.6}$$

But you are already familiar with this result from elementary calculus! It is usually stated there in the following terms:

- $P(x, y)\, dx + Q(x, y)\, dy$ is an **exact differential** if and only if $\partial P/\partial y = \partial Q/\partial x$.

Area formula: Green's Theorem also provides us with an explicit formula for the area S enclosed by a simple closed contour C in the plane. Consider the vector field

$$\mathbf{u}(\mathbf{r}) = -y\, \mathbf{e}_x + x\, \mathbf{e}_y. \tag{8.7}$$

Therefore $\partial u_y/\partial x = -\partial u_x/\partial y = 1$. Equation (8.4) then gives the area formula

$$\boxed{\int_S dS = S = \tfrac{1}{2} \oint_C (x\, dy - y\, dx).} \tag{8.8}$$

In other words, the evaluation of an *area*, which is given by a double integral, has been reduced to a single integration. There is a simple way to understand this formula. In general, if $\mathbf{r}$ is the position vector of a point in space, the area element representing the infinitesimal triangle formed by the three vectors $\mathbf{r}$, $d\mathbf{r}$, and $\mathbf{r} + d\mathbf{r}$ is given by $\frac{1}{2}(\mathbf{r} \times d\mathbf{r})$. And in two dimensions, the counterpart of the cross product $(\mathbf{r} \times d\mathbf{r})$ is given by $\epsilon_{ij}\, x_i\, dx_j = x\, dy - y\, dx$, since $\epsilon_{12} = -\epsilon_{21} = 1$.

8.1.4 A Topological Restriction; "Exact" Versus "Closed"

A very important condition must be kept in mind concerning the applicability of Green's Theorem (more generally, of Stokes' Theorem). This is best understood with the help of a simple example.

The azimuthal angle in plane polar coordinates is given by $\varphi = \tan^{-1}(y/x)$. Therefore

$$d\varphi = \frac{-y}{x^2 + y^2}\, dx + \frac{x}{x^2 + y^2}\, dy. \tag{8.9}$$

Hence $d\varphi$ is of the form $P\, dx + Q\, dy$, with

$$\frac{\partial P}{\partial y} = \frac{\partial Q}{\partial x} = \frac{y^2 - x^2}{(x^2 + y^2)^2}. \tag{8.10}$$

A blind application of Green's Theorem (Eq. (8.4)) would, therefore, demand that $\oint_C d\varphi = \oint_C \nabla\varphi \cdot d\boldsymbol{\ell}$ be equal to zero, for any closed contour C. And this is indeed so, *except* when the contour C encircles the origin of coordinates a nonzero number of times. In the latter case, of course, $\oint_C d\varphi = 2n\pi \neq 0$, where the integer n is the number of times that C winds around the origin. Here n is positive (respectively, negative) for encirclement in the anticlockwise (respectively, clockwise) sense. Thus, there is an apparent violation of Stokes' Theorem (or Green's Theorem, in this case).

The resolution of this difficulty lies in the fact that the vector field $\nabla\varphi$ is *singular* at the origin. (In fact, the azimuthal angle φ is indeterminate at the origin, as you know.) Recall that we have already encountered this field in the form $K\,\nabla\varphi = (K/\varrho)\,\mathbf{e}_\varphi$ (where K is a constant) in Chap. 6, Sect. 6.2.3, as a model of the velocity field in fluid flow around a *vortex* at the origin.

- Stokes' Theorem (Eq. (8.2)), or Green's Theorem (Eq. (8.4)) is not valid when the vector field $\mathbf{u}$ has singularities on the surface S or on the contour C bounding the surface.

There is another way out. We could try to eliminate the singularity at the origin by a mathematical device. What if we *removed* the origin of coordinates from the space, and considered the *punctured* plane, denoted by $\mathbb{R}^2 - \{(0,0)\}$? The curl of the vector field

$$\mathbf{u}(\mathbf{r}) = \frac{-y}{x^2+y^2}\,\mathbf{e}_x + \frac{x}{x^2+y^2}\,\mathbf{e}_y \tag{8.11}$$

vanishes at all points in this punctured space. And, in accordance with Stokes' Theorem, the circulation $\oint_C \mathbf{u}\cdot d\boldsymbol{\ell}$ of this vector field around any closed contour C also vanishes, *as long as the winding number of C around the origin is zero*. Figure 8.3a shows a closed contour over which the circulation of $\mathbf{u}$ vanishes. The contour in this case is *deformable* to the contours C_1 and C_2 shown in Fig. 8.3b. Observe that it comprises *two* closed paths, each of which encircles the origin once, but in *opposite* senses. In contrast, the circulation of $\mathbf{u}$ around the closed path C in Fig. 8.3c does *not* vanish. This again appears to be a violation of Stokes' Theorem. But there is, in fact, no contradiction here, for the following reason:

- Stokes' Theorem requires that the surface S be *simply-connected*: that is, any closed path in it must be continuously shrinkable to a point, without leaving the

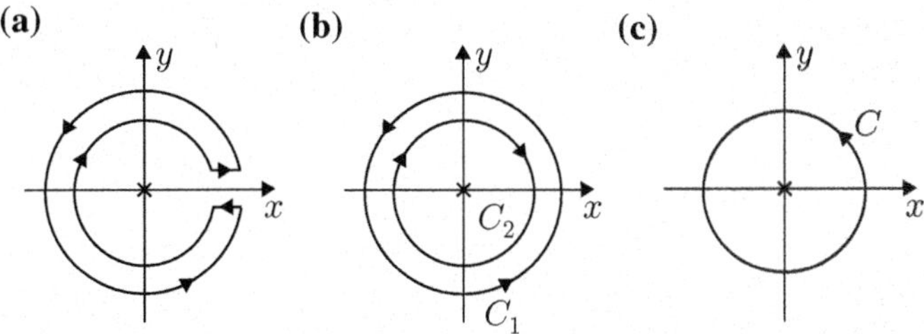

Fig. 8.3 Contours of integration in the punctured plane. The origin is indicated by a cross to show that it does not belong to the space. **a** The closed contour shown does *not* enclose the origin. It is equivalent to the pair of contours C_1 and C_2 shown in (b). The closed contour in (c) *does* enclose the origin

surface. (More precisely, the space must be "contractible" in a certain precise mathematical sense.)

- The punctured plane, on the other hand, is not simply-connected. Any closed path that has a nonzero winding number around the origin (i.e., the puncture or deleted point) *cannot* be smoothly and continuously shrunk to a point without leaving the space.

Thus, there is a *topological* restriction on the applicability of Stokes' Theorem: the requirement of simple connectivity on the surface concerned.

The example we have just considered is the simplest one that helps us understand a fundamental aspect of differential geometry: the distinction between what is known as a **closed differential form** and an **exact differential form**. Because of its importance, I will mention it here, although I have stated more than once that differential geometry is outside the scope of this book.

- In the punctured plane, the one-form $d\varphi$ is *closed* but *not exact*.
- An exact form is closed, but the converse is only true *locally*, and not necessarily globally, precisely because the topology may be "nontrivial".

The reason I mention this (albeit without further explanation) is because it is the starting point of some far-reaching developments in mathematics.

8.1.5 Gauss's Theorem

Gauss's Theorem increases once again the dimensionality of the integrals in the fundamental theorem of the calculus. For a *three*-dimensional compact region of volume V, enclosed by the *two*-dimensional closed surface S, we have

$$\boxed{\int_V (\nabla \cdot \mathbf{u})\, dV = \oint_S \mathbf{u} \cdot d\mathbf{S}.} \tag{8.12}$$

The symbol $\oint_S$ denotes an integral over a *closed* surface S. Recall that the unit normal of each area element on S is the outward normal. As in the case of the previous theorems, it is assumed that the vector field $\mathbf{u}(\mathbf{r})$ is well-behaved at all points in V and on S. More specifically, is assumed that V is compact, its surface S is piecewise smooth, and that $\mathbf{u}$ is continuously differentiable in the region concerned.

The most familiar physical application of Gauss's Theorem, of course, is encountered in electrostatics. **Gauss's Law** relates the flux of the electrostatic field over a closed surface to the total charge enclosed by the surface. This law follows directly from the application of Gauss's Theorem to a vector field with an inverse-square law fall-off.

Several special cases of Gauss's Theorem follow readily. For instance, if $\phi(\mathbf{r})$ is a scalar field, then

$$\int_V (\nabla \phi)\, dV = \oint_S \phi\, d\mathbf{S}. \tag{8.13}$$

Similarly, if $\mathbf{v}(\mathbf{r})$ is a vector field,

$$\int_V (\nabla \times \mathbf{v})\, dV = \oint_S d\mathbf{S} \times \mathbf{v}. \tag{8.14}$$

More generally, for two vector fields $\mathbf{v}(\mathbf{r})$ and $\mathbf{w}(\mathbf{r})$,

$$\int_V [\mathbf{w} \cdot (\nabla \times \mathbf{v}) - \mathbf{v} \cdot (\nabla \times \mathbf{w})]\, dV = \oint_S (\mathbf{v} \times \mathbf{w}) \cdot d\mathbf{S}. \tag{8.15}$$

★ **2.** Establish (a) Eq. (8.13) (b) Eq. (8.14) (c) Eq. (8.15).

★ **3.** Gauss's Theorem can be quite useful in evaluating certain integrals.

(a) Let S be a closed surface enclosing a volume V in three-dimensional space. Show that

$$\oint_S \nabla (r^2) \cdot d\mathbf{S} = 6V.$$

(b) Let S be the surface of a toroid that is centered at the origin, with radius a and cross-sectional radius b. Show that

$$\oint_S \mathbf{r} \cdot d\mathbf{S} = 6\pi^2 a b^2.$$

(c) Consider the vector field $\mathbf{u}(\mathbf{r}) = \mathbf{a}\, e^{i\mathbf{k}\cdot\mathbf{r}}$ where $\mathbf{a}$ and $\mathbf{k}$ are constant vectors. Let V and S denote the volume and surface of a sphere of radius R centered at the origin. Show that

$$\oint_S \mathbf{u} \cdot d\mathbf{S} = 4\pi R^3 (i\mathbf{k} \cdot \mathbf{a}) \left\{ \frac{\sin(kR) - kR \cos(kR)}{(kR)^3} \right\}.$$

8.1.6 Green's Identities and Reciprocity Relation

Gauss's Theorem leads to two identities that are quite useful in applications.

Let S be a closed surface enclosing a volume V, and let ϕ and ψ be two scalar fields that are regular in V and on S. Then, applying Gauss's Theorem to the vector field $\psi\, \nabla \phi$, it follows that

$$\boxed{\int_V (\psi \, \nabla^2 \phi + \nabla\phi \cdot \nabla\psi) \, dV = \oint_S \psi \, \nabla\phi \cdot d\mathbf{S} = \oint_S \psi \, \frac{\partial \phi}{\partial n} \, dS ,} \tag{8.16}$$

where $\partial\phi/\partial n$ denotes the rate of change of ϕ in the direction of the unit normal $\mathbf{n}$ to the surface element $d\mathbf{S}$. This is **Green's first identity**.

Interchanging ϕ and ψ in the above and subtracting the result from Eq. (8.16), the $\nabla\phi \cdot \nabla\psi$ term cancels out. We get

$$\boxed{\int_V (\psi \, \nabla^2 \phi - \phi \, \nabla^2 \psi) \, dV = \oint_S \Big(\psi \, \frac{\partial \phi}{\partial n} - \phi \, \frac{\partial \psi}{\partial n} \Big) \, dS.} \tag{8.17}$$

This is **Green's second identity**. It is used, for instance, in electrostatics and in scattering theory in quantum mechanics.

Green's reciprocity relation in electrostatics should be familiar to you. Let $\phi(\mathbf{r})$ be the electrostatic potential due to a static charge distribution $\rho(\mathbf{r})$ that is confined to a finite region of space, so that ϕ vanishes at spatial infinity. Similarly, let $\phi'(\mathbf{r})$ be the electrostatic potential due to a finite charge distribution $\rho'(\mathbf{r})$. Then

$$\boxed{\int \phi(\mathbf{r}) \, \rho'(\mathbf{r}) \, dV = \int \phi'(\mathbf{r}) \, \rho(\mathbf{r}) \, dV,} \tag{8.18}$$

where the volume integrals extend over all space.

★ **4.** Derive Eq. (8.18).

8.1.7 Comment on the Generalized Stokes' Theorem

I have already mentioned that all the theorems in the foregoing are particular cases of the fundamental theorem of calculus on manifolds, called (the generalized) Stokes' Theorem. In order to pique your curiosity, I merely state this theorem, with an admittedly very cursory (and heuristic) explanation of the terms involved.[1]

Differential geometry is concerned with differentiable manifolds. Smooth curves and smooth surfaces are, respectively, examples of one-dimensional and two-dimensional manifolds. Their generalizations in higher dimensions are manifolds in three or more dimensions. "Differentiable" essentially means that you can treat a small neighborhood of any point in the manifold as a patch of Euclidean space, and

[1]Ideally, an elementary account of differential geometry ought to have been included in this book. It would have made the discussion of several other topics included here more unified and compact. I have, however, decided against including some differential geometry for several reasons. Among these are the following. A self-contained account would take up too much space; and the subject is not (yet) in the standard curriculum for most students of physics and engineering, although this situation is changing rapidly.

carry out the familiar operations of calculus in it. If the (right-handed, say) Cartesian coordinate system that you start with at any point can be moved around everywhere on the manifold and brought back to the starting point without undergoing a change of orientation (or "handedness"), the manifold is an oriented differentiable manifold.

On such a manifold, one defines so-called differential forms. A **zero-form** is a function (of the local coordinates). An operation called **exterior differentiation** is defined, with the help of which higher order forms can be constructed. A **one-form** is the differential of a function. Higher order forms are the analogs of area elements, volume elements, and so on. They are constructed with the help of a generalization of the idea of a cross product, called the **wedge product**. Scalar, vector, and tensor fields are most naturally expressed in the language of differential geometry.

The generalized Stokes' Theorem then says

- If ω is an $(n-1)$-form with compact support on a smooth oriented n-dimensional manifold M with boundary ∂M, and d denotes the exterior derivative, then

$$\int_M \mathrm{d}\omega = \oint_{\partial M} \omega. \tag{8.19}$$

In other words, the integral of the n-form $\mathrm{d}\omega$ over M is equal to the integral of the $(n-1)$-form ω over its boundary ∂M.

For further details, you must refer to a text on differential geometry.

8.2 Harmonic Functions

Laplace's equation, $\nabla^2\phi = 0$, is one of the most basic equations of mathematical physics. It appears in a wide variety of situations. To give just two instances, the electrostatic potential in a charge-free region satisfies this equation; so do the velocity potential and stream function of an incompressible fluid in irrotational flow, as you have seen already in Eqs. (7.22) and (7.23) of Chap. 7, Sect. 7.2.4.

8.2.1 Mean Value Property

The solutions of Laplace's equation in a given domain are harmonic functions in that domain. These functions have many interesting and important properties. In one dimension, harmonic functions are trivially found, because $d^2\phi/dx^2 = 0$ simply implies that ϕ is a linear function of x, i.e., $\phi(x) = ax + b$. But this also means that *the value of the function at any point x is the arithmetic average of its values at points symmetrically placed on either side of x*, i.e.,

$$\phi(x) = \tfrac{1}{2}\big[\phi(x+\epsilon) + \phi(x-\epsilon)\big]. \tag{8.20}$$

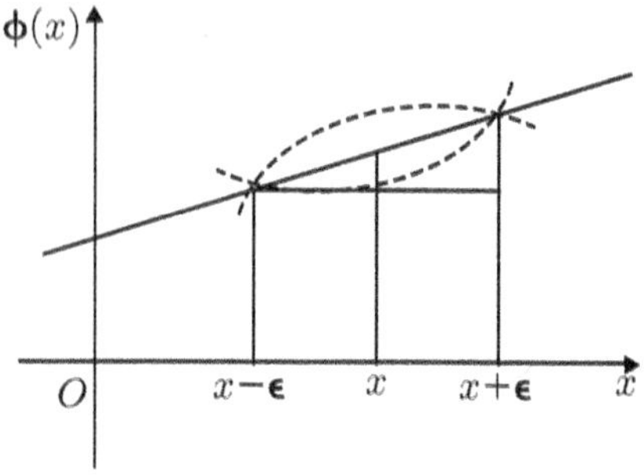

Fig. 8.4 A linear function satisfies the mean value property of Eq. (8.20). Functions that have curvature (shown by dotted lines) do not satisfy this property

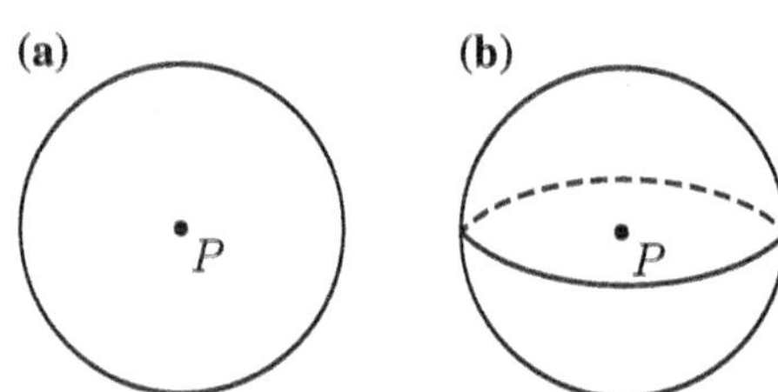

Fig. 8.5 The value of a harmonic function at the point P is the arithmetic mean of its values on **a** a circle centered at P, in two dimensions; **b** a sphere centered at P, in three dimensions

The lack of curvature of a straight line guarantees that this remains true for *any* value of ϵ, not necessarily infinitesimal. Figure 8.4 illustrates this property. Rewriting Eq. (8.20) as $[\phi(x+\epsilon)-\phi(x)]-[\phi(x)-\phi(x-\epsilon)]=0$, dividing this by ϵ^2, and taking the limit $\epsilon \to 0$ takes us back to Laplace's equation $d^2\phi/dx^2 = 0$.

The same average or mean value property leads to Laplace's equation in any number of dimensions. In two dimensions, for instance, suppose we impose (again, for any value of ϵ) the requirement

$$\phi(x, y) = \tfrac{1}{4}\big[\phi(x+\epsilon, y) + \phi(x-\epsilon, y) + \phi(x, y+\epsilon) + \phi(x, y-\epsilon)\big]. \qquad (8.21)$$

Dividing the equation by ϵ^2 and passing to the limit $\epsilon \to 0$ leads to $\partial^2\phi/\partial x^2 + \partial^2\phi/\partial y^2 = 0$. In fact, the arguments of the functions on the right-hand side in Eq. (8.21) can be those of *any* four points symmetrically located about (x, y). That is, they could be located on the corners of any square whose center is at (x, y). The sides of the square need not be parallel to the coordinate axes. This would still lead to $\nabla^2\phi = 0$, because the Laplacian operator is a scalar. It is invariant under rotations about the origin. Going a step further, this means that $\phi(x, y)$ is actually the mean of the values at all points on a circle of radius ϵ centered at (x, y).

The extension of this discussion to higher dimensions is straightforward. Figures 8.5 a and b illustrate this property in two and three dimensions, respectively. Imposing the mean value property leads to Laplace's equation and hence implies a harmonic function. Conversely, we may start with a harmonic function (i.e., a solution of Laplace's equation), and ask whether the mean value property is satisfied. This is called the **mean value theorem** for harmonic functions. In an arbitrary number d of dimensions, a harmonic function $\phi(\mathbf{r})$ satisfies

$$\phi(\mathbf{r}) = \frac{1}{V_d}\int_{\text{ball}} dV'\,\phi(\mathbf{r}') = \frac{1}{S_d}\int_{\text{sphere}} dS'\,\phi(\mathbf{r}'). \tag{8.22}$$

Here, "ball" denotes a hypersphere of arbitrary radius R centered at $\mathbf{r}$, and "sphere" stands for its hypersurface. Recall that the "volume" V_d of the ball, and the "area" S_d of its "surface", have already been recorded in Eqs. (6.18) and (6.21), respectively, of Chap. 6, Sect. 6.1.3. Repeating them for ready reference,

$$V_d(R) = \frac{\pi^{d/2} R^d}{\Gamma\left(1 + \frac{1}{2}d\right)} \quad \text{and} \quad S_d(R) = \frac{2\pi^{d/2} R^{d-1}}{\Gamma\left(\frac{1}{2}d\right)}. \tag{8.23}$$

★ **5.** Let $\phi(\mathbf{r})$ be a harmonic function in some region. If V is a volume in this region bounded by the closed surface S, show that

$$\text{(a)} \oint_S \frac{\partial \phi}{\partial n}\, dS = 0 \quad \text{and} \quad \text{(b)} \int_V |\nabla \phi|^2\, dV = \oint_S \phi\, \frac{\partial \phi}{\partial n}\, dS\,.$$

8.2.2 *Harmonic Functions Have No Absolute Maxima or Minima*

Let us return to Laplace's equation. It is immediately clear that its solutions in dimensions $d \geq 2$ need not be restricted to functions that are linear in the Cartesian coordinates. The simple reason is that it is the *sum* of second derivatives, $\sum_{i=1}^{d} \partial^2\phi/\partial x_i^2$, that is required to vanish—and not the individual second derivatives themselves.

- The set of harmonic functions is quite extensive in $d \geq 2$, and has a rich structure.

In Chap. 7, Sect. 7.2.4, I have mentioned that the velocity potential $\phi(x, y)$ and stream function $\psi(x, y)$ in the two-dimensional irrotational flow of an incompressible fluid satisfy the Cauchy–Riemann conditions (7.24), and are harmonic functions. The real and imaginary parts of *any* analytic function of a complex variable $z = x + iy$ satisfy these conditions, and are harmonic functions in specified regions of the xy-plane. Analytic functions of a complex variable will be considered in Chaps. 22–27. In Chap. 16, Sect. 16.4.7, we will discuss another very useful class of harmonic functions in three-dimensional space—namely, **spherical harmonics**.

Another important property of harmonic functions emerges readily. Consider a function $\phi(x_1, \ldots, x_d)$ of the coordinates in d-dimensional space. As you know, at an absolute maximum of such a function, *each* of the second derivatives $\partial^2\phi/\partial x_j^2$ (where $1 \leq j \leq d$) must be positive. Similarly, at an absolute minimum, each $\partial^2\phi/\partial x_j^2$ must be negative. Hence their sum cannot add up to zero in either case. That is, ϕ cannot be a harmonic function if it has absolute maxima or minima.

- A harmonic function *cannot* have an absolute maximum or minimum at any point in the *interior* of a region in which it is harmonic.

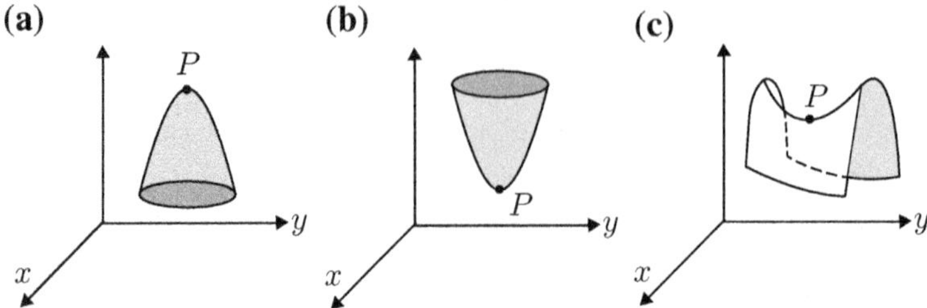

Fig. 8.6 Surfaces representing a function of two variables with **a** an absolute maximum, **b** an absolute minimum, and **c** a saddle point

Thus, a harmonic function can have maxima or minima, if any, *only* at points on the *boundary* of the region in which it is harmonic. But it can, of course, have **critical points** where the first derivative $\partial\phi/\partial x_j = 0$ for *every* j, at points within the region. In the calculus of functions of several variables, the critical points of a function $\phi(x_1, \dots, x_d)$ are the points at which its gradient $\nabla\phi$ vanishes. In the generic case, these must necessarily be points where there is a local minimum with respect to some of the coordinates, and a local maximum with respect to some others. In other words, these must be **saddle points**. Figures 8.6 a–c depict the distinction between a maximum, minimum, and saddle point in the two-dimensional case. Simple examples are provided by the functions $-x^2 - y^2$, $x^2 + y^2$, and $x^2 - y^2$, respectively. Each of these functions has a critical point at (0, 0). The first two are not harmonic functions, while the last one is.

The absence of absolute minima in a harmonic function has an interesting physical consequence: There can be no point of *stable* equilibrium for a test charge placed in the electrostatic field of a given set of charges. The electrostatic potential at any point not occupied by any of the given set of charges satisfies Laplace's equation. Hence it cannot have a minimum at any such point. Equivalently:

- A set of charges cannot be held in *stable* equilibrium by electrostatic forces alone. This is known as **Earnshaw's Theorem** in electrostatics.
- All equilibrium points in electrostatics are, therefore, of the saddle-point type, and are unstable.

8.2.3 What Is the Significance of the Laplacian?

In Sect. 6.2.6, I stated that the Laplacian and its generalizations "are of deep and profound significance" from the mathematical point of view. I now explain briefly why this is so, without really going into the technical details (which are beyond the scope of this book).

The solutions of Laplace's equation, $\nabla^2\phi = 0$, may be regarded as the eigenfunctions of the Laplacian operator[2] corresponding to the eigenvalue 0. More generally, we have the eigenvalue equation

$$\boxed{-\nabla^2\phi = \lambda\phi.} \tag{8.24}$$

This is just the Helmholtz equation. In order to identify the distinct solutions of this differential equation (i.e., the eigenfunctions of the operator $-\nabla^2$) and the corresponding eigenvalues (the spectrum of the operator), we must specify boundary conditions. Typically, one considers the equation in some *bounded* region $\mathcal{R}$ of d-dimensional Euclidean space, with the condition $\phi = 0$ at all points on the boundary of $\mathcal{R}$. These are called **Dirichlet boundary conditions**. Some far-reaching mathematical results can then be deduced:

- The eigenvalues of $-\nabla^2$ comprise an *infinite* set of positive real numbers λ_n $(n = 1, 2, \ldots)$, such that

$$0 < \lambda_1 \leq \lambda_2 \leq \lambda_3 \leq \ldots, \quad \text{where } \lambda_n \to \infty \text{ as } n \to \infty. \tag{8.25}$$

- These eigenvalues, called the Dirichlet eigenvalues of the Laplacian, are proportional to the squares of the frequencies of the normal modes of vibration of the region concerned.

Observe that 0 is *not* one of the eigenvalues in this set. A harmonic function (other than the trivial function that is identically equal to zero) cannot vanish at all points on the boundary of a finite region, i.e., it cannot satisfy the Dirichlet boundary condition. The positivity of the eigenvalues is the reason why it is more convenient to consider the operator $-\nabla^2$ rather than ∇^2 itself.

- Remarkably enough, a knowledge of the spectrum $\{\lambda_n\}$ of the Laplacian operator in $\mathcal{R}$ is tantamount to a detailed knowledge of the geometry and topology of $\mathcal{R}$ itself.

These statements can be extended to curved spaces called **Riemannian manifolds**. The generalization of the Laplacian operator to that case is called the **Laplace–Beltrami operator**.

The extent to which the spectrum determines the domain is a topic on which a vast mathematical literature exists. A particular aspect of the problem is famously[3] represented by the question, "*Can you hear the shape of a drum*?" In other words, if we know all the natural frequencies of vibration of a drumhead, can we work backwards to deduce the exact shape of the drumhead uniquely (up to an *isometry* or congruence)? The answer, it turns out, is "no", in general. Although the "shape of a drum" question refers to the situation in two dimensions, the general problem pertains

---

[2]Some of these terms may be unfamiliar to you, but they will become clear after the discussion of linear vector spaces and operators in Chaps. 10–15.

[3]Made famous by the mathematical physicist Mark Kac (1910–1984).

to any number of dimensions $d \geq 2$. There exist regions that are not isometric to each other, and yet have exactly the same set of Dirichlet eigenvalues. Such shapes are said to be *isospectral* domains.

Returning to the spectrum itself, an early result due to H. Weyl is the following. Let V be the "volume" of a compact d-dimensional manifold $\mathcal{R}$, and S its "surface area". Then, if $N(b)$ is the number of eigenvalues of $-\nabla^2$ that are less than a prescribed positive number b,

$$N(b) = \frac{V}{(2\pi)^d} b^{d/2} + c(d)\, S\, b^{(d-1)/2} + \text{terms of lower order in } b, \tag{8.26}$$

where $c(d)$ is a known constant that depends on the dimensionality of the manifold. The coefficients of the lower order terms depend on other properties of the domain such as its connectivity.

Similar information is carried by the *generating function* $G(t)$ defined as

$$G(t) = \sum_{n=1}^{\infty} e^{-\lambda_n t}, \tag{8.27}$$

where t is a positive variable. The properties of $G(t)$ as a function of t are important. Clearly, when t is very large, $G(t)$ is dominated by its leading term, $e^{-\lambda_1 t}$. Now, a great deal is known about the leading eigenvalue λ_1 . In particular, it is the solution to a minimization problem. Consider all functions $f(\mathbf{r})$ that are square-integrable in $\mathcal{R}$, i.e., for which the integral $\int_V dV\, |f|^2$ is finite. Then the eigenfunction corresponding to λ_1 is the function for which the integral of $\int_V dV\, |\nabla f|^2$ is the least. More precisely,

$$\boxed{\lambda_1 = \text{lim inf} \left\{ \frac{\int_V dV\, |\nabla f|^2}{\int_V dV\, |f|^2} \right\}.} \tag{8.28}$$

Here "lim inf" denotes, as usual, the greatest lower bound of the quantity concerned, over all square-integrable functions in $\mathcal{R}$. Equation (8.28) is often used in variational problems in several contexts.

The behavior of $G(t)$ for small values of t is far more intricate and interesting. It is obvious that $G(t) \to \infty$ as $t \to 0$, because each term in the infinite sum tends to unity. The precise manner in which $G(t)$ diverges is informative. It turns out that, as $t \to 0$, $G(t)$ is given by an *asymptotic series* in powers of $t^{1/2}$. In three dimensions, for instance, this series takes the form

$$G(t) \sim \frac{V}{(2\pi)^{3/2}} t^{-3/2} - \frac{S}{16\pi} t^{-1} + (\cdots) t^{-1/2} + (\cdots) t^0 + (\cdots) t^{1/2} + \cdots . \tag{8.29}$$

Once again, the coefficients of the successive terms in this series carry both geometric and topological information about the region concerned. Formulas of this kind have deep generalizations, and they have applications in several parts of theoretical physics such as statistical physics and quantum field theory, among others.

Even more generally, the Laplacian can be extended to the case of differential forms of arbitrary order. It is then called the **Laplace–de Rham operator**. This operator is intimately related to the Laplace–Beltrami operator, and shares many of its properties. The Laplacian is thus of fundamental importance. Mathematicians strive continually for extensions and generalizations. The Laplacian itself is prototypical of a class of differential operators called **elliptic operators**, whose properties continue to be studied in depth.

In Chap. 29, the fundamental Green function corresponding to the Laplacian operator in d-dimensional Euclidean space will be derived, in connection with the solution of Poisson's equation.

8.3 Singularities of Planar Vector Fields

In all the theorems of vector calculus discussed above, it has been assumed that the vector fields involved are not singular at any point in the regions of interest. The study of the singularities of vector fields is a subject in its own right. Going into it here will take us too far afield. But it is useful to understand at least the simplest of cases, namely, point singularities of vector fields in a two-dimensional plane. Among other applications, an important one is to the analysis of two-dimensional dynamical systems.

8.3.1 Critical Points and the Poincaré Index

The planar vector field

$$\mathbf{u}(\mathbf{r}) = u(x, y)\, \mathbf{e}_x + v(x, y)\, \mathbf{e}_y \tag{8.30}$$

is obviously singular at points where either u, or v, or both become infinite. More interestingly, the vector field is also singular at points where it *vanishes*, i.e., at points that are the common roots of the simultaneous equations

$$u(x, y) = 0 \quad \text{and} \quad v(x, y) = 0. \tag{8.31}$$

Such points are called the **critical points** of the vector field. I use this term in a generic sense. As mentioned in Sect. 8.2.2, the critical points of a scalar function $\phi(x_1, x_2, \dots, x_d)$ are the points where $\nabla\phi = 0$.

An important characterizer of a critical point is an integer called its **Poincaré index**. It is defined as

$$\boxed{n \stackrel{\text{def.}}{=} \frac{1}{2\pi}\oint_C \frac{(u\,dv - v\,du)}{(u^2+v^2)},} \tag{8.32}$$

where C encircles the singularity once in the positive (or anticlockwise) sense, and does not encircle any other singularity of the vector field. The expression on the right-hand side of Eq. (8.32) is guaranteed to be an integer, no matter what $u(x\,,\,y)$ and $v(x\,,\,y)$ are, as along as these functions are single-valued. It is not difficult to see why this is so, and where the formula for n comes from. Just as we can combine x and y to form the complex variable $z = x + iy$, we can combine u and v to form the complex number $\zeta = u + iv$. This complex number can be written in polar form as $\zeta = R\,e^{i\psi}$. Making a complete circuit around the singularity of the vector field must change the argument ψ by an integer multiple of 2π, since u and v are supposed to be single-valued at all nonsingular points. Hence $\oint_C d\psi = 2\pi n$. Equation (8.32) follows on using the fact that

$$\psi = \tan^{-1}(v/u) \implies d\psi = \frac{u\,dv - v\,du}{u^2+v^2}. \tag{8.33}$$

The Poincaré index n, also called the **winding number**, is a property of the *singularity* concerned. It is a **topological invariant**: *All* contours that encircle the singularity once in the positive sense yield the same value of n. The contour integral on the right-hand side of Eq. (8.32) is an explicit analytical expression. For this reason, n is called an *analytical* topological invariant.

- If C does not enclose any singularity of the vector field at all, then ψ returns to its original value when the complete circuit is traversed, and $n = 0$.

The converse is *not* true: if the winding number of a contour C is zero, it does not follow that there is no singularity enclosed by the contour, as you will see shorly.

The simplest singularities of a planar vector field are easily understood with the help of a set of representative examples. Consider the eight planar vector fields listed below, each with a singularity at the origin of coordinates:

$$\left.\begin{array}{llll}
\text{(a)} & x\,\mathbf{e}_x + y\,\mathbf{e}_y & \text{(b)} & y\,\mathbf{e}_x - x\,\mathbf{e}_y \\[2ex]
\text{(c)} & x\,\mathbf{e}_x - y\,\mathbf{e}_y & \text{(d)} & y\,\mathbf{e}_x + x\,\mathbf{e}_y \\[2ex]
\text{(e)} & -x\,\mathbf{e}_x - y\,\mathbf{e}_y & \text{(f)} & -y\,\mathbf{e}_x + x\,\mathbf{e}_y \\[2ex]
\text{(g)} & -x\,\mathbf{e}_x + y\,\mathbf{e}_y & \text{(h)} & -y\,\mathbf{e}_x - x\,\mathbf{e}_y\,.
\end{array}\right\} \tag{8.34}$$

★ **6.** Sketch the field lines, and find the Poincaré index of the singularity at (0, 0) of each of the vector fields in (8.34)(a)–(h).

These examples help us classify the elementary singularities of a planar vector field. The origin is a **node** in cases (a) and (e). The former is like a *source*, while the latter is like a *sink*. Note that both of these critical points have the *same* Poincaré

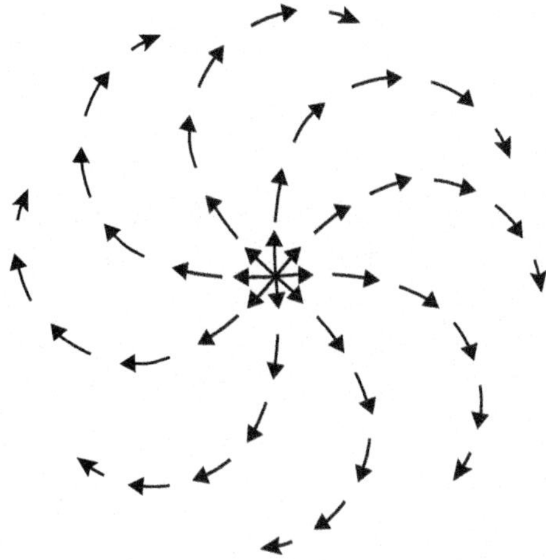

Fig. 8.7 The field lines in the vicinity of a node can be continuously deformed to look like those around a center

index, contrary to what you might guess off-hand. In (b) and (f), we have an *elliptic* critical point, called a **center**. Once again, contrary to what one might guess, the Poincaré index does not depend on whether the field lines encircle the critical point in the clockwise sense or anticlockwise sense. In all the foregoing cases, the index is given by $n = +1$. In cases (c), (d), (g), and (h), the origin is a *hyperbolic* critical point, called a **saddle point**. In these cases, $n = -1$.

What is the significance of the fact that the index ($n = +1$) at a node (with radial field lines) is the same as that of a center (concentric circles as field lines)? It means that either one of the two field patterns can be deformed smoothly and continuously to the other pattern. For instance, radial field lines emanating from the origin can be gradually bent over as we move away from the origin, till the field lines look tangential to circles at a sufficiently large distance from the critical point. This is illustrated in Fig. 8.7. Similarly, the fact that the Poincaré index of a saddle point is different from that of a node or center means that the field lines in the two cases *cannot* be deformed smoothly and continuously into each other. The two cases are *topologically* distinct from each other.

The vector fields in (8.34) are the simplest ones illustrating the singularities described above. Now consider a general critical point located at some point $(\bar{x}, \bar{y})$, say. That is, $u(\bar{x}.\bar{y}) = 0$ and $v(\bar{x}.\bar{y}) = 0$. In the neighborhood of the critical point, we may expand the functions u and v in Taylor series in powers of $(x - \bar{x})$ and $(y - \bar{y})$. Thus

$$\left.\begin{aligned} u(x, y) &\simeq (x - \bar{x})\,(\partial u/\partial x)_{(\bar{x},\bar{y})} + (y - \bar{y})\,(\partial u/\partial y)_{(\bar{x},\bar{y})} + \cdots, \\ v(x, y) &\simeq (x - \bar{x})\,(\partial v/\partial x)_{(\bar{x},\bar{y})} + (y - \bar{y})\,(\partial v/\partial y)_{(\bar{x},\bar{y})} + \cdots, \end{aligned}\right\} \qquad (8.35)$$

where the dots stand for quadratic and higher order terms in $(x - \bar{x})$ and $(y - \bar{y})$. The subscripts outside the brackets indicate that the partial derivatives are to be evaluated at $(\bar{x}, \bar{y})$. The critical point at $(\bar{x}, \bar{y})$ is *simple* or **nondegenerate** if the determinant of the Jacobian matrix L of partial derivatives at that point does not vanish. That is, if

$$\det L = \begin{vmatrix} (\partial u/\partial x)_{(\bar{x},\bar{y})} & (\partial u/\partial y)_{(\bar{x},\bar{y})} \\ (\partial v/\partial x)_{(\bar{x},\bar{y})} & (\partial v/\partial y)_{(\bar{x},\bar{y})} \end{vmatrix} \neq 0. \tag{8.36}$$

This is the standard, or generic, situation. It means that the components of the vector field $\mathbf{u}$ are well approximated, in the vicinity of the critical point, by terms that are *linear* in the shifted coordinates $(x - \bar{x})$ and $(y - \bar{y})$.

★ **7.** Show that, for a simple critical point, the Poincaré index

$$n = \begin{cases} +1 \text{ if } \det L > 0 \\ -1 \text{ if } \det L < 0. \end{cases}$$

★ **8.** Show that $\det L < 0$ if and only if the two eigenvalues of the Jacobian matrix L are real and opposite in sign. This is precisely the case in which the critical point is a saddle point.

Classification of simple critical points: For the sake of completeness, here is the classification of simple critical points of planar vector fields, in brief. Recall that examples of a node, a center and a saddle point have already been encountered.

(i) A node, if the two eigenvalues are real and have the same sign.
(ii) A saddle point, if the two eigenvalues are real and have opposite signs.
(iii) A center, if the two eigenvalues are pure imaginary.
(iv) A **spiral point** (also called a **focus**), if the eigenvalues are a complex conjugate pair with a nonzero imaginary part.

All other instances are sub-cases of the above.

8.3.2 Degenerate Critical Points and Unfolding Singularities

In general, a planar vector field may have several critical points, as the functions $u(x, y)$ and $v(x, y)$ may be nonlinear. The simultaneous equations $u = 0,\ v = 0$ may then have more than one set of real solutions in x and y.

Suppose the contour C on the right-hand side of Eq. (8.32) encloses more than one critical point of the vector field $\mathbf{u}$, as in Fig. 8.8a. It can then be deformed into the contour in Fig. 8.8b, and finally to a sum of disjoint pieces as in Fig. 8.8c, each of which encloses just one singularity. The value of the original contour integral does not change during this deformation process. An important consequence follows immediately:

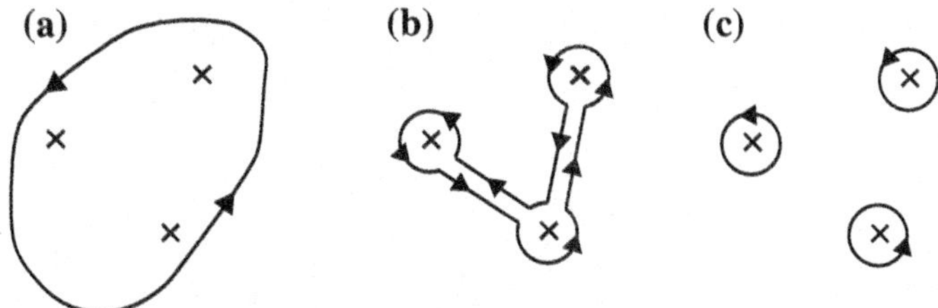

Fig. 8.8 **a** A contour enclosing three critical points of a vector field **u** (marked by crosses). **b**, **c** Its successive deformations

- If C encloses more than one singularity of the vector field, the net value of the integer n is the algebraic sum of the Poincaré indices of the individual singularities.[4]
- If C does not enclose any singularity of the vector field, the integral over C is zero. But the converse is not necessarily true.

For instance, if C encloses a node as well as a saddle point, the net value of $n = 1 - 1 = 0$. Here is an illustrative example of what happens in this situation. Consider the vector field

$$\mathbf{u}(\mathbf{r}) = x(x-a)\,\mathbf{e}_x - y\,\mathbf{e}_y, \tag{8.37}$$

where a is a positive constant. It is clear that the field has two critical points, at $(0, 0)$ and $(a, 0)$, respectively. An inspection of the Jacobian matrix L at each of these points shows that $(0, 0)$ is a node, while $(a, 0)$ is a saddle point.

★ **9.** Consider the vector field given by Eq. (8.37).

(a) Schematically sketch the field lines of $\mathbf{u}(\mathbf{r})$.

(b) Show that the contour integral $\dfrac{1}{2\pi}\displaystyle\oint_C \frac{u\,dv - v\,du}{u^2+v^2}$ is equal to

(i) $+1$, if C only encircles $(0, 0)$ once in the positive sense;
(ii) -1, if C only encircles $(a, 0)$ once in the positive sense;
(iii) 0, if C encircles both critical points once in the positive sense.

(c) Sketch the field lines of $\mathbf{u}(\mathbf{r})$ in the limit $a \to 0$.

In the limit $a \to 0$, the vector field in Eq. (8.37) becomes

$$\mathbf{u}(\mathbf{r}) = x^2\,\mathbf{e}_x - y\,\mathbf{e}_y. \tag{8.38}$$

This field has a single singularity, located at the origin. However, it is a higher order or **degenerate** critical point. The component u of the vector field is no longer *generic*,

[4]This result should remind you of Cauchy's residue theorem for a contour integral in which the contour encloses several poles of the integrand. See Eq. (23.23) of Chap. 23, Sect. 23.3.2.

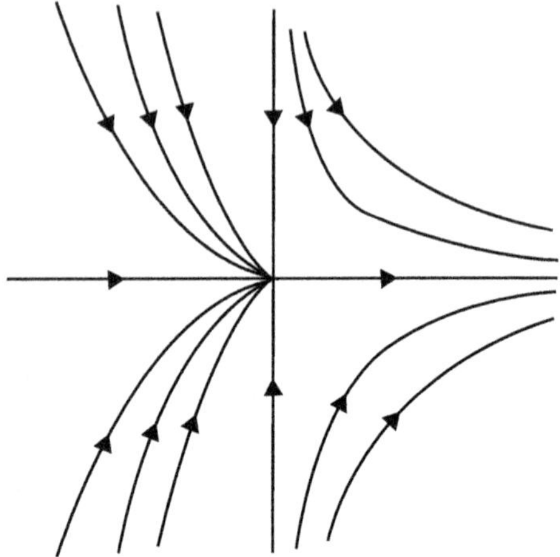

Fig. 8.9 Field lines in the vicinity of the saddle-node singularity of the vector field in Eq. (8.38)

in the sense that it has no terms that are *linear* in the coordinates. In the linear approximation, we would have to set this component equal to zero identically. It is therefore clear that naive linearization *fails* in such instances. This situation is typical of degenerate critical points. On the other hand, the example above shows how the critical point arises from the *coalescence* of two simple critical points, namely, the node at $(0, 0)$ and the saddle point at $(a, 0)$, as $a \to 0$ from above. Hence the critical point is called a **saddle node**. Figure 8.9 shows the field pattern of the vector field in Eq. (8.38) in the vicinity of the saddle node at the origin.

This example suggests a general technique. When we are faced with a degenerate critical point, with vector field components that are intrinsically nonlinear in its vicinity, we must convert the nonlinear functions to limits of products of linear factors. This reverse process is called **unfolding**. There is a well-defined mathematical procedure that classifies the different distinct kinds of nonlinearities (for regular, differentiable functions) and their unfolding. This procedure is a part of what is called **catastrophe theory**, which falls under the purview of **bifurcation theory** in the study of dynamical systems, and of **singularity theory** and **Morse theory** in algebra. A bifurcation is said to occur in a dynamical system if there is an abrupt *qualitative* change in its behavior, when a parameter is changed by an infinitesimal amount across a threshold value.

★ **10.** Consider the following planar vector fields:

$$\text{(i) } \mathbf{u}(\mathbf{r}) = (x^2 - y^2)\,\mathbf{e}_x + 2xy\,\mathbf{e}_y \quad \text{(ii) } \mathbf{u}(\mathbf{r}) = x(x^2 - 3y^2)\,\mathbf{e}_x + y(3x^2 - y^2)\,\mathbf{e}_y.$$

(a) Sketch the field lines in each case. Note that $(0, 0)$ is the only critical point in both cases.
(b) Show that the Poincaré index of the critical point is $n = 2$ in case (i), and $n = 3$ in case (ii).
(c) Unfold the singularity in case (i) by replacing z^2 with $z(z - \epsilon)$ where ϵ is a small positive number (say). The vector field is then given by

$$\mathbf{u}(\mathbf{r}) = [x(x-\epsilon) - y^2]\,\mathbf{e}_x + (2x-\epsilon)y\,\mathbf{e}_y.$$

Show that the critical points of this vector field are nodes at $(0, 0)$ and $(\epsilon, 0)$.

8.3.3 Singularities of Three-Vector Fields

I conclude this chapter with a few lines on the singularities of vector fields in three-dimensional Euclidean space. The classification of these singularities is considerably more complicated than it is for planar vector fields. It is immediately evident that we could now have both *point* singularities as well as *line* singularities. (More accurately, it is possible to have "wall" singularities as well, just as it is possible to have line singularities that act as walls in a plane.) The remarks that follow are restricted to point singularities.

A general vector field can now be written as

$$\mathbf{u}(\mathbf{r}) = u(x, y, z)\,\mathbf{e}_x + v(x, y, z)\,\mathbf{e}_y + w(x, y, z)\,\mathbf{e}_z. \tag{8.39}$$

Critical points correspond to the isolated roots of the simultaneous equations $u = 0,\; v = 0,\; w = 0$. As in the case of planar vector fields, we may linearize the functions u, v and w in the vicinity of a critical point, and examine the nature of the field lines in generic cases. Once again, this is determined by the three eigenvalues of the Jacobian matrix L of partial derivatives evaluated at the critical point. When all three eigenvalues are real and positive, the critical point is a source (a node); when they are all negative, we have a sink (also a node). When only two of the eigenvalues have the same sign, we have the three-dimensional analog of a saddle point. When there is a pair of complex conjugate eigenvalues and the third eigenvalue is positive, the critical point is a **saddle focus**, and so on. As I have already mentioned, the systematic study of the singularities (of scalar, vector, and tensor fields and of more general objects) is a well-developed part of mathematics. The subject also finds numerous applications in physics—for instance, in fluid dynamics, in classical and quantum field theory, in the study of defects in condensed matter, in plasma physics, and in astrophysics and cosmology.

8.4 Solutions

1. Set $\mathbf{u}(\mathbf{r}) = \mathbf{a}\,\phi(\mathbf{r})$ (where $\mathbf{a}$ is a constant vector) in Eq. (8.2). Use a vector identity for the curl of the product $\mathbf{a}\,\phi(\mathbf{r})$. ▶

2. (a) Set $\mathbf{u}(\mathbf{r}) = \mathbf{a}\,\phi(\mathbf{r})$ (where $\mathbf{a}$ is a constant vector) in Eq. (8.12). Note that if $\mathbf{a}\cdot\mathbf{X} = \mathbf{a}\cdot\mathbf{Y}$ for every vector $\mathbf{a}$, then the vectors $\mathbf{X}$ and $\mathbf{Y}$ must be equal to each other. (b) Set $\mathbf{u} = \mathbf{a}\times\mathbf{v}$ where $\mathbf{a}$ is a constant vector. You need to make the same argument as in (a) above, when $\mathbf{a}\times\mathbf{X} = \mathbf{a}\times\mathbf{Y}$ for arbitrary $\mathbf{a}$. ▶

3. (a) $\nabla \cdot \nabla(r^2) = \nabla^2(r^2)$ is trivially found in Cartesian coordinates to be equal to 6. The result quoted follows at once.
(b) The volume of the toroid is easily found, using a general result for the volume of a surface of revolution that you would have encountered in an elementary course on the calculus (the Theorem of Pappus). The torus is a surface of revolution obtained by taking the cross-sectional circle of radius b with center at a distance a from the origin, and moving it in a circle of radius a. The volume swept out is, therefore, the product of the area πb^2 and the distance $2\pi a$ moved by the center of the circle.
(c) Use the fact that $\nabla \cdot (\mathbf{a}\, e^{i\mathbf{k}\cdot\mathbf{r}}) = (i\mathbf{k}\cdot\mathbf{a})\, e^{i\mathbf{k}\cdot\mathbf{r}}$. You have to evaluate the integral $\int_V e^{i\mathbf{k}\cdot\mathbf{r}}\, dV$ over the volume of a sphere of radius R centered at the origin. It is therefore natural to use spherical polar coordinates for the integration over $\mathbf{r}$. The value of the integral is a scalar, i.e., it is rotationally invariant. Hence it cannot depend on the *direction* of the vector $\mathbf{k}$. The latter can be chosen to lie along the polar or z-axis—or rather, the polar axis can be chosen to lie along the direction of $\mathbf{k}$. This greatly simplifies the calculation, because $\mathbf{k}\cdot\mathbf{r}$ now becomes $k\, r\, \cos\theta$, which is independent of the azimuthal angle φ. Hence the integral over φ is done trivially, to get a factor 2π. Carry out the integration over θ and r to obtain the result quoted. ▶

4. Use Poisson's equation for the electrostatic potential (Eq. (4.31) of Chap. 4, Sect. 4.2.6) in reverse, i.e., write $\rho(\mathbf{r}) = -\epsilon_0\, \nabla^2\phi(\mathbf{r})$, and use Green's second identity. The finite extent of the charge distributions $\rho(\mathbf{r})$ and $\rho'(\mathbf{r})$ guarantees that the right-hand side of the identity vanishes in the limit when the volume integral extends over all space. ▶

5. Use Green's first identity, Eq. (8.16). The choice of ψ in each case is obvious. ▶

6. Cases (a)–(d) correspond to distinctly different field configurations. Cases (e)–(h) are, respectively, the same configurations as in (a)–(d), with all arrows on the field lines reversed in sense.

The vector field (a) is just the two-dimensional position vector $\mathbf{r}$ itself. Its field lines are therefore radial lines emanating from the origin. The Poincaré index $n = +1$ for (a) and (e).

The field lines in cases (b) and (f) are concentric circles centered at the origin, directed clockwise and anticlockwise, respectively. $n = +1$ in *both* cases.

The field lines in (c) and (g) are rectangular hyperbolas with the coordinate axes as the asymptotes. Similarly, the field lines in (d) and (h) are also rectangular hyperbolas, but with the lines $y = \pm x$ as the asymptotes. $n = -1$ in all these cases.

Figures. 8.10a–d show the field lines of the vector fields in (a)–(d). ▶

7. Let $\xi = (x - \bar{x})$ and $\eta = (y - \bar{y})$. You can regard the linear approximation of Eq. (8.35) as a coordinate transformation from the pair (ξ, η) to the pair $(u\,,\, v)$. This is a linear transformation that is orientation-preserving if $\det L > 0$, and orientation-changing if $\det L < 0$. That is, the right- or left-handedness of the original coordinate system remains unaltered if $\det L > 0$, and is flipped if $\det L < 0$. When the closed contour C around the critical point is traversed once in the anticlockwise sense, the

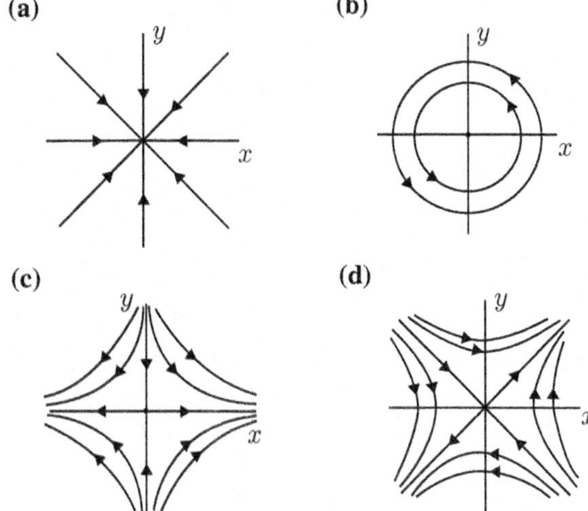

Fig. 8.10 Field lines of a vector field in the vicinity of **a** a node, **b** a center, **c** and **d** saddle points

argument $\tan^{-1}(\eta/\xi)$ changes by 2π. If the transformation to $(u\,,\ v)$ is orientation-preserving [respectively, orientation-changing], the argument $\tan^{-1}(v/u)$ changes by 2π [respectively, -2π]. Hence $n = 1$ [respectively, -1]. ▶

8. $\det L$ is just the product of the two eigenvalues of L. The latter must either be a pair of real numbers, or a complex conjugate pair. ▶

9. (b) In cases (i) and (ii), you can linearize the vector field in the vicinity of the critical point concerned, and take the contour to be a small circle around that point. In case (iii), you can expand the contour *outwards* to make it a large circle of radius R, and take the limit $R \to \infty$.

Remark When $a = 0$, the two critical points coincide. Note that the contour integral above remains equal to 0, even though it now encircles just one singularity of the vector field. ▶

10. (a) First express the field components in plane polar coordinates.
(b) Introduce the complex variable $z = x + iy$. It is then evident that u and v are just the real and imaginary parts of z^2 in case (i), and of z^3 in case (ii). The Poincaré indices follow immediately: a single circuit around the origin in the z-plane increases the argument of z by 2π, and hence that of z^n by $2n\pi$.

Remark The critical point in case (i) above may be termed a **dipole**. As you already know from electrostatics, a point dipole is obtained as a limiting case of a positive

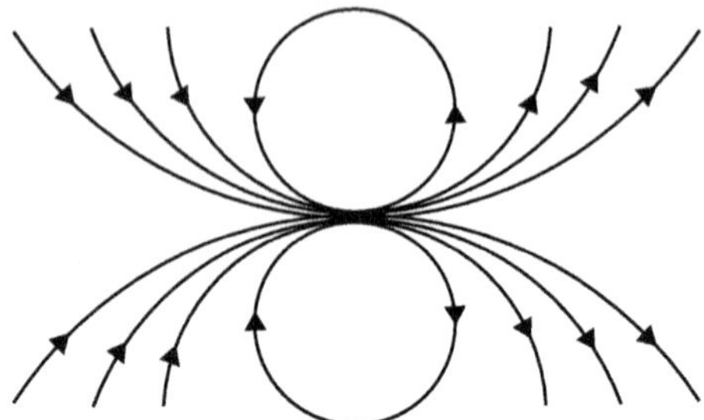

Fig. 8.11 Field lines of a vector field in the vicinity of a dipole critical point

charge (or source) and a negative charge (or sink), when the separation tends to zero while the magnitude of the charge tends to infinity, such that the product of the two remains finite. In the unfolded form, the node at $(0, 0)$ is a sink, while that at $(\epsilon, 0)$ is a source. The coalescence of these points as $\epsilon \to 0$ leads to the dipole singularity. Figure 8.11 illustrates the field lines in the vicinity of a dipole. ▶

Chapter 9
A Bit of Electromagnetism and Special Relativity

In Chap. 7, we have seen how vector calculus is useful in fluid dynamics. Let us now take a brief look at the other major applications of this tool in classical physics, namely, electromagnetism. I shall assume that you are already familiar with the elements of this subject. Its physical aspects will, therefore, be mentioned only very cursorily. The emphasis here will be on seeing how vector calculus provides a natural language for the description of **electromagnetic fields** (EM fields, for short). In the process, I will discuss the important concept of **gauge invariance** and some relevant aspects of **special relativity**, **four-vectors**, and **Lorentz-invariance**. But I shall not explicitly introduce the **metric tensor**, **contravariant** and **covariant vectors**, etc. Important as these concepts are, they would involve too lengthy a digression.

9.1 Classical Electromagnetism

9.1.1 Maxwell's Field Equations

We shall use SI units throughout. Recall that **Maxwell's equations** for the electric field $\mathbf{E}(\mathbf{r}, t)$ and magnetic field $\mathbf{B}(\mathbf{r}, t)$ in free space are

$$\boxed{\nabla \cdot \mathbf{B} = 0, \quad (\nabla \times \mathbf{E}) + \partial \mathbf{B}/\partial t = 0,} \tag{9.1}$$

and

$$\boxed{\nabla \cdot \mathbf{E} = \rho/\epsilon_0\,, \quad (\nabla \times \mathbf{B}) - \mu_0\epsilon_0\, \partial \mathbf{E}/\partial t = \mu_0\, \mathbf{j}.} \tag{9.2}$$

Here ϵ_0 and μ_0 denote, respectively, the permittivity and permeability of free space. As you know, the combination $(\mu_0\epsilon_0)^{-1/2}$ turns out to be the speed c of electromagnetic waves in free space. It is very important to bear in mind the following basic facts regarding these field equations:

V. Balakrishnan, *Mathematical Physics*,
https://doi.org/10.1007/978-3-030-39680-0_9

(i) The EM fields are induced by the charge density $\rho(\mathbf{r}, t)$ and the current density $\mathbf{j}(\mathbf{r}, t)$. The latter are the *sources* of the fields, and are supposed to be given quantities. Maxwell's equations are valid in general, and not merely for static charges or steady currents.
(ii) The equations specify the divergence and the curl of the EM fields. These quantities suffice to specify the fields completely, *in a form that is independent of any particular choice of coordinates*: recall, at this stage, the discussion in Chap. 6, Sect. 6.2.7 on the significance of specifying the divergence and curl of a vector field.
(iii) The equations are partial differential equations of the first order in the spatial coordinates as well as the time. This uniformity is related to the fact that the equations are *relativistic* in nature: they are invariant under Lorentz transformations, as we shall see below.
(iv) Together with appropriate initial conditions and boundary conditions, the field equations determine the EM fields uniquely in any situation, once the sources ρ and $\mathbf{j}$ are specified.

Writing out Eqs. (9.1) and (9.2) in component form, we find that there are actually eight equations. But the total number of unknowns is just six, namely, the components of the two vector fields $\mathbf{E}$ and $\mathbf{B}$. Does this mean that the equations over-determine the fields? That would be bad, because when you have more equations than there are unknowns, there are no nontrivial solutions, in general. But this is not the case with Maxwell's equations. The equation of continuity

$$\frac{\partial \rho}{\partial t} + \nabla \cdot \mathbf{j} = 0 \tag{9.3}$$

relates the sources ρ and $\mathbf{j}$. *The conservation of electric charge follows from the equation of continuity*. Similarly, the fact that the divergence of $\mathbf{B}$ vanishes identically implies that magnetic monopoles do not exist, and is again a statement about the sources of the EM fields (see the remarks that follow in the next paragraph). The upshot of these two conditions on the sources is that one has just the right number of independent equations (namely, six) to determine the fields completely and uniquely, given appropriate initial and boundary conditions.

A remark on magnetic monopoles is in order here. The existence of magnetic monopoles would make Maxwell's equations more symmetrical in $\mathbf{E}$ and $\mathbf{B}$. The right-hand sides of the two equations in (9.1) would then be nonzero. They would be proportional, respectively, to the magnetic monopole density ρ_{m} and the negative of the magnetic current density $\mathbf{j}_{\mathrm{m}}$. These sources would then be related by the continuity equation $\partial \rho_{\mathrm{m}}/\partial t + \nabla \cdot \mathbf{j}_{\mathrm{m}} = 0$. It turns out, however, that in nature ρ_{m} and $\mathbf{j}_{\mathrm{m}}$ are identically equal to zero. (We are not discussing electroweak unification or the early universe here!)

Before we go on to the analysis of the field equations, there are a couple of other obvious but noteworthy points. I mention these because they are sometimes overlooked or misunderstood.

- The electric field is *not* a conservative vector field in general, because its curl is not identically equal to zero.

E becomes a conservative vector field only when there is no *time-varying* magnetic field. It can then be expressed as the gradient of a scalar field. Electrostatic fields fall in this category.

- The magnetic field, too, is not a conservative vector field in general. Its curl is not zero even if all the currents present are steady currents rather than time-varying currents.

Therefore, even a static magnetic field cannot be expressed as the gradient of some scalar field. This is why magnetostatics is a little more involved, mathematically, than electrostatics: For example, **Coulomb's Law** for the electrostatic field due to a static charge is certainly less complicated than the **Biot–Savart Law** for a magnetostatic field due to a steady current.

9.1.2 The Scalar and Vector Potentials

There is a fundamental difference between Eqs. (9.1), the first pair of Maxwell's equations, on the one hand, and Eqs. (9.2), the second pair, on the other.

- Equations (9.1) are *homogeneous* equations. They do *not* involve the sources ρ and $\mathbf{j}$ of the EM fields. Hence the information they carry is valid no matter what the sources are.
- In contrast, Eqs. (9.2) are *inhomogeneous* equations in the fields. They involve the sources ρ and $\mathbf{j}$ of the EM fields.

Since div $\mathbf{B} = 0$, it follows that $\mathbf{B}$ is *always* expressible as the curl of another vector field, conventionally denoted by $\mathbf{A}$:

$$\nabla \cdot \mathbf{B} = 0 \quad \Rightarrow \quad \boxed{\mathbf{B}(\mathbf{r}, t) = \nabla \times \mathbf{A}(\mathbf{r}, t).} \tag{9.4}$$

$\mathbf{A}$ is called the **vector potential**. The second homogeneous equation then becomes

$$\nabla \times \left(\mathbf{E} + \frac{\partial \mathbf{A}}{\partial t}\right) = 0. \tag{9.5}$$

Since the curl of a gradient is identically zero, it follows that the vector in the brackets in the equation above can *always* be written as the gradient of a scalar field:

$$\mathbf{E} + \frac{\partial \mathbf{A}}{\partial t} = -\nabla\phi, \quad \text{or} \quad \boxed{\mathbf{E}(\mathbf{r}, t) = -\frac{\partial}{\partial t}\mathbf{A}(\mathbf{r}, t) - \nabla\phi(\mathbf{r}, t).} \tag{9.6}$$

ϕ is called the **scalar potential**. The minus sign in its definition is a matter of convention. It ensures that, in the special case of an electro*static* field, ϕ is the electrostatic potential, rather than $-\phi$. The homogeneous Maxwell equations thus yield the convenient *representations* in Eqs. (9.4) and (9.6) for the magnetic and electric fields in terms of the vector and scalar potentials. Once these representations are written down, the content of Eqs. (9.1), the homogeneous pair of Maxwell's equations, is exhausted.

Inserting these representations for **E** and **B** in Eqs. (9.2), the inhomogeneous pair of Maxwell's equations, we get

$$\frac{\partial}{\partial t}(\nabla \cdot \mathbf{A}) + \nabla^2 \phi = -\frac{\rho}{\epsilon_0}, \tag{9.7}$$

and

$$\nabla\left(\frac{1}{c^2}\frac{\partial \phi}{\partial t} + \nabla \cdot \mathbf{A}\right) + \frac{1}{c^2}\frac{\partial^2 \mathbf{A}}{\partial t^2} - \nabla^2 \mathbf{A} = \mu_0\, \mathbf{j}\,. \tag{9.8}$$

Equations (9.7) and (9.8) comprise four equations for four quantities (the scalar ϕ and the three components of the vector **A**).

★ **1.** What happens to the relationship between ρ and **j** implied by the equation of continuity?

Equations (9.7) and (9.8) are fairly involved equations—in particular, they are *coupled* equations, rather than separate equations for ϕ and **A**. To proceed, it is helpful to simplify them first by using the freedom that is available to us in the choice of the potentials.

9.1.3 Gauge Invariance and Choice of Gauge

The curl of a gradient is identically zero. Therefore Eq. (9.4) also shows that the magnetic field **B**, a physical quantity, does not change if the gradient of an arbitrary (but differentiable) scalar function $\chi(\mathbf{r}, t)$ is added to the vector potential. The electric field would change under this modification, but it can be made invariant by subtracting the partial time derivative of $\chi(\mathbf{r}, t)$ from the scalar potential.

- It is easy to see from Eqs. (9.4) and (9.6) that the physical EM fields do not get affected if the potentials **A** and ϕ are replaced by $\mathbf{A}'$ and ϕ', respectively, where

$$\boxed{\mathbf{A}' = \mathbf{A} + \nabla\chi \quad \text{and} \quad \phi' = \phi - \frac{\partial \chi}{\partial t}\,.} \tag{9.9}$$

  These comprise a **gauge transformation** of the EM potentials. The fact that **E** and **B** remain unaltered is called the **gauge invariance** of the EM fields.

The arbitrariness in **A** and ϕ implied by the foregoing is called **gauge freedom**. The choice of any specific function χ "fixes the gauge". Maxwell's field equations are obviously gauge invariant, because they involve the fields **E** and **B** directly, rather than the potentials.

- All physical or measurable quantities pertaining to the EM fields, such as the energy of the field, its momentum, angular momentum, and so on, *must* be expressible in terms of these fields (rather than the potentials alone), and must therefore be gauge invariant as well.

This statement implies that the potentials themselves are auxiliary mathematical quantities rather than physical observables. This is certainly true in *classical* electrodynamics, but not so in *quantum physics*. Explaining why would take us too far afield. I merely mention the following:

- Observable effects such as the **Aharonov–Bohm effect** confirm the physical nature of the electromagnetic potentials in quantum mechanics.

But I hasten to add, in order to avoid any misunderstanding, that the gauge invariance of electromagnetism continues to hold good in quantum mechanics and in quantum field theory.

Gauge fixing is done by selecting a suitable *scalar* function $\chi(\mathbf{r}, t)$. Therefore any particular choice of gauge must involve a single *scalar* condition or equation to be satisfied by $\mathbf{A}$, or ϕ, or both. In particular, a *vector* condition will not be satisfied in general, although it may be possible to satisfy it in particular instances. For example, in electrostatics, we are concerned with static charges alone, with no currents present. There is only an electrostatic field, and no magnetic field at all. In this case it is clearly possible to have a vanishing vector potential at all points and for all t (i.e., to satisfy the *vector* equation $\mathbf{A} = 0$). But this is obviously not possible in more general situations. When a magnetic field is present, $\mathbf{A}$ cannot be identically equal to 0.

In contrast, the condition $\phi = 0$ is a scalar condition. Therefore it appears to be acceptable as a gauge condition. The gauge in which the scalar potential vanishes identically is called the **Weyl gauge**. Here is how it can be implemented. Suppose we are given a scalar potential $\phi(\mathbf{r}, t)$ that is *not* identically equal to zero. Then, by selecting the gauge-fixing function χ as a solution of the equation $\phi - \partial\chi/\partial t = 0$, we have a new scalar potential ϕ' that *does* vanish identically. Of course we would simultaneously have to change over from the original vector potential $\mathbf{A}$ to the new vector potential $\mathbf{A}' = \mathbf{A} + \nabla\chi$. But this change does not affect the physical fields $\mathbf{E}$ and $\mathbf{B}$.

Among the infinite number of possible scalar conditions for fixing the gauge, a few stand out as the most *useful* choices, in practice. Magnetostatics (in which the sources are restricted to steady currents alone), for instance, is conveniently studied in the Weyl gauge. Magnetostatics is such a special case that we can set $\phi = 0$ and still have the freedom to choose $\mathbf{A}$ appropriately, as will be seen shortly. It is even possible (but not very sensible!) to do *electro*statics without a scalar potential. For instance, suppose the fields are given by $\mathbf{E} = E_0\,\mathbf{e}_x,\;\; \mathbf{B} = 0$ (where E_0 is a constant). These fields are easily obtained from the potentials $\phi = 0,\;\; \mathbf{A} = -E_0\,\mathbf{e}_x\, t$.

Two particular choices of gauge are especially important, and are the ones used most frequently. They are discussed below.

9.1.4 The Coulomb Gauge

It is *always* possible to choose the vector potential such that

$$\boxed{\nabla \cdot \mathbf{A} = 0.} \tag{9.10}$$

This is called the **Coulomb gauge**. Here is an argument to show how this choice is always possible. Suppose we are given a vector potential such that $\nabla \cdot \mathbf{A}$ is equal to some (scalar) function $f(\mathbf{r}, t)$ that does *not* vanish identically. The relation $\mathbf{A}' = \mathbf{A} + \nabla\chi$ then gives $\nabla \cdot \mathbf{A}' = f + \nabla^2\chi$. All we need to do to ensure that $\nabla \cdot \mathbf{A}' = 0$ is to choose χ to be a solution of the Poisson equation

$$\nabla^2\chi(\mathbf{r}, t) = -f(\mathbf{r}, t). \tag{9.11}$$

This is a well-defined and standard mathematical problem that has a solution under fairly general conditions. (In Chap. 29, we will derive the so-called fundamental solution of Poisson's equation.) It follows that we can always make a gauge transformation to a vector potential that is solenoidal.

Even after we impose the requirement that $\nabla \cdot \mathbf{A}$ be equal to zero (we may as well drop the prime), there is still a good deal of arbitrariness in the vector potential $\mathbf{A}$. You can of course add any constant vector to $\mathbf{A}$. But you can also add $\nabla\psi$, where ψ is any scalar function that satisfies $\nabla^2\psi = 0$, without changing the physical fields $\mathbf{E}$ and $\mathbf{B}$. The new vector potential thus obtained will continue to remain solenoidal, so that we are still in the Coulomb gauge. Therefore, when we choose the Coulomb gauge in any given situation, we are really choosing a *family* of vector potentials, each of them a solenoidal vector field, rather than a single unique vector potential. The boundary conditions on the fields, however, may restrict this freedom considerably. They may even make the potentials in the Coulomb gauge unique, in some instances.

On setting $\nabla \cdot \mathbf{A} = 0$, Eq. (9.7) for ϕ gets simplified. The vector potential $\mathbf{A}$ disappears from this equation, which now reduces to a well-studied equation of mathematical physics—namely, Poisson's equation:

$$\boxed{\nabla^2\phi\,(\mathbf{r}, t) = -\frac{\rho\,(\mathbf{r}, t)}{\epsilon_0}.} \tag{9.12}$$

I have written out the possible t-dependence of ρ (and hence that of ϕ) explicitly in this equation, in order to emphasize that it is valid *in general*, once we choose to work in the Coulomb gauge. It is *not* restricted to the case of a static charge distribution, i.e., it is not restricted to electrostatics alone.

- The great advantage of the Coulomb gauge is that the vector potential is decoupled from the equation obeyed by the scalar potential. Moreover, the latter is the standard Poisson equation.

In Eq. (5.57) of Chap. 5, Sect. 5.3.5, I have already written down the fundamental solution to Poisson's equation—the solution that satisfies the "natural" boundary condition $\phi = 0$ for $r \to \infty$. In other words, the potential vanishes at spatial infinity in all directions. This solution is based on the assumption that the charge distribution ρ (the source of the field) is confined to a finite region of space. Moreover, it pertains to a static charge distribution. But the *same* formal solution also applies to the time-dependent case, Eq. (9.12): t is just a parameter (an "idle spectator", if you like) that appears on both sides of this equation. The fundamental solution to the equation can therefore be written down at once. It is

$$\phi(\mathbf{r}, t) = \frac{1}{4\pi\epsilon_0} \int d^3r' \, \frac{\rho(\mathbf{r}', t)}{|\mathbf{r} - \mathbf{r}'|} . \tag{9.13}$$

But Eq. (9.13) implies that, if the charge distribution at any point $\mathbf{r}'$ changes with time, the scalar potential $\phi(\mathbf{r}, t)$ at any other point $\mathbf{r}$ changes *instantaneously*, no matter how far apart $\mathbf{r}'$ and $\mathbf{r}$ are. Hence $-\nabla\phi$ also changes instantaneously. Does this mean that the electric field $\mathbf{E}$, a physical measurable, also changes instantaneously? Not at all. The field cannot change instantaneously. That would violate the limitation posed by special relativity—namely, that no signal can propagate at a speed greater than c. The other term in $\mathbf{E}$, namely, $-\partial\mathbf{A}/\partial t$, *must* therefore exhibit such a time dependence that any violation of special relativity is canceled out in the sum $-\partial\mathbf{A}/\partial t - \nabla\phi = \mathbf{E}$. Incidentally, this point also serves as a reminder that it is the electric *field* that is the physical quantity, while the potentials are auxiliary quantities (at the classical level, at any rate).

Once the solution to the Poisson equation (9.12) is obtained, we may substitute it in Eq. (9.8) for the vector potential, and move the corresponding term to the right-hand side of that equation. The equation for the vector potential in the Coulomb gauge then becomes

$$\frac{1}{c^2}\frac{\partial^2\mathbf{A}}{\partial t^2} - \nabla^2\mathbf{A} = \mu_0\,\mathbf{j} - \frac{1}{c^2}\nabla\frac{\partial\phi}{\partial t} . \tag{9.14}$$

The right-hand side comprises known quantities, and acts as a source term for $\mathbf{A}$. Equation (9.14) is again a standard equation of mathematical physics, the inhomogeneous **wave equation**. The solutions of this equation are also well-studied. (We will discuss the fundamental solution of the wave equation in Chap. 31.) Once ϕ and $\mathbf{A}$ are known, the fields $\mathbf{E}$ and $\mathbf{B}$ are easily computed.

- In principle, therefore, the Coulomb gauge reduces the complete solution of Maxwell's equations to the solution of Poisson's equation for ϕ and the inhomogeneous wave equation for $\mathbf{A}$.

9.1.5 Electrostatics

In the special case of electrostatics, we are concerned with a single static field $\mathbf{E}(\mathbf{r})$ that satisfies the restricted field equations

$$\boxed{\nabla \cdot \mathbf{E}(\mathbf{r}) = \frac{\rho(\mathbf{r})}{\epsilon_0} \quad \text{and} \quad \nabla \times \mathbf{E}(\mathbf{r}) = 0.} \tag{9.15}$$

The source of the field is a static charge density $\rho(\mathbf{r})$. Since $\mathbf{E}(\mathbf{r})$ is an irrotational vector field in this case, it follows at once that we can write $\mathbf{E}(\mathbf{r}) = -\nabla\phi(\mathbf{r})$, where

$$\boxed{\nabla^2 \phi\,(\mathbf{r}) = -\frac{\rho\,(\mathbf{r})}{\epsilon_0}.} \tag{9.16}$$

Thus,

- any problem in electrostatics can be reduced to the solution of Poisson's equation with appropriate boundary conditions.

Recall that the solution for $\phi(\mathbf{r})$ in the simple case of natural boundary conditions, $\lim_{r\to\infty} \phi = 0$, has already been written down in Eq. (5.57) of Chap. 5, Sect. 5.3.5. Taking the gradient with respect to $\mathbf{r}$ on both sides of this equation, the solution for $\mathbf{E}$ is

$$\boxed{\mathbf{E}(\mathbf{r}) = \frac{1}{4\pi\epsilon_0} \int d^3r\,'\, \frac{\rho(\mathbf{r}\,')(\mathbf{r} - \mathbf{r}\,')}{|\mathbf{r} - \mathbf{r}\,'|^3},} \tag{9.17}$$

as expected. In other cases, the boundary conditions lead to more intricate solutions. Any complexity in electrostatics arises essentially because of the boundary conditions.

9.1.6 *Magnetostatics*

In magnetostatics, too, there is no time dependence, and we are concerned with the special case of a single static vector field $\mathbf{B}(\mathbf{r})$. The restricted field equations in this case are

$$\boxed{\nabla \cdot \mathbf{B}(\mathbf{r}) = 0 \quad \text{and} \quad \nabla \times \mathbf{B}(\mathbf{r}) = \mu_0\, \mathbf{j}(\mathbf{r}).} \tag{9.18}$$

The source of the field is a steady current density $\mathbf{j}(\mathbf{r})$. Since $\mathbf{B}$ is a solenoidal vector field, it follows at once that $\mathbf{B}(\mathbf{r}) = \nabla \times \mathbf{A}(\mathbf{r})$. Now use the standard identity for $\nabla \times (\nabla \times \mathbf{A})$, and work in the Coulomb gauge, so that $\nabla \cdot \mathbf{A} = 0$. We then have

$$\boxed{\nabla^2 \mathbf{A}(\mathbf{r}) = -\mu_0\, \mathbf{j}(\mathbf{r}).} \tag{9.19}$$

But this is just Poisson's equation once again, this time for a vector field rather than a scalar field. Each Cartesian component of $\mathbf{j}$ acts as the source for the corresponding component of $\mathbf{A}$. As before, if the current density is confined to a finite region of space, and we assume natural boundary conditions ($\mathbf{A}$ vanishes at spatial infinity in all directions), the solution can be written down exactly as in the case of the scalar potential in electrostatics. Thus

$$\mathbf{A}(\mathbf{r}) = \frac{\mu_0}{4\pi}\int d^3r\,'\,\frac{\mathbf{j}(\mathbf{r}\,')}{|\mathbf{r}-\mathbf{r}\,'|}\,. \tag{9.20}$$

It remains to find the magnetic field $\mathbf{B}(\mathbf{r}) = \nabla \times \mathbf{A}(\mathbf{r})$. The result is

$$\boxed{\mathbf{B}(\mathbf{r}) = \frac{\mu_0}{4\pi}\int d^3r\,'\,\frac{\mathbf{j}(\mathbf{r}\,')\times(\mathbf{r}-\mathbf{r}\,')}{|\mathbf{r}-\mathbf{r}\,'|^3}\,.} \tag{9.21}$$

This is the general form of the **Biot–Savart Law** for the magnetostatic field due to a steady current density.

★ **2.** Derive Eq. (9.21) from (9.20).
In the special case of a line (a thin wire) carrying a steady current I, we have $d^3r\,'\,\mathbf{j}(\mathbf{r}\,') \to I\,d\boldsymbol{\ell}$, where $d\boldsymbol{\ell}$ is the line element along the wire. The Biot–Savart Law then reduces to the familiar form that we learn in high school:

$$\mathbf{B}(\mathbf{r}) = \frac{\mu_0 I}{4\pi}\int \frac{d\boldsymbol{\ell}\times\mathbf{R}}{R^3}\,, \tag{9.22}$$

where the integral is along the length of the wire, and $\mathbf{R}$ is the vector *from* the line element $d\boldsymbol{\ell}$ *to* the field point $\mathbf{r}$.

9.1.7 The Lorenz Gauge

Let us return to the general Eqs. (9.7) and (9.8) for the scalar and vector potentials. You have seen that the choice of the Coulomb gauge condition $\nabla \cdot \mathbf{A} = 0$ eliminates $\mathbf{A}$ from Eq. (9.7), and reduces it to Poisson's equation for ϕ. Similarly, the condition

$$\boxed{\frac{1}{c^2}\frac{\partial \phi}{\partial t} + \nabla \cdot \mathbf{A} = 0} \tag{9.23}$$

decouples ϕ from Eq. (9.8), and reduces it to the inhomogeneous wave equation for $\mathbf{A}$ (Eq. (9.24) below). A choice of potentials such that Eq. (9.23) is satisfied is called the **Lorenz gauge**.

★ **3.** Use an argument similar to that given in the case of the Coulomb gauge to show that the Lorenz gauge can *always* be implemented.

Once again, when we choose the Lorenz gauge, what we really have is a whole *family* of potentials, all of which satisfy Eq. (9.23). They differ from each other by gauge functions ψ that are solutions of the *homogeneous* wave equation $(1/c^2)\,(\partial^2\psi/\partial t^2) - \nabla^2\psi = 0$. These could be nontrivial functions, so that the different potentials obtained could look quite different from each other—and yet lead to the same EM fields.

The vector potential in the Lorenz gauge satisfies the wave equation with the current density as the source:

$$\frac{1}{c^2}\frac{\partial^2 \mathbf{A}}{\partial t^2} - \nabla^2 \mathbf{A} = \mu_0\,\mathbf{j}\,. \tag{9.24}$$

Further, using the relation $\nabla \cdot \mathbf{A} = -(1/c^2)\,(\partial\phi/\partial t)$ in Eq. (9.7), we get

$$\frac{1}{c^2}\frac{\partial^2 \phi}{\partial t^2} - \nabla^2 \phi = \frac{\rho}{\epsilon_0}\,. \tag{9.25}$$

Thus, in the Lorenz gauge, the scalar potential *also* satisfies the wave equation, with the charge density as the source.

- In principle, therefore, the Lorenz gauge reduces the complete solution of Maxwell's equations to the solution of inhomogeneous wave equations for both ϕ and $\mathbf{A}$.

This is not a coincidence! Rather, it is the consequence of a deeper symmetry of the electromagnetic field equations, which I will discuss in Sect. 9.3 below. This symmetry, or **relativistic invariance**, is a fundamental reason why the Lorenz gauge is of special interest.

Let us go back, for a moment, to the Coulomb gauge condition $\nabla \cdot \mathbf{A} = 0$. The left-hand side of this equation is a scalar, which means that it is unchanged under rotations of the coordinate axes. Suppose the Coulomb gauge is chosen in a given frame of reference. Now transform to a new frame of reference whose coordinate axes are rotated with respect to the original axes. It is then guaranteed that we continue to remain in the Coulomb gauge. This is clearly a convenient property. However, if we transform to a new frame of reference that is *moving* at a uniform velocity with respect to the original frame, this is no longer true. This is because $\nabla \cdot \mathbf{A}$ is *not* a scalar under such "velocity transformations" (or **boosts**). We have to make *another* gauge transformation in the moving frame to get back into the Coulomb gauge; and this can always be done, of course.

In contrast to this situation, the combination $(1/c^2)(\partial\phi/\partial t) + \nabla \cdot \mathbf{A}$ turns out to be unchanged under rotations of the coordinate axes *as well as* boosts to uniformly moving frames of reference—i.e., under the full set of Lorentz transformations. Hence this combination it is not only an "ordinary" scalar, but also a **Lorentz scalar**. Therefore the condition $(1/c^2)(\partial\phi/\partial t) + \nabla \cdot \mathbf{A} = 0$, if satisfied in a given frame of reference, remains satisfied in all frames that are inertial with respect to the original frame. In other words,

- the Lorenz gauge condition is Lorentz-invariant.[1]

[1]The Lorenz gauge is named after the Danish mathematician and physicist, L. V. Lorenz (1829–1891). Lorentz transformations are named after the Dutch physicist H. A. Lorentz (1853–1928).

This property of relativistic invariance or **Lorentz-invariance** of the Lorenz gauge condition makes it very useful in dealing with electromagnetic radiation phenomena and relativistic electrodynamics, as well as quantum electrodynamics and quantum field theory.

This brings us to a consideration of the properties of the EM fields and the Maxwell equations under Lorentz transformations. A brief digression into special relativity helps set the stage. Once again, I assume that you are familiar with the elements of the subject. What follows is merely a quick recapitulation of some salient features, and not a detailed account of special relativity. As already stated at the beginning of this chapter, I shall not explicitly use the metric tensor, contravariant and covariant vectors, etc., owing to limitations of space.

9.2 Special Relativity

9.2.1 The Principle and the Postulate of Relativity

Special Relativity is based, as you know, on (i) a general *principle* and (ii) a physical *postulate*.

- The **Principle of Relativity** says that the laws of physical phenomena are unchanged in form (*form-invariant*) for all mutually inertial observers, i.e., in all frames of reference related to each other by Lorentz transformations.

The term **Lorentz transformation** is often used in elementary discussions to mean just a transformation to a frame of reference moving uniformly with respect to the original frame. This is sometimes called a velocity transformation, the technical term being a boost. Lorentz transformations actually comprise all possible rotations of the spatial axes as well as boosts in all possible directions. More precisely: these constitute the set of *homogeneous, proper* Lorentz transformations, the so-called *special* Lorentz transformations. Such transformations comprise the **Lorentz group**, denoted by $SO(3, 1)$. *Inhomogeneous* Lorentz transformations also includes shifts of the origin of the spacetime coordinates by constant amounts. Inhomogeneous Lorentz transformations also form a group, called the inhomogeneous Lorentz group or the **Poincaré group**. The principle of relativity stated above applies to this extended set of transformations. But we will not consider these here. There are also *improper* transformations such as parity and time reversal.

- The **Postulate of Relativity** says that there exists a fundamental limiting speed in nature, that is the same in all mutually inertial frames of reference.

Light propagates in a vacuum with this limiting speed, denoted by c. So does any particle whose rest mass happens to be exactly zero.

The question arises as to what happens in different sets of mutually inertial frames of reference, which may be *accelerating* with respect to each other. Without going into details, I merely mention that, strictly speaking, the principle of (special) relativity stated above is only valid in "flat" spacetime, i.e., spacetime in the absence of any curvature or gravitational fields. Gravitation enters the picture because of the **Principle of Equivalence**, which says, broadly speaking, that any acceleration is locally equivalent to the effect of an appropriate gravitational field. The latter, in turn, is a manifestation of the curvature of spacetime. There is a specific criterion to determine whether any given region of spacetime is flat or not. It takes fairly intense gravitational fields to produce significant curvature in a region of spacetime, so that the latter may be taken to be flat to a good approximation even in the presence of mild gravitational fields. This is why special relativity, rather than general relativity, suffices to handle all situations except those involving the effects very high gravitational fields, such as gravitational waves. But the validity of this simplification is of course contingent upon the level of accuracy desired in any given situation.

9.2.2 *Boost Formulas*

Consider a frame of reference S, and another frame of reference S$'$ moving at a uniform velocity $\mathbf{v}$ with respect to it (Fig. 9.1). Let $(\mathbf{r}, t)$ and $(\mathbf{r}', t')$ be the respective spacetime coordinates in S and S$'$. We assume (for simplicity) that the origins and the Cartesian axes of the two frames coincide at $t = 0$. You are no doubt familiar with the Lorentz transformation formulas (or boost formulas) in the special case when $\mathbf{v} = v\,\mathbf{e}_x$. Let us be a little different, and write down the transformation rules for a boost velocity $\mathbf{v}$ in any *arbitrary* direction.

Resolve the coordinate vector $\mathbf{r}$ (in S) into components along $\mathbf{v}$ and transverse to it. (Recall, if necessary, Eq. (5.22) of Chap. 5, Sect. 5.1.4.) That is,

$$\mathbf{r} = \mathbf{r}_{\parallel} + \mathbf{r}_{\perp} = \frac{\mathbf{r}\cdot\mathbf{v}}{v^2}\,\mathbf{v} + \left(\mathbf{r} - \frac{\mathbf{r}\cdot\mathbf{v}}{v^2}\,\mathbf{v}\right). \tag{9.26}$$

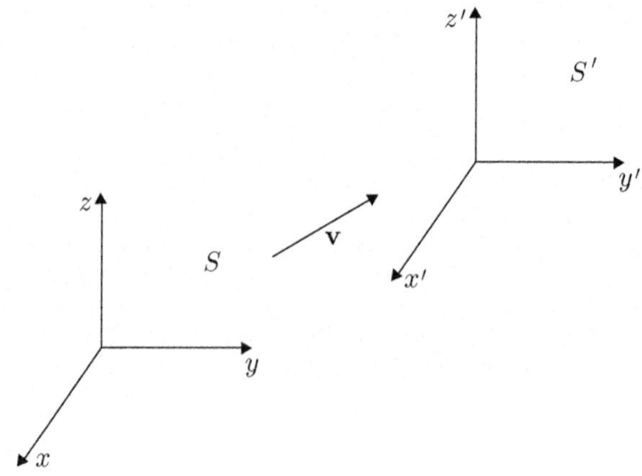

Fig. 9.1 Mutually inertial frames S and S$'$

The transverse component $\mathbf{r}_\perp$ is not affected by the boost, as you might expect. The longitudinal component undergoes the familiar transformation. Let

$$\gamma_v \stackrel{\text{def.}}{=} 1/\sqrt{1-(v/c)^2}\,. \tag{9.27}$$

Then the Lorentz transformation formulas corresponding to the boost are

$$\boxed{\begin{aligned} ct' &= \gamma_v\left(ct - \frac{\mathbf{r}\cdot\mathbf{v}}{c}\right),\\ \mathbf{r}' &= \gamma_v\left(\frac{\mathbf{r}\cdot\mathbf{v}}{v^2}\mathbf{v} - \mathbf{v}t\right) + \left(\mathbf{r} - \frac{\mathbf{r}\cdot\mathbf{v}}{v^2}\mathbf{v}\right).\end{aligned}} \tag{9.28}$$

In three-dimensional Euclidean space, the square of the distance to any point, $r^2 = x_i\,x_i$, is preserved under rotations of the coordinate axes about the origin. In the same way, what is preserved under Lorentz transformations is the square of the interval from the origin to any point in spacetime, defined to be $c^2\,t^2 - r^2$. The surface $r^2 =$ constant is a sphere in space. The hypersurface $c^2\,t^2 - r^2 =$ constant is a *hyperboloid* in spacetime.

★ **4.** Given Eqs. (9.28), verify that $c^2\,t'^2 - r'^2 = c^2\,t^2 - r^2$.

9.2.3 *Collinear Boosts: Velocity Addition Rule*

The familiar special case is the one in which the frame S$'$ moves with a velocity $\mathbf{v} = v\,\mathbf{e}_x$ along the x-axis of S. The formulas in Eqs. (9.28) then reduce to the familiar ones for the spacetime coordinates in S$'$, namely,

$$ct' = \gamma_v\left(ct - \frac{x\,v}{c}\right),\quad x' = \gamma_v(x - vt),\quad y' = y,\quad z' = z. \tag{9.29}$$

Now suppose a third frame of reference S$''$ is moving at a uniform velocity $u\,\mathbf{e}_x$ with respect to S$'$ (Fig. 9.2). The spacetime coordinates in this frame are therefore given by

$$ct'' = \gamma_u\left(ct' - \frac{x'\,u}{c}\right),\quad x'' = \gamma_u(x' - ut'),\quad y'' = y',\quad z'' = z', \tag{9.30}$$

where $\gamma_u = 1/\sqrt{1-(u/c)^2}$. Putting in the expressions for the primed variables from Eq. (9.29), we find that the spacetime coordinates in S$''$ are related to those of the original frame S by a *single* boost $w\,\mathbf{e}_x$, according to

$$ct'' = \gamma_w\left(ct - \frac{xw}{c}\right),\quad x'' = \gamma_w(x - wt),\quad y'' = y,\quad z'' = z, \tag{9.31}$$

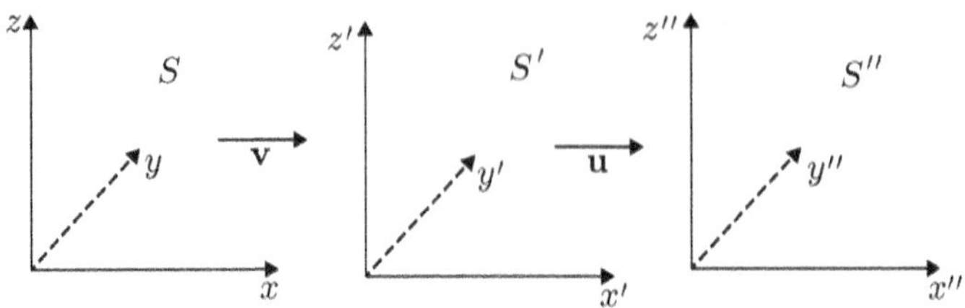

Fig. 9.2 Three collinearly moving frames

where $\gamma_w = 1/\sqrt{1-(w/c)^2}$, and

$$\boxed{w = \frac{v+u}{1+(vu/c^2)}}\,. \tag{9.32}$$

As you know, the striking difference between the nonrelativistic and relativistic situations is the change in the rule for the addition of velocities implied by the expression for w in Eq. (9.32).

★ **5.** Verify Eqs. (9.31) and (9.32).

The foregoing establishes an important result:

- The resultant of two successive boosts in the same direction is again a boost in the same direction.
- The resultant boost is *nonlinear* in the individual boosts, and explicitly involves the fundamental velocity c.

And now for a surprise: This conclusion does *not* hold good if the two boosts are in *different* directions! Instead, this is what happens:

- Two successive boosts in different directions are equivalent to a single boost *together with* a rotation.

Why is it that two boost velocity vectors $\mathbf{v}$ and $\mathbf{u}$ do not just add up to produce a resultant boost velocity $(\mathbf{v}+\mathbf{u})$, modulated by some "correction factor" involving c, as in Eq. (9.32)? A physical way of understanding the reason why is as follows. The second equation in (9.28), the formula for $\mathbf{r}'$, indicates the way in which the components of *any* three-vector transform under a boost. It shows that the component of the vector *along* the direction of the boost, and the part *normal* to the boost, transform in *different* ways. Now, when a boost $\mathbf{v}$ is followed by a boost $\mathbf{u}$, the latter acts on not only the original coordinate $\mathbf{r}$, but also on the original boost velocity vector $\mathbf{v}$, because $\mathbf{r}'$ involves both $\mathbf{r}$ and $\mathbf{v}$. The part of $\mathbf{v}$ that is directed along $\mathbf{u}$ and the part that is normal to $\mathbf{u}$ get transformed in different ways. This produces a kind of "twist", whose effect shows up as a *rotation* of the axes.

This rotation is called a **Wigner rotation**. It is responsible for the phenomenon of **Thomas precession**. An important consequence of the foregoing is that, while the set of all possible rotations constitutes a *subgroup* of the group of Lorentz transformations, the set of all possible boosts does not.

9.2.4 Rapidity

As I have just stated, a second boost $\mathbf{u}$ transforms not only the original coordinate vector $\mathbf{r}$, but also the first boost velocity vector $\mathbf{v}$. But this is true, of course, even when $\mathbf{u}$ is along the *same* direction as $\mathbf{v}$. The effective resultant of the velocities $v\,\mathbf{e}_x$ and $u\,\mathbf{e}_x$ is $w\,\mathbf{e}_x$, where w is given by Eq. (9.32). Here v and u lie in the range $(-c\,,\,c)$.

The velocity addition rule can also be interpreted as follows. A physical object at rest in the frame S$''$ moves with a velocity $u\,\mathbf{e}_x$ in the frame S$'$, which is itself moving with a velocity $v\,\mathbf{e}_x$ with respect to a frame S. Then the velocity of this object as measured in S is $w\,\mathbf{e}_x$, where w is given by Eq. (9.32).

- The existence of a limiting velocity c necessarily makes the law of addition of velocities *nonlinear* in the individual velocities.

Note that $|w| \to c$ as $|u| \to c$, consistent with the postulate of relativity. For all $|v| < c$ and $|u| < c$, we have $|w| < c$ as well, as expected.

The moving object could be light itself, in a fluid medium of refractive index μ. Then $u = c/\mu$, which is less than c. Let the fluid move with velocity v with respect to the lab frame in the same direction as the light beam. Then, using Eq. (9.32), the speed of light as measured in the lab frame is given by

$$w = c\left(\frac{\mu v + c}{\mu c + v}\right). \tag{9.33}$$

This is precisely the result that is exploited in **Fizeau's interferometer**, used in the measurement of the speed of light in a moving medium. In practice, the fluid velocity v is generally much smaller than c. To first order in v/c, Eq. (9.33) reduces to

$$w \simeq \frac{c}{\mu} + \left(1 - \frac{1}{\mu^2}\right) v. \tag{9.34}$$

In his experiment, Fizeau used this expression for w, which was at that time derived on the basis of the erroneous, pre-relativity assumption that light was "dragged" by a moving medium. But we know now that (fortunately!) it coincides with the correct answer to leading order in v/c.

Although the formula for the resultant velocity w in Eq. (9.32) is a nonlinear function of v and u, there does exist a specific function of the speed, in terms of which the composition law for collinear velocities becomes simple algebraic addition. This quantity is the **rapidity** corresponding to the one-dimensional velocity v, defined as

$$\boxed{\xi_v \stackrel{\text{def.}}{=} \tanh^{-1}(v/c).} \tag{9.35}$$

Equation (9.32) then reduces to

$$\boxed{\xi_w = \xi_v + \xi_u\,.} \tag{9.36}$$

In other words:

- When the velocities are collinear, *rapidities*, rather than velocities, add up.

For $v \ll c$, the rapidity $\xi_v \simeq v/c$ to leading order. As $v \to c,\ \xi_v \to \infty$. Observe also that the quantities γ_v and ξ_v are related according to

$$\gamma_v = 1/\sqrt{1-(v/c)^2} = \cosh\,\xi_v\,. \tag{9.37}$$

The utility of the rapidity variable will become clear when the transformation matrix corresponding to a boost is written down in Sect. 9.2.6 (see Eq. (9.47) below).

Finally, for completeness, I mention that the rule for the addition of noncollinear velocities is somewhat more complicated than the one for collinear velocities. The rule is *non-commutative*, in the sense that the answers are different if the roles of **v** and **u** are interchanged in the foregoing. And when more than two velocities are to be "added", the rule is *non-associative* as well.

9.2.5 *Lorentz Scalars and Four-Vectors*

The spacetime coordinates $(ct\,,\,\mathbf{r})$ form a **four-vector**, which will be denoted by $\underline{x}$. Analogous to the case of three-vectors in Euclidean space, any other set of four quantities comprises a four-vector $\underline{a}$ if it transforms, under *Lorentz* transformations, exactly as $\underline{x}$ does. The first component of $\underline{a}$ is the **time-like component** of the four-vector, while the Cartesian components of **a** are its **space-like components**. If E and **p** denote, respectively, the energy and linear momentum of a particle, then $\underline{p} = (E/c\,,\,\mathbf{p})$ is the **four-momentum** of the particle. Obviously, all the four components of a four-vector must have the same physical dimensions. This is achieved with the help of a suitable factor involving the fundamental constant c (as in the case of ct or E/c, for example). Of immediate relevance to us here are the **four-vector current density** $\underline{j}$ and the **four-vector potential** $\underline{A}$, defined respectively as

$$\boxed{\underline{j} = (c\rho\,,\,\mathbf{j})\quad\text{and}\quad\underline{A} = (\phi/c\,,\,\mathbf{A}).} \tag{9.38}$$

As already stated in Sect. 9.1.7, a Lorentz scalar is a quantity that remains invariant under Lorentz transformations. A crucial feature of special relativity is incorporated in the way the scalar product of two four-vectors $\underline{a}$ and $\underline{b}$ is *defined* so as to produce a Lorentz scalar. Recall that the square of the interval from the origin to any point in spacetime, $c^2t^2 - r^2$, is preserved under Lorentz transformations. This means that the scalar product of $\underline{x} = (ct\,,\,\mathbf{r})$ with itself must be defined as

$$\underline{x} \cdot \underline{x} \stackrel{\text{def.}}{=} c^2t^2 - r^2. \tag{9.39}$$

Similarly, the "square" of the four-momentum of a particle must be a Lorentz scalar. For a free particle, it is given by

$$\boxed{\underline{p} \cdot \underline{p} = (E^2/c^2) - \mathbf{p}^2 = m^2c^2,} \tag{9.40}$$

where the constant m is the **rest mass** of the particle. For physical particles, $m \geq 0$. The relative minus sign between the squares of the time-like and space-like components is all-important. More generally, the scalar product of any two four-vectors is defined as the product of their time-like components *minus* the ordinary dot product of the three-vectors representing their space components. This minus sign emerges automatically when we define an appropriate metric tensor, and introduce contravariant and covariant indices.

The *four-dimensional gradient operator* is defined as

$$\underline{\partial} \stackrel{\text{def.}}{=} \left(\frac{1}{c}\frac{\partial}{\partial t}\,,\, -\nabla\right). \tag{9.41}$$

The minus sign in the space-like components in the definition above ensures that the four-dimensional divergence (or four-divergence) of $\underline{x}$ is correctly given by

$$\underline{\partial} \cdot \underline{x} = \frac{1}{c}\frac{\partial}{\partial t}ct - (-\nabla \cdot \mathbf{r}) = 1 + 3 = 4, \tag{9.42}$$

the dimensionality of spacetime. The equation of continuity is, in this notation,

$$\partial\rho/\partial t + \nabla \cdot \mathbf{j} = \underline{\partial} \cdot \underline{j} = 0. \tag{9.43}$$

Hence

- the equation of continuity is the statement that the four-divergence of the four-current density is zero.

But $\underline{\partial} \cdot \underline{j}$ is a Lorentz scalar, so that it remains equal to zero in *all* mutually inertial frames of reference. This is as it should be, because the physical content of the equation of continuity—namely, the **conservation of electric charge**—is expected to remain valid for all mutually inertial observers.

We have seen that we can construct, from the del operator ∇, the Laplacian operator $\nabla \cdot \nabla = \nabla^2$ that is a scalar under rotations of the coordinate axes. The relativistic analog of the Laplacian operator ∇^2 is the **d'Alembertian**

$$\boxed{\square \stackrel{\text{def.}}{=} \underline{\partial} \cdot \underline{\partial} = \frac{1}{c^2} \frac{\partial^2}{\partial t^2} - \nabla^2 \, .} \tag{9.44}$$

The d'Alembertian, also called the **box operator** or the **wave operator**, is (by construction) a Lorentz scalar.

9.2.6 Matrices Representing Lorentz Transformations

In Eq. (5.2) of Chap. 5, Sect. 5.1.2, we have written down the (3×3) orthogonal matrix $R(\alpha)$ that represents a rotation in the xy-plane through an angle α, in three-dimensional space. It was also established that a general proper rotation is represented by an orthogonal (3×3) matrix with unit determinant (and real elements). Such matrices comprise the proper rotation group in three dimensions, the special orthogonal group $SO(3)$.

What is the analogous representation of a proper Lorentz transformation in $(3 + 1)$-dimensional spacetime, if we write the spacetime coordinate $\underline{x}$ in the form of a (4×1) column vector with elements (ct, x, y, z)? Under a Lorentz transformation,

$$\underline{x} \mapsto \underline{x}' = \Lambda \, \underline{x}, \tag{9.45}$$

where Λ is the (4×4) matrix representing the transformation. Recall that Lorentz transformations comprise rotations of the spatial coordinate axes as well as boosts. The former are easy to represent, as spatial rotations do not affect the time coordinate at all. Hence a rotation is represented simply by a (4×4) matrix of the form

$$\Lambda \text{ (for a rotation of the spatial axes)} = \begin{pmatrix} 1 & 0 & 0 & 0 \\ 0 & R_{11} & R_{12} & R_{13} \\ 0 & R_{21} & R_{22} & R_{23} \\ 0 & R_{31} & R_{32} & R_{33} \end{pmatrix}, \tag{9.46}$$

where R_{ij} is the (ij)th matrix element of a (3×3) rotation matrix R. (That is, R is an orthogonal matrix with unit determinant.) A *boost*, however, mixes up the spatial and time coordinates. Consider, for definiteness, a boost from a frame S to a frame S$'$ moving with a velocity $\mathbf{v} = v\,\mathbf{e}_x$ along the x-axis of S. The spacetime coordinates in the two frames are related by the transformation formulas (9.29). These relations may be written in a very convenient form using the rapidity variable ξ_v corresponding to the boost. We note that $\cosh \xi_v = \gamma_v$, as already pointed out in Eq. (9.37), and therefore $\sinh \xi_v = (v/c)\,\gamma_v$. Then the transformed spacetime coordinate $\underline{x}'$ in S$'$ corresponding to the spacetime coordinate $\underline{x}$ in S is given by $\underline{x}' = \Lambda \, \underline{x}$, where

$$\Lambda \text{ (for a boost in the } x\text{-direction)} = \begin{pmatrix} \cosh \xi_v & -\sinh \xi_v & 0 & 0 \\ -\sinh \xi_v & \cosh \xi_v & 0 & 0 \\ 0 & 0 & 1 & 0 \\ 0 & 0 & 0 & 1 \end{pmatrix}. \tag{9.47}$$

Observe the rough similarity with the matrix $R(\alpha)$. Instead of of the cosine and sine of α, we now have hyperbolic cosine and hyperbolic sine of ξ_v. Unlike R, however, the matrix Λ is *not* an orthogonal matrix. Recall that the orthogonality of a rotation matrix follows from the fact that $r'^2 = r^2$. Under a Lorentz transformation, however, we have the more general requirement $c^2 t'^2 - r'^2 = c^2 t^2 - r^2$. This requirement imposes a condition on every Lorentz transformation matrix. Define the "metric matrix"

$$g \overset{\text{def.}}{=} \operatorname{diag}(1, -1, -1, -1) = \begin{pmatrix} 1 & 0 & 0 & 0 \\ 0 & -1 & 0 & 0 \\ 0 & 0 & -1 & 0 \\ 0 & 0 & 0 & -1 \end{pmatrix}. \tag{9.48}$$

It can then be shown that any general Lorentz transformation matrix Λ obeys the following **pseudo-orthogonality** condition

$$\boxed{\Lambda^{\mathrm{T}} g \, \Lambda = g.} \tag{9.49}$$

Equating the corresponding determinants, it follows that $(\det \Lambda)^2 = 1$, so that

$$\det \Lambda = \pm 1. \tag{9.50}$$

As in the case of rotations, transformations with $\det \Lambda = 1$ comprise the set of *proper* Lorentz transformations. These form a group, the **special pseudo-orthogonal group** $SO(3, 1)$. It also follows from Eq. (9.49) that

$$\Lambda^{\mathrm{T}} = g \, \Lambda^{-1} g, \quad \Lambda^{-1} = g \, \Lambda^{\mathrm{T}} g, \quad \text{and} \quad \Lambda \, g \, \Lambda^{\mathrm{T}} = g. \tag{9.51}$$

★ **6.** The pseudo-orthogonality condition (9.49) and several relations that follow from it are established quite easily.

(a) Show that the Lorentz-invariance of $c^2 t^2 - r^2$ implies that the transformation matrix Λ satisfies Eq. (9.49).
(b) Establish the relations in (9.51).
(c) Show that the Lorentz transformations (equivalently, the corresponding transformation matrices) form a group.
(d) Show that the matrix Λ corresponding to a rotation of the coordinate axes, Eq. (9.46), also satisfies the pseudo-orthogonality condition (9.49).

9.3 Relativistic Invariance of Electromagnetism

9.3.1 Covariant Form of the Field Equations

The Lorenz gauge condition is now seen to take on a very compact form in terms of four-vectors. We have

$$\frac{1}{c^2}\frac{\partial\phi}{\partial t} + \nabla\cdot\mathbf{A} = \underline{\partial}\cdot\underline{A} = 0. \tag{9.52}$$

In other words, the Lorenz gauge condition is just the requirement that the four-divergence of the four-vector potential be zero. The analogy with the Coulomb gauge $\nabla\cdot\mathbf{A} = 0$ is now obvious.

- The Lorenz gauge is the relativistic generalization of the Coulomb gauge.

The great advantage of the Lorenz gauge is that it remains unchanged under Lorentz transformations, as I have already emphasized. This is now obvious by inspection, because $\underline{\partial}\cdot\underline{A}$ is a Lorentz scalar. Many other Lorentz-invariant gauges are possible, of course: for instance, the gauge in which $\underline{x}\cdot\underline{A} = 0$.

In the Lorenz gauge, the vector and scalar potentials of EM satisfy the wave equations (9.24) and (9.25), respectively. Since $(\phi/c, \mathbf{A})$ and $(c\rho, \mathbf{j})$ are in fact four-vectors, these equations may be combined into the single compact equation

$$\boxed{\Box\underline{A} = \underline{j}, \quad \text{with the gauge condition } \underline{\partial}\cdot\underline{A} = 0.} \tag{9.53}$$

Thus Maxwell's equations in free space are reduced to the wave equation for the four-vector potential in the Lorenz gauge. Both the gauge condition (9.52) and the wave equation (9.53) are *manifestly covariant*, i.e., they are form-invariant under Lorentz transformations.

But this is still an indirect way of exhibiting the relativistic invariance of electromagnetism. The physical electric and magnetic fields **E** and **B** are themselves just three-vectors. Moreover, the equations connecting them to the scalar and vector potentials do not appear to be Lorentz-covariant (although they are, as you will see shortly). How does one reconcile these facts with Lorentz-invariance? This task requires the introduction of the electromagnetic field tensor.

9.3.2 The Electromagnetic Field Tensor

Although the relations $\mathbf{B} = \nabla\times\mathbf{A}$ and $\mathbf{E} = -\partial\mathbf{A}/\partial t - \nabla\phi$ do not at all appear to be form-invariant under Lorentz transformations, this is not so. The relation between the fields and the four-vector potential is indeed Lorentz-covariant. In order to write it down, however, we need to introduce the metric tensor, and extend the index notation appropriately. As stated at the beginning of this chapter, I shall not do so here. Instead, I will state the relevant facts in words.

Taken together, the electric and magnetic fields have six Cartesian components. Hence they cannot represent either a Lorentz scalar or a four-vector. What about tensors of higher order? In four dimensions, the number of components of a tensor of rank 2 is $4^2 = 16$. But this is reduced to six independent components if the tensor is an *antisymmetric* one. Now consider the relation $\mathbf{B} = \nabla \times \mathbf{A}$. The components of the magnetic field, $\partial_i A_j - \partial_j A_i$, already suggest an antisymmetric tensor. And they are, indeed, the space–space components of an antisymmetric tensor in $(3+1)$-dimensional spacetime—the *four-dimensional analog* of the curl of the four-vector $\underline{\mathrm{A}}$. The electric field $\mathbf{E} = -\partial \mathbf{A}/\partial t - \nabla\phi$ appears to be of a different form altogether, but this is not so. Recall that $(1/c)(\partial/\partial t)$ and ϕ/c are the time-like components of the four-vectors $\underline{\partial}$ and $\underline{\mathrm{A}}$, respectively. It is then easy to see that the components of $\mathbf{E}$, too, are the time–space components of this four-dimensional curl. This tensor is called the **electromagnetic field tensor**. Written in matrix form (with a certain choice of the metric for the spacetime coordinates), it looks like this:

$$\mathsf{F} = \begin{pmatrix} 0 & -E_x/c & -E_y/c & -E_z/c \\ E_x/c & 0 & -B_z & B_y \\ E_y/c & B_z & 0 & -B_x \\ E_z/c & -B_y & B_x & 0 \end{pmatrix}. \tag{9.54}$$

In other words, the electric and magnetic fields *together* comprise an antisymmetric tensor of rank 2, in a spacetime of $(3+1)$ dimensions. This tensor has definite transformation properties under Lorentz transformations (see Eq. (9.58) below). Maxwell's equations, too, can be written in a manifestly covariant form in terms of the EM field tensor and the four-current $\underline{\mathrm{j}}$.

9.3.3 Transformation Properties of E *and* B

What is of immediate interest to us is the way **E** and **B** change under Lorentz transformations. We already know what they do under rotations of the coordinate axes, of course. Since **E** and **B** are three-vectors, they transform exactly like the spatial coordinate **r** does under proper rotations of the axes. In particular, the components of **E** do not get mixed up with those of **B**, and vice versa. Under a *boost*, however, we may immediately expect such a mix-up to happen. As you know, a moving charge (i.e., a current) generates a magnetic field, and a static charge will look like a moving charge to an observer who is moving with respect to it. The fact that **E** and **B** are actually components of the same field tensor also shows that they will get mixed up when we boost from one inertial frame to another.

I now write down (without proof) the transformation rules for **E** and **B** under a boost from a frame S to a frame S′, which is moving with an arbitrary velocity **v** $(v < c)$ with respect to S. As usual, unprimed and primed quantities denote variables in S and S′, respectively. Further, let the subscripts $\parallel$ and $\perp$ denote components respectively along the direction of the boost **v** and perpendicular to it. Then:

$$\boxed{E'_{\parallel} = E_{\parallel}\,, \quad \mathbf{E}'_{\perp} = \gamma_v \left\{\mathbf{E}_{\perp} + (\mathbf{v} \times \mathbf{B}_{\perp})\right\}} \tag{9.55}$$

and

$$\boxed{B'_{\parallel} = B_{\parallel}\,, \quad \mathbf{B}'_{\perp} = \gamma_v \left\{\mathbf{B}_{\perp} - \frac{(\mathbf{v} \times \mathbf{E}_{\perp})}{c^2}\right\}.} \tag{9.56}$$

Here $\gamma_v = 1/\sqrt{1-(v/c)^2}$, as usual. Several noteworthy points follow from these relations.

- The components of the EM fields *along* the direction of the boost are unaffected by the boost.
- It is the transverse components $\mathbf{E}_{\perp}$ and $\mathbf{B}_{\perp}$ that get mixed up with each other as a consequence of the boost.
- For sufficiently small boosts, such that v^2/c^2 (note the *square*) is negligible compared to unity, we have

$$\mathbf{E}' \simeq \mathbf{E} + (\mathbf{v} \times \mathbf{B}) \quad \text{and} \quad \mathbf{B}' \simeq \mathbf{B} - \frac{(\mathbf{v} \times \mathbf{E})}{c^2}\,. \tag{9.57}$$

- In hindsight, these relations suggest how the Lorentz force on a moving charge arises—or, from another point of view, how *the magnetic field itself is a natural consequence of charges in motion*!

Although the Lorentz transformation rules for the EM fields have been simply written down in the foregoing, they can in fact be guessed by extrapolating from the transformation rule for a second-rank tensor T in three-dimensional space, under a rotation R of the coordinate axes. Recall that this rule has been written down in Chap. 5, Sect. 5.1.2. We need here the first of Eqs. (5.7), which implies that $T' = R\,T\,R^{\mathrm{T}}$. It is not surprising, then, that the transformation rule for the EM field tensor F under Lorentz transformations is precisely

$$\boxed{\mathsf{F}' = \Lambda\,\mathsf{F}\,\Lambda^{\mathrm{T}}.} \tag{9.58}$$

In the earlier instance, $T' = R\,T\,R^{\mathrm{T}}$ could also be written as $T' = R\,T\,R^{-1}$ (as in Eq. (5.8)), because $R^{\mathrm{T}} = R^{-1}$. It is important to note that we cannot do this in the present instance, because $\Lambda^{\mathrm{T}} \neq \Lambda^{-1}$.

★ **7.** Apply the transformation rule (9.58) to the case of a boost in the x-direction, for which Λ is given by Eq. (9.47), to find the transformed field tensor F'. Hence deduce that, under such a boost,

$$E'_x = E_x\,, \quad E'_y = \gamma_v\left(E_y - v\,B_z\right), \quad E'_z = \gamma_v\left(E_z + v\,B_y\right)$$

and

$$B'_x = B_x\,, \quad B'_y = \gamma_v\left(B_y + \frac{v\,E_z}{c^2}\right), \quad B'_z = \gamma_v\left(B_z - \frac{v\,E_y}{c^2}\right).$$

These expressions are evidently consistent with the general formulas in Eqs. (9.55) and (9.56).

9.3.4 Lorentz Invariants of the Electromagnetic Field

The combination $(\mathbf{E}^2 - c^2\,\mathbf{B}^2)$ is obviously a scalar under rotations of the coordinate axes. It turns out that it is also invariant under boosts, so that it is actually a Lorentz scalar. This quantity (apart from a constant factor) is the **Lagrangian density** of the electromagnetic field. Similarly, the quantity $\mathbf{E}\cdot\mathbf{B}$ is a scalar under proper rotations, and changes sign under the parity transformation. (Recall that $\mathbf{E}$ is a polar vector, while $\mathbf{B}$ is an axial vector.) Hence $(\mathbf{E}\cdot\mathbf{B})^2$ is a scalar under all rotations, proper and improper. It turns out to be invariant under boosts as well. That is,

$$\mathbf{E}^2 - c^2\,\mathbf{B}^2 = \mathbf{E}'^{\,2} - c^2\,\mathbf{B}'^{\,2}, \quad (\mathbf{E}\cdot\mathbf{B})^2 = (\mathbf{E}'\cdot\mathbf{B}')^2. \tag{9.59}$$

These two quantities are the only two independent Lorentz scalars that can be formed from the EM fields. They are called the *invariants* of the electromagnetic field.

★ **8.** Using the transformation formulas (9.55) and (9.56), verify Eqs. (9.59).

An interesting consequence of the Lorentz-invariance of $(\mathbf{E}\cdot\mathbf{B})^2$ is as follows. If $\mathbf{E}\cdot\mathbf{B} = 0$ in one frame of reference, it remains so for all frames obtained from it by Lorentz transformations.

- As a result, transverse electromagnetic waves remain transverse electromagnetic waves for all mutually inertial observers. That is, light remains light in all inertial frames.

This is only to be expected, given the postulate of relativity that we started out with!

More generally, since $\mathbf{E}\cdot\mathbf{B} = \mathbf{E}'\cdot\mathbf{B}'$, we have $EB\,\cos\theta = E'B'\,\cos\theta'$, where θ and θ' are the angles between the electric and magnetic fields in the frame S and the boosted frame S′, respectively. It follows immediately that $\cos\theta$ and $\cos\theta'$ must have the same sign. That is, if $\mathbf{E}$ and $\mathbf{B}$ make an acute (respectively, right and obtuse) angle with each other in a given frame, they continue to make such an angle in any Lorentz-transformed frame.

★ **9.** Let $\mathbf{E}$ and $\mathbf{B}$ be constant, uniform fields making an arbitrary acute angle θ with each other, in a frame of reference S.

(a) Show that it is always possible to find a boosted frame S′ such that $\mathbf{E}'$ is parallel to $\mathbf{B}'$.
(b) Find an expression for the boost velocity required to go from S to S′.

9.3.5 Energy Density and the Poynting Vector

The **energy density** of the EM field is given by

$$W = \tfrac{1}{2}\,\epsilon_0\,(\mathbf{E}^2 + c^2\mathbf{B}^2). \tag{9.60}$$

W is *not* a Lorentz scalar, in contrast to the Lagrangian density of the EM field. The energy flux density of the EM field (i.e., the amount of energy crossing unit area per unit time) is given by the **Poynting vector**

$$\mathbf{S} = \frac{1}{\mu_0}\,(\mathbf{E} \times \mathbf{B}). \tag{9.61}$$

In view of their physical meanings, we may expect W and $\mathbf{S}$ to be related by a continuity equation. This equation is

$$\frac{\partial W}{\partial t} + \nabla \cdot \mathbf{S} = -(\mathbf{E} \cdot \mathbf{j}). \tag{9.62}$$

★ **10.** From Maxwell's equations and the definitions of W and $\mathbf{S}$, show that Eq. (9.62) is satisfied.

The right-hand side of Eq. (9.62) is just the rate of Ohmic dissipation, as you might expect. In the absence of sources ($\rho = 0$, $\mathbf{j} = 0$), i.e., for a pure radiation field, the quantity $\partial W/\partial t + \nabla \cdot \mathbf{S} = 0$ in all mutually inertial frames of reference.

***The central role of gauge invariance in fundamental physics*:** A final remark is in order. In the very brief recapitulation of the elementary aspects of gauge invariance given in the foregoing, I have not gone into several aspects of gauge transformations that play a central role in modern physics. These include singular gauges, gauge transformations in quantum mechanics and quantum field theory, non-abelian gauges, gauge fields and non-integrable phases, and the natural relationship between gauge fields and fiber bundles. Arguably, the single most striking discovery of physics itself, to date, is the following. It is so fundamental that it merits being written in boldface!

- **Local gauge invariance underlies the profound connection between gauge fields and the fundamental forces of nature.**

9.4 Solutions

1. Treating (9.7) and (9.8) as equations for ρ and $\mathbf{j}$ for a moment, verify that the equation of continuity, $\partial\rho/\partial t + \nabla \cdot \mathbf{j} = 0$, is *automatically* satisfied. ▶

2. Use the identity given in Chap. 6, Sect. 6.2.8 for the curl of the product of a scalar and a vector, namely, $\nabla \times (\phi\,\mathbf{u}) = \phi\,(\nabla \times \mathbf{u}) + (\nabla\,\phi) \times \mathbf{u}$. Remember that the curl involves derivatives with respect to the components of $\mathbf{r}$ (and not $\mathbf{r}'$). ▶

3. Suppose the left-hand side of Eq. (9.23) is not zero, but some function $g(\mathbf{r}, t) \not\equiv 0$. Make a gauge transformation, choosing the gauge function $\chi(\mathbf{r}, t)$ to be a solution of the equation

$$\frac{1}{c^2}\frac{\partial^2 \chi}{\partial t^2} - \nabla^2 \chi = g.$$

This is again a standard equation, the inhomogeneous wave equation. ▶

6. (a) You have merely to note that, if $\underline{x}$ is written as a (4×1) column vector with elements (ct, x, y, z), then

$$c^2 t^2 - r^2 = \underline{x}^{\mathrm{T}}\, g\, \underline{x}.$$

Hence, if $\underline{x}' = \Lambda\, \underline{x}$, we have

$$c^2\, t'^2 - r'^2 = \underline{x}'^{\mathrm{T}}\, g\, \underline{x}' = (\Lambda\, \underline{x})^{\mathrm{T}}\, g\, (\Lambda\, \underline{x}) = \underline{x}^{\mathrm{T}}\, (\Lambda^{\mathrm{T}}\, g\, \Lambda)\, \underline{x}.$$

Since this is required to be equal to $\underline{x}^{\mathrm{T}}\, g\, \underline{x}$ for all $\underline{x}$, we must have $\Lambda^{\mathrm{T}}\, g\, \Lambda = g$.
(b) Use the fact that $g^2 = I$, so that $g^{-1} = g$. The relations concerned follow in a straightforward manner.
(c) The set of transformation matrices includes the identity matrix, and each Λ has an inverse. Further, if Λ_1 and Λ_2 are Lorentz transformation matrices,

$$(\Lambda_1\, \Lambda_2)^{\mathrm{T}}\, g\, (\Lambda_1\, \Lambda_2) = \Lambda_2^{\mathrm{T}}\, \Lambda_1^{\mathrm{T}}\, g\, \Lambda_1\, \Lambda_2 = \Lambda_2^{\mathrm{T}}\, g\, \Lambda_2 = g.$$

Hence $\Lambda_1\, \Lambda_2$ also represents a Lorentz transformation.
(d) Use the fact that R itself satisfies the condition $R^{\mathrm{T}}\, R = I$. ▶

9. (a) Let E denote the magnitude of $\mathbf{E}$, and similarly for the other vectors. Using the fact that $\mathbf{E} \cdot \mathbf{B}$ and $\mathbf{E}^2 - c^2\, \mathbf{B}^2$ are Lorentz-invariant, the following simultaneous equations must be satisfied in a frame in which $\mathbf{E}'$ is parallel to $\mathbf{B}'$:

$$E'\, B' = E\, B\, \cos\theta, \quad E'^2 - c^2\, B'^2 = E^2 - c^2\, B^2.$$

For any value of θ such that $0 < \cos\theta < 1$, these equations yield positive definite solutions for E' and B', given by

$$\left.\begin{aligned} E' &= (1/\sqrt{2})\left\{E^2 - c^2\, B^2 + \left(E^4 + 2c^2 E^2 B^2 \cos 2\theta + c^4 B^4\right)^{1/2}\right\}^{1/2}, \\ B' &= (E\, B\, \cos\theta)/E'. \end{aligned}\right\}$$

Hence such a frame definitely exists.
(b) In order to find the boost required, the argument goes as follows. The frame S$'$ (in which $\mathbf{E}'$ is parallel to $\mathbf{B}'$) certainly cannot be unique, by the following reasoning. Suppose we find any one such frame. Then the fields will not change in any frame boosted with respect to it along the (common) direction of $\mathbf{E}'$ and $\mathbf{B}'$. Hence they will remain parallel to each other in all these other frames as well.

Let us now try to find a boost $\mathbf{v}$ that is *normal* to *both* $\mathbf{E}$ and $\mathbf{B}$, i.e., it is parallel (or antiparallel) to $\mathbf{E} \times \mathbf{B}$. Since there is no $B_{\|}$ or $E_{\|}$ in this case, the transformation laws (9.55) and (9.56) for the fields read, in this case,

$$\mathbf{E}' = \gamma_v \left\{\mathbf{E} + (\mathbf{v} \times \mathbf{B})\right\}, \quad \mathbf{B}' = \gamma_v \left\{\mathbf{B} - \frac{(\mathbf{v} \times \mathbf{E})}{c^2}\right\}.$$

But we must have $\mathbf{E}' \times \mathbf{B}' = 0$, since these fields are parallel to each other. Imposing this condition and simplifying, we find

$$[\mathbf{v} \cdot (\mathbf{E} \times \mathbf{B})]\,\mathbf{v} - (E^2 + c^2 B^2)\,\mathbf{v} + c^2(\mathbf{E} \times \mathbf{B}) = 0.$$

Suppose $\mathbf{v}$ is along $\mathbf{E} \times \mathbf{B}$ (a similar argument can be given if it is directed opposite to $\mathbf{E} \times \mathbf{B}$). Taking the dot product of the last equation with $\mathbf{v}$ then yields a quadratic equation for the magnitude v of the boost. The solution is

$$v = \frac{E^2 + c^2 B^2 \pm \left(E^4 + 2c^2 E^2 B^2 \cos 2\theta + c^4 B^4\right)^{1/2}}{2EB \sin\theta},$$

where the sign must be chosen. So as to obtain a positive root. ▶

10. Use the vector identity $\nabla \cdot (\mathbf{u} \times \mathbf{v}) = \mathbf{v} \cdot (\nabla \times \mathbf{u}) - \mathbf{u} \cdot (\nabla \times \mathbf{v})$. ▶

Chapter 10
Linear Vector Spaces

Linear vector spaces comprise a basic topic in mathematics, besides occurring in many forms in a very large variety of physical applications. Foremost among these is quantum mechanics, for which linear vector spaces provide the natural language. It is therefore helpful to use Dirac notation right from the beginning.

In view of the importance of linear vector spaces in applications, I devote a number of chapters to various aspects of the subject. Matrices appear frequently in connection with linear vector spaces. We will discuss many aspects of matrix algebra in Chaps. 11 and 12. But I shall assume that you are already familiar with some of the most elementary aspects of matrices such as the basic terminology of the subject, the rules for matrix addition and multiplication, etc. Infinite-dimensional linear spaces and operators will be considered in Chaps. 13–15. Even the standard topic of orthogonal polynomials and related special functions will be treated in Chap. 16 essentially from the point of view of their role as basis sets in function spaces, as this provides a natural and unified perspective on this topic.

As always, the emphasis will be on applying the results of various theorems, rather than on formal statements of the theorems themselves and their rigorous proofs.

10.1 Definitions and Basic Properties

10.1.1 Definition of a Linear Vector Space

A **linear vector space** (for which I shall use the abbreviation LVS), denoted by $\mathbb{V}$, is a set of elements $|\psi\rangle, |\phi\rangle, |\chi\rangle, \ldots$ called *vectors*, with an operation called *addition* satisfying the following properties:

(i) $|\phi\rangle + |\psi\rangle = |\psi\rangle + |\phi\rangle \in \mathbb{V}$ for every $|\phi\rangle, |\psi\rangle \in \mathbb{V}$.
(ii) $|\phi\rangle + \big(|\psi\rangle + |\chi\rangle\big) = \big(|\phi\rangle + |\psi\rangle\big) + |\chi\rangle$.

V. Balakrishnan, *Mathematical Physics*,
https://doi.org/10.1007/978-3-030-39680-0_10

(iii) $\exists$ a unique *null vector* $|\Omega\rangle \in \mathbb{V}$ such that $|\phi\rangle + |\Omega\rangle = |\phi\rangle, \ \forall\, |\phi\rangle \in \mathbb{V}$.
(iv) $\exists$ a unique vector $-|\phi\rangle \in \mathbb{V}$ for every $|\phi\rangle$ such that $|\phi\rangle + (-|\phi\rangle) = |\Omega\rangle$.

The null vector $|\Omega\rangle$ must not be confused with the symbol $|\,0\,\rangle$, which is often used to denote some specific *non*-null vector such as the ground state of a quantum mechanical system.

Over and above the addition of vectors, an LVS also has an operation of *multiplication by scalars* $a, b, \ldots$ belonging to some *field*. The word "field" is used here in the mathematical sense. But I will not formally define this object here since the only ones we are going to use are the fields of real numbers ($\mathbb{R}$) and complex numbers ($\mathbb{C}$), whose basic properties you know already. This operation of multiplication by scalars has the following postulated properties:

(v) $a\,|\phi\rangle \in \mathbb{V}, \ \forall a \in \mathbb{R}$ or $\mathbb{C}$ and $|\phi\rangle \in \mathbb{V}$.
(vi) $a\,(b\,|\psi\rangle) = (ab)\,|\psi\rangle$.
(vii) $a\,\big(|\psi\rangle + |\phi\rangle\big) = a\,|\psi\rangle + a\,|\phi\rangle$.
(viii) $(a+b)\,|\psi\rangle = a\,|\psi\rangle + b\,|\psi\rangle$.
(ix) $1\,|\psi\rangle = |\psi\rangle$.
(x) $0\,|\psi\rangle = 0$.

Owing to property (x) above, we may as well use the usual symbol 0 both for the null vector $|\Omega\rangle$ and for the usual scalar zero. That is why you will often come across equations in which the left-hand side is a vector (or an element of an LVS), and the right-hand side is just 0, although consistency of notation demands that it should also be a vector (the null vector).

Other terms used for an LVS are a **linear space** or a **vector space**. The elements $|\phi\rangle, \ |\psi\rangle$, etc. are also called **ket vectors**. If the scalars $a, b, \ldots$ are restricted to the real numbers, the LVS is a real linear space; if the scalars are complex numbers, we have a complex linear space. An asterisk will be used to denote the complex conjugate of a complex number.

10.1.2 The Dual of a Linear Space

Every LVS $\mathbb{V}$ has a **dual space** $\widetilde{\mathbb{V}}$ that is also an LVS. The notation used for the elements of $\widetilde{\mathbb{V}}$ is $\langle\phi|, \ \langle\psi| \ \ldots$. These vectors are called **bra vectors** to distinguish them from the elements of $\mathbb{V}$ (which are ket vectors). The dual of the dual space $\widetilde{\mathbb{V}}$ is again $\mathbb{V}$ itself, i.e., $\widetilde{\widetilde{\mathbb{V}}} \equiv \mathbb{V}$.

It is very helpful to keep in mind the concrete example of n-dimensional Euclidean space for any finite $n \geq 2$, in which the ket vectors can be represented by $(n \times 1)$ matrices or **column vectors**. The dual space in this instance is again n-dimensional Euclidean space, and bra vectors may be represented by $(1 \times n)$ matrices or **row vectors**.

I must mention here that the proper definition of the dual space $\widetilde{\mathbb{V}}$ is actually a little more involved. It is the space of *linear functionals* formed from the elements of $\mathbb{V}$ with certain prescribed properties. It turns out that these linear functionals are essentially the inner products to be introduced shortly. Further, the LVS formed by these linear functionals is isomorphic to the space of the bra vectors. We shall therefore regard the latter space itself as the dual space $\widetilde{\mathbb{V}}$ without further elaboration.

10.1.3 The Inner Product of Two Vectors

The quantity $\langle\phi|\psi\rangle$ may now be regarded as formed by a bilinear combination of a bra and a ket, and is called an **inner product**. It is a scalar (in general, complex)—hence the alternative name, "scalar product". The general properties of the inner product are as follows:

(i) $\langle\phi|\big(|\psi\rangle + |\chi\rangle\big) = \langle\phi|\,\psi\rangle + \langle\phi|\,\chi\rangle.$
(ii) $\big(\langle\psi| + \langle\chi|\big)\,|\phi\rangle = \langle\psi|\,\phi\rangle + \langle\chi|\,\phi\rangle.$
(iii) $\langle\phi|a\,\psi\rangle = a\langle\phi|\psi\rangle.$
(iv) $\langle a\,\phi|\psi\rangle = a^*\langle\phi|\psi\rangle.$
(v) $\langle\phi|\psi\rangle = \langle\psi|\phi\rangle^*.$

Note, in particular, the complex conjugation in (iv) and (v) above. If $|\psi\rangle$ is represented by an $(n \times 1)$ column vector with complex elements $x_1,\ x_2,\ \ldots,\ x_n$, then $\langle\psi|$ is given by the $(1 \times n)$ row vector $(x_1^*\ x_2^*\ \cdots\ x_n^*)$. The vectors $|\phi\rangle$ and $|\psi\rangle$ are orthogonal to each other if and only if $\langle\phi|\psi\rangle = 0$.

The inner product of a vector with itself helps define the **norm** of a vector. This is a generalization of the idea of the length or magnitude of the usual kind of vector in Euclidean space. The norm of $|\psi\rangle$ is defined as

$$\boxed{\|\psi\| \stackrel{\text{def.}}{=} \langle\psi|\psi\rangle^{1/2}.} \tag{10.1}$$

The norm of a vector is a positive number, in general. It vanishes if and only if $|\psi\rangle = |\Omega\rangle$, the null vector.

I mention at this point that it is possible to have linear vector spaces in which an inner product of two different vectors is not defined, but the norm of a vector is defined in some specific manner (not as in Eq. (10.1), of course). I shall not consider such spaces in this book. See also the remarks at the end of Sect. 13.3.1 in Chap. 13.

10.1.4 Basis Sets and Dimensionality

A **basis set** or a **basis** in an LVS is a set of vectors $\{|\psi_k\rangle,\ k = 1, 2, \ldots\}$ in the LVS satisfying the following two requirements:

(a) ***Linear independence:*** They must be **linearly independent** of each other. That is, none of the vectors of the set should be expressible as a linear combination of the others. An equation of the form $\sum_k a_k\,|\psi_k\rangle = 0$ must have no solution other than the trivial one $a_k = 0$ for every k.

(b) ***Spanning the space:*** They must **span** the LVS, i.e., *every* vector $|\chi\rangle \in \mathbb{V}$ must be expressible as a linear combination of the form $|\chi\rangle = \sum_k b_k \, |\psi_k\rangle$.

The concept of the span of a vector (or a set of vectors) is easily understood with an example. In three-dimensional Euclidean space, for instance, the span of $\mathbf{e}_x$ is the whole of the x-axis; the span of the pair $(\mathbf{e}_x, \mathbf{e}_y)$ or the pair $(\mathbf{e}_x + \mathbf{e}_y, \; 2\mathbf{e}_x - 3\mathbf{e}_y)$ is the xy-plane; and so on.

- It is very important for you to appreciate the fact that the requirements (a) and (b) above are quite *distinct* from each other.

For example, in three-dimensional Euclidean space, the vectors $(\mathbf{e}_x, \mathbf{e}_y)$ are linearly independent, but do not span the space. On the other hand, the vectors $(\mathbf{e}_x, \mathbf{e}_y, \mathbf{e}_z, \mathbf{e}_x + \mathbf{e}_y + \mathbf{e}_z)$ do span the space, but are not linearly independent because any one of these four vectors can be written as a linear combination of the other three.

The dimensionality of an LVS: While a basis in an LVS is not at all unique, the *number* of vectors constituting the basis is unique to the LVS. This number is called the **dimensionality** of the LVS.

- Every basis in an n-dimensional LVS has exactly n vectors.
- No collection of r vectors in an n-dimensional LVS can be linearly independent if $r > n$.
- No collection of r vectors in an n-dimensional LVS can span the space if $r < n$.

A fundamental theorem that simplifies the study of any *finite*-dimensional LVS is the following:

- Every n-dimensional LVS is isomorphic to $\mathbb{R}^n$, i.e., to the set of "ordinary vectors" in n-dimensional Euclidean space. (Strictly speaking, $\mathbb{R}^n$ must be "endowed with an Euclidean metric" before we can call it Euclidean space.)

Therefore, the familiar and elementary geometrical insight and experience we have with regard to vectors in two- and three-dimensional space suffice, with a straightforward extension to higher values of n, to study finite-dimensional vector spaces. In particular, numerous properties of matrices of finite order can be readily understood along these lines.

★ **1.** Check whether the following sets of elements form an LVS over the field of real numbers or the field of complex numbers, as the case may be. If they do, find the dimensionality of the LVS.

(a) The set of all $(n \times n)$ matrices with complex entries.
(b) The set of all tensors of rank 2 with real elements in three dimensions.
(c) The set of all antisymmetric tensors of rank 2 with complex elements in three dimensions.
(d) The set of all (2×2) matrices whose trace is zero.
(e) The set of all real solutions of the differential equation $y'' - 3y' + 2y = 0$, where y is a function of the real variable x, and a prime denotes the derivative with respect to x.

(f) The set of all $(n \times n)$ unitary matrices. (U is unitary iff $U^\dagger U = UU^\dagger = I$.)
(g) The set of all $(n \times n)$ Hermitian matrices (with multiplication by real scalars).

★ **2.** In the three-dimensional LVS with basis vectors

$$|\phi_1\rangle = \begin{pmatrix} 1 \\ 0 \\ 0 \end{pmatrix}, \quad |\phi_2\rangle = \begin{pmatrix} 0 \\ 1 \\ 0 \end{pmatrix}, \quad |\phi_3\rangle = \begin{pmatrix} 0 \\ 0 \\ 1 \end{pmatrix},$$

find a vector $|\psi\rangle$ such that $\langle\phi_i\,|\psi\rangle = 1$ for $i = 1, 2$, and 3.

★ **3.** In a real n-dimensional LVS, consider the vectors $|\psi_k\rangle$ $(1 \le k \le n)$ given by

$$|\psi_1\rangle = \begin{pmatrix} 1 \\ 0 \\ 0 \\ 0 \\ \vdots \\ 0 \end{pmatrix}, \quad |\psi_2\rangle = \frac{1}{\sqrt{2}}\begin{pmatrix} 1 \\ 1 \\ 0 \\ 0 \\ \vdots \\ 0 \end{pmatrix}, \quad |\psi_3\rangle = \frac{1}{\sqrt{3}}\begin{pmatrix} 1 \\ 1 \\ 1 \\ 0 \\ \vdots \\ 0 \end{pmatrix}, \quad \cdots, \quad |\psi_n\rangle = \frac{1}{\sqrt{n}}\begin{pmatrix} 1 \\ 1 \\ 1 \\ 1 \\ \vdots \\ 1 \end{pmatrix}.$$

(a) Does the set $\{|\psi_k\rangle\ (1 \le k \le n)\}$ form a basis in the space?
(b) Construct a vector $|\chi\rangle$ such that $\langle\psi_k|\chi\rangle = 1$ for every k $(1 \le k \le n)$.

Infinite-dimensional linear vector spaces also occur very frequently in all kinds of applications.

- An infinite-dimensional LVS is one which does not have a finite basis, i.e., it does not have a basis set of n vectors for any finite value of n.

An example of an infinite-dimensional LVS is the set of all polynomials (of all orders) of a complex variable z. While most properties of finite-dimensional vector spaces remain valid for infinite-dimensional ones (under suitable conditions), there are others that are no longer necessarily valid. What follows in the rest of this chapter is formally applicable to both finite- and infinite- dimensional spaces—in the latter case, with certain convergence conditions tacitly assumed. Some of the subtleties that are specific to infinite-dimensional spaces will be touched upon in Chap. 13.

10.2 Orthonormal Basis Sets

10.2.1 Gram–Schmidt Orthonormalization

From elementary coordinate geometry, we know that points in space can be specified either in terms of Cartesian coordinates, or in terms of *oblique* coordinates. The former system has numerous advantages over the latter. In the same way, it is often advantageous to use an **orthonormal basis** $\{|\phi_k\rangle\}$ in an LVS. The elements of such a basis satisfy the relations

$$\langle\phi_j|\phi_k\rangle = 0 \text{ for } j \neq k, \quad \text{and} \quad \langle\phi_k|\phi_k\rangle = 1 \text{ for every } k. \tag{10.2}$$

That is, we have the orthonormality condition

$$\boxed{\langle\phi_j|\phi_k\rangle = \delta_{jk} \quad \textbf{(orthonormality)}.} \tag{10.3}$$

The other crucial property of such a basis is the completeness relation, given by

$$\boxed{\sum_j |\phi_j\rangle\langle\phi_j| = I \quad \textbf{(completeness)}.} \tag{10.4}$$

Here I is the **unit operator** or **identity operator** in the LVS, that is, when it acts on any vector in the LVS, the vector remains unchanged. The summation over j in Eq. (10.4) runs over all members of the basis set. It is thus an infinite sum in an infinite-dimensional LVS.

- It is important to note that orthonormality is a *scalar* condition while completeness is an *operator* relation.

Operators will be discussed at length in subsequent chapters. But it is useful to note the following right here. (I will return to this point in Chap. 12, Sect. 12.1.1.) A symbol like $\langle\phi\,|\psi\rangle$ is a complex *number*, in general—think of it as the product of a $(1 \times n)$ row matrix and an $(n \times 1)$ column matrix. On the other hand, a symbol like $|\phi\rangle\langle\psi|$ is an *operator*—think of it as the product of an $(n \times 1)$ column matrix and a $(1 \times n)$ row matrix. The result is an $(n \times n)$ matrix. This can "act" on ket vectors from the left, for instance, to produce other ket vectors.

An arbitrary basis $\{\,|\psi_1\rangle,\ |\psi_2\rangle,\ \ldots\}$ can be transformed into an orthonormal basis $\{\,|\phi_1\rangle,\ |\phi_2\rangle,\ \ldots\}$ using the **Gram–Schmidt orthonormalization** procedure, which goes as follows:

(i) Start with $|\psi_1\rangle$ and normalize it to get

$$|\phi_1\rangle = \frac{|\psi_1\rangle}{\|\,\psi_1\,\|} = \frac{|\psi_1\rangle}{\langle\psi_1|\psi_1\rangle^{1/2}}\,. \tag{10.5}$$

(ii) Now take the next vector, $|\psi_2\rangle$, and subtract from it the *projection* of this ket along the direction of $|\phi_1\rangle$. In Eq. (5.21) of Chap. 5, Sect. 5.1.4, you have seen that $(\mathbf{e}_b \cdot \mathbf{a})\,\mathbf{e}_b$ is the projection of a vector $\mathbf{a}$ along the direction of another vector $\mathbf{b}$. What we need here is the analog or generalization of this result to vectors in an arbitrary LVS. The required projection is just $\langle\phi_1|\psi_2\rangle\,|\phi_1\rangle$. Normalizing the result after subtraction from $|\psi_2\rangle$ yields

$$|\phi_2\rangle = \frac{|\psi_2\rangle - \langle\phi_1|\psi_2\rangle\,|\phi_1\rangle}{\left[\langle\psi_2|\psi_2\rangle - |\langle\phi_1|\psi_2\rangle|^2\right]^{1/2}}. \tag{10.6}$$

(iii) Next, subtract from $|\psi_3\rangle$ the projections of this ket along both $|\phi_1\rangle$ and $|\phi_2\rangle$, and normalize the result to get $|\phi_3\rangle$, and so on. Continuing this procedure yields the orthonormal required. The general element of the set is given by

$$|\phi_k\rangle = \frac{|\psi_k\rangle - \sum_{i=1}^{k-1}\langle\phi_i|\psi_k\rangle\,|\phi_i\rangle}{\left[\langle\psi_k|\psi_k\rangle - \sum_{i=1}^{k-1}|\langle\phi_i|\psi_k\rangle|^2\right]^{1/2}}, \quad k \geq 2. \tag{10.7}$$

★ **4.** In a three-dimensional LVS, consider the three vectors

$$|\psi_1\rangle = \begin{pmatrix}1\\1\\1\end{pmatrix}, \quad |\psi_2\rangle = \begin{pmatrix}1\\1\\0\end{pmatrix}, \quad |\psi_3\rangle = \begin{pmatrix}1\\0\\1\end{pmatrix}.$$

(a) Show that they are linearly independent.
(b) Use the Gram–Schmidt procedure to construct an orthonormal basis $\{|\phi_k\rangle\}$ from these three vectors.

The result of the Gram–Schmidt orthogonalization procedure can be written compactly, and in a suggestive manner, in terms of **projection operators**. These operators will be discussed in Chap. 12, Sect. 12.1.2. It will be seen there that the expression in Eq. (10.7) can be rewritten in a more suggestive form, as in Eq. (12.11).

10.2.2 Expansion of an Arbitrary Vector

Once you have a basis in an LVS, any element of the LVS can be written as a linear combination of the basis vectors. When the basis is an orthonormal one, some simple but crucial properties follow.

Let $\{|\phi_k\rangle\}$ be an orthonormal basis in an LVS. Then any arbitrary element $|\psi\rangle$ of the LVS can be expanded in the form

$$|\psi\rangle = \sum_j c_j\,|\phi_j\rangle, \qquad \text{(expansion formula)} \tag{10.8}$$

where the coefficients are given by

$$\boxed{c_k = \langle \phi_k | \psi \rangle \qquad \text{(inversion formula).}} \tag{10.9}$$

The inner product $\langle \phi_k | \psi \rangle$ is a measure of the "overlap" between the two states concerned: it is a measure of "how much of the vector $|\psi\rangle$ is along the vector $|\phi_k\rangle$". The expansion (10.8) is *unique*, in the sense that each given vector $|\psi\rangle$ has a unique set of coefficients $\{c_k\}$. Conversely, given the full set of coefficients c_k, the vector ψ is determined uniquely. Incidentally, this is precisely the reason why, in elementary vector algebra, we often refer to the *components* (a_1, a_2, a_3) of a vector $\mathbf{a}$ as the vector itself. It is obvious that a specific choice of basis is implicit when we do so.

Change of basis: There is nothing unique about an orthonormal basis in an LVS. The formulas connecting the expansion of any vector in two different basis sets are simple and straightforward. Let $\{|\phi_k\rangle\}$ and $\{|\chi_j\rangle\}$ be two sets of orthonormal basis vector in the LVS. Let the expansions of any vector in the two basis sets be given by

$$|\psi\rangle = \sum_k c_k \, |\phi_k\rangle = \sum_j d_j \, |\chi_j\rangle. \tag{10.10}$$

The respective expansion coefficients are then related according to

$$c_k = \sum_j d_j \, \langle \phi_k | \chi_j \rangle \quad \text{and} \quad d_j = \sum_k c_k \, \langle \chi_j | \phi_k \rangle. \tag{10.11}$$

Recall that the inner products $\langle \phi_k | \chi_j \rangle$ and $\langle \chi_j | \phi_k \rangle$ are complex conjugates of each other.

In Chap. 13, Sect. 13.2.2, we will encounter the notion of a *continuous* basis, that is, a basis set labeled by a continuous variable, instead of the discrete indices k and j above. Formulas analogous to those in Eqs. (10.11) hold good for a change of basis from a discrete to a continuous basis, and vice versa.

10.2.3 Basis Independence of the Inner Product

The inner product of any two elements of an LVS can also be "expanded" with the help of an orthonormal basis. When this is done in two different basis sets and the results are compared, you get a relation between the corresponding expansion coefficients.

Let $|\psi\rangle$ and $|\Psi\rangle$ be two vectors in an LVS, with expansions in two different orthonormal basis sets $\{|\phi_k\rangle\}$ and $\{|\chi_j\rangle\}$ given, respectively, by

$$|\psi\rangle = \sum_k c_k \, |\phi_k\rangle = \sum_j d_j \, |\chi_j\rangle \quad \text{and} \quad |\Psi\rangle = \sum_k C_k \, |\phi_k\rangle = \sum_j D_j \, |\chi_j\rangle. \tag{10.12}$$

It follows from the orthonormality of each basis that

$$\langle\psi|\Psi\rangle = \sum_k c_k^* \, C_k = \sum_j d_j^* \, D_j \, . \tag{10.13}$$

In particular, the norm of any vector is independent of the basis in terms of which you may choose to expand the vector. Setting $|\Psi\rangle = |\psi\rangle$ in Eq. (10.13), we get

$$\|\psi\|^2 = \sum_k |c_k|^2 = \sum_j |d_j|^2. \tag{10.14}$$

The identity (10.13), and its special case (10.14), are examples of **Parseval's Theorem**, also called Parseval's formula, or Parseval's identity. This name is generally associated with the corresponding result connecting square-integrable functions[1] and their Fourier transforms, but it is actually a more general result. The counterpart of the theorem in the case of orthogonal polynomials will be encountered in Chap. 16, Sect. 16.1.3. The versions involving Fourier series and Fourier integrals will appear in Chap. 17, Sect. 17.1.4, and Chap. 18, Sect. 18.1.2, respectively.

★ **5.** As always, it is instructive to work out the steps in the foregoing explicitly.

(a) In order to establish Eqs. (10.11), start with $c_k = \langle\phi_k|\,\psi\rangle = \langle\phi_k|\,I\,|\psi\rangle$ and use the completeness relations $\sum_l |\phi_l\rangle\langle\phi_l| = I = \sum_j |\chi_j\rangle\langle\chi_j|$ for the identity operator; similarly for d_j .
(b) Verify Eq. (10.13).

10.3 Some Important Inequalities

10.3.1 The Cauchy–Schwarz Inequality

In elementary vector algebra, we learn that the inner (or scalar) product of two vectors **a** and **b** in Euclidean space is $\mathbf{a} \cdot \mathbf{b} = a\,b\,\cos\,\theta$, where a, b are the respective magnitudes of the two vectors, and θ is the angle between them. Since $0 \le |\cos\,\theta| \le 1$, it follows that $|\mathbf{a} \cdot \mathbf{b}| \le a\,b$. Moreover, the equality sign applies only when $|\cos\,\theta| = 1$ or θ is either 0 or π, i.e., *if and only if the two vectors are collinear.*

The **Cauchy–Schwarz inequality** is just the generalization of these simple properties to any LVS. It asserts that

$$\boxed{|\langle\phi\,|\psi\rangle| \le \|\,\phi\,\|\,\|\,\psi\,\|,} \tag{10.15}$$

[1] Square-integrable functions will be defined in Chap. 13, Sect. 13.2.1.

the equality holding if and only if $|\phi\rangle$ and $|\psi\rangle$ are linearly dependent. This result is easily established as follows.

Proof Consider the inner product

$$\langle \phi + a\,\psi \,|\, \phi + a\,\psi \rangle \tag{10.16}$$

where $|\phi\rangle,\ |\psi\rangle \in \mathbb{V}$, and a is an arbitrary complex number. This quantity is nonnegative, because it is the square of the norm of a vector in the LVS. It is zero if and only if $|\phi\rangle + a|\psi\rangle$ is the null vector, in which case $|\phi\rangle = -a|\psi\rangle$, which means that $|\phi\rangle$ and $|\psi\rangle$ are linearly dependent. Expanding the inner product in (10.16), we have

$$\langle \phi|\phi\rangle + a^*\,\langle \psi|\phi\rangle + a\,\langle \phi|\psi\rangle + |a|^2\,\langle \psi|\psi\rangle \geq 0. \tag{10.17}$$

This inequality is valid for *any* complex number a. In particular, we may choose it to be given by

$$a = -\,\frac{\langle \psi|\phi\rangle}{\langle \psi|\psi\rangle}, \quad \text{so that} \quad a^* = -\,\frac{\langle \phi|\psi\rangle}{\langle \psi|\psi\rangle}. \tag{10.18}$$

Simplifying the resulting expression, we get

$$\langle \phi|\phi\rangle\,\langle \psi|\psi\rangle \geq \langle \phi|\psi\rangle\langle \psi|\phi\rangle = |\langle \phi|\psi\rangle|^2. \tag{10.19}$$

Taking square roots on both sides immediately yields the Cauchy–Schwarz inequality, $|\langle \phi\,|\psi\rangle| \leq \|\,\phi\,\|\,\|\,\psi\,\|$. The equality sign applies if and only if $|\phi\rangle$ and $|\psi\rangle$ are linearly dependent.

The generalized Uncertainty Principle: Among the numerous applications of the Cauchy–Schwarz inequality, the derivation of the generalized **Uncertainty Principle** in quantum mechanics is noteworthy. I state this principle here without going into the definitions of the various terms therein. (These terms will become clear after we consider operators in subsequent chapters.) Let A and B be the Hermitian (more precisely, self-adjoint) operators representing any two physical observables pertaining to a quantum mechanical system, and let $[A,\,B] = iC$ be their commutator. (The factor i ensures that C is also a Hermitian operator.) Then, if ΔA and ΔB are the standard deviations of the two observables in any state of the system, and $\langle C\rangle$ is the expectation value of C in that state, we must have

$$\boxed{(\Delta A)\,(\Delta B) \geq \tfrac{1}{2}\,|\langle C\rangle|.} \tag{10.20}$$

★ **6.** Let $|\psi\rangle$ and $|\phi\rangle$ be two linearly independent vectors in a linear vector space.

(a) Suppose the space is a *real* LVS. Find the value of the (real) scalar α that makes $\|\,\psi - \alpha\,\phi\,\|$ a minimum, and find this minimum value.
(b) Suppose the space is a *complex* LVS. Find the value of the (complex) scalar α that makes $\|\,\psi - \alpha\,\phi\,\|$ a minimum, and find this minimum value.

(c) Use the result of (b) to answer the following question: In the foregoing proof of the Cauchy–Schwarz inequality (Eqs. (10.16)–(10.19)), can there be a better choice of the scalar a than the one in Eq. (10.18), that will result in an improved or more stringent version of the inequality?

10.3.2 The Triangle Inequality

In plane geometry, the sum of the lengths of any two sides of a triangle is greater than the length of the third side. In fact, this condition is used as a fundamental property of the distance function or metric in more complicated spaces as well.

In an LVS, the Cauchy–Schwarz inequality leads to the corresponding **triangle inequality**: For any two vectors $|\phi\rangle,\ |\psi\rangle \in \mathbb{V}$,

$$\boxed{\parallel \phi + \psi \parallel \leq \parallel \phi \parallel + \parallel \psi \parallel .} \tag{10.21}$$

★ **7.** Use the Cauchy–Schwarz inequality to establish the triangle inequality (10.21).

Requiring that the triangle inequality holds good in infinite-dimensional spaces imposes some conditions on the elements of such spaces, as you will see in Chap. 13, Sect. 13.1 in the case of square-summable infinite sequences. When applied to certain function spaces, the triangle inequality is called the **Minkowski inequality**.

10.3.3 The Gram Determinant Inequality

The Cauchy–Schwarz inequality is itself a special case of an inequality involving an arbitrary number of elements of an LVS. As we saw in Chap. 5, Sect. 5.1.5, the Gram determinant of three vectors in three-dimensional Euclidean space is always greater than or equal to zero. It vanishes if and only if the vectors are linearly dependent. This is a special case of a more general result.

Let $|\psi_1\rangle,\ |\psi_2\rangle,\ \ldots,\ |\psi_k\rangle$ be vectors in an LVS. The Gram determinant of this set of vectors, $G(\psi_1,\ \ldots,\ \psi_k)$, satisfies the inequality

$$G(\psi_1,\ \ldots,\ \psi_k) \stackrel{\text{def.}}{=} \begin{vmatrix} \langle\psi_1|\psi_1\rangle & \langle\psi_1|\psi_2\rangle & \cdots & \langle\psi_1|\psi_k\rangle \\ \langle\psi_2|\psi_1\rangle & \langle\psi_2|\psi_2\rangle & \cdots & \langle\psi_2|\psi_k\rangle \\ \vdots & \vdots & \ddots & \vdots \\ \langle\psi_k|\psi_1\rangle & \langle\psi_k|\psi_2\rangle & \cdots & \langle\psi_k|\psi_k\rangle \end{vmatrix} \geq 0. \tag{10.22}$$

The $>$ sign applies if and only if all k vectors are linearly independent. If any of the vectors is linearly dependent on the rest, the Gram determinant of the set vanishes. It is obvious that the Cauchy–Schwarz inequality is the special case of this result, corresponding to $k = 2$. In this sense, the inequality in (10.22) is a generalization of the Cauchy–Schwarz inequality. Note also that, in an LVS of n dimensions, you cannot have more than n linearly independent vectors. Three important conclusions follow:

- The Gram determinant of any k vector in an LVS of n dimensions vanishes identically for all $k > n$.
- For $k \leq n$, the Gram determinant vanishes if and only if the vectors are not linearly independent.
- Hence the Gram determinant of a set of vectors in an LVS provides us with a test to determine whether they are linearly independent.

★ **8.** Show that, if $|\psi_1\rangle$, $|\psi_2\rangle$, and $|\psi_3\rangle$ are linearly independent vectors in an LVS, then

$$\| \psi_1 \|^2 \| \psi_2 \|^2 \| \psi_3 \|^2 > \{ \| \psi_1 \|^2 |\langle\psi_2|\psi_3\rangle|^2 + \| \psi_2 \|^2 |\langle\psi_3|\psi_1\rangle|^2 + \| \psi_3 \|^2 |\langle\psi_1|\psi_2\rangle|^2 \} - 2\,\mathrm{Re}\,\{\langle\psi_1|\psi_2\rangle\langle\psi_2|\psi_3\rangle\langle\psi_3|\psi_1\rangle\}.$$

10.4 Solutions

1. (a) An n^2-dimensional complex LVS. (b) A nine-dimensional real LVS. (c) A three-dimensional complex LVS. (d) A three-dimensional complex LVS. (e) A two-dimensional real LVS (comprising all functions of the form $y = a\,e^x + b\,e^{2x}$ where a and b are real numbers). (f) Not an LVS (the sum of two unitary matrices is not unitary in general). (g) An n^2-dimensional real LVS. ▶

2. $|\psi\rangle$ is a (3×1) column vector with each element equal to unity.

Remark The geometrical interpretation of this fact is elementary. The three unit vectors given may be represented by the Cartesian unit vectors $\mathbf{e}_x$, $\mathbf{e}_y$, and $\mathbf{e}_z$ in three-dimensional space. $|\psi\rangle$ is then represented by the vector $(\mathbf{e}_x + \mathbf{e}_y + \mathbf{e}_z)$, which is the (directed) body diagonal of a unit cube placed with a corner at the origin and its sides along the positive coordinate axes. ▶

3. (a) Yes. (b) $|\chi\rangle$ is an $(n \times 1)$ column vector with elements $x_j = \sqrt{j} - \sqrt{j-1}$, where $1 \leq j \leq n$. ▶

4. (a) None of the vectors can be written as a linear combination of the other two. It is easy to see that the equation $a|\psi_1\rangle + b|\psi_2\rangle + c|\psi_3\rangle = 0$ implies that $a = b = c = 0$. Note, incidentally, that the given vectors can be represented in Cartesian coordinates as $\mathbf{e}_x + \mathbf{e}_y + \mathbf{e}_z$, $\mathbf{e}_x + \mathbf{e}_y$ and $\mathbf{e}_x + \mathbf{e}_z$, respectively.

(b) Applying the Gram–Schmidt procedure, we find

$$|\phi_1\rangle = \frac{1}{\sqrt{3}}\begin{pmatrix}1\\1\\1\end{pmatrix}, \quad |\phi_2\rangle = \frac{1}{\sqrt{6}}\begin{pmatrix}1\\1\\-2\end{pmatrix}, \quad |\phi_3\rangle = \frac{1}{\sqrt{2}}\begin{pmatrix}-1\\1\\0\end{pmatrix}.$$

You can represent these vectors in Cartesian coordinates by the unit vectors

$$(\mathbf{e}_x + \mathbf{e}_y + \mathbf{e}_z)/\sqrt{3}, \quad (\mathbf{e}_x + \mathbf{e}_y - 2\mathbf{e}_z)/\sqrt{6} \text{ and } (-\mathbf{e}_x + \mathbf{e}_y)/\sqrt{2},$$

and check out that they form a right-handed triad of unit vectors. You will find it instructive to draw a figure. ▶

6. (a) The norm $\| \psi - \alpha \phi \|$ is the positive square root of the quantity

$$N(\alpha) = \| \psi - \alpha \phi \|^2 = \langle \psi - \alpha \phi \,|\, \psi - \alpha \phi \rangle$$

as a function of the real variable α. Minimizing N is equivalent to minimizing $\| \psi - \alpha \phi \|$. Imposing the conditions for a minimum, namely, $dN/d\alpha = 0$ and $d^2N/d\alpha^2 > 0$, it is easily seen that a minimum occurs at $\alpha = [\text{Re}\,\langle \psi|\phi\rangle]/\langle \phi|\phi\rangle$. The minimum value of N is found to be

$$\min_{\alpha} \| \psi - \alpha \phi \|^2 = \left[\| \psi \|^2 \| \phi \|^2 - \left(\text{Re}\,\langle \phi \,|\psi\rangle\right)^2 \right] / \| \phi \|^2 .$$

(b) $\| \psi - \alpha \phi \|^2 = N(\alpha, \alpha^*)$ is now a function of both α and α^*. Remember that any complex number α and its complex conjugate α^* are linearly independent! To find the minimum of N, you must write it as a function of the pair of real variables α_1 and α_2, where $\alpha = \alpha_1 + i\alpha_2$. Imposing the conditions for a minimum, namely,

$$\frac{\partial N}{\partial \alpha_1} = 0, \quad \frac{\partial N}{\partial \alpha_2} = 0,$$

and further

$$\frac{\partial^2 N}{\partial \alpha_1^2} > 0, \quad \frac{\partial^2 N}{\partial \alpha_2^2} > 0, \quad \frac{\partial^2 N}{\partial \alpha_1^2}\frac{\partial^2 N}{\partial \alpha_2^2} > \left(\frac{\partial^2 N}{\partial \alpha_1\, \partial \alpha_2}\right)^2,$$

we find that they are satisfied at the value $\alpha = \langle \phi \,|\psi\rangle / \langle \phi|\phi\rangle$. The minimum value of N is given by

$$\min_{\alpha,\, \alpha^*} \| \psi - \alpha \phi \|^2 = \left[\| \psi \|^2 \| \phi \|^2 - |\langle \phi \,|\psi\rangle|^2 \right] / \| \phi \|^2 .$$

(c) You would have noticed the close similarity between the solution in (b) and the derivation of the Cauchy–Schwarz inequality in Eqs. (10.16)–(10.19) (with the obvious identification $a = -\alpha$). The minimization above should tell you that the

Cauchy–Schwarz inequality is the *best* possible one for the quantities appearing in it. That is, there is no better choice of a in Eq. (10.18) that will yield an improved or more stringent inequality in the general case. ▶

7. Start with

$$\| \phi + \psi \|^2 = \langle \phi + \psi \,|\, \phi + \psi \rangle = \| \phi \|^2 + \| \psi \|^2 + 2\,\mathrm{Re}\, \langle \phi \,|\psi \rangle .$$

Use the fact that $\mathrm{Re}\, \langle \phi \,|\psi \rangle \leq |\mathrm{Re}\, \langle \phi \,|\psi \rangle| \leq |\langle \phi \,|\psi \rangle|$. ▶

8. Use (10.22) in the case $k = 3$ and simplify. ▶

Chapter 11
A Look at Matrices

In this chapter, we continue our discussion of linear vector spaces and linear algebra. I shall assume that you are familiar with the most basic concepts and operations of matrix algebra, specifically those pertaining to square matrices of finite order: the addition and multiplication of matrices, the concepts of eigenvalues and eigenvectors, and so on. Occasionally, I shall also use terms such as *Hermitian* matrix and *unitary* matrix, under the assumption that you are familiar with the corresponding definitions. In some cases, I shall return to these terms at appropriate junctures and define them formally, merely for the sake of completeness.

Loosely speaking, one can look upon a matrix in two different ways. The first, as an element of some LVS: it is easy to check that all $(n \times n)$ matrices form an LVS, for each given positive integer value of n. The second, as the representation of an operator: it acts on the elements (or vectors) of some LVS, to produce other vectors. Which interpretation is involved in any particular case will be clear from the context. Both aspects will be involved in what follows.

11.1 Pauli Matrices

11.1.1 Expansion of a (2×2) Matrix

Let us begin with the simplest case, that of (2×2) matrices. For this purpose, it is most convenient to introduce and study the properties of certain special (2×2) matrices, called **Pauli matrices**. In quantum mechanics, they are closely associated with spin-$\frac{1}{2}$ particles such as electrons, protons, and neutrons. Numerous situations, both in quantum mechanics as well as in a variety of other contexts, can be reduced to so-called **two-level systems** in which the Pauli matrices play a significant role.

V. Balakrishnan, *Mathematical Physics*,
https://doi.org/10.1007/978-3-030-39680-0_11

(In quantum mechanics, this refers to a system whose states are ket vectors in a two-dimensional LVS.) These matrices are therefore of fundamental importance.

Consider a general (2×2) matrix $M = \begin{pmatrix} a & b \\ c & d \end{pmatrix}$. Treated as an element of the LVS of (2×2) matrices, this is shorthand for the linear combination (or expansion)

$$M = a \begin{pmatrix} 1 & 0 \\ 0 & 0 \end{pmatrix} + b \begin{pmatrix} 0 & 1 \\ 0 & 0 \end{pmatrix} + c \begin{pmatrix} 0 & 0 \\ 1 & 0 \end{pmatrix} + d \begin{pmatrix} 0 & 0 \\ 0 & 1 \end{pmatrix}. \tag{11.1}$$

The four matrices on the right-hand side constitute the **natural basis** in the LVS of (2×2) matrices. But there is another, very useful, basis that consists of the (2×2) unit matrix I and the three Pauli matrices. The standard definition of the latter set of matrices is

$$\boxed{\sigma_1 \stackrel{\text{def.}}{=} \begin{pmatrix} 0 & 1 \\ 1 & 0 \end{pmatrix}, \quad \sigma_2 \stackrel{\text{def.}}{=} \begin{pmatrix} 0 & -i \\ i & 0 \end{pmatrix}, \quad \sigma_3 \stackrel{\text{def.}}{=} \begin{pmatrix} 1 & 0 \\ 0 & -1 \end{pmatrix}.} \tag{11.2}$$

Then, *any* (2×2) matrix M can be written as a linear combination

$$M = \begin{pmatrix} a & b \\ c & d \end{pmatrix} = \alpha_0 \, I + \alpha_1 \, \sigma_1 + \alpha_2 \, \sigma_2 + \alpha_3 \, \sigma_3 \, . \tag{11.3}$$

★ **1.** Show that the (2×2) unit matrix I and the three Pauli matrices form a basis in the LVS of all (2×2) matrices, that is, the matrix elements (or expansion coefficients) $(a, \, b, \, c \, , d)$ in the natural basis (11.1) *uniquely* determine the expansion coefficients $(\alpha_0, \, \alpha_1, \, \alpha_2, \, \alpha_3)$ in (11.3), and *vice versa*.

The basis in (11.3) has numerous advantages over the natural basis. These follow from the basic properties of the Pauli matrices, which are listed next.

11.1.2 Basic Properties of the Pauli Matrices

Some of the basic properties of the Pauli matrices σ_i $(i = 1, 2, 3)$ are as follows:

(i) Each σ_i is Hermitian $(\sigma_i = \sigma_i^\dagger)$ and traceless, with determinant equal to -1.
(ii) Each $\sigma_i^2 = I$, where I is the (2×2) unit matrix. Hence $\sigma_i^{-1} = \sigma_i$.
(iii) Each σ_i has eigenvalues 1 and -1 , with normalized eigenvectors $\begin{pmatrix} 1 \\ 0 \end{pmatrix}$ and $\begin{pmatrix} 0 \\ 1 \end{pmatrix}$, respectively.
(iv) The product of any two different Pauli matrices is proportional to the remaining one:

$$\sigma_1 \, \sigma_2 = i \ \sigma_3 \, , \quad \sigma_2 \, \sigma_3 = i \ \sigma_1 \, , \quad \sigma_3 \, \sigma_1 = i \ \sigma_2 \, . \tag{11.4}$$

(v) The preceding statement, combined with the fact that the square of any Pauli matrix is the unit matrix, implies that

$$\sigma_k \, \sigma_l = \delta_{kl} \, I + i \, \epsilon_{klm} \, \sigma_m \, . \tag{11.5}$$

(vi) Hence the **commutator** of any two different Pauli matrices is a constant times the remaining Pauli matrix:

$$[\sigma_k \, , \, \sigma_l] \overset{\text{def.}}{=} \sigma_k \, \sigma_l - \sigma_l \, \sigma_k = 2 \, i \, \epsilon_{klm} \, \sigma_m \, . \tag{11.6}$$

(vii) In particular, the matrices

$$J_k = \tfrac{1}{2} \sigma_k \quad (k = 1, 2, 3) \tag{11.7}$$

satisfy the **commutation relations**

$$\boxed{[J_k \, , \, J_l] = i \, \epsilon_{klm} \, J_m \, .} \tag{11.8}$$

This is an important fact. Equation (11.8) is precisely the so-called **angular momentum algebra**—or the Lie algebra of the **infinitesimal generators** of the rotation group in three-dimensional space. Thus, there is a close connection between the Pauli matrices and rotations in three-dimensional space.[1]

(viii) Similarly, the **anticommutator**[2] of two distinct Pauli matrices is identically zero. Combining this with the fact that the square of any Pauli matrix is the unit matrix, we have

$$[\sigma_k \, , \, \sigma_l]_+ \overset{\text{def.}}{=} \sigma_k \, \sigma_l + \sigma_l \, \sigma_k = 2 \, \delta_{kl} \, I. \tag{11.9}$$

It is customary to denote the set of Pauli matrices $(\sigma_1 \, , \, \sigma_2 \, , \, \sigma_3)$ by the "vector" $\boldsymbol{\sigma}$. This is more than just a matter of notation; there is a proper justification for it, because of the way the Pauli matrices transform under a rotation of the coordinate system: they transform precisely like the components of a vector. More will be said about this in Sect. 11.3.2 below, and in Chap. 15, Sect. 15.3.1. Here are some useful identities involving $\boldsymbol{\sigma}$. Let $\mathbf{a}$ and $\mathbf{b}$ denote ordinary vectors in three-dimensional Euclidean space. Then

$$[\boldsymbol{\sigma} \, , (\mathbf{a} \cdot \boldsymbol{\sigma})] = 2i \, (\mathbf{a} \times \boldsymbol{\sigma}) \tag{11.10}$$

$$(\mathbf{a} \cdot \boldsymbol{\sigma}) \, (\mathbf{b} \cdot \boldsymbol{\sigma}) = (\mathbf{a} \cdot \mathbf{b}) \, I + i \, (\mathbf{a} \times \mathbf{b}) \cdot \boldsymbol{\sigma} \tag{11.11}$$

$$[(\mathbf{a} \cdot \boldsymbol{\sigma}) \, , (\mathbf{b} \cdot \boldsymbol{\sigma})] = 2i \, (\mathbf{a} \times \mathbf{b}) \cdot \boldsymbol{\sigma} \tag{11.12}$$

---

[1] I have already stated in Chap. 5, Sect. 5.2 that we will come back to this topic in Sect. 11.3.1, as well as in Chap. 12, Sect. 12.4.2 and Chap. 15, Sects. 15.3.1 and 15.3.3.

[2] Sometimes the notation $\{ \, , \, \}$ is used for the anticommutator, but I prefer the notation $[\, , \,]_+$ in order to avoid confusion with other quantities such as the Poisson bracket.

★ **2.** Use the properties of the Pauli matrices to establish the identities in Eqs. (11.10)–(11.12).

★ **3.** Let $\mathbf{n} = (n_1\,,\, n_2\,,\, n_3)$ be any unit vector in three-dimensional Euclidean space.

(a) Show that the eigenvalues of the matrix $\boldsymbol{\sigma}\cdot\mathbf{n}$ remain equal to 1 and -1, independent of the direction of the unit vector $\mathbf{n}$. (Remember that $n_1^2 + n_2^2 + n_3^2 = 1$.) Find the corresponding normalized eigenvectors.

(b) Show that $(\boldsymbol{\sigma}\times\mathbf{n})\,(\boldsymbol{\sigma}\cdot\mathbf{n}) = i\,\boldsymbol{\sigma} - i\,(\boldsymbol{\sigma}\cdot\mathbf{n})\,\mathbf{n}$.

(c) Show that $(\boldsymbol{\sigma}\cdot\mathbf{n})\,(\boldsymbol{\sigma}\times\mathbf{n}) = i\,(\boldsymbol{\sigma}\cdot\mathbf{n})\mathbf{n} - i\,\boldsymbol{\sigma}$.

Remark (a) The spin operator of a spin-$\frac{1}{2}$ particle is given by $\mathbf{S} = \frac{1}{2}\hbar\boldsymbol{\sigma}$. Hence the eigenvalues of *any* component (along any arbitrary direction, not restricted to the z-axis, or any of the axes in Cartesian coordinates!) of the spin operator of such a particle are given by $\pm\frac{1}{2}\hbar$.

The identities in (b) and (c), and some of those to follow, are particularly useful—for instance, in the study of the behavior of the intrinsic magnetic moment of an electron in an applied magnetic field.

11.2 The Exponential of a Matrix

11.2.1 Occurrence and Definition

The exponential of a square matrix or, more generally, the exponential of an *operator*, occurs frequently in physical applications. Indeed, it is ubiquitous in physics. In a very broad sense, there is good reason to regard *the exponentiation of an operator* as the central problem of mathematical physics itself. Here are two important instances:

- ***Quantum mechanics***: Given a system with a time-independent Hamiltonian H, the time evolution of the physical observables pertaining to the system is governed by its time-development operator $e^{-iHt/\hbar}$.
- ***Equilibrium statistical mechanics***: In the canonical ensemble, all statistical averages of physical quantities pertaining to the system are governed by its density operator $e^{-H/(k_B T)}$.

In both these subjects, therefore, the basic problem is to find the exponential of the Hamiltonian. Even differential operators such as d/dx, d^2/dx^2, ∇^2, etc., can be exponentiated, and a specific meaning can be given to the resulting operators. You have already come across an example in elementary calculus. The **Taylor expansion**

$$f(x+a) = f(x) + af\,'(x) + \frac{a^2}{2!}f\,''(x) + \cdots \tag{11.13}$$

can be written symbolically as

$$\boxed{f(x+a) = e^{a(d/dx)}\, f(x).} \tag{11.14}$$

This tells you that the effect of exponentiating the operator $a\,(d/dx)$ is a *translation* of the argument x of the function it acts on by the amount a.

For the moment, however, we are only concerned with the exponential of an $(n \times n)$ square matrix M, for a finite value of n. This is defined as

$$\boxed{\exp M \stackrel{\text{def.}}{=} I + M + \frac{M^2}{2!} + \cdots = \sum_{n=0}^{\infty} \frac{M^n}{n!},} \tag{11.15}$$

exactly analogous to the definition of e^z for any complex number z. Here I is the unit matrix that is of the same order as the matrix M. Just as the power series for e^z is absolutely convergent for all finite values of $|z|$, the series for e^M is convergent for all matrices with finite elements. (This statement can be made more precise, but it suffices for the present.)

★ **4.** Show that, if σ_k is any one of the three Pauli matrices and α is any complex number, then

$$e^{i\alpha\sigma_k} = (\cos\,a)\,I + i\,(\sin\,\alpha)\,\sigma_k\,.$$

This formula is a special case of the identity (11.16) below.

11.2.2 The Exponential of an Arbitrary (2 × 2) *Matrix*

Expanding an arbitrary (2 × 2) matrix in terms of the unit matrix and the Pauli matrices enables us to derive a very useful formula for the exponential of the matrix.

Let $\mathbf{a}$ be an ordinary vector with Cartesian components $(a_1\,,\, a_2\,,\, a_3)$, and let $(\mathbf{a}\cdot\boldsymbol{\sigma})$ stand for the matrix $a_1\,\sigma_1 + a_2\,\sigma_2 + a_3\,\sigma_3$, as usual. Then

$$\boxed{e^{i\,\mathbf{a}\cdot\boldsymbol{\sigma}} = I\,\cos\,a + i\,\frac{(\mathbf{a}\cdot\boldsymbol{\sigma})}{a}\,\sin\,a, \quad \text{where} \quad a = (a_1^2 + a_2^2 + a_3^2)^{1/2}.} \tag{11.16}$$

This formula is the (2 × 2) matrix analog of the familiar Euler identity $e^{i\theta} = \cos\,\theta + i\,\sin\,\theta$ for any real number θ.

★ **5.** Establish the identity in Eq. (11.16).

Equation (11.16) is readily extended to the case of $e^{\boldsymbol{\alpha}\cdot\boldsymbol{\sigma}}$, where $\boldsymbol{\alpha}\cdot\boldsymbol{\sigma}$ stands for $\alpha_1\,\sigma_1 + \alpha_2\,\sigma_2 + \alpha_3\,\sigma_3$, and $(\alpha_1\,,\, \alpha_2\,,\, \alpha_3)$ are real numbers. Simply set $i\,\mathbf{a} = \boldsymbol{\alpha}$, i.e., $a_1 = -i\,\alpha_1\,,\; a_2 = -i\,\alpha_2\,,\; a_3 = -i\,\alpha_3$ in the formula to get

$$\boxed{e^{\alpha\cdot\sigma} = I\,\cosh\alpha + \frac{(\boldsymbol{\alpha}\cdot\boldsymbol{\sigma})}{\alpha}\,\sinh\alpha, \quad \text{where} \quad \alpha = (\alpha_1^2+\alpha_2^2+\alpha_3^2)^{1/2}.} \quad (11.17)$$

What has actually been done here is an analytic continuation of the original formula from real values of the quantities (a_1, a_2, a_3) to pure imaginary values of these variables.[3] Equations (11.16) and (11.17) are completely equivalent to each other, and are actually valid for arbitrary *complex* numbers $(\alpha_1, \alpha_2, \alpha_3)$ or (a_1, a_2, a_3). (You must then remember that α or a need not be real quantities, in general.)

We know that an arbitrary (2×2) matrix M can be expanded as

$$M = \alpha_0\, I + \alpha_1\,\sigma_1 + \alpha_2\,\sigma_2 + \alpha_3\,\sigma_3 = \alpha_0\, I + \boldsymbol{\alpha}\cdot\boldsymbol{\sigma}, \quad (11.18)$$

and that the unit matrix commutes with all matrices. It follows from (11.17) that we have the closed-form expression

$$\boxed{e^{M} = e^{\alpha_0}\Big(I\,\cosh\alpha + \frac{(\boldsymbol{\alpha}\cdot\boldsymbol{\sigma})}{\alpha}\,\sinh\alpha\Big)} \quad (11.19)$$

for the exponential of *any* (2×2) matrix M. Recall that $\alpha_0 = \frac{1}{2}\,\mathrm{Tr}\,M$.

Significantly enough,

- there is no counterpart of such a simple closed-form expression for the exponential of a general $(n \times n)$ matrix for $n > 2$.

In some special cases, however, such higher order matrices may be exponentiated quite easily.

★ **6.** Here are two examples of matrices that can be exponentiated readily.

(a) If $M = \begin{pmatrix} 0 & 1 & 0 \\ 0 & 0 & 1 \\ 0 & 0 & 0 \end{pmatrix}$, find e^M. Hence write down the eigenvalues of e^M.

(b) Let M be the $(n \times n)$ with each element $M_{ij} = 1$ for $1 \le i \le n$ and $1 \le j \le n$. Find e^M.

In the next section, we shall encounter a family of (3×3) matrices of physical significance that can be exponentiated easily.

---

[3] Analytic continuation is a crucial property of analytic functions of complex variables. We will consider it in some detail in Chap. 25.

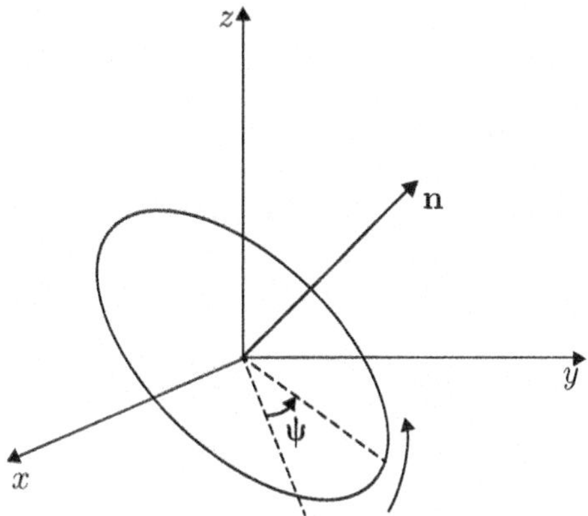

Fig. 11.1 Specifying a rotation of the coordinate axes about an axis along a unit vector **n**, through an angle ψ

11.3 Rotation Matrices in Three Dimensions

11.3.1 Generators of Infinitesimal Rotations and Their Algebra

Let us return to rotations in three-dimensional space, and consider the (3×3) orthogonal matrices that describe these rotations. Recall, at this stage, the remarks made at the end of Chap. 5, Sect. 5.1.2, regarding the representation of rotations by (3×3) matrices. There are many ways of parametrizing rotations in three dimensions. A very useful way in applications is via three **Euler angles**. These are used, for instance, in studying the rotational motion of a rigid body. Any given orientation of the triad of coordinate axes may be reached from an initial reference orientation by a succession of three rotations about a prescribed set of three different axes. There are, in fact, 12 different conventions for defining Euler angles. I will not digress into these. Our objective here is somewhat different: we are interested in the rotation matrices *per se* and in their algebraic properties. We are also interested in finding the *explicit* transformation formula of a vector under an arbitrary rotation of the coordinate axes. As shown below, there is quite an easy way to arrive at the exact answer without any tedious algebra.

Euler's rotation theorem: A convenient way of parametrizing any given rotation is to specify the *axis* of rotation, i.e., the *direction* in space about which the triad of Cartesian coordinate axes is rotated, and the *amount* or angle of rotation about this axis. (This is essentially **Euler's rotation theorem**.) We may therefore denote a general rotation matrix by $R(\mathbf{n}, \psi)$, where $\mathbf{n}$ is the unit vector along the axis of rotation, and ψ is the angle of rotation about this axis (see Fig. 11.1). It appears to be quite obvious that the range of ψ is $0 \leq \psi < 2\pi$. We may take it to be so, for the present. (Somewhat surprisingly, this point requires deeper examination. I will return to it in Chap. 15, Sect. 15.3.5.) All you need in order to write down the rotation matrices corresponding to the special case of rotations about the three Cartesian axes themselves is to recall a result from elementary coordinate geometry. A rotation by an angle ψ of the coordinate axes about the origin in the xy-plane gives the new coordinates $x' = x \cos\psi + y \sin\psi$ and $y' = -x \sin\psi + y \cos\psi$. The z coordinate is left unchanged. Therefore

$$R(\mathbf{e}_z, \psi) = \begin{pmatrix} \cos\psi & \sin\psi & 0 \\ -\sin\psi & \cos\psi & 0 \\ 0 & 0 & 1 \end{pmatrix}. \tag{11.20}$$

By cyclic permutation of xyz, we may write down the other two matrices

$$R(\mathbf{e}_x, \psi) = \begin{pmatrix} 1 & 0 & 0 \\ 0 & \cos\psi & \sin\psi \\ 0 & -\sin\psi & \cos\psi \end{pmatrix}, \quad R(\mathbf{e}_y, \psi) = \begin{pmatrix} \cos\psi & 0 & -\sin\psi \\ 0 & 1 & 0 \\ \sin\psi & 0 & \cos\psi \end{pmatrix}. \tag{11.21}$$

It is easily checked that each of these matrices is orthogonal, and has a determinant equal to $+1$. Hence each of them can be built up from the identity matrix *by a succession of infinitesimal rotations* about the axis concerned. We can work backwards from Eqs. (11.20) and (11.21) to see how this is done.

Consider, for definiteness, $R(\mathbf{e}_z, \psi)$. We could implement such a rotation by n successive rotations about the z-axis, each through an infinitesimal angle $\delta\psi$, such that $n\,\delta\psi = \psi$. The matrix $R(\mathbf{e}_z, \delta\psi)$ is easily written down: use the fact that $\sin\delta\psi \simeq \delta\psi$ and $\cos\delta\psi \simeq 1$ to first order in $\delta\psi$. Separating out the (3×3) unit matrix, which corresponds to the identity transformation (or zero rotation), we get

$$R(\mathbf{e}_z, \delta\psi) = I + i\,(\delta\psi)\,J_3\,, \quad \text{where } J_3 = \begin{pmatrix} 0 & -i & 0 \\ i & 0 & 0 \\ 0 & 0 & 0 \end{pmatrix}. \tag{11.22}$$

The parameter $\delta\psi$ has been factored out in the expression above. This makes the elements of the matrix J_3 pure numbers that are independent of the angle of rotation. The reason for separating out the factor i in the definition of J_3 is to ensure that J_3 is a *Hermitian* matrix.[4] The finite-angle rotation matrix $R(\mathbf{e}_z, \psi)$ is then given by

$$R(\mathbf{e}_z, \psi) = \underbrace{R(\mathbf{e}_z, \delta\psi)\cdots R(\mathbf{e}_z, \delta\psi)}_{n \text{ factors}} = [R(\mathbf{e}_z, \delta\psi)]^n = [I + i\,(\delta\psi)\,J_3]^n. \tag{11.23}$$

Setting $\delta\psi = \psi/n$ and passing to the limit $n \to \infty$,

[4]That is, it is equal to its complex conjugate transpose, or $J_3 = J_3^\dagger$. Observe that I have used the same symbol (J_3) for another matrix earlier in this chapter, in Eq. (11.7): namely, the (2×2) matrix $\frac{1}{2}\sigma_3$. This is deliberate, and the reason will become clear shortly.

$$R(\mathbf{e}_z, \psi) = \lim_{n\to\infty} \left(I + \frac{i\,\psi\,J_3}{n}\right)^n = e^{i\,\psi\,J_3}. \tag{11.24}$$

Repeat the procedure above for the matrices $R(\mathbf{e}_x, \psi)$ and $R(\mathbf{e}_y, \psi)$, to get

$$R(\mathbf{e}_x, \psi) = e^{i\,\psi\,J_1} \quad \text{and} \quad R(\mathbf{e}_y, \psi) = e^{i\,\psi\,J_2}, \tag{11.25}$$

where

$$J_1 = \begin{pmatrix} 0 & 0 & 0 \\ 0 & 0 & -i \\ 0 & i & 0 \end{pmatrix} \quad \text{and} \quad J_2 = \begin{pmatrix} 0 & 0 & i \\ 0 & 0 & 0 \\ -i & 0 & 0 \end{pmatrix}. \tag{11.26}$$

To sum up

- The form $R(\mathbf{e}_z, \delta\psi) = I + i\,(\delta\psi)\,J_3$ makes it quite clear why the matrix J_3 is called the **generator** of an infinitesimal rotation about the z-axis (i.e., a rotation about the origin, in the xy-plane). Similarly, J_1 and J_2 may be identified as the generators of infinitesimal rotations about the x and y axes (i.e., about the origin, in the yz-plane and zx-plane), respectively.
- The matrix corresponding to rotation by a finite angle is obtained by exponentiating the corresponding generator. This is a general feature of groups of transformations (more generally, of Lie groups).
- The matrices J_1, J_2, and J_3 are Hermitian, and they satisfy the commutation relations

$$\boxed{[J_k\,,\; J_l] = i\epsilon_{klm}\,J_m\,.} \tag{11.27}$$

This is the same angular momentum algebra that is satisfied by the matrices $\frac{1}{2}\,\sigma_k$, as you have seen already in Eqs. (11.7) and (11.8).

★ **7.** It is instructive to check out the statements made above.

(a) Start with Eqs. (11.21) and write down the corresponding matrices for an infinitesimal rotation angle $\delta\psi$ about the directions $\mathbf{e}_x$ and $\mathbf{e}_y$, respectively. Express these as $[I + i\,(\delta\psi)\,J_1]$ and $[I + i\,(\delta\psi)\,J_2]$, respectively, to verify that J_1 and J_2 are the matrices written down in Eqs. (11.26).

(b) Using the expressions given in Eqs. (11.26) and (11.22) for the matrices J_k, directly calculate the exponentials $\exp\,(i\psi\,J_k)$ for $k = 1, 2$, and 3 by summing the corresponding exponential series. Verify that you recover Eqs. (11.21) and (11.20) for the finite-angle rotation matrices $R(\mathbf{e}_x, \psi)$, $R(\mathbf{e}_y, \psi)$, and $R(\mathbf{e}_z, \psi)$.

(c) Verify that the generators J_k satisfy the commutation relations in Eqs. (11.27).

Here is an important observation:

- The fact the (3×3) matrices J_k in Eqs. (11.26) and (11.22) satisfy the same commutation relations as the (2×2) matrices $\frac{1}{2}\,\sigma_k$ means that these two sets of matrices are just two different representations of the same Lie algebra, namely, the angular momentum algebra.

That is the reason why I have used the same symbols for both sets of matrices. We shall consider the angular momentum algebra further in Chap. 12, Sect. 12.4.2 and in Chap. 15, Sects. 15.3.1–15.3.3.

11.3.2 The General Rotation Matrix

Equations (11.24) and (11.25) show that the matrices corresponding to rotations about the three Cartesian axes can be written as exponentials of the corresponding infinitesimal generators. What about a *general* rotation by an angle ψ, about an axis $\mathbf{n}$ pointing in an arbitrary direction in space?

I have mentioned in Sect. 11.1.2 that the Pauli matrices $(\sigma_1\,,\,\sigma_2\,,\,\sigma_3)$ transform under rotations like the Cartesian components of a vector, which is why it makes sense to denote the triplet collectively by $\boldsymbol{\sigma}$. This is actually a special case of the following more general statement (that will be proved in Chap. 15, Sect. 15.3.1):

- The three generators of infinitesimal rotations in three-dimensional space, $(J_1\,,\,J_2\,,\,J_3)$, themselves transform under rotations like the components of a vector.

It is therefore natural to denote the triplet by the vector symbol $\mathbf{J}$. Then, if the components of the direction vector $\mathbf{n}$ are given by $(n_1\,,\,n_2\,,\,n_3)$, we are guaranteed that

$$\boxed{R(\mathbf{n},\,\psi) = e^{i\,(J_1\,n_1+J_2\,n_2+J_3\,n_3)\,\psi} = e^{i\,(\mathbf{J}\cdot\mathbf{n})\,\psi}.} \tag{11.28}$$

Since the different matrices J_k do not commute with each other, however, the right-hand side of (11.28) is *not* equal to the product of exponentials, i.e.,

$$e^{i\,(J_1\,n_1+J_2\,n_2+J_3\,n_3)\,\psi} \neq e^{i\,J_1\,n_1\,\psi}\,e^{i\,J_2\,n_2\,\psi}\,e^{i\,J_3\,n_3\,\psi}.$$

In spite of this problem, it turns out to be possible to compute the exponential of the (3×3) matrix $i\,(\mathbf{J}\cdot\mathbf{n})\,\psi$ exactly, and in closed form. Here is how this is done.

We want to find $e^{M\psi}$, where $M = i\,(\mathbf{J}\cdot\mathbf{n})$. Using the definitions of the matrices J_k above, we have

$$M = i\,(\mathbf{J}\cdot\mathbf{n}) = \begin{pmatrix} 0 & n_3 & -n_2 \\ -n_3 & 0 & n_1 \\ n_2 & -n_1 & 0 \end{pmatrix}. \tag{11.29}$$

In order to find the powers of M explicitly, it is helpful to note that the general element of M is given by

$$M_{ij} = \epsilon_{ijk}\,n_k\,. \tag{11.30}$$

Now use the fundamental relation between the Levi-Civita symbol and the Kronecker delta, Eq. (5.13) of Chap. 5, Sect. 5.1.3. It follows readily that

$$(M^2)_{ij} = n_i\, n_j - \delta_{ij}\,, \quad \text{and hence} \quad (M^3)_{ij} = -\epsilon_{ijk}\, n_k = -M_{ij}\,. \tag{11.31}$$

The fact that $M^3 = -M$ immediately enables us to simplify the exponential $e^{M\psi}$. The final answer for the matrix elements of the rotation matrix $R(\mathbf{n}, \psi)$ is both simple and elegant. It reads

$$\boxed{R_{ij}(\mathbf{n}, \psi) = \delta_{ij} \cos \psi + n_i\, n_j\, (1 - \cos \psi) + \epsilon_{ijk}\, n_k \sin \psi.} \tag{11.32}$$

Even more explicitly, if the spherical polar angles of the unit vector $\mathbf{n}$ are θ and φ, the direction cosines are given by

$$n_1 = \sin \theta \cos \varphi, \quad n_2 = \sin \theta \sin \varphi, \quad n_3 = \cos \theta. \tag{11.33}$$

Substituting these expressions in Eq. (11.32), you can write down the complete rotation matrix for an arbitrary rotation $R(\mathbf{n}, \psi)$.

★ **8.** Once again, you will find it instructive to work through the algebra to arrive at the results above.

(a) First verify Eqs. (11.29)–(11.31).
(b) Hence derive Eq. (11.32) for the elements of the general rotation matrix $R(\mathbf{n}, \psi)$.
(c) Write down the rotation matrix $R(\mathbf{n}, \psi)$ explicitly, using spherical polar coordinates for $\mathbf{n}$ as in Eqs. (11.33).

★ **9.** Show that the eigenvalues of the general rotation matrix $R(\mathbf{n}, \psi)$ are 1, $e^{i\psi}$ and $e^{-i\psi}$.

The matrix R has real elements, but it is not symmetric. If it had been so, all its eigenvalues would have been real. (The eigenvalues of a real symmetric matrix are real.) Rather, it is an *orthogonal* matrix with *real* elements. It is therefore a special case of a *unitary* matrix. As we shall see in Chap. 12, Sect. 12.2.1, all the eigenvalues of a unitary matrix must be complex numbers with magnitude equal to unity, i.e., they must lie on the unit circle in the complex plane.

What can be said about the eigen*vectors* of $R(\mathbf{n}, \psi)$? It should be obvious on physical grounds that $\mathbf{n}$ itself, i.e., the column vector with elements $(n_1\,, n_2\,, n_3)$, is the (normalized) eigenvector corresponding to the eigenvalue 1. This is just the statement that a rotation about an axis directed along $\mathbf{n}$ leaves the coordinates of all points on that axis unchanged. The other two eigenvectors, corresponding to the eigenvalues $e^{\pm i\psi}$, cannot be vectors with real elements. If that were so, it would mean that a rotation about $\mathbf{n}$ also leaves unchanged some two other directions in physical space. But this is obviously not true.

Rotation matrices in rigid-body dynamics: Rotation matrices have an obvious physical application in the dynamics of rigid bodies. One introduces a *space-fixed* coordinate frame with an origin $\mathrm{O_s}$, and a *body-fixed* coordinate frame that co-moves with the rigid body (see Fig. 11.2), with its origin $\mathrm{O_b}$ at the center-of-mass of the body (say). The instantaneous configuration of the rigid body is then given by the position

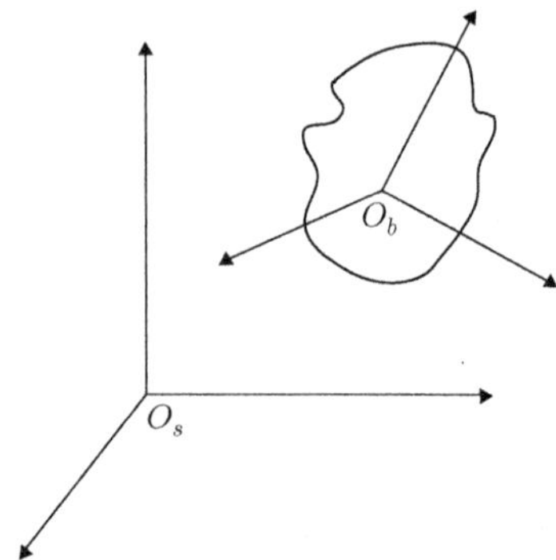

Fig. 11.2 Space-fixed and body-fixed coordinate frames

vector of O_b (with respect to O_s) as well as and the orientation of the body-fixed frame with respect to the space-fixed frame. The latter is evidently given by some rotation matrix. The time evolution of this rotation matrix then describes the orientational dynamics of the rigid body.

Suppose we are given the *numerical values* of the elements of a rotation matrix that describes the orientation of a rigid body. Can we deduce the angle ψ through which the body-fixed frame has been rotated with respect to the space-fixed frame, without a knowledge of the rotation axis $\mathbf{n}$? Recall that the *trace* of $R(\mathbf{n}, \psi)$ is just the sum of its eigenvalues, so that

$$\text{Tr}\, R(\mathbf{n}, \psi) = 1 + e^{i\psi} + e^{-i\psi} = 1 + 2\cos\psi, \tag{11.34}$$

independent of the axis $\mathbf{n}$. Therefore $\frac{1}{2}\big(\text{Tr}\, R(\mathbf{n}, \psi) - 1\big)$ gives the numerical value of $\cos\psi$, regardless of the direction of $\mathbf{n}$. Extracting ψ itself (given that $\cos\psi$ is double-valued in $[0, 2\pi]$) involves further technicalities that I shall not go into here.

Finally, one may ask: what does the general rotation matrix look like when the generators are represented by the Pauli matrices, i.e., when $J_k = \frac{1}{2}\sigma_k$? This question will be answered in Chap. 15, Sect. 15.3.3. A related aspect is suggested by the following observation.

A simple but important observation: As mentioned at the beginning, although we expect the angle of rotation ψ to take values in the range $0 \le \psi < 2\pi$, this point requires further examination. A hint is provided by the following simple observation. If we set $\psi = \pi$ in the general expression (11.32) for $R_{ij}(\mathbf{n}, \psi)$, we get

$$R_{ij}(\mathbf{n}, \pi) = -\delta_{ij} + 2n_i\, n_j\,. \tag{11.35}$$

Note that the term linear in $\mathbf{n}$ has vanished. As a consequence, a rotation by an angle π about the axis $-\mathbf{n}$ is given by exactly the same expression! That is,

$$\boxed{R_{ij}(\mathbf{n}, \pi) = R_{ij}(-\mathbf{n}, \pi), \quad \text{or} \quad R(\mathbf{n}, \pi) = R(-\mathbf{n}, \pi).} \tag{11.36}$$

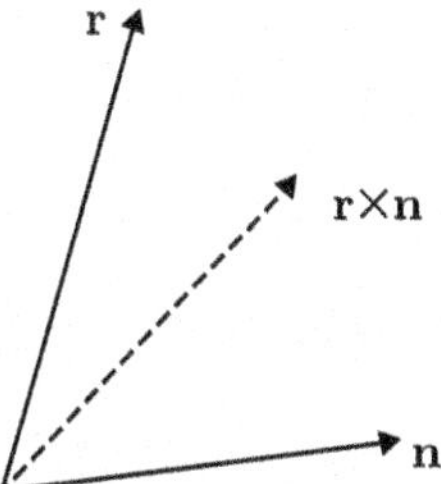

Fig. 11.3 Under the rotation $R(\mathbf{n}, \psi)$, the transformed vector $\mathbf{r}'$ has components along $\mathbf{r}$, $\mathbf{n}$ and $\mathbf{r} \times \mathbf{n}$

In other words:

- A rotation through π about an axis pointing in any direction $\mathbf{n}$ is exactly the same thing as a rotation through π about an axis pointing in the opposite direction $-\mathbf{n}$.

This seemingly trivial observation has profound consequences, as you will see in Chap. 15, Sect. 15.3.5.

11.3.3 The Finite Rotation Formula for a Vector

Once we have $R(\mathbf{n}, \psi)$ explicitly, it is straightforward to apply it to an arbitrary position vector $\mathbf{r}$. We have

$$x'_i = R_{ij}\, x_j = x_i \cos\psi + n_i\, x_j\, n_j\, (1 - \cos\psi) + \epsilon_{ijk}\, x_j\, n_k \sin\psi. \qquad (11.37)$$

Expressing this formula back in terms of the vectors $\mathbf{r}$ and $\mathbf{n}$ helps us understand it in physical terms:

$$\boxed{\mathbf{r}' = (\cos\psi)\, \mathbf{r} + (1 - \cos\psi)\, (\mathbf{r} \cdot \mathbf{n})\, \mathbf{n} + (\sin\psi)\, (\mathbf{r} \times \mathbf{n}).} \qquad (11.38)$$

Equation (11.38) is sometimes called the **finite rotation formula** for a vector. Needless to say, it also tells us precisely how *any* vector transforms under the rotation $R(\mathbf{n}, \psi)$, by the very definition of a vector. In other words, by its very definition, any vector $\mathbf{A}$ transforms under the rotation according to

$$\boxed{\mathbf{A} \to \mathbf{A}' = (\cos\psi)\, \mathbf{A} + (1 - \cos\psi)\, (\mathbf{A} \cdot \mathbf{n})\, \mathbf{n} + (\sin\psi)\, (\mathbf{A} \times \mathbf{n}).} \qquad (11.39)$$

The formula is almost completely deducible with the help of simple general arguments, as follows. A rotation is a linear, homogeneous transformation. Given the vectors $\mathbf{r}$ and $\mathbf{n}$, what can the vector $\mathbf{r}'$ possibly be? It must be linear and homogeneous (and of first degree) in $\mathbf{r}$. As a vector in three-dimensional space, it must be expressible as a linear combination of the two given vectors, $\mathbf{r}$ and $\mathbf{n}$, and the vector $(\mathbf{r} \times \mathbf{n})$ that is normal to the plane formed by the two. (See Fig. 11.3.) Thus, linearity and homogeneity already tell us that $\mathbf{r}'$ must be a linear combination of $\mathbf{r}$, $(\mathbf{r}\cdot\mathbf{n})\,\mathbf{n}$, and $(\mathbf{r}\times\mathbf{n})$. This exhausts the dependence on $\mathbf{r}$ and $\mathbf{n}$. Hence the coefficients in the expansion can only be scalars that depend on the angle ψ. Let us call these $f_1(\psi)$, $f_2(\psi)$, and $f_3(\psi)$, respectively. Thus, we must have

$$\mathbf{r}' = f_1(\psi)\,\mathbf{r} + f_2(\psi)\,(\mathbf{r}\cdot\mathbf{n})\,\mathbf{n} + f_3(\psi)\,(\mathbf{r}\times\mathbf{n}). \tag{11.40}$$

It is obvious that a rotation by a multiple of 2π brings us back to the original coordinate axes. Hence each $f_i(\psi)$ must be a periodic function of ψ, with period 2π. The identity transformation corresponds to $\psi = 0$, and it must take every point to itself. Hence $f_1(0) = 1$ and $f_2(0) = f_3(0) = 0$. In the particular case when $\mathbf{r}$ is collinear with $\mathbf{n}$, we have $\mathbf{r} = \pm r\mathbf{n}$, so that $(\mathbf{r}\cdot\mathbf{n})\,\mathbf{n} = \mathbf{r}$ and also $\mathbf{r}\times\mathbf{n} = 0$. But points on the axis of rotation must remain unchanged, so that $\mathbf{r}'$ must reduce to $\mathbf{r}$ itself. This condition gives $f_1(\psi) + f_2(\psi) = 1$. Squaring both sides of Eq. (11.40) and setting $f_2 = 1 - f_1$ gives

$$r'^2 = (f_1^2 + f_3^2)\,r^2 + (1 - f_1^2 - f_3^2)\,(\mathbf{r}\cdot\mathbf{n})^2. \tag{11.41}$$

But $r'^2 = r^2$ under a rotation, whatever be $\mathbf{n}$. Therefore $f_1^2(\psi) + f_3^2(\psi)$ must be identically equal to 1. The simplest possible solutions that immediately suggest themselves, taking into account the other conditions on these functions, are $f_1(\psi) = \cos\,\psi$ and $f_3(\psi) = \sin\,\psi$. Proving properly that this is so takes a little more work: as you might expect, it once again involves the building up of a finite rotation as a continuous sequence of infinitesimal rotations.

11.4 The Eigenvalue Spectrum of a Matrix

11.4.1 The Characteristic Equation

The **eigenvalues** of an $(n\times n)$ matrix M are the roots of its **characteristic equation** (sometimes called the **secular equation**)

$$\boxed{P_M(\lambda) \stackrel{\text{def.}}{=} \det\,\big(\lambda I - M\big) = 0.} \tag{11.42}$$

$P_M(\lambda)$ is a polynomial of order n in λ, called the **characteristic polynomial** of M. The equation determining the eigenvalues is of the form

$$P_M(\lambda) = \lambda^n + c_1\,\lambda^{n-1} + c_2\,\lambda^{n-2} + \cdots + c_n = 0. \tag{11.43}$$

The fundamental theorem of algebra guarantees that this equation has exactly n roots $(\lambda_1\,,\,\lambda_2\,,\,\ldots\,,\,\lambda_n)$ in the field of complex numbers. That is, Eq. (11.43) can be written as

$$P_M(\lambda) = \prod_{i=1}^{n} (\lambda - \lambda_i) = 0. \tag{11.44}$$

Some of the roots may be repeated. The set of eigenvalues is the **spectrum** of the matrix M. (More generally, the spectrum of an operator is the set of its eigenvalues.)

The coefficients c_i in the characteristic equation are determined, of course, by the elements of the matrix M. It follows from elementary algebra that

$$\left.\begin{aligned} \lambda_1 + \lambda_2 + \cdots + \lambda_n &= -c_1\,, \\ \lambda_1\lambda_2 + \lambda_1\lambda_3 + \cdots + \lambda_{n-1}\lambda_n &= c_2\,, \\ \cdots\cdots\cdots\cdots &= \cdots \\ \lambda_1\,\lambda_2\,\cdots\,\lambda_n &= (-1)^n c_n\,. \end{aligned}\right\} \tag{11.45}$$

What is noteworthy is that all the coefficients c_n be expressed in terms of certain *invariant* quantities, i.e., quantities that are unchanged under similarity transformations on M:

- The eigenvalues of an $(n \times n)$ matrix M can be expressed in terms of the invariant quantities $\text{Tr}\, M,\ \text{Tr}\,(M^2),\ \ldots,\ \text{Tr}\,(M^n)$, where Tr denotes the **trace** (i.e., the sum of the diagonal elements of the matrix concerned).

In the simple cases $n = 2$ and $n = 3$, for instance, the characteristic equations can be written in the form

$$\lambda^2 - T_1\,\lambda + \tfrac{1}{2}\,(T_1^2 - T_2) = 0 \tag{11.46}$$

and

$$\lambda^3 - T_1\,\lambda^2 + \tfrac{1}{2}\,(T_1^2 - T_2)\,\lambda - \tfrac{1}{6}\,T_1^3 + \tfrac{1}{2}\,T_1 T_2 - \tfrac{1}{3}\,T_3 = 0, \tag{11.47}$$

where $T_r = \text{Tr}\,(M^r)$.

★ **10.** Establish Eqs. (11.46) and (11.47).

11.4.2 Gershgorin's Circle Theorem

In numerical analysis, it is helpful to be able to put some bounds on the eigenvalues of a matrix. A very useful result in this regard is **Gershgorin's Circle Theorem**. Let M_{ii} denote the diagonal element in the ith row of an $(n \times n)$ matrix M. Let

$$r_i = \sum_{\substack{j=1 \\ j \neq i}}^{n} |M_{ij}| \tag{11.48}$$

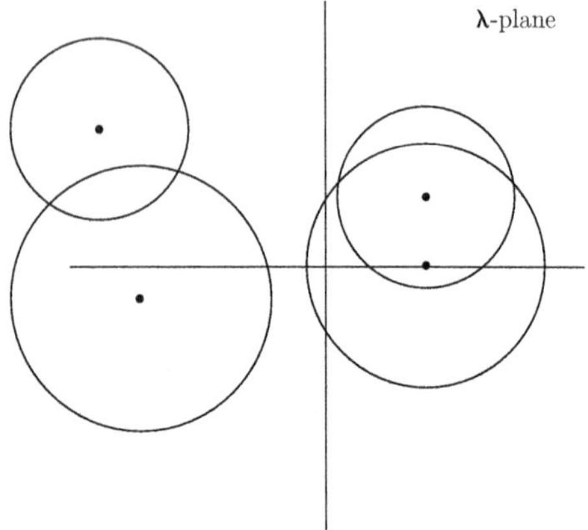

Fig. 11.4 All the eigenvalues of any $(n \times n)$ matrix M lie in the union of the n Gershgorin disks of M in the complex plane. The dots indicate the centers of the disks shown

be the sum of the magnitudes of the nondiagonal elements of the ith row. Draw a circle in the complex plane, with M_{ii} as the center and r_i as the radius. This is called a **Gershgorin disk**. The disk includes its boundary circle. Carry out this construction for each of the rows of the matrix, from $i = 1$ to $i = n$. (Figure 11.4 depicts some of the Gershgorin disks of a matrix M.) Then the theorem states that:

- Every eigenvalue of M lies in one or more of the Gershgorin disks. Hence all the eigenvalues lie in the union of the disks.
- The statement remains true if the columns of M are used instead of the rows, in defining the radii of the disks.

Proof: Let λ be an eigenvalue of the matrix M, with eigenvector $\mathbf{u}$, so that $M\mathbf{u} = \lambda\mathbf{u}$. Suppose u_k is the element of the column vector $\mathbf{u}$ with the largest magnitude among all its elements. Then

$$\sum_{j=1}^{n} M_{kj}\, u_j = \lambda\, u_k\,, \quad \text{or} \quad \sum_{\substack{j=1 \\ j\neq k}}^{n} M_{kj}\, u_j = (\lambda - M_{kk})\, u_k\,. \tag{11.49}$$

Take absolute values on both sides, and divide out by $|u_k|$ (which is guaranteed to be nonzero, because $\mathbf{u}$ is not the null vector). Use the fact that the absolute value of a sum is less than or equal to the sum of the absolute values of its individual terms. It follows that

$$|\lambda - M_{kk}| \leq r_k\,, \tag{11.50}$$

where r_k is defined as in Eq. (11.48), with k replacing i. Hence the theorem.

★ **11.** Work through the steps above to establish Gershgorin's Circle Theorem. Check out that the result goes through if the radius r_i is defined as the sum of the magnitudes of the nondiagonal elements of the ith column, instead of the ith row.

A specific application of the theorem occurs in the context of certain random (Markov) processes. This will be discussed in Chap. 21, Sect. 21.2.3.

There are several straightforward extensions of the basic result derived above. Here is an example:

- If one of the Gershgorin disks is disjoint from the rest, then exactly one eigenvalue lies in this disk.

The foregoing results are general ones, applicable to all square matrices. But more can be said regarding the eigenvalues of special kinds of matrices (e.g., Hermitian, unitary, etc.), as you will see in Chap. 12, Sect. 12.2.

11.4.3 The Cayley–Hamilton Theorem

The Cayley–Hamilton Theorem is a simple-looking, but nontrivial and important result in matrix analysis. It has generalizations in abstract algebra. The theorem states that

- every $(n \times n)$ matrix M satisfies its own characteristic equation.

That is, M satisfies the *matrix* equation

$$P_M(M) = M^n + c_1\, M^{n-1} + c_2\, M^{n-2} + \cdots + c_n\, I = \prod_{i=1}^{n} (M - \lambda_i\, I) = 0, \tag{11.51}$$

where the coefficients c_i are those occurring in the characteristic equation, Eq. (11.43). As a consequence of this theorem, we are guaranteed that the nth power of any $(n \times n)$ matrix M, namely, the matrix M^n, can be written as a linear combination of the matrices I , M , M^2 , ..., M^{n-1}. Hence all higher powers M^{n+k} can also be expressed as such linear combinations. In principle, therefore, e^M can be expressed as a linear combination of the matrices I , M , M^2 , ..., M^{n-1}. Finding the coefficients of such an expansion, however, may not be an easy task: in general, each coefficient would itself be an infinite series.

Some simplification can occur in certain special cases such as that of **triangular matrices**. A matrix in which all the elements below [respectively, above] the principal diagonal are zero is called an **upper triangular matrix** [respectively, **lower triangular matrix**]. It follows by inspection that

- the eigenvalues of a triangular matrix are just the elements on its principal diagonal. (Think of how you would write out the determinant $\det(\lambda I - M)$ on the left-hand side of the characteristic equation.)

It is also easy to see that

- any power of an upper [respectively, lower] triangular matrix is an upper [respectively, lower] triangular matrix.

★ **12.** Consider the $(n \times n)$ upper triangular matrix M whose elements are equal to unity on the diagonal just above the principal diagonal, all other elements being equal to zero. That is,

$$M = \begin{pmatrix} 0 & 1 & 0 & 0 & \cdots & 0 & 0 \\ 0 & 0 & 1 & 0 & \cdots & 0 & 0 \\ 0 & 0 & 0 & 1 & \cdots & 0 & 0 \\ \vdots & \vdots & \vdots & \vdots & \ddots & \vdots & \vdots \\ 0 & 0 & 0 & 0 & \cdots & 0 & 1 \\ 0 & 0 & 0 & 0 & \cdots & 0 & 0 \end{pmatrix}, \quad \text{or} \quad M_{ij} = \delta_{i+1\,,\,j}\,, \quad \text{where } 1 \le i \le (n-1).$$

Show that $\exp M$ is also an upper triangular matrix, with elements given by

$$\left(e^{M}\right)_{i\,,\,i+k} = 1/k!$$

where, for each k in the range $1 \le k \le (n-1)$, the index i takes values in the range $1 \le i \le (n-k)$.

A natural question that arises is the following: Suppose we find, quite independently of a knowledge of its eigenvalues, that an $(n \times n)$ matrix M satisfies an nth order polynomial equation of the form

$$M^{n} + b_1\, M^{n-1} + b_2\, M^{n-2} + \cdots + b_n\, I = 0. \tag{11.52}$$

Can we then conclude that the polynomial

$$\lambda^{n} + b_1\, \lambda^{n-1} + b_2\, \lambda^{n-2} + \cdots + b_{n-1}\, \lambda + b_n \tag{11.53}$$

must *necessarily* be the characteristic polynomial $P_M(\lambda)$ of the matrix M? If so, that would in some sense be a *converse* of the Cayley–Hamilton Theorem. The answer to the question is "no". I will return to this matter in Chap. 12, Sect. 12.3.5, after we discuss the idea of the minimal polynomial of a matrix.

11.4.4 The Resolvent of a Matrix

Let M be any $(n \times n)$ matrix. An important quantity associated with it is its **resolvent**, defined as the *inverse* of the matrix $(zI - M)$, where z is a complex variable. That is,

$$\boxed{R_M(z) \stackrel{\text{def.}}{=} (z\,I - M)^{-1}.} \tag{11.54}$$

The resolvent $R_M(z)$ does not exist for a given value of z if and only if z belongs to the set of eigenvalues of M, i.e., if z lies in the spectrum of M. A convenient representation of the resolvent of a matrix will be given in Eq. (12.68) of Chap. 12, Sect. 12.5.1.

The definition of the resolvent and the statements just made are not restricted to matrices in a finite-dimensional Euclidean space. They are valid for any general linear operator in an LVS. In Chap. 32, Sect. 32.1.2, we will encounter the resolvent of an integral operator, in the context of integral equations.

★ **13.** The eigenvalues of a (3×3) matrix M are given to be 0, $i\,\alpha$ and $-i\,\alpha$, where α is a positive constant.

(a) Using the Cayley–Hamilton Theorem, show that

$$e^{iM} = I + \frac{i\,\sinh\,\alpha}{\alpha}\,M + \frac{(1 - \cosh\,\alpha)}{\alpha^2}\,M^2.$$

(b) Similarly, show that the resolvent of M reduces to a polynomial in M, namely,

$$R_M(z) = \frac{I}{z} + \frac{M z}{z^2 + \alpha^2} + \frac{M^2}{z^2 + \alpha^2}.$$

11.5 A Generalization of the Gaussian Integral

The n-dimensional generalization of the standard one-dimensional Gaussian integral occurs quite often in applications. In Chap. 2, Sect. 2.1.2, we have already considered a couple of simple special cases. The general n-dimensional Gaussian integral is given by

$$\int_{-\infty}^{\infty} dx_1 \cdots \int_{-\infty}^{\infty} dx_n\, \exp\Big\{-\sum_{i=1}^{n}\sum_{j=1}^{n} A_{ij}\, x_i\, x_j\Big\}, \tag{11.55}$$

where the coefficients A_{ij} are the elements of an $(n \times n)$ real symmetric matrix A whose eigenvalues are all positive. If we regard $(x_1\,, x_2\,, \ldots, x_n)$ as the components of a column vector $\mathbf{x}$, then its transpose $\mathbf{x}^{\mathrm{T}}$ is a row vector with components $(x_1\,, x_2\,, \ldots, x_n)$. The integrand in Eq. (11.55) can be written more compactly as $\exp\,(-\mathbf{x}^{\mathrm{T}} A\, \mathbf{x})$. Then

$$\boxed{\int_{-\infty}^{\infty} dx_1 \cdots \int_{-\infty}^{\infty} dx_n\, e^{-\mathbf{x}^{\mathrm{T}} A\,\mathbf{x}} \equiv \int d^n x\, e^{-\mathbf{x}^{\mathrm{T}} A\,\mathbf{x}} = \sqrt{\frac{\pi^n}{\det\, A}}.} \tag{11.56}$$

This is a very important result.

★ **14.** Establish Eq. (11.56).

From the general formula (11.56), we can read off the special cases considered in Chap. 2, Sect. 2.1.2, namely, Eqs. (2.6) and (2.9). I repeat them for ready reference. Let $-1 < \mu < 1$ and $-\frac{1}{2} < \nu < 1$. Then

$$\int_{-\infty}^{\infty} dx \int_{-\infty}^{\infty} dy \; e^{-(x^2+2\mu\, x\, y+y^2)} = \frac{\pi}{\sqrt{1-\mu^2}} \tag{11.57}$$

and

$$\int_{-\infty}^{\infty} dx \int_{-\infty}^{\infty} dy \int_{-\infty}^{\infty} dz \; e^{-(x^2+y^2+z^2+2\nu\, x\, y+2\nu\, y\, z+2\nu\, z\, x)} = \frac{\pi^{3/2}}{(1-\nu)\sqrt{1+2\nu}}\,. \tag{11.58}$$

It is easily seen that det $A = (1-\mu^2)$ and $(1-\nu)^2(1+2\nu)$, respectively, in the two cases. In the second case, the fact that the eigenvalue $1-\nu$ of the matrix A is a two-fold *repeated* eigenvalue is responsible for the *stronger* divergence of the integral at $\nu = 1$ than the inverse square root divergence at $\nu = -\frac{1}{2}$. This is the answer to the question posed at the end of Sect. 2.1.2 in Chap. 2.

11.6 Inner Product in the Linear Space of Matrices

I have stated (and used the fact) that the set of all $(n \times n)$ matrices with complex entries forms an LVS ($\mathbb{V}$, say). The dual space $\widetilde{\mathbb{V}}$ is the space of all the Hermitian conjugates of the matrices in $\mathbb{V}$, which is again $\mathbb{V}$ itself. This LVS is therefore self-dual. The interesting question is how one defines the inner product of two elements in this case.

Let A and B be two elements in this LVS, i.e., two $(n \times n)$ matrices.[5] One way of defining the inner product $(A\,,\,B)$ of the two elements is

$$(A\,,\,B) \stackrel{\text{def.}}{=} \text{Tr}\,(A^\dagger\, B)\,, \tag{11.59}$$

where $A^\dagger$ denotes the Hermitian conjugate of A.

★ **15.** Use the definition of the inner product given above to verify that

(a) $(B\,,\,A) = (A\,,\,B)^*$ (where $*$ denotes the complex conjugate).
(b) $(A\,,\,A) \geq 0$, the equality sign being applicable only when A is the null matrix.

★ **16.** Let A be an $(n \times n)$ Hermitian matrix that is not a multiple of the unit matrix. Show that

$$\text{Tr}\,(A^2) > (1/n)\,\big(\text{Tr}\,A\big)^2.$$

★ **17.** If A is an arbitrary $(n \times n)$ matrix with elements A_{ij} and U is any unitary $(n \times n)$ matrix, show that

$$\sum_{i=1}^{n}\sum_{j=1}^{n} |A_{ij}|^2 \geq (1/n)\,\big|\,\text{Tr}\,(U^\dagger\, A)\,\big|^2.$$

---

[5] We should really use ket vectors to denote these elements. But as these objects are matrices, I shall stick to plain A, B, etc., here, in order to avoid confusion.

Inequalities such as these find application in quantum mechanics and statistical mechanics, especially in connection with **density matrices**.

11.7 Solutions

1. It is trivially seen that the relationship between the sets of coefficients is a linear one, with unique solution sets. Thus

$$a = \alpha_0 + \alpha_3\,,\quad b = \alpha_1 - i\alpha_2\,,\quad c = \alpha_1 + i\alpha_2\,,\quad d = \alpha_0 - \alpha_3\,.$$

Hence

$$\alpha_0 = \tfrac{1}{2}(a+d),\ \ \alpha_1 = \tfrac{1}{2}(b+c),\ \ \alpha_2 = \tfrac{1}{2}i(b-c),\ \ \alpha_3 = \tfrac{1}{2}(a-d).$$

Remark Note that the coefficient of the unit matrix in the expansion (11.3) is $\alpha_0 = \frac{1}{2}\,\mathrm{Tr}\,M$. It follows that any *traceless* (2×2) matrix can be expanded uniquely as a linear combination of the three Pauli matrices. ▶

2. Use the index notation and the summation convention. For Eq. (11.10), for instance, you have to evaluate the commutator $[\sigma_i\,,\ a_j\,\sigma_j] = a_j\,[\sigma_i\,,\ \sigma_j]$. ▶

5. Use the fact that $\sigma_i^2 = I$ for each i, and the fact that the anticommutator of any two different Pauli matrices is the null matrix. ▶

6. (a) Observe that M^2 is a (3×3) matrix with $(M^2)_{13} = 1$, and all other elements equal to zero. The matrix M^3 is the null matrix, implying that M^n is the null matrix for all $n \geq 3$. Hence $e^M = I + M + M^2/2!$ in this case, and we get

$$e^M = \begin{pmatrix} 1 & 1 & \frac{1}{2} \\ 0 & 1 & 1 \\ 0 & 0 & 1 \end{pmatrix}.$$

This is an upper triangular matrix (all the elements below the principal diagonal are zero). Hence the eigenvalues of the matrix e^M are the diagonal elements themselves, i.e., 1, 1, and 1.

Remark The eigenvalues of an upper or lower triangular matrix are just its diagonal elements (see Sect. 11.4.3). It is easy to see how this comes about. Consider the secular equation $\det\,(\lambda I - A) = 0$ whose roots give the eigenvalues of any square matrix A. Think about how the determinant is evaluated.

(b) Observe that $M^2 = nM$ in this case. The power series for e^M can therefore be summed easily, to get

$$e^M = I + \frac{(e^n - 1)}{n}\, M.$$

The matrix M with all $M_{ij} = 1$ is an interesting one, and I will return to some other properties of this matrix in Chap. 12, Sect. 12.3.6. ▶

8. Write out the power series for $\exp(M\psi)$. Use the fact that $M^3 = -M$ to sum the series. The result is just a linear combination of the three matrices I, M, and M^2. ▶

9. The eigenvalues of M must be determined first: if λ is an eigenvalue of the matrix M, then $e^{\lambda\psi}$ is an eigenvalue of he rotation matrix $R(\mathbf{n}, \psi) = e^{M\psi}$. It is straightforward to find the eigenvalues of M. They are 0, i, and $-i$. Hence the eigenvalues of $R(\mathbf{n}, \psi) = e^{M\psi}$ are 1, $e^{i\psi}$, and $e^{-i\psi}$.

Remark In the present instance, the eigenvalues of M can be written down at once from the fact that $M^3 = -M$. Anticipating what will be said in connection with the characteristic polynomial of a matrix (Sect. 11.4.1), the Cayley–Hamilton Theorem (Sect. 11.4.3), and the minimal polynomial of a matrix (Chap. 12, Sect. 12.3.5), I mention here that the cubic equation $M^3 + M = 0$ is the lowest order polynomial equation that the (3×3) matrix M satisfies. Hence $M^3 + M$ must be both the minimal polynomial of M as well as its characteristic polynomial. It follows that its eigenvalues are the roots of the equation $\lambda^3 + \lambda = 0$, namely, 0 and $\pm i$. ▶

10. You need to use the following property: If $\lambda_1, \ldots, \lambda_n$ are the eigenvalues of M, then $\lambda_1^r + \lambda_2^r + \cdots + \lambda_n^r = T_r$ for each integer r from 1 to n. This statement is easily seen to be true when M can be diagonalized by a similarity transformation S—that is, when we can find a matrix S such that $S^{-1}MS = D$, where D is a diagonal matrix with the eigenvalues as its diagonal elements. But it remains valid even when M cannot be diagonalized. We will return to this point in Chap. 12, Sect. 12.3.7. ▶

11. Use the fact that M and its transpose M^{T} have the same set of eigenvalues. ▶

12. This problem is the generalization of the (3×3) case considered in Sect. 11.2.2. The matrix M^2 has elements equal to unity on the diagonal immediately above the one on which M has nonzero elements, all its other elements being equal to zero. In M^3, this is pushed up further to the next diagonal, and so on. Finally, the matrix M^n has a single nonzero element, $(M^n)_{1n} = 1$, and M^r is the null matrix for all $r > n$. ▶

13. The characteristic equation is $\lambda(\lambda^2 + \alpha^2) = 0$. Hence, by the Cayley–Hamilton Theorem, $M^3 = -\alpha^2 M$. This helps you write both the exponential series for $\exp(iM)$ and the binomial expansion of $(zI - M)^{-1}$ as linear combinations of the matrices I, M, and M^2. Collect the coefficients of M and M^2 in each case and re-sum the series concerned, to obtain the expressions quoted above.

Remark In the case of the binomial expansion of the resolvent $R_M(z) = (1/z)(I - M/z)^{-1}$ in powers of M/z, you must assume that z is kept in the region $|z| > \alpha$, so that the series converges absolutely. After re-summation of the coefficients of M and M^2, the closed-form expressions that result are applicable by analytic continuation to the region $|z| \leq \alpha$ as well. Observe that $R_M(z)$ diverges at the eigenvalues of M, namely, at the points $z = 0$, $i\alpha$, and $-i\alpha$, as expected. These statements will be better understood after a study of functions of a complex variable and analytic continuation in Chaps. 22–26. ▶

14. Assume that the real symmetric matrix A can be diagonalized by an orthogonal transformation. (This is discussed in Chap. 12, Sect. 12.3.4.) Thus $A = S^{-1} D S = S^{\mathrm{T}} D S$, where S is an orthogonal matrix and D is a diagonal matrix with the eigenvalues of A as the diagonal elements. If $S\mathbf{x} = \mathbf{y}$, the quantity $\mathbf{x}^{\mathrm{T}} A\, \mathbf{x}$ reduces to $\sum_j \lambda_j\, y_j^2$. The change of variables of integration from $(x_1, \ldots, x_n)$ to $(y_1, \ldots, y_n)$ involves the determinant of the Jacobian matrix. But you can write this down using the orthogonality property of S. We then have a product of decoupled Gaussian integrals in the components of $\mathbf{y}$. It only remains to note that the product of the eigenvalues of A is equal to $\det A$. ▶

15. (a) Using the summation convention for repeated indices,

$$(A\,,\, B) = \mathrm{Tr}\,(A^{\dagger}\, B) = (A^{\dagger} B)_{ii} = (A^{\dagger})_{ij}\, B_{ji} = A^{*}_{ji}\, B_{ji}\,.$$

On the other hand,

$$(B\,,\, A) = \mathrm{Tr}\,(B^{\dagger}\, A) = (B^{\dagger} A)_{ii} = (B^{\dagger})_{ij}\, A_{ji} = B^{*}_{ji}\, A_{ji}\,,$$

which is obviously the complex conjugate of $(A\,,\, B)$.
(b) Setting $B = A$, we have $(A\,,\, A) = |A_{ji}|^2$, which is positive definite unless each $A_{ji} = 0$. ▶

16. Use the Cauchy–Schwarz inequality $|(A\,,\, B)|^2 < (A\,,\, A)(B\,,\, B)$, with B set equal to I, the $(n \times n)$ identity matrix. ▶

17. Use the Cauchy–Schwarz inequality for the elements U and A of the LVS of $(n \times n)$ matrices. ▶

Chapter 12
More About Matrices

12.1 Matrices as Operators in a Linear Space

12.1.1 *Representation of Operators*

As I have mentioned already, we may also view matrices as the representations of **operators** acting on the elements of an LVS. In an n-dimensional Euclidean space (denoted by $\mathbb{R}^n$), the natural basis is given by the ket vectors represented by the column matrices

$$|\phi_1\rangle = \begin{pmatrix} 1 \\ 0 \\ 0 \\ \vdots \\ 0 \end{pmatrix}, \quad |\phi_2\rangle = \begin{pmatrix} 0 \\ 1 \\ 0 \\ \vdots \\ 0 \end{pmatrix}, \quad \ldots, \quad |\phi_n\rangle = \begin{pmatrix} 0 \\ 0 \\ \vdots \\ 0 \\ 1 \end{pmatrix}. \tag{12.1}$$

This space is self-dual. The natural basis in the dual space is formed by the bra vectors $\langle\phi_i|$. These are represented by the row matrices that are the Hermitian conjugates of the column matrices above:

$$\langle\phi_1| = \begin{pmatrix} 1 \; 0 \cdots 0 \end{pmatrix}, \; \langle\phi_2| = \begin{pmatrix} 0 \; 1 \; 0 \cdots 0 \end{pmatrix}, \; \ldots, \; \langle\phi_n| = \begin{pmatrix} 0 \cdots 0 \; 1 \end{pmatrix}. \tag{12.2}$$

We know that a quantity like $\langle\phi_i|\phi_j\rangle$ is a pure number, i.e., a scalar. What sort of quantity is an object like $|\phi_i\rangle\langle\phi_j|$? It is an *operator*, as already pointed out in Chap. 10, Sect. 10.2.1. From Eqs. (12.1) and (12.2), it is obvious that this operator is represented by an $(n \times n)$ matrix whose (ij)th element is unity, all its other elements being zero. This simple property immediately suggests the following interpretation of *any* general $(n \times n)$ matrix A:

V. Balakrishnan, *Mathematical Physics*,
https://doi.org/10.1007/978-3-030-39680-0_12

- Any $(n \times n)$ matrix A with elements a_{ij} can be regarded as the *representation* of an abstract operator A given by

$$\boxed{\mathsf{A} = \sum_{i=1}^{n} \sum_{j=1}^{n} a_{ij} \, |\phi_i\rangle\langle\phi_j|.} \tag{12.3}$$

- The n^2 operators $|\phi_i\rangle\langle\phi_j|$, where the indices i and j run from 1 to n, comprise the natural basis for all *operators* acting on the vectors of the LVS.

The orthonormality of the natural basis immediately yields

$$\boxed{a_{ij} = \langle\phi_i|\,\mathsf{A}\,|\phi_j\rangle.} \tag{12.4}$$

This should immediately tell you why an object like $\langle\phi_i|\,\mathsf{A}\,|\phi_j\rangle$ is called (what else!) a **matrix element** in quantum mechanics. This terminology is retained even when the LVS is infinite-dimensional, and even when the basis set itself is a non-denumerable or "continuous" basis.

- I shall exploit this fact to use the same symbol for both the abstract operator and its matrix representation, that is, the symbol A will be used both for A and for the matrix representing it (in the natural basis, unless otherwise specified).

Which of the two is meant will be clear from the context, and no confusion should arise. This is an abuse of notation, but it aids notational simplicity. I have already made use of this convention in the case of the unit operator and the unit matrix, denoting both objects by I.

★ **1.** In a two-dimensional linear space, let $\{\, |\phi_1\rangle,\ |\phi_2\rangle \,\}$ constitute an orthonormal basis. Consider the operator

$$H = a\big(\, |\phi_1\rangle\langle\phi_1| - |\phi_2\rangle\langle\phi_2| + |\phi_2\rangle\langle\phi_1| + |\phi_1\rangle\langle\phi_2| \,\big),$$

where a is a real constant. Find the eigenvalues and eigenvectors of H, and express the latter as linear combinations of $|\phi_1\rangle$ and $|\phi_2\rangle$.

12.1.2 Projection Operators

Given an orthonormal basis $\{|\phi_j\rangle\}$ in an LVS, we have just seen that the set of operators $\{|\phi_i\rangle\langle\phi_j|\}$ forms a basis for the operators acting on the vectors in the LVS. Of these, the n w "diagonal" members of the set are special. The operator

$$\boxed{P_j \stackrel{\text{def.}}{=} |\phi_j\rangle\langle\phi_j|} \tag{12.5}$$

is a **projection operator**: when it acts on any vector in the LVS, it projects out the part of the vector "along" the unit vector $|\phi_j\rangle$. Since $\langle\phi_j|\phi_j\rangle = 1$ for each j, we have (with no summation over the repeated index j)

$$P_j^2 \equiv P_j\, P_j = P_j\,. \tag{12.6}$$

Hence, for each j,

$$P_j\,(I - P_j) = 0 \text{ (the null operator)}. \tag{12.7}$$

The projection operator P_j is a Hermitian (actually, self-adjoint) operator,[1] i.e.,

$$P_j = P_j^\dagger\,. \tag{12.8}$$

The properties in Eqs. (12.6) and (12.8) actually serve to *define* projection operators, in general. Note also that

$$P_i\, P_j = 0 \quad \text{if} \quad i \neq j. \tag{12.9}$$

The geometrical meaning of this relation is simple: if you project any vector along a given basis vector $|\phi_j\rangle$, the subsequent projection of the result along any other orthogonal direction is obviously zero. The completeness relation $\sum_j |\phi_j\rangle\langle\phi_j| = I$ (Eq. (10.4) of Chap. 10, Sect. 10.2.1) is just the statement that

$$\sum_j P_j = I. \tag{12.10}$$

Again, this is intuitively obvious.

I have mentioned in Chap. 10, Sect. 10.2.1 that the result of Gram–Schmidt orthonormalization can be presented in a suggestive form using projection operators. Here it is: Eq. (10.7) for the orthonormalized ket $|\phi_k\rangle$ can be written as

$$|\phi_k\rangle = \frac{\left(I - \sum_{i=1}^{k-1} P_i\right)|\psi_k\rangle}{\langle\psi_k|\left(I - \sum_{i=1}^{k-1} P_i\right)|\psi_k\rangle^{1/2}}\,. \tag{12.11}$$

As you ought to expect, the denominator in the right-hand side of Eq. (12.11) is just the norm of the ket vector in the numerator. To establish this, you need to use the self-adjointness property of projection operators, Eq. (12.8).

★ **2.** Verify that Eq. (10.7) can be re-expressed as in (12.11).

---

[1] I have yet to define the Hermitian conjugate of an operator, and what is meant by saying that it is Hermitian or self-adjoint (and the subtle distinction between the two). That will be done in Chap. 14, Sect. 14.2.3. For the present, take it to be that operator whose matrix representation is the Hermitian conjugate (or complex conjugate transpose) of the matrix representing the operator itself.

12.2 Hermitian, Unitary, and Positive Definite Matrices

12.2.1 Definitions and Eigenvalues

Here is a quick recapitulation of a number of definitions, properties, and standard results that are very useful in applications. We restrict ourselves to square matrices of general finite order $(n \times n)$.

(i) A matrix R is **orthogonal** if $R\,R^{\mathrm{T}} = I$, where the superscript T denotes the transpose. Hence $R^{\mathrm{T}} = R^{-1}$ for an orthogonal matrix.
(ii) A matrix H is **Hermitian** if $H = H^{\dagger}$. (Recall that the superscript † stands for the complex conjugate transpose.)

It follows that the matrix element $\langle\phi|H|\phi\rangle$ is real for any ket vector $|\phi\rangle$.

(iii) A Hermitian matrix P is **positive definite** [respectively, positive semi-definite] if, for every $|\phi\rangle$, the matrix element $\langle\phi|P|\phi\rangle > 0$ [respectively, ≥ 0].

The concept of a positive semi-definite matrix extends to non-Hermitian matrices as well, but I shall not go into this here.

(iv) If M is any matrix, the matrices $MM^{\dagger}$ and $M^{\dagger}M$ are positive semi-definite.
(v) If P is a Hermitian positive semi-definite matrix, there exists a *unique* Hermitian positive semi-definite matrix S such that $S^2 = P$, and conversely.

Thus, among all the matrices whose square equals a given positive semi-definite matrix P, there is a unique positive semi-definite "square root" of P. The assertions in (iv) and (v) extend to the more general case of positive semi-definite operators in any LVS, both finite- and infinite-dimensional.

(vi) A matrix M is **normal** if it commutes with its Hermitian conjugate, i.e., if $MM^{\dagger} = M^{\dagger}M$.
(vii) A matrix U is **unitary** if $U\,U^{\dagger} = I$, the unit matrix. Hence $U^{\dagger} = U^{-1}$ for a unitary matrix (see below).
(viii) A matrix M and its transpose M^{T} have the same set of eigenvalues. Further, if λ is an eigenvalue of M, then λ^* is an eigenvalue of $M^{\dagger}$.

It is often helpful to think of Hermitian matrices and unitary matrices as (roughly) the matrix analogs of real numbers and complex numbers of unit modulus, respectively. Their respective eigenvalues are, in fact, such numbers:

(ix) The eigenvalues of a Hermitian matrix are real. A special case of this result is that a symmetric matrix with real elements has real eigenvalues.
(x) The eigenvalues of a Hermitian positive semi-definite [respectively, positive definite] matrix are all nonnegative [respectively, positive].
(xi) Every eigenvalue of a unitary matrix lies on the unit circle in the complex plane, i.e., it is of the form $e^{i\theta}$, where θ is a real number.

Some additional remarks are in order. For a matrix of *finite* order, the left and right inverses are identical. Hence the orthogonality condition $R\,R^{\mathrm{T}} = I$ implies automatically that $R^{\mathrm{T}}\,R = I$. Similarly, the unitarity condition $U\,U^{\dagger} = I$ implies that $U^{\dagger}\,U = I$. This is not necessarily so for matrices of infinite order (and for operators in general). A given infinite-dimensional matrix may have a left inverse but not a right inverse, or *vice versa*. The conditions for unitarity are then $U\,U^{\dagger} = I$ *as well as* $U^{\dagger}\,U = I$. If only one of these conditions is satisfied, we have a **partial isometry**. If both conditions are satisfied, we have a **total isometry**, which is the same thing as unitarity.

12.2.2 The Eigenvalues of a Rotation Matrix in d Dimensions

A proper rotation about the origin in d-dimensional Euclidean space (where $d \geq 2$) can be represented by an $(d \times d)$ orthogonal matrix with real elements and determinant equal to 1. Such matrices form a group, the special orthogonal group $SO(d)$. Now, it is obvious that a real orthogonal matrix is a special case of a unitary matrix. Hence all its eigenvalues must lie on the unit circle in the complex plane. Recall, for instance, the result found in Chap. 11, Sect. 11.3.2 for a rotation matrix in three dimensions: the eigenvalues of the (3×3) rotation matrix $R(\mathbf{n}, \psi)$, which is orthogonal and has real elements, are 1, $e^{i\psi}$, and $e^{-i\psi}$. The eigenvector corresponding to the eigenvalue 1 is obviously $\mathbf{n}$, the axis of rotation: every point on this axis remains unchanged under the rotation.

- The existence of such a direction for each R is what enables us to identify a specific *axis* of rotation with every rotation in three dimensions.

We are so accustomed to this property that its validity seems to be "intuitively obvious" for rotations in any number of dimensions, but that is not so. Consider, as the simplest counter-example, rotations about the origin in two-dimensional space: there is no "axis of rotation" for a rotation in two dimensions! A rotation (about the origin) of the coordinate axes through an angle α is represented by the (2×2) orthogonal matrix

$$R(\alpha) = \begin{pmatrix} \cos\,\alpha & \sin\,\alpha \\ \sin\,\alpha & \cos\,\alpha \end{pmatrix}. \tag{12.12}$$

The eigenvalues of $R(\alpha)$ are $e^{\pm i\alpha}$. There is no eigenvalue equal to 1, and it is obvious that no direction in the plane is left unchanged by the rotation.

These considerations can be extended to a proper rotation matrix R in an arbitrary number of dimensions $d \geq 2$ as follows. As stated above, all the eigenvalues of R must lie on the unit circle in the complex plane. Moreover, the product of all the eigenvalues must be equal to the determinant of R, i.e., $+1$. Every complex eigenvalue $e^{i\psi}$ must occur along with its complex conjugate $e^{-i\psi}$, since all the coefficients in the characteristic equation are real. The product of these two quantities is 1. It is obvious that the eigenvalue -1, if it occurs, must occur an even number of times.

It follows at once that, if d is *even*, $+1$ may or may not be an eigenvalue of R. Hence there may or may not be a definite direction in the space along which points are left unchanged by the rotation. On the other hand, when n is *odd*, it is clear that the eigenvalue $+1$ must *necessarily occur at least once*, since the other $(d-1)$ eigenvalues pair off, the product of each pair being 1. This means that there is always at least one direction along which points are left unchanged by any rotation in a d-dimensional space, when d is an odd integer ≥ 3.

The foregoing discussion should serve as an indication that the rotation groups in even- and odd-dimensional spaces differ significantly in their properties. Much more can be said in this regard, but I shall not do so here. I reiterate that

- the existence of a *unique* axis of rotation associated with every rotation is an exclusive property of three-dimensional space.

12.2.3 The General Form of a (2×2) Unitary Matrix

Sets of special kinds of matrices form **groups** under matrix multiplication, and these groups are of great importance in physics. Foremost among these are orthogonal matrices and unitary matrices. Thus all $(n \times n)$ orthogonal matrices with real elements form the **orthogonal group** $O(n)$, while all $(n \times n)$ unitary matrices with complex elements form the **unitary group** $U(n)$. This group, and one of its subgroups, $SU(n)$, have turned out to be of fundamental importance in atomic, nuclear, and subnuclear physics, among other areas. For the present, let us restrict ourselves to some properties of unitary (2×2) matrices, as these occur most frequently in numerous applications.

What is the general form of a unitary (2×2) matrix? Consider an arbitrary (2×2) matrix

$$U = \begin{pmatrix} \alpha & \beta \\ \gamma & \delta \end{pmatrix}, \quad \text{so that} \quad U^{\dagger} = \begin{pmatrix} \alpha^* & \gamma^* \\ \beta^* & \delta^* \end{pmatrix}. \tag{12.13}$$

The elements α, β, γ, and δ are complex numbers.[2] Hence eight real parameters are required to specify a given matrix. Requiring that U be unitary, i.e., imposing the condition $UU^{\dagger} = I$ yields relationships between these parameters and reduces the number of independent ones. Note, further, that the determinant of a unitary matrix of any finite order *must* have a modulus equal to unity: since $\det UU^{\dagger} = \det I = 1$, we have

$$\det (U\,U^{\dagger}) = (\det U)(\det U^{\dagger}) = (\det U)(\det U^{\mathrm{T}})^* = |\det U|^2 = 1. \tag{12.14}$$

Hence $|\det U| = 1$, which implies that

[2] Here I have used the symbol U to denote the *matrix* representing a general element of the *group* $U(2)$, but this should not cause any confusion.

$$\det U = e^{i\theta} \tag{12.15}$$

in general, where θ is any real number. As you will see shortly, the general form of a unitary (2×2) matrix is then given by

$$U = \begin{pmatrix} \alpha & \beta \\ -e^{i\theta}\,\beta^* & e^{i\theta}\,\alpha^* \end{pmatrix}, \quad \text{where} \quad |\alpha|^2 + |\beta|^2 = 1. \tag{12.16}$$

Hence, if α_1, α_2 [respectively, β_1, β_2] are the real and imaginary parts of α [respectively, β], we have

$$\alpha_1^2 + \alpha_2^2 + \beta_1^2 + \beta_2^2 = 1. \tag{12.17}$$

This provides one condition among four real variables, leaving three independent parameters. Together with θ, these comprise the four independent parameters needed to specify a general (2×2) unitary matrix.

★ **3.** Show that a general (2×2) unitary matrix has the form given in Eq. (12.16), and that such matrices form a group.

The special unitary group $SU(2)$**:** In the special case when the matrices are also *unimodular*, i.e., when the determinant is equal to $+1$, the parameter $\theta = 0$. The matrix in Eq. (12.16) then reduces further to

$$\boxed{U = \begin{pmatrix} \alpha & \beta \\ -\beta^* & \alpha^* \end{pmatrix}, \quad \text{where} \quad |\alpha|^2 + |\beta|^2 = 1.} \tag{12.18}$$

Such unimodular unitary matrices also form a group by themselves, the **special unitary group** $SU(2)$. This is a subgroup of $U(2)$.

- The group $SU(2)$ is of fundamental importance in physics.
- It turns out that $SU(2)$ is intimately related to $SO(3)$, the group of rotations in three dimensions. There is a 2-to-1 correspondence (or **homomorphism**) between the two groups: there are two distinct elements of $SU(2)$ corresponding to each element of $SO(3)$. I will discuss this relationship in Chap. 15, Sect. 15.3.3.

Number of independent parameters in an $(n \times n)$ ***unitary matrix***: A general $(n \times n)$ matrix with complex elements has $2n^2$ independent real parameters. Requiring that the matrix be unitary reduces this number to n^2. Thus $U(n)$ is an n^2-parameter group. That is, the dimensionality of the **parameter space** of the group $U(n)$ is n^2.

The determinant of any element of $U(n)$ is a complex number of unit modulus, i.e., is of the form $e^{i\theta}$, where θ is a real number. If, further, we require that the matrices be unimodular, then $\theta = 0$ and the number of independent real parameters decreases by unity to $n^2 - 1$. Such matrices comprise the special or unimodular unitary group $SU(n)$, which is a subgroup of $U(n)$. Thus the dimensionality of the parameter space of the group $SU(n)$ is $n^2 - 1$.

In physical applications (e.g., in quantum mechanics), certain ways of writing an arbitrary unitary matrix are very useful. Among these are

$$U = e^{iH} \quad \text{and} \quad U = (I + iH)^{-1}(I - iH), \tag{12.19}$$

where H is Hermitian. Such representations are valid more generally for unitary and Hermitian *operators* as well, as you might expect. In particular, consider the Schrödinger equation for the state vector of a quantum mechanical system governed by a time-independent, Hermitian **Hamiltonian** H:

$$i\hbar \frac{d|\Psi(t)\rangle}{dt} = H|\Psi(t)\rangle. \tag{12.20}$$

The formal solution of this equation is

$$|\Psi(t)\rangle = U(t, t_0)\,|\Psi(t_0)\rangle, \quad \text{where} \quad U(t, t_0) = e^{-i(t-t_0)H/\hbar}. \tag{12.21}$$

The **time-development operator** $U(t, t_0)$ is a *unitary* operator.

Digression: An important generalization arises when the Hamiltonian has an explicit dependence on time, although it remains Hermitian. The time-development operator U is then no longer given by the simple exponential form in (12.21). Rather, it is a so-called **time-ordered exponential**, given by an infinite series:

$$\begin{aligned} U(t, t_0) &= \mathcal{T}\left\{\exp\left[\frac{-i}{\hbar}\int_{t_0}^{t} dt'\, H(t')\right]\right\} \\ &\stackrel{\text{def.}}{=} I + \sum_{n=1}^{\infty}\left(\frac{-i}{\hbar}\right)^n \int_{t_0}^{t} dt_1 \int_{t_0}^{t_1} dt_2 \cdots \int_{t_0}^{t_{n-1}} dt_n\, H(t_1)\, H(t_2) \cdots H(t_n). \end{aligned} \tag{12.22}$$

But U remains a unitary operator, preserving the inner product $\langle\Psi(t)|\Psi(t)\rangle$, i.e., maintaining the conservation of total probability.

12.3 Diagonalization of a Matrix and all That

If a square matrix M is in a diagonal form, the entries on the principal diagonal are, of course, its eigenvalues. But it is a common misconception to imagine that (i) you need to be able to reduce M to diagonal form in order to find its eigenvalues, or that (ii) M can be diagonalized in all cases. *Neither* of these statements is true. Matrices play such a prominent role in so many applications (including quantum mechanics), that it is worth taking a quick look at some relevant features from linear algebra. For details, refer to any text on linear algebra and matrix analysis.

Much of what follows is customarily discussed in terms of general rectangular matrices and, even more generally, in the context of linear operators. But let us restrict ourselves here to square matrices of finite order.

12.3.1 Eigenvectors, Nullspace, and Nullity

Recall the meaning of an eigenvector and eigenvalue of an arbitrary $(n \times n)$ matrix M. If the homogeneous equations

$$M\,\mathbf{u} = \lambda\,\mathbf{u} \quad \text{and} \quad \mathbf{v}^{\mathrm{T}}\,M = \lambda\,\mathbf{v}^{\mathrm{T}} \tag{12.23}$$

have nontrivial (i.e., nonzero) solutions for the column vector $\mathbf{u}$ and the row vector $\mathbf{v}^{\mathrm{T}}$ for some particular value of λ, then $\mathbf{u}$ is a **right eigenvector** and $\mathbf{v}^{\mathrm{T}}$ is a **left eigenvector** of M corresponding to the eigenvalue λ.[3] Each $\mathbf{u}$ is equal to the corresponding $\mathbf{v}$ if and only if M is a symmetric matrix. The condition for the homogeneous equations (12.23) to have nontrivial solutions is precisely the characteristic equation $\det(\lambda I - M) = 0$ (Eq. (11.42) of Chap. 11, Sect. 11.4.1). As noted there, this equation is guaranteed to have n roots (some of which may be repeated or multiple roots) in the field of complex numbers. These roots comprise the spectrum of the matrix M.

- The problem of finding the spectrum of a matrix is therefore just a matter of finding the roots of an nth order polynomial equation in λ. It has nothing to do with diagonalizing M.

It is obvious from the characteristic equation that 0 is an eigenvalue of M if and only if $\det M = 0$. In that case M has at least one nontrivial right eigenvector $\mathbf{u}$ that is "annihilated" by M—that is, M acts upon the column vector $\mathbf{u}$ to produce the null vector, according to $M\,\mathbf{u} = 0$. As you know, the matrix M is then **singular**, and the matrix inverse M^{-1} does not exist—i.e., the matrix M is not invertible.

The equation $M\,\mathbf{u} = 0$ may have more than one solution. That is, there may be more than one eigenvector corresponding to the eigenvalue 0 of the matrix M. The span of these eigenvectors is called the **right nullspace** of the matrix. The dimensionality of the nullspace is called the **nullity** of the matrix, and may have any integer value from 0 to n. As I have mentioned, M can also be regarded as a linear operator in the LVS of column vectors of order n, because it acts on column vectors to produce other column vectors in the same space. In this view, M is a **linear map**. The set of vectors that it maps to the null vector is precisely the set of its eigenvectors corresponding to eigenvalue 0. In this context, this set is called the **kernel** of the linear operator (or map).

There is another way to understand the right nullspace of an $(n \times n)$ matrix M. Each row of the matrix is an n-component object, like a row vector. Let us therefore write

$$\mathbf{w}^{(i)} = (M_{i1}\; M_{i2}\; \ldots\; M_{in}) \tag{12.24}$$

[3] I reiterate that we are considering finite-dimensional matrices here. In the case of infinite-dimensional matrices, it is possible for a matrix to have right eigenvectors but no left eigenvectors, or *vice versa*.

for the row vector whose components are the elements of the ith row of M. The vectors $\{\mathbf{w}^{(i)} \mid 1 \le i \le n\}$ span the **row space** of the matrix. If there is a column vector $\mathbf{u}$ satisfying $M\,\mathbf{u} = 0$, we have

$$M_{ij}\, u_j = 0, \quad \text{or} \quad \mathbf{w}^{(i)} \cdot \mathbf{u} = 0 \quad \text{for each } i. \tag{12.25}$$

Therefore the vectors of the right nullspace are *orthogonal* to the vectors spanning the row space.

- The right nullspace is the **orthogonal complement** of the row space of a matrix.

Similarly, we have the **column space** of M, which is the space spanned by its columns.

- The **left nullspace** of M is the orthogonal complement of its column space.
- As you might guess, the left and right nullspaces of a (square) matrix have the same dimensionality.

Note that $\det M = 0$ implies that M cannot have n linearly independent rows.[4] The usual operations used in elementary algebra to simplify determinants involve adding or subtracting multiples of any row to any other row. These operations do not change the value of the determinant. If, by these operations, we are able to reduce all the elements of any row to zero, the determinant vanishes. But this is exactly the same thing as saying that the row concerned is linearly dependent on other rows.

In contrast, if $\det M \neq 0$, then the matrix M is invertible. It has no eigenvalue that is equal to zero. Its n rows are linearly independent of each other. They span the n-dimensional LVS of row vectors. Hence they form a basis in that LVS.

***Caution*:** Saying that an $(n \times n)$ matrix has n linearly independent rows (or columns) is *not* the same thing as saying that the matrix has n linearly independent *eigenvectors*!

12.3.2 The Rank of a Matrix and the Rank-Nullity Theorem

The **rank** of a matrix is a fundamental concept in matrix analysis. There are several equivalent definitions of the rank of an $(n \times n)$ square matrix M.

(i) The row rank of M is the maximal number of linearly independent rows in M. Hence it is the dimensionality of the row space of M.
(ii) The column rank of M is the maximal number of linearly independent columns in M. Hence it is the dimensionality of the column space of M.
(iii) The row rank and column rank of a (square) matrix are always equal to each other. Hence we may simply call it the rank of the matrix.
(iv) The rank of M is the order of the largest nonvanishing minor in M.

---

[4]Or columns. The statements that follow are valid if "row" is replaced with "column". Recall also that the value of a determinant does not change if its rows and columns are interchanged.

(v) Treating M as a linear operator in the LVS of column matrices, the rank of M is the dimensionality of the **image** of the map.

The rank of M can be any integer from 1 to n. (Only the null matrix has rank 0.)

- If M is of full rank n, then it has n linearly independent rows (or columns), $\det M \neq 0$, and it is invertible and *vice versa.*
- If $\det M = 0$, then M has less than n linearly independent rows (or columns), and has a rank between 1 and $(n - 1)$. It is a **rank-deficient matrix**.

The rank-nullity theorem is an important relationship connecting the nullity of an $(n \times n)$ matrix and its rank:

$$\boxed{\text{The rank of } M + \text{ the nullity of } M = n.} \tag{12.26}$$

it is easy to see heuristically why this should be so: the rank is the dimensionality of the row space of a matrix, while the nullity is the dimensionality of the orthogonal complement of the row space. In mathematics, they like to write the theorem as

$$\dim(\text{im } M) + \dim(\ker M) = n, \tag{12.27}$$

where "im" and "ker" stand for the image and kernel, respectively, of the linear map M. Note that the theorem corroborates the fact that the dimensionality of the left nullspace of a matrix is equal to that of its right nullspace, since its row rank is equal to its column rank.

Specific examples illustrating these concepts will follow shortly, in Sect. 12.3.6.

12.3.3 Degenerate Eigenvalues and Defective Matrices

Simple roots of the characteristic equation $\det(M - \lambda I) = 0$ correspond to **nondegenerate eigenvalues**. For every nondegenerate eigenvalue there is an eigenvector, and these eigenvectors are linearly independent of each other.

Multiple or repeated roots of the characteristic equation are **degenerate eigenvalues**, and require special attention. The number of times a degenerate eigenvalue μ occurs, say r_μ, is called its **algebraic multiplicity**. The crucial point is the following:

- There *may* not exist r_μ linearly independent eigenvectors corresponding to the eigenvalue μ, but only s_μ of them, where $1 \leq s_\mu \leq r_\mu$. This number s_μ is the **geometric multiplicity** of the eigenvalue μ.

If an $(n \times n)$ matrix has a full set of n linearly independent eigenvectors,[5] we could use them to form a basis (in which to represent both column vectors in the LVS as well as other matrices or operators). But when we have a degenerate eigenvalue

[5]It is important to note that I am now referring to the linearly independent *eigenvectors* of the matrix, and *not* to its linearly independent rows or columns.

with $s_\mu < r_\mu$, this is no longer possible. In such a case, we must also find the so-called **generalized eigenvectors** of M (see below), and use them along with the other eigenvectors to form a basis.

- An $(n \times n)$ matrix that has fewer than n linearly independent eigenvectors is called a **defective matrix**.
- Defective matrices always have less than n distinct eigenvalues, i.e., some eigenvalues are repeated.
- *The converse of the statement just made is not necessarily true*! A matrix that has less than n distinct eigenvalues, i.e., a matrix that has some repeated eigenvalues, may still have n distinct eigenvectors. An example will be given shortly, in Sect. 12.3.6.

Exactly *how* the generalized eigenvectors are to be found, is a separate question. For instance, if $s_\mu = 1$, there is just one eigenvector corresponding to the eigenvalue μ. This eigenvector is the solution of $(M - \mu I)\,\mathbf{u} = 0$, while the generalized eigenvectors are the nontrivial solutions of the equations $(M - \mu I)^j\,\mathbf{u} = 0$ for $j = 2, 3, \ldots, r_\mu$. I do not go into this aspect any further, as it is not our main concern here.

12.3.4 When Can a Matrix Be Diagonalized?

We are ready, now, to turn to a basic question in matrix analysis:

- When can an $(n \times n)$ matrix M be diagonalized by a **similarity transformation**?

That is, under what conditions can a nonsingular $(n \times n)$ matrix S be found such that $S^{-1}MS = \Lambda$, where Λ is a diagonal matrix with the eigenvalues of M as its diagonal elements?

The answer is actually a special case of a whole class of results that go under the general name of **matrix decomposition**. This term refers to the possible ways in which a matrix can be "factorized", e.g., written as a product of matrices with special properties such as symmetry, unitarity, etc.—analogous, for instance, to the way any complex number z can be written as the product of a nonnegative real number r and a complex number of unit modulus, $e^{i\theta}$.

- The diagonalization of a matrix by a similarity transformation is one such form of factorization.

I will not list the various decompositions and the conditions under which they are possible. Instead, I quote a number of results that are useful in physical applications, followed by some illustrative examples in Sect. 12.3.6. Some comments regarding other forms of matrix decomposition are made in Sect. 12.3.8.

(a) Let us start with a *sufficiency* condition for a matrix M to be diagonalizable.

- An $(n \times n)$ matrix M can be diagonalized if it has n distinct eigenvalues.

- In other words, all the roots of its characteristic equation must be simple roots.

Remember that the above is a *sufficiency* condition, not a *necessary* one. In one of the examples given below, you will encounter a matrix whose characteristic equation has a repeated root, and yet the matrix is diagonalizable.
(b) Next, let us consider a *necessary and sufficient* condition for diagonalization by a *special* kind of transformation, namely, a **unitary transformation**. This is important in several physical applications, notably in quantum mechanics. Recall that a matrix M is normal if $M\,M^\dagger = M^\dagger\,M$. Examples of normal matrices are

- Hermitian matrices ($M^\dagger = M$),
- skew-Hermitian matrices ($M^\dagger = -M$),
- unitary matrices ($M^\dagger M = M M^\dagger = I$),
- and the counterparts of the above among matrices with real elements—namely, real symmetric matrices, real antisymmetric matrices, and real orthogonal matrices, respectively.

A normal matrix enjoys the following property:

- Every normal matrix M can be diagonalized by a unitary transformation, $U^\dagger M U$, where $U^\dagger = U^{-1}$.

A corollary of this result is that

- every real symmetric matrix M can be diagonalized by an *orthogonal* transformation, $R^{\mathrm{T}} M R$, where $R^{\mathrm{T}} = R^{-1}$.

Owing to the "necessary and sufficient" nature of the condition under discussion, the converse result is also valid:

- Given a matrix M, if there is a unitary matrix U such that $U^\dagger M U$ is a diagonal matrix, then M must be a normal matrix.

(c) Now let us turn to a *necessary and sufficient* condition in the general case.

- An $(n \times n)$ matrix M is diagonalizable by a similarity transformation *if, and only if,* it has n linearly independent eigenvectors. In other words, M must not be defective.

A defective matrix cannot be diagonalized by a similarity transformation.

12.3.5 The Minimal Polynomial of a Matrix

The condition for diagonalizability can be phrased in another way. This involves the concept of the **minimal polynomial** of a matrix.

- Every $(n \times n)$ matrix M has a *unique* **minimal polynomial** $p_M(\lambda)$ of some degree $m \le n$, such that the polynomial equation $p_M(M) = 0$ is satisfied.

No polynomial in M that is of order lower than m can vanish identically. The minimal polynomial $p_M(\lambda)$ of a matrix M is distinct from its characteristic polynomial $P_M(\lambda)$ in some cases, while in others it is the characteristic polynomial itself.

- A matrix is diagonalizable by a similarity transformation *if and only if* all the roots of its minimal polynomial are *simple* roots, i.e., the equation $p_M(\lambda) = 0$ has no repeated roots.

I shall not go into the way the minimal polynomial of an $(n \times n)$ matrix is determined. It is obvious that its degree cannot be greater than n, in any case. A noteworthy property is that $p_M(\lambda)$ is always a factor of the characteristic polynomial $P_M(\lambda)$. Hence all the roots of the equation $p_M(\lambda) = 0$ are *included* in the set of eigenvalues of M. In fact, a stronger result can be established: the roots of $p_M(\lambda) = 0$ *exhaust* the set of eigenvalues of M, but (obviously) with lower algebraic multiplicities, in general. This implies at once that, if none of the eigenvalues of a matrix is a repeated eigenvalue, then the minimal and characteristic polynomials of that matrix are the same, apart from an overall sign. The converse is not necessarily true: a matrix may have repeated eigenvalues, and yet its minimal and characteristic polynomials may be essentially the same. We will come across examples of this possibility shortly.

Recall, now, the question posed in Chap. 11, Sect. 11.4.3, after the Cayley–Hamilton Theorem was stated. Suppose an $(n \times n)$ matrix M is found to satisfy an nth order polynomial equation as in Eq. (11.52). Does this equation, with M replaced by λ as in Eq. (11.53), necessarily have to be the characteristic equation of M? The answer is now obvious: of course not! Since $p_M(M) \equiv 0$, left-multiplication of both sides by any arbitrary polynomial in M will still yield zero. In particular, you can left-multiply by any polynomial in M of order $(n - m)$, and produce an nth order polynomial equation satisfied by M. There is therefore no "converse" of the Cayley–Hamilton Theorem in this sense.

On the other hand, if you *know* that the eigenvalues of M are distinct, then the minimal polynomial p_M is the characteristic polynomial P_M itself (apart from an overall sign), as already stated. In *that* case, given an nth order polynomial equation for M like Eq. (11.52), you can assert that Eq. (11.53) is indeed the characteristic equation whose roots yield the eigenvalues of M. This is why I asserted that the eigenvalues of the (3×3) matrix $M = i\,\mathbf{J} \cdot \mathbf{n}$ encountered in Chap. 11, Sect. 11.3.2, were 0, i, and $-i$, based on the fact that $M^3 + M = 0$ in that case.

12.3.6 Simple Illustrative Examples

I now apply the foregoing to some simple but very instructive examples, spelling out the details.

Example (i): Consider the matrix

$$M = \begin{pmatrix} 0 & 1 \\ 0 & 0 \end{pmatrix}. \tag{12.28}$$

The eigenvalues are 0 and 0. However, there is only one eigenvector corresponding to this repeated eigenvalue, namely, $\binom{1}{0}$ (apart from a constant factor). The span of this eigenvector is one-dimensional (the x-axis in $\mathbb{R}^2$, for instance). Therefore the nullity of M is 1. Its rank is also 1, because the row space is just the span of the row vector $(0\ 1)$. (The other row vector is just the null vector.) Alternatively, we may note that the order of the largest nonvanishing minor is 1, so that the rank is 1. Hence the rank-nullity theorem reads, in this case, $1 + 1 = 2$.

M has only one eigenvector, and is therefore a defective matrix. It cannot be diagonalized by a similarity transformation. The minimal polynomial in this case is the characteristic polynomial itself: $p(M) = P(M) = M^2$. The equation $\lambda^2 = 0$ has a double root, corroborating the fact that the matrix cannot be diagonalized.

Example (ii): Consider the matrix

$$M = \begin{pmatrix} 1 & 1 \\ 0 & 1 \end{pmatrix}. \tag{12.29}$$

The matrix is upper triangular, so that the eigenvalues are its diagonal elements, 1 and 1. The rows vectors $(1\ 1)$ and $(0\ 1)$ are obviously linearly independent. Hence the rank of M is 2. There is no zero eigenvalue, so that the nullity is 0. Hence the rank-nullity theorem is satisfied, $2 + 0 = 2$.

M has only one eigenvector, which is proportional to $\binom{1}{0}$. It is therefore a defective matrix, and hence not diagonalizable. Since M is not a multiple of the unit matrix, it cannot satisfy a linear equation of the form $aM + bI = 0$. Once again, the minimal polynomial is the characteristic polynomial itself, $p(M) = P(M) = (M - I)^2$. The fact that the equation $(\lambda - 1)^2 = 0$ has a double root corroborates the conclusion that the matrix cannot be diagonalized.

★ **4.** Show that the conclusions drawn in Example (ii) above extend, for every $n \geq 2$, to the following case: an $(n \times n)$ upper triangular matrix in which all the elements on the principal diagonal and above it are equal to unity, and all other elements are zero. That is,

$$M = \begin{pmatrix} 1 & 1 & 0 & 0 & \cdots & 0 & 0 \\ 0 & 1 & 1 & 0 & \cdots & 0 & 0 \\ 0 & 0 & 1 & 1 & \cdots & 0 & 0 \\ \vdots & \vdots & \vdots & \vdots & \ddots & \vdots & \vdots \\ 0 & 0 & 0 & 0 & \cdots & 1 & 1 \\ 0 & 0 & 0 & 0 & \cdots & 0 & 1 \end{pmatrix}.$$

Example (iii): Consider the matrix

$$M = \begin{pmatrix} 1 & 0 & 0 & 1 \\ 1 & 1 & 0 & 0 \\ 0 & 1 & 1 & 0 \\ 0 & 0 & 1 & 1 \end{pmatrix}. \tag{12.30}$$

It is easily shown that det $M = 0$. The characteristic equation

$$(\lambda - 1)^4 - 1 = 0$$

immediately yields the eigenvalues $2,\ 1+i,\ 1-i$, and 0. The corresponding eigenvectors are obtained quite easily. They are proportional to

$$\begin{pmatrix}1\\1\\1\\1\end{pmatrix}, \quad \begin{pmatrix}1\\-i\\-1\\i\end{pmatrix}, \quad \begin{pmatrix}1\\i\\-1\\-i\end{pmatrix} \quad \text{and} \quad \begin{pmatrix}1\\-1\\1\\-1\end{pmatrix}, \tag{12.31}$$

respectively. It is obvious that the nullity is 1 (there is a single nontrivial eigenvector corresponding to the eigenvalue 0). Labeling the row vectors of M as α, β, γ, and δ, respectively, it is easy to see that $\alpha - \beta + \gamma - \delta = 0$. Therefore any one of the four row vectors can be written down as a liner combination of the other three, i.e., there are three linearly independent rows. Hence the rank of M is 3. The rank-nullity theorem is satisfied as $3 + 1 = 4$. There are four linearly independent eigenvectors, and the matrix is not defective. It can be diagonalized by a similarity transformation.

Observe the way in which the elements of the first row of the matrix in Eq. (12.30) "go around" one step at a time to produce the other rows. This is an example of a **circulant matrix**. Circulant matrices occur frequently in applications, and comprise an important class of matrices. I shall return to them in Sect. 12.3.9.

★ **5.** Let S be the matrix formed by the eigenvectors listed in Eq. (12.31), i.e.,

$$S = \begin{pmatrix}1 & 1 & 1 & 1\\ 1 & -i & i & -1\\ 1 & -1 & -1 & 1\\ 1 & i & -i & -1\end{pmatrix}.$$

Show explicitly that the matrix M given in Eq. (12.30) is diagonalized by the similarity transformation $S^{-1}MS$.

Example (iv): This is an important example that should help dispel some common misconceptions. Consider the $(n \times n)$ Hermitian matrix M with *every* element equal to unity, i.e.,

$$M = \begin{pmatrix}1 & 1 & \cdots & 1\\ 1 & 1 & \cdots & 1\\ \vdots & \vdots & \ddots & \vdots\\ 1 & 1 & \cdots & 1\end{pmatrix}. \tag{12.32}$$

You have already encountered this matrix in Chap. 11, Sect. 11.2.2, where its exponential was evaluated.

M is a real symmetric matrix, which suffices to ensure that it can be diagonalized by an orthogonal transformation. Equivalently, it has n linearly independent eigenvectors. All the rows of M being identical to each other, there is only one independent row vector, $(1\ 1\ \ldots\ 1)$. Hence the rank of M is 1. (The order of the largest nonvanishing minor is also easily seen to be just 1.) Its nullity must therefore be $(n-1)$. In other words, 0 is an $(n-1)$-fold repeated eigenvalue of M, with an equal number of linearly independent eigenvectors. The characteristic polynomial of the matrix is

$$P_M(\lambda) = \lambda^{n-1}(\lambda - n). \tag{12.33}$$

It follows that the remaining eigenvalue is just n. We could have written this down by inspection, by observing that $M^2 = nM$. Hence the minimal polynomial is the quadratic

$$p_M(\lambda) = \lambda(\lambda - n), \tag{12.34}$$

whose roots are 0 and n. These are simple roots, corroborating the fact that M can be diagonalized. So we have, here, a matrix that is highly rank-deficient, and has $(n-1)$ eigenvalues equal to zero. And yet it is *not* a defective matrix.

★ **6.** Consider the matrix M given by Eq. (12.32).

(a) Find n linearly independent eigenvectors corresponding to the n eigenvalues of M.
(b) Apply the Gram–Schmidt orthonormalization procedure given in Chap. 10, Sect. 10.2.1, to obtain an orthonormal set of eigenvectors of M.

★ **7.** Consider the Hermitian matrix

$$H = \begin{pmatrix} 1 & 0 & 0 \\ 0 & \frac{3}{2} & -\frac{1}{2} \\ 0 & -\frac{1}{2} & \frac{3}{2} \end{pmatrix}.$$

(a) Find the rank, nullity, eigenvalues, and eigenvectors of H. Construct an orthonormal basis using the eigenvectors.
(b) Obtain the similarity transformation which diagonalizes H. Is the transformation matrix orthogonal?

12.3.7 Jordan Normal Form

For the sake of completeness, I state a general result that is applicable whether or not a matrix can be diagonalized by a similarity transformation.

- Every $(n \times n)$ matrix can always be brought to the so-called **Jordan normal form** (also called the Jordan canonical form) by a similarity transformation.

This form is an upper triangular matrix in which the eigenvalues occur along the diagonal, and the super-diagonal just above the principal diagonal has entries that are either 0 or 1. The matrix is in *block-diagonal* form, composed of **Jordan blocks.** Each Jordan block is a square upper triangular matrix, with a particular eigenvalue as its diagonal elements, unity as each element of the super-diagonal, and zero everywhere else.

Here is an example. Consider a (6×6) matrix with a two-fold degenerate eigenvalue λ_1, a simple or nondegenerate eigenvalue λ_2, and a three-fold degenerate eigenvalue λ_3. Suppose there is only one eigenvector for each eigenvalue. By a similarity transformation, such a matrix can always be cast (up to a permutation of the three Jordan blocks) in the Jordan normal form

$$\left(\begin{array}{cc|c|ccc} \lambda_1 & 1 & 0 & 0 & 0 & 0 \\ 0 & \lambda_1 & 0 & 0 & 0 & 0 \\ \hline 0 & 0 & \lambda_2 & 0 & 0 & 0 \\ \hline 0 & 0 & 0 & \lambda_3 & 1 & 0 \\ 0 & 0 & 0 & 0 & \lambda_3 & 1 \\ 0 & 0 & 0 & 0 & 0 & \lambda_3 \end{array}\right). \tag{12.35}$$

This form comprises three Jordan blocks: a (2×2) block corresponding to the eigenvalue λ_1, a (1×1) block for the eigenvalue λ_2, and a (3×3) block for the eigenvalue λ_3.

In general, a given eigenvalue may have more than one Jordan block associated with it. The number of such blocks is the *geometric* multiplicity of the eigenvalue. The size of each block is determined by a more detailed procedure which will not be described here. But the sum of the sizes of *all* the Jordan blocks corresponding to a given eigenvalue is equal to the *algebraic* multiplicity of the eigenvalue. This provides us with yet another way of stating the condition for the diagonalizability of an $(n \times n)$ matrix:

- A matrix can be diagonalized by a similarity transformation if and only if the geometric multiplicity of each eigenvalue is equal to its algebraic multiplicity.

When that happens, all the Jordan blocks reduce to (1×1) blocks, and the Jordan normal form becomes a diagonal matrix.

- A diagonal matrix is thus a special case of the Jordan normal form.

Note, finally, that the diagonal elements of the rth power of a matrix in Jordan normal form has the rth power of the eigenvalues on its diagonal elements. This means, in effect, that the trace of the rth power of any matrix is just the sum of the rth powers of its eigenvalues. This establishes the assertion made in Chap. 11, Sect. 11.4.1: namely, that the coefficients of the characteristic equation of any $(n \times n)$ matrix can be expressed in terms of the traces $T_1 = \text{Tr}\, M,\ T_2 = \text{Tr}\,(M^2),\ \ldots,\ T_n = \text{Tr}\,(M^n)$.

12.3.8 Other Matrix Decompositions

I have already stated that diagonalization by a similarity transformation is just one of several possible matrix decompositions. Some other important matrix decompositions that are possible for a general $(n \times n)$ matrix are as follows.

***Decomposition into triangular matrices*:** Any $(n \times n)$ matrix M can be written as the product of a lower triangular matrix T_L and an upper triangular matrix T_U, i.e.,

$$M = T_L\, T_U\,. \tag{12.36}$$

***Schur decomposition*:** Any $(n \times n)$ matrix can be written in the form

$$M = U^\dagger T U, \tag{12.37}$$

where U is a unitary matrix and T is an upper triangular matrix. The diagonal elements of T are the eigenvalues of M.

***Singular value decomposition*:** Any $(n \times n)$ matrix can be written in the form

$$M = U_2^\dagger\, D\, U_1, \tag{12.38}$$

where U_1 and U_2 are unitary matrices, and D is a diagonal matrix. The elements of D are *not* the eigenvalues of the matrix, but its **singular values**. These are the positive square roots of the eigenvalues of $M^\dagger M$. The matrix $M^\dagger M$ is Hermitian and positive semi-definite, that is, all its eigenvalues are real, nonnegative numbers. Its positive eigenvalues are also eigenvalues of the matrix $MM^\dagger$.

12.3.9 Circulant Matrices

As I have mentioned already, the matrix in Eq. (12.30) of Sect. 12.3.6 is a circulant matrix. These matrices have very interesting and well-studied properties.

Given a set of n numbers $a_0, a_1, \ldots, a_{n-1}$, a general $(n \times n)$ circulant matrix is of the form

$$M = \begin{pmatrix} a_0 & a_1 & a_2 & \cdots & a_{n-2} & a_{n-1} \\ a_{n-1} & a_0 & a_1 & \cdots & a_{n-3} & a_{n-2} \\ a_{n-2} & a_{n-1} & a_0 & \cdots & a_{n-4} & a_{n-3} \\ \vdots & \vdots & \vdots & \ddots & \vdots & \vdots \\ a_2 & a_3 & a_4 & \cdots & a_0 & a_1 \\ a_1 & a_2 & a_3 & \cdots & a_{n-1} & a_0 \end{pmatrix}, \tag{12.39}$$

or

$$M_{ij} = \begin{cases} a_{j-i} & \text{for } i \leq j, \\ a_{j-i+n} & \text{for } i > j. \end{cases} \tag{12.40}$$

Let

$$\omega_k = e^{2\pi i k/n}, \quad k = 0, 1, \ldots, (n-1) \tag{12.41}$$

denote the nth roots of unity. Then the eigenvalues of M are $\lambda_0, \lambda_1, \ldots, \lambda_{n-1}$, where

$$\lambda_k = \sum_{r=0}^{n-1} a_r\, \omega_k^r \quad (k = 0, 1, \ldots, n-1). \tag{12.42}$$

It is obvious that all the powers ω_k^r of each ω_k can be expressed in terms of the set $\{\omega_l \,|\, l = 0, 1, \ldots, (n-1)\}$ itself. The first eigenvalue is just the sum of the elements of the first (or any other) row, i.e., $\lambda_0 = a_0 + \cdots + a_{n-1}$. The right eigenvector corresponding to the eigenvalue λ_k is given by the transpose of the row vector $\left(1 \; \omega_k \; \omega_k^2 \; \cdots \; \omega_k^{n-1}\right)$. Since $\omega_0 = 1$, the eigenvector corresponding to λ_0 is just the uniform column vector with each element equal to 1.

Some other important properties of circulant matrices (of a given order) are as follows.

- The product of two circulant matrices is again a circulant matrix.
- Any two circulant matrices commute with each other.
- If M is a circulant matrix, so is $M^\dagger$.
- Hence $MM^\dagger = M^\dagger M$, so that every circulant matrix is a normal matrix. Therefore it can be diagonalized by a similarity transformation (in fact, by a unitary transformation).
- Since the eigenvectors of a circulant matrix do not depend on the values of its elements, all circulant matrices (of a given order order n) can be diagonalized by the *same* similarity transformation. Thus $SMS^{-1} = \Lambda \equiv \text{diag}\,(\lambda_0, \lambda_,, \ldots, \lambda_{n-1})$, where

$$S = \begin{pmatrix} 1 & 1 & 1 & \cdots & 1 \\ 1 & \omega_1 & \omega_2 & \cdots & \omega_{n-1} \\ 1 & \omega_1^2 & \omega_2^2 & \cdots & \omega_{n-1}^2 \\ \vdots & \vdots & \vdots & \ddots & \vdots \\ 1 & \omega_1^{n-1} & \omega_2^{n-1} & \cdots & \omega_{n-1}^{n-1} \end{pmatrix}. \tag{12.43}$$

★ **8.** Let $0 < p < 1$, and $q = 1 - p$. The $(n \times n)$ tridiagonal circulant matrix

$$M = \begin{pmatrix} -1 & q & 0 & \cdots & 0 & p \\ p & -1 & q & \cdots & 0 & 0 \\ 0 & p & -1 & \cdots & 0 & 0 \\ \vdots & \vdots & \vdots & \ddots & \vdots & \vdots \\ 0 & 0 & 0 & \cdots & -1 & q \\ q & 0 & 0 & \cdots & p & -1 \end{pmatrix}$$

is encountered in certain applications, e.g., in the problem of a biased random walk on a chain of n sites arranged in a ring. (A specific instance is given in Sect. 12.3.10 below.) Find the rank of M, its nullity, and its set of eigenvalues.

12.3.10 A Simple Illustration: A 3-state Random Walk

Here is an illustration involving a very simple instance of a random walk that uses the properties of a circulant matrix. I will consider random walks in greater detail in the sequel, especially in Chap. 21, Sect. 21.5.2. But the example that follows can be understood on its own. Moreover, this example and its generalizations serve as useful models in diverse physical situations.

Imagine a system that can be in any one of three states, labeled by j ($= 1, 2$ or 3). The system jumps randomly from any state to either of the other two states, with the following probabilities: the *a priori* probability of a jump from state 1 to 2, or 2 to 3, or 3 to 1, is given to be p, where $0 < p < 1$. The *a priori* probability of a jump in the reverse direction, i.e., from 3 to 2, or 2 to 1, or 1 to 3, is $q = 1 - p$. It helps to picture the states $1, 2$, and 3 as sites on a ring, and the jumps from one state to another as a random walk between the sites (Fig. 12.1).

Let $P(j, t)$ denote the probability that the system is in state j at time t. Let λ be the transition rate, i.e., the probability, *per unit time*, that the system jumps from any state to any other state. We now assume that the jumps from one state to another constitute a **stationary Markov process**. (This class of random processes will be discussed in Chap. 21, Sect. 21.2.) The probabilities $P(j, t)$ then satisfy the following set of rate equations or **master equations**:

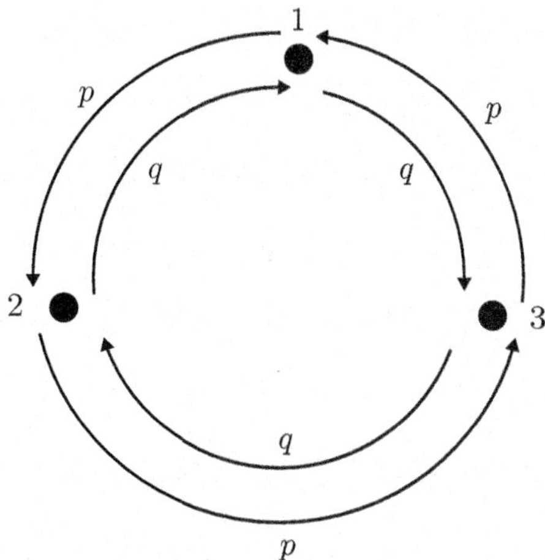

Fig. 12.1 Biased random walk on a ring with 3 sites. p and $q = 1 - p$ denote the site-to-site jump probabilities

$$\left.\begin{aligned}\frac{d}{dt}P(1,t) &= \lambda\big[-P(1,t) + q\,P(2,t) + p\,P(3,t)\big],\\ \frac{d}{dt}P(2,t) &= \lambda\big[p\,P(1,t) - P(2,t) + q\,P(3,t)\big],\\ \frac{d}{dt}P(3,t) &= \lambda\big[q\,P(1,t) + p\,P(2,t) - P(3,t)\big].\end{aligned}\right\} \qquad (12.44)$$

The positive and negative terms on the right-hand side of each equation in (12.44) can be understood as "gain" and "loss" terms contributing to the rate of change of the probability concerned. I will return to this aspect in Chap. 21, Sect. 21.2.2, to explain how such rate equations arise.

In order to solve Eq. (12.44), it is convenient to let $\mathsf{P}(t)$ denote the column vector whose jth row is $P(j,t)$. Then Eq. (12.44) can be written in the compact form

$$\frac{d\mathsf{P}(t)}{dt} = W\mathsf{P}(t), \quad \text{where the matrix} \quad W = \lambda\begin{pmatrix}-1 & q & p\\ p & -1 & q\\ q & p & -1\end{pmatrix}. \qquad (12.45)$$

W is called the **transition matrix**. The formal solution to this equation, for any given initial probability distribution $\mathsf{P}(0)$, is given by

$$\mathsf{P}(t) = e^{Wt}\,\mathsf{P}(0). \qquad (12.46)$$

The problem thus reduces to finding the exponential of the transition matrix.

We can do this explicitly. Observe that W is a circulant matrix. Hence it is a normal matrix, and can be diagonalized. The eigenvalues of W are

$$\lambda_0 = 0,\;\; \lambda_1 = (-1 + q\omega^* + p\omega)\lambda,\;\; \lambda_2 = (-1 + q\omega + p\omega^*)\lambda, \qquad (12.47)$$

where $\omega = e^{2\pi i/3}$ is a cube root of unity. The matrix comprising the corresponding right eigenvectors of W as its columns is

$$S = \frac{1}{\sqrt{3}}\begin{pmatrix}1 & 1 & 1\\ 1 & \omega^* & \omega\\ 1 & \omega & \omega^*\end{pmatrix}. \qquad (12.48)$$

S diagonalizes W by a similarity transformation, i.e.,

$$S^{-1}WS = \text{diag}\,(\lambda_0,\, \lambda_1,\, \lambda_2). \qquad (12.49)$$

It follows that

$$S^{-1}W^nS = \text{diag}\,(\lambda_0^n,\, \lambda_1^n,\, \lambda_2^n). \qquad (12.50)$$

Hence

$$S^{-1}\,e^{Wt}\,S = \text{diag}\,\big(e^{\lambda_0 t},\; e^{\lambda_1 t},\; e^{\lambda_2 t}\big). \qquad (12.51)$$

Therefore

$$e^{Wt} = S \begin{pmatrix} e^{\lambda_0 t} & 0 & 0 \\ 0 & e^{\lambda_1 t} & 0 \\ 0 & 0 & e^{\lambda_2 t} \end{pmatrix} S^{-1}. \tag{12.52}$$

For definiteness, let us take the system to be in state 1 (say) at $t = 0$. Therefore $\mathsf{P}(0) = \begin{pmatrix}1 & 0 & 0\end{pmatrix}^{\mathrm{T}}$. Then, at any time $t \geq 0$, we find

$$\left.\begin{aligned} P(1,t) &= \tfrac{1}{3} + \tfrac{2}{3} e^{-3\lambda t/2} \cos\left[\tfrac{\sqrt{3}}{2}(p-q)t\right], \\ P(2,t) &= \tfrac{1}{3} + \tfrac{2}{3} e^{-3\lambda t/2} \cos\left[\tfrac{\sqrt{3}}{2}(p-q)t - \tfrac{2}{3}\pi\right], \\ P(3,t) &= \tfrac{1}{3} + \tfrac{2}{3} e^{-3\lambda t/2} \cos\left[\tfrac{\sqrt{3}}{2}(p-q)t + \tfrac{2}{3}\pi\right]. \end{aligned}\right\} \tag{12.53}$$

It is easily checked that $P(1,0) = 1, \;\; P(2,0) = P(3,0) = 0$ as required. Moreover, as $t \to \infty$, every $P(j,t)$ exponentially approaches the stationary or equilibrium value $\frac{1}{3}$, with a **relaxation time** $\frac{2}{3}\lambda^{-1}$. The *directional bias*, characterized by the difference $(p-q)$, leads to an oscillatory approach of the probabilities (as functions of time) to this common stationary value. In the absence of such a bias (i.e., when $p = q = \frac{1}{2}$), each probability tends monotonically to its equilibrium value.

★ **9.** Work out all the steps leading from Eq. (12.44) to the solutions given by Eq. (12.53).

12.4 Commutators of Matrices

12.4.1 Mutually Commuting Matrices in Quantum Mechanics

The question of matrix diagonalization has a bearing on quantum mechanics, where, as you know, physical quantities are represented by operators. The latter are often represented by matrices.

Let us consider the case of finite-dimensional matrices. Let A, B, and C be three $(n \times n)$ matrices. Given that the commutators $[A, B] = 0$ and $[B, C] = 0$, it does *not* necessarily follow that $[A, C] = 0$. In other words, the commutation property is *not* transitive. It is of interest in quantum mechanics to consider sets of mutually commuting matrices (more generally, mutually commuting operators acting on the vectors of some LVS).

- If the members of a set of $(n \times n)$ Hermitian matrices commute with each other, they can be diagonalized simultaneously, i.e., by the same unitary transformation matrix U.

This is a special case of a property that is of primary importance in quantum mechanics.

- Given a quantum mechanical system, one looks for a *maximal set* of mutually commuting self-adjoint operators[6] representing the physical observables pertaining to the system.
- This set is guaranteed to have a *complete set of common eigenvectors* that forms a basis in the LVS of the states of the system.
- It is advantageous to make sure that the Hamiltonian operator H of the system belongs to this maximal commuting set, because H governs the time evolution of the system. The basis states are then *stationary states*, i.e., eigenstates of H.

★ **10.** You will find it most instructive to work out the following exercise right up to the last step. Consider the two Hermitian matrices

$$A = \begin{pmatrix} 1 & 0 & 1 \\ 0 & 0 & 0 \\ 1 & 0 & 1 \end{pmatrix} \quad \text{and} \quad B = \begin{pmatrix} 2 & 1 & 1 \\ 1 & 0 & -1 \\ 1 & -1 & 2 \end{pmatrix}.$$

(a) Verify that they can be simultaneously diagonalized.
(b) Find the common eigenvectors of A and B.
(c) Construct the unitary transformation S that diagonalizes the two matrices, such that $S^{-1}AS$ and $S^{-1}BS$ are diagonal matrices.

Two further points ought to be noted, as they are directly relevant to quantum mechanics. They also help dispel some common misconceptions. I make one of these remarks here.

- When two operators (or matrices) do *not* commute with each other, it is still possible for them to have a common eigenvector or a number of common eigenvectors. But these common eigenvectors cannot comprise a *complete* set in the space concerned.

An explicit example of this possibility will be given in Chap. 14, Sect. 14.4.2. The second remark is also deferred to the same place.

12.4.2 The Lie Algebra of $(n \times n)$ Matrices

This is an appropriate place to introduce the idea of a Lie algebra. Recall that this concept has already been mentioned in Chap. 5, Sect. 5.1.4, in connection with the cross product of two vectors in three-dimensional Euclidean space. It has also been indicated in Chap. 11, Sects. 11.1.2 and 11.3.1, that the generators J_i of infinitesimal rotations in three dimensions form a Lie algebra. The importance of Lie algebras lies in the fact that they *generate* Lie groups, which underlie various continuous symmetries and invariances in physics. This process involves the exponentiation of

---

[6]As mentioned earlier, in the case of finite-dimensional matrices (which is what we are concerned with here), "self-adjoint" is the same as 'Hermitian'. The distinction between Hermitian and self-adjoint operators in the general case will be explained in Chap. 14, Sect. 14.2.3.

the so-called infinitesimal generators of the group. You have already seen an example of such an exponential form in the case of the rotation group, as discussed in Chap. 11, Sects. 11.3.1 and 11.3.2 (see, in particular, Eq. (11.28)).

A Lie algebra is a special kind of linear vector space. Its elements not only form an LVS but also possess a *binary operation* (called the **Lie bracket**) involving any two elements of the set, such that the result of the operation is also an element of the set. Further, this operation[7] has a specific set of properties, listed below. As always, I consider only the simplest cases, omitting formal mathematical definitions and generalizations. The linear vector spaces we are concerned are those in which the scalar multipliers of the elements of the LVS are either real numbers or complex numbers. It is convenient, for the moment, to denote the elements of the LVS by $\alpha,\ \beta,\ \gamma,\ \ldots$. The binary operation will be denoted by $\circ$. Thus, $(\alpha \circ \beta)$ is also an element of the set, and so on. The special properties demanded of this operation are as follows:

(i) Anti-symmetry: The operation is antisymmetric under the interchange of the pair of elements. That is,

$$(\alpha \circ \beta) = -(\beta \circ \alpha). \tag{12.54}$$

(ii) Associativity: The operation is associative. If $a,\ b$, and c are scalar multipliers of elements in the LVS,

$$\alpha \circ (b\beta + c\gamma) = b\,(\alpha \circ \beta) + c\,(\alpha \circ \gamma), \quad (a\alpha + b\beta) \circ \gamma = a\,(\alpha \circ \gamma) + b\,(\beta \circ \gamma). \tag{12.55}$$

(iii) Jacobi identity: The Jacobi identity is satisfied:

$$\alpha \circ (\beta \circ \gamma) + \beta \circ (\gamma \circ \alpha) + \gamma \circ (\alpha \circ \beta) = 0. \tag{12.56}$$

The most familiar example of a Lie algebra is provided, as I have already mentioned in Chap. 5, Sect. 5.1.4, by "ordinary" vectors in three-dimensional Euclidean space, $\mathbb{R}^3$. The cross product of any two such vectors **a** and **b** satisfies the anti-symmetry property $(\mathbf{a} \times \mathbf{b}) = -(\mathbf{b} \times \mathbf{a})$, as well as the triple cross product identity, Eq. (5.20). The binary operation in this Lie algebra is therefore the cross product. The *generators* of this Lie algebra are, as you might guess, the three unit vectors $(\mathbf{e}_x,\ \mathbf{e}_y,\ \mathbf{e}_z)$.

The second important example of a Lie algebra is provided by square matrices of any given order. As you know, the set of $(n \times n)$ matrices forms an LVS. Let $A,\ B,\ C$ be $(n \times n)$ matrices. The commutator of two matrices clearly satisfies the anti-symmetry property

$$[A,\ B] = AB - BA = -[B,\ A]. \tag{12.57}$$

The associativity property (12.55) is also obviously satisfied. It is easily checked that

[7]This binary operation is not to be confused with the inner product of two elements of the LVS.

$$[A,\, BC] = [A,\, B]\, C + B\, [A,\, C], \quad [AB,\, C] = A\, [B,\, C] + [A,\, C]\, B. \quad (12.58)$$

Using these properties, it follows at once that the Jacobi identity is satisfied, namely,

$$[A,\, [B,\, C]] + [B,\, [C,\, A]] + [C,\, [A,\, B]] = 0. \quad (12.59)$$

Thus, $(n \times n)$ matrices form a Lie algebra, with the *commutator* of two matrices as the Lie bracket. This Lie algebra is denoted by $\mathfrak{gl}(n)$. It is associated with the Lie group formed by the set of all nonsingular $(n \times n)$ matrices, called the **general linear group** $GL(n)$ —more precisely, $GL(n, \mathbb{R})$ or $GL(n, \mathbb{C})$, depending on whether the matrix elements are restricted to real numbers or can take complex values.

★ **11.** Verify Eqs. (12.58) and (12.59).

Various subsets of matrices also form interesting and physically relevant Lie algebras. For example:

- The set of *traceless* $(n \times n)$ *matrices* forms the Lie algebra $\mathfrak{sl}(n)$, associated with the special linear group $SL(n)$ of nonsingular, unimodular $(n \times n)$ matrices.
- The set of $(d \times d)$ *antisymmetric matrices with real elements* comprises a Lie algebra. The associated Lie group, the special orthogonal group $SO(d)$, is the group of proper rotations in d-dimensional Euclidean space.[8]
- The set of *skew-Hermitian* $(n \times n)$ *matrices* forms the Lie algebra $\mathfrak{u}(n)$ associated with the Lie group $U(n)$ of unitary $(n \times n)$ matrices.

★ **12.** Verify that the following subsets of $(n \times n)$ matrices form linear vector spaces, and that in each case, the commutator of two matrices is again a matrix of the same kind:

(a) traceless matrices,
(b) real antisymmetric matrices, and
(c) skew-Hermitian matrices.

***A Lie algebra in Hamiltonian dynamics*:** Here is another important example of a Lie algebra. Consider a Hamiltonian system in classical dynamics, with n degrees of freedom. Let the generalized coordinates be $(q_1,\, \ldots,\, q_n)$, and let the corresponding canonically conjugate momenta be $(p_1,\, \ldots,\, p_n)$. As you know, the **Poisson bracket** of any two differentiable functions F and G of the dynamical variables is given by

$$\{F,\, G\} \stackrel{\text{def.}}{=} \sum_{i=1}^{n} \left(\frac{\partial F}{\partial q_i} \frac{\partial G}{\partial p_i} - \frac{\partial F}{\partial p_i} \frac{\partial G}{\partial q_i} \right). \quad (12.60)$$

It is straightforward to verify that all such differentiable functions of the dynamical variables form a linear vector space. One can go further:

[8] You might then ask why the generators J_i of rotations in three dimensions are Hermitian rather than antisymmetric. This has to do with real versus complex Lie algebras.

– Differentiable functions of the generalized coordinates and momenta of a Hamiltonian system with n degrees of freedom form a Lie algebra, with the Poisson bracket as the Lie bracket.

★ **13.** Verify this statement.

This Lie algebra is related to the **symplectic group** $Sp(2n, \mathbb{R})$, which is the group of **canonical transformations** of the phase space variables of a Hamiltonian system with n degrees of freedom. Other examples of Lie algebras, involving operators (or infinite-dimensional matrices), will be encountered in Chap. 15, Sect. 15.1.1.

12.5 Spectral Representation of a Matrix

We have seen how an $(n \times n)$ matrix is really an expansion in the natural basis. The natural question that arises is: What does a matrix (or, more generally, an operator) look like in terms of its eigenvectors and eigenvalues? The answer is provided by the **spectral decomposition** of a matrix. For simplicity, let us consider the straightforward case of a diagonalizable $(n \times n)$ matrix M.

12.5.1 Right and Left Eigenvectors of a Matrix

If S is the transformation matrix that diagonalizes M, we have

$$S^{-1}MS = \Lambda, \quad \text{or} \quad M = S\Lambda S^{-1}, \tag{12.61}$$

where Λ is a diagonal matrix with the eigenvalues of M as its elements. This representation of M is essentially a spectral decomposition, as you will see. Its great advantage is that it also provides, automatically, a representation for any sufficiently regular *function* of the matrix concerned. Equation (12.61) implies that any positive integral power of M is given by

$$M^k = S\Lambda^k S^{-1}, \tag{12.62}$$

where the elements of the diagonal matrix Λ^k are just the powers λ_i^k of the eigenvalues of M. Therefore, for any function f that is a convergent power series of its argument, we have

$$\boxed{f(M) = Sf(\Lambda)S^{-1}, \quad \text{where} \quad f(\Lambda) = \text{diag}\,\big(f(\lambda_1), f(\lambda_2), \ldots, f(\lambda_n)\big).} \tag{12.63}$$

What follows is essentially valid not only for matrices but also for operators in an LVS. Let us therefore switch to bra and ket notation, as this makes matters more transparent. The first point to note is that a general $(n \times n)$ matrix M is not symmetric. This complicates matters a bit. As stated in Sect. 12.3.1, such a matrix has *two* sets of eigenvectors, namely, right eigenvectors $\{|\psi_i\rangle\}$ and left eigenvectors $\{\langle\chi_i|\}$, which share the same set of eigenvalues $\{\lambda_i\}$. Thus

$$M\,|\psi_i\rangle = \lambda_i\,|\psi_i\rangle \quad \text{and} \quad \langle\chi_i|\,M = \lambda_i\,\langle\chi_i|\,. \tag{12.64}$$

The point is that $|\psi_i\rangle$ and $\langle\chi_i|$ are *not*, in general, adjoints of each other.[9] They can be chosen so as to form, in general, a **bi-orthogonal set** of eigenvectors. To keep matters simple, let us assume that the eigenvalues are nondegenerate. We can arrange the normalization of these vectors such that the orthonormality and completeness relations are

$$\left.\begin{aligned} \langle\chi_i|\psi_j\rangle &= \delta_{ij} \quad \text{(orthonormality)} \\ \sum_i |\psi_i\rangle\langle\chi_i| &= I \quad \text{(completeness).} \end{aligned}\right\} \tag{12.65}$$

Then the spectral representation[10] of M is just

$$\boxed{M = \sum_i \lambda_i\,|\psi_i\rangle\langle\chi_i|.} \tag{12.66}$$

It follows from the orthonormality property above that

$$M^k = \sum_i (\lambda_i)^k\,|\psi_i\rangle\langle\chi_i|, \quad \text{and} \quad f(M) = \sum_i f(\lambda_i)\,|\psi_i\rangle\langle\chi_i|. \tag{12.67}$$

***Representation of the resolvent*:** Recall the definition of the resolvent of M in Eq. (11.54) of Chap. 11, Sect. 11.4.4: $R(z) = (zI - M)^{-1}$. We may now write this in the form

$$\boxed{R(z) = \sum_i \frac{|\psi_i\rangle\langle\chi_i|}{z - \lambda_i}\,.} \tag{12.68}$$

I have already mentioned that $R(z)$ exists for all complex numbers z except those belonging to the eigenvalue spectrum of M. The factor $(z - \lambda_i)^{-1}$ in Eq. (12.68) makes it obvious why this is so.

---

[9] I have used the notation $|\psi_i\rangle$ and $\langle\chi_i|$ for the right and left eigenvectors, so as to avoid any possible confusion with the natural basis, for which I have used the symbols $|\phi_i\rangle$ and $\langle\phi_i|$.

[10] Note that the term "spectral decomposition" is used sometimes for the representation $S^{-1}MS = \Lambda$ of a diagonalizable matrix.

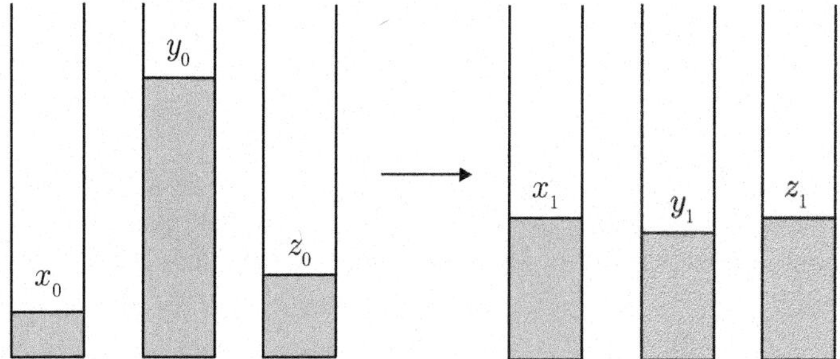

Fig. 12.2 Example to illustrate the spectral representation of a (3 × 3) matrix

12.5.2 An Illustration

As an illustration of the use of the spectral representation of a matrix, consider the following problem.

Three identical cylindrical glasses of uniform cross-section, containing unequal amounts of a beverage, are placed on a table (Fig. 12.2). The initial levels in the three glasses are given by the positive numbers x_0, y_0, and z_0, respectively. In the first step, the levels of glasses 1 and 2 are first equalized,[11] and then the levels of glasses 2 and 3 are equalized, and finally the levels of glasses 3 and 1 are equalized. This cyclic procedure is repeated over and over again. What are the levels x_k, y_k, and z_k in the respective glasses after k iterations of the procedure? What are the limiting values of these levels as $k \to \infty$?

It is clear that, for any positive integer k, we must have

$$x_k + y_k + z_k = x_0 + y_0 + z_0, \tag{12.69}$$

because no liquid is lost. We expect intuitively that, as $k \to \infty$, each of the quantities x_k, y_k, and z_k will approach the common limiting value $\frac{1}{3}(x_0 + y_0 + z_0)$. The task is to find the exact solution for *any* finite k.

Let $\mathbf{x}_k$ be the column vector with elements (x_k, y_k, z_k). It is evident that the levels x_1, y_1, and z_1 at the end of the first cycle are certain linear combinations of x_0, y_0, and z_0. A little calculation shows that

$$\mathbf{x}_1 = M\,\mathbf{x}_0 \quad \text{where} \quad M = \begin{pmatrix} \frac{3}{8} & \frac{3}{8} & \frac{1}{4} \\ \frac{1}{4} & \frac{1}{4} & \frac{1}{2} \\ \frac{3}{8} & \frac{3}{8} & \frac{1}{4} \end{pmatrix}. \tag{12.70}$$

As the cycle repeats itself, it is clear that

$$\mathbf{x}_k = M\,\mathbf{x}_{k-1} = M^k\,\mathbf{x}_0\,. \tag{12.71}$$

[11]That is, the level in each glass now becomes $\frac{1}{2}(x_0 + y_0)$.

All we have to do, therefore, is to evaluate the kth power of the matrix M. We could, of course, try to diagonalize M and hence read off M^k. But M is not symmetric, and the algebra would be tedious. Here is an equivalent, but far less tedious, way to arrive at the answer.

Note that two columns (or rows) of M are identical to each other. Hence $\det M = 0$, so that 0 is an eigenvalue. Further, the elements of M are positive numbers, and each row adds up to unity. M is therefore a so-called **stochastic matrix.**[12] It follows that $(1\;1\;1)^{\mathrm{T}}$ is a right eigenvector, corresponding to the eigenvalue 1. Finally, from the coefficient of λ^2 in the eigenvalue equation $\det(\lambda I - M) = 0$ (or from the fact that the sum of the eigenvalues is equal to $\frac{7}{8}$, the trace of M), you can deduce that the remaining eigenvalue is $-\frac{1}{8}$. Let us label these eigenvalues $\lambda_1 = 0$, $\lambda_2 = -\frac{1}{8}$, and $\lambda_3 = 1$.

The right and left eigenvectors of M corresponding to its three eigenvalues are easily found. They are

$$\left.\begin{aligned} |\psi_1\rangle &= \begin{pmatrix} 1 \\ -1 \\ 0 \end{pmatrix}, \quad \langle\chi_1| = (1\;0\;-1)\,; \\ |\psi_2\rangle &= \begin{pmatrix} 1 \\ -2 \\ 1 \end{pmatrix}, \quad \langle\chi_2| = \tfrac{1}{3}\,(-1\;-1\;2)\,; \\ |\psi_3\rangle &= \begin{pmatrix} 1 \\ 1 \\ 1 \end{pmatrix}, \quad \langle\chi_3| = \tfrac{1}{3}\,(1\;1\;1)\,. \end{aligned}\right\} \tag{12.72}$$

I have chosen the overall multiple of each left eigenvector such that $\langle\chi_i|\psi_i\rangle = 1$ for each i. It is easy to check that $\langle\chi_i|\psi_j\rangle = 0$ for $i \neq j$. Hence the orthonormality condition (the first equation in (12.65)) is satisfied.

Writing out the three matrices $|\psi_i\rangle\langle\chi_i|$ $(i = 1, 2, 3)$, it is readily verified that

$$\sum_{i=1}^{3} |\psi_i\rangle\langle\chi_i| = I \quad \text{and} \quad \sum_{i=1}^{3} \lambda_i\, |\psi_i\rangle\langle\chi_i| = M. \tag{12.73}$$

It is now a very simple matter to use the fact that

$$M^k = \sum_{i=1}^{3} (\lambda_i)^k\, |\psi_i\rangle\langle\chi_i| \tag{12.74}$$

[12]Each *column* of M also adds up to unity. Hence M is, in fact, a **doubly stochastic** matrix. This last property is no longer valid in the general case, i.e., when the process is conducted with $n > 3$ glasses, as you will see subsequently.

to write down the matrix M^k explicitly. Substitution in Eq. (12.71) leads at once to explicit solutions for x_k, y_k, and z_k. For every positive integer k, we have

$$x_k = z_k = \frac{(x_0 + y_0 + z_0)}{3} + \frac{(-1)^k(2z_0 - x_0 - y_0)}{(3 \times 8^k)},$$
$$y_k = \frac{(x_0 + y_0 + z_0)}{3} - \frac{2(-1)^k(2z_0 - x_0 - y_0)}{(3 \times 8^k)}. \tag{12.75}$$

The conservation condition (12.69) is obviously satisfied by these solutions.

★ **14.** Derive Eq. (12.70), and work out the subsequent steps to arrive at the solutions given by Eq. (12.75).

***An example of relaxation to equilibrium*:** The problem just discussed provides as example of the phenomenon of *relaxation to an equilibrium state*. Each of the quantities x_k, y_k, and x_k "relaxes" to the same limit $\frac{1}{3}(x_0 + y_0 + z_0)$ as $k \to \infty$. This process occurs exponentially rapidly, with a characteristic **relaxation time** that can be read off from the identity $8^{-k} = e^{-k \ln 8}$. The relaxation time is thus equal to $1/(\ln 8)$, in the time units we have chosen (i.e., unit time for each full cycle of the process). The approach to the limiting value is oscillatory, as each of the levels alternately overshoots and undershoots the asymptotic value with decreasing amplitude.

Can some kind of physical meaning be attached to the three eigenvalues of M in this case? The eigenvalue $\lambda_3 = 1$ is associated with the *equilibrium* or asymptotic solution as $k \to \infty$. The eigenvalue $\lambda_1 = 0$ does not appear to contribute directly to M^k at all. But it is indirectly present, of course, *via* the completeness relation. Its role in the present problem is to maintain the equality of x_k and z_k for all $k \geq 1$. This leaves the remaining eigenvalue, $\lambda_2 = -\frac{1}{8}$, to control the time evolution of the three levels to their common limiting value.

The kind of analysis carried out above often occurs in the study of **Markov chains** in discrete time (although the present example involved a deterministic process rather than a random one). You will encounter another important instance of a Markov chain (the simple random walk) in Chap. 19, Sect. 19.4.1. The more general case of **Markov processes** will be considered in several places in the sequel, in Chaps. 20, 21, and 30.

★ **15.** Let us generalize the example just considered to the case of n glasses placed in a ring. Label the glasses by j, where $1 \leq j \leq n$ and the label $(n + 1) \equiv 1$. A complete cycle is the equalization of levels successively between the pairs of glasses $(j, j + 1)$ from $j = 1$ to $j = n$.

(a) Write down the corresponding "transfer matrix" M.
(b) Show that 0 and 1 are nondegenerate eigenvalues of M, and find the corresponding eigenvectors.

12.6 Solutions

1. The matrix representation of the operator H in the basis $\big(\,|\phi_1\rangle,\;|\phi_2\rangle\,\big)$ is

$$H = \begin{pmatrix} a & a \\ a & -a \end{pmatrix}.$$

The eigenvalues of this matrix are $a\sqrt{2}$ and $-a\sqrt{2}$. The corresponding (unnormalized) eigenvectors are

$$\begin{pmatrix} 1 \\ \sqrt{2}-1 \end{pmatrix} \quad \text{and} \quad \begin{pmatrix} 1-\sqrt{2} \\ 1 \end{pmatrix}$$

respectively. Since the basis $\big\{\,|\phi_1\rangle,\;|\phi_2\rangle\,\big\}$ corresponds to representing these vectors by $\begin{pmatrix} 1 \\ 0 \end{pmatrix}$ and $\begin{pmatrix} 0 \\ 1 \end{pmatrix}$, respectively, the eigenvectors of H may be identified as

$$|\phi_1\rangle + (\sqrt{2}-1)\,|\phi_2\rangle \quad \text{and} \quad (1-\sqrt{2})\,|\phi_1\rangle + |\phi_2\rangle,$$

respectively. ▶

2. Start with a general matrix U as in Eq. (12.13). Imposing the condition $U\,U^\dagger = I$ gives the equations

$$|\alpha|^2 + |\beta|^2 = 1, \quad |\gamma|^2 + |\delta|^2 = 1, \quad \alpha\,\gamma^* + \beta\,\delta^* = 0, \quad \alpha^*\,\gamma + \beta^*\,\delta = 0.$$

These comprise four independent conditions. The last of these gives $\delta = -\alpha^*\,\gamma/\beta^*$. But the determinant condition (12.15) implies that

$$\alpha\,\delta - \beta\,\gamma = e^{i\theta},$$

where θ is some real number. Substitute for δ in this equation, and use the condition $|\alpha|^2 + |\beta|^2 = 1$. We immediately get

$$\gamma = -\beta^*\,e^{i\theta} \quad \text{and hence} \quad \delta = \alpha^*\,e^{i\theta}.$$

This establishes Eq. (12.16).

The identity matrix is obviously unitary. If U is unitary, so is $U^\dagger$. And every unitary matrix has an inverse, since $U^{-1} = U^\dagger$. Finally, the product of two unitary matrices U and V of the same order, U and V, say, is also unitary: $(UV)\,(UV)^\dagger = UVV^\dagger U^\dagger = U\,U^\dagger = I$. Hence all unitary $(n \times n)$ matrices form a group, denoted by $U(n)$. For $n = 2$, the group is $U(2)$. ▶

3. The n row vectors are clearly linearly independent. M is an upper triangular matrix, so that its diagonal elements are its eigenvalues. Each of these is equal to unity. Hence, for all $n \geq 2$, the rank of M is n, and the nullity is 0. There is only one right eigenvector, namely, the transpose of the row vector

$$\begin{pmatrix}1\ 0 \cdots 0\end{pmatrix}^{\mathrm{T}}.$$

Hence the matrix is a defective one, and cannot be diagonalized by a similarity transformation.

The minimal polynomial of this matrix is its characteristic polynomial itself: $p(M) = P(M) = (M - I)^n$. The equation $(\lambda - 1)^n = 0$ has only a multiple root, corroborating the fact that M cannot be diagonalized. ▶

4. (a) The eigenvectors can be written down by inspection. To start with, it is obvious that the column vector with every element equal to 1 is the right eigenvector corresponding to $\lambda_1 = n$. The rest of the column vectors, corresponding to the repeated eigenvalue 0, can also be guessed readily. In order to save some space, let us work with the transposes of the column vectors concerned, i.e., the corresponding row vectors. As M is a real symmetric matrix, it follows that these row vectors are precisely the *left* eigenvectors of M.

Here is a rather obvious choice for the (left) eigenvectors:

$$\begin{aligned}
\langle\psi_1| &= \begin{pmatrix}1\ 1\ 1 \cdots 1\end{pmatrix} \\
\langle\psi_2| &= \begin{pmatrix}1\ -1\ 0 \cdots 0\end{pmatrix}, \\
\langle\psi_3| &= \begin{pmatrix}1\ 0\ -1 \cdots 0\end{pmatrix}, \\
\cdots &= \cdots\cdots\cdots\cdots\cdots\cdots, \\
\langle\psi_n| &= \begin{pmatrix}1\ 0\ 0 \cdots -1\end{pmatrix}.
\end{aligned}$$

Here $\langle\psi_1|$ corresponds to the eigenvalue n, while $\langle\psi_2|, \ldots, \langle\psi_n|$ correspond to the $(n-1)$–fold repeated eigenvalue 0.

(b) It remains to orthonormalize this set of eigenvectors using the Gram–Schmidt procedure. Observe that $\langle\psi_1|$ is already orthogonal to all the other eigenvectors $\langle\psi_i|,\ 2 \leq i \leq n$. The latter eigenvectors, however, are not orthogonal to each other. It is easily checked that this leads to somewhat involved expressions when orthonormalization is carried out. But *any* set of $n - 1$ linear combinations of the eigenvectors $\langle\psi_2|, \ldots, \langle\psi_n|$ that span the subspace spanned by these vectors would serve just as well as the eigenvectors for the eigenvalue 0. A somewhat more convenient starting point in this regard is the choice

$$\langle\psi_2| = \begin{pmatrix}1\ (1-n)\ 1 \cdots 1\end{pmatrix}, \quad \ldots, \quad \langle\psi_n| = \begin{pmatrix}1\ 1\ 1 \cdots (1-n)\end{pmatrix}.$$

The orthogonalization procedure then yields the set of normalized left eigenvectors

$$\begin{aligned}
\langle\phi_1| &= \begin{pmatrix}1\ 1\ 1\ 1\ 1 \cdots 1\end{pmatrix} / \sqrt{n}, \\
\langle\phi_2| &= \begin{pmatrix}1\ (1-n)\ 1\ 1\ 1 \cdots 1\end{pmatrix} / \sqrt{n(n-1)},
\end{aligned}$$

$$\langle\phi_3| = \big(1\;0\;(2-n)\;1\;1\;\cdots\;1\big)\big/\sqrt{(n-1)(n-2)},$$
$$\langle\phi_4| = \big(1\;0\;0\;(3-n)\;1\;\cdots\;1\big)\big/\sqrt{(n-2)(n-3)},$$
$$\cdots = \cdots\cdots\cdots\cdots\cdots\cdots\cdots$$
$$\langle\phi_n| = \big(1\;0\;0\;0\;0\;\cdots\;-1\big)\big/\sqrt{(2)(1)}.$$

The orthonormalized right eigenvectors of M are just the ket vectors corresponding to the bra vectors above. ▶

5. (a) The rows of H are linearly independent, so that its rank is 3. Hence its nullity must be zero. The roots of the characteristic equation $\det(\lambda I - H) = 0$ are trivially found. The eigenvalues are $\lambda_1 = \lambda_2 = 1,\ \lambda_3 = 2$. Hence the characteristic equation has a multiple root. But H is a real symmetric matrix. Therefore it must be diagonalizable by a similarity transformation $S^{-1}HS$, where S is an orthogonal matrix.

The right eigenvector $\mathbf{u}$ corresponding to any eigenvalue λ is found, of course, by writing out the equation $H\mathbf{u} = \lambda\mathbf{u}$ in components and solving the resulting simultaneous equations for the components u_i. This example serves to illustrate the typical situation when a repeated eigenvalue has more than one eigenvector. The equation $H\mathbf{u} = \mathbf{u}$ yields $u_1 = u_1,\ u_2 = u_3$, and $u_3 = u_2$. This means that the eigenvectors corresponding to λ_1 and λ_2 are of the form $\big(u_1\ u_2\ u_2\big)^{\mathrm{T}}$, where u_1 and u_2 are arbitrary. Clearly, there are two linearly independent possibilities, represented by the standard choices $u_1 = 1,\ u_2 = 0$, and $u_1 = 0,\ u_2 = 1$, respectively. The equation $H\mathbf{u} = 2\mathbf{u}$ for the eigenvalue λ_3 gives $u_1 = 0,\ u_2 = -u_3$. This eigenvector is therefore proportional to $\big(0\ 1\ -1\big)^{\mathrm{T}}$. Normalizing the eigenvectors, we arrive at the orthonormal set of eigenvectors

$$|\phi_1\rangle = \begin{pmatrix}1\\0\\0\end{pmatrix},\quad |\phi_2\rangle = \frac{1}{\sqrt{2}}\begin{pmatrix}0\\1\\1\end{pmatrix}\quad \text{and}\quad |\phi_3\rangle = \frac{1}{\sqrt{2}}\begin{pmatrix}0\\1\\-1\end{pmatrix},$$

corresponding to the eigenvalues 1, 1, and 2, respectively.

(b) Writing the column vectors $|\phi_i\rangle$ next to each other, we have the matrix

$$S = \frac{1}{\sqrt{2}}\begin{pmatrix}\sqrt{2} & 0 & 0\\ 0 & 1 & 1\\ 0 & 1 & -1\end{pmatrix}.$$

S is a symmetric matrix, and further, $S^2 = I$. Hence $S^{-1} = S = S^{\mathrm{T}}$, i.e., S is orthogonal. It is easily verified that $S^{-1}HS = \operatorname{diag}(1, 1, 2)$. Thus the real symmetric matrix H is diagonalized by the orthogonal matrix S, as expected. ▶

6. The elements of each row (or column) add up to zero. Hence any column (or row) of M is the negative of the sum of the rest of the columns (or rows). Moreover, 0 is an eigenvalue of M, with the column vector $(1\ 1\ \cdots\ 1)^{\mathrm{T}}$ as the corresponding right eigenvector. The rank of M is therefore $(n-1)$, and its nullity is 1.

Since the only nonzero elements of the first row of M are $a_0 = -1$, $a_1 = q$, and $a_{n-1} = p$, with $p + q = 1$, the eigenvalues follow at once from the general formula in Eq. (12.42). After some simplification, we get

$$\lambda_0 = 0,$$
$$\lambda_k = -1 + \cos(2\pi k/n) + i(q-p)\sin(2\pi k/n), \quad \text{where } k = 1, 2, \cdots, n-1.$$

Remark In accordance with Gershgorin's Circle Theorem (Chap. 11, Sect. 11.4.2), all the eigenvalues lie within or on a circle of radius 1 centered at -1 in the complex plane. ▶

7. You will find it helpful to use the identities $\omega^* = \omega^{-1} = \omega^2$. To write down the inverse of S, note that W is a circulant matrix and hence a normal matrix. It can therefore be diagonalized by a *unitary* transformation. Thus $S^{-1} = S^{\dagger}$. ▶

8. (a) As A and B are Hermitian matrices, it is guaranteed that each of them can be diagonalized by a *unitary* transformation. It is easily verified that $AB = BA\ (= 3A)$. Since A and B commute with each other, it follows that they can be diagonalized simultaneously.

(b) Solving the equation $\det(\lambda I - A) = 0$ yields the eigenvalues of A, namely, 0, 0, and 2. Note that A has only one independent nontrivial row, namely, $(1\ 0\ 1)$. Hence its rank is 1, so that its nullity must be 2, i.e., there must exist two linearly independent eigenvectors corresponding to the eigenvalue 0. Moreover, any linear combination of these is also obviously an eigenvector with eigenvalue 0. But not all such combinations can be expected to be eigenvectors of B as well.

The eigenvalues of B are $-1, 2$, and 3, respectively. The rank of B is 3, and its nullity is 0. *Because B has no repeated eigenvalues, it is more convenient to find the eigenvectors of B, and then to check that they are also eigenvectors of A*. Solve the equation $B\mathbf{u} = \lambda\mathbf{u}$ for the components of $\mathbf{u}$, with λ set equal to each of the eigenvalues of B in turn. It is straightforward to obtain the eigenvectors

$$\begin{pmatrix} a \\ -2a \\ -a \end{pmatrix}, \quad \begin{pmatrix} b \\ b \\ -b \end{pmatrix}, \quad \begin{pmatrix} c \\ 0 \\ c \end{pmatrix} \quad \text{(where } a, b, c \text{ are constants)}$$

corresponding to the eigenvalues -1, 2, and 3, respectively. It is easily checked that they are also eigenvectors of A, corresponding to the eigenvalues 0, 0, and 2, respectively.

(c) Write down the three column vectors above as a (3×3) matrix S, and hence its Hermitian conjugate $S^\dagger$. (Take the constants a, b, and c to be real, in order to avoid unnecessary complication.) This yields

$$SS^\dagger = \begin{pmatrix} a^2+b^2+c^2 & -2a^2+b^2 & -a^2-b^2+c^2 \\ -2a^2+b^2 & 4a^2+b^2 & 2a^2-b^2 \\ -a^2-b^2+c^2 & 2a^2-b^2 & a^2+b^2+c^2 \end{pmatrix}.$$

Now equate $SS^\dagger$ to the unit matrix, and read off the consistent solutions obtained for a, b, and c. Up to an overall sign, $a = 1/\sqrt{6},\ b = a\sqrt{2},\ c = a\sqrt{3}$. We thus arrive at the unitary matrix

$$S = \frac{1}{\sqrt{6}}\begin{pmatrix} 1 & \sqrt{2} & \sqrt{3} \\ -2 & \sqrt{2} & 0 \\ -1 & -\sqrt{2} & \sqrt{3} \end{pmatrix}, \text{ hence } S^{-1} = S^\dagger = \frac{1}{\sqrt{6}}\begin{pmatrix} 1 & -2 & -1 \\ \sqrt{2} & \sqrt{2} & -\sqrt{2} \\ \sqrt{3} & 0 & \sqrt{3} \end{pmatrix}.$$

It is straightforward to check that S diagonalizes both A and B according to $S^{-1}AS = S^\dagger AS = \text{diag}\,(0,\ 0,\ 2)$ and $S^{-1}BS = S^\dagger BS = \text{diag}\,(-1,\ 2,\ 3)$.

Remark Note the order in which S^{-1} and S appear in the above. The matrices SAS^{-1} and SBS^{-1} are *not* diagonal matrices. This has to do with the manner in which the transformation matrix S has been constructed in all the examples considered: its *columns* are the *right* eigenvectors of the matrix being diagonalized.

You could have proceeded by finding the common set of *left* eigenvectors of A and B, and writing these down as the *rows* of a transformation matrix T. The diagonalization of A and B is then achieved by the similarity transformations $TAT^{-1} = \text{diag}\,(0,\ 0,\ 2)$ and $TBT^{-1} = \text{diag}\,(-1,\ 2,\ 3)$. But then T turns out to be precisely the same as the matrix S^{-1} found above. Check this out! ▶

9. The anti-symmetry of the Poisson bracket is obvious. You have to check that the relations (12.58) and (12.59), with commutators replaced by Poisson brackets, are satisfied by functions of the phase space variables. ▶

10. (a) Let the levels in the n glasses after k iterations of the procedure be given by the components of the column vector

$$\mathbf{x}(k) = \big(x_1(k)\ x_2(k)\ \ldots\ x_n(k)\big)^{\mathrm{T}}.$$

A little calculation shows that $\mathbf{x}(1) = M\,\mathbf{x}(0)$, where the elements of the $(n \times n)$ transfer matrix M are as follows:

$$M_{1k} = M_{nk} = \begin{cases} 2^{-2} + 2^{-n} & \text{for } k = 1, 2 \\ 2^{-n-2+k} & \text{for } 3 \le k \le n. \end{cases}$$

For $2 \le j \le n-1$,

$$M_{jk} = \begin{cases} 2^{-j} & \text{for } k = 1, 2 \\ 2^{-2-j+k} & \text{for } 3 \le k \le j+1 \\ 0 & \text{for } k > j+1. \end{cases}$$

Writing out the matrix explicitly,

$$M = \begin{pmatrix} 2^{-2}+2^{-n} & 2^{-2}+2^{-n} & 2^{-n+1} & 2^{-n+2} & \cdots & 2^{-3} & 2^{-2} \\ 2^{-2} & 2^{-2} & 2^{-1} & 0 & \cdots & 0 & 0 \\ 2^{-3} & 2^{-3} & 2^{-2} & 2^{-1} & \cdots & 0 & 0 \\ 2^{-4} & 2^{-4} & 2^{-3} & 2^{-2} & \cdots & 0 & 0 \\ \vdots & \vdots & \vdots & \vdots & \ddots & \vdots & \vdots \\ 2^{-n+1} & 2^{-n+1} & 2^{-n+2} & 2^{-n+3} & \cdots & 2^{-2} & 2^{-1} \\ 2^{-2}+2^{-n} & 2^{-2}+2^{-n} & 2^{-n+1} & 2^{-n+2} & \cdots & 2^{-3} & 2^{-2} \end{pmatrix}.$$

As in the case $n = 3$ worked out earlier, we have

$$\mathbf{x}(k) = M\,\mathbf{x}(k-1) = M^k\,\mathbf{x}(0) \quad \text{for every positive integer } k.$$

(b) The matrix M has only two distinct Gershgorin disks (Chap. 11, Sect. 11.4.2) in the complex λ-plane. The first of these is centered at $\frac{1}{4} + 2^{-n}$, and has a radius $\frac{3}{4} - 2^{n}$. Its rightmost point therefore passes through 1. The second disk is centered at $\frac{1}{4}$, and has a radius $\frac{3}{4}$. It, too, passes through the point 1. The first disk is contained within the second, except for the common point of tangency at 1. All the eigenvalues of M must therefore lie on or within the second disk.

Each row sum of M is equal to unity, so that M is a stochastic matrix. (It is *doubly* stochastic only in the case $n = 3$.) Hence the uniform column vector $\begin{pmatrix}1 & 1 & \cdots & 1\end{pmatrix}^{\mathrm{T}}$ is a right eigenvector of M, corresponding to the eigenvalue 1. This eigenvector is essentially the asymptotic or equilibrium distribution, in which the levels in all the glasses become equal to each other.

The first $(n-1)$ rows of M are linearly independent of each other, while the last row of M is identical to the first row. Hence det $M = 0$, and 0 is an eigenvalue of M. This reflects the fact that $x_1(k) = x_n(k)$ for all k. The eigenvector corresponding to the eigenvalue 0 can be found by solving the equation $M\,\mathbf{u} = M_{jk}u_k = 0$. We have

$$M_{2k}u_k = 0 \;\Rightarrow\; \tfrac{1}{4}(u_1 + u_2) + \tfrac{1}{2}u_3 = 0.$$

Using this in the successive equations for $j = 3,\ 4,\ \ldots,\ (n-1)$, it follows that

$$u_4 = u_5 = \cdots = u_n = 0.$$

Now setting $j = 1$ (or $j = n$), we find that

$$\tfrac{1}{4}(u_1 + u_2) + \tfrac{1}{2}M_{n-1,k}u_k = 0, \quad \text{or} \quad u_1 + u_2 = 0.$$

This implies that $u_3 = 0$, and further, that the eigenvector is proportional to $(1\ {-1}\ 0 \cdots 0)^{\mathrm{T}}$. You could, of course, have guessed this result by noting that the first two elements of each row of M are equal. ▶

Chapter 13
Infinite-Dimensional Vector Spaces

We have seen that matrices provide representations of operators in linear vector spaces of a finite number of dimensions. In physical applications (e.g., in quantum mechanics), however, infinite-dimensional spaces occur frequently. We then have to deal with infinite-dimensional matrices and other kinds of operators, such as differential or integral operators. Many interesting features and subtleties arise in infinite-dimensional spaces, that are not present in infinite-dimensional spaces. I will therefore begin with a couple of infinite-dimensional spaces that are of great importance in physical applications, especially in quantum mechanics. We will then go on to some of the basic properties of linear operators that act on the elements of an LVS.

13.1 The Space ℓ_2 of Square-Summable Sequences

The most natural infinite-dimensional generalization of the Euclidean spaces $\mathbb{R}^d$ $(d = 1, 2, \dots)$ is the space of **square-summable sequences**, ℓ_2. It comprises all *infinite* sequences $(x_1, x_2, \dots)$ such that

$$\sum_{n=1}^{\infty} |x_n|^2 < \infty \,. \tag{13.1}$$

Note that this requires $|x_n|^2$ to tend to zero as $n \to \infty$ more rapidly than n^{-1}, i.e., $|x_n|$ must vanish faster than $n^{-1/2}$. This convergence condition ensures that the inner product of two elements in the LVS (to be defined below) is always finite, and that the usual properties of vectors in an LVS, such as the Cauchy–Schwarz and triangle inequalities (Eqs. (10.15) and (10.21) in Chap. 10, Sects. 10.3.1 and 10.3.2,

V. Balakrishnan, *Mathematical Physics*,
https://doi.org/10.1007/978-3-030-39680-0_13

respectively) are satisfied. Note that the components x_n may be complex numbers, in general.

The ket vectors of ℓ_2 may be represented by infinite-dimensional column matrices,

$$|\psi\rangle = \begin{pmatrix} x_1 \\ x_2 \\ \vdots \end{pmatrix}, \quad |\chi\rangle = \begin{pmatrix} y_1 \\ y_2 \\ \vdots \end{pmatrix}, \ldots. \tag{13.2}$$

The space ℓ_2 is self-dual. The bra vectors that are elements of the dual LVS may be represented by infinite-dimensional row matrices. Thus the adjoint of the ket $|\psi\rangle$ above is the bra

$$\langle\psi| = \begin{pmatrix} x_1^* \; x_2^* \cdots \end{pmatrix}. \tag{13.3}$$

The inner product is defined as

$$\boxed{\langle\chi|\psi\rangle = \sum_{n=1}^{\infty} y_n^* \, x_n \,.} \tag{13.4}$$

The square of the norm of a vector is thus

$$\|\psi\|^2 = \langle\psi|\psi\rangle = \sum_{n=1}^{\infty} |x_n|^2 < \infty\,. \tag{13.5}$$

It is evident that the ℓ_2-norm $\|\psi\|$ vanishes if and only if $|\psi\rangle$ is the null vector, *all* of whose components are zero.

★ **1.** Consider the infinite sequence $(x_1, x_2, \ldots)$ whose nth element x_n is given by

(a) $(-1)^n (\ln n)/n$ (b) $n!/(2n)!$ (c) $(n!)^2/(2n)!$ (d) $(2/n)^n$
(e) $1/(n^2+1)^{1/2}$ (f) $(2n+1)/(3n+4)^2$ (g) e^n/n^n (h) $2^{-n/2}$
(i) $n!/[(2n)!]^{1/2}$ (j) $1/\sqrt{n}\,\ln(n+1)$ (k) $2^n n!/[(2n)!]^{1/2}$.

Which of the sequences above belong to ℓ_2?

For completeness, I mention that ℓ_2 is actually a member of an infinite *family* of linear vector spaces of sequences. The space ℓ_p, where p is any number ≥ 1, is defined as comprising all infinite sequences that satisfy the condition

$$\left(\sum_{n=1}^{\infty} |x_n|^p\right)^{1/p} < \infty\,. \tag{13.6}$$

The left-hand side in (13.6) is called the ℓ_p-norm of the vector concerned. It turns out that the dual of the space ℓ_p is the space ℓ_q, where p and q are related according to

$$(1/p) + (1/q) = 1. \tag{13.7}$$

It is clear, therefore, that ℓ_2 is special because it is self-dual ($p = q = 2$). The norm, too, then becomes precisely what we would naturally regard as the length of the vector in Euclidean geometry. ℓ_2 is also the space that is most often relevant in applications.

13.2 The Space $\mathcal{L}_2$ of Square-Integrable Functions

13.2.1 Definition of $\mathcal{L}_2$

Functions belonging to some specified class (e.g., differentiable functions) often constitute a linear vector space (a **function space**). Among these, an important one is the LVS $\mathcal{L}_2(-\infty, \infty)$ of **square-integrable functions** of a real variable x, namely, all functions that satisfy the condition

$$\boxed{\int_{-\infty}^{\infty} dx\, |f(x)|^2 < \infty.} \tag{13.8}$$

It is straightforward to check out that such functions satisfy all the conditions required of the elements of an LVS. Note that, while x is real, $f(x)$ itself could be a complex-valued function, in general. One can also speak of $\mathcal{L}_2$ functions that are defined in some interval $[a, b]$, that is, functions that satisfy

$$\int_a^b dx\, |f(x)|^2 < \infty, \quad \text{so that} \quad f(x) \in \mathcal{L}_2[a\,,\, b]. \tag{13.9}$$

If no range is specified, it is understood that we are dealing with $\mathcal{L}_2(-\infty\,,\,\infty)$. The left-hand side in (13.8) defines the square of the norm of an element of this space, as we will see shortly.

13.2.2 Continuous Basis

It is intuitively clear that a function space is obtained from a space of sequences (or ordered sets of numbers) when the *index* j in the sequence

$$(x_1, x_2, \ldots, x_j, \ldots) \tag{13.10}$$

becomes a *continuous* variable. In this sense the value of $f(x)$ for *each* given value of the argument x represents a "component" of an abstract vector. The latter is conveniently and suggestively denoted by $| f \rangle$. The idea is that there is a **continuous basis** $\{| x \rangle\}$ in the LVS, labeled by the values of x. The elements of this basis satisfy the orthonormality and completeness relations

$$\langle x' | x \rangle = \delta(x' - x) \quad \textbf{(orthonormality)} \tag{13.11}$$

and

$$\int_{-\infty}^{\infty} dx \, | x \rangle \langle x | = I \quad \textbf{(completeness)}, \tag{13.12}$$

where I is the unit operator in the space.[1] Then

$$| f \rangle = I | f \rangle = \int_{-\infty}^{\infty} dx \, | x \rangle \langle x | | f \rangle = \int_{-\infty}^{\infty} dx \, \big(\langle x | f \rangle\big) | x \rangle \equiv \int_{-\infty}^{\infty} dx \, f(x) \, | x \rangle. \tag{13.13}$$

We thus arrive at an important conclusion:

- The function $f(x)$ is just shorthand for the expansion coefficient of a vector in the $| x \rangle$-basis, i.e.,

$$\boxed{f(x) \equiv \langle x | f \rangle, \quad \text{and hence} \quad \langle f | x \rangle = f^*(x).} \tag{13.14}$$

This way of looking upon functions of a continuous variable is very useful.

Thus $\mathcal{L}_2$ may be regarded as the natural continuum analog of the space ℓ_2. It is also a self-dual LVS. The inner product of any two vectors $| f \rangle$ and $| g \rangle$ then reduces to

$$\begin{aligned} \langle f | g \rangle \equiv \langle f | \Big(\int_{-\infty}^{\infty} dx \, | x \rangle \langle x | \Big) | g \rangle &= \int_{-\infty}^{\infty} dx \, \langle f | x \rangle \langle x | g \rangle \\ &= \int_{-\infty}^{\infty} dx \, f^*(x) \, g(x). \end{aligned} \tag{13.15}$$

This inner product is often written as (f, g), but it is helpful in physical applications to retain the Dirac notation $\langle f | g \rangle$ for function spaces as well.

It is evident that the square of the norm of an element $f \in \mathcal{L}_2(-\infty, \infty)$ is given by

$$\boxed{\| f \|^2 = \langle f | f \rangle = \int_{-\infty}^{\infty} dx \, | f(x) |^2.} \tag{13.16}$$

Recall the Cauchy–Schwarz inequality for vectors in an LVS, $|\langle f | g \rangle|^2 \le \| f \|^2 \, \| g \|^2$. As applied to $\mathcal{L}_2(-\infty, \infty)$, this inequality reads

[1] Compare these equations with the corresponding ones for a discrete basis, Eqs. (10.3) and (10.4) of Sect. 10.2.1.

$$\left|\int_{-\infty}^{\infty} dx\, f^*(x)\, g(x)\right|^2 \leq \left(\int_{-\infty}^{\infty} dx\, |f(x)|^2\right) \left(\int_{-\infty}^{\infty} dx\, |g(x)|^2\right). \tag{13.17}$$

The inequality reduces to an equality if and only if $f(x)$ is a constant multiple of $g(x)$. All the foregoing statements apply to any $\mathcal{L}_2[a\,,\,b]$ as well.

★ **2.** Identify the functions $f(x)$ that belong to $\mathcal{L}_2(-\infty\,,\,\infty)$ among the functions listed below. Wherever $f(x)$ is indeterminate, assume that it is defined by continuity.

(a) $(x^2+1)^{-1/2}$ (b) $(\sin x)/x$ (c) $(\sin x)^2/x^2$ (d) $x^3\, e^{-x^2}$
(e) $(\tanh x)/x$ (f) $e^{-x}\cos x$ (g) e^{-1/x^2} (h) $(x^2-1)^{-1}$.

★ **3.** Use the Cauchy–Schwarz inequality in $\mathcal{L}_2(-\infty\,,\,\infty)$ to show that

$$\int_{-\infty}^{\infty} \frac{e^{-x^2/2}\, dx}{(x^2+1)^{1/2}} < \pi^{3/4}.$$

Compare this upper bound with the numerical value of the integral obtained using any convenient integration routine.

13.2.3 Weight Functions: A Generalization of $\mathcal{L}_2$

We have defined an element of $\mathcal{L}_2(-\infty\,,\,\infty)$ as any function $f(x)$ that satisfies (13.8), namely, $\int_{-\infty}^{\infty} dx\, |f(x)|^2 < \infty$. The convergence of the integral immediately imposes a *necessary* (but not sufficient) condition upon $f(x)$: it must tend to zero more rapidly than $|x|^{-1/2}$ as $|x| \to \infty$. This rules out polynomials, for instance, as members of $\mathcal{L}_2(-\infty\,,\,\infty)$, because the corresponding integrals diverge. On the other hand, as we shall see in Chap. 16, specific families of polynomials are used to construct natural basis sets in various function spaces. These spaces include those with a semi-infinite domain (e.g., $[0,\,\infty)$) and an infinite domain $(-\infty,\,\infty)$. How is the divergence problem overcome?

This is done by introducing a convergence factor or **weight function** $\rho(x)$ into the definition of the inner product in the function space. The function $\rho(x)$ is positive in the region of integration, and vanishes sufficiently rapidly as $|x| \to \infty$. Thus, instead of the condition (13.8) for the square integrability of a function, we impose the condition

$$\int_{-\infty}^{\infty} dx\, \rho(x)\, |f(x)|^2 < \infty. \tag{13.18}$$

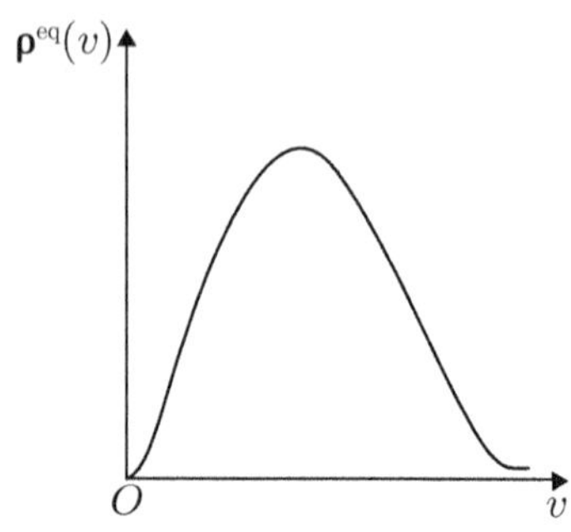

Fig. 13.1 Maxwellian distribution of the speed of a molecule in a classical ideal gas (Eq. (13.19)) (schematic)

The weight $\rho(x)$ is chosen such that the convergence of the integral is secured for the class of functions of interest.[2] For example, if we choose $\rho(x) = e^{-x^2}$, it is clear that $f(x)$ could be any polynomial of arbitrarily high degree. A similar generalization is applicable to $\mathcal{L}_2[a\,,\,b]$ in any finite interval $[a,\,b]$. The weight $\rho(x)$ is then positive in $(a,\,b)$, and may or may not vanish at the end points a and b. I will return to these matters in greater detail in Chap. 16, when we discuss families of orthogonal polynomials.

The Maxwellian distribution of molecular speeds provides a physical example of the use of a weight factor in function space. Consider a classical ideal gas of particles of mass m in thermal equilibrium at a temperature T. According to equilibrium statistical mechanics, the normalized probability density $\rho^{\text{eq}}(v)$ of the speed v of a molecule is given by

$$\rho^{\text{eq}}(v) = \left(\frac{m}{2\pi k_B T}\right)^{3/2} 4\pi v^2 \exp\left\{-\frac{mv^2}{2k_B T}\right\}, \tag{13.19}$$

where k_B is Boltzmann's constant, and $0 \leq v < \infty$ (see Fig. 13.1.) The rth moment of the speed is given by

$$\langle v^r \rangle = \int_0^\infty dv\, \rho^{\text{eq}}(v)\, v^r. \tag{13.20}$$

Note that there is no denominator on the right-hand side, because the PDF is normalized to unity, i.e., $\int_0^\infty dv\, \rho^{\text{eq}}(v) = 1$. For any positive value of r, there is no problem of convergence at the upper limit of integration, because of the Gaussian factor in $\rho^{\text{eq}}(v)$. At the lower limit 0, the behavior of the integrand in Eq. (13.20) is $\sim v^{r+2}$ because of the extra factor of v^2 supplied by $\rho^{\text{eq}}(v)$. Therefore the integral is convergent not only for all positive values of r, but also for negative values as long as $r + 2 > -1$, i.e., $r > -3$. In particular, the quantities $\langle v^{-1} \rangle$ and $\langle v^{-2} \rangle$ are finite.

---

[2]More generally, what is involved is a so-called **integration measure** $d\mu(x)$. The form $d\mu(x) = \rho(x)\,dx$ is a special case, corresponding to the existence of a weight or density $\rho(x)$. I will comment further on this in Chap. 16, Sect. 16.1.2.

★ **4.** Consider functions of v in the function space $\mathcal{L}_2\,[0, \infty)$, with a Gaussian weight function given by $\rho^{\text{eq}}(v)$, Eq. (13.19).

(a) Use the Cauchy–Schwarz inequality to show that the average value of the reciprocal of the speed of a molecule is greater than the reciprocal of the average speed. That is,

$$\langle v^{-1}\rangle > \langle v\rangle^{-1}.$$

(b) Use an analogous argument to show that

$$\langle v^{-2}\rangle > \langle v\rangle^{-2}.$$

(c) Using the formula of Eq. (3.16) of Chap. 3, Sect. 3.1.5 for the general Gaussian integral, write down the four average values $\langle v\rangle$, $\langle v^{-1}\rangle$, $\langle v^2\rangle$, and $\langle v^{-2}\rangle$, and verify that both the inequalities above are indeed satisfied.

(d) If α is any real number in the range $-3 < \alpha < 3$, show that

$$\langle v^{\alpha}\rangle\langle v^{-\alpha}\rangle = (1-\alpha^2)\sec\left(\tfrac{1}{2}\pi\alpha\right).$$

We will have the occasion to return to the Maxwellian distribution of velocities in Chap. 20, Sect. 20.2.3, and again in Chap. 21, Sect. 21.7.3.

13.2.4 $\mathcal{L}_2(-\infty\,,\,\infty)$ Functions and Fourier Transforms

The connection between $\mathcal{L}_2(-\infty\,,\,\infty)$ functions and Fourier transforms will become clearer after we discuss Fourier transforms in Chap. 18. But it is both helpful and relevant to note the following points right here:

- The Fourier transform $\widetilde{f}(k)$ of a square-integrable function $f(x)$ is also square-integrable. In other words, if $f(x) \in \mathcal{L}_2(-\infty\,,\,\infty)$, then $\widetilde{f}(k)$ also belongs to $\mathcal{L}_2(-\infty\,,\,\infty)$. As you will see in Chap. 18, Sect. 18.1.2, this is yet another manifestation of Parseval's Theorem (encountered in Chap. 10, Sect. 10.2.3).
- We may therefore look upon a square-integrable function and its Fourier transform as two different, but *equivalent*, ways of representing the same vector or element of the LVS of square-integrable functions.
- From this point of view, the Fourier transform operation in $\mathcal{L}_2(-\infty\,,\,\infty)$ is nothing but a change of basis (from the $\{|\,x\,\rangle\}$ basis to the $\{|\,k\rangle\}$ basis).

These facts have an immediate implication for quantum mechanics:

(i) The state vector of a quantum mechanical system such as a particle or a collection of particles may be represented either by its position-space wave function or by its momentum-space wave function. The two wave functions comprise a Fourier transform pair.

(ii) If the position-space wave function of a particle is normalizable, so is its momentum-space wave function.

13.2.5 The Wave Function of a Particle

The generalization of the space $\mathcal{L}_2$ to square-integrable functions in more than one variable is straightforward. In three dimensions, for instance, we have the space of functions of $\mathbf{r}$ that are square-integrable, namely,

$$\int d^3r\,|f(\mathbf{r})|^2 < \infty\,. \tag{13.21}$$

The relevance of such $\mathcal{L}_2$ spaces to quantum mechanics is now obvious. Recall that the wave function of a particle satisfies, when the state of the particle is a bound state in some potential, the normalization condition

$$\int d^3r\,|\psi(\mathbf{r},t)|^2 = 1\,. \tag{13.22}$$

This condition follows from the interpretation of $|\psi(\mathbf{r},t)|^2$ as the probability density of the particle in position space.

- This is how the function space $\mathcal{L}_2$ of square-integrable functions enters quantum mechanics naturally.

From the mathematical point of view,

- the wave function $\psi(\mathbf{r},t)$ of a quantum mechanical particle actually is nothing but the coefficient in the expansion, in the **position basis**, of the abstract state vector $|\Psi(t)\rangle$ of the particle.

The state vector is an element of a certain LVS called a Hilbert space (see Sect. 13.3.1 below). The position basis is the continuous basis formed by the position eigenstates $|\mathbf{r}\rangle$ of the particle. The orthonormality and completeness relations for this basis are

$$\boxed{\langle\mathbf{r}'|\mathbf{r}\rangle = \delta^{(3)}(\mathbf{r}'-\mathbf{r}) \quad\text{and}\quad \int d^3r\,|\mathbf{r}\rangle\langle\mathbf{r}| = I,} \tag{13.23}$$

where I denotes the unit operator in the Hilbert space. Then

$$|\Psi(t)\rangle = \int d^3r\,|\mathbf{r}\rangle\langle\mathbf{r}|\Psi(t)\rangle = \int d^3r\,\big(\langle\mathbf{r}|\Psi(t)\rangle\big)\,|\mathbf{r}\rangle \equiv \int d^3r\,\psi(\mathbf{r},t)\,|\mathbf{r}\rangle, \tag{13.24}$$

so that

$$\boxed{\psi(\mathbf{r},t) \overset{\text{def.}}{=} \langle\mathbf{r}|\Psi(t)\rangle\,.} \tag{13.25}$$

It is obvious that Eq. (13.25) is a three-dimensional version of Eq. (13.14).

Exactly the same sort of relationships hold good in the **momentum basis**. This is the continuous basis formed by the momentum eigenstates of the particle. The orthonormality and completeness relations for this basis are

$$\boxed{\langle \mathbf{p}' \,|\mathbf{p}\,\rangle = \delta^{(3)}(\mathbf{p}' - \mathbf{p}) \quad \text{and} \quad \int d^3p \,|\mathbf{p}\rangle\,\langle\mathbf{p}| = I.} \tag{13.26}$$

Further,

$$|\Psi\,(t)\rangle = \int d^3p \,|\mathbf{p}\rangle\,\langle\mathbf{p}\,|\Psi\,(t)\rangle = \int d^3p \,\big(\langle\mathbf{p}|\Psi\,(t)\rangle\big)\,|\mathbf{p}\,\rangle \equiv \int d^3p \;\widetilde{\psi}(\mathbf{p}, t)\,|\mathbf{p}\,\rangle, \tag{13.27}$$

so that

$$\boxed{\widetilde{\psi}(\mathbf{p}, t) \stackrel{\text{def.}}{=} \langle\,\mathbf{p}\,|\Psi\,(t)\rangle\,.} \tag{13.28}$$

Once again, the wave function $\widetilde{\psi}(\mathbf{p}, t)$ is nothing but the coefficient in the expansion of the state vector $|\Psi\,(t)\rangle$ in the momentum basis. The relationship between the expansion coefficients (or wave functions) $\psi(\mathbf{r}, t)$ and $\widetilde{\psi}(\mathbf{p}, t)$ follows readily (recall the change-of-basis relations (10.11) of Chap. 10, Sect. 10.2.2):

$$\widetilde{\psi}(\mathbf{p}, t) = \langle\mathbf{p}|\Psi\,(t)\rangle = \int d^3r \;\psi(\mathbf{r}, t)\,\langle\,\mathbf{p}\,|\,\mathbf{r}\,\rangle\,. \tag{13.29}$$

The discussion so far has been purely mathematical. The key *physical* input comes in via the specification of the scalar product $\langle\,\mathbf{p}\,|\,\mathbf{r}\,\rangle$ representing the "overlap" between a position eigenstate and a momentum eigenstate of a particle.

- This overlap is determined by the fundamental **canonical commutation relation** $[x_j,\, p_k] = i\hbar\,\delta_{jk}\, I$ between the position and momentum operators of the particle. (Here the indices j and k label Cartesian components of the vectors concerned.)

$\langle\,\mathbf{p}\,|\,\mathbf{r}\,\rangle$ turns out to be proportional to $e^{-i\mathbf{p}\cdot\mathbf{r}/\hbar}$. With the standard proportionality factor (or normalization constant) that is used in this regard, we have

$$\boxed{\widetilde{\psi}(\mathbf{p}, t) = \frac{1}{(2\pi\hbar)^{3/2}} \int d^3r \; e^{-i\mathbf{p}\cdot\mathbf{r}/\hbar}\,\psi(\mathbf{r}, t).} \tag{13.30}$$

In other words, $\widetilde{\psi}(\mathbf{p}, t)$ is essentially the (three-dimensional) Fourier transform of $\psi(\mathbf{r}, t)$. Similarly,

$$\psi(\mathbf{r}, t) = \langle\mathbf{r}|\Psi\,(t)\rangle = \int d^3p \;\widetilde{\psi}(\mathbf{p}, t)\,\langle\,\mathbf{r}\,|\,\mathbf{p}\,\rangle\,. \tag{13.31}$$

Since $\langle\,\mathbf{r}\,|\,\mathbf{p}\,\rangle = \langle\,\mathbf{p}\,|\,\mathbf{r}\,\rangle^* = e^{i\mathbf{p}\cdot\mathbf{r}/\hbar}$,

$$\boxed{\psi(\mathbf{r}, t) = \frac{1}{(2\pi\hbar)^{3/2}} \int d^3p \; e^{i\mathbf{p}\cdot\mathbf{r}/\hbar}\,\widetilde{\psi}(\mathbf{p}, t).} \tag{13.32}$$

That is, $\psi(\mathbf{r}, t)$ is the inverse Fourier transform of $\widetilde{\psi}(\mathbf{p}, t)$, as you would expect.

13.3 Hilbert Space and Subspaces

In Sects. 13.3.1 and 13.3.2, I list some essential definitions and terminologies, in order to make the discussion as self-contained as possible. The reader is reminded once again that this is not intended to be a substitute for more detailed formal treatments of the topics concerned.

13.3.1 Hilbert Space

When we deal with infinite-dimensional linear vector spaces, we must be careful about questions of convergence. In particular, we must consider the convergence of infinite sums or linear combinations of the form $\sum_{n=1}^{\infty} c_n \, |\phi_n\rangle$. The concept of a **Cauchy sequence** of vectors is fundamental in this regard.

- A sequence of vectors $|\phi_n\rangle$, where $n = 1, 2, \ldots$ *ad inf.*, is a Cauchy sequence if the difference vector $|\phi_n\rangle - |\phi_m\rangle$ tends to the null vector $|\Omega\rangle$ as both n and $m \to \infty$.
- Cauchy sequences of vectors are guaranteed to tend to definite limits. Thus, $\lim_{n\to\infty} |\phi_n\rangle$ is some definite vector $|\phi\rangle$ if and only if $|\phi_n\rangle$ is a member of a Cauchy sequence.

It is clear that, in manipulating vectors in an infinite-dimensional space, it will be convenient if all such limit vectors are *also* elements of the LVS. This leads naturally to the next definition:

- An LVS which includes all limit vectors of Cauchy sequences among its elements is said to be a **complete linear space**.
- An LVS in which an inner product is defined, and which is complete in the sense just described, is called a **Hilbert space**.

Hilbert spaces can be finite-dimensional or infinite-dimensional. As you know, a primary reason for the importance of Hilbert spaces is that

- the state vectors of a quantum mechanical system are elements of a Hilbert space.

We have spoken of continuous basis sets. It is important to know whether a given infinite-dimensional Hilbert space has *only* such basis sets, or whether it also has *countably* infinite, *discrete* basis sets as well. If it does, then it is called a **separable Hilbert space**. In most simple applications of quantum mechanics, we only deal with separable Hilbert spaces. This circumstance helps avoid many technical complications related to continuous basis sets, which I have ignored in the heuristic discussion given here.

More generally, we could have an LVS that is complete, and on which the *norm* of a vector is defined, but not an inner product of any arbitrary pair of vectors. Such a space is called a **Banach space**. Every Hilbert space is a Banach space, but the converse is not necessarily true. Although we shall not deal with Banach spaces in this book, I mention this because they play a significant role in functional analysis and operator theory.

13.3.2 Linear Manifolds and Subspaces

In applications, one is often concerned with just a part of a linear vector space, and this portion appears to be "self-contained" in some sense. The concept of a subspace arises naturally in this regard.

Linear manifold: A subset $\mathbb{U}$ of an LVS $\mathbb{V}$ is a **linear manifold** if the following property is satisfied: Given any pair of elements $|\phi\rangle, |\chi\rangle \in \mathbb{U}$, any arbitrary linear combination $\alpha\,|\phi\rangle + \beta\,|\chi\rangle$ (where α and β are scalars) is also an element of $\mathbb{U}$. It is immediately obvious that the idea of a linear manifold can be extended to linear combinations of more than two vectors. In principle, such combinations could also involve *infinite* sums over vectors, raising the question of convergence to limit vectors. This leads naturally to the next definition.

- A linear manifold $\mathbb{U}$ is a **subspace** of the LVS $\mathbb{V}$ if it is complete.

That is, all the limit vectors of all Cauchy sequences of vectors belonging to $\mathbb{U}$ also lie in $\mathbb{U}$. In practice, a subset $\mathbb{U}$ of vectors in an LVS $\mathbb{V}$ is a subspace if the following conditions are met:

(i) $\mathbb{U}$ must contain the null vector.
(ii) The sum of any two vectors in $\mathbb{U}$ must lie in $\mathbb{U}$.
(iii) Any scalar multiple of any vector in $\mathbb{U}$ must lie in $\mathbb{U}$.

The essential difference between a linear manifold and a subspace of an LVS is as follows:

- A subspace of an LVS is also an LVS, with the same operations of addition and scalar multiplication as the original LVS.

Moreover:

- If $\mathbb{U}_1$ and $\mathbb{U}_2$ are subspaces of an LVS, then so is their sum (or union) $\mathbb{U}_1 \cup \mathbb{U}_2$, as well as their intersection $\mathbb{U}_1 \cap \mathbb{U}_2$.
- The dimensions of these subspaces are related according to

$$\dim\,(\mathbb{U}_1 \cup \mathbb{U}_2) = \dim\,\mathbb{U}_1 + \dim\,\mathbb{U}_2 - \dim\,(\mathbb{U}_1 \cap \mathbb{U}_2). \tag{13.33}$$

The kind of "inclusion-exclusion" formula in Eq. (13.33) may be familiar to you in more than one context—e.g., in probability theory. Figure 13.2 illustrates the subspaces concerned. It helps us understand roughly why the contribution of the "double-counted" intersection must be subtracted out.

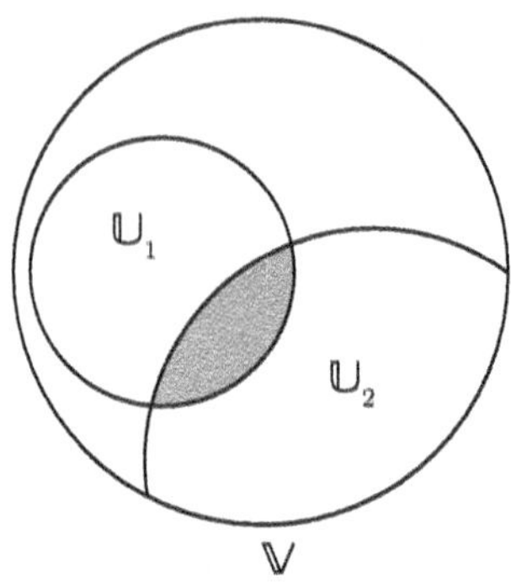

Fig. 13.2 The dimensionality of the union of two subspaces $\mathbb{U}_1$ and $\mathbb{U}_2$ is the sum of the dimensionalities of the individual subspaces, minus that of their intersection (shaded)

13.4 Solutions

1. All that you have to check out in each case is whether $|x_n|$ vanishes faster than $1/\sqrt{n}$ as $n \to \infty$. Wherever factorials appear, use Stirling's formula (Eq. (2.11) of Chap. 2, Sect. 2.2). The sequences from (a) to (i) are all square-summable, and belong to ℓ_2. The sequences in (j) and (k) are not square-summable. As $N \to \infty$, the sums $\sum^N |x_n|^2$ *diverge* in these cases like ln (ln N) and $N^{3/2}$, respectively. ▶

2. You have to check that $\langle f \mid f \rangle$ exists in each case. This requires $|f(x)|^2$ to approach zero more rapidly than $1/|x|$ as $x \to \pm\infty$, and further, to have no nonintegrable singularity in $(-\infty, \infty)$. The functions in (a) to (e) satisfy these requirements, and are elements of $\mathcal{L}_2(-\infty\,,\,\infty)$. In cases (b), (c), and (e), the apparent singularity at $x = 0$ is removed by the vanishing of the numerator, and $f(x) = 1$ (by continuity) at $x = 0$. The functions in (f), (g), and (h) are not square-integrable. In (f), the factor e^{-x} diverges as $x \to -\infty$. In (g), $f(x)$ tends to a nonzero constant (unity) as $x \to \pm\infty$. In (h), $f(x)$ has nonintegrable singularities at the points $x = \pm 1$ in the range of integration. ▶

3. In $\mathcal{L}_2(-\infty\,,\,\infty)$, consider the elements

$$f(x) = e^{-x^2/2} \quad \text{and} \quad g(x) = (x^2+1)^{-1/2},$$

and apply the Cauchy–Schwarz inequality. Since $\langle f \mid f \rangle = \pi^{1/2}$ and $\langle g \mid g \rangle = \pi$, the desired inequality follows at once. This is not the best upper bound on the value of the integral, of course. To give you an idea of how good a bound it is, the numerical value of the integral is $1.9793\cdots$, while $\pi^{3/4} = 2.3597\cdots$. ▶

4. The rth moment of the speed, $\langle v^r \rangle$, is given by Eq. (13.20) for all $r > -3$.
(a) You need to show that $\langle v \rangle \langle v^{-1} \rangle > 1$. You can make the identifications $\langle v \rangle = \|f\|^2$ and $\langle v^{-1} \rangle = \|g\|^2$, if you set $f(v) = v^{1/2}$ and $g(v) = v^{-1/2}$. Then $\langle f | g \rangle = 1$ because $\rho^{\text{eq}}(v)$ is normalized to unity. The required inequality follows at once from the Cauchy–Schwarz inequality.

(b) Similarly, setting $f(v) = v$ and $g(v) = v^{-1}$, we have the identifications $\langle v^2 \rangle = \|f\|^2$ and $\langle v^{-2} \rangle = \|g\|^2$. Once again, $\langle f | g \rangle = 1$ because $\int_0^\infty dv\, \rho^{\text{eq}}(v) = 1$. Hence the Cauchy–Schwarz inequality implies that the product $\langle v^2 \rangle \langle v^{-2} \rangle > 1$.
(c) Using the general formula of Eq. (3.16) for Gaussian integrals, we find

$$\langle v \rangle = \left(\frac{8k_B T}{m\pi}\right)^{1/2} \quad \text{and} \quad \langle v^{-1} \rangle = \left(\frac{2m}{\pi k_B T}\right)^{1/2}, \quad \text{so that} \quad \langle v \rangle \langle v^{-1} \rangle = 4/\pi,$$

which is greater than unity. Similarly,

$$\langle v^2 \rangle = \frac{3k_B T}{m} \quad \text{and} \quad \langle v^{-2} \rangle = \frac{m}{k_B T}, \quad \text{so that} \quad \langle v^2 \rangle \langle v^{-2} \rangle = 3,$$

which again is greater than unity.
(d) Using the Gaussian integral (3.16), it is easy to see that

$$\langle v^\alpha \rangle = 2\pi^{-1/2} \Gamma\left(\tfrac{1}{2}(3+\alpha)\right) (2k_B T/m)^{\alpha/2}, \quad \text{as long as} \quad \alpha > -3.$$

Hence

$$\langle v^{-\alpha} \rangle = 2\pi^{-1/2} \Gamma\left(\tfrac{1}{2}(3-\alpha)\right) (2k_B T/m)^{-\alpha/2}, \quad \text{as long as} \quad \alpha < 3.$$

Therefore

$$\langle v^\alpha \rangle \langle v^{-\alpha} \rangle = 4\pi^{-1} \Gamma\left(\tfrac{1}{2}(3+\alpha)\right) \Gamma\left(\tfrac{1}{2}(3-\alpha)\right), \quad \text{provided} \quad -3 < \alpha < 3.$$

Now use the functional equation $\Gamma(x+1) = x\,\Gamma(x)$ (Eq. (3.14) of Chap. 3, Sect. 3.1.4) to get

$$\langle v^\alpha \rangle \langle v^{-\alpha} \rangle = (1-\alpha^2) \Gamma\left(\tfrac{1}{2}(1+\alpha)\right) \Gamma\left(\tfrac{1}{2}(1-\alpha)\right) \quad (-3 < \alpha < 3).$$

At this stage, you need the so-called reflection formula[3] satisfied by the gamma function, namely, $\Gamma(x)\,\Gamma(1-x) = \pi \operatorname{cosec} \pi x$. Using this formula, we get

$$\langle v^\alpha \rangle \langle v^{-\alpha} \rangle = (1-\alpha^2) \sec\left(\tfrac{1}{2}\pi\alpha\right) \quad (-3 < \alpha < 3).$$

It is easily checked that the results obtained earlier for $\alpha = 1$ and $\alpha = 2$ are recovered from this general expression. In the former case, since $(1-\alpha^2)$ vanishes and $\sec\left(\frac{1}{2}\pi\alpha\right)$ becomes infinite at $\alpha = 1$, you must pass to the limit, as $\alpha \to 1$, of the product of these two factors. ▶

[3]In Chap. 25, Sect. 25.2.7, we shall see how this formula, Eq. (25.48), comes about.

Chapter 14
Linear Operators on a Vector Space

We turn now to operators on an LVS. To start with, we need a number of simple definitions. Most of these are obvious in the case of finite-dimensional vector spaces. But the case of infinite-dimensional spaces is nontrivial, and due care must be exercised. As always, I shall merely describe and list some relevant results in brief, without getting into technical details and proofs of assertions, or the formal mathematics. These details and formal proofs are, however, particularly important at least for some of the topics in this chapter. It is therefore advisable to supplement the discussion given here with more extended treatments to be found in standard texts on operator algebra and functional analysis (see, e.g., the Bibliography).

14.1 Some Basic Notions

14.1.1 Domain, Range, and Inverse

An **operator** A in an LVS $\mathbb{V}$ is an entity that "acts" on a vector $|\phi\rangle$ of the LVS to produce an element $|\psi\rangle$ of the LVS: this is written as

$$A\,|\phi\rangle = |\psi\rangle. \tag{14.1}$$

Domain: A moment's thought shows that the foregoing statement needs to be qualified. The point is that when a given operator A acts on an *arbitrary* element of $\mathbb{V}$, it may not necessarily produce a vector that remains an element of $\mathbb{V}$. That is, A may take some vectors of $\mathbb{V}$ "out" of $\mathbb{V}$. In general, $A\,|\phi\rangle$ will remain inside $\mathbb{V}$ only when $|\phi\rangle$ belongs to some *subset* of the elements of $\mathbb{V}$. This subset is called the **domain** $\mathbb{D}_A$ of the operator A.

V. Balakrishnan, *Mathematical Physics*,
https://doi.org/10.1007/978-3-030-39680-0_14

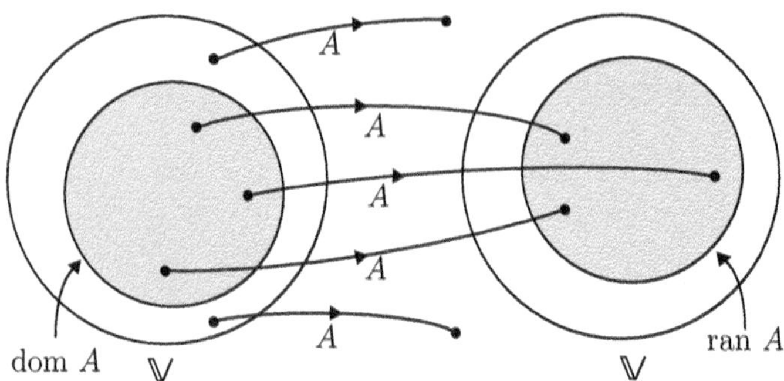

Fig. 14.1 Illustrating the domain and range of an operator A on an LVS $\mathbb{V}$. The dots denote vectors. A may be regarded as a map that maps vectors to other vectors. Vectors in $\mathbb{D}_A$, the domain of A (shaded area on the left), are mapped to vectors in Δ_A, the range of A (shaded area on the right). Vectors in $\mathbb{V}$ that are not in $\mathbb{D}_A$ are mapped to vectors that do not lie in $\mathbb{V}$

Range: As $|\phi\rangle$ runs through all the elements of $\mathbb{D}_A$, the vectors $A\,|\phi\rangle = |\psi\rangle$ that result constitute a set Δ_A (that may either be all of $\mathbb{V}$ itself, or a subset of $\mathbb{V}$). This set Δ_A is called the **range** of the operator A. If A is regarded as a map, Δ_A is the *image* of the map. Recall that this term has already been introduced in Chap. 12, Sect. 12.3.2, in connection with the rank of a matrix. Figure 14.1 illustrates these concepts pictorially.

The domain plays an important role in what is meant by the equality of two operators.

- Two operators are equal *if and only if* they have (i) the same domain, and (ii) the same action on each given vector in their common domain.

Inverse: When does an operator have an **inverse**? If an operator A maps *each* pair of *different* elements of $\mathbb{D}_A$ into a pair of *different* elements of Δ_A, then A has an inverse A^{-1}. This operator maps the elements of Δ_A into the elements of $\mathbb{D}_A$. Alternatively

- If, for *every* $|\psi\rangle \in \Delta_A$, there is a *unique* $|\phi\rangle \in \mathbb{D}_A$ such that $A\,|\phi\rangle = |\psi\rangle$, then A has an inverse. We then have

$$A^{-1}\,|\psi\rangle = |\phi\rangle \;\text{ if and only if }\; A\,|\phi\rangle = |\psi\rangle. \tag{14.2}$$

14.1.2 Linear Operators, Norm, and Bounded Operators

We are ready, now, to define a **linear operator** on an LVS. A is a linear operator if the two conditions below are satisfied:

(i) The domain $\mathbb{D}_A$ is a linear manifold. (You will find it helpful, at this point, to review the definition of a linear manifold given in Chap. 13, Sect. 13.3.2.)
(ii) For any two vectors $|\phi\rangle$ and $|\chi\rangle$ in $\mathbb{D}_A$, we have

$$A\,(\alpha\,|\phi\rangle + \beta\,|\chi\rangle) = \alpha\,A|\phi\rangle + \beta\,A|\chi\rangle, \tag{14.3}$$

where α and β are arbitrary scalars.

In this book, we shall be concerned only with linear operators.

The norm of a linear operator: The norm of a vector provides a measure of its "length". We now need a measure of how "large" an operator A is. Such a measure is provided by the **norm** of the operator. Take each vector $|\phi\rangle \in \mathbb{D}_A$, and find the norm of the vector $A\,|\phi\rangle = |\psi\rangle$. The value of this quantity depends, of course, on the norm of $|\phi\rangle$ itself. To eliminate this trivial dependence, divide by the norm of $|\phi\rangle$. The largest of the quantities thus obtained is a plausible measure of how "large" the operator A itself is. More precisely, the norm of an operator A is defined as

$$\boxed{\|A\| \stackrel{\text{def.}}{=} \sup_{|\phi\rangle\,\in\,\mathbb{D}_A} \frac{\|A\,\phi\|}{\|\phi\|} = \sup_{|\phi\rangle\,\in\,\mathbb{D}_A} \frac{\langle A\,\phi\,|\,A\,\phi\,\rangle^{1/2}}{\langle\,\phi\,|\,\phi\,\rangle^{1/2}}.} \tag{14.4}$$

Here "sup" stands for the *supremum* or the least upper bound. If a is any scalar, then the norm of the operator $a\,A$ is given by

$$\|a\,A\| = |a|\,\|A\|. \tag{14.5}$$

Bounded and unbounded operators: A linear operator A is a **bounded operator** if $\|A\| < \infty$. If the supremum in Eq. (14.4) is infinite, then A is an **unbounded operator**.

- In general, *differential operators over function spaces are unbounded operators.*

Several identities and relationships that hold good for finite-dimensional matrices are not necessarily valid for unbounded operators. Caution must therefore be exercised when dealing with such operators, which occur quite frequently in quantum mechanics. Here is an example. If A and B are $(n \times n)$ matrices, then $\text{Tr}\,(AB) = \text{Tr}\,(BA)$, whether or not A and B commute with each other. This property, called the **cyclic invariance of the trace**, is not necessarily valid when the matrices are infinite-dimensional.

This statement is illustrated by the following well-known example. A conjugate position–momentum pair in quantum mechanics satisfies the canonical commutation relation $qp - pq = i\hbar I$, where I is the unit operator. Take the trace of both sides of this relation. Setting $\text{Tr}\,(qp) = \text{Tr}\,(pq)$ immediately leads to a fallacy: the left-hand side vanishes while the right-hand side does not. The point is that q and p are *unbounded* operators, and *cannot* be represented by *finite*-dimensional matrices. We will find appropriate representations for these operators as differential operators and infinite-dimensional matrices, respectively, in Sects. 14.3.1 and 14.4.4.

Linear operators can be combined to produce other linear operators. Let A and B be linear operators in $\mathbb{V}$. The following important properties hold good:

(i) Any linear combination $C = aA + bB$ (where a and b are scalars) is also a linear operator, with domain $\mathbb{D}_C = \mathbb{D}_A \cap \mathbb{D}_B$.
(ii) The operators AB and BA are also linear operators.
(iii) If A and B are bounded linear operators in all of $\mathbb{V}$, then so are AB and BA.
(iv) The norm of a product of two operators satisfies the inequality

$$\|AB\| \leq \|A|\ \|B\|. \tag{14.6}$$

★ **1.** If A and B are linear operators in $\mathbb{V}$ with domains $\mathbb{D}_A$, $\mathbb{D}_B$ and ranges Δ_A, Δ_B, respectively, what are the domains of the operators AB and BA?

★ **2.** In $\mathcal{L}_2(-\infty, \infty)$, consider the linear operators A_1, A_2, A_3, and A_4 whose actions on any element $f(x)$ are as follows:

$$\text{(a) } A_1\, f(x) = \frac{d}{dx}\, f(x) \qquad \text{(b) } A_2\, f(x) = x^n\, f(x),\ n = 1, 2, \ldots$$
$$\text{(c) } A_3\, f(x) = e^{-x^2/2}\, f(x) \qquad \text{(d) } A_4\, f(x) = \frac{1}{(x^2+1)}\, f(x).$$

Which of the operators A_i are bounded operators?

★ **3.** Let $|\phi\rangle = (x_1, x_2, x_3, \ldots)$ be an arbitrary element of ℓ_2. Consider the operators A_i whose actions on $|\phi\rangle$ are as follows:

$$\begin{aligned}
\text{(i)}\quad & A_1|\phi\rangle = (2x_1, x_2, x_3, \ldots),\\
\text{(ii)}\quad & A_2|\phi\rangle = (0, x_1, x_2, x_3, \ldots),\\
\text{(iii)}\quad & A_3|\phi\rangle = (x_2, x_3, x_4, \ldots),\\
\text{(iv)}\quad & A_4|\phi\rangle = (x_1/1!, x_2/2!, x_3/3!, \ldots),\\
\text{(v)}\quad & A_5|\phi\rangle = (1!\,x_1, 2!\,x_2, 3!\,x_3, \ldots),\\
\text{(vi)}\quad & A_6|\phi\rangle = (x_1, x_1 + x_2, x_1 + x_2 + x_3, \ldots),\\
\text{(vii)}\quad & A_7|\phi\rangle = (1, x_1, x_2, x_3, \ldots).
\end{aligned}$$

(a) Identify the linear operators among the A_i.
(b) Write down matrix representations (in the natural basis) for each of the linear operators.
(c) What are the domains of the linear operators?
(d) Which of the linear operators are bounded? And what are their norms?
(e) Which of the linear operators have inverses?

14.2 The Adjoint of an Operator

Next, we turn to the important concept of the *adjoint* of a linear operator.

14.2.1 Densely Defined Operators

From the mathematical point of view, operators in finite-dimensional linear spaces can be handled in a straightforward manner. Very broadly speaking, this statement extends to the case of *bounded* operators in infinite-dimensional spaces. It is the case of *unbounded* operators that requires careful specification of many conditions and caveats. These involve going into functional analysis at a level beyond the scope of this book.

A fundamental aspect of an unbounded operator is that its domain may be a *subset* of the LVS, rather than all of it. Many technical complications follow as a consequence. As the intention is to keep things as simple as possible, let us confine our attention to the restricted situation in which two conditions are met. These restrictions enable us to make some rigorous statements that are also quite straightforward. Moreover, this situation is precisely the one that is of direct interest in physical applications, especially in quantum mechanics. The restrictions are as follows.

(i) The LVS is assumed to be a Hilbert space, rather than a general linear space. Recall that a Hilbert space is a complete linear vector space (the limits of all Cauchy sequences of vectors are elements of the space), on which an inner product of vectors is defined.
(ii) The second assumption is that the domain of any operator we consider is *dense* in the space: in other words, in the neighborhood of any element of the space, there is at least one vector belonging to the domain of the operator. Such an operator is said to be **densely defined** on the space concerned. Many of the technical difficulties associated with unbounded operators are not present for densely defined operators.

14.2.2 Definition of the Adjoint Operator

Let A be a linear operator acting on the vectors of a linear space $\mathbb{V}$. The inner product of any two arbitrary vectors is denoted by $\langle\chi|\,\psi\,\rangle$, as usual. Suppose A first acts on the vector $|\chi\rangle$, and then we take the inner product of the resulting vector with $|\psi\rangle$, to get $\langle A\,\chi|\psi\rangle$. The question is: Can we first act on the other vector $|\psi\rangle$ with some operator B, and then take the inner product with $\langle\chi|$, to get the same answer as before? This is the basic idea behind the adjoint of an operator.

Adjoint of an operator: If, for every pair of vectors $|\psi\rangle$ and $|\chi\rangle$ (in the domain of A), there exists an operator B such that

$$\langle\chi|\,A\,\psi\,\rangle = \langle\,B\,\chi\,|\psi\rangle, \tag{14.7}$$

then the operator B is said to be the **adjoint** of the operator A. In order to show that it is related to the particular operator A, it will be denoted by $A^{\dagger}$. The *defining* equation for the adjoint $A^\dagger$ of an operator A is therefore the relation[1]

$$\boxed{\langle\chi|\,A\,\psi\rangle = \langle A^\dagger\,\chi|\psi\rangle} \tag{14.8}$$

that must be satisfied for every pair of vectors in the domain of A. It follows that the adjoint of the equation $A\,|\psi\rangle = |\phi\rangle$ is $\langle\psi|\,A^\dagger = \langle\phi|$.

(i) In the case of a *finite-dimensional* LVS of n dimensions, in which operators are representable by $(n \times n)$ matrices, the adjoint $A^\dagger$ is nothing but the *Hermitian conjugate* (i.e., complex conjugate transpose) of the matrix A. (Based on the case of $(n \times n)$ matrices, the adjoint of a general operator is often loosely called its Hermitian conjugate in the physics literature.)
(ii) For operators that are not represented by finite-dimensional matrices, the adjoint must be found, in principle, from the defining relation, Eq. (14.8).
(iii) When A is an unbounded operator, the existence of its adjoint operator $A^\dagger$ is guaranteed, provided A is densely defined.
(iv) The adjoint of the adjoint of an operator is the operator itself, $(A^\dagger)^\dagger = A$.
(v) If c is a scalar, $(cA)^\dagger = c^*\,A^\dagger$, where c^* is the complex conjugate of c.
(vi) Most importantly, in general,

$$\boxed{\mathbb{D}_{A^\dagger} \supseteq \mathbb{D}_A.} \tag{14.9}$$

That is, the domain of $A^\dagger$ is generally larger than that of A, and contains the latter. Explicit examples of this fact will be given shortly, involving the derivative operator on a function space.

Note: The more conventional definition used in mathematics for the adjoint of an operator A is as follows: if, for every $|\psi\rangle,\ |\chi\rangle \in \mathbb{D}_A$, there exists an operator B such that

$$\langle A\,\chi\,|\psi\rangle = \langle\chi|\,B\,\psi\rangle, \tag{14.10}$$

then B is the adjoint of A. (Compare this with Eq. (14.7).) The identifying equation for $A^\dagger$ is therefore

$$\langle A\,\chi|\psi\rangle = \langle\chi|A^\dagger\,\psi\rangle, \tag{14.11}$$

rather than Eq. (14.8). As long as A acts in a finite-dimensional LVS, and is represented by an $(n \times n)$ matrix, it makes no difference whether we use (14.8) or (14.11) as the defining equation for the adjoint. It does make a difference, however, when A is an unbounded operator, and is represented (for instance) by a differential operator. Now, the definition in Eqs. (14.10)–(14.11) is consistent with the manner in which the inner product $\langle f\,|\,g\rangle$ is defined in the mathematics literature, namely, $\langle f\,|\,g\rangle = \int_{-\infty}^{\infty} dx\; f(x)\,g^*(x)$. In the physics literature, however, it is customary to

[1] There is an important matter of notation involved here. See the ***Note*** that follows below, after Eq. (14.9).

define this inner product as $\langle f \mid g \rangle = \int_{-\infty}^{\infty} dx\, f^*(x)\, g(x)$, as we have already done in Eq. (13.15) of Chap. 13, Sect. 13.2.2. With this definition, the appropriate definition of the adjoint is as given by Eqs. (14.7)–(14.8), and we shall stick to it.

14.2.3 Symmetric, Hermitian, and Self-adjoint Operators

An operator A is **self-adjoint** if $A = A^\dagger$. It is obvious that this is a natural definition of self-adjointness. But is important to recall here what is meant by the equality of two operators. As stated in Sect. 14.1.1, the two operators must have (i) the same domain, and (ii) the same action on each given vector in their common domain (see below).

For an operator represented by a finite-dimensional matrix, it is trivial to verify whether A is self-adjoint or not. All you have to do is to check whether the matrix A is Hermitian. In this case, "self-adjoint" is the same thing as "Hermitian". It is important to remember that this is *not* true for more general operators. Hermitian does *not* necessarily imply self-adjoint when the LVS is infinite-dimensional, and the operators are not representable as finite-dimensional matrices, and may not even be bounded operators.

In general, an operator A is self-adjoint if, and only if, *both* of the conditions listed below are satisfied:

(i) The operator $A^\dagger$ as identified from the relation $\langle A\,\chi|\psi\rangle = \langle\chi|\,A^\dagger\,\psi\rangle$ must be the same as the operator A. The operator is then said to be **symmetric**.

If a symmetric operator is also a *bounded* operator, then it is said to be a **Hermitian operator**. In physics, however, this distinction is often not made, and any symmetric operator is (loosely) called Hermitian.

(ii) Further, the *domains* of the two operators must be identical, i.e., we must have $\mathbb{D}_A = \mathbb{D}_{A^\dagger}$—whereas, in general, the domain $\mathbb{D}_A$ is contained in the larger domain $\mathbb{D}_{A^\dagger}$.

Only if both (i) and (ii) are satisfied can we say that A is self-adjoint, and write $A = A^\dagger$ as an *operator* equality.

Why is it necessary to make this distinction between symmetric (or "Hermitian") operators and self-adjoint operators? Apart from mathematical correctness, there is an important implication for quantum mechanics. It can be shown that all the eigenvalues of a self-adjoint operator are real. In elementary treatments of quantum mechanics, we learn that real physical observables are represented by Hermitian operators, and that these have only real eigenvalues. The more accurate statement is the following:

- In quantum mechanics, (real-valued) physical observables are represented by *self-adjoint* operators, and these have only real eigenvalues.

This distinction between the properties of symmetry and self-adjointness of linear operators (the latter is a stronger requirement) becomes relevant because many of the basic physical quantities in quantum mechanics, such as the position and momentum operators of particles, are unbounded operators. The momentum (or derivative) operator on a finite interval provides an explicit example of an operator that is symmetric but not self-adjoint, as you will see in Sect. 14.3.3. Another example, to be discussed in Sect. 14.3.6, is the radial component of the linear momentum of a particle moving in d-dimensional space, where $d \geq 2$.

14.3 The Derivative Operator in $\mathcal{L}_2$

14.3.1 The Momentum Operator of a Quantum Particle

In Chap. 13, Sect. 13.2.5, you have seen how the function space $\mathcal{L}_2$ makes its appearance in quantum mechanics: normalizable wave functions are $\mathcal{L}_2$ functions. The derivative operator, that acts on these wave functions, also appears naturally. It does so as a consequence of the fundamental canonical commutation relation between the self-adjoint operators representing a conjugate coordinate–momentum pair. For just a moment, let us indicate operators with an overhead caret, in order to make things clear. (This symbol will be omitted subsequently, because the occurrence of an operator will be clear from the context.) As I have already pointed out in Sect. 14.1.2, the fundamental commutation relation

$$\boxed{[\hat{x},\, \hat{p}] = \hat{x}\,\hat{p} - \hat{p}\,\hat{x} = i\hbar\,\hat{I},} \tag{14.12}$$

where $\hat{I}$ is the unit operator, immediately *rules out* the possibility of representing the position and momentum operators $\hat{x}$ and $\hat{p}$ of a particle by $(n \times n)$ matrices for any *finite* value of n. On the other hand, two representations that are consistent with the commutation relation are the following:

(i) Let the state vector $|\Psi\,(t)\rangle$ of a particle moving in one dimension be represented in the position basis by the position-space wave function $\psi(x\,,\, t) = \langle\, x\,|\Psi\,(t)\rangle$. Then, the position operator acting on $|\Psi\,(t)\rangle$ is represented by the variable x multiplying $\psi(x,\, t)$, while the momentum operator acting on $|\Psi\,(t)\rangle$ is represented by the derivative operator $-i\hbar\,\partial/\partial x$ acting on $\psi(x,\, t)$. That is,

$$\boxed{\langle\, x\,|\,\hat{x}\,|\,\Psi(t)\,\rangle = x\,\langle\, x\,|\,\Psi(t)\,\rangle = x\,\psi(x,\, t)} \tag{14.13}$$

and

$$\boxed{\langle\, x\,|\,\hat{p}\,|\Psi\,(t)\rangle = -\,i\hbar\,\frac{\partial}{\partial x}\,\langle\, x\,|\Psi\,(t)\rangle = -\,i\hbar\,\frac{\partial}{\partial x}\,\psi(x,\, t).} \tag{14.14}$$

(I have used a partial derivative with respect to x even in this one-dimensional case because $\psi(x, t)$ is a function of both x and t.)

(ii) On the other hand, let the state vector $|\Psi(t)\rangle$ be represented in the momentum basis by the momentum-space wave function $\widetilde{\psi}(p, t) = \langle p | \Psi(t)\rangle$. It is now the momentum operator acting on $|\Psi(t)\rangle$ that is represented by p multiplying $\widetilde{\psi}(p, t)$. The position operator acting on $|\Psi(t)\rangle$ is then represented by the derivative operator $+i\hbar\,\partial/\partial p$ acting on $\widetilde{\psi}(p, t)$. That is,

$$\boxed{\langle p \,|\, \hat{p} \,|\Psi(t)\rangle = p\,\langle p \,|\Psi(t)\rangle = p\,\widetilde{\psi}(p, t)} \tag{14.15}$$

and

$$\boxed{\langle p \,|\, \hat{x} \,|\Psi(t)\rangle = +i\hbar\,\frac{\partial}{\partial p}\,\langle p \,|\Psi(t)\rangle = +i\hbar\,\frac{\partial}{\partial p}\,\widetilde{\psi}(p, t).} \tag{14.16}$$

Note the plus sign in Eq. (14.16), in contrast to the minus sign in Eq. (14.14). This difference in sign arises from, and is necessary for consistency with, the commutation relation (14.12).

For a particle moving in two or more spatial dimensions, the partial derivatives are replaced by the corresponding gradient operators. Symbolically, therefore,

$$\boxed{\hat{\mathbf{r}} \rightarrow \mathbf{r}, \quad \hat{\mathbf{p}} \rightarrow -i\hbar\,\nabla_{\mathbf{r}} \quad \text{(position basis)},} \tag{14.17}$$

and

$$\boxed{\hat{\mathbf{p}} \rightarrow \mathbf{p}, \quad \hat{\mathbf{r}} \rightarrow +i\hbar\,\nabla_{\mathbf{p}} \quad \text{(momentum basis)}.} \tag{14.18}$$

In each case, these are *unique* representations of the operators concerned.

A proper understanding of Eqs. (14.13)–(14.18) is necessary in order to begin to understand quantum mechanics!

14.3.2 The Adjoint of the Derivative Operator in $\mathcal{L}_2(-\infty, \infty)$

The next question that arises is: What is the adjoint of the derivative operator d/dx in the linear space $\mathcal{L}_2(-\infty, \infty)$? We could take recourse to physics, and argue as follows: the momentum operator[2] of a particle moving in one dimension must be self-adjoint. But this operator is represented in the position basis by $-i\hbar\, d/dx$. Since i changes to $-i$ when the adjoint is taken, d/dx must change to $-d/dx$ so that p itself remains equal to its adjoint. Hence the adjoint of the operator d/dx must be $-d/dx$.

This heuristic reasoning is easily made more precise as follows. Consider any two vectors $|\,f\,\rangle$ and $|\,g\,\rangle$, represented in the function space $\mathcal{L}_2(-\infty, \infty)$ by the

[2] From now on, I will drop the overhead caret that we have used to denote operators. You can identify them from the context.

square-integrable functions $f(x)$ and $g(x)$, respectively. Equation (14.8) reads, in this case,

$$\boxed{\int_{-\infty}^{\infty} dx\, f^*(x)\,\big(A\,g(x)\big) = \int_{-\infty}^{\infty} dx\,\big(A^\dagger f(x)\big)^*\, g(x).} \tag{14.19}$$

In the case when $A = d/dx$, we have

$$\begin{aligned}\langle f\,|\,A\,g\rangle &= \int_{-\infty}^{\infty} dx\, f^*(x)\left(\frac{d}{dx}g(x)\right) \\ &= f^*(x)\,g(x)\Big|_{-\infty}^{\infty} - \int_{-\infty}^{\infty} dx\left(\frac{d}{dx}f^*(x)\right)g(x),\end{aligned} \tag{14.20}$$

on integrating by parts. But both $g(x)$ and $f(x)$ (and hence $f^*(x)$) must vanish as $x \to \pm\infty$, because they belong to $\mathcal{L}_2(-\infty\,,\,\infty)$. Therefore

$$\langle f\,|\,A\,g\rangle = \int_{-\infty}^{\infty} dx\left(-\frac{d}{dx}f(x)\right)^* g(x) \equiv \langle A^\dagger f\,|\,g\rangle. \tag{14.21}$$

This suggests that the adjoint we seek is in fact just $-d/dx$. It is trivial to see that the domain of d/dx is the same as that of $-d/dx$. In both cases, the domain is that subset of functions $f(x) \in \mathcal{L}_2(-\infty\,,\,\infty)$ whose first derivatives are also square-integrable, i.e., $f\,'(x) \in \mathcal{L}_2(-\infty\,,\,\infty)$. Hence we may read off the result

$$A = \frac{d}{dx} \quad\Rightarrow\quad A^\dagger = -\frac{d}{dx}\,. \tag{14.22}$$

Multiplying by $\pm i$, it follows that $\pm i\,d/dx$ is a self-adjoint operator in $\mathcal{L}_2(-\infty\,,\,\infty)$. Further, on functions that are differentiable n times,

$$\boxed{\left(\frac{d^n}{dx^n}\right)^\dagger = (-1)^n\left(\frac{d^n}{dx^n}\right).} \tag{14.23}$$

★ **4.** In the vector space $\mathcal{L}_2(-\infty\,,\,\infty)$ of square-integrable functions of a real variable x, show that the adjoint of the operator A is as indicated in each case:

(a) $A = x\,\dfrac{d}{dx} \quad\Rightarrow\quad A^\dagger = -\dfrac{d}{dx}\,x$

(b) $A = x^2 + \dfrac{d^2}{dx^2} \quad\Rightarrow\quad A^\dagger = x^2 + \dfrac{d^2}{dx^2}$

(c) $A = e^{i\,\xi x}$ ($\xi =$ real constant) $\quad\Rightarrow\quad A^\dagger = e^{-i\,\xi x}$

(d) $A = e^{i\,\eta\, d/dx}$ ($\eta =$ real constant) $\quad\Rightarrow\quad A^\dagger = e^{i\,\eta\, d/dx}$

(e) $A = x^m\,\dfrac{d^n}{dx^n}$ ($m,\,n =$ positive integers) $\quad\Rightarrow\quad A^\dagger = (-1)^n\,\dfrac{d^n}{dx^n}\,x^m$.

14.3.3 When Is $-i\,(d/dx)$ Self-adjoint in $\mathcal{L}_2[a\,,\,b]$?

We have seen that the operator $-i\,(d/dx)$ is self-adjoint in $\mathcal{L}_2(-\infty\,,\,\infty)$. Its domain is the set of square-integrable functions whose first derivatives are also square-integrable. In the context of quantum mechanics, the x-component of the momentum operator of a particle is thus represented, in the position basis, by the operator $-i\hbar\,(d/dx)$. What happens if the position is restricted to an interval $[a, b]$ where a and b are *finite* real numbers?

We must now work in the space $\mathcal{L}_2[a\,,\,b]$. Let $A = -i\,(d/dx)$. Consider, first, any two functions $f(x)$ and $g(x)$ belonging to $\mathcal{L}_2[a\,,\,b]$ whose first derivatives are also square-integrable in $[a,\,b]$. Then, integrating by parts (as usual),

$$\begin{aligned}\langle f\,|A\,g\rangle &= \int_a^b dx\; f^*(x)\Big(-i\frac{d}{dx}g(x)\Big)\\ &= \Delta(a,b) + \int_a^b dx\,\Big(-i\frac{d}{dx}f(x)\Big)^*\,g(x)\\ &= \Delta(a,b) + \langle A\,f\,|\,g\,\rangle,\end{aligned} \tag{14.24}$$

where

$$\Delta(a,b) = i\big[f^*(a)g(a) - f^*(b)g(b)\big]. \tag{14.25}$$

Therefore $\langle f\,|A\,g\,\rangle = \langle A\,f\,|\,g\,\rangle$ if, and only if, $\Delta(a,b)$ vanishes identically. This is ensured if we restrict ourselves to functions that vanish at the boundaries a and b, i.e., functions that satisfy the conditions $g(a) = g(b) = 0$ and $f(a) = f(b) = 0$. Then, over the domain

$$\mathbb{D}_A = \Big\{g(x) \in \mathcal{L}_2[a\,,\,b]\,\Big|\,g' \in \mathcal{L}_2[a\,,\,b],\ \text{and } g(a) = g(b) = 0\Big\}, \tag{14.26}$$

the operator A is *symmetric*.

And now for the crucial observation: note that it is *not necessary* to impose the additional boundary conditions $f(a) = f(b) = 0$ in order to derive this result! It *suffices* to stipulate the conditions $g(a) = g(b) = 0$. The immediate conclusion is that the domain of the adjoint $A^\dagger$ is larger than that of the operator A: it comprises *all* functions $f(x) \in \mathcal{L}_2[a\,,\,b]$ such that $f' \in \mathcal{L}_2[a\,,\,b]$, and is *not* restricted to those that further satisfy $f(a) = f(b) = 0$. Thus,

$$\mathbb{D}_{A^\dagger} = \Big\{f(x) \in \mathcal{L}_2[a\,,\,b]\,\Big|\,f' \in \mathcal{L}_2[a\,,\,b]\Big\} \supset \mathbb{D}_A. \tag{14.27}$$

We may therefore conclude that

- the operator $-i(d/dx)$ is symmetric, but not self-adjoint, over $\mathcal{L}_2[a\,,\,b]$.

14.3.4 Self-adjoint Extensions of Operators

Is there a way to make the operator $A = -i\,(d/dx)$ self-adjoint in some suitable subset of $\mathcal{L}_2[a\,,\,b]$? In order to achieve this, we must not only show that A is symmetric over some domain, but also ensure that its adjoint has the *same* domain. It is intuitively clear that, if this is at all possible, it must be done as follows: $\mathbb{D}_A$ must be *enlarged* beyond the overly-restrictive domain in (14.26), and $\mathbb{D}_{A^\dagger}$ must be *shrunk* from the overly-broad domain in (14.27), till the two of them match. As you might expect, the procedure involves manipulating the boundary conditions appropriately.

Consider, instead of the boundary conditions $g(a) = g(b) = 0$, the condition

$$g(b) = e^{i\theta}\,g(a), \tag{14.28}$$

where θ is an arbitrary real constant. The quantity $\Delta(a, b)$ defined in Eq. (14.25) then reduces to

$$\Delta(a, b) = i g(a)\big[f^*(a) - e^{i\theta} f^*(b)\big]. \tag{14.29}$$

This quantity vanishes identically if $e^{i\theta} f^*(b) = f^*(a)$, or

$$f(b) = e^{i\theta}\,f(a). \tag{14.30}$$

Note that we are *compelled* to select this boundary condition for $f(x)$, there being no freedom left in its choice. But Eq. (14.30) is precisely the boundary condition satisfied by $g(x)$, Eq. (14.28). The domains $\mathbb{D}_A$ and $\mathbb{D}_{A^\dagger}$ are therefore the same. Hence

- For every given value of the real constant θ, the operator $A = -i(d/dx)$ is self-adjoint over the domain

$$\boxed{\mathbb{D}_A = \mathbb{D}_{A^\dagger} = \Big\{h(x) = \mathcal{L}_2[a\,,\,b]\,\Big|\,h' \in \mathcal{L}_2[a\,,\,b],\ \text{and}\ h(b) = e^{i\theta}\,h(a)\Big\},} \tag{14.31}$$

for every value of the real parameter θ.

What we now have is a **self-adjoint extension** of the operator $-i\,(d/dx)$ over $\mathcal{L}_2[a\,,\,b]$. In fact, we have a whole *family* of such extensions, its members labeled by the value of the continuous real parameter θ.

14.3.5 Deficiency Indices

The boundary condition (14.28) is not the outcome of mere guesswork. The extension of a symmetric operator to a self-adjoint operator is an important problem in functional analysis, and there is a systematic approach to it. I will only touch upon it very cursorily and heuristically here, merely quoting the relevant results, without going into how they arise.

Let A be a symmetric operator over some domain. Whether it has a self-adjoint extension or not depends on the so-called **deficiency indices** n_+ and n_- associated with it. These are defined as the respective numbers of linearly independent solutions of the eigenvalue equations

$$A\,|\phi_+\rangle = i\,|\phi_+\rangle \quad \text{and} \quad A\,|\phi_-\rangle = -i\,|\phi_-\rangle. \tag{14.32}$$

Three possibilities then arise:

(i) If $n_+ = n_- = 0$, A is either self-adjoint or has a *unique* self-adjoint extension. In the latter case, A is said to be **essentially self-adjoint**.

(ii) If $n_+ = n_- = n$, where $n \neq 0$, there exist self-adjoint extensions of A over an appropriate domain. In fact, there exists an n-parameter *family* of such extensions.

(iii) If $n_+ \neq n_-$, there can be *no* self-adjoint extension of A.

I have already stated that no explanation will be given here of how these results come about. But it is worth adding a line indicating what is involved. The analysis hinges on the so-called **Cayley transform** of the operator A, defined as $(A - iI)(A + iI)^{-1}$. This is how the particular operator combinations in Eq. (14.32) arise.

★ **5.** The example that follows will help you understand how the idea of deficiency indices works. Use the foregoing to check out the self-adjointness properties of the operator $A = -i\,(d/dx)$ over the spaces

(i) $\mathcal{L}_2(-\infty\,,\,\infty)$ (the infinite line);
(ii) $\mathcal{L}_2[a\,,\,b]$, where a and b are finite (a finite interval);
(iii) $\mathcal{L}_2[0, \infty)$ (the semi-infinite line).

★ **6.** Consider the anticommutator

$$A_r = [x^r\,,\,p]_+ = x^r p + p x^r$$

over $\mathcal{L}_2(-\infty\,,\,\infty)$, where $p = -i\,(d/dx)$ is the momentum operator (setting $\hbar = 1$ for convenience) and r is a positive integer. Show that

(a) A_r is symmetric for every $r \geq 1$;
(b) A_1 is self-adjoint;
(c) A_r is self-adjoint for all even values of $r \geq 2$; and
(d) A_r has no self-adjoint extension for all odd values of $r \geq 3$.

14.3.6 The Radial Momentum Operator in $d \geq 2$ Dimensions

The radial component of the momentum operator of a quantum mechanical particle provides an instructive physical application of the foregoing. We set $\hbar = 1$ in this section for simplicity of notation.

Consider a particle moving in d-dimensional space, where $d \geq 2$. Results corresponding to the physically interesting cases $d = 2$ and $d = 3$ can be read off from the general formula to be derived below. As we have seen in Eq. (14.17), the linear momentum of the particle is represented in the position basis by $-i\,\nabla_{\mathbf{r}}$. Can we then conclude that the *radial* component of the momentum, p_r, is represented by $-i\,\partial/\partial r$, since the radial component of the gradient operator is just $\partial/\partial r$?

The answer is "no", because there is a problem of noncommutativity involved here. *Classically*, the radial component of the momentum is given by $p_r = \mathbf{e}_r \cdot \mathbf{p} = \mathbf{p} \cdot \mathbf{e}_r$, where $\mathbf{e}_r = \mathbf{r}/r$. In quantum mechanics, however, $\mathbf{e}_r \cdot \mathbf{p} \neq \mathbf{p} \cdot \mathbf{e}_r$, because the Cartesian components of $\mathbf{p}$ do not commute with the corresponding components of $\mathbf{r}/r$. A linear combination of the form

$$p_r = \alpha\,(\mathbf{e}_r \cdot \mathbf{p}) + (1 - \alpha)\,(\mathbf{p} \cdot \mathbf{e}_r)$$

suggests itself as a possible definition of p_r. The requirement that p_r be a symmetric operator then implies that α must be equal to $\frac{1}{2}$, leading to the definition

$$\boxed{p_r \stackrel{\text{def.}}{=} \tfrac{1}{2}\left(\frac{\mathbf{r}}{r} \cdot \mathbf{p} + \mathbf{p} \cdot \frac{\mathbf{r}}{r}\right)} \tag{14.33}$$

for the radial momentum operator. Using the position-space representation $\mathbf{p} \to -i\,\nabla_{\mathbf{r}}$ for the momentum operator, the position-space representation of the radial momentum operator is then given by

$$\boxed{p_r \to -i\left(\frac{d-1}{2r} + \frac{\partial}{\partial r}\right).} \tag{14.34}$$

The range of r is $0 \leq r < \infty$, and the measure of integration over r is $r^{d-1}\,dr$. We are therefore concerned with the space of functions $h(r) \in \mathcal{L}_2[0, \infty)$ such that

$$\int_0^\infty dr\, r^{d-1}\, |h(r)|^2 < \infty. \tag{14.35}$$

The radial component of the momentum is a real physical observable. The operator p_r must therefore either be self-adjoint, or essentially self-adjoint, or at least have a self-adjoint extension (or extensions).

★ **7.** p_r turns out to be an essentially self-adjoint operator over $\mathcal{L}_2[0, \infty)$. The weight factor r^{d-1} plays a crucial role in establishing this result, as you will see.

(a) Show that $-i\,(\partial/\partial r)$ is *not* a symmetric operator.

(b) Verify that Eq. (14.33) yields Eq. (14.34) for p_r, on using $\mathbf{p} \to -i\,\nabla_{\mathbf{r}}$.
(c) Show that p_r is a symmetric operator in a certain domain.
(d) Show that there is a unique self-adjoint extension of p_r, so that it is an essentially self-adjoint operator.
(e) Corroborate this result by showing that the deficiency indices of p_r are $n_+ = 0$ and $n_- = 0$.

14.4 Nonsymmetric Operators

We know that physical observables correspond to self-adjoint operators (or their self-adjoint extensions) in quantum mechanics. But this does not mean that other operators are without any use at all. They occur quite frequently, and have interesting properties. Remember that the eigenvalues of a non-self-adjoint operator need not be real, in general. The example that follows shows some of the possibilities in this regard.

14.4.1 The Operators $x \pm ip$

A simple but important example of non-self-adjoint (in fact, nonsymmetric and hence non-Hermitian) operators already occurs in the elementary quantum mechanics of a particle moving in one dimension. Let us work in suitable units such that both the position x and the momentum p have the physical dimensions of $\sqrt{\hbar}$. It is tedious to carry around the factor $(i\hbar)$ in the canonical commutation relation (14.12) between x and p. This factor is easily eliminated by defining the linear combinations

$$\boxed{a \stackrel{\text{def.}}{=} \frac{x + ip}{\sqrt{2\hbar}} \quad \text{and (hence)} \quad a^{\dagger} \stackrel{\text{def.}}{=} \frac{x - ip}{\sqrt{2\hbar}}\,.} \tag{14.36}$$

You would very likely have come across the operators a and $a^{\dagger}$ in the context of the quantum mechanical linear harmonic oscillator. But these linear combinations of x and p can be defined in general, and are not restricted to the particular problem of the oscillator.[3] In terms of these nonsymmetric operators, the fundamental canonical commutation relation becomes

$$\boxed{[a,\, a^{\dagger}] = I.} \tag{14.37}$$

[3] What the particular problem of the harmonic oscillator does is to provide two "dimensionful" parameters—the mass m and the natural frequency ω of the oscillator. These parameters, in combination with $\hbar$, provide natural scales of length and momentum in the oscillator problem, proportional to $(\hbar/m\omega)^{1/2}$ and $(m\hbar\omega)^{1/2}$, respectively.

In the position basis, the representations for the operators a and $a^\dagger$ are of course

$$a = \frac{1}{\sqrt{2\hbar}}\left(x + \hbar\,\frac{d}{dx}\right) \quad \text{and} \quad a^\dagger = \frac{1}{\sqrt{2\hbar}}\left(x - \hbar\,\frac{d}{dx}\right), \tag{14.38}$$

respectively. It is easy to derive the following results:

(i) The operator a has a continuous infinity of complex eigenvalues and corresponding normalizable eigenfunctions.
(ii) In contrast, its adjoint $a^\dagger$ has no normalizable eigenfunctions at all.

★ **8.** The representations in Eqs. (14.38) can be used to establish these results.

(a) Show that a has an eigenvalue 0, with a normalizable eigenvector $|\Phi_0\rangle$. Let $\langle x \mid \Phi_0\rangle = \Phi_0(x)$ be the corresponding eigenfunction in the position basis. The eigenvalue equation $a\,|\Phi_0\rangle = 0$ is given in this basis by the differential equation

$$\frac{1}{\sqrt{2\hbar}}\left(x + \hbar\frac{d}{dx}\right)\Phi_0(x) = 0.$$

Show that this equation has the normalized Gaussian solution

$$\Phi_0(x) = (\pi\hbar)^{-1/4}\,e^{-x^2/(2\hbar)}.$$

(b) The operator $a^\dagger$ does not have have a similar property, i.e., it has no normalizable eigenfunction with 0 as the eigenvalue. Working in the position basis, show that the differential equation

$$\frac{1}{\sqrt{2\hbar}}\left(x - \hbar\frac{d}{dx}\right) f(x) = 0$$

does not have a solution satisfying the condition $\int_{-\infty}^{\infty} dx\,|f(x)|^2 < \infty$.

(c) Remarkably enough, *every* complex number z is an eigenvalue (!) of the operator a. Show that, for any complex number z, the eigenvalue equation

$$\frac{1}{\sqrt{2\hbar}}\left(x + \hbar\,\frac{d}{dx}\right)\psi_z(x) = z\,\psi_z(x)$$

has a normalizable solution

$$\psi_z(x) = C(z)\,\exp\left(-\frac{x^2}{2\hbar} + \sqrt{\frac{2}{\hbar}}\,z\,x\right),$$

where $C(z)$ is a normalization constant. More will be said about these eigenfunctions subsequently.

(d) Now consider the adjoint operator $a^\dagger$. Suppose it has an eigenvalue λ. Working in the position basis, the corresponding eigenfunction $f_\lambda(x)$ must satisfy the differential equation

$$\frac{1}{\sqrt{2\hbar}}\left(x - \hbar\,\frac{d}{dx}\right) f_\lambda(x) = \lambda\, f_\lambda(x).$$

Show that $a^\dagger$ has *no* normalizable eigenfunctions at all.

(e) Check out that exactly the same conclusions as above are reached if you use the momentum basis, in which

$$a = \frac{i}{\sqrt{2\hbar}}\left(p + \hbar\,\frac{d}{dp}\right) \quad \text{and} \quad a^\dagger = \frac{i}{\sqrt{2\hbar}}\left(-\,p + \hbar\,\frac{d}{dp}\right).$$

14.4.2 Oscillator Ladder Operators and Coherent States

The results just derived may be familiar to you in the context of the operator formalism for the quantum mechanical linear harmonic oscillator. But I reiterate that they are based solely on the canonical commutation relation $[x,\ p] = i\hbar\, I$. In the units used here (corresponding to setting $m = 1$ and $\omega = 1$, where m is the mass of the oscillator and ω is its natural frequency), the Hamiltonian of the oscillator is

$$\boxed{H = \tfrac{1}{2}\,(x^2 + p^2) = \hbar\,\big(a^\dagger\, a + \tfrac{1}{2} I\big).} \tag{14.39}$$

In this context, a and $a^\dagger$ are the well-known lowering and raising operators, or **ladder operators** for the linear harmonic oscillator. In the position basis and the momentum basis, respectively, the oscillator Hamiltonian is represented by

$$H = \frac{1}{2}\left(x^2 - \hbar^2\,\frac{d^2}{dx^2}\right) \quad \text{and} \quad H = \frac{1}{2}\left(p^2 - \hbar^2\,\frac{d^2}{dp^2}\right). \tag{14.40}$$

The self-adjoint operator

$$\boxed{N \stackrel{\text{def.}}{=} a^\dagger\, a} \tag{14.41}$$

is easily shown to have the infinite spectrum of discrete eigenvalues $0,\ 1,\ 2,\ \ldots,$ if we insist that the corresponding eigenstates $|\,0\,\rangle,\ |1\rangle,\ |2\rangle,\ \ldots$ have finite norms. That is,

$$\boxed{N\,|n\rangle = n\,|n\rangle, \quad \text{where} \quad n = 0, 1, 2, \ldots.} \tag{14.42}$$

Orthonormalizing the eigenstates, we have

$$\langle n\,|\,n'\,\rangle = \delta_{nn'}\,. \tag{14.43}$$

It is obvious that each $|n\rangle$ is also an eigenstate of the oscillator Hamiltonian H, corresponding to the eigenvalue

$$E_n = \hbar\left(n + \tfrac{1}{2}\right). \tag{14.44}$$

(Remember that we have chosen units such that $\omega = 1$.) The ground state energy $E_0 = \frac{1}{2}\hbar$ is, of course, the familiar zero-point energy of the oscillator. As you know, the action of a and $a^\dagger$ on the normalized eigenstate $|n\rangle$ is given by

$$\boxed{a\,|0\rangle = 0, \quad a\,|n\rangle = \sqrt{n}\,|n-1\rangle, \quad a^\dagger\,|n\rangle = \sqrt{(n+1)}\,|n+1\rangle.} \tag{14.45}$$

This is why a and $a^\dagger$ are called lowering and raising operators.

There is a close link between the Hamiltonian of the linear harmonic oscillator and that of a single-mode radiation field. Although we will not enter into a discussion of quantum field theory here, it is helpful to mention that this connection arises when the electromagnetic field is *quantized*.

- As a consequence of field quantization, $|n\rangle$ can also be interpreted as a state in which there are n photons (of a given frequency and polarization).

For this reason, N is called the (photon) **number operator**. a and $a^\dagger$ are the (photon) **annihilation operator** and **creation operator**, respectively. The state $|n\rangle$ is called a **Fock state**. The ground state $|0\rangle$ then corresponds to the zero-photon or **vacuum state** of the radiation field. I will use this terminology on occasion. But note the following: in the context of radiation, the self-adjoint linear combinations

$$x = \sqrt{(\hbar/2)}\,(a + a^\dagger) \quad \text{and} \quad p = i\sqrt{(\hbar/2)}\,(a^\dagger - a) \tag{14.46}$$

do *not* represent the "position" or "linear momentum" of a photon. Rather, they represent certain physical observables pertaining to the radiation field, and go by the general name of **quadratures**.

The eigenfunction $\Phi_0(x)$ found earlier, namely,

$$\boxed{\Phi_0(x) = (\pi\hbar)^{-1/4}\, e^{-x^2/(2\hbar)},} \tag{14.47}$$

is just the wave function in the position basis of the ground state $|0\rangle$ of the oscillator. In other words,

$$\Phi_0(x) \stackrel{\text{def.}}{=} \langle x\,|\,0\rangle. \tag{14.48}$$

The eigenfunctions of the operator a found in the foregoing, namely,

$$\psi_z(x) = C(z)\,\exp\left(-\frac{x^2}{2\hbar} + \sqrt{\frac{2}{\hbar}}\,z\,x\right), \tag{14.49}$$

where z is any complex number, are the position-space wave functions corresponding to the so-called **coherent states** of the harmonic oscillator (or of the radiation field), in the x-basis. That is,

$$\psi_z(x) \stackrel{\text{def.}}{=} \langle x \,|\, z \rangle, \tag{14.50}$$

where the coherent state $|\, z \,\rangle$ is an eigenstates of a with (complex) eigenvalue z:

$$\boxed{a \,|\, z \rangle = z \,|\, z \rangle, \quad z \in \mathbb{C}.} \tag{14.51}$$

With one exception, the states $|\, z \,\rangle$ are *not* eigenstates of the oscillator Hamiltonian H. (The eigenstates of H are, of course, the Fock states $\{|\, n \rangle\}$.) The only exception is the case $z = 0$: the functions $\Phi_0(x)$ and $\psi_0(x)$ are the same, i.e.,

$$\Phi_0(x) \equiv \psi_0(x). \tag{14.52}$$

(Remember that the subscript 0 in Φ_0 refers to $n = 0$, while the same subscript in ψ_0 refers to $z = 0$!) Coherent states (and their generalizations) play a prominent role in quantum optics. We will return to them in Chap. 15, Sect. 15.4.1. The normalization factor $C(z)$ in Eq. (14.49) is determined, and the full wave function $\psi_z(x)$ is written down explicitly, in Eqs. (16.52)–(16.54) of Chap. 16, Sect. 16.2.4. I will consider a different aspect of the coherent state $|\, z \,\rangle$, related to its role as a state describing coherent radiation, in Chap. 19, Sect. 19.2.2.

To summarize:

(i) The operators x and $p = -i\hbar\,(d/dx)$ are unbounded, self-adjoint operators over $\mathcal{L}_2(-\infty\,,\,\infty)$. They have no eigenfunctions in this space.

(ii) The operator $a = (x + ip)/\sqrt{2\hbar}$ is an unbounded nonsymmetric operator over $\mathcal{L}_2(-\infty\,,\,\infty)$. It has a *continuous complex spectrum*, namely, all complex numbers z with real and imaginary parts lying in $(-\infty,\,\infty)$. The corresponding eigenfunctions (which have finite $\mathcal{L}_2$ norms, of course) are given by Eq. (14.49) in the x-basis.

(iii) Its adjoint, the operator $a^\dagger = (x - ip)/\sqrt{2\hbar}$, is also an unbounded nonsymmetric operator over $\mathcal{L}_2(-\infty\,,\,\infty)$. But it has no eigenvectors of finite norm, and hence an empty spectrum.

(iv) The number operator $N = a^\dagger a$ is an unbounded self-adjoint operator over $\mathcal{L}_2(-\infty\,,\,\infty)$. It has a *denumerably infinite, real, nonnegative spectrum* comprising the integers $0,\,1,\,\ldots$. The corresponding eigenvectors, normalized to unity, comprise the set $\{|\, n \,\rangle\}$ of Fock states. This set provides a denumerable, orthonormal basis in the Hilbert space. The latter is therefore a separable Hilbert space. The general state $|n\rangle$ has the position-space wave function $\Phi_n(x) = \langle\, x \,|\, n \,\rangle$. It will be written down in full in Eq. (16.47) of Chap. 16, Sect. 16.2.5, after we discuss Hermite polynomials. The extension of the foregoing statements to the linear harmonic oscillator Hamiltonian $H = \hbar\left(N + \frac{1}{2}I\right)$ is trivial, because the term $\frac{1}{2}I$ merely adds $\frac{1}{2}$ to the eigenvalues of N.

(v) Although a and N do *not* commute with each other, they share a common eigenstate, the zero-eigenvalue state $|\, 0 \,\rangle$. In the language of wave functions, this corresponds to the common eigenfunction $\Phi_0(x) = \psi_0(x)$. As I have pointed out in Chap. 12, Sect. 12.4.1, there is nothing to forbid this from happening. But a and N certainly cannot have a *complete set* of common eigenstates, of course. In fact, they have no common eigenstate other than $|0\rangle$.

14.4.3 Eigenvalues and Non-normalizable Eigenstates of *x* and *p*

The position and momentum operators x and p have no normalizable eigenstates, i.e., they have no eigenfunctions in $\mathcal{L}_2(-\infty\,,\,\infty)$. Each of these operators has, however, a continuous spectrum of eigenvalues. This is just the set of all real numbers $(-\infty, \infty)$. The corresponding sets of eigenstates $\{|\,x\,\rangle\}$ and $\{|\,p\,\rangle\}$ constitute *continuous* basis sets in the Hilbert space, as discussed in Chap. 13, Sects. 13.2.2 and 13.2.5. What follows below is the one-dimensional counterpart of the relations that have already been written down, or are implicit, in Chap. 13, Sect. 13.2.5. I repeat some of this material because it is necessary to understand it thoroughly. In order to make matters as clear as possible, the subscripts "pos" and "mom" have been used in Eqs. (14.54)–(14.57).

The eigenstates $\{|\,x\,\rangle\}$ and $\{|\,p\,\rangle\}$ satisfy δ-function orthonormality conditions, the one-dimensional versions of the relevant relations in Eqs. (13.23) and (13.26):

$$\langle x\,|\,x\,'\rangle = \delta(x - x\,') \quad \text{and} \quad \langle p\,|\,p\,'\rangle = \delta(p - p\,'). \tag{14.53}$$

It is obvious that these eigenstates do not have a finite $\mathcal{L}_2$ norm. The position-space wave function corresponding to a position eigenvalue x_0 with eigenstate $|x_0\rangle$ is a δ-function,

$$\psi_{\text{pos}}(x) \stackrel{\text{def.}}{=} \langle x\,|\,x_0\rangle = \delta(x - x_0), \tag{14.54}$$

while the associated momentum-space wave function is a plane wave:

$$\widetilde{\psi}_{\text{pos}}(p) \stackrel{\text{def.}}{=} \langle p\,|\,x_0\rangle = (2\pi\hbar)^{-1/2}\, e^{-ipx_0}. \tag{14.55}$$

Similarly, the momentum-space wave function corresponding to a momentum eigenvalue p_0 with eigenstate $|\,p_0\rangle$ is a δ-function,

$$\widetilde{\psi}_{\text{mom}}(p) \stackrel{\text{def.}}{=} \langle p\,|\,p_0\rangle = \delta(p - p_0), \tag{14.56}$$

while the associated position-space wave function is a plane wave:

$$\psi_{\text{mom}}(x) \stackrel{\text{def.}}{=} \langle x\,|\,p_0\rangle = (2\pi\hbar)^{-1/2}\, e^{ixp_0}. \tag{14.57}$$

Now, the square of a δ-function does not exist, and the integrals

$$\int_{-\infty}^{\infty} dp\, |\widetilde{\psi}_{\text{pos}}(p)|^2 \quad \text{and} \quad \int_{-\infty}^{\infty} dx\, |\psi_{\text{mom}}(x)|^2$$

are obviously divergent. Therefore, neither a δ-function nor a plane wave is an element of $\mathcal{L}_2(-\infty, \infty)$. But we can superpose each of these types of wave functions to form normalizable **wave packets** in position space or in momentum space that are centered about some given value x_0 or p_0, respectively. The most commonly used forms are **Gaussian wave packets**, i.e., packets whose shape is given by a Gaussian function. In Chap. 18, Sect. 18.1.5, we will consider an interesting property of these wave packets. They will be studied in greater detail in Chap. 30, Sect. 30.4.2, in connection with the spreading of a quantum mechanical wave packet as a function of time.

The foregoing illustrates an important fact concerning operators in general:

- The eigenvalue spectrum of an operator is dependent on the *class* of eigenfunctions considered admissible.

Finally, let us turn to the second of the two comments promised in Chap. 12, at the end of Sect. 12.4.1: the assertion is that, if two (self-adjoint) operators commute with each other, they share a common complete set of eigenstates (in some specified LVS).

- But this does not mean that *every* eigenstate of either one of two commuting operators is necessarily an eigenstate of the other!

Here is a simple example to clarify the point. Consider the linear harmonic oscillator Hamiltonian H, given by Eq. (14.39). It is invariant under the parity transformation $x \mapsto -x$. In other words: let P be the parity operator, whose action on any position-space wave function is given by

$$P\,\psi(x, t) = \psi(-x, t). \tag{14.58}$$

By the statement, "H is invariant under the parity transformation", we mean that

$$P^{-1}\,H\,P = H \tag{14.59}$$

itself. That is,

$$H\,P = P\,H, \quad \text{i.e., the commutator } [H,\ P] = 0. \tag{14.60}$$

The eigenfunctions of H, the position-space wave functions $\Phi_n(x) \equiv \langle x\,|\,n\rangle$, form a complete orthonormal set in $\mathcal{L}_2(-\infty, \infty)$. As you know from elementary quantum mechanics, the function $\Phi_n(x)$ is essentially a Hermite polynomial of order n, multiplied by a Gaussian factor.[4] Each of these wave functions is *also* an eigenfunction of the parity operator, because it is either an even function or an odd function of x.

---

[4]As I have mentioned earlier, the exact expression for $\Phi_n(x)$ is given in Eq. (16.47) of Chap. 16, Sect. 16.2.5.

On the other hand it is obvious that, although any *arbitrary* even function or odd function of x is an eigenfunction of P, it is not necessarily an eigenfunction of H.

But then, you might argue, the space of admissible functions must be specified first. In this case it is $\mathcal{L}_2(-\infty\,,\,\infty)$. Even so, *all* square-integrable functions of a definite parity are certainly not eigenfunctions of H. For instance, $\Phi_0(x) + \Phi_2(x)$ is an even function, but it is not an eigenfunction of H. All that can be asserted is that H and P share a common complete set of eigenfunctions in $\mathcal{L}_2(-\infty\,,\,\infty)$. In this instance, this complete set is the set of eigenfunctions of H. Moreover, every even function belonging to the space $\mathcal{L}_2(-\infty\,,\,\infty)$ can be expanded uniquely[5] in terms of the set $\{\Phi_{2n}(x)\}$; similarly, every odd function in this space can be expanded uniquely in terms of the set $\{\Phi_{2n+1}(x)\}$, where $n = 0,\ 1,\ 2,\ \ldots$.

14.4.4 Matrix Representations for Unbounded Operators

We have seen that the commutation relation $[x,\ p] = i\hbar I$ cannot be satisfied by finite-dimensional matrix representations of x and p. But these operators *can* be represented by *infinite*-dimensional matrices. Consider, for instance, the Hermitian matrices

$$x = \sqrt{\frac{\hbar}{2}} \begin{pmatrix} 0 & \sqrt{1} & 0 & 0 & \cdot\cdot \\ \sqrt{1} & 0 & \sqrt{2} & 0 & \cdot\cdot \\ 0 & \sqrt{2} & 0 & \sqrt{3} & \cdot\cdot \\ 0 & 0 & \sqrt{3} & 0 & \cdot\cdot \\ \cdot & \cdot & \cdot & \cdot & \cdot\cdot \end{pmatrix} \tag{14.61}$$

and

$$p = -i\sqrt{\frac{\hbar}{2}} \begin{pmatrix} 0 & \sqrt{1} & 0 & 0 & \cdot\cdot \\ -\sqrt{1} & 0 & \sqrt{2} & 0 & \cdot\cdot \\ 0 & -\sqrt{2} & 0 & \sqrt{3} & \cdot\cdot \\ 0 & 0 & -\sqrt{3} & 0 & \cdot\cdot \\ \cdot & \cdot & \cdot & \cdot & \cdot\cdot \end{pmatrix}. \tag{14.62}$$

Equivalently, the matrices corresponding to a and $a^\dagger$ are

$$a = \begin{pmatrix} 0 & \sqrt{1} & 0 & 0 & \cdots \\ 0 & 0 & \sqrt{2} & 0 & \cdots \\ 0 & 0 & 0 & \sqrt{3} & \cdots \\ \cdot & \cdot & \cdot & \cdot & \cdots \\ \cdot & \cdot & \cdot & \cdot & \cdots \end{pmatrix}, \quad a^\dagger = \begin{pmatrix} 0 & 0 & 0 & \cdots\cdot \\ \sqrt{1} & 0 & 0 & \cdots\cdot \\ 0 & \sqrt{2} & 0 & \cdots\cdot \\ 0 & 0 & \sqrt{3} & \cdots\cdot \\ \cdot & \cdot & \cdot & \cdots\cdot \end{pmatrix}. \tag{14.63}$$

[5] As a linear combination of the eigenfunctions concerned, with unique expansion coefficients.

★ **9.** Verify that the commutation relations $[x,\ p] = i\hbar\, I$ and $[a,\ a^\dagger] = I$ are satisfied by the infinite-dimensional matrices in Eqs. (14.61), (14.62) and (14.63), respectively. What is the *basis* in which the foregoing explicit matrix representations have been written down?

Finally, note how the same operator can have very different-looking representations when it acts on elements of different linear spaces. In the position basis, the action of the position operator is just multiplication by x, while the momentum operator is the derivative operator $-i\hbar\, d/dx$. They are self-adjoint operators over $\mathcal{L}_2(-\infty\,,\ \infty)$. And now we find that, over ℓ_2, the very same operators are represented by infinite-dimensional matrices!

14.5 Solutions

1. $\mathbb{D}_{AB} = \mathbb{D}_A \cap \Delta_B,\quad \mathbb{D}_{BA} = \mathbb{D}_B \cap \Delta_A.$ ▶

2. (a), (b) It is clear that the domains of A_1 and A_2 cannot be the whole of the LVS $\mathcal{L}_2(-\infty\,,\ \infty)$. Rather, the domains are those subsets of the LVS for which $f\,'(x)$ and $x^n\, f(x)$ are themselves square-integrable. The corresponding norms squared are given, respectively, by the suprema of

$$\frac{\int_{-\infty}^{\infty} dx\, |f\,'(x)|^2}{\int_{-\infty}^{\infty} dx\, |f(x)|^2} \quad \text{and} \quad \frac{\int_{-\infty}^{\infty} dx\, x^{2n}\, |f(x)|^2}{\int_{-\infty}^{\infty} dx\, |f(x)|^2},$$

as $f(x)$ runs over the domains of the operators concerned. These quantities can be as large as one pleases. Hence A_1 and A_2 must be unbounded operators.
(c), (d) The squared norms of A_3 and A_4 are given by the suprema of

$$\frac{\int_{-\infty}^{\infty} dx\, e^{-x^2}\, |f(x)|^2}{\int_{-\infty}^{\infty} dx\, |f(x)|^2} \quad \text{and} \quad \frac{\int_{-\infty}^{\infty} dx\, (x^2+1)^{-2}\, |f(x)|^2}{\int_{-\infty}^{\infty} dx\, |f(x)|^2},$$

respectively. Since both e^{-x^2} and $(x^2+1)^{-2}$ are positive definite functions that decay monotonically, as $|x|$ increases, from a maximum value of 1 at $x = 0$, it is clear that each of these squared norms is bounded by unity in any case. A_3 and A_4 are therefore bounded operators. ▶

3. (a) Let $|\phi\rangle = (x_1,\ x_2, \ldots)$ and $|\chi\rangle = (y_1,\ y_2, \ldots)$ be any two vectors in ℓ_2. You have to ensure that the condition Eq. (14.3) is satisfied in each case, i.e., that $A_i\,(\alpha\,|\phi\rangle + \beta\,|\chi\rangle) = \alpha\, A_i|\phi\rangle + \beta\, A_i|\chi\rangle$. It is straightforward to verify that all the operators except A_7 satisfy this condition. A_7 is not a *linear* operator!
(b) The matrices representing the operators A_1 to A_6 are (obviously) infinite-dimensional. Let a_{jk} denote the (jk)th matrix element of the matrix, in each case. Then

(i) $a_{11} = 2,\ a_{jj} = 1$ for $j \geq 2,\ a_{jk} = 0$ for $j \neq k$ (diagonal matrix).

(ii) $a_{jk} = \delta_{j,k+1}$. Each element in the diagonal just below the principal diagonal is equal to 1. Every other element is 0.
(iii) $a_{jk} = \delta_{j+1,k}$. Each element in the diagonal just above the principal diagonal is equal to 1. Every other element is 0.
(iv) $a_{jk} = \delta_{jk}/j!$ (diagonal matrix).
(v) $a_{jk} = j!\,\delta_{jk}$ (diagonal matrix).
(vi) $a_{jk} = 1$ for all $j \geq k$, and $a_{jk} = 0$ for all $j < k$. The matrix is a lower triangular one, with every element equal to 1 on and below the principal diagonal.

(c) In cases (i), (ii), (iii), and (iv), $\mathbb{D}_{A_i}$ is all of ℓ_2. (v) Since $A_5|\phi\rangle$ must remain in ℓ_2, it is clear that $\mathbb{D}_{A_5}$ is restricted to those sequences in ℓ_2 that fall off rapidly enough to ensure the convergence of $\sum_{n=1}^{\infty}(n!)^2\,|x_n|^2$. That is, $|x_n|$ must vanish as $n \to \infty$ more rapidly than $1/(n!\sqrt{n})$. (vi) It is clear that $\mathbb{D}_{A_6}$ comprises only those sequences of ℓ_2 for which the modulus of the sum $(x_1 + \cdots + x_n)$ itself vanishes faster than $n^{-1/2}$ as $n \to \infty$. This means, among other things, that there must be cancelations between the different terms of the sum.
(d) Cases (i), (ii), (iii), and (iv) are bounded operators, with $\|A_i\| = 1$. Cases (v) and (vi) are unbounded operators, since $\langle A_i\,\phi \mid A_i\,\phi\rangle$ can be arbitrarily large.
(e) (i) A_1 has an inverse. The matrix representing A_1^{-1} is also a diagonal matrix, with its first element equal to $\frac{1}{2}$ and all the other diagonal elements equal to 1. We have $A_1^{-1}(y_1,\, y_2,\, \ldots) = \left(\frac{1}{2}y_1,\, y_2,\, \ldots\right)$. (ii), (iii) A_2 and A_3 do not have inverses. (iv), (v) A_4 and A_5 are inverses of each other. (vi) It is evident that A_6^{-1} must be that operator whose action on any vector is given by $A_6\,(y_1,\, y_2,\, y_3,\, \ldots) = (y_1,\, y_2 - y_1,\, y_3 - y_2,\, \ldots)$. Thus A_6^{-1} is represented by a lower triangular matrix with each diagonal element equal to 1, each element on the diagonal below the principal diagonal equal to -1, and all other elements equal to 0. ▶

4. Use the definition of the adjoint as given by Eq. (14.8). In $\mathcal{L}_2(-\infty\,,\,\infty)$, this means that you must identify $A^\dagger$ using the relation (14.19). In each case, integrate the left-hand side by parts the required number of times, to bring it to the form given on the right-hand side. ▶

5. Equations (14.32) are, in this case,

$$-i\,(d/dx)\phi_\pm(x) = \pm i\,\phi_\pm(x).$$

The respective solutions are $\phi_+(x) \propto e^{-x}$ and $\phi_-(x) \propto e^{x}$.
(i) It is obvious that neither e^x nor e^{-x} is square-integrable in $(-\infty, \infty)$. Therefore $n_+ = n_- = 0$ in this case. Hence, as you know already,

- $-i\,(d/dx)$ is self-adjoint over $\mathcal{L}_2(-\infty\,,\,\infty)$.

(ii) In $\mathcal{L}_2[a\,,\,b]$ (where a and b are finite), each of the solutions e^{-x} and e^x is square-integrable. Hence $n_+ = n_- = 1$. Therefore, as we have just seen,

- $-i\,(d/dx)$ has a one-parameter family of self-adjoint extensions over $\mathcal{L}_2[a\,,\,b]$. Each value of the real parameter θ in Eq. (14.31) represents a member of the family.

(iii) The solution e^{-x} is square-integrable in $[0, \infty)$, but the solution e^{x} is not. Hence $n_+ = 1$ and $n_- = 0$.

- Since $n_+ \neq n_-$, $-i\,(d/dx)$ has no self-adjoint extension over $\mathcal{L}_2[0, \infty)$.

This statement is borne out by a direct calculation. Going back to Eqs. (14.24) and (14.25) with $a = 0$ and $b = \infty$, we must have $g(\infty) = 0$, because $\int_0^\infty dx\,|g(x)|^2$ must be finite, since $g(x)$ belongs to $\mathcal{L}_2[0, \infty)$. That leaves $\Delta(0, \infty) = if^*(0)g(0)$. But this quantity vanishes identically when $g(0) = 0$, leaving no freedom to select the boundary condition to be satisfied by $f(x)$ at $x = 0$. In other words, there is no possibility of a self-adjoint extension of the operator $-i\,(d/dx)$ in the function space $\mathcal{L}_2[0, \infty)$. ▶

6. (a) is easily established in a straightforward manner.
(b) The eigenvalue equations (14.32) for the operator A_r are given by

$$\frac{d\phi_+}{dx} = -\Big(\frac{r}{2x} + \frac{1}{2x^r}\Big)\,\phi_+ \quad\text{and}\quad \frac{d\phi_-}{dx} = -\Big(\frac{r}{2x} - \frac{1}{2x^r}\Big)\,\phi_-,$$

respectively. For $r = 1$, these reduce to

$$d\phi_+/dx = -\phi_+/x \quad\text{and}\quad d\phi_-/dx = 0,$$

with solutions $\phi_+ \propto x$ and $\phi_- =$ constant. As neither of these is square-integrable in $(-\infty, \infty)$, we have $n_+ = 0$, $n_- = 0$. Hence A_1 is self-adjoint.
(c), (d) When $r \geq 2$, the eigenvalue equations for $\phi_\pm$ have solutions

$$\phi_+(x) \propto x^{-r/2} \exp\big[1/\{2(r-1)x^{r-1}\}\big]$$

and

$$\phi_-(x) \propto x^{-r/2} \exp\big[-1/\{2(r-1)x^{r-1}\}\big],$$

respectively. But $\phi_+(x)$ diverges strongly as $x \to 0$ from the positive side, and is not square-integrable for any for any $r \geq 2$. Hence $n_+ = 0$. Similarly, when $r - 1$ is an odd integer (i.e., when r is an even integer), f_- diverges strongly when $x \to 0$ from the negative side, and hence is not square-integrable. Therefore n_- is also equal to 0. This implies that A_r is essentially self-adjoint for even values of $r \geq 2$. On the other hand, when $r - 1$ is an even integer, i.e., when r is an *odd* integer ≥ 3, f_- is well-behaved (in fact, vanishes rapidly) as $x \to 0$ both from above and from below. Hence $n_- = 1$ for odd values of $r \geq 3$. Since $n_+ \neq n_-$ in this case, the operator A_r has no self-adjoint extension when $r = 3, 5, \ldots$. ▶

7. The first step is to put bounds on the possible behavior of the class of functions involved, based on the square-integrability condition (14.35). Simple power counting places limits on how slowly any function $h(r)$ belonging to the function space can decay as $r \to \infty$, and how singular it can possibly be at $r = 0$. If $|h(r)| \sim 1/r^\alpha$ as $r \to \infty$, then the convergence of the integral $\int^\infty dr\, r^{d-1}/r^{2\alpha}$ requires that

$$2\alpha - d + 1 > 1, \quad \text{or} \quad \alpha > \tfrac{1}{2}d.$$

In other words, $h(r)$ must vanish as $r \to \infty$ *more rapidly* than $1/r^{d/2}$. Similarly, if $|h(r)| \sim 1/r^{\beta}$ as $r \to 0$, then the convergence of the integral $\int_0 dr\, r^{d-1}/r^{2\beta}$ requires that

$$2\beta - d + 1 < 1, \quad \text{or} \quad \beta < \tfrac{1}{2}d.$$

That is, $h(r)$ must be *less singular* than $1/r^{d/2}$ at the origin. (It can, of course, be finite at that point.) So we start with the class of functions meeting these requirements. Let us denote this function space by $\mathcal{V}$ for brevity.

(a) Consider the quantity

$$-i \int_0^{\infty} dr\, r^{d-1}\, f^*(r)\, \frac{\partial g(r)}{\partial r}.$$

Integrating by parts, it is immediately clear that the presence of the weight factor r^{d-1} makes it impossible for the operator $-i\,(\partial/\partial r)$ to be symmetric.

(b) Use the formula for the divergence of a scalar times a vector, and the fact that $\nabla \cdot \mathbf{r} = d$.

(c) Consider the integral

$$\int_0^{\infty} dr\, r^{d-1}\, f^*(r)\, [p_r\, g(r)] = \int_0^{\infty} dr\, r^{d-1}\, f^*(r)\Big[-i\,\Big(\frac{d-1}{2r} + \frac{\partial}{\partial r}\Big) g(r)\Big].$$

After integration by parts and a little simplification, this quantity becomes equal to

$$\Delta + \int_0^{\infty} dr\, r^{d-1}\Big[-i\,\Big(\frac{d-1}{2r} + \frac{\partial}{\partial r}\Big) f(r)\Big]^* g(r) = \Delta + \int_0^{\infty} dr\, r^{d-1}\, [p_r\, f(r)]^*\, g(r),$$

where

$$\Delta = -i\Big[r^{d-1}\, f^*(r)\, g(r)\Big]_{r=0}^{\infty}.$$

Therefore, *provided* Δ vanishes identically, we have

$$\int_0^{\infty} dr\, r^{d-1}\, f^*(r)\, [p_r\, g(r)] = \int_0^{\infty} dr\, r^{d-1}\, [p_r\, f(r)]^*\, g(r),$$

establishing that p_r is a symmetric operator in a certain domain. Now, since both f and g vanish more rapidly than $r^{-d/2}$ as $r \to \infty$, the contribution to Δ from the upper limit ($r = \infty$) is zero. This leaves us with

$$\Delta = i \lim_{r \to 0} \Big[r^{d-1}\, f^*(r)\, g(r)\Big].$$

But f and g are elements of $\mathcal{V}$, which permits them to be as singular as $r^{-(d/2)+\epsilon}$ at the origin, where ϵ is a positive number that can be arbitrarily small (but not zero). In such cases Δ does not vanish. On the other hand, by restricting $g(r)$ to functions that are not so singular at $r = 0$ as those allowed in $\mathcal{V}$, we can make Δ vanish, ensuring that p_r is a symmetric operator. Specifically, $g(r)$ must be less singular than $r^{1-d/2}$ at the origin. The domain of p_r, however, is now a subset of $\mathcal{V}$ while that of its adjoint is $\mathcal{V}$, since the function $f(r)$ can be any function in all of $\mathcal{V}$. Thus, p_r is symmetric, but not self-adjoint.

(d) The expression for Δ suggests how this quantity can be made to vanish while ensuring that the domains of p_r and its adjoint are the *same* subset of $\mathcal{V}$. All that needs to be done is to impose the *same* restriction on $g(r)$ and $f(r)$ as $r \to 0$, such that $\Delta = 0$. Obviously, this is ensured if the $r \to 0$ behavior of these functions is less singular than $r^{-(d-1)/2}$. This subset of $\mathcal{V}$ is then the common domain of the symmetric operator p_r and its adjoint, and we have a unique self-adjoint extension: p_r is *essentially self-adjoint.*

(e) The deficiency indices of the operator p_r given by Eq. (14.34) are easily found. The solution to

$$p_r\,\phi_-(r) = -i\phi_-(r)$$

is

$$\phi_-(r) \propto e^r/r^{(d-1)/2}.$$

But this function does not satisfy the square-integrability condition (14.35) because it diverges as $r \to \infty$. Hence $n_- = 0$.

On the other hand, the solution to

$$p_r\,\phi_+(r) = +i\phi_+(r)$$

is

$$\phi_+(r) \propto e^{-r}/r^{(d-1)/2}.$$

This function satisfies the square-integrability condition (14.35). But it is as singular as $r^{-(d-1)/2}$ at the origin. The common domain of p_r and its adjoint, however, comprises functions that must be *less* singular than $r^{-(d-1)/2}$ at $r = 0$, as we have seen above. Therefore n_+ is also equal to 0. We are then guaranteed that p_r, if not already self-adjoint (it is not), is essentially self-adjoint. ▶

8. Each part of this problem involves the solution of a simple first-order differential equation of the general form

$$\psi'(x) \pm (k + x)\psi(x) = 0,$$

where k is a constant. This equation can be integrated in a straightforward manner. The solution either involves a factor $e^{-x^2/2}$, which makes it normalizable, or it involves a factor $e^{x^2/2}$, which makes it non-normalizable. ▶

9. Consider the corresponding matrix representation for the self-adjoint operator N. Note that it is a diagonal matrix, that is,

$$N = a^\dagger a = \text{diag}\,(0,\ 1,\ 2,\ \ldots).$$

The basis in the LVS must therefore be the set of eigenvectors of N. Observe that this is just the natural basis. In other words, we are working in a basis in which the Fock states have the representations

$$|\,0\,\rangle = \big(1\;0\;0\cdots\big)^{\mathrm{T}},\quad |\,1\,\rangle = \big(0\;1\;0\cdots\big)^{\mathrm{T}},\quad \cdots.$$

Remark: The operator $a\,a^\dagger$ is just $a^\dagger\,a + I$. Hence it is also given by a diagonal matrix in this basis: $a\,a^\dagger = \text{diag}\,(1,\ 2,\ 3,\ \ldots)$. ▶

Chapter 15
Operator Algebras and Identities

Sets of linear operators on a vector space are often related to each other by an interesting mathematical structure: they comprise an *algebra*, such as the Lie algebra of the generators of rotations (the angular momentum algebra) that has already been encountered in Chap. 11, Sect. 11.3.1. We have also touched upon Lie algebras briefly in Chap. 12, Sect. 12.4.2. In this chapter, I begin with a discussion of the basic algebra satisfied by the ladder operators a and $a^\dagger$ introduced in Chap. 14, Sect. 14.4.1, and some other algebras associated with higher powers of these operators. I shall then consider a number of operator identities that are extremely useful in calculations involving non-commuting operators. Following that, I turn to a few physical applications of these results, and discuss some of the relevant properties of the groups $SO(2)$, $SO(3)$, $SU(2)$ and $SU(1, 1)$ that are intimately connected to these applications.

15.1 Operator Algebras

15.1.1 The Heisenberg Algebra

What is remarkable is the following fact:

- All the results and conclusions arrived at in Chap. 14, from the beginning of Sect. 14.4 onwards, are consequences of a single operator relation: namely, the commutation relation $[x,\ p] = i\hbar\, I$, or, equivalently, $[a,\ a^\dagger] = I$.

When the commutators of a set of operators are linear combinations of operators belonging to the same set, we have an **operator algebra**. You have already encountered examples of Lie algebras in Chap. 12, Sect. 12.4.2. And now we have another example, generated by the operators $a,\ a^\dagger$ and the unit operator I, the binary operation being the commutator of two operators. The commutation relations in this instance are

V. Balakrishnan, *Mathematical Physics*,
https://doi.org/10.1007/978-3-030-39680-0_15

$$\boxed{[a,\,a^\dagger] = I,\ [a,\,I] = 0,\ [a^\dagger,\,I] = 0.} \tag{15.1}$$

It is easily checked that these operators form a Lie algebra, the so-called **canonical commutator algebra**. More generally, let q_j and p_j (where $1 \le j \le n$) be the conjugate position–momentum pairs of a system with n degrees of freedom. Then the canonical commutator algebra is

$$[q_j,\,q_k] = 0,\ [p_j,\,p_k] = 0,\ [q_j,\,p_k] = i\hbar\,\delta_{jk}\,I,\ [q_j,\,I] = 0,\ [p_j,\,I] = 0. \tag{15.2}$$

As stated in Sect. 12.4.2, Lie groups are associated with Lie algebras; the elements of the group are obtained by exponentiating the generators of the algebra. In the present instance, the general element of the group concerned is of the form $\exp\,(t_1\,a + t_2\,a^\dagger + t_3\,I)$, where the t_i are scalar constants. We will encounter exponentials of this kind in Sect. 15.4.1, in connection with coherent states.

The algebra in Eq. (15.1) is an example of a **Heisenberg algebra**. The term "Heisenberg algebra" is actually used in a somewhat wider sense in mathematics. In its simplest form, it refers to a Lie algebra with three basic elements or generators $\alpha,\ \beta$, and γ, and a binary operation denoted by $\circ$, satisfying the following relations:

$$\alpha \circ \beta = \gamma,\ \alpha \circ \gamma = 0,\ \beta \circ \gamma = 0. \tag{15.3}$$

In order to see what an explicit representation of the algebra (15.3) looks like, identify $\alpha,\ \beta$, and γ with the three matrices

$$T_1 = \begin{pmatrix} 0 & 1 & 0 \\ 0 & 0 & 0 \\ 0 & 0 & 0 \end{pmatrix},\quad T_2 = \begin{pmatrix} 0 & 0 & 0 \\ 0 & 0 & 1 \\ 0 & 0 & 0 \end{pmatrix} \quad\text{and}\quad T_3 = \begin{pmatrix} 0 & 0 & 1 \\ 0 & 0 & 0 \\ 0 & 0 & 0 \end{pmatrix}, \tag{15.4}$$

respectively, and the binary operation $\circ$ with the commutator. Then the relations corresponding to (15.3) are

$$[T_1,\,T_2] = T_3,\ [T_1,\,T_3] = 0,\ [T_2,\,T_3] = 0. \tag{15.5}$$

The matrices T_i generate the Lie algebra of strictly upper triangular (3×3) matrices, with the commutator of two matrices as the binary operation. (Strictly upper triangular means that all the elements on and below the principal diagonal are identically zero.) The elements of the corresponding group are obtained by exponentiating the linear combination $\sum_{i=1}^{3} t_i\,T_i$, where the t_i are scalar parameters. This yields the general element

$$g(t_1,\,t_2,\,t_3) = e^{t_1\,T_1 + t_2\,T_2 + t_3\,T_3} = \begin{pmatrix} 1 & t_1 & \frac{1}{2}t_1\,t_2 + t_3 \\ 0 & 1 & t_2 \\ 0 & 0 & 1 \end{pmatrix}. \tag{15.6}$$

We see that $\det\, g = 1$. The inverse of g is given by

$$g^{-1}(t_1, t_2, t_3) = \begin{pmatrix} 1 & -t_1 & \frac{1}{2}t_1 t_2 - t_3 \\ 0 & 1 & -t_2 \\ 0 & 0 & 1 \end{pmatrix}. \qquad (15.7)$$

Note that the elements of the Lie *algebra* are strictly upper triangular matrices (all diagonal elements equal to zero), while those of the Lie *group* are triangular matrices with each diagonal element equal to unity.

Upper triangular matrices of the form (15.6) constitute the **Heisenberg group** H_3 under matrix multiplication. The subscript is meant to indicate that it comprises (3×3) matrices. There are generalizations to higher odd dimensions, denoted by H_{2n+1}, and other generalizations as well. All these objects are of interest in representation theory and quantum mechanics.

★ **1.** Consider the Heisenberg group of (3×3) upper triangular matrices.

(a) Start with the three matrices T_i in Eq. (15.4), and verify the commutation relations (15.5). Check that the set $\{T_i\}$ forms a Lie algebra.
(b) Show that $(t_1 T_1 + t_2 T_2 + t_3 T_3)^3 \equiv 0$ (the null matrix) for all values of the parameters t_i. Hence verify that the general matrix $g = e^{t_1 T_1 + t_2 T_2 + t_3 T_3}$ reduces to the form given by the final equation in (15.6).
(c) Let $\mathbf{t}$ and $\mathbf{t}'$ stand for the sets of parameters (t_1, t_2, t_3) and (t'_1, t'_2, t'_3), respectively. Show that the product of any two upper triangular matrices $g(\mathbf{t})$ and $g(\mathbf{t}')$ is again an upper triangular matrix of the same form.
(d) Is the Heisenberg group commutative,[1] i.e., do its elements satisfy $g(\mathbf{t})\, g(\mathbf{t}') = g(\mathbf{t}')\, g(\mathbf{t})$?
(e) Verify that the inverse of the matrix $g(\mathbf{t})$ is given by Eq. (15.7).
(f) What is the nullity of the general matrix $g(\mathbf{t})$ in Eq. (15.6), and what is its rank?
(g) How many independent eigenvectors does $g(\mathbf{t})$ have? Is it a defective matrix, or can it be diagonalized by a similarity transformation?

★ **2.** Let us return to the ladder operators a and $a^\dagger$ satisfying the fundamental commutation relation $[a, a^\dagger] = I$.

(a) Find the commutators $[a^n, a^\dagger]$ and $[a, a^{\dagger n}]$, where n is any positive integer.
(b) Hence show that $e^a\, a^\dagger\, e^{-a} = a^\dagger + I$ and $e^{-a^\dagger}\, a\, e^{a^\dagger} = a + I$.

15.1.2 Some Other Basic Operator Algebras

Besides the Heisenberg algebra (15.1) involving a, $a^\dagger$, and I, there are other algebras associated with these operators.

★ **3.** Here are a couple of examples.

---

[1]Let g_1 and g_2 be any two distinct elements of a group. The group is commutative or **abelian** if the elements $g_1 \circ g_2$ and $g_2 \circ g_1$ are the same.

(a) Show that the operators $a,\ a^{\dagger},\ N\ (= a^{\dagger}\, a)$, and I form a closed algebra under commutation. (The operator $a\, a^{\dagger}$ is not an independent one, as it is just $N + I$.)
(b) Hence find the commutators $[a^n,\ N]$, $[e^a,\ N]$, $[N,\ a^{\dagger\, n}]$, and $[N,\ e^{a^{\dagger}}]$.
(c) Remarkably enough, operators that are *quadratic* in a and $a^{\dagger}$ also form an algebra. Show that the operators $a^2,\ a^{\dagger\, 2}$, and $N + \frac{1}{2}I$ form a closed algebra under commutation. This algebra is of considerable significance, as you will see shortly.

The Lie algebra $\mathfrak{su}(2)$: Recall the algebra of the generators $J_1,\ J_2$, and J_3 of rotations (i.e., the angular momentum algebra) in three-dimensional space, given by Eq. (11.8) of Chap. 11, Sect. 11.1.2. Writing out the commutators explicitly,

$$[J_1,\, J_2] = i\, J_3,\ \ [J_2,\, J_3] = i\, J_1,\ \ [J_3,\, J_1] = i\, J_2\,. \tag{15.8}$$

As usual, a general element of the rotation group $SO(3)$ is given by the exponential of a linear combination of these three generators: you have already seen, in Chap. 11, Sect. 11.3.2, that a rotation about any direction $\mathbf{n}$ by an angle ψ is given by $R(\mathbf{n},\, \psi) = e^{i\, (\mathbf{J}\cdot\mathbf{n})\, \psi}$. See Eqs. (11.28) and (11.32).

I have mentioned earlier that the angular momentum algebra (15.8) is also the Lie algebra $\mathfrak{su}(2)$ of the generators of the Lie group $SU(2)$, the group of unimodular, unitary (2×2) matrices. The general form of a matrix belonging to $SU(2)$ has been deduced earlier, in Eq. (12.18) of Chap. 12, Sect. 12.2.3. We will see how such a matrix also represents a general rotation in three-dimensional space, and connect up the groups $SO(3)$ and $SU(2)$, in Sects. 15.3.2 and 15.3.3 below.

As you may know from the quantum mechanical theory of angular momentum, it is useful to define ladder operators that are the analogs of the raising and lowering operators a and $a^{\dagger}$, according to

$$\boxed{J_{\pm} \stackrel{\text{def.}}{=} J_1 \pm i\, J_2\,.} \tag{15.9}$$

The angular momentum algebra (15.8) may then be rewritten in the form

$$\boxed{[J_+,\, J_-] = 2J_3,\ \ [J_+,\, J_3] = -J_+,\ \ [J_-,\, J_3] = J_-\,.} \tag{15.10}$$

This alternative way of expressing the algebra of the generators is very useful, for instance, in determining the eigenvalues and eigenstates of J_i and $\mathbf{J}^2$.

The Lie algebra $\mathfrak{su}(1,\, 1)$: Now consider the three operators

$$\boxed{K_+ = \tfrac{1}{2}a^2,\ \ K_- = \tfrac{1}{2}a^{\dagger\, 2},\ \ K_3 = \tfrac{1}{2}\big(a^{\dagger}\, a + \tfrac{1}{2}I\big).} \tag{15.11}$$

From the results derived in the foregoing, we see that these operators satisfy the commutation relations

$$\boxed{[K_+,\, K_-] = 2K_3,\ \ [K_+,\, K_3] = K_+,\ \ [K_-,\, K_3] = -K_-\,.} \tag{15.12}$$

Compare these relations with those in Eq. (15.10). The two sets of commutators only differ in the sign of the right-hand side in the second and third commutators. The algebra in (15.12) is the Lie algebra $\mathfrak{su}(1, 1)$, and the operators $K_\pm$ and K_3 are the generators of the **pseudo-unitary Lie group** $SU(1, 1)$. This group occurs, for instance, in classical mechanics, optics, and also in quantum optics, in connection with **squeezed states**. We will consider a class of squeezed states briefly in Sect. 15.4.2. More will be said about $SU(1, 1)$ and $\mathfrak{su}(1, 1)$ in Sects. 15.4.4 and 15.4.5. The group $SU(1, 1)$ will be encountered once again in Chap. 27, Sect. 27.4.3.

***An infinite-dimensional algebra*:** The first powers of the operators a and $a^\dagger$ (together with I) form an algebra, and so do their second powers. It is natural to ask: does this remain true for higher powers of a and $a^\dagger$?

Interestingly, a closed algebra with a *finite* number of generators is no longer possible once we go beyond the second power of the operators. When operators of the form $a^n\, a^{\dagger\, m}$ (where $n + m \geq 3$) are included, increasingly higher powers of these operators occur in the commutators. Consider, for instance, the commutator $[a^n,\, a^{\dagger\, n}]$ for a general positive integer $n \geq 2$. The answer can be expressed as a polynomial in the number operator N, given by

$$\boxed{[a^n,\, a^{\dagger\, n}] = n!\, I + \sum_{r=1}^{n-1} \frac{(n!)^2}{(r!)^2 (n-r)!}\, N(N-I)(N-2I)\cdots\big(N-(r-1)I\big).} \tag{15.13}$$

As you have already seen, when $n = 2$ we have $[a^2,\, a^{\dagger\, 2}] = 4(N + \frac{1}{2}I)$. When $n = 3$,

$$[a^3,\, a^{\dagger\, 3}] = 9N^2 + 9N + 6I. \tag{15.14}$$

The first term on the right-hand side is of the *fourth* order in the operators a and $a^\dagger$. In fact, once $n + m \geq 3$, the algebra satisfied by the operators $a^n\, a^{\dagger\, m}$ involves the entire set of operators for *all* nonnegative integer values of n and m. This makes the algebra infinite-dimensional.

★ **4.** Verify Eq. (15.13) using the method of induction.

15.2 Useful Operator Identities

Let A and B be linear operators in an LVS with a common domain. In general, A and B may not commute with each other, that is, $[A, B] \equiv AB - BA \neq 0$. There exist several operator identities that are useful in this situation. Some of these follow. In order to avoid having to specify the domains of the individual operators that arise, I shall assume that all the operators involved are elements of some Lie algebra. This is the situation that occurs most commonly in physical applications of the identities considered.

15.2.1 Perturbation Series for an Inverse Operator

Frequently, we encounter the problem of finding the inverse of an operator of the form $(A + \epsilon\, B)$, where the inverse A^{-1} of A exists and is known, and ϵ is a scalar of sufficiently small magnitude such that $\|\epsilon\, B\| \ll \|A\|$, or $|\epsilon| \ll \|A\|/\|B\|$. Here, $\|A\|$ and $\|B\|$ are the norms of the operators A and B.[2] We can then write a so-called **perturbation series** for the inverse $(A + \epsilon\, B)^{-1}$. It is a convergent infinite series in powers of ϵ:

$$\boxed{(A + \epsilon\, B)^{-1} = A^{-1} - \epsilon\, A^{-1}\, B\, A^{-1} + \epsilon^2\, A^{-1}\, B\, A^{-1}\, B\, A^{-1} + \cdots .} \tag{15.15}$$

Note that the inverse of B itself does not appear in this formula, *nor is it required to exist*. We will come across an example of such a formula in Chap. 32, Sect. 32.1.5, in the iterative solution to an integral equation. Essentially the same idea is used to write down, in Sect. 32.2.5, the perturbative solution for the scattering amplitude in the problem of the scattering of a quantum mechanical particle by a potential.

★ **5.** Derive the expansion in Eq. (15.15).

15.2.2 Hadamard's Lemma

Hadamard's lemma is a remarkable and extremely useful formula that expresses the operator $e^{\lambda A}\, B\, e^{-\lambda A}$, where λ is a scalar parameter (real or complex), as a power series in λ, with coefficients that are multiple nested commutators of A with B. The formula is

$$\boxed{e^{\lambda A}\, B\, e^{-\lambda A} = B + \lambda\, [A,\, B] + \frac{\lambda^2}{2!}\, [A,\, [A,\, B]] + \frac{\lambda^3}{3!} \Big[A,\, \big[A,\, [A,\, B]\big]\Big] + \cdots} \tag{15.16}$$

If any of the multiple commutators on the right-hand side should happen to be zero, the series terminates.

You can derive Eq. (15.16) as follows. Consider the operator

$$F(\lambda) = e^{\lambda A}\, B\, e^{-\lambda A} \tag{15.17}$$

as a function of λ. Differentiate both sides with respect to λ, paying attention to the fact that A and F do not commute with each other, in general. We have

$$F'(\lambda) = A\, F(\lambda) - F(\lambda)\, A. \tag{15.18}$$

[2] The norm of a linear operator has been defined in Eq. (14.4) of Chap. 14, Sect. 14.1.2.

Now, $F(\lambda)$ is a regular function of λ that can be expanded in a Taylor series in powers of λ, with coefficients that are operators, of course. Thus

$$F(\lambda) = F(0) + \lambda\, F'(0) + \frac{\lambda^2}{2!}\, F''(0) + \cdots . \tag{15.19}$$

The existence of such a power series expansion for $F(\lambda)$ implies the existence of an analogous series for its derivative $F'(\lambda)$ as well.[3] Hence

$$F'(\lambda) = F'(0) + \lambda\, F''(0) + \frac{\lambda^2}{2!}\, F'''(0) + \cdots . \tag{15.20}$$

★ **6.** Use these expressions to derive Eq. (15.16).

***Special case* (i):** $\big[A, [A, B]\big] = 0$. Hadamard's lemma simplifies considerably in the important special case when A commutes with the commutator $[A, B]$. The infinite series in Eq. (15.16) then terminates after the first two terms, and we have

$$\boxed{e^{\lambda A}\, B\, e^{-\lambda A} = B + \lambda\, C, \quad \text{if } [A, C] = 0 \text{ where } C = [A, B].} \tag{15.21}$$

The identity $e^{a}\, a^{\dagger}\, e^{-a} = a^{\dagger} + I$ obtained in Sect. 15.1.1 follows at once from Eq. (15.21), because $[a, a^{\dagger}]$ is just the unit operator, with which all operators commute.

***Special case* (ii):** $[A, B] = \alpha A$, where α is a scalar. This case is obviously just a subcase of the one above. We have, in this instance,

$$\boxed{e^{\lambda A}\, B\, e^{-\lambda A} = B + \lambda \alpha A.} \tag{15.22}$$

An example is provided by $A = a,\ B = a^{\dagger} a$, where $[a, a^{\dagger}] = I$ (in this case $\alpha = 1$). Another trivial example (in the space of infinitely differentiable functions of x) is $A = e^{x},\ B = d/dx$ (in this case $\alpha = -1$).

***Special case* (iii):** $[A, B] = \beta B$, where β is a scalar. Once again, the right-hand side of Hadamard's lemma simplifies to yield

$$\boxed{e^{\lambda A}\, B\, e^{-\lambda A} = e^{\lambda \beta}\, B, \quad \text{if } [A, B] = \beta B.} \tag{15.23}$$

★ **7.** These and similar special cases may be established directly, without using the Hadamard formula itself. Deduce (a) Eq. (15.21) and (b) Eq. (15.23) directly.

---

[3]This assertion follows from the fact that $F(\lambda)$ is an *analytic* function of the complex variable λ—in this instance, for all finite values of $|\lambda|$. The point here is that the derivative of an analytic function is also an analytic function. We will discuss analytic functions of a complex variable in Chaps. 22–27.

15.2.3 Weyl Form of the Canonical Commutation Relation

Going back to the general formula (15.16), replace B by e^B in it. Then

$$e^{\lambda A}\, e^B\, e^{-\lambda A} = e^B + \lambda\,[A,\, e^B] + \frac{\lambda^2}{2!}\,[A,\,[A,\, e^B]] + \frac{\lambda^3}{3!}\,[A,\,[A,\,[A,\, e^B]]] + \cdots \tag{15.24}$$

This formula leads to a very useful result in the special case when *both* A and B commute with their commutator. This is

$$e^{\lambda A}\, e^B\, e^{-\lambda A} = e^B\, e^{\lambda C}, \quad \text{if } [A,\, C] = [B,\, C] = 0 \text{ where } C = [A,\, B]. \tag{15.25}$$

★ **8.** Establish Eq. (15.25).

It follows from (15.25) that

$$\boxed{e^A\, e^B = e^C\, e^B\, e^A, \quad \text{if } [A,\, C] = [B,\, C] = 0 \text{ where } C = [A,\, B].} \tag{15.26}$$

An important application of the identity (15.26) is the case $A = i\,\xi\,x,\;\; B = i\,\eta\,p$ where ξ and η are real scalars and $x,\; p$ are the conjugate position and momentum operators in one dimension. We then have

$$\boxed{e^{i\,\xi\,x}\, e^{i\,\eta\,p} = e^{-i\hbar\xi\eta}\, e^{i\,\eta\,p}\, e^{i\,\xi\,x}.} \tag{15.27}$$

This is called the **Weyl form** of the canonical commutation relation. Note the factor $e^{-i\hbar\xi\eta}$ on the right-hand side: The appearance of such a "phase factor" when a product of operators is written in a different order is a feature typical of quantum mechanics.

15.2.4 The Zassenhaus Formula

The exponentials of operators occur everywhere. It is often required to find the exponential of the sum (or a linear combination) of operators that do not commute with each other. For instance, the elements of Lie groups are generally exponentials of linear combinations of the generators of its Lie algebra, as in the case of the matrix $e^{t_k\, T_k}$ in Eq. (15.6), corresponding to an element of the Heisenberg group (repeated indices are to be summed over). Another example that you have encountered already is the operator $e^{i\,(\mathbf{J}\cdot\mathbf{n})\,\psi} = e^{i\,J_k\,n_k\,\psi}$ specifying a finite rotation in three-dimensional space. One of the most common instances arises in quantum mechanics, when a system has a Hamiltonian H that is the sum of two or more non-commuting parts. The task is then to determine the time-development operator of the system, namely, the exponential $e^{-iHt/\hbar}$.

The problem is that, when two operators A and B do not commute with each other, e^{A+B} *cannot* be written as the product $e^A\,e^B$. The general expression is of the form $e^{A+B} = e^A\,e^B\,\cdots$, where the dots stand for a complicated infinite product of the exponentials of operators (as you will see shortly in Eqs. (15.29) and (15.30) below). This is called the **Zassenhaus formula.**

But there is a special case in which the answer is actually quite simple, as in the case of Hadamard's lemma. This is the case when both A and B commute with their commutator, i.e., $[A,\ C] = 0$ and $[B,\ C] = 0$ where $C = [A,\ B]$. Sometimes this happens because C is just a constant multiple of the unit operator—for example, when $A = x$ and $B = p$, or when $A = a$ and $B = a^\dagger$. But the result is valid whenever A and B commute with C. It reads

$$\boxed{e^{A+B} = e^A\,e^B\,e^{-\frac{1}{2}C},\quad \text{provided}\ [A,\ C] = [B,\ C] = 0.} \tag{15.28}$$

★ **9.** Derive the formula (15.28). As in the case of Hadamard's lemma, you can do this by first defining the operator

$$G(\lambda) = e^{\lambda A}\,e^{\lambda B},$$

where λ is a scalar parameter (i.e., it is not an operator). Derive and solve a first-order differential equation for $G(\lambda)$, using the boundary condition $G(0) = I$.

Returning to the general case, the Zassenhaus formula says that

$$e^{A+B} = e^A\,e^B\,e^{Z_1}\,e^{Z_2}\,e^{Z_3}\,\ldots \tag{15.29}$$

where

$$\left.\begin{aligned} Z_1 &= -\tfrac{1}{2}\,C \\ Z_2 &= \tfrac{1}{6}[A,\ C] + \tfrac{1}{3}[B,\ C] \\ Z_3 &= -\tfrac{1}{24}\big[A,\ [A,\ C]\big] - \tfrac{1}{8}\big[A,\ [B,\ C]\big] - \tfrac{1}{8}\big[B,\ [B,\ C]\big] \\ &\cdots\cdots\ \cdots\ \cdots \end{aligned}\right\} \tag{15.30}$$

The subsequent terms in the product involve higher order multiple commutators. They can be determined by a systematic but quite tedious iterative procedure.

15.2.5 The Baker–Campbell–Hausdorff Formula

The **Baker–Campbell–Hausdorff formula** is the complement of the Zassenhaus formula. It expresses the *product* of exponentials, $e^A\,e^B$, in terms of the exponential of an infinite *sum* of operators involving multiple commutators. Once again, in the special case in which both A and B commute with their commutator C, the formula reduces to a simple form that is just Eq. (15.28) rewritten as

$$\boxed{e^A\, e^B = e^{A+B+\frac{1}{2}C}, \quad \text{provided } [A,\, C] = [B,\, C] = 0.} \tag{15.31}$$

In the general case, the formula is

$$e^A\, e^B = e^{A+B+\frac{1}{2}C+\frac{1}{12}[A,\,C]-\frac{1}{12}[B,\,C]-\frac{1}{24}[B,\,[A,\,C]]+\cdots}, \tag{15.32}$$

where $\cdots$ in the exponent on the right-hand side stands for an infinite sum of higher order multiple commutators. As in the case of the Zassenhaus formula, the successive terms in the sum can be found by a systematic procedure.

A final remark: Hadamard's lemma is the operator identity underlying all these formulas, and several others as well.

15.3 Some Physical Applications

In this final section of the present chapter, I consider three physical examples that use and illustrate some of the operator identities discussed in the foregoing. The first of these involves the rotation generators in three-dimensional space, i.e., the angular momentum operators $(J_1,\, J_2,\, J_3)$. It will be shown that these transform like a 3-vector under rotations of the coordinate axes, justifying their identification as the Cartesian components of the operator-valued vector $\mathbf{J}$. The second and third examples are relevant to quantum optics. They involve the displacement operator and the squeezing operator that are constructed from the ladder operators a and $a^\dagger$. These operators act on the vacuum state $|\,0\,\rangle$ to yield coherent states and squeezed states, respectively.

These applications also provide us an opportunity to discuss some of the pertinent properties of the Lie algebras $\mathfrak{su}(2)$ and $\mathfrak{su}(1,\,1)$, and the corresponding Lie groups $SU(2)$ and $SU(1,\,1)$.

15.3.1 Angular Momentum Operators

In Eqs. (11.7) and (11.8) of Chap. 11, Sect. 11.1.2, we found that the Pauli matrices (multiplied by $\frac{1}{2}$) satisfy the so-called angular momentum commutation relations, or the angular momentum algebra (repeated in Eq. (15.8)). Further on, in Sect. 11.3.1, we found a set of three (3×3) matrices that also satisfy the relations (15.8). I stated there that these are just different *representations* of the generators $(J_1,\, J_2,\, J_3)$ of rotations in three dimensions; and further, that this triplet of generators transforms like the components of a vector $\mathbf{J}$ under rotations of the coordinate axes.

It is important to understand that $\mathbf{J}$ is an *operator*-valued vector: it is a triplet of operators $(J_1,\, J_2,\, J_3)$ whose components transform among themselves like the components of a vector under rotations of the coordinate axes. That is,

- the generators of rotations in three-dimensional space themselves transform like (the components of) a vector under rotations.
- The actual *representation* of each J_k depends on what it is required to act on.

The matrices $\frac{1}{2}\sigma_k$ (where σ_k denotes the Pauli matrices) provide a (2×2) matrix representation for these operators. The three matrices in Eqs. (11.26) and (11.22) provide a (3×3) matrix representation of the operators. But these operators have an infinite number of other representations, including infinite-dimensional ones—for example, when they act on state vectors in infinite-dimensional Hilbert spaces in quantum mechanics. Let us therefore denote by $\mathsf{R}(\mathbf{n}, \psi)$ the rotation transformation per se, i.e., the transformation that rotates the coordinate axes in three-dimensional space by an angle ψ about the direction $\mathbf{n}$. What we have so far denoted by $R(\mathbf{n}, \psi)$ is actually the (3×3) real orthogonal matrix that *represents* the abstract transformation $\mathsf{R}(\mathbf{n}, \psi)$. Such a transformation of the coordinate axes *induces* a transformation in the Hilbert space of interest. This is a **unitary transformation**, given by $U\big(\mathsf{R}(\mathbf{n}, \psi)\big) = e^{i\,(\mathbf{J}\cdot\mathbf{n})\,\psi}$, *where the generators* $\mathbf{J}$ *now have the representation appropriate to the Hilbert space.*

We now have the machinery to demonstrate that $\mathbf{J}$ is a vector under rotations of the coordinate axes. Note the following points:

(i) This must be shown to be true *independent of any particular representation* for the quantities $J_1,\ J_2$, and J_3.
(ii) This means that the only input we can use is the *algebra* satisfied by the three operators, namely, $[J_k,\ J_l] = i\,\epsilon_{klm}\,J_m$.

Let us directly consider the general case of a rotation about an arbitrary axis $\mathbf{n}$ through an angle ψ. (All special cases can then be read off from it.) The task is to show that $\mathbf{J}$ satisfies the finite rotation formula, which is

$$e^{-i\,(\mathbf{J}\cdot\mathbf{n})\,\psi}\,J_i\,e^{i\,(\mathbf{J}\cdot\mathbf{n})\,\psi} = (\cos\,\psi)\,J_i + (1 - \cos\,\psi)\,(J_j\,n_j)\,n_i + (\sin\,\psi)\,\epsilon_{ijk}\,J_j\,n_k\,. \tag{15.33}$$

It is important to note that the transformation rule for the operator $\mathbf{J}$ is not just $\mathbf{J}' = R(\mathbf{n}, \psi)\,\mathbf{J}$. This is because $\mathbf{J}$ is not an ordinary vector, but an *operator*-valued vector. I reiterate that the components of $\mathbf{J}$ are not necessarily (3×3) matrices, nor do we care what the actual representation is. In vector form, Eq. (15.33) reads

$$\boxed{e^{-i\,(\mathbf{J}\cdot\mathbf{n})\,\psi}\,\mathbf{J}\,e^{i\,(\mathbf{J}\cdot\mathbf{n})\,\psi} = (\cos\,\psi)\,\mathbf{J} + (1 - \cos\,\psi)\,(\mathbf{J}\cdot\mathbf{n})\,\mathbf{n} + (\sin\,\psi)\,(\mathbf{J}\times\mathbf{n}).} \tag{15.34}$$

Compare this with the finite rotation formula for the coordinate $\mathbf{r}$ of a point in three-dimensional space, given in Eq. (11.38) of Chap. 11, Sect. 11.3.3. The right-hand sides of the two equations are identical in form. Therefore, once Eq. (15.33) is proved, we may assert that $\mathbf{J}$ itself transforms like a vector under rotations.

In order to establish Eq. (15.33), apply Hadamard's lemma (Eq. (15.16)) to the problem, with the identifications

$$\lambda = -i\psi, \quad A = \mathbf{J}\cdot\mathbf{n} = J_k\, n_k, \quad B = J_i\,. \tag{15.35}$$

For ease of notation, let us write the r-fold multiple commutator in Hadamard's lemma as C_r. Then

$$[A,\, C_r] = C_{r+1}, \quad \text{where} \quad C_1 = [A,\, B] \;\text{ and }\; r = 1, 2, \ldots\,. \tag{15.36}$$

Using the commutator algebra $[J_k,\, J_l] = i\,\epsilon_{klm}\, J_m$, it is straightforward to show that

$$C_1 = i\,\epsilon_{ilk}\, J_l\, n_k \quad \text{and} \quad C_2 = [A,\, C_1] = J_i - n_i\, n_k\, J_k\,. \tag{15.37}$$

Further,

$$C_3 = [A,\, C_2] = i\,\epsilon_{ilk}\, J_l\, n_k = C_1, \quad \text{and hence} \quad C_4 = C_2, \;\; C_5 = C_1, \;\ldots\,. \tag{15.38}$$

Thus, $C_{2r+1} = C_1$ and $C_{2r} = C_2$. The infinite series in Hadamard's lemma (15.16) can now be summed easily. Equations (15.33) and (15.34) follow.

★ **10.** Verify Eqs. (15.37) and (15.38). Insert these in Hadamard's lemma (15.16), and complete the steps to derive Eqs. (15.33) and (15.34).

There is a small subtlety involved in the left-hand sides of Eqs. (15.33) and (15.34). The factor $e^{i\,(\mathbf{J}\cdot\mathbf{n})\,\psi}$, which we had identified with $R(\mathbf{n},\, \psi)$ in Chap. 11, Sect. 11.3.2, appears on the *right* of J_i, while its adjoint $e^{-i\,(\mathbf{J}\cdot\mathbf{n})\,\psi}$ appears on its left. You might have expected the reverse, based perhaps on the transformation rule for a second-rank tensor, Eq. (5.8) of Chap. 5, Sect. 5.1.2. But the order given here is correct. It is related to the fact that $\mathbf{J}$ itself is an operator (e.g., in a Hilbert space), and this is the transformation rule for an operator, as opposed to that for a state vector.

15.3.2 Representation of Rotations by $SU(2)$ Matrices

What is the consequence of the fact that the Pauli matrices also provide a representation of the generators of rotations in three-dimensional space?

To recapitulate: Consider a proper rotation of the coordinate axes in three-dimensional space, about an axis along the unit vector $\mathbf{n}$, and through an angle ψ. The corresponding element of the rotation group is $\mathsf{R}(\mathbf{n},\, \psi) = e^{i\,(\mathbf{J}\cdot\mathbf{n})\,\psi}$, where $\mathbf{J} = (J_1,\, J_2,\, J_3)$ and the components J_k satisfy the angular momentum algebra $[J_k,\, J_l] = i\,\epsilon_{klm}\, J_m$. When the position vector of a point is written as a column vector whose elements are the Cartesian components of $\mathbf{r}$, the generators J_k are represented by the (3×3) matrices given in Eqs. (11.26) and (11.22) of Chap. 11, Sect. 11.3.1. $\mathsf{R}(\mathbf{n},\, \psi)$ is then represented by the (3×3) orthogonal matrix $R(\mathbf{n},\, \psi)$, an element of the group $SO(3)$. This matrix is obtained by calculating the exponential of the (3×3) matrix $i\,(\mathbf{J}\cdot\mathbf{n})\,\psi$. We have done so in Chap. 11, Sect. 11.3.2, and written down the general matrix element of the rotation matrix in Eq. (11.32).

On the other hand, the generators J_k can also be represented in terms of the Pauli matrices according to $J_k = \frac{1}{2}\sigma_k$. The matrix representing the rotation $\mathsf{R}(\mathbf{n}, \psi)$ is then the (2×2) matrix $U(\mathbf{n}, \psi) = e^{i\,(\boldsymbol{\sigma}\cdot\mathbf{n})\,\psi/2}$. (I have used the symbol U for this matrix in anticipation of the fact that it is a unitary matrix.) $U(\mathbf{n}, \psi)$ can be written down using the identity in Eq. (11.16) of Chap. 11, Sect. 11.2.2 for the exponential of the matrix $i\,(\mathbf{a}\cdot\boldsymbol{\sigma})$, where $\mathbf{a}$ is any ordinary 3-vector. Setting $\mathbf{a} = \frac{1}{2}\mathbf{n}\psi$ in that formula, we get

$$\boxed{U(\mathbf{n}, \psi) = e^{i\,(\mathbf{n}\cdot\boldsymbol{\sigma})\,\psi/2} = I\,\cos\tfrac{1}{2}\psi + i\,(\boldsymbol{\sigma}\cdot\mathbf{n})\,\sin\tfrac{1}{2}\psi.} \tag{15.39}$$

Since the Pauli matrices are Hermitian, it follows that

$$\boxed{U^{\dagger}(\mathbf{n}, \psi) = e^{-i\,(\mathbf{n}\cdot\boldsymbol{\sigma})\,\psi/2} = I\,\cos\tfrac{1}{2}\psi - i\,(\boldsymbol{\sigma}\cdot\mathbf{n})\,\sin\tfrac{1}{2}\psi.} \tag{15.40}$$

If the unit vector $\mathbf{n} = (n_1, n_2, n_3)$, the explicit form of $U(\mathbf{n}, \psi)$ is given by

$$U(\mathbf{n}, \psi) = \begin{pmatrix} \cos\frac{1}{2}\psi + i n_3 \sin\frac{1}{2}\psi & (n_2 + i n_1)\sin\frac{1}{2}\psi \\ (-n_2 + i n_1)\sin\frac{1}{2}\psi & \cos\frac{1}{2}\psi - i n_3 \sin\frac{1}{2}\psi \end{pmatrix}. \tag{15.41}$$

Recall, now, the general form of a (2×2) unitary matrix with determinant equal to 1, i.e., an element of the group $SU(2)$. As you have seen in Eq. (12.18) of Chap. 12, Sect. 12.2.3, such a matrix must necessarily be of the form

$$U = \begin{pmatrix} \alpha & \beta \\ -\beta^* & \alpha^* \end{pmatrix}, \quad \text{where} \quad |\alpha|^2 + |\beta|^2 = 1. \tag{15.42}$$

It is easily checked that the matrix $U(\mathbf{n}, \psi)$ in Eq. (15.41) is precisely of this form. (Remember that $n_1^2 + n_2^2 + n_3^2 = 1$.) This becomes even more obvious if we write the components of $\mathbf{n}$ in terms of its spherical polar coordinates $(1, \theta, \varphi)$ (Eq. (11.33) of Chap. 11, Sect. 11.3.2):

$$n_1 = \sin\theta\,\cos\varphi, \quad n_2 = \sin\theta\,\sin\varphi, \quad n_3 = \cos\theta. \tag{15.43}$$

Then

$$\boxed{U(\mathbf{n}, \psi) = \begin{pmatrix} \cos\frac{1}{2}\psi + i\cos\theta\,\sin\frac{1}{2}\psi & i\,e^{-i\varphi}\sin\theta\,\sin\frac{1}{2}\psi \\ i\,e^{i\varphi}\sin\theta\,\sin\frac{1}{2}\psi & \cos\frac{1}{2}\psi - i\cos\theta\,\sin\frac{1}{2}\psi \end{pmatrix}.} \tag{15.44}$$

Further, $U^{\dagger}(\mathbf{n}, \psi) = U^{-1}(\mathbf{n}, \psi) = U(\mathbf{n}, -\psi)$. (The last relation is obvious on physical grounds.) We may conclude, then, that an arbitrary proper rotation in three-dimensional space can be represented by an element of $SU(2)$.

The next question is that if a rotation in three-dimensional space is represented by a (2×2) unitary matrix, rather than a (3×3) orthogonal matrix, how does it act on the coordinate of any point? The answer is as follows. Instead of writing the Cartesian coordinates of any point $\mathbf{r}$ as a column vector, we use them to form the (2×2) matrix

$$\mathbf{r} \cdot \boldsymbol{\sigma} = x_1 \sigma_1 + x_2 \sigma_2 + x_3 \sigma_3 = \begin{pmatrix} x_3 & x_1 - i x_2 \\ x_1 + i x_2 & -x_3 \end{pmatrix}. \tag{15.45}$$

Then, if $\mathbf{r} \mapsto \mathbf{r}'$ under the rotation transformation, we have

$$\boxed{\mathbf{r} \cdot \boldsymbol{\sigma} \mapsto \mathbf{r}' \cdot \boldsymbol{\sigma} = U(\mathbf{n}, \psi)\, (\mathbf{r} \cdot \boldsymbol{\sigma}) U^\dagger(\mathbf{n}, \psi),} \tag{15.46}$$

where the unitary matrix $U(\mathbf{n}, \psi)$ is given by Eq. (15.41) or (15.44).

★ **11.** You must make sure that Eq. (15.46) is consistent with the transformation rule for the coordinates under a rotation of the axes as given by the finite rotation formula, Eq. (11.37) of Chap. 11, Sect. 11.3.3.

(a) Using Eq. (15.41) for $U(\mathbf{n}, \psi)$, work out the right-hand side of Eq. (15.46) explicitly.

(b) Verify that the transformation rules for the Cartesian components of $\mathbf{r}$ are the same as those given by the finite rotation formula

$$x'_i = x_i \cos \psi + n_i\, x_j\, n_j\, (1 - \cos \psi) + \epsilon_{ijk}\, x_j\, n_k\, \sin \psi.$$

(c) Use Eqs. (15.39) and (15.40) for U and $U^\dagger$, respectively, in Eq. (15.46). Simplify the expression obtained, to get

$$\mathbf{r}' \cdot \boldsymbol{\sigma} = (\cos \psi)\, (\mathbf{r} \cdot \boldsymbol{\sigma}) + (1 - \cos \psi)\, (\mathbf{r} \cdot \mathbf{n})\, (\mathbf{n} \cdot \boldsymbol{\sigma}) + (\sin \psi)\, (\mathbf{r} \times \mathbf{n}) \cdot \boldsymbol{\sigma}$$

This is exactly the relation we get on taking the dot product of both sides of the rotation formula for the position vector (Eq. (11.38) of Chap. 11, Sect. 11.3.3) with $\boldsymbol{\sigma}$.

15.3.3 Connection Between the Groups $SO(3)$ and $SU(2)$

We are ready, now, to discuss the close connection between the special orthogonal group $SO(3)$ and the special unitary group $SU(2)$.

Equation (15.46) immediately reveals an important property of the representation of rotations by matrices belonging to $SU(2)$. The left-hand-side $\mathbf{r}' \cdot \boldsymbol{\sigma}$, and hence $\mathbf{r}'$ itself, is unchanged if the matrix $U(\mathbf{n}, \psi)$ is multiplied by an arbitrary phase factor $e^{i\alpha}$, because this will cancel against the factor $e^{-i\alpha}$ coming from $U^\dagger(\mathbf{n}, \psi)$. But the requirement $\det U = 1$ imposes the condition $e^{2i\alpha} = 1$. Hence the parameter α can only take the values 0 and π, so that $e^{i\alpha} = \pm 1$. We conclude that there are *two* different matrices, $U(\mathbf{n}, \psi)$ and $-U(\mathbf{n}, \psi)$, that correspond to every rotation matrix

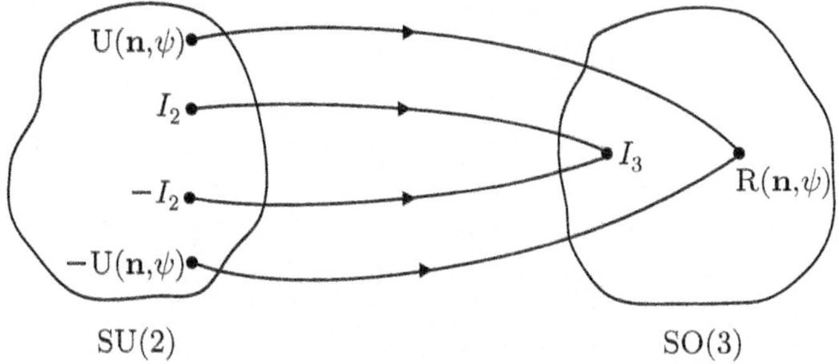

Fig. 15.1 Schematic illustration of the 2-to-1 group homomorphism from $SU(2)$ to $SO(3)$

$R(\mathbf{n}, \psi)$. For convenience let us denote, for the moment, the (2×2) unit matrix by I_2 and the (3×3) unit matrix by I_3. In mathematical terms:

- The correspondence between the elements of $SU(2)$ and those of $SO(3)$ is a mapping from one group to another, called a **group homomorphism**.
- For each given $\mathbf{n}$ and ψ, the elements $U(\mathbf{n}, \psi)$ and $-U(\mathbf{n}, \psi)$ of $SU(2)$ are both mapped to the element $R(\mathbf{n}, \psi)$ of $SO(3)$. The mapping from $SU(2)$ to $SO(3)$ is thus a 2-to-1 homomorphism.
- In particular, the identity element I_2 and the element $-I_2$ of $SU(2)$ are both mapped to the identity element I_3 of $SO(3)$. The set $\{I_2, -I_2\}$ is the **kernel** of the homomorphism. (This is the set of elements whose image under the mapping is the identity element in the target space.)
- The elements I_2 and $-I_2$ form a group by themselves. This is the cyclic group of order 2. It is isomorphic to (i.e., it is in 1-to-1 correspondence with) $\mathbb{Z}_2$, the additive group of integers modulo 2.

Figure 15.1 shows, *very* schematically, some features of this mapping. The actual parameter spaces of $SU(2)$ and $SO(3)$ are compact three-dimensional spaces with different topologies. I shall touch upon this aspect in Sects. 15.3.4 and 15.3.5 below.

We now consider a set whose general element is the *pair* $\{U(\mathbf{n}, \psi), \; -U(\mathbf{n}, \psi)\}$. It is not hard to show that such a set is, in fact, a group. It is called the **quotient group** $SU(2)/\mathbb{Z}_2$, which is read as "$SU(2)$ modulo $\mathbb{Z}_2$". Its identity element is, naturally, the pair $\{I_2, -I_2\}$.

- There is a **group isomorphism** (or 1-to-1 correspondence) between this quotient group and the rotation group $SO(3)$. This is written as

$$\boxed{SU(2)/\mathbb{Z}_2 \cong SO(3).} \tag{15.47}$$

It turns out that $SU(2)$ is the so-called **universal covering group** of the rotation group $SO(3)$. The matters touched upon in the foregoing underlie truly deep and profound physical consequences. Among these are

- The eigenvalue spectrum $\{j(j+1) \mid j = 0, \frac{1}{2}, 1, \frac{3}{2}, 2, \dots\}$ of the operator $\mathbf{J}^2$.

- The double connectivity of the parameter space of $SO(3)$, in contrast to the simply-connected parameter space of $SU(2)$.
- The existence of "single-valued" or tensor representations of the rotation group (labeled by integer values of j), and "double-valued" or spinor representations (labeled by half-odd-integer values of j).
- The existence of **bosons**, with integer values of the spin quantum number, and **fermions**, with half-odd-integer values of this quantum number.
- The **spin-statistics theorem** of quantum field theory, as a result of which a collection of identical bosons satisfies Bose–Einstein statistics, while a collection of identical fermions satisfies Fermi–Dirac statistics.

To reiterate: the groups $SU(2)$ and $SO(3)$ are related by a 2-to-1 homomorphism. For every element of $SO(3)$, or, equivalently, for every proper rotation $R(\mathbf{n}, \psi)$ of the coordinate axes in three-dimensional space, there are two matrices belonging to the group $SU(2)$. The question that now arises naturally is: What are the respective *topological spaces* of the parameters specifying the elements of these groups? What follows is a brief account of this aspect, for the sake of completeness.

15.3.4 The Parameter Space of SU(2)

A general element of the special unitary group $SU(2)$ has the form given by Eq. (12.18) of Chap. 12, Sect. 12.2.3, namely,

$$U \in SU(2) \Longrightarrow U = \begin{pmatrix} \alpha & \beta \\ -\beta^* & \alpha^* \end{pmatrix}, \quad \text{where} \quad |\alpha|^2 + |\beta|^2 = 1. \tag{15.48}$$

The set of such (2×2) matrices actually forms the so-called **fundamental representation** of the abstract $SU(2)$ group. Setting $\alpha = \alpha_1 + i\alpha_2$ and $\beta = \beta_1 + i\beta_2$, an element of $SU(2)$ is therefore specified by four real parameters satisfying the condition

$$\alpha_1^2 + \alpha_2^2 + \beta_1^2 + \beta_2^2 = 1. \tag{15.49}$$

But this is the equation specifying the "surface" of a "sphere" of unit radius embedded in a four-dimensional Euclidean space. More precisely, the parameter space of $SU(2)$ is a three-dimensional space, namely, the three-dimensional unit sphere S^3. All the four parameters α_1, α_2, β_1, and β_2 are bounded (in fact, none of them can exceed unity in magnitude). As a result, $SU(2)$ is a **compact group**. The parameter space is also **simply connected**, as I mentioned in passing in Sect. 15.3.3: any closed loop in the parameter space can be shrunk continuously to a point, without any part of the loop leaving the space.

As already stated, you would need at least four Euclidean dimensions to have a "concrete" model of S^3. On the other hand, an analogy helps us visualize the space. Consider the way we might construct the 2-sphere S^2, as follows. Take two disks,

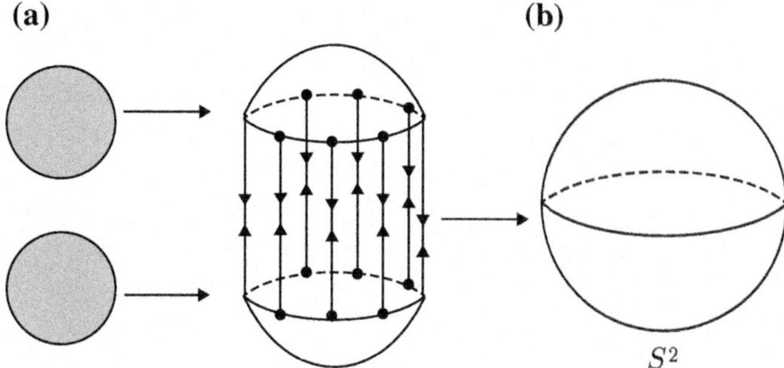

Fig. 15.2 **a** Two closed disks are deformed to hemispherical shells, and corresponding points on their boundary circles (S^1 and S^1) are identified (indicated by lines with two-way arrows). **b** Gluing the two boundaries together, the 2-sphere S^2 is obtained

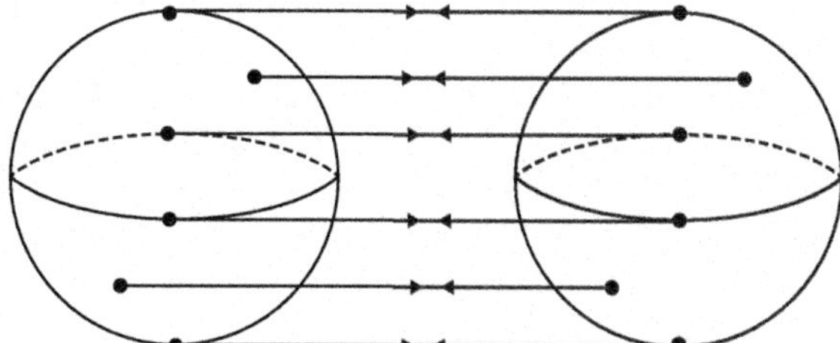

Fig. 15.3 Two closed solid spheres, each with a boundary S^2. When corresponding points on their boundaries are identified with each other (indicated by lines with two-way arrows), a 3-sphere S^3 is obtained

each with a boundary that is a circle, i.e., a 1-sphere S^1. Each disk can be deformed continuously into the shape of a hemispherical shell. Now identify corresponding points on their boundaries, as shown by the dotted lines in Fig. 15.2a. Glue the boundaries together, as in Fig. 15.2b, and the outcome is the 2-sphere S^2. This is a compact two-dimensional manifold with no boundary. Any closed path drawn on the space can be shrunk continuously to a point without any part of it leaving the space. S^2 is a simply-connected space.

Similarly, let us start with two solid spheres, each with a boundary (or bounding surface) S^2. Now imagine gluing together corresponding points on their boundaries, as indicated in Fig. 15.3. Although you cannot do this while remaining in three-dimensional space (you need at least four dimensions to do so explicitly), this point-wise identification is mathematically well-defined. The outcome is the 3-sphere S^3, a compact three-dimensional manifold with no boundary. It is also a simply-connected space.

***The Poincaré conjecture*:** This is an appropriate place to mention a profound result in mathematics (more specifically, in topology) that is relatively easy to state, but enormously difficult to establish. In somewhat loose terms, the **Poincaré conjecture** asserts that every closed, simply-connected three-dimensional manifold is topologically equivalent ("homeomorphic") to the 3-sphere. Enunciated by H. Poincaré in 1904, this proposition remained a conjecture for nearly a hundred years in spite of intensive efforts to either prove or disprove it. It was finally shown to be true by G. Perelman in 2002–03 using, in particular, earlier results of R. Hamilton. Remarkably enough, the proof relies heavily on advances in the mathematics of nonlinear partial differential equations. Moreover, the "conjecture" itself is a special case of a much broader result called the **geometrization conjecture**. Investigations into this broad class of problems has catalyzed advances in several related areas of modern mathematics for more than a century now, as deep connections between seemingly unrelated topics continue to be discovered.[4]

15.3.5 The Parameter Space of $SO(3)$

Let us turn, now, to the parameter space of $SO(3)$. According to Euler's rotation theorem (Chap. 11, Sect. 11.3.1), a rotation of the coordinate axes in three-dimensional space can be parametrized by an axis of rotation specified by a unit vector $\mathbf{n}$, and an angle ψ of rotation about this axis. Since the components of the unit vector $\mathbf{n}$ satisfy the condition $n_1^2 + n_2^2 + n_3^2 = 1$, there are two independent parameters in $\mathbf{n}$. The set $\{\mathbf{n}, \psi\}$ therefore comprises three independent parameters. The parameter space of $SO(3)$ is therefore three-dimensional, just as the parameter space of $SU(2)$ is.

Now, $\mathbf{n}$ and ψ can be combined to form a single three-dimensional vector of magnitude ψ and direction $\mathbf{n}$, according to

$$\boldsymbol{\Psi} = \psi\,\mathbf{n}. \tag{15.50}$$

Let the "tail" of each vector $\boldsymbol{\Psi}$ be placed at the origin. Then, taking ψ to lie in the range $0 \le \psi < 2\pi$, the space of $\boldsymbol{\Psi}$ appears to be a solid ball of radius 2π centered at the origin. Each vector $\boldsymbol{\Psi}$, i.e., each member of $SO(3)$, would then be represented by a point in the *interior* of this ball: the direction of the radius vector from the origin to that point specifies $\mathbf{n}$, and the magnitude of the radius vector specifies ψ. The center O of the ball represents the (3×3) identity matrix that corresponds to no rotation at all.

But this conclusion is incorrect. First of all, the angle ψ need not run up to 2π. The range $0 \le \psi \le \pi$ suffices. To see why this is so, consider the expression obtained in Eq. (11.32) of Chap. 11, Sect. 11.3.2 for the (3×3) rotation matrix $R(\mathbf{n}, \psi)$: we have

$$R_{ij}(\mathbf{n}, \psi) = \delta_{ij}\, \cos\, \psi + n_i\, n_j\, (1 - \cos\, \psi) + \epsilon_{ijk}\, n_k\, \sin\, \psi. \tag{15.51}$$

Suppose ψ is in the range $\pi \le \psi < 2\pi$, and let $\psi' = 2\pi - \psi$. Then ψ' lies in the range $0 < \psi' \le \pi$, Further,

[4] For a semi-popular exposition of the Poincaré conjecture, see D. O'Shea, *The Poincaré Conjecture*, Walker & Co., NY, 2008. For a succinct technical review of the subject, see T. Tao, arXiv:math/0610903v1 [math.DG].

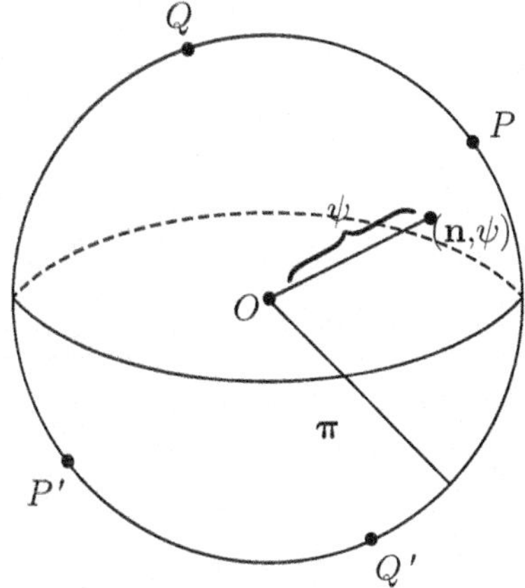

Fig. 15.4 A depiction of the parameter space of the group $SO(3)$. It comprises all the points inside and on a solid ball of radius π. Further, every pair of antipodal points on the surface of the ball represents a single point of the space. Thus P and P' are supposed to be the same point, as are Q and Q'. A typical point in the space is also shown

$$\begin{aligned} R_{ij}(\mathbf{n}, \psi) &= R(\mathbf{n}, 2\pi - \psi') \\ &= \delta_{ij} \cos \psi' + n_i\, n_j\, (1 - \cos \psi') - \epsilon_{ijk}\, n_k \sin \psi' \\ &= R_{ij}(-\mathbf{n}, \psi'). \end{aligned} \quad (15.52)$$

In other words, a rotation about any axis $\mathbf{n}$ through an angle greater than π (and less than 2π) is already accounted for, as a rotation about the antipodal direction $-\mathbf{n}$ through an angle less than π. As a consequence, the parameter space of $\boldsymbol{\Psi}$ need only be a ball of radius π, rather than 2π. Further, the surface of this ball (corresponding to a rotation angle of π) is also part of the parameter space.

But there is more to it. As you have already seen in Eq. (11.36) of Chap. 11, Sect. 11.3.2, $R(\mathbf{n}, \pi) = R(-\mathbf{n}, \pi)$. That is, a rotation by π about $\mathbf{n}$ is exactly the same thing as a rotation by π about $-\mathbf{n}$. This means that any point on the surface of the ball in parameter space must be *identical* to its antipodal point on that surface! This makes the topology of the parameter space nontrivial. A space of this sort cannot be embedded in three-dimensional Euclidean space. You cannot, therefore, hold a completely accurate model of this space in your hand, or envision it, except in mathematical terms. Figure 15.4 is a partial depiction of the space, with an (imagined) line joining the pair (P, P'), and another imagined line joining the pair (Q, Q') of antipodal points, to remind us of the fact that each of them actually represents a single point in the space. This must be done for *every* pair of antipodal points on the surface of the ball. Such a space is called a **projective space**. In the present instance, there is a group isomorphism between $SO(3)$ and the quotient group $SU(2)/\mathbb{Z}_2$ (Eq. (15.47) of Sect. 15.3.3), Further, the parameter space of $SU(2)$ is the 3-sphere S^3. The parameter space of $SO(3)$ is therefore the real projective space $S^3/\mathbb{Z}_2$, and is denoted by $\mathbb{R}P^3$. This three-dimensional space cannot be embedded in the four-dimensional Euclidean space $\mathbb{R}^4$, unlike the parameter space S^3 of the group $SU(2)$. It turns out that the lowest dimensional Euclidean space in which $\mathbb{R}P^3$ can be embedded is the five-dimensional space $\mathbb{R}^5$.

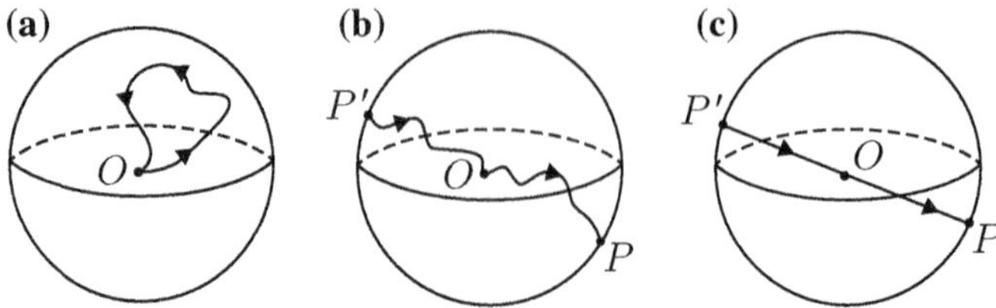

Fig. 15.5 Examples of the two classes of closed paths or loops in the parameter space of the group $SO(3)$. The loop in **a** can be shrunk continuously to a point without leaving the space, while that in **b** cannot. The closed path in **b** can be straightened out, and is therefore equivalent, to that in **c**

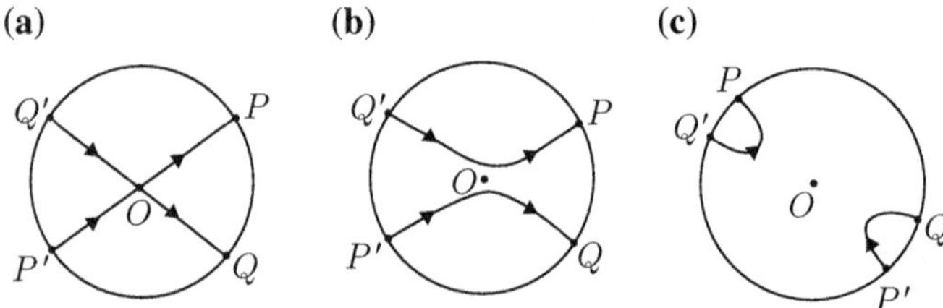

Fig. 15.6 How a pair of closed paths in the parameter space of $SO(3)$ can be deformed continuously to a single point

A crucial feature of the parameter space $\mathbb{R}P^3$ is the following. There are two completely different ("homotopically distinct") classes of closed paths or loops in it. Figure 15.5a and b show, respectively, representative examples of these paths. The loop in Fig. 15.5a has the conventional shape of a loop, and it can be shrunk to a point continuously and without leaving the space. The "loop" in Fig. 15.5b, on the other hand, does not look like a closed path, but it is. It starts from the center O of the ball, which corresponds to no rotation at all (the identity element of $SO(3)$) and goes out to P, restarts from its equivalent antipodal point P', and returns to O. Since P and P' are identified as a single point, this path is actually a *closed* path in the parameter space! It can, at best, be straightened out to form a diameter of the ball, passing through the origin (the identity element of $SO(3)$), as in Fig. 15.5c. It cannot be deformed continuously to a closed path of the type shown in Fig. 15.5a.

It can be shown that there are no other classes of closed paths in the parameter space of $SO(3)$, other than the two classes described above. This enables us to conclude that this space is **doubly connected**. The profound physical consequences listed in Sect. 15.3.3, following Eq. (15.47), are intimately related to this property. In particular, the existence of single-valued or tensor representations of the rotation group, corresponding to $j = 0, 1, 2, \ldots$, is connected to the equivalence class of closed paths as in Fig. 15.5a. The existence of double-valued or spinor representations of the rotation group, corresponding to $j = \frac{1}{2}, \frac{3}{2}, \ldots$, is associated with the equivalence class of closed paths as in Fig. 15.5b or c.

It is interesting to see how the latter kind of closed path, if traced out *twice* in succession, can be deformed to a point. Figure 15.6a shows such a path, traversed in the sequence

$$O \to P = P' \to O \to Q = Q' \to O. \tag{15.53}$$

Since this is a closed path, we may start anywhere on it to traverse the whole path and return to the starting point. Consider, therefore, traversing it in the sequence

$$P' \to O \to Q = Q' \to O \to P = P'. \tag{15.54}$$

Now deform this path slightly, so as to avoid the crossing at O, as shown in Fig. 15.6b. Observe that P is now joined to Q', while Q is joined to P'. Continue to deform the path by moving P toward Q' along the surface of the ball. This process automatically moves the antipodal point P' toward Q, as in Fig. 15.6c. In the end, each of these line segments can be shrunk to a point, to leave a single pair of antipodal points on the surface of the ball. But these two points count as a single point. Thus, *two* full rotations make this class of closed paths equivalent to no rotation at all. This is the property underlying the statement that a spinor changes sign under a rotation of the coordinate axes by 2π, but returns to its original value under a rotation by 4π.

The parameter space of $SO(d)$, $d \neq 3$**:** It is natural, at this stage, to ask: what is the dimensionality of the parameter space of the rotation group $SO(d)$ for values of d other than 3? What is the connectivity of this space?

A general element of the proper rotation group $SO(d)$, where $d \geq 2$, is specified by $\frac{1}{2}d(d-1)$ parameters. This quantity is therefore the dimensionality of the corresponding parameter space. It turns out that

- the parameter space of $SO(d)$ is a doubly connected space for every $d > 3$, just as it is in the case $d = 3$.

I merely mention this fact here, as a more detailed discussion of this aspect would take us too far afield.

15.3.6 *The Parameter Space of* $SO(2)$

Remarkably enough, the situation in two dimensions is very different from the foregoing. You have seen (e.g., in Eq. (12.12) of Chap. 12, Sect. 12.2.2) that a proper rotation of the coordinate axes about the origin in two-dimensional space, through an angle α, can be represented by the orthogonal matrix

$$R(\alpha) = \begin{pmatrix} \cos\alpha & \sin\alpha \\ \sin\alpha & \cos\alpha \end{pmatrix}. \tag{15.55}$$

The angle α runs from 0 to 2π, so that its "space" is just the set of points on a circle (of unit radius, say). The parameter space of the group $SO(2)$ of proper rotations

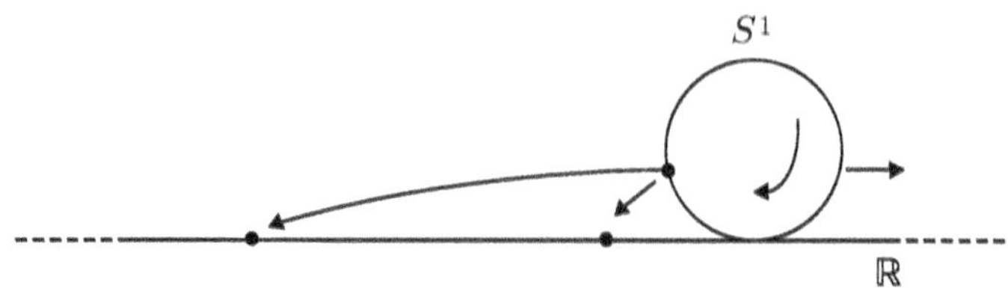

Fig. 15.7 Rolling out a circle onto a line maps the points of S^1 to points of $\mathbb{R}$. Each point on S^1 is mapped to a denumerable infinity of points on $\mathbb{R}$. The dots denote one such point in S^1 and two of its images in $\mathbb{R}$

in two dimensions is therefore the circle, or 1-sphere, S^1. This space is connected, but *not* simply connected. You can have closed paths on it that wind around S^1 an arbitrary integer number of times, in either the positive or the negative sense. Each such class of paths is labeled by this **winding number** n, which can take on all possible integer values (that is, $n \in \mathbb{Z}$). Each class is distinct from the others, since a path with a winding number n cannot be smoothly deformed to another path with a winding number m, where $n \neq m$. As a consequence, the parameter space S^1 has *infinite* connectivity.

What would be the universal covering group of $SO(2)$? A heuristic way of arriving at the answer to this question would be to ask how the parameter space S^1 can be mapped to a simply connected space. "Rolling out" the circle onto a line suggests itself: in the process, the space S^1 gets mapped to the space $\mathbb{R}$—repeatedly so, an infinite number of times, as shown in Fig. 15.7. Each point on S^1 (taken to have unit radius, say) is mapped to an infinite set of points on the line, separated by a distance 2π from its neighbors on either side of it. Now, $\mathbb{R}$ is obviously simply connected. It is also a Lie group as well: real numbers comprise an abelian Lie group, with addition as the group composition law. These facts suggest that there is a group homomorphism between $\mathbb{R}$ and $SO(2)$, and that $\mathbb{R}$ is the universal covering group of $SO(2)$. Analogous to Eq. (15.47) for $SO(3)$, we now have the group isomorphism

$$\boxed{\mathbb{R}/2\pi\mathbb{Z} \cong SO(2).} \tag{15.56}$$

The factor 2π can be eliminated by a trivial rescaling.

Comparison of the cases $d = 3$ and $d = 2$: At this stage, it is helpful to pause to make a comparison of the three- and two-dimensional cases. In Sects. 15.3.3, 15.3.4, and 15.3.5, you have seen the following:

(i) The parameter space of $SO(3)$ has double connectivity.
(ii) $SU(2)$ is the universal covering group of $SO(3)$.
(iii) There is a 2-to-1 homomorphism between $SU(2)$ and $SO(3)$.
(iv) The parameter space of $SU(2)$ is the 3-sphere S^3, which is a compact, simply connected space.
(v) There is a group isomorphism $SO(3) \cong SU(2)/\mathbb{Z}_2$.

The corresponding facts in the case of $SO(2)$ are as follows:

(i) The parameter space of $SO(2)$ has infinite connectivity.
(ii) $\mathbb{R}$ is the universal covering group of $SO(2)$.
(iii) There is a many-to-one homomorphism between $\mathbb{R}$ and $SO(2)$.
(iv) The parameter space of $\mathbb{R}$ is simply connected, but noncompact.
(v) There is a group isomorphism $SO(2) \cong \mathbb{R}/\mathbb{Z}$.

These properties have deep implications for two-dimensional systems in condensed matter physics and quantum field theory.

***Isomorphism between** $SO(2)$ **and** $U(1)$:* We have seen that the three-dimensional rotation group has a two-dimensional representation, by virtue of the relation between $SO(3)$ and $SU(2)$. That is, rotations in three-dimensional space can be represented not only by (3×3) orthogonal matrices, but also by (2×2) unitary matrices. The latter comprise the lowest dimensional nontrivial representation of rotations in three-dimensional space. Likewise, rotations in two-dimensional space also have a *one-dimensional representation*! Instead of the (2×2) orthogonal matrix in Eq. (15.55), a rotation through an angle α can be represented simply by the complex number $e^{i\alpha}$. It is obvious that the abelian group property

$$R(\alpha)\, R(\alpha') = R(\alpha')\, R(\alpha) = R(\alpha + \alpha') \tag{15.57}$$

is satisfied if we represent $R(\alpha)$ by $e^{i\alpha}$. Further, $R(0) = 1$ (the identity element) and $R^{-1}(\alpha) = R^{\dagger}(\alpha)$. But $e^{i\alpha}$ may be regarded as a (1×1) *unitary* matrix, i.e., as an element of the unitary group $U(1)$.[5] In fact, the groups $SO(2)$ and $U(1)$ are related by a group isomorphism:

$$U(1) \cong SO(2). \tag{15.58}$$

15.4 Some More Physical Applications

15.4.1 The Displacement Operator and Coherent States

Let us return to the ladder operators a and $a^{\dagger}$ related to the linear harmonic oscillator that were introduced in Chap. 14, Sects. 14.4.1 and 14.4.2. The coherent state $|\, z\,\rangle$ was defined in Eq. (14.51) as the eigenstate of the lowering operator a corresponding to the complex eigenvalue z. Its position-space wave function was written down, up to a normalization constant $C(z)$, in Eq. (14.49).

We discussed the Heisenberg algebra formed by a, $a^{\dagger}$, and I, Eq. (15.1), in Sect. 15.1.1. Exponentiating linear combinations of the generators a and $a^{\dagger}$, we get unitary operators that are elements of the Heisenberg group. When these operators act on the vacuum state $|\, 0\,\rangle$, we get coherent states. Consider the operator

[5] It is obvious that the *special* unitary group $SU(1)$ is just the trivial group with a single element, namely, 1.

$$\boxed{D(z) \stackrel{\text{def.}}{=} e^{z\,a^\dagger - z^*\,a},} \tag{15.59}$$

where z is an arbitrary complex number. $D(z)$ is called the **displacement operator** for a reason that will become clear shortly. We should really write $D(z, z^*)$ because D depends on z as well as its complex conjugate z^*, and these are linearly independent quantities. But I will stick to the customary notation $D(z)$ that is adopted for the sake of notational simplicity.

The exponential operator in Eq. (15.59) can be expressed in alternative forms. Since a and $a^\dagger$ commute with their commutator (which is just the unit operator), the simplified version (15.28) of the Zassenhaus formula is applicable. We get

$$\boxed{D(z) = e^{-\frac{1}{2}|z|^2}\, e^{z\,a^\dagger}\, e^{-z^*\,a}} \tag{15.60}$$

and also

$$D(z) = e^{\frac{1}{2}|z|^2}\, e^{-z^*\,a}\, e^{z\,a^\dagger}. \tag{15.61}$$

The first of these, Eq. (15.60), is the **normal-ordered form** of the operator $D(z)$. The *normal ordering* of any expression involving a and $a^\dagger$ means that all powers of $a^\dagger$ have been brought to the left of all powers of a, by repeated use of the canonical commutation relation $[a, a^\dagger] = I$. For example, the normal-ordered form of $a^2\, a^{\dagger\,2}$ is $a^{\dagger\,2}\, a^2 + 4a^\dagger\, a + 2I$. Normal ordering is of great help in simplifying calculations in quantum optics and in field theory in general, because of relations such as $a^n |\,0\,\rangle = 0$ and $\langle\,0\,|\,a^{\dagger\,n} = 0$ for all positive integer values of n. Equation (15.61) is the *anti*normal-ordered form of $D(z)$, in which all powers of a have been brought to the left of all powers of $a^\dagger$. This kind of ordering is also useful on occasion.

Using Eq. (15.60), it is easy to find the state that is obtained when $D(z)$ acts on the ground state $|\,0\,\rangle$. We get

$$D(z)\,|\,0\,\rangle = e^{-\frac{1}{2}|z|^2} \sum_{n=0}^{\infty} \frac{z^n}{\sqrt{n!}}\,|\,n\,\rangle. \tag{15.62}$$

Applying the operator a to each side of Eq. (15.62) gives, after a little simplification,

$$a\,D(z)\,|\,0\,\rangle = z\,D(z)\,|\,0\,\rangle. \tag{15.63}$$

Therefore $D(z)\,|\,0\,\rangle$ is precisely the coherent state $|\,z\,\rangle$, i.e.,

$$\boxed{|\,z\,\rangle = D(z)\,|\,0\,\rangle = e^{z\,a^\dagger - z^*\,a}|\,0\,\rangle = e^{-\frac{1}{2}|z|^2} \sum_{n=0}^{\infty} \frac{z^n}{\sqrt{n!}}\,|\,n\,\rangle.} \tag{15.64}$$

Note that the state $|\,z\,\rangle$ is normalized to unity, i.e., $\langle z\,|\,z\,\rangle = 1$.

★ **12.** Starting with the definition (15.59), establish Eqs. (15.60)–(15.63).

The multiplication rule for the operator $D(z)$ is of interest. It is, of course, the composition law for the elements of the Heisenberg group generated by $a, a^\dagger$, and I. If z and w are any two complex numbers, we find

$$\boxed{D(z)\,D(w) = e^{\frac{1}{2}(zw^* - z^* w)}\,D(z+w).} \tag{15.65}$$

Further,

$$\boxed{D(z)\,D(w) = e^{(zw^* - z^* w)}\,D(w)\,D(z).} \tag{15.66}$$

Note that the exponentials in Eqs. (15.65) and (15.66) are pure phase factors, since $(zw^* - z^* w) = 2i\,\mathrm{Im}\,(zw^*)$.

★ **13.** Derive Eqs. (15.65) and (15.66).

Equation (15.66) is just the Weyl form of the canonical commutation relation, Eq. (15.27), on making the identifications $z = i\xi\,(\hbar/2)^{1/2}$ and $w = -\eta\,(\hbar/2)^{1/2}$.

The unitarity of $D(z)$**:** Note that $D(0) = I$. An important property of $D(z)$ can be deduced from Eq. (15.65). Setting $w = -z$, we have $D(z)\,D(-z) = I$. Therefore $D(-z) = D^{-1}(z)$. On the other hand, it follows from the definition (15.59) that $D^\dagger(z) = e^{z^* a - z a^\dagger} = D(-z)$. Hence

$$\boxed{D^\dagger(z) = D^{-1}(z), \text{ so that } D(z) \text{ is a \textit{unitary} operator.}} \tag{15.67}$$

Recall that $D(z)$ is an element of the Heisenberg group. Hence we have, here, a representation of the Heisenberg group that is unitary. In contrast, Eq. (15.6) exhibits a representation of an element of this group by a (3×3) upper triangular matrix $g(\mathbf{t})$ that is *not* unitary. This is consistent with the fact that

- the Heisenberg group does not have any *finite*-dimensional *unitary* representations.

This property is an example of a general result that will be stated at the end of Sect. 15.4.5.

Why is $D(z)$ called the *displacement* operator? Applying the unitary transformation $D(z)$ to a and $a^\dagger$, we have (using Hadamard's lemma)

$$D^\dagger(z)\,a\,D(z) = e^{z^* a - z a^\dagger}\,a\,e^{z a^\dagger - z^* a} = a + zI. \tag{15.68}$$

Taking the adjoint of this equation,

$$D^\dagger(z)\,a^\dagger\,D(z) = e^{z^* a - z a^\dagger}\,a^\dagger\,e^{z a^\dagger - z^* a} = a^\dagger + z^* I. \tag{15.69}$$

In other words, the action of $D(z)$ is to *displace* the raising and lowering operators by zI and $z^* I$, respectively. Correspondingly, the operators x and p are also displaced, according to

$$\left.\begin{aligned} D^\dagger(z)\,x\,D(z) &= x + \sqrt{(2\hbar)}\,z_1\,I \\ D^\dagger(z)\,p\,D(z) &= p + \sqrt{(2\hbar)}\,z_2\,I, \end{aligned}\right\} \tag{15.70}$$

where z_1 and z_2 are, respectively, the real and imaginary parts of z. In Eqs. (16.54) and (16.55) of Chap. 16, Sect. 16.2.6, we shall write down the x-space and p-space wave functions corresponding to the coherent state $|\,z\,\rangle$. The corresponding probability densities are Gaussians centered at $x = \sqrt{(2\hbar)}\,z_1$ and $p = \sqrt{(2\hbar)}\,z_2$, respectively. Equation (15.70) should help you understand why this is so, and also why the state $|\,z\,\rangle = D(z)\,|\,0\,\rangle$ is also known as a **displaced vacuum state**.

Using the unitarity of $D(z)$, it also follows from Eqs. (15.68) and (15.69) that

$$\begin{aligned} D^\dagger(z)\,a^\dagger\,a\,D(z) &= D^\dagger(z)\,a^\dagger\,D(z)\,D^\dagger(z)\,a\,D(z) \\ &= (a^\dagger + z^* I)(a + z I). \end{aligned} \tag{15.71}$$

Therefore the mean value of the number operator in any coherent state $|\,z\,\rangle$ is given by

$$\boxed{\langle\,z\,|N|\,z\,\rangle = \langle\,0\,|D^\dagger(z)\,a^\dagger\,a\,D(z)|\,0\,\rangle = |z|^2.} \tag{15.72}$$

In the context of radiation, this means that the mean number of photons (of a given frequency and state of polarization) in a coherent state $|\,z\,\rangle$ is $|z|^2$. This is an important result. I will return to it in Chap. 19, Sect. 19.2.2, when we discuss the photon number statistics of ideal, single-mode laser light.

***Minimum uncertainty states*:** Coherent states have yet another important property: They are **minimum uncertainty states**.

- The product of the uncertainties in the canonically conjugate pair of quadratures x and p attains its lowest possible value, $\frac{1}{2}\hbar$, in any coherent state $|\,z\,\rangle$.

The expectation values of the quadratures x and p in a normalized coherent state $|\,z\,\rangle$ are easily found. They are

$$\langle\,z\,|\,x\,|\,z\,\rangle = \sqrt{(2\hbar)}\,z_1, \quad \langle\,z\,|\,p\,|\,z\,\rangle = \sqrt{(2\hbar)}\,z_2\,. \tag{15.73}$$

Similarly, one finds

$$\langle\,z\,|\,x^2\,|\,z\,\rangle = 2\hbar\,z_1^2 + \tfrac{1}{2}\hbar, \quad \langle\,z\,|\,p^2\,|\,z\,\rangle = 2\hbar\,z_2^2 + \tfrac{1}{2}\hbar. \tag{15.74}$$

Therefore each of the variances $(\Delta x)^2 \stackrel{\text{def.}}{=} \langle x^2\rangle - \langle x\rangle^2$ and $(\Delta p)^2 \stackrel{\text{def.}}{=} \langle p^2\rangle - \langle p\rangle^2$ is equal to $\frac{1}{2}\hbar$, and

$$\boxed{\Delta x\,\Delta p = \tfrac{1}{2}\hbar \quad \text{in any coherent state}\ |\,z\,\rangle.} \tag{15.75}$$

★ **14.** Derive Eqs. (15.73) and (15.74), and hence Eq. (15.75).

Note that coherent states comprise a whole *family* of states that share the minimum uncertainty property with the oscillator ground state $|\,0\,\rangle$. Moreover, all these states have "symmetric" uncertainties in the two quadratures, i.e.,

$$\boxed{\Delta x = \Delta p = \sqrt{\tfrac{1}{2}\hbar} \quad \text{for every coherent state } |\,z\,\rangle.} \tag{15.76}$$

15.4.2 The Squeezing Operator and the Squeezed Vacuum

There are several generalizations of coherent states. Notable among them are squeezed states, mentioned earlier in Sect. 15.1.2. Here is a short account of some aspects of the **squeezing operator** $S(z)$ and the states it produces when it acts on the vacuum state $|\,0\,\rangle$.

Analogous to the definition of the displacement operator, the definition of the squeezing operator is

$$\boxed{S(z) \stackrel{\text{def.}}{=} e^{\frac{1}{2}(za^{\dagger 2} - z^* a^2)} = e^{zK_- - z^* K_+}, \quad \text{where} \quad z \in \mathbb{C}.} \tag{15.77}$$

Recall that the operators $K_\pm$ and K_3 have been defined in Eq. (15.11) of Sect. 15.1.2. As in the case of the displacement operator, we should write the squeezing operator as $S(z, z^*)$, but it is written as $S(z)$ for notational simplicity. The reason for the name "squeezing operator" will become clear shortly.

It follows from the definition (15.77) that $S^\dagger(z) = e^{z^* K_+ - zK_-} = S(-z)$. On the other hand, $S(z)S(-z) = S(0) = I$, because the exponents in $S(z)$ and $S(-z)$ differ only by sign, and therefore commute with each other. It follows that $S(-z) = S^{-1}(z)$. Therefore

$$\boxed{S^\dagger(z) = S^{-1}(z), \ \text{so that } S(z) \ \text{is a } \textit{unitary} \text{ operator.}} \tag{15.78}$$

When $S(z)$ acts on the vacuum state, it produces the so-called **squeezed vacuum state**

$$\boxed{|\sigma(z)\rangle \stackrel{\text{def.}}{=} S(z)|\,0\,\rangle, \quad z \in \mathbb{C}.} \tag{15.79}$$

This class of states has many interesting properties, depending on the value of the complex parameter z. It is convenient to set $z = re^{i\theta}$.

To start with, a unitary transformation by $S(z)$ upon a and $a^\dagger$ leads to

$$\boxed{\left.\begin{aligned} S^\dagger(z)\, a\, S(z) &= (\cosh r)\, a + (e^{i\theta} \sinh r)\, a^\dagger, \\ S^\dagger(z)\, a^\dagger\, S(z) &= (e^{-i\theta} \sinh r)\, a + (\cosh r)\, a^\dagger. \end{aligned}\right\}} \tag{15.80}$$

These expressions are derived (see below) by using Hadamard's lemma, Eq. (15.16). The unitary transformation by the squeezing operator $S(z)$ takes each of the operators a and $a^\dagger$ to linear combinations of these operators. The transformed versions of a and $a^\dagger$ satisfy the same commutation relation as the original operators do. Equation (15.80) constitute a **Bogoliubov transformation** of a and $a^\dagger$. It follows that $\langle 0\,|S^\dagger(z)\,a\,S(z)|\,0\,\rangle = \langle\sigma(z)|\,a\,|\sigma(z)\rangle = 0$, and hence $\langle\sigma(z)|\,a^\dagger\,|\sigma(z)\rangle = 0$. Therefore the expectation values of the quadratures x and p in the state $|\sigma(z)\rangle$ also vanish, i.e.,

$$\langle\sigma(z)|\,x\,|\sigma(z)\rangle = 0, \quad \langle\sigma(z)|\,p\,|\sigma(z)\rangle = 0. \tag{15.81}$$

We further find

$$\langle\sigma(z)|\,a^2\,|\sigma(z)\rangle = \tfrac{1}{2}e^{i\theta}\,\sinh 2r, \quad \langle\sigma(z)|\,a^{\dagger 2}\,|\sigma(z)\rangle = \tfrac{1}{2}e^{-i\theta}\,\sinh 2r, \tag{15.82}$$

and

$$\boxed{\langle\sigma(z)|\,a^\dagger\,a\,|\sigma(z)\rangle = \langle\sigma(z)|\,N\,|\sigma(z)\rangle = \sinh^2 r.} \tag{15.83}$$

Equation (15.83) gives the mean photon number in the squeezed vacuum state. It follows from the results above that the mean values of x and p are zero, while their variances are

$$\left.\begin{aligned}(\Delta x)^2 &= \langle\sigma(z)|\,x^2\,|\sigma(z)\rangle = \tfrac{1}{2}\hbar\,(\cosh 2r + \cos\theta\,\sinh 2r),\\ (\Delta p)^2 &= \langle\sigma(z)|\,p^2\,|\sigma(z)\rangle = \tfrac{1}{2}\hbar\,(\cosh 2r - \cos\theta\,\sinh 2r).\end{aligned}\right\} \tag{15.84}$$

The uncertainty product therefore becomes

$$(\Delta x)(\Delta p) = \tfrac{1}{2}\hbar\,\cosh\,(2r)\,\big(1 - \cos^2\theta\,\tanh^2 2r\big)^{1/2}. \tag{15.85}$$

Note that the variances of x and p, and the uncertainty product above, are real nonnegative quantities, as required. Note also that, the uncertainties Δx and Δp remain the same for the pair of angles θ and $2\pi - \theta$. As far as these quantities are concerned it suffices, therefore, to consider values of θ in the range 0–π. I will return to the discussion of these uncertainties shortly.

The variance of the number operator N in a squeezed vacuum state can also be computed readily. We find

$$\langle\sigma(z)|\,N^2\,|\sigma(z)\rangle = \tfrac{1}{2}\sinh^2(2r) + \sinh^4 r. \tag{15.86}$$

Using Eq. (15.83) for the mean value of N, we get

$$\text{Var}\,(N) = \tfrac{1}{2}\sinh^2(2r). \tag{15.87}$$

★ **15.** Work through the steps involved to obtain Eqs. (15.80)–(15.87).

For the sake of completeness, I record here (without proof) the expansion of the normalized squeezed vacuum state $|\sigma(z)\rangle$ in terms of Fock states. It can be shown that

$$|\sigma(z)\rangle = \frac{1}{\sqrt{\cosh r}} \sum_{n=0}^{\infty} \frac{\sqrt{(2n)!}}{2^n\, n!} \left(e^{i\theta} \tanh r\right)^n |\, 2n\,\rangle. \tag{15.88}$$

Since the squeezing operator $S(z)$ only involves *even* powers of a and $a^\dagger$, it is not surprising that only the even Fock states $\{|\, 2n\,\rangle\}$ occur in the expansion (15.88). Using this expansion, the expectation values of N and all its higher powers, as well as the probability distribution of the number of photons in the squeezed vacuum state, can be found readily. I will return to these matters in Chap. 19, Sect. 19.2.3.

15.4.3 Values of z That Produce Squeezing in x or p

The following conclusions follow immediately from Eqs. (15.84) and (15.85):
(i) If the parameter z is real positive, i.e., if $\theta = 0$ and $z = r > 0$, we have

$$\Delta x = (\hbar/2)^{1/2}\, e^r \quad \text{and} \quad \Delta p = (\hbar/2)^{1/2}\, e^{-r}. \tag{15.89}$$

The product $(\Delta x)(\Delta p)$ equals its lowest possible value $(= \frac{1}{2}\hbar)$, the same value that it has in the vacuum state $|\, 0\,\rangle$. But the uncertainty in the p-quadrature is squeezed to a value *below* that in the vacuum state, i.e., below $(\hbar/2)^{1/2}$. This is the origin of the term "squeezing".
(ii) If the parameter z is real negative, i.e., $\theta = \pi$ and $z = -r < 0$, the situation is reversed: we now have

$$\Delta x = (\hbar/2)^{1/2}\, e^{-r} \quad \text{and} \quad \Delta p = (\hbar/2)^{1/2}\, e^{r}. \tag{15.90}$$

It is now the x-quadrature that is squeezed.

Figure 15.8 depicts, in the $(\Delta x, \Delta p)$ plane, the minimum uncertainty curve $\Delta x\, \Delta p = \frac{1}{2}\hbar$. This is one branch of a rectangular hyperbola. The shaded regions correspond to states that are squeezed in either the x-quadrature or the p-quadrature.

Although $S(z)|\, 0\,\rangle$ is called a "squeezed state" for any $z \in \mathbb{C}$, it is not true that squeezing in either quadrature occurs for all values of the complex parameter z. What values z produce states that are squeezed in one of the two quadratures?

Consider Δp first. It follows from the second of Eq. (15.84) that

$$\Delta p < (\hbar/2)^{1/2} \Rightarrow \cosh 2r - \cos\theta \sinh 2r < 1. \tag{15.91}$$

For any given value of r, the upper bound on θ below which squeezing in p occurs is given by

$$\cosh 2r - \cos\theta_{\text{max}} \sinh 2r = 1, \quad \text{or} \quad \cos\theta_{\text{max}} = \tanh r. \tag{15.92}$$

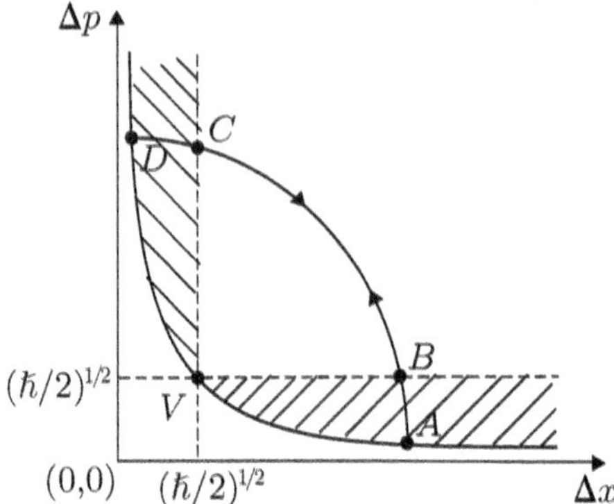

Fig. 15.8 The $(\Delta x, \Delta p)$ plane, showing the hyperbola $(\Delta x)(\Delta p) = \frac{1}{2}\hbar$. Points on this line correspond to minimum uncertainty states. The shaded regions between the hyperbola and the lines $\Delta x = (\hbar/2)^{1/2}$ and $\Delta p = (\hbar/2)^{1/2}$, respectively, correspond to states that exhibit squeezing in the x and p quadratures. V denotes the vacuum state $|\,0\,\rangle$. The points A, B, C, and D lie on the arc of a circle, as explained in the text below. They correspond to states with the same value of r, but different values of θ

In other words, for a given value of r, the argument θ must lie in the range

$$0 \le \theta < \theta_{\text{max}}(r), \quad \text{where} \quad \theta_{\text{max}}(r) = \cos^{-1}(\tanh r). \tag{15.93}$$

The same conclusion applies, of course, when θ lies in the range

$$2\pi - \theta_{\text{max}}(r) < \theta \le 2\pi. \tag{15.94}$$

The full range of θ for which squeezing in p occurs is therefore given by

$$2\pi - \theta_{\text{max}}(r) < \theta < \theta_{\text{max}}(r). \tag{15.95}$$

Likewise, the condition for the squeezing of the x-quadrature is

$$\Delta x < (\hbar/2)^{1/2} \Rightarrow \cosh 2r + \cos\theta \sinh 2r < 1. \tag{15.96}$$

Once again, simplification leads to the condition

$$\Delta x < (\hbar/2)^{1/2} \Rightarrow \pi - \theta_{\text{max}}(r) < \theta < \pi + \theta_{\text{max}}(r). \tag{15.97}$$

Figure 15.9 shows, schematically, the regions of the complex plane in which z must be located in order to produce states $S(z)|\,0\,\rangle$ that are squeezed either in p or in x. The hairpin-shaped curves on the right and left half-planes are polar plots of the equations

$$\cos\theta = \tanh r \quad \text{and} \quad \cos(\pi - \theta) = \tanh r,$$

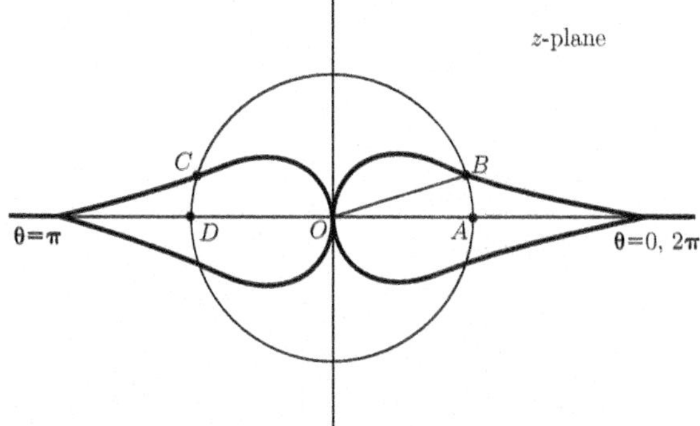

Fig. 15.9 The complex z-plane, showing the regions in which the state $S(z)|\,0\,\rangle$ is squeezed in p (the interior of the "hairpin" curve straddling the positive real axis) or in x (the interior of the curve straddling the negative real axis). The circle depicts the locus of the point $z = r\,e^{i\theta}$ as θ runs from 0 to 2π, for a fixed value of r. The angle AOB is equal to θ_{max} for that particular value of r

respectively. Any value of z lying inside these lobes leads to a state $S(z)|\,0\,\rangle$ that is squeezed, respectively, in the p and x quadratures. It is clear that $\theta_{\text{max}} \to 0$ as $r \to \infty$. The points A, B, C, D in Fig. 15.9 correspond, respectively, to

$$\theta = 0,\ \theta_{\text{max}}(r),\ \pi - \theta_{\text{max}}(r)\ \text{ and }\ \pi,$$

respectively. Referring back to Fig. 15.8, the $(\Delta x,\, \Delta p)$ values for the corresponding squeezed states are again denoted by A, B. C, and D for ready identification. Note that these points lie on the arc of a circle, because it follows trivially from Eq. (15.84) that

$$(\Delta x)^2 + (\Delta p)^2 = \hbar \cosh 2r. \tag{15.98}$$

As θ increases further from π to 2π, the path in the $(\Delta x,\, \Delta p)$ plane is retraced from D through C and B to A.

15.4.4 The Squeezing Operator and the Group SU(1, 1)

The squeezing operator $S(z)$ offers us an opportunity to discuss some basic aspects of the important Lie group $SU(1, 1)$ and its Lie algebra $\mathfrak{su}(1, 1)$.

$S(z)$ is obtained by exponentiating a linear combination of the operators K_+ and K_-, which are two of the three generators of the Lie algebra, $\mathfrak{su}(1, 1)$. Hence $S(z)$ is an element of the Lie group $SU(1, 1)$ generated by the Lie algebra. $SU(1, 1)$ is the group made up of unimodular (2×2) matrices that are "pseudo-unitary" in the sense to be specified shortly.

Before we discuss $SU(1, 1)$, here is how the squeezing operator $S(z)$ is directly connected to a (2×2) matrix $\mathsf{S}(z)$ belonging $SU(1, 1)$. Write the *operators* a and $a^\dagger$ as the elements of a (2×1) column vector. The Bogoliubov transformation given by Eq. (15.80) may then be written as

$$\begin{pmatrix} a \\ a^\dagger \end{pmatrix} \to \mathsf{S}(z) \begin{pmatrix} a \\ a^\dagger \end{pmatrix} \quad \text{where} \quad \mathsf{S}(z) = \begin{pmatrix} \cosh r & e^{i\theta} \sinh r \\ e^{-i\theta} \sinh r & \cosh r \end{pmatrix}. \tag{15.99}$$

The matrix $\mathsf{S}(z)$ is an element of the group $SU(1, 1)$, as you will see by comparing it with the general form of an $SU(1, 1)$ matrix given in Eq. (15.101) below.

The special **pseudo-unitary group** $SU(1, 1)$ is the group of (2×2) matrices with unit determinant that satisfy the *pseudo-unitarity* condition

$$U^\dagger g U = g, \quad \text{where} \quad g = \begin{pmatrix} 1 & 0 \\ 0 & -1 \end{pmatrix} \tag{15.100}$$

is the **metric matrix**. This condition can also be written as $U^{-1} = g\, U^\dagger g$ since $g^{-1} = g$. Proceeding as we did in the case of $SU(2)$ in Chap. 12, Sect. 12.2.3, we find that any element of $SU(1, 1)$ must have one of *two* possible general forms. These are given, respectively, by

$$U_+ = \begin{pmatrix} \alpha & \beta \\ \beta^* & \alpha^* \end{pmatrix}, \quad \text{where} \quad |\alpha|^2 - |\beta|^2 = 1, \tag{15.101}$$

and

$$U_- = \begin{pmatrix} \alpha & \beta \\ -\beta^* & -\alpha^* \end{pmatrix}, \quad \text{where} \quad -|\alpha|^2 + |\beta|^2 = 1. \tag{15.102}$$

Such matrices comprise the fundamental representation of the pseudo-unitary group $SU(1, 1)$. Because the expression $|\alpha|^2 - |\beta|^2$ can have either sign, the group is also called the *indefinite* special unitary group $SU(1, 1)$.

For completeness, I mention that the indefinite special unitary group $SU(p, q)$ (where p and q are positive integers) consists of all $(p + q) \times (p + q)$ matrices satisfying the conditions $\det U = 1$ and $U^\dagger g\, U = g$, where the metric matrix $g = \text{diag}\,(\underbrace{1, 1, \ldots, 1}_{p \text{ terms}}, \underbrace{-1, -1, \ldots, -1}_{q \text{ terms}})$.

★ **16.** Show that any element of $SU(1, 1)$ must be of the form given by Eq. (15.101) or Eq. (15.102).

The parameter space of $SU(1, 1)$ thus comprises the hyperboloids

$$\alpha_1^2 + \alpha_2^2 - \beta_1^2 - \beta_2^2 = 1 \quad \text{and} \quad \alpha_1^2 + \alpha_2^2 - \beta_1^2 - \beta_2^2 = -1. \tag{15.103}$$

It is obvious that the parameters can take on values that are arbitrarily large in magnitude. As a consequence, $SU(1, 1)$ is a **noncompact group**. Observe that the parameter space is not *connected*, in the sense that it is made up of two disjoint pieces.[6] The component that is *connected to the identity*, i.e., the set of matrices that can be reached by a continuous path in parameter space from the point representing the identity matrix, is the set of matrices of the U_+ type. (The identity matrix obviously corresponds to $\alpha_1 = 1,\ \alpha_2 = \beta_1 = \beta_2 = 0$.) As already stated, you can get the set of matrices of the U_- type by multiplying the metric matrix g and a matrix of the U_+ type.

We will encounter the group $SU(1, 1)$ once again in Chap. 27, Sect. 27.4.3, in connection with a group of transformations of the complex plane that leaves the unit circle unchanged.

15.4.5 $SU(1, 1)$ *Generators in Terms of Pauli Matrices*

I have already stated that the Lie algebra $\mathfrak{su}(1, 1)$ is specified by the generators $K_\pm$ and K_3 that satisfy the commutation relations (15.12) of Sect. 15.1.2. Let us write this in a form that can be compared readily with the commutation relations (15.8) satisfied by the generators J_1, J_2, and J_3 of the $\mathfrak{su}(2)$ algebra. Let $K_\pm = K_1 \pm i K_2$, so that $K_1 = \frac{1}{2}(K_+ + K_-)$ and $K_2 = \frac{1}{2}i(K_- - K_+)$. The commutation relations are then

$$\boxed{[K_1,\ K_2] = iK_3,\quad [K_2,\ K_3] = -iK_1,\quad [K_3,\ K_1] = -iK_2\,.} \tag{15.104}$$

It follows from Eq. (15.11) that these operators can be represented in terms of the ladder operators a and $a^\dagger$ as

$$\boxed{K_1 = \tfrac{1}{4}\big(a^{\dagger 2} + a^2\big),\quad K_2 = \tfrac{1}{4}i\big(a^{\dagger 2} - a^2\big),\quad K_3 = \tfrac{1}{2}\big(a^\dagger a + \tfrac{1}{2}I\big).} \tag{15.105}$$

Note that each K_i in (15.105) is represented by a self-adjoint operator. Equivalently, $K_+ = K_-^\dagger$ in this representation. Bear in mind, too, that any matrix representation of $a,\ a^\dagger$, and their higher powers must necessarily involve *infinite*-dimensional matrices. The implication of these facts will become clear shortly.

---

[6]A simpler example in three-dimensional Euclidean space will help you visualize at least some cross-sections of the hypersurface. The hyperboloid $x^2 + y^2 - z^2 = 1$ is a single connected surface of revolution about the z-axis, whose central cross-section (in the plane $z = 0$) is the unit circle $x^2 + y^2 = 1$. The hyperboloid $x^2 + y^2 - z^2 = -1$ has two disjoint pieces, an "upper" surface of revolution that has a minimum at $(0, 0, 1)$, and a "lower" surface of revolution that has a maximum at $(0, 0, -1)$.

But it is also possible to find a representation for the generators K_1, K_2, and K_3 in terms of (2×2) matrices! We can work backwards from the form of the general element of $SU(1,1)$ to find an explicit (2×2) matrix representations for the generators K_i. The result is

$$K_1 = \frac{1}{2}\begin{pmatrix} 0 & -i \\ -i & 0 \end{pmatrix}, \quad K_2 = \frac{1}{2}\begin{pmatrix} 0 & 1 \\ -1 & 0 \end{pmatrix}, \quad K_3 = \frac{1}{2}\begin{pmatrix} 1 & 0 \\ 0 & -1 \end{pmatrix}. \tag{15.106}$$

We thus have, in terms of the Pauli matrices, the representations

$$\boxed{K_1 = -\tfrac{1}{2}\, i\sigma_1, \quad K_2 = \tfrac{1}{2}\, i\sigma_2, \quad K_3 = \tfrac{1}{2}\,\sigma_3\,.} \tag{15.107}$$

★ **17.** Starting with a general (2×2) matrix belonging to $SU(1,1)$, show that the generators have the representation (15.107).

***A general property of noncompact groups*:** Observe, now, that K_1 and K_2 are now represented by *non*Hermitian matrices. Equivalently, the corresponding representations of the operators K_+ and K_- are

$$K_- = \begin{pmatrix} 0 & -i \\ 0 & 0 \end{pmatrix} \quad \text{and} \quad K_+ = \begin{pmatrix} 0 & 0 \\ -i & 0 \end{pmatrix} \neq K_-^{\dagger}. \tag{15.108}$$

This is in contrast to the self-adjoint operators representing the generators in Eq. (15.105).

This difference illustrates a general feature. In physical applications, a Lie group often arises as a group of symmetry transformations of some system. The generators of the group may correspond to physical observables, in which case one must associate self-adjoint operators with them. (For example, the generators of the rotation group $SO(3)$ or the special unitary group $SU(2)$ are the components J_i of the angular momentum operator.) The group elements corresponding to the transformations are then exponentials of linear combinations of the generators. If these are symmetry transformations, they must be represented by unitary operators. And, as you know, if T is a Hermitian matrix (or a self-adjoint operator) and θ is a real parameter, then the exponential $e^{i\theta T}$ is a unitary matrix (respectively, a unitary operator). On the other hand, it turns out that

- a *noncompact* group like $SU(1,1)$ cannot have any *finite*-dimensional unitary representations, although it does have *infinite*-dimensional representations that are unitary.

The representations of the generators in Eqs. (15.107) and (15.105) are consistent with this general result. Recall that we have already encountered this feature in the case of the Heisenberg group (which is also a noncompact group), in Sect. 15.4.1.

15.5 Solutions

1. (a) Each of the eight matrices

$$T_1^2,\ T_2^2,\ T_3^2,\ T_2\,T_1,\ T_2\,T_3,\ T_3\,T_2,\ T_1\,T_3 \text{ and } T_3\,T_1$$

reduces to the null matrix. The only nonzero matrix that is quadratic in the T_i is the matrix $T_1\,T_2 = T_3$. The commutation relations (15.5) follow at once. The relations defining a Lie algebra, Eqs. (12.54)–(12.56) of Chap. 12, Sect. 12.4.2, are also easily verified, with the commutator of two matrices as the binary operation.
(b) From the foregoing, it follows that

$$(t_1\,T_1 + t_2\,T_2 + t_3\,T_3)^2 = t_1\,t_2\,T_1\,T_2 = t_1\,t_2\,T_3\,.$$

This yields $(t_1\,T_1 + t_2\,T_2 + t_3\,T_3)^3 = 0$. The exponential required reduces to the form given by (15.6).
(c), (d) We find $g(\mathbf{t})\,g(\mathbf{t}') = g(\mathbf{s})$, where $\mathbf{s} = (s_1,\, s_2,\, s_3)$ with

$$s_1 = t_1 + t'_1,\ s_1 = t_2 + t'_2,\ s_3 = t_3 + t'_3 + \tfrac{1}{2}(t_1\,t'_2 - t_2\,t'_1).$$

Hence $g(\mathbf{t})\,g(\mathbf{t}')$ has the same form as an element of the group. Note carefully, however, that $\mathbf{s} \neq \mathbf{t} + \mathbf{t}'$, because of the extra final term in s_3. We have

$$g(\mathbf{t})\,g(\mathbf{t}') = g(\mathbf{t} + \mathbf{t}') + \tfrac{1}{2}(t_1\,t'_2 - t_2\,t'_1)\,T_3\,.$$

The presence of this term also implies that $g(\mathbf{t})\,g(\mathbf{t}') \neq g(\mathbf{t}')\,g(\mathbf{t})$: the Heisenberg group is not a commutative group. The commutator of two general elements is given by

$$[g(\mathbf{t}),\, g(\mathbf{t}')] = (t_1\,t'_2 - t_2\,t'_1)\,T_3\,.$$

Although the Heisenberg group is *not* a commutative group, it comes close to being one. It is an example of what is known as a **nilpotent** Lie group.[7]
(e) Setting $\mathbf{t}' = -\mathbf{t}$, it is easy to see that $g(\mathbf{t})\,g(-\mathbf{t}) = g(\mathbf{0}) = I$, the identity matrix. It follows that $g^{-1}(\mathbf{t}) = g(-\mathbf{t})$, which yields Eq. (15.7).
(f) Inspection shows that all three rows of the general matrix $g(\mathbf{t})$ are linearly independent. The eigenvalues of this upper triangular matrix are its diagonal elements, namely, 1, 1, and 1. There are no zero eigenvalues. Hence the rank of $g(\mathbf{t})$ is 3, and its nullity is 0.
(g) The only right eigenvector of $g(\mathbf{t})$ corresponding to the three-fold degenerate eigenvalue 1 is proportional to $\begin{pmatrix}1 & 0 & 0\end{pmatrix}^{\mathrm{T}}$. Hence it is a defective matrix, and cannot be diagonalized by a similarity transformation. ▶

[7] For further details on what this means, you must refer to a text on group theory.

2. (a) It is trivially shown that $[a^n,\, a^\dagger] = n\, a^{n-1}$ and $[a,\, a^{\dagger\, n}] = n\, a^{\dagger\, n-1}$.
(b) In general, for any function $f(a)$ that can be expanded in a power series in a, we have $[f(a),\, a^\dagger] = f\,'(a)$, where the prime denotes the derivative. Therefore $[e^a,\, a^\dagger] = e^a$. This gives $e^a\, a^\dagger = a^\dagger\, e^a + e^a$. Multiplying by e^{-a} from the right, we get $e^a\, a^\dagger\, e^{-a} = a^\dagger + I$. Taking the adjoint of each side of this relation, $e^{-a^\dagger}\, a\, e^{a^\dagger} = a + I$.

Remark (i) Similarly, for any function $g(a^\dagger)$ that can be expanded in a power series in $a^\dagger$, we have $[a,\, g(a^\dagger)] = g\,'(a^\dagger)$.
(ii) The identities in (b) above follow immediately from Hadamard's lemma, to be discussed in Sect. 15.2.2. ▶

3. (a) $[a,\, a^\dagger] = I,\ [a,\, N] = a,\ [N,\, a^\dagger] = a^\dagger$, and of course all the operators concerned commute with I.
(b) $[a^n,\, N] = na^n\,;\ \ [e^a,\, N] = a\, e^a\,;\ \ [N,\, a^{\dagger\, n}] = na^{\dagger\, n}\,;\ \ [N,\, e^{a^\dagger}] = a^\dagger\, e^{a^\dagger}$.
(c) $[a^2,\, a^{\dagger\, 2}]{=}4(N + \frac{1}{2}I);\ \ [a^2,\, (N + \frac{1}{2}I)]{=}2a^2;\ \ [a^{\dagger\, 2},\, (N + \frac{1}{2}I)]{=} - 2a^{\dagger\, 2}$. ▶

5. Start with the identity $A + \epsilon\, B = A(I + \epsilon\, A^{-1} B)$, take the inverse of each side of the equation, and expand the resolvent operator $(I + \epsilon\, A^{-1} B)^{-1}$ in powers of ϵ. ▶

6. Express each side of Eq. (15.18) as a power series in λ using Eqs. (15.20) and (15.19), and equate the coefficients of like powers of λ. Note that $F(0) = B$. Determine the successive derivatives $F^{(n)}(0)$ recursively, to arrive at the formula (15.16) for $F(\lambda)$. ▶

7. (a) Let $C = [A,\, B]$, as usual. Note that, if $[A,\, C] = 0$, then $[A^n,\, B] = n A^{n-1}\, C$. Hence

$$[e^{\lambda A},\, B] = \lambda\, C\, e^{\lambda A}, \quad \text{so that} \quad e^{\lambda A}\, B = B\, e^{\lambda A} + \lambda\, C\, e^{\lambda A}.$$

Right-multiplying both sides of this equation by $e^{-\lambda A}$, we get Eq. (15.21).
(b) Here is a very useful trick to establish such operator identities: work with suitable *representations* of the operators concerned, if these can be found, and then revert to the abstract operators themselves. In the present instance, consider the space of infinitely differentiable functions of x. Then, if $A = d/dx$ and $B = e^{\beta x}$, we have $[A,\, B] = \beta B$. Now, for any infinitely differentiable function $f(x)$,

$$e^{\lambda d/dx}\, e^{\beta x}\, e^{-\lambda d/dx}\, f(x) = e^{\lambda d/dx} \left[e^{\beta x}\, f(x - \lambda)\right] = e^{\beta(x+\lambda)}\, f(x) = e^{\lambda\beta}\, e^{\beta x}\, f(x).$$

Reverting to the general operators A and B, Eq. (15.23) follows. ▶

8. If $[B,\, C] = 0$, then $[A,\, B^n] = n B^{n-1}\, C$, and hence $[A,\, e^B] = e^B\, C$. Insert this in (15.24) and simplify, using also the fact that $[A,\, C] = 0$. ▶

9. Differentiating $G(\lambda)$ with respect to λ,

$$\begin{aligned}G'(\lambda) &= A\,e^{\lambda A}\,e^{\lambda B} + e^{\lambda A}\,B\,e^{\lambda B} = A\,e^{\lambda A}\,e^{\lambda B} + e^{\lambda A}\,B\,e^{-\lambda A}\,e^{\lambda A}\,e^{\lambda B}\\ &= [A + e^{\lambda A}\,B\,e^{-\lambda A}]\,G(\lambda).\end{aligned}$$

Using Hadamard's lemma in the special case when A commutes with the commutator $C = [A,\,B]$ (Eq. (15.21)), we have

$$G'(\lambda) = (A + B + \lambda C)\,G(\lambda).$$

This equation for $G'(\lambda)$ can be integrated over λ, *because the two operators* $(A + B)$ *and* C *commute with each other*. The result is

$$G(\lambda) = e^{\lambda(A+B)+\frac{1}{2}\lambda^2 C}\,G(0) = e^{\lambda(A+B)+\frac{1}{2}\lambda^2 C},$$

using the boundary condition $G(0) = I$. Setting $\lambda = 1$,

$$G(1) = e^A\,e^B = e^{A+B+\frac{1}{2}C} = e^{A+B}\,e^{\frac{1}{2}C},$$

again because $A + B$ commutes with C. Right-multiplying by $e^{-\frac{1}{2}C}$, we arrive at the identity (15.28). ▶

11. (b) It suffices to check out the special cases in which $\mathbf{r}$ is respectively equal to $\mathbf{e}_x = (1, 0, 0)$, $\mathbf{e}_y = (0, 1, 0)$, and $\mathbf{e}_z = (0, 0, 1)$.
(c) Use the identities involving $\boldsymbol{\sigma}$ in Eqs. (11.11) and (11.12) of Chap. 11, Sect. 11.1.2. ▶

12. The Zassenhaus formula immediately yields Eq. (15.60) (and, similarly, Eq. (15.61)) from Eq. (15.59), because $[a, a^\dagger] = I$. It follows from $a\,|\,0\,\rangle = 0$ that $e^{-z^*a}|\,0\,\rangle = |\,0\,\rangle$. Expand the exponential $e^{z\,a^\dagger}$ in a power series and repeatedly use the relation $a^\dagger\,|\,k\,\rangle = \sqrt{(k+1)}\,|\,k+1\,\rangle$, where k is any nonnegative integer. Equation (15.63) follows on using the relation $a\,|\,n\,\rangle = \sqrt{n}\,|\,n-1\,\rangle$ and simplifying. ▶

13. The simplified version (15.31) of the Baker–Campbell–Hausdorff formula that is applicable here yields Eq. (15.65). Interchange z and w to write down $D(w)\,D(z)$. Equation (15.66) follows at once. ▶

14. Equations (15.73) and (15.74) are easily derived using Eq. (14.46) and the fact that $a\,|\,z\,\rangle = z\,|\,z\,\rangle$, $\langle\,z\,|\,a^\dagger = z^*\,\langle\,z\,|$. ▶

15. Use Hadamard's lemma, Eq. (15.16), to derive the first of Eq. (15.80). $S^\dagger\,a^\dagger\,S$ is just the adjoint of $S^\dagger\,a\,S$. Write $S^\dagger\,a^2\,S = S^\dagger\,a\,S\,S^\dagger\,a\,S$, etc., and make repeated use of Eq. (15.80). ▶

16. Let

$$U = \begin{pmatrix} \alpha & \beta \\ \gamma & \delta \end{pmatrix}, \quad \text{so that} \quad U^\dagger g\, U = \begin{pmatrix} |\alpha|^2 - |\gamma|^2 & \alpha^*\beta - \gamma^*\delta \\ \alpha\beta^* - \gamma\delta^* & |\beta|^2 - |\delta|^2 \end{pmatrix}.$$

Setting $U^\dagger g\, U = g$ gives the relations

$$|\alpha|^2 - |\gamma|^2 = 1, \quad |\beta|^2 - |\delta|^2 = -1, \quad \alpha^*\beta = \gamma^*\delta.$$

Using the third relation in the first two leads to $|\alpha| = |\delta|$ and $|\beta| = |\gamma|$. Therefore we may set $\delta = \alpha^*\, e^{i\theta}$ and $\gamma = \beta^*\, e^{i\phi}$, where ϕ and θ are real numbers. The condition $\alpha^*\beta = \gamma^*\delta$ then gives $e^{i(\theta-\phi)} = 1$, or $\theta - \phi = 2\pi$. Hence U must be of the form

$$U = \begin{pmatrix} \alpha & \beta \\ e^{i\phi}\beta^* & e^{i\phi}\alpha^* \end{pmatrix}.$$

Since $\det\, U = e^{i\phi}(|\alpha|^2 - |\beta|^2)$ must be real (and equal to unity), ϕ can only be 0 or π. Correspondingly, U must have the general form U_+ or U_- as given by Eqs. (15.101) and (15.102), respectively.

Remark If $U^\dagger g\, U = g$, then $(gU)^\dagger g\,(gU) = g$ as well, because $g^2 = I$. The matrix U_- is just $g\, U_+$. ▶

17. For this purpose, you must start with a matrix of the form U_+ in the connected component of $SU(1, 1)$, that is infinitesimally close to the identity element. This is a matrix in which $\alpha_1 = 1$ and the other three parameters are infinitesimals. Write it in the form

$$\begin{pmatrix} 1 + i\,\delta\alpha_2 & \delta\beta_1 + i\,\delta\beta_2 \\ \delta\beta_1 - i\,\delta\beta_2 & 1 - i\,\delta\alpha_2 \end{pmatrix} = \begin{pmatrix} 1 & 0 \\ 0 & 1 \end{pmatrix} + i\,(\delta\beta_1)\begin{pmatrix} 0 & -i \\ -i & 0 \end{pmatrix} + i\,(\delta\beta_2)\begin{pmatrix} 0 & 1 \\ -1 & 0 \end{pmatrix}$$
$$+\, i\,(\delta\alpha_2)\begin{pmatrix} 1 & 0 \\ 0 & -1 \end{pmatrix}.$$

The three matrices multiplying the infinitesimal parameters on the right-hand side, multiplied by $\frac{1}{2}$, satisfy the commutation relations (15.104) obeyed by K_1, K_2, and K_3. Hence they represent the infinitesimal generators of the group. It is evident that they are, respectively, $-\frac{1}{2}\,i\sigma_1$, $\frac{1}{2}\,i\sigma_2$, and $\frac{1}{2}\,\sigma_3$. ▶

Chapter 16
Orthogonal Polynomials

16.1 General Formalism

16.1.1 Introduction

Families of **orthogonal polynomials** play a very important role in numerous applications in mathematical physics, as part of the class of so-called **special functions**. This is a classical topic in mathematics and has been the subject of investigation for centuries now. Many of these special functions were first introduced into mathematics as solutions of specific second-order differential equations. Second-order ordinary differential equations comprise one of the important topics that I do not discuss in this book owing to limitations of space, given that you are very likely to be already familiar with the rudiments of this topic. In any case, in applications we are interested primarily in the properties of the *solutions* of these equations. Between this chapter and those on analytic functions of a complex variable, there will be a fair amount of discussion of these properties.

Special functions also have a large number of other interesting properties that have deep implications. For instance, they

- satisfy difference equations, i.e., recursion relations;
- obey various "addition theorems";
- have diverse series and integral representations;
- are closely related to group representations;
- have analytic continuations to complex argument as well as order;

and so on. The list of their properties and relationships is truly staggering, and there exist several well-known sets of volumes devoted to this subject.

In order to present a concise summary of so vast a subject, the following plan has been adopted. I begin with a listing of some *general* properties of sets or families of orthogonal polynomials. Some of these are not usually considered in elementary treatments of the subject. But they should help you gain a perspective on the subject.

V. Balakrishnan, *Mathematical Physics*,
https://doi.org/10.1007/978-3-030-39680-0_16

Next, I emphasize and use the fact that *there are basically three distinct families of orthogonal polynomials* (in one variable). The accompanying exercises will focus on the specific cases of the most commonly occurring polynomials, such as Hermite and Legendre polynomials. Later on, when we study analytic functions of a complex variable, I will return to more properties of some of these functions as well as some other special functions (such as Bessel functions).

I have already mentioned that the literature on the subject is vast. Most texts on mathematical physics have chapters devoted to special functions and orthogonal polynomials. The emphasis differs considerably from one book to another. But there is another minor, and yet annoying, problem with this particular topic: the considerable variation in notational conventions, normalization factors, etc. It becomes necessary, therefore, to stick to a uniform and well-specified notation. The monograph by A. F. Nikiforov and V. B. Uvarov, *Special Functions of Mathematical Physics*, is a modern, rigorous, and exhaustive discussion of the subject (see the Bibliography). In the summary account given in this chapter, I have followed the treatment and notation of Chaps. I and II of this book (referred to as NU), wherever relevant. I have also listed a few of the classics in the Bibliography. Some of them were first published about a century ago, but have remained in print ever since!

Keeping in mind the discussion of linear vector spaces in the preceding chapters, I find it most convenient to introduce orthogonal polynomials from the following point of view:

- Orthogonal polynomials constitute basis sets in suitable linear spaces of functions of a real variable x in a specified interval $(a,\ b)$.

This is the point of view that is emphasized in this chapter. Function spaces are generally infinite-dimensional. Hence a basis in such a space consists of an infinite set of polynomials. We restrict ourselves to real vector spaces, and the basis sets will comprise only real functions of x. In order to keep open the possibility of a potential application to complex-valued functions, however, in the next section I shall formally retain the complex conjugation symbol wherever it occurs. The abstract notation for the set of basis vectors will be $\{|\phi_n\rangle\}$ as usual. In a given function space, their explicit representation will be denoted by the functions $\phi_n(x),\ n = 0,\ 1,\ \ldots$. That is, in the basis $\{|\,x\,\rangle\}$ of "position" eigenkets, we have (recalling Eq. (13.14) of Chap. 13, Sect. 13.2.2)

$$\langle\, x\,|\phi_n\,\rangle \equiv \phi_n(x), \quad \text{and hence} \quad \langle\phi_n|\,x\,\rangle = \phi_n^*(x)\,. \tag{16.1}$$

16.1.2 Orthogonality and Completeness

We are interested here in families of *polynomials* in x. Clearly, if the range $(a,\ b)$ is infinite or semi-infinite, polynomials will not be integrable in the range concerned—the integrals will diverge as $|x| \to \infty$. This difficulty has been mentioned earlier, in Chap. 13, Sect. 13.2.3, in connection with the space $\mathcal{L}_2(-\infty\,,\ \infty)$. We have also seen how the difficulty may be overcome. The way to ensure that inner products remain finite is to introduce a suitable convergence factor or weight function $\rho(x)$, such as e^{-x^2}, whenever an integration over x is involved. The orthonormality relation must be generalized to incorporate such a weight function.[1] Moreover, the normalization of the functions $\phi_n(x)$ may be such that there is also an n-dependent factor on the right-hand side, multiplying the Kronecker delta function. (Normalization is generally a matter of convention.) Taking these features into account, we proceed as follows. All the steps have been spelt out in order to make things as clear as possible.

Let us start with the orthogonality condition among the basis vectors,

$$\langle \phi_n | \phi_m \rangle = A_n \, \delta_{nm} \quad \text{(no summation over } n\text{)}, \tag{16.2}$$

where the positive constant A_n is the normalization constant. When $A_n = 1$, orthogonality becomes orthonormality. This is just a matter of terminology, since we can always take the basis vector to be $A_n^{-1/2} \, |\phi_n\rangle$ instead of $|\phi_n\rangle$. But all the formulas that follow will then have to be changed appropriately. It is helpful to retain the normalization constant A_n explicitly, because it helps in matching the standard conventions used in the definitions of different families of orthogonal polynomials.

Next, we use the completeness relation satisfied by the continuous basis $\{| \, x \,\rangle\}$. Instead of the usual relation $\int_a^b dx \, | \, x \,\rangle \langle \, x \, | = I$, we now have

$$\int_a^b dx \, \rho(x) \, | \, x \,\rangle \langle \, x \, | = I, \tag{16.3}$$

where the weight function $\rho(x)$ is strictly positive for $a < x < b$. It may remain finite, or vanish, or diverge at either or both of the end-points a and b. The case $\rho\,(x) = 1$ corresponds to the so-called **Lebesgue measure**—in simple terms, it means that we just integrate any function after multiplying it by dx. When $\rho\,(x)$ is not unity but some positive function of x in $(a, \, b)$, we say that the integration is with respect to an **integration measure** $\rho\,(x)\,dx$ that is different from the Lebesgue measure.[2] Inserting the expansion (16.3) for the unit operator in the relation

$$\langle \phi_n | \phi_m \rangle = \langle \phi_n | I \, | \phi_m \rangle = A_n \, \delta_{nm}, \tag{16.4}$$

[1]This is just one of the motivating reasons for the generalization. There are other reasons as well. For instance, orthogonal polynomials are related to group representations, and weight functions occur naturally in that context, as so-called *invariant measures over group manifolds*.

[2]It is customary to write $\rho(x)\,dx \equiv d\mu(x)$ for the measure of integration. If the indefinite integral (or *primitive*) of the weight $\rho\,(x)$ exists in explicit form, then we would have $\mu(x) = \int \rho\,(x)\,dx$. But $\mu(x)$ may not even exist as an elementary function, or even in explicit form, for a general $\rho\,(x)$—nor is it *required* to do so. For example, there is no elementary function $\mu(x)$ such that $d\mu(x)/dx = e^{-x^2}$. To keep things simple, I shall just write out the measure as $\rho\,(x)\,dx$.

it follows that

$$\boxed{\int_a^b dx\,\rho\,(x)\,\phi_n^*(x)\,\phi_m(x) = A_n\,\delta_{nm} \qquad \textbf{(orthogonality)}.} \tag{16.5}$$

Setting $n = m$, we have

$$A_n = \int_a^b dx\,\rho(x)\,|\phi_n(x)|^2\,. \tag{16.6}$$

Another consequence of the introduction of the integration measure $\rho(x)\,dx$ must be noted. Take the definition of any function $f(x)$ where $x \in [a,\, b]$, namely, $f(x) \equiv \langle\, x\,|f\,\rangle$. Insert the unit operator on the right-hand side, so that $\langle\, x\,|f\,\rangle = \langle\, x\,|I\,|\,f\,\rangle$. Using the completeness relation for the position eigenkets, we have

$$f(x) = \int_a^b dx\,'\,\rho(x\,')\,\langle\, x\,|\,x\,'\,\rangle\,\langle\, x\,'\,|\,f\,\rangle = \int_a^b dx\,'\,\rho(x\,')\,\langle\, x\,|\,x\,'\,\rangle\,f(x\,'). \tag{16.7}$$

But $f(x) = \int_a^b dx\,\delta(x - x\,')\,f(x\,')$. Comparing the two expressions, we conclude that

- it is $\rho(x\,')\,\langle\, x\,|\,x\,'\,\rangle$, rather than just $\langle\, x\,|\,x\,'\,\rangle$, that now acts like the Dirac δ-function $\delta(x - x\,')$.

Effectively, the inner product $\langle\, x\,|\,x\,'\,\rangle$ becomes a *weighted* δ-function:

$$\langle\, x\,|\,x\,'\,\rangle = \frac{\delta(x - x\,')}{\rho(x\,')} = \frac{\delta(x - x\,')}{\rho(x)} = \frac{\delta(x - x\,')}{\sqrt{\rho(x)\,\rho(x\,')}}\,. \tag{16.8}$$

The last two equations in (16.8) follow because the δ-function allows us to replace x by $x\,'$, and vice versa, in all other factors. The final form above is convenient because it is symmetric in x and $x\,'$, just like $\delta(x - x\,')$ itself.

Now consider the completeness relation of the basis set $\{|\phi_n\rangle\}$, namely,

$$\sum_{n=0}^{\infty} \frac{|\phi_n\rangle}{\sqrt{A_n}}\,\frac{\langle\phi_n|}{\sqrt{A_n}} = I. \tag{16.9}$$

Applying $\langle\, x\,|$ from the left and $|\,x\,'\,\rangle$ from the right, we get

$$\sum_{n=0}^{\infty} \frac{\phi_n(x)\,\phi_n^*(x\,')}{A_n} = \langle\, x\,|\,x\,'\,\rangle = \frac{\delta(x - x\,')}{\sqrt{\rho(x)\,\rho(x\,')}}\,. \tag{16.10}$$

Thus the completeness relation reads

$$\boxed{\sum_{n=0}^{\infty} \frac{1}{A_n} \sqrt{\rho(x)\,\rho(x')}\,\phi_n(x)\,\phi_n^*(x') = \delta(x-x') \qquad \textbf{(completeness)}.} \tag{16.11}$$

16.1.3 Expansion and Inversion Formulas

Let $\{\phi_n(x)\}$ be a family of *real-valued* orthogonal polynomials defined in $x \in [a, b]$. An arbitrary[3] function $f(x)$ in the function space can be expanded in terms of the complete set of basis functions as

$$f(x) = \sum_{n=0}^{\infty} f_n\,\phi_n(x) \qquad \textbf{(expansion formula)}. \tag{16.12}$$

It follows from the orthogonality relation (16.5) satisfied by the polynomials that the general coefficient f_n is given by

$$\boxed{f_n = \frac{1}{A_n} \int_a^b dx\, \rho(x)\,\phi_n(x)\, f(x) \qquad \textbf{(inversion formula)}.} \tag{16.13}$$

The expansion is unique: $f(x)$ is uniquely determined by the set of coefficients $\{f_n\}$, and vice versa. In using such expansions, however, we need to be more precise. The *class* of functions $f(x)$ that can be expanded as above must be specified.

In general, the expansion for $f(x)$ is valid "almost everywhere", i.e., at all points except a set of measure zero. The sense in which it converges to $f(x)$ must also be stated. Similarly, given the coefficients $\{f_n\}$, the function $f(x)$ is determined everywhere except on a set of measure zero. I do not go into these technicalities here, as our focus is on the orthogonal polynomials themselves. When necessary (e.g., in connection with the expansion of a function in terms of Legendre polynomials), these details will be specified.

It is clear that each $\phi_n(x)$ is orthogonal, not only to every other $\phi_m(x)$, but also to every polynomial of degree less than n. This result is made explicit by the equation

$$\int_a^b dx\, \rho(x)\, x^m\, \phi_n(x) = 0 \quad \text{for all} \quad m < n, \tag{16.14}$$

for any given n.

★ **1.** Establish Eq. (16.14).

[3]But see the remarks that follow shortly.

***Parseval's Theorem*:** Let $f(x)$ be a real-valued square-integrable function in (a, b). If its expansion in terms of the set of orthogonal polynomials $\{\phi_n(x)\}$ is given by

$$f(x) = \sum_{n=0}^{\infty} f_n\, \phi_n(x), \tag{16.15}$$

then

$$\sum_{n=0}^{\infty} A_n\, f_n^2 = \int_a^b dx\, \rho(x)\, f^2(x). \tag{16.16}$$

This is yet another version of Parseval's Theorem (or Parseval's formula), which we encountered first in Chap. 10, Sect. 10.2.3. As I have already mentioned there, the term is often reserved for the special case of this result as applied to a Fourier series expansion or a Fourier integral representation. But the underlying formula is more generally applicable, and depends only on the fact that $f(x) \in \mathcal{L}_2(a, b)$.

★ **2.** Derive Eq. (16.16).

16.1.4 Uniqueness and Explicit Representation

The following is a crucial and quite remarkable result:

- Once the interval (a, b) and the weight $\rho(x)$ are specified, the family of polynomials $\{\phi_n(x)\}$ satisfying the orthonormality relation is essentially uniquely determined, up to a multiplicative normalization constant.

This constant may be chosen in more than one way. One often finds that different authors use slightly different definitions for various standard orthogonal polynomials, based on different choices of the normalization constant. I shall follow the most frequently adopted definitions in this regard.

How, precisely, does a knowledge of the interval (a, b) and the weight $\rho(x)$ determine the set of polynomials $\{\phi_n(x)\}$? The zeroth-order polynomial is, of course, a constant:

$$\phi_0(x) = C_0\,. \tag{16.17}$$

For $n \geq 1$, we have the following explicit formula.[4] Let M_k be the kth moment of the weight function, i.e.,

$$M_k = \int_a^b dx\, \rho(x)\, x^k, \quad \text{where} \quad k = 0, 1, \ldots\,. \tag{16.18}$$

[4]NU, p. 34: Eq. (6) of Chap. II, Sect. 6.2.

Note that each M_k is a constant (independent of x). Then $\phi_n(x)$ is expressible as the determinant of an $(n+1) \times (n+1)$ matrix, according to

$$\phi_n(x) = (-1)^n \, C_n \begin{vmatrix} 1 & x & x^2 & \cdots & x^n \\ M_0 & M_1 & M_2 & \cdots & M_n \\ M_1 & M_2 & M_3 & \cdots & M_{n+1} \\ \vdots & \vdots & \vdots & \ddots & \vdots \\ M_{n-1} & M_n & M_{n+1} & \cdots & M_{2n-1} \end{vmatrix} \quad (n \geq 1), \tag{16.19}$$

where C_n is a normalizing constant.

★ **3.** Using the representation of Eq. (16.19) for $\phi_n(x)$, prove Eq. (16.14).

16.1.5 Recursion Relation

Orthogonal polynomials obey a number of **recursion relations**, i.e., difference equations in the order n. I merely quote[5] the basic relation from which several other recursion relations may be deduced by using various properties of the polynomials. Let a_n and b_n denote the respective coefficients of x^n and x^{n-1} in the polynomial $\phi_n(x)$, i.e.,

$$\phi_n(x) = a_n \, x^n + b_n \, x^{n-1} + \cdots . \tag{16.20}$$

Then, for all $n \geq 1$, $\phi_n(x)$ satisfies the recursion relation

$$\frac{a_n}{a_{n+1}} \, \phi_{n+1}(x) + \Big(\frac{b_n}{a_n} - \frac{b_{n+1}}{a_{n+1}} - x\Big)\phi_n(x) + \frac{a_{n-1}}{a_n} \frac{A_n}{A_{n-1}} \, \phi_{n-1}(x) = 0 . \tag{16.21}$$

Here a_{n+1} and b_{n+1} denote, respectively, the coefficients of x^{n+1} and x^n in the polynomial $\phi_{n+1}(x)$. Similarly, a_{n-1} is the coefficient of x^{n-1} in the polynomial $\phi_{n-1}(x)$. Note that the recursion relation is a "three-step" recursion relation, i.e., it connects any three consecutive polynomials ϕ_{n-1}, ϕ_n and ϕ_{n+1}. Such a relation is the discrete analog of a *second*-order ordinary differential equation.

16.2 The Classical Orthogonal Polynomials

16.2.1 Polynomials of the Hypergeometric Type

Ordinary differential equations and special functions are properly discussed in terms of an independent variable z that is a *complex* variable, rather than just a real vari-

[5]NU, pp. 36 and 37: Eqs. (7) and (11) of Chap. II, Sect. 6.3.

able x. This brings out the true nature and scope of their analytic properties. We are, however, primarily interested here in orthogonal polynomials in a real function space. I shall therefore continue to use x instead of z, except for a brief interlude in Sect. 16.2.2 below.

In Sect. 16.1, you have had a glimpse of the approach to orthogonal polynomials as basis sets in function spaces, constructed by specifying the interval (a, b) and the weight function $\rho(x)$. The subject can be developed from another, seemingly totally different, angle. In this approach, a fundamental role is played by the **hypergeometric differential equation**. This is a second-order ordinary differential equation of the form[6]

$$\sigma(x)\,\frac{d^2\phi}{dx^2}+\tau(x)\,\frac{d\phi}{dx}+\lambda\,\phi=0, \tag{16.22}$$

where $\sigma(x)$, $\tau(x)$ and λ are polynomials of order at most 2, 1, and 0, respectively. That is,

$$\sigma(x)=\sigma_2\,x^2+\sigma_1\,x+\sigma_0,\quad \tau(x)=\tau_1\,x+\tau_0, \tag{16.23}$$

where $\sigma_2,\ \sigma_1,\ \sigma_0,\ \tau_1,\ \tau_0$, and λ are constants. We are only interested here in the relationship of the equation with orthogonal polynomials. The discussion is therefore restricted to what is relevant to our immediate purposes. (But see the brief digression that follows in Sect. 16.2.2.) The crucial points are as follows:

(i) The hypergeometric equation (16.22) has a *polynomial* solution $\phi_n(x)$ whenever the constant λ takes on certain special values λ_n, where

$$\lambda_n=-\big[n(n-1)\sigma_2+n\,\tau_1\big]\ \text{ and }\ n=0,1,\ldots. \tag{16.24}$$

Such solutions are called **polynomials of the hypergeometric type.**

(ii) All the so-called **classical orthogonal polynomials** are polynomials of the hypergeometric type, satisfying the following additional conditions on the weight function $\rho(x)$:

$$\big[\sigma(x)\,\rho(x)\big]_{x=a}=0\ \text{ and }\ \big[\sigma(x)\,\rho(x)\big]_{x=b}=0. \tag{16.25}$$

(iii) There are essentially only three distinct possible choices for the interval (a, b), namely: infinite, semi-infinite, or finite. By suitable linear changes of the independent variable (i.e., by letting $x\mapsto c_1\,x+c_2$), we can take the interval to be

$$\text{I. } (a, b)=(-\infty, \infty)\qquad \text{II. } [a, b)=[0, \infty)\qquad \text{III. } [a, b]=[-1, 1].$$

We shall assume henceforth that this has been done in all cases.

[6]NU, p. 2: Eq. (9) of Chap. I, Sect. 1.

(iv) Corresponding to the three intervals above, there are three distinct classes of orthogonal polynomials, broadly classified into the **Hermite, generalized Laguerre**, and **Jacobi types**, respectively. After suitable changes of variables, they can be brought to the so-called canonical or standard forms, which are given below.

The hypergeometric equation is so fundamental that a brief digression to mention a few salient points is worthwhile, before we resume the discussion of orthogonal polynomials.

16.2.2 The Hypergeometric Differential Equation

There is a very elaborate theory of the hypergeometric differential equation and its solutions that forms the backbone of the subject of second-order ordinary differential equations. As already mentioned, the general second-order linear differential equation for a scalar function ϕ must really be studied in the *complex* domain, i.e., the independent variable x should be replaced by a complex variable z. For our present purposes, however, I will merely state some of the relevant results. In the process, it will be necessary to use some terminology from the theory of analytic functions of a complex variable, anticipating the discussion in Chaps. 22–26.

The most general linear second-order differential equation is of the form

$$A(z)\,\frac{d^2\phi}{dz^2} + B(z)\,\frac{d\phi}{dz} + C(z)\,\phi = 0. \tag{16.26}$$

The point $z = a$ is a **regular singular point** of the equation if the function $B(z)/A(z)$ has a simple pole, and/or the function $C(z)/A(z)$ has at most a double pole, at that point. Any second-order ordinary differential equation with three regular singular points can be brought, by suitable transformations, to the hypergeometric form (16.22), namely,

$$\sigma(z)\,\frac{d^2\phi}{dz^2} + \tau(z)\,\frac{d\phi}{dz} + \lambda\,\phi = 0. \tag{16.27}$$

(In an abuse of notation, ϕ and z have been used to denote the transformed variables as well.) Such an equation has two linearly independent solutions. Our interest here is in possible *polynomial* solutions.

The hypergeometric function: The general hypergeometric equation (16.27) can be converted by elementary transformations to the standard form in which this equation is usually written. Re-labeling the independent variable (after the transformation) as z once again, this standard form (which you will find in textbooks) is

$$\boxed{z(1-z)\,\frac{d^2\phi}{dz^2} + [c-(a+b+1)z]\,\frac{d\phi}{dz} - ab\,\phi = 0,} \tag{16.28}$$

where a, b, and c are constants (or parameters). The regular singular points of this equation are at $z = 0$, 1, and ∞. The two linearly independent solutions can be written in terms of **hypergeometric functions**. In the neighborhood of $z = 0$, these solutions are given (when $c \neq$ integer) by

$$ {}_2F_1(a, b; c; z) \quad \text{and} \quad z^{1-c}\, {}_2F_1(a + 1 - c, b + 1 - c; 2 - c; z), \tag{16.29}$$

respectively. The hypergeometric function ${}_2F_1(a, b; c; z)$ is defined, for $|z| < 1$ and $c \neq 0, -1, -2, \ldots$, by the infinite series

$$\boxed{{}_2F_1(a, b; c; z) = \sum_{n=0}^{\infty} \frac{\Gamma(n + a)}{\Gamma(a)} \frac{\Gamma(n + b)}{\Gamma(b)} \frac{\Gamma(c)}{\Gamma(n + c)} \frac{z^n}{n!}.} \tag{16.30}$$

This power series converges absolutely for $|z| < 1$. The first of the solutions in (16.29) is the regular solution (at $z = 0$) of the differential equation. The other solution is the so-called singular solution, because of the presence of the factor z^{1-c} (remember that c is not an integer, in general). As indicated earlier, the hypergeometric function is actually an analytic function of the complex variable z. It is defined for all values of z by the process of analytic continuation.[7] In fact, the proper way to understand the beautifully intricate analytic properties of functions such as ${}_2F_1(a, b; c; z)$ is to study it as a function of several complex variables—in this case, z *as well as* the parameters a, b, and c. (This point of view was already appreciated and espoused by Gauss himself.)

To return to the aspect of immediate interest to us: the condition (16.24) under which the regular solution of the hypergeometric equation reduces to a polynomial of finite order can be stated more simply in terms of the parameters appearing in ${}_2F_1(a, b; c; z)$. We assume that c is not a negative integer or zero. This is the generic case. The special case $c = -n$ can be handled separately, and is not of interest to us here.

- The infinite series in Eq. (16.30) terminates, and ${}_2F_1(a, b; c; z)$ reduces to a polynomial of order n, when either a or b is equal to $-n$, where $n = 0, 1, 2, \ldots$.

The confluent hypergeometric function: Some of the orthogonal polynomials to be considered below are actually special cases of the **confluent hypergeometric function** ${}_1F_1(a; c; z)$, rather than the hypergeometric function. This function is the regular solution (at $z = 0$) of the **confluent hypergeometric equation**

$$\boxed{z\,\frac{d^2\phi}{dz^2} + (c - z)\,\frac{d\phi}{dz} - a\,\phi = 0,} \tag{16.31}$$

where a and c are constant parameters. This equation arises as a limiting case when the regular singular points at 1 and ∞ of the hypergeometric equation coalesce to

[7] We will discuss analytic continuation in Chap. 25.

form an irregular singular point at ∞ (hence the adjective, "confluent"). The function ${}_1F_1(a;\ c;\ z)$ is given, for $c \neq 0, -1, -2, \ldots$, by the infinite series

$$\boxed{{}_1F_1(a;\ c;\ z) = \sum_{n=0}^{\infty} \frac{\Gamma(n+a)}{\Gamma(a)}\ \frac{\Gamma(c)}{\Gamma(n+c)}\ \frac{z^n}{n!}.} \tag{16.32}$$

This power series for ${}_1F_1(a;\ c;\ z)$ is absolutely convergent for *all* finite values of $|z|$ (i.e., for all $|z| < \infty$). In contrast, the power series (16.30) for the hypergeometric function ${}_2F_1(a,\ b;\ c;\ z)$ is absolutely convergent only if $|z| < 1$, as already stated. You can readily verify these statements using the **ratio test**.[8] Use Stirling's formula to express the large-n behavior of the gamma functions in both series. The essential result required is that, if α and β are independent of n, then the ratio

$$\frac{\Gamma(n+\alpha)}{\Gamma(n+\beta)} \xrightarrow[n\to\infty]{} n^{\alpha-\beta}. \tag{16.33}$$

★ **4.** Use Eq. (16.33) and the ratio test to verify that the series (16.30) is absolutely convergent when $|z| < 1$, whereas the series (16.32) is absolutely convergent for all finite values of $|z|$.

The general hypergeometric function ${}_pF_q$: For the sake of completeness, I mention here that the general hypergeometric function ${}_pF_q$ is defined in the neighborhood of $z = 0$ by a power series in z that involves p "numerator" parameters $\{a_i\}$ and q "denominator" parameters $\{c_i\}$. Let the symbol $[a]_n$ stand for the ratio $\Gamma(n + a)/\Gamma(a)$. Then

$${}_pF_q(a_1,\ a_2,\ \ldots,\ a_p\ ;\ c_1,\ c_2,\ \ldots,\ c_q\ ;\ z) = \sum_{n=0}^{\infty} \frac{[a_1]_n \cdots [a_p]_n}{[c_1]_n \cdots [c_q]_n}\ \frac{z^n}{n!}. \tag{16.34}$$

The radius of convergence (in the complex z-plane) of the series in Eq, (16.34) is

(i) zero when $p \geq q + 2$;
(ii) nonzero and finite when $p = q + 1$; and
(iii) infinite when $p \leq q$.

A very large number of functions, both elementary functions as well as special functions, are actually hypergeometric functions corresponding to various values of the parameters. For instance, it is seen easily that

$${}_0F_0(z) = {}_1F_1(a;\ a;\ z) = e^z, \quad {}_1F_0(1;\ z) = (1-z)^{-1}, \tag{16.35}$$

---

[8] It is assumed that the reader is familiar with this test to determine the absolute convergence or otherwise of an infinite series. We will discuss the convergence of power series (including the ratio test) in a little more detail in Chap. 22, Sect. 22.5.

and so on. I will indicate the explicit connection in the case of each of the families of orthogonal polynomials to be discussed below. These will turn out to be either ${}_2F_1$ or ${}_1F_1$ functions, with a numerator parameter that is a negative integer. This ensures that the series terminates after a finite number of terms, yielding a polynomial.

16.2.3 Rodrigues Formula and Generating Function

For each of the three types of polynomials listed in Sect. 16.2.1, there exists a representation called the **Rodrigues formula**. This formula expresses each nth order polynomial $\phi_n(x)$ as the nth derivative of a specific function of x, as follows:

$$\phi_n(x) = \frac{k_n}{\rho(x)}\frac{d^n}{dx^n}\left[\sigma^n(x)\,\rho(x)\right], \quad \text{(Rodrigues formula)} \tag{16.36}$$

where k_n is an n-dependent constant. Recall that $\rho(x)$ is the weight, and $\sigma(x)$ is the coefficient of the second derivative term $\phi''(x)$ in the hypergeometric differential equation (16.22). In practice, the Rodrigues formula provides one of the most convenient ways of obtaining the polynomials explicitly, perhaps even more so than the evaluation of the determinant involved in the explicit representation of Eq. (16.19).

The generating function: Where does the Rodrigues formula come from? It originates from the **generating function** for each family of orthogonal polynomials. Consider a function $F(x,t)$ of two variables x and t that is in the form of a power series in t, in which the coefficient of t^n is the polynomial $\phi_n(x)$, apart from a possible n-dependent constant c_n. That is,

$$F(x,t) = \sum_{n=0}^{\infty} c_n\,\phi_n(x)\,t^n. \tag{16.37}$$

In its region of convergence, such a power series is just the Taylor series, about the point $t=0$, of $F(x,t)$ regarded as a function of t for any given value of x. Hence the coefficient of t^n is $(1/n!)$ times the nth derivative of F with respect to t at $t=0$. Therefore

$$\phi_n(x) = \frac{1}{n!\,c_n}\left[\frac{\partial^n F(x,t)}{\partial t^n}\right]_{t=0}. \tag{16.38}$$

Suppose, now, that we are able to find $F(x,t)$ *in closed form*. (Evidently, this is the reason for allowing for the set of constants $\{c_n\}$.) Then $F(x,t)$ is a generating function for the set of polynomials $\{\phi_n(x)\}$, and Eq. (16.38) is a Rodrigues formula for the set. The examples below will show explicitly how this works.

As I have mentioned earlier, we will study power series as analytic functions of a complex variable in Chap. 22, Sect. 22.5. A few points about power series, however, must be mentioned here, as they are relevant to our immediate purposes.

A general infinite series in nonnegative integer powers of t converges absolutely inside some circle of convergence centered at the origin in the complex t-plane. In this region, it defines an analytic function in the sense of analytic functions of a complex variable. Term by term differentiation and integration of the power series are permissible as long as one remains within the circle of convergence. Moreover, the relationship between the generating function and its power series expansion is to be regarded as one between analytic functions. (In fact, the generating function $F(z, t)$ is best regarded as an analytic function of the complex variables z and t.) The closed-form expression for the generating function could be regarded as the **analytic continuation** of the function represented by the power series to all values of t, including those outside the region of convergence.

The second point is that, if two absolutely convergent power series in t, say $\sum_n \alpha_n t^n$ and $\sum_n \beta_n t^n$, are equal to each other in some common region of the complex t-plane, then $\alpha_n = \beta_n$ for each value of n. This is because the difference between the series vanishes identically, implying that the analytic function represented by this difference vanishes identically. Hence each of its derivatives must also be zero; and $\alpha_n - \beta_n$ is proportional to its nth derivative at $t = 0$. We will frequently make use of this fact in what follows.

Finally, note that generating functions may also involve negative integer powers of t in the case of families of special functions that are *not* polynomials, such as Bessel functions. The corresponding power series are more general than Taylor series, and are called **Laurent series**. We will discuss the properties of such series in Chap. 23, Sect. 23.2.4.

Let us now consider each of the three classes of polynomials in succession.

16.2.4 Class I. Hermite Polynomials

The range of x in this case is the whole of the x-axis. The weight factor is a Gaussian, as you might expect. Specifically,

$$\boxed{(a, b) = (-\infty, \infty), \quad \rho(x) = e^{-x^2}, \quad \sigma(x) = 1.} \quad \text{(Hermite)} \tag{16.39}$$

The generating function of the Hermite polynomials $H_n(x)$ is given by

$$\boxed{F^{\text{her}}(x, t) = e^{2tx - t^2} = \sum_{n=0}^{\infty} \frac{H_n(x)}{n!}\, t^n\,.} \tag{16.40}$$

The radius of convergence of the series on the right-hand side in (16.40) is infinite, i.e., the series converges absolutely for all $|t| < \infty$ in the complex t-plane. We will use this fact in Chap. 18, Sect. 18.2.2, when we consider the eigenfunctions of the Fourier transform operator.

The Rodrigues formula for the Hermite polynomials $H_n(x)$ is

$$\boxed{H_n(x) = (-1)^n\, e^{x^2}\, \frac{d^n}{dx^n}\, e^{-x^2}.} \tag{16.41}$$

The orthonormality relation is

$$\int_{-\infty}^{\infty} dx\, e^{-x^2}\, H_n(x)\, H_m(x) = \sqrt{\pi}\, 2^n\, n!\, \delta_{nm}, \tag{16.42}$$

while the completeness relation is

$$\sum_{n=0}^{\infty} \frac{e^{-(x^2+x'^2)/2}}{\sqrt{\pi}\, 2^n\, n!}\, H_n(x)\, H_n(x') = \delta(x - x'). \tag{16.43}$$

$H_n(x)$ is the regular solution of Hermite's differential equation

$$\boxed{\frac{d^2\phi}{dx^2} - 2x\,\frac{d\phi}{dx} + 2n\,\phi = 0.} \tag{16.44}$$

The Hermite polynomials $H_n(x)$ are related to the confluent hypergeometric function by

$$\left.\begin{aligned} H_{2n}(x) &= \frac{(-1)^n (2n)!}{n!}\, {}_1F_1(-n;\, \tfrac{1}{2};\, x^2), \\ H_{2n+1}(x) &= \frac{(-1)^n (2n+1)!}{n!}\, {}_1F_1(-n;\, \tfrac{3}{2};\, x^2)\, x. \end{aligned}\right\} \tag{16.45}$$

The first few Hermite polynomials are

$$H_0(x) = 1,\;\; H_1(x) = 2x,\;\; H_2(x) = 4x^2 - 2,\;\; H_3(x) = 8x^3 - 12x. \tag{16.46}$$

★ **5.** A number of properties of the Hermite polynomials can be established with the help of the generating function.

(a) Show that $H_{2n+1}(0) = 0$ and $H_{2n}(0) = \dfrac{(-1)^n\,(2n)!}{n!}$.
(b) Establish the Rodrigues formula (16.41) for Hermite polynomials.
(c) Show that $H'_n(x) = 2n\, H_{n-1}(x) \quad (n \geq 1)$.
(d) Show that $H_{n+1}(x) - 2x\, H_n(x) + 2n\, H_{n-1}(x) = 0 \quad (n \geq 1)$.
(e) Hence show that $H_n(x)$ satisfies Hermite's differential equation (16.44).
(f) Derive the **addition theorem** for Hermite polynomials,

$$H_n(x + y) = \frac{1}{2^{n/2}} \sum_{k=0}^{n} \binom{n}{k} H_{n-k}(\sqrt{2}\,x)\, H_k(\sqrt{2}\,y).$$

Similarly, other special functions also satisfy such addition theorems. The case of generalized Laguerre polynomials will be considered in Sect. 16.2.7. The important case of spherical harmonics, Eq. (16.138), will be dealt with in Sect. 16.4.8.

16.2.5 Linear Harmonic Oscillator Eigenfunctions

Hermite polynomials occur, for instance, in the problem of the quantum mechanical linear harmonic oscillator.

The Hamiltonian of the oscillator, and its representations in the position basis and momentum basis, has been written down in Eqs. (14.39) and (14.40) of Chap. 14, Sect. 14.4.2. We learn in basic quantum mechanics that the position-space wave functions corresponding to the eigenstates of the Hamiltonian are precisely the Hermite polynomials, multiplied by a Gaussian factor. We have chosen units such that the mass and the natural frequency of the oscillator are set equal to unity. Both x and p then have the physical dimensions of $\sqrt{\hbar}$. In these units, the position-space wave functions are given by

$$\boxed{\Phi_n(x) = \frac{1}{\left(2^n n! \sqrt{\pi\hbar}\right)^{1/2}}\, e^{-x^2/(2\hbar)}\, H_n(x/\sqrt{\hbar}), \quad n = 0, 1, \ldots .} \tag{16.47}$$

The n-dependent constant factor in $\Phi_n(x)$ ensures the normalization of the wave function according to $\int_{-\infty}^{\infty} dx\, |\Phi_n(x)|^2 = 1$, i.e., that $\Phi_n(x) \in \mathcal{L}_2(-\infty\,,\,\infty)$.

What about the momentum-space wave functions $\widetilde{\Phi}_n(p)$ corresponding to the eigenstates of the Hamiltonian? The relation between $\widetilde{\Phi}_n(p)$ and $\Phi_n(x)$ is, of course, the one-dimensional analog of Eq. (13.30) of Chap. 13, Sect. 13.2.5. In the present case, this relation reads

$$\widetilde{\Phi}_n(p) = \frac{1}{(2\pi\hbar)^{1/2}} \int_{-\infty}^{\infty} dx\, e^{-ipx/\hbar}\, \Phi_n(x). \tag{16.48}$$

One may try to evaluate this integral directly, after putting in the expression in Eq. (16.47) for $\Phi_n(x)$. But this is not necessary. The integral can be computed using a little trick that again involves the generating function of the Hermite polynomials. I will do so in Chap. 18, Sect. 18.2.2, when we consider the eigenfunctions of the Fourier transform operator. Meanwhile, the answer can (almost) be guessed. The harmonic oscillator Hamiltonian is very special, because of its obvious symmetry under the interchange of the position and momentum variables x and p. As a consequence, in the symmetric units we are using (in which both x and p are in units of $\sqrt{\hbar}$), we may expect $\widetilde{\Phi}_n(p)$ to be exactly the same as $\Phi_n(x)$, *apart from a possible phase factor*. This does indeed turn out to be the case. The exact answer is

$$\boxed{\widetilde{\Phi}_n(p) = \frac{(-i)^n}{\left(2^n n! \sqrt{\pi\hbar}\right)^{1/2}} \, e^{-p^2/(2\hbar)} \, H_n(p/\sqrt{\hbar}) \equiv (-i)^n \, \Phi_n(p).} \tag{16.49}$$

This is an illustration of yet another marvelous property of the functions $\Phi_n(x)$, as we will see in Chap. 18, Sect. 18.2.2: $\Phi_n(x)$ is also an eigenfunction of the Fourier transform operator, and the factor $(-i)^n$ in the second equation in (16.49) is essentially the corresponding eigenvalue.

16.2.6 Oscillator Coherent State Wave Functions

The explicit expression in Eq. (16.47) for $\Phi_n(x)$ enables us to find the oscillator coherent state wave functions as well, quite easily. Recall from Chap. 14, Sect. 14.4.2 that *every* complex number z is an eigenvalue of the ladder operator a. The corresponding eigenstate $|\, z \,\rangle$ has a normalizable position-space wave function $\langle\, x \,|\, z \,\rangle = \psi_z(x)$. We found $\psi_z(x)$ (up to a normalization constant $C(z)$) in Eq. (14.49), by solving the eigenvalue equation $a \,|\, z \,\rangle = z \,|\, z \,\rangle$ as a first-order differential equation in the position basis. Let us now see how this can be done using the generating function for the Hermite polynomials. In the process, the normalization constant $C(z)$ will also get determined explicitly.

The input we need for this purpose is an expansion of the coherent state $|\, z \,\rangle$ in the basis $\{|\, n \,\rangle\}$ of the eigenstates of the number operator $a^\dagger a$. This expansion has already been written down in Eq. (15.64) of Chap. 15, Sect. 15.4.1. Repeating it for ready reference,

$$|\, z \,\rangle = e^{-\frac{1}{2}|z|^2} \sum_{n=0}^{\infty} \frac{z^n}{\sqrt{n!}} \,|\, n \,\rangle, \quad \text{where} \quad z \in \mathbb{C}. \tag{16.50}$$

The factor $e^{-\frac{1}{2}|z|^2}$ ensures that $|\, z \,\rangle$ is normalized to unity according to $\langle\, z \,|\, z \,\rangle = 1$. This expansion is convergent for all finite values of the complex number z. It follows immediately that

$$\psi_z(x) \stackrel{\text{def.}}{=} \langle\, x \,|\, z \,\rangle = e^{-\frac{1}{2}|z|^2} \sum_{n=0}^{\infty} \frac{z^n}{\sqrt{n!}} \langle\, x \,|\, n \,\rangle = e^{-\frac{1}{2}|z|^2} \sum_{n=0}^{\infty} \frac{z^n}{\sqrt{n!}} \Phi_n(x). \tag{16.51}$$

Now use the expression for $\Phi_n(x)$ given in Eq. (16.47). The sum over n can be read off, by comparing it with the generating function formula in Eq. (16.40). The result is

$$\psi_z(x) = \frac{1}{(\pi\hbar)^{1/4}} \exp\left\{ -\frac{1}{2}\Big(\frac{x^2}{\hbar} + \frac{2\sqrt{2}\,zx}{\sqrt{\hbar}} - z^2 - |z|^2 \Big) \right\}, \quad z \in \mathbb{C}. \tag{16.52}$$

Hence the normalization constant $C(z)$ introduced in Eq. (14.49) is given by

$$C(z) = (\pi\hbar)^{-1/4}\, e^{-\frac{1}{2}(z^2+|z|^2)} = (\pi\hbar)^{-1/4}\, e^{-z_1(z_1+iz_2)}, \tag{16.53}$$

where $z = z_1 + iz_2$.

The real and imaginary parts of the eigenvalue z corresponding to the coherent state $|\, z\,\rangle$ have a specific significance. The ground state wave function $\Phi_0(x)$ of the oscillator is a Gaussian centered at $x = 0$ (Eq. (14.47) of Chap. 14, Sect. 14.4.2). Moreover, this state is itself a coherent state, corresponding to the special case $z = 0$ (recall Eq. (14.52), $\Phi_0(x) = \psi_0(x)$). What is the shape of the function $\psi_z(x)$ for a general value of z? Set $z = z_1 + iz_2$. Equation (16.52) then becomes, after a bit of simplification,

$$\boxed{\psi_z(x) = (\pi\hbar)^{-1/4}\, e^{-(x-\sqrt{2\hbar}\,z_1)^2/(2\hbar)}\; e^{iz_2\,(\sqrt{\hbar}\,z_1-\sqrt{2}\,x)/\sqrt{\hbar}}.} \tag{16.54}$$

The wave function is therefore a *displaced* Gaussian, modulated by a phase factor (the second exponential factor on the right-hand side).

- The center of the Gaussian in x is displaced from the origin by the amount $\sqrt{(2\hbar)}\, z_1$.

A similar calculation shows that the momentum-space wave function corresponding to the coherent state $|\, z\,\rangle$ is given by

$$\boxed{\widetilde{\psi}_z(p) \overset{\text{def.}}{=} \langle\, p\,|\, z\,\rangle = (\pi\hbar)^{-1/4}\, e^{-(p-\sqrt{2\hbar}\,z_2)^2/(2\hbar)}\; e^{iz_1\,(\sqrt{2}\,p-\sqrt{\hbar}\,z_2)/\sqrt{\hbar}}.} \tag{16.55}$$

As you might expect, this is again a displaced Gaussian modulated by a phase factor.

- The center of the Gaussian in p is displaced from the origin by the amount $\sqrt{(2\hbar)}\, z_2$.

Refer, now, to Eq. (15.70) of Chap. 15, Sect. 15.4.1. These equations show that the unitary transformation $D(z)$ shifts the operators x and p precisely by $\sqrt{(2\hbar)}\, z_1\, I$ and $\sqrt{(2\hbar)}\, z_2\, I$, respectively. These facts, together with the results just derived, should make it fully clear why an oscillator coherent state is also called a displaced vacuum state, as already stated.

★ **6.** Starting from Eq. (16.51), work out the steps to arrive at Eqs. (16.54) and (16.55).

As I have mentioned earlier, we will return to oscillator coherent states briefly in Chap. 19, Sect. 19.2.2.

16.2.7 Class II. Generalized Laguerre Polynomials

The range in this case is semi-infinite, $0 \le x < \infty$. We might then expect an exponentially decaying weight factor, e^{-x}. But this factor can be multiplied by any power

x^α where the parameter α is not necessarily an integer, without affecting the convergence properties of the integrals concerned as $x \to \infty$ or at $x = 0$ (in the latter case, as long as $\alpha > -1$). We thus have, in this case,

$$\boxed{[a,\, b) = [0,\, \infty), \quad \rho(x) = x^\alpha\, e^{-x}, \quad \sigma(x) = x.} \quad \text{(Genlzd. Laguerre)} \tag{16.56}$$

The generating function of the **generalized Laguerre polynomials** $L_n^\alpha(x)$ is given by

$$\boxed{F^{\text{lag}}(x,t) = \frac{1}{(1-t)^{\alpha+1}} \exp\left(\frac{tx}{t-1}\right) = \sum_{n=0}^{\infty} L_n^\alpha(x)\, t^n.} \tag{16.57}$$

The power series in t converges absolutely for $|t| < 1$. When the parameter $\alpha = 0$, corresponding to the weight function e^{-x}, the generalized Laguerre polynomial reduces to the (ordinary) **Laguerre polynomial**, denoted by $L_n(x)$:

$$L_n(x) \stackrel{\text{def.}}{=} L_n^0(x). \tag{16.58}$$

Let us consider the general case $\alpha \neq 0$. The corresponding results for the ordinary Laguerre polynomials are easily obtained by setting $\alpha = 0$ in the various formulas.

The Rodrigues formula for the generalized Laguerre polynomials is

$$\boxed{L_n^\alpha(x) = \frac{1}{n!}\, x^{-\alpha}\, e^x\, \frac{d^n}{dx^n}\, (x^{n+\alpha}\, e^{-x}).} \tag{16.59}$$

The orthonormality relation is

$$\int_0^\infty dx\, x^\alpha\, e^{-x}\, L_n^\alpha(x)\, L_m^\alpha(x) = \frac{\Gamma(\alpha+n+1)}{n!}\, \delta_{nm}\,. \tag{16.60}$$

Note that the normalization constant A_n reduces to unity when $\alpha = 0$.

The completeness relation is

$$\sum_{n=0}^{\infty} \frac{n!}{\Gamma(\alpha+n+1)}\, (x\, x')^{\alpha/2}\, e^{-(x+x')}\, L_n^\alpha(x)\, L_n^\alpha(x') = \delta(x - x'). \tag{16.61}$$

The generalized Laguerre polynomial satisfies the differential equation

$$x\, \frac{d^2\phi}{dx^2} + (\alpha + 1 - x)\, \frac{d\phi}{dx} + n\, \phi = 0\,. \tag{16.62}$$

$L_n^\alpha(x)$ is related to the confluent hypergeometric function according to

$$L_n^{\alpha}(x) = \frac{\Gamma(\alpha + n + 1)}{\Gamma(\alpha + 1)\, n!}\, {}_1F_1(-n;\ \alpha + 1;\ x). \tag{16.63}$$

★ **7.** Once again, the generating function and the Rodrigues formula help us derive various formulas pertaining to the generalized Laguerre polynomials.

(a) Show that $L_n^{\alpha}(0) = \dfrac{\Gamma\,(n + \alpha + 1)}{\Gamma(\alpha + 1)\, n!}$.

(b) Show that $L_0^{\alpha}(x) = 1,\ L_1^{\alpha}(x) = \alpha + 1 - x,\ L_2^{\alpha}(x) = \frac{1}{2}(\alpha + 2 - x)^2 - \frac{1}{2}(\alpha + 2)$.

(c) Show that $L_n^{\alpha+1}(x) = \sum_{k=0}^{n} L_k^{\alpha}(x)$.

(d) Establish the following addition theorem for the generalized Laguerre polynomial:

$$L_n^{\alpha+\beta+1}(x + y) = \sum_{k=0}^{n} L_{n-k}^{\alpha}(x)\, L_k^{\beta}(y) = \sum_{k=0}^{n} L_k^{\alpha}(x)\, L_{n-k}^{\beta}(y).$$

Connection with the hydrogen atom radial wave function: You will recall that generalized Laguerre polynomials occur in the radial wave functions of the bound states of a charged quantum mechanical particle of mass m in an attractive Coulomb potential $V(r) = -K/r$, where K is a positive constant: this is the hydrogen atom problem. The energy levels are proportional to $-1/n^2$, where n $(= 1, 2, \ldots)$ is the principal quantum number. Let $\kappa_n = mK/(n\hbar^2)$. Its reciprocal κ_n^{-1} is n times the Bohr radius $\hbar^2/(mK)$. This is the length scale associated with the set of bound states belonging to the quantum number n. The radial wave function corresponding to principal quantum number n and orbital angular momentum quantum number l is

$$R_{nl}(r) \propto e^{-\kappa_n r}\, (2\kappa_n r)^l\, L_{n-l-1}^{2l+1}(2\kappa_n r). \tag{16.64}$$

The exponential factor $e^{-\kappa_n r}$ ensures the normalizability of the wave function at the upper end $(r \to \infty)$. As $r \to 0$, the leading behavior of the radial wave function is $\sim r^l$.

16.2.8 Class III. Jacobi Polynomials

Finally, we turn to the case of a finite range, which can be taken to be $-1 \le x \le 1$, as already stated. The Jacobi polynomials correspond to

$$\boxed{[a,\, b] = [-1,\, 1],\quad \rho(x) = (1 - x)^{\alpha}\, (1 + x)^{\beta},\quad \sigma(x) = 1 - x^2.} \quad \text{(Jacobi)} \tag{16.65}$$

The polynomials are denoted by $P_n^{(\alpha,\,\beta)}(x)$, and depend on two parameters, α and β. The conditions $\alpha > -1,\ \beta > -1$ are needed to ensure the convergence of the integrals representing inner products in the function space concerned. I list below a few important properties of Jacobi polynomials,[9] merely for completeness, before going on to the special cases of these polynomials (such as Legendre and associated Legendre polynomials) that appear more frequently in physical applications.

The Rodrigues formula for Jacobi polynomials is

$$P_n^{(\alpha,\,\beta)}(x) = \frac{(-1)^n}{2^n\,n!}\,(1-x)^{-\alpha}(1+x)^{-\beta}\,\frac{d^n}{dx^n}\,[(1-x)^{n+\alpha}(1+x)^{n+\beta}\,]. \quad (16.66)$$

The orthonormality condition is given by

$$\int_{-1}^{1} dx\,(1-x)^{\alpha}\,(1+x)^{\beta}\,P_n^{(\alpha,\,\beta)}(x)\,P_m^{(\alpha,\,\beta)}(x) = \frac{2^{\alpha+\beta+1}\Gamma(\alpha+n+1)\Gamma(\beta+n+1)}{(\alpha+\beta+2n+1)\Gamma(\alpha+\beta+n+1)\,n!}\,\delta_{nm}\,. \quad (16.67)$$

The Jacobi polynomial $P_n^{(\alpha,\,\beta)}(x)$ satisfies the differential equation

$$(1-x^2)\,\frac{d^2\phi}{dx^2} + [\beta-\alpha-(\alpha+\beta+2)\,x]\,\frac{d\phi}{dx} + n(n+\alpha+\beta+1)\,\phi = 0\,. \quad (16.68)$$

$P_n^{(\alpha,\,\beta)}(x)$ is related to the hypergeometric function according to

$$P_n^{(\alpha,\,\beta)}(x) = \frac{\Gamma(\alpha+n+1)}{\Gamma(\alpha+1)\,n!}\;{}_2F_1\big(-n,\ n+\alpha+\beta+1;\ \alpha+1;\ \tfrac{1}{2}(1-x)\big). \quad (16.69)$$

In their general form, the Jacobi polynomials are not symmetric in the parameters α and β. But when $\alpha = \beta$, the Jacobi polynomials reduce to more commonly occurring polynomials that are very important and useful in physical applications. Let us therefore consider them in a little more detail.

16.3 Gegenbauer Polynomials

16.3.1 Ultraspherical Harmonics

When $\alpha = \beta$, the Jacobi polynomials reduce to the **Gegenbauer polynomials**. It is conventional to set

$$\alpha = \beta = \lambda - \tfrac{1}{2}, \quad \text{where} \quad \lambda > -\tfrac{1}{2}\,. \quad (16.70)$$

[9] Nikiforov and Uvarov, *op. cit.*

Thus, in this case we have

$$\boxed{[a,\, b] = [-1,\, 1], \quad \rho(x) = (1-x^2)^{\lambda-\frac{1}{2}}, \quad \sigma(x) = 1 - x^2.} \quad \text{(Gegenbauer)} \tag{16.71}$$

The Gegenbauer polynomial, also called the **ultraspherical polynomial**, is denoted by $C_n^\lambda(x)$. Its generating function is

$$\boxed{F^{\text{geg}}(x,t) = \frac{1}{(1-2tx+t^2)^\lambda} = \sum_{n=0}^{\infty} C_n^\lambda(x)\, t^n.} \tag{16.72}$$

$C_n^\lambda(x)$ is related to a Jacobi polynomial and a hypergeometric function as follows:

$$\begin{aligned} C_n^\lambda(x) &= \frac{\Gamma(n+2\lambda)\,\Gamma(\lambda+\frac{1}{2})}{\Gamma(2\lambda)\,\Gamma(n+\lambda+\frac{1}{2})}\, P_n^{(\lambda-\frac{1}{2},\,\lambda-\frac{1}{2})}(x) \\ &= \frac{\Gamma(2\lambda+n)}{\Gamma(2\lambda)\, n!}\; {}_2F_1\big(-n,\, n+2\lambda;\, \lambda+\tfrac{1}{2};\, \tfrac{1}{2}(1-x)\big). \end{aligned} \tag{16.73}$$

Some care is needed in defining $C_n^\lambda(x)$ in the case $\lambda = 0$. I will not go into this here, but see Sect. 16.3.2 below.

$C_n^\lambda(x)$ is the regular, polynomial solution of the differential equation

$$(1-x^2)\,\frac{d^2\phi}{dx^2} - (2\lambda+1)\,x\,\frac{d\phi}{dx} + n\,(n+2\lambda)\,\phi = 0\,. \tag{16.74}$$

Among other reasons, the physical significance of the Gegenbauer polynomials arises from the following fact:

- When λ takes on the value $\frac{1}{2}d - 1$ for any integer $d \geq 3$, the Gegenbauer polynomials $C_n^\lambda(x)$ represent the so-called spherical harmonics in a space of d dimensions. For $d > 3$, these are called **ultraspherical harmonics**.

I will say more about this after we discuss Legendre polynomials and spherical harmonics.

For certain special values of λ, the Gegenbauer polynomials reduce further to the Chebyshev and Legendre polynomials, as follows.

16.3.2 Chebyshev Polynomials of the 1st Kind

When $\alpha = \beta = -\frac{1}{2}$ (so that $\lambda = 0$), the Jacobi polynomial reduces to the **Chebyshev polynomial of the first kind**, denoted by $T_n(x)$. We have in this case

$$\boxed{[a,\, b] = [-1,\, 1], \quad \rho(x) = (1-x^2)^{-\frac{1}{2}}, \quad \sigma(x) = 1 - x^2.} \quad \text{(Chebyshev I)} \tag{16.75}$$

The generating function for $T_n(x)$ is given by

$$\boxed{F^{\text{chel}}(x,t) = \frac{1-xt}{1-2xt+t^2} = \sum_{n=0}^{\infty} T_n(x)\, t^n.} \tag{16.76}$$

There is an alternative generating function that is also useful. This is

$$G^{\text{chel}}(x,t) = \frac{1-t^2}{1-2xt+t^2} = T_0(x) + 2\sum_{n=1}^{\infty} T_n(x)\, t^n. \tag{16.77}$$

The Rodrigues formula for $T_n(x)$ is

$$T_n(x) = \frac{(-1)^n\, 2^n\, n!}{(2n)!}\,(1-x^2)^{\frac{1}{2}}\,\frac{d^n}{dx^n}\,[(1-x^2)^{n-\frac{1}{2}}]. \tag{16.78}$$

The orthonormality relation is

$$\int_{-1}^{1}\frac{dx}{\sqrt{1-x^2}}\,T_n(x)\,T_m(x) = A_n\,\delta_{nm}, \text{ where } A_0 = \pi,\ \ A_n = \tfrac{1}{2}\pi\ (n \geq 1). \tag{16.79}$$

$T_n(x)$ is the regular, polynomial solution of the differential equation

$$(1-x^2)\,\frac{d^2\phi}{dx^2} - x\,\frac{d\phi}{dx} + n^2\,\phi = 0\,. \tag{16.80}$$

When $\lambda = 0$, care must be exercised in simplifying the formulas in (16.73). The correct expressions for $T_n(x)$ in terms of a Jacobi polynomial and a hypergeometric function are

$$T_n(x) = \frac{2^{2n}\,(n!)^2}{(2n)!}\,P_n^{(-\frac{1}{2},\,-\frac{1}{2})}(x) = {}_2F_1\big(-n,\, n;\ \tfrac{1}{2};\ \tfrac{1}{2}(1-x)\big). \tag{16.81}$$

The first few polynomials are

$$T_0(x) = 1,\ \ T_1(x) = x,\ \ T_2(x) = 2x^2 - 1,\ \ T_3(x) = 4x^3 - 3x,\ \ldots\,. \tag{16.82}$$

Once again, since $F^{\text{chel}}(-x,-t) = F^{\text{chel}}(x,t)$, it is easy to see that the polynomials $T_n(x)$ are parity eigenfunctions, satisfying

$$T_n(-x) = (-1)^n\,T_n(x). \tag{16.83}$$

The Chebyshev polynomial $T_n(x)$ has a very simple and useful representation as a trigonometric function. Since $x \in [-1,\ 1]$, we may write $x \equiv \cos\theta$. Then:

- $T_n(x)$ has the very simple form

$$\boxed{T_n(\cos\,\theta) = \cos\,n\theta\,.} \tag{16.84}$$

It follows trivially that $T_n(1) = 1$ and $T_n(-1) = (-1)^n$. From Eq. (16.84), it is easy to see that

- $T_n(x)$ satisfies a functional equation that exhibits the remarkable "nesting" property

$$\boxed{T_m\big(T_m(x)\big) = T_{mn}(x).} \tag{16.85}$$

★ **8.** Establish the following properties of $T_n(x)$:

(a) $T_n(0) = 0$ when n is odd; $T_n(0) = (-1)^{n/2}$ when n is even.
(b) $T\,'_n(0) = 0$ when n is even; $T\,'_n(-1)^{(n-1)/2}$ when n is odd.
(c) The recursion relation $T_{n+1}(x) - 2x\,T_n(x) + T_{n-1}(x) = 0$, for every $n \geq 1$.
(d) The functional equation (16.85).

16.3.3 Chebyshev Polynomials of the Second Kind

When $\alpha = \beta = \frac{1}{2}$, i.e., $\lambda = 1$, the Jacobi polynomial reduces to the **Chebyshev polynomial of the second kind**, denoted by $U_n(x)$. We now have

$$\boxed{[a,\,b] = [-1,\,1], \quad \rho(x) = (1-x^2)^{\frac{1}{2}}, \quad \sigma(x) = 1 - x^2.} \quad \text{(Chebyshev II)} \tag{16.86}$$

The generating function for $U_n(x)$ is given by

$$\boxed{F^{\text{che2}}(x,t) = \frac{1}{1 - 2xt + t^2} = \sum_{n=0}^{\infty} U_n(x)\,t^n.} \tag{16.87}$$

The Rodrigues formula for $U_n(x)$ is

$$U_n(x) = \frac{(-1)^n\,2^n\,(n+1)!}{(2n+1)!}\,(1-x^2)^{-\frac{1}{2}}\,\frac{d^n}{dx^n}\,[(1-x^2)^{n+\frac{1}{2}}]. \tag{16.88}$$

The orthonormality relation is

$$\int_{-1}^{1} dx\,(1-x^2)^{1/2}\,U_n(x)\,U_m(x) = \tfrac{1}{2}\pi\,\delta_{nm}\,. \tag{16.89}$$

$U_n(x)$ is the regular, polynomial solution of the differential equation

$$(1-x^2)\,\frac{d^2\phi}{dx^2} - 3x\,\frac{d\phi}{dx} + n\,(n+2)\,\phi = 0\,. \tag{16.90}$$

In terms of a Jacobi polynomial and the hypergeometric function,

$$U_n(x) = \frac{2^{2n}\, n!\,(n+1)!}{(2n+1)!}\, P_n^{(\frac{1}{2},\,\frac{1}{2})}(x) = (n+1)\,{}_2F_1\big(-n,\, n+2;\, \tfrac{3}{2};\, \tfrac{1}{2}(1-x)\big). \tag{16.91}$$

The first few polynomials are

$$U_0(x) = 1,\;\; U_1(x) = 2x,\;\; U_2(x) = 4x^2 - 1,\;\; U_3(x) = 8x^3 - 4x,\;\; \ldots. \tag{16.92}$$

As in the case of $T_n(x)$, we have $F^{\text{che2}}(-x, -t) = F^{\text{che2}}(x, t)$. Hence the polynomials $U_n(x)$ are also parity eigenfunctions, satisfying

$$U_n(-x) = (-1)^n\, U_n(x). \tag{16.93}$$

Once again, setting $x = \cos\theta$, a representation is obtained for $U_n(x)$ as a trigonometric function:

- $U_n(x)$ is given by the simple form

$$\boxed{U_n(\cos\theta) = \frac{\sin\,(n+1)\theta}{\sin\,\theta}.} \tag{16.94}$$

It follows trivially that $U_n(1) = (n+1)$ and $U_n(-1) = (-1)^n(n+1)$.

★ **9.** Establish the following properties of $U_n(x)$:

(a) $U_n(x) = \dfrac{1}{(n+1)}\,\dfrac{d\,T_{n+1}(x)}{dx}$.
(b) $U_n(0) = 0$ when n is odd; $U_n(0) = (-1)^{n/2}/(n+1)$ when n is even.
(c) $U'_n(0) = 0$ when n is even; $U'_n(0) = (-1)^{(n-1)/2}\,(n+1)$ when n is odd.
(d) The recursion relation $U_{n+1}(x) - 2x\,U_n(x) + U_{n-1}(x) = 0$, for every $n \geq 1$.
(e) $U_n(x) - x\,U_{n-1}(x) = T_n(x),\;\; n \geq 1$.
(f) $U_n(x) - U_{n-2}(x) = 2\,T_n(x),\;\; n \geq 2$.

Chebyshev polynomials have many other interesting properties, which are of considerable use in numerical analysis and the approximation of functions by polynomials.

16.4 Legendre Polynomials

16.4.1 *Basic Properties*

Finally, let us consider the Jacobi polynomial $P_n^{(\alpha,\,\beta)}(x)$ in the case when $\alpha = \beta = 0$, i.e., the Gegenbauer polynomial $C_n^{\lambda}(x)$ when $\lambda = \frac{1}{2}$. The polynomial in this case is the very familiar **Legendre polynomial**, denoted of course by $P_n(x)$. We now have

$$\boxed{[a,\, b] = [-1,\, 1], \quad \rho(x) = 1, \quad \sigma(x) = 1 - x^2.} \quad \text{(Legendre)} \tag{16.95}$$

The generating function for $P_n(x)$ is given by

$$\boxed{F^{\text{leg}}(x, t) = \frac{1}{(1 - 2xt + t^2)^{1/2}} = \sum_{n=0}^{\infty} P_n(x)\, t^n.} \tag{16.96}$$

For $x \in [-1, 1]$, the power series in t converges absolutely for $|t| < 1$, i.e., inside the unit circle in the complex t-plane.

$P_n(x)$ is a polynomial of order n. The first few Legendre polynomials are

$$P_0(x) = 1, \quad P_1(x) = x, \quad P_2(x) = \tfrac{1}{2}(3x^2 - 1), \quad P_3(x) = \tfrac{1}{2}(5x^3 - 3x), \quad \ldots. \tag{16.97}$$

The Rodrigues formula for $P_n(x)$ is

$$\boxed{P_n(x) = \frac{1}{2^n\, n!}\, \frac{d^n}{dx^n}\, (x^2 - 1)^n.} \tag{16.98}$$

The polynomials are even functions for even n, and odd functions for odd n:

$$P_n(-x) = (-1)^n\, P_n(x). \tag{16.99}$$

The orthonormality relation is

$$\boxed{\int_{-1}^{1} dx\, P_n(x)\, P_m(x) = \frac{2}{2n+1}\, \delta_{nm}\,.} \tag{16.100}$$

★ **10.** Show that, if $(m - n)$ is even,

$$\int_0^1 dx\, P_m(x)\, P_n(x) = \begin{cases} 0 & \text{if } m \neq n \\ 1/(2n+1) & \text{if } m = n\,. \end{cases}$$

$P_n(x)$ is the regular, polynomial solution of Legendre's differential equation,

$$\boxed{(1 - x^2)\, \frac{d^2\phi}{dx^2} - 2x\, \frac{d\phi}{dx} + n\,(n+1)\,\phi = 0.} \tag{16.101}$$

The other linearly independent solution of Eq. (16.101), called the **Legendre function of the second kind** and denoted by $Q_n(x)$, will be discussed briefly in Sect. 16.4.5 below. More will be said about Legendre functions of both the first and second kinds in Chap. 25, Sect. 25.1.2, and again in Chap. 26, Sect. 26.2.5.

★ **11.** Write the differential equation (16.101) in the form of the hypergeometric equation (16.28), and identify $P_n(x)$ with a hypergeometric function.

The relation of $P_n(x)$ with Jacobi polynomials, Gegenbauer polynomials and the hypergeometric function is

$$P_n(x) = P_n^{(0,\,0)}(x) = C_n^{1/2}(x) = {}_2F_1\big(-n,\, n+1;\ 1;\ \tfrac{1}{2}(1-x)\big). \qquad (16.102)$$

Note how the Legendre polynomial is a very special case, the progression being

hypergeometric → Jacobi → Gegenbauer → Legendre.

Observe that Legendre's differential equation, Eq. (16.101), can be written in the form

$$\frac{d}{dx}\left\{(1-x^2)\,\frac{d}{dx}\right\}\phi = -n(n+1)\phi. \qquad (16.103)$$

This is an eigenvalue equation for the differential operator $(d/dx)(1-x^2)(d/dx)$. The latter is a self-adjoint operator in the space of functions defined on $[-1,\ 1]$ with suitable boundary conditions. Legendre's differential equation is a prime example of the **Sturm–Liouville eigenvalue problem** that is as important aspect of the theory of second-order differential equations. It also has a direct application to the problem of a quantum mechanical particle moving in a potential.

★ **12.** Once again, numerous properties of Legendre polynomials can be established with the help of the generating function, given in Eq. (16.96). In what follows, a prime denotes the derivative with respect to x.

(a) Establish the recursion relation

$$n\,P_n(x) - (2n-1)\,x\,P_{n-1}(x) + (n-1)\,P_{n-2}(x) = 0.$$

(b) Show that $x\,P'_n(x) - P'_{n-1}(x) = n\,P_n(x)$.
(c) Show that $P'_n(x) - x\,P'_{n-1}(x) = n\,P_{n-1}(x)$.
(d) Hence deduce the following relations:

(i) $P'_{n+1}(x) - P'_{n-1}(x) = (2n+1)\,P_n(x)$.

(ii) $(x^2-1)\,P'_n(x) = n\,x\,P_n(x) - n\,P_{n-1}(x)$.

(iii) $(x^2-1)\,P'_n(x) = (n+1)\,P_{n+1}(x) - (n+1)\,x\,P_n(x)$.

(e) Use the foregoing to show that $P_n(x)$ satisfies Legendre's differential equation, Eq. (16.101).
(f) Show that $P_n(1) = 1$ and $P_n(-1) = (-1)^n$.
(g) Show that $P_{2n+1}(0) = 0$ and $P_{2n}(0) = \dfrac{(-1)^n\,(2n)!}{2^{2n}\,(n!)^2}$.

Legendre polynomials of negative integer order: Observe that the parameter n appears in the combination $n(n+1)$ in the differential equation (16.101) (or Eq. (16.103)) satisfied by the Legendre polynomials. But $n(n+1) = (n+\frac{1}{2})^2 - \frac{1}{4}$, which is a symmetric function of $n+\frac{1}{2}$. Hence the differential equation is invariant under the interchange $n+\frac{1}{2} \leftrightarrow -n-\frac{1}{2}$, or $n \leftrightarrow -n-1$. It turns out that the solution $P_n(x)$ of the equation is also invariant under this exchange. In other words, Legendre polynomials of negative integer order exist, and may in fact be defined via the relation

$$\boxed{P_{-n-1}(x) \stackrel{\text{def.}}{=} P_n(x), \quad n = 0,\, 1,\, \ldots .} \tag{16.104}$$

The identity (16.104) is actually a special case of the relation $P_{-\nu-1}(z) = P_\nu(z)$ (Eq. (25.4) of Chap. 25, Sect. 25.1.2) that is satisfied by the Legendre function of the first kind for all complex values of both its argument z and its order ν.

16.4.2 $P_n(x)$ by Gram–Schmidt Orthonormalization

Here is a very natural way in which the Legendre polynomials arise. Consider the infinite set of monomials $\{x^0,\, x^1,\, x^2,\, x^3, \ldots\}$ in the function space $\mathcal{L}_2[-1,\, 1]$. These functions are linearly independent of each other, and they span the space. Hence they constitute a basis set in the space. But they are not mutually orthogonal, since the inner product $\int_{-1}^{1} dx\, x^m\, x^n \neq 0$ when m and n are both even or both odd. But these functions can, however, be transformed to such a basis by the Gram–Schmidt orthonormalization procedure of Chap. 10, Sect. 10.2.1. Since the weight $\rho(x) = 1$ in the case of Legendre polynomials, the inner product is defined as

$$\langle \chi | \psi \rangle = \int_{-1}^{1} dx\, \chi^*(x)\, \psi(x). \tag{16.105}$$

The result of the procedure is precisely the set of Legendre polynomials, $\{P_n(x)\}$, if the orthonormality condition (16.100) is imposed.

★ **13.** Start with the monomials

$$\psi_0(x) = x^0 = 1,\ \psi_1(x) = x^1,\ \psi_2(x) = x^2,\ \psi_3(x) = x^3,$$

and apply the Gram–Schmidt procedure to obtain an orthogonal set of functions from these. Use the orthonormality condition (16.100), and show that the functions obtained are

$$P_0(x) = 1,\ \ P_1(x) = x,\ \ P_2(x) = \tfrac{1}{2}(3x^2 - 1),\ \ P_3(x) = \tfrac{1}{2}(5x^3 - 3x),$$

if the normalization

$$\left(n+\tfrac{1}{2}\right)\int_{-1}^{1} dx\, P_n^2(x) = 1$$

is used.

16.4.3 Expansion in Legendre Polynomials

Any *continuous, integrable* function $f(x)$ where $x \in [-1,\ 1]$ can be expanded in a series of Legendre polynomials, according to

$$\boxed{f(x) = \sum_{n=0}^{\infty} (2n+1)\, f_n\, P_n(x).} \tag{16.106}$$

Note the extra factor $(2n + 1)$ in the summand on the right-hand side—this is the usual convention adopted. The coefficients f_n in the expansion are then given by

$$\boxed{f_n = \tfrac{1}{2}\int_{-1}^{1} dx\, P_n(x)\, f(x).} \tag{16.107}$$

If, further, $f(x) \in \mathcal{L}_2[-1,\ 1]$, then Parseval's Theorem gives

$$\sum_{n=0}^{\infty} (2n+1)\, |f_n|^2 = \tfrac{1}{2}\int_{-1}^{1} dx\, |f(x)|^2. \tag{16.108}$$

But the right-hand side of this equation is finite, by definition. Hence, in order to ensure the convergence of the series on the left-hand side, the coefficient $|f_n|$ must tend to zero as $n \to \infty$ more rapidly than n^{-1}. Conversely, suppose we encounter an expansion in Legendre polynomials where $|f_n|$ decays to zero less rapidly than n^{-1} as $n \to \infty$. We may then deduce that the corresponding function $f(x)$ is not *square*-integrable in $[-1,\ 1]$.

Once again, it is often convenient to set $x = \cos\theta$, where $\theta \in [0,\ \pi]$. Any continuous integrable function $g(\theta)$ of the angular variable θ can be expanded as

$$\boxed{g(\theta) = \sum_{n=0}^{\infty} (2n+1)\, g_n\, P_n(\cos\theta).} \tag{16.109}$$

The corresponding inversion formula for the coefficients is

$$\boxed{g_n = \tfrac{1}{2}\int_{0}^{\pi} d\theta\, \sin\theta\, P_n(\cos\theta)\, g(\theta).} \tag{16.110}$$

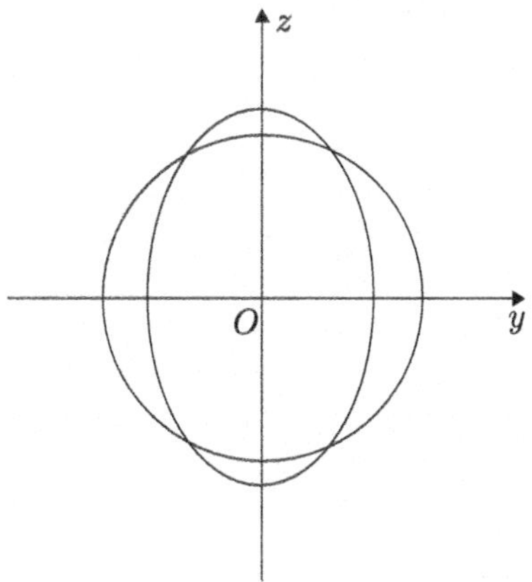

Fig. 16.1 Longitudinal section of a slightly deformed sphere (not to scale). The circle is a cross-section of the undeformed sphere

★ **14.** The equation of the surface of a sphere of radius a centered at the origin is $r = a$, in spherical polar coordinates. Suppose, now, that the sphere is deformed to the shape of a figure of revolution like a spheroid. The spherical polar coordinates of any point on surface of the deformed sphere satisfy the equation

$$r(\theta) = a\ [1 + \epsilon\, P_2(\cos\,\theta)]\,.$$

For small values of ϵ ($\ll 1$), a longitudinal section of the solid is as shown in Fig. 16.1.

(a) Show that its volume is given by $V = \frac{4}{3}\,\pi a^3\,\left(1 + \frac{3}{5}\,\epsilon^2 + \frac{2}{35}\,\epsilon^3\right)$.
(b) Show that its surface area is given by $S = 4\pi\, a^2\,\left(1 + \frac{1}{5}\,\epsilon^2\right)$.

16.4.4 Expansion of x^n in Legendre Polynomials

In Sect. 16.4.2, you have seen how Legendre polynomials arise as linear combinations of the monomials $x^0,\ x^1,\ x^2, \ldots$, on applying the Gram–Schmidt orthonormalization procedure. We now seek the converse result, namely, the expansion of x^n as a linear combination of Legendre polynomials. Such an expansion is needed in many applications—notably, in the **partial wave analysis** of the scattering amplitude in the quantum theory of scattering of a particle from a central potential. (This topic will be discussed in Chap. 32, Sect. 32.3.)

Let n be a nonnegative integer, and let

$$x^n = \sum_{l=0}^{\infty}(2l + 1)c_{nl}\, P_l(x). \tag{16.111}$$

Then

$$c_{nl} = \tfrac{1}{2} \int_{-1}^{1} dx\, x^n\, P_l(x). \tag{16.112}$$

It is obvious that the expansion of x^n cannot involve any Legendre polynomial $P_l(x)$ of order $l > n$. Hence the sum in Eq. (16.111) is cut off at $l = n$. Moreover, $P_l(x)$ is an even or odd function of x according as l is even or odd. It is clear, therefore, that c_{nl} vanishes unless n and l have the same parity (i.e., $n - l$ must be an even integer). To evaluate c_{nl} for $l \le n$, use the Rodrigues formula (16.98) for $P_l(x)$ in Eq. (16.112) for c_{nl}, and integrate by parts l times. All the boundary terms vanish, because the rth derivative of $(x^2 - 1)^l$ vanishes at $x = \pm 1$ for all $r < l$. Hence

$$c_{nl} = \frac{n!}{2^{l+1}\, l!\, (n-l)!} \int_{-1}^{1} dx\, x^{n-l}\, (1 - x^2)^l. \tag{16.113}$$

Use the fact that $n - l$ is even, set $x = \cos\theta$, and apply the formula in Eq. (3.20) of Chap. 3, Sect. 3.1.5 to the resulting trigonometric integral. The result is

$$c_{nl} = \frac{n!}{2^{l+1}\, (n-l)!}\, \frac{\Gamma\big(\frac{1}{2}(n - l + 1)\big)}{\Gamma\big(\frac{1}{2}(n + l + 3)\big)}. \tag{16.114}$$

But the ratio of gamma functions can be simplified further with the help of the formula for the gamma function of a half-odd integer (the first of Eq. (3.17) of Chap. 3, Sect. 3.1.5). The final expression for c_{nl} is

$$c_{nl} = \frac{2^l\, n!\, \big(\frac{1}{2}(n + l)\big)!}{(n + l + 1)!\, \big(\frac{1}{2}(n - l)\big)!} \quad (0 \le l \le n, \quad n - l \text{ even}). \tag{16.115}$$

★ **15.** Starting from Eq. (16.111), work out all the steps described above to derive Eq. (16.115).

The result just derived will be used in Chap. 32, Sect. 32.3.2, when we discuss the quantum theory of scattering. The expansion of the function $e^{ikr\cos\theta}$ in the set of Legendre polynomials $\{P_\ell(\cos\theta)\}$ will be required there. In turn, this involves the expansion of $(\cos\theta)^n$ in terms of Legendre polynomials. The outcome is the expansion in Eq. (32.70) of the function $e^{ikr\cos\theta}$.

16.4.5 Legendre Function of the Second Kind

The Legendre polynomial $P_n(x)$ is only one of the two linearly independent solutions of Legendre's differential equation (16.101). The other solution is $Q_n(x)$, the Legendre function of the second kind. $Q_n(x)$ is not a polynomial even when $n = 0, 1, 2, \ldots$, as it involves polynomials as well as logarithmic functions. For instance,

$$Q_0(x) = \frac{1}{2} \ln \left(\frac{1+x}{1-x}\right), \quad Q_1(x) = \frac{x}{2} \ln \left(\frac{1+x}{1-x}\right) - 1, \quad \cdots . \tag{16.116}$$

For positive integer values of n, the general form of $Q_n(x)$ is given by

$$Q_n(x) = \frac{1}{2} P_n(x) \ln \left(\frac{1+x}{1-x}\right) - R_{n-1}(x), \tag{16.117}$$

where $R_{n-1}(x)$ is a polynomial of degree $(n-1)$. A general formula relating Q_n and P_n will be given in Eq. (26.30) of Chap. 26, Sect. 26.2.5.

As pointed out earlier in connection with the hypergeometric function, we must really regard both the independent variable x and the parameter n in Legendre's differential equation as complex variables (z and ν, in customary notation). $P_\nu(z)$ and $Q_\nu(z)$ are Legendre functions of the first and second kinds, respectively. They have very interesting analytic properties as functions of their argument z and index ν. I will consider some of these briefly in Chap. 25, Sect. 25.1.2, and in a little more detail in Chap. 26, Sect. 26.2.5.

$Q_n(x)$ and $P_n(x)$ satisfy the same differential equation, in which the parameter n appears in the combination $n(n+1)$. It is therefore natural to ask whether Q_n also has the same symmetry property as P_n (Eq. (16.104)) under the interchange $n \leftrightarrow -n-1$. The answer is *no*, it does not! In fact, the analytic function $Q_\nu(z)$ has a *singularity* (as a function of the complex variable ν) whenever ν is equal to a negative integer, as you will see in Eq. (25.7) of in Chap. 25, Sect. 25.1.2.

16.4.6 Associated Legendre Functions

In the solution of Laplace's equation in spherical polar coordinates, there occur certain functions that are closely related to the Legendre polynomials. These functions satisfy the differential equation

$$(1-x^2)\,\frac{d^2\phi}{dx^2} - 2x\,\frac{d\phi}{dx} + \left[n\,(n+1) - \frac{m^2}{1-x^2}\right]\phi = 0\,. \tag{16.118}$$

The regular solution of this equation is the **associated Legendre function** of the first kind, denoted by $P_n^m(x)$. When m is an integer satisfying $0 \le m \le n$, this solution has no singularity for $x \in (-1,\ 1)$, and is given by

$$P_n^m(x) = (-1)^m\,(1-x^2)^{m/2}\,\frac{d^m}{dx^m}\,P_n(x) \quad (0 \le m \le n). \tag{16.119}$$

It is obvious that $P_n^0(x) \equiv P_n(x)$. The presence of the factor $(1-x^2)^{m/2}$ in Eq. (16.119) makes it clear that $P_n^m(x)$ is a *polynomial* in x only when m is an even integer.

Insert the Rodrigues formula (16.98) for $P_n(x)$ in Eq. (16.119) to get

$$P_n^m(x) = \frac{(-1)^{m+n}}{2^n\, n!}\,(1-x^2)^{m/2}\,\frac{d^{m+n}}{dx^{m+n}}\,(1-x^2)^n. \tag{16.120}$$

It is immediately clear from the last equation that we can actually extend the definition of $P_n^m(x)$ to *negative* integer values of m, in the range $-n \le m < 0$. Therefore, for every integer value of $n \ge 0$, the associated Legendre function $P_n^m(x)$ is defined by Eq. (16.120) for all integer values of m from $-n$ to n. That is,

$$\boxed{P_n^m(x) = \frac{(-1)^{m+n}}{2^n\, n!}\,(1-x^2)^{m/2}\,\frac{d^{m+n}}{dx^{m+n}}\,(1-x^2)^n,\quad n \ge 0,\ -n \le m \le n.} \tag{16.121}$$

This fact enables us to deduce an important symmetry property of $P_n^m(x)$, which is

$$P_n^{-m}(x) = (-1)^m\,\frac{(n-m)!}{(n+m)!}\,P_n^m(x). \tag{16.122}$$

When n is a nonnegative integer and m is an integer, the associated Legendre function $P_n^m(x)$ vanishes for $|m| > n$.

★ **16.** Establish Eq. (16.122).

The orthonormality condition for $P_n^m(x)$ is

$$\int_{-1}^{1} dx\, P_n^m(x)\, P_l^m(x) = \frac{2}{(2n+1)}\,\frac{(n-m)!}{(n+m)!}\,\delta_{nl},\quad \text{where}\quad 0 \le m \le n. \tag{16.123}$$

Once again, since $x \in [-1,\ 1]$, may set $x = \cos\theta$ where $\theta \in [0,\ \pi]$. The associated Legendre function is then given by

$$P_n^m(\cos\theta) = \frac{(-1)^{m+n}}{2^n\, n!}\,(\sin\theta)^m\,\frac{d^{m+n}}{d(\cos\theta)^{m+n}}\,(\sin\theta)^{2n}. \tag{16.124}$$

16.4.7 Spherical Harmonics

The set of Legendre polynomials is an orthonormal basis in the space of functions of an angle $\theta \in [0,\ \pi]$. The angle θ may be regarded as the polar angle in spherical polar coordinates in three-dimensional space. What is the corresponding basis in the space of functions of *both* the angular coordinates, θ and φ, where $0 \le \varphi < 2\pi$?

Such a basis is constructed by combining the associated Legendre functions with the periodic functions $\cos(m\varphi)$ and $\sin(m\varphi)$ (or, equivalently, $e^{\pm im\varphi}$), where m takes on integer values. The **spherical harmonic** $Y_{nm}(\theta, \varphi)$ is defined as

$$\boxed{Y_{nm}(\theta, \varphi) = \left[\frac{(2n+1)}{4\pi}\frac{(n-m)!}{(n+m)!}\right]^{1/2} P_n^m(\cos\theta)\, e^{im\varphi},} \tag{16.125}$$

where the associated Legendre function $P_n^m(\cos\theta)$ is given by Eq. (16.124). As is often the case with special functions, there are several alternative definitions, differing from each other by normalization constants and phase factors. I have adopted the one that is used most often in physics. With this definition of the spherical harmonics, the orthonormalization condition for these functions is

$$\boxed{\int d\Omega\, Y_{nm}^*(\theta, \varphi)\, Y_{n'm'}(\theta, \varphi) = \delta_{nn'}\, \delta_{mm'}\,.} \tag{16.126}$$

Here $d\Omega = \sin\theta\, d\theta\, d\varphi$, and the integration runs over all directions in space, i.e., from 0 to 2π in φ, and from 0 to π in θ.

Setting $n = 0,\ m = 0$ in the definition (16.125) of $Y_{nm}(\theta, \varphi)$ gives

$$Y_{00}(\theta, \varphi) = \frac{1}{\sqrt{4\pi}}\,. \tag{16.127}$$

It then follows from the orthonormality relation (16.126) that

$$\int d\Omega\, Y_{nm}^*(\theta, \varphi) = \int d\Omega\, Y_{nm}(\theta, \varphi) = \sqrt{4\pi}\, \delta_{n,0}\, \delta_{m,0}\,. \tag{16.128}$$

The symmetry property (16.122) of $P_n^m(\cos\theta)$ implies that

$$\boxed{Y_{nm}^*(\theta, \varphi) = (-1)^m\, Y_{n,-m}(\theta, \varphi).} \tag{16.129}$$

★ **17.** Verify the symmetry property given by Eq. (16.129).

Expansion of functions defined on the surface of a sphere: The spherical harmonics form an orthonormal basis for the expansion of continuous, integrable functions defined on the surface of a sphere in three dimensions. The expansion reads

$$\boxed{f(\theta, \varphi) = \sum_{n=0}^{\infty}\sum_{m=-n}^{n} f_{nm}\, Y_{nm}(\theta, \varphi).} \tag{16.130}$$

The inversion formula for the coefficients in the expansion is

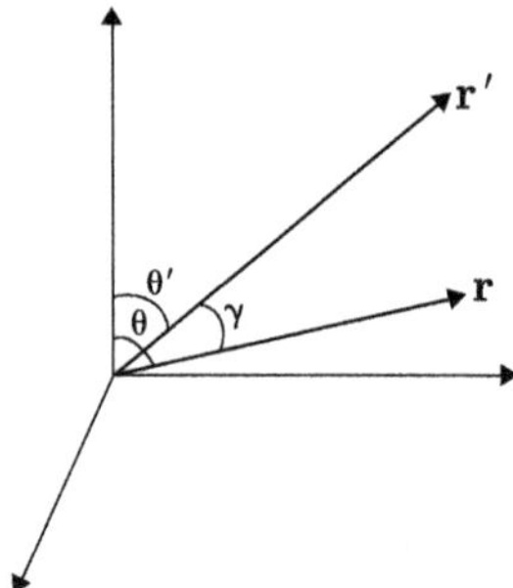

Fig. 16.2 Two arbitrary vectors **r** and **r**′, and the angle γ between them

$$\boxed{f_{nm} = \int d\Omega\, Y^{*}_{nm}(\theta,\, \varphi)\, f(\theta,\, \varphi).} \tag{16.131}$$

In particular, if $f(\theta,\, \varphi)$ is a square-integrable function on the sphere, Parseval's Theorem gives

$$\int d\Omega\, |f(\theta,\, \varphi)|^2 = \sum_{n=0}^{\infty} \sum_{m=-n}^{n} |f_{nm}|^2. \tag{16.132}$$

16.4.8 Expansion of the Coulomb Kernel

Perhaps the most common occurrence of an expansion in Legendre polynomials (and spherical harmonics) is in the expansion of the "$1/r$ potential" or Coulomb potential—or, more precisely, the expansion of the **Coulomb kernel** $1/|\mathbf{r} - \mathbf{r}'|$ associated with the electrostatic or gravitational potential in three dimensions. (Recall, for instance, Eq. (5.57) of Chap. 5, Sect. 5.3.5, for the solution of Poisson's equation in electrostatics under natural boundary conditions.) This is a reflection of the fact that, in three dimensions, this kernel is the **Green function** corresponding to the ∇^2 operator. More precisely, the solution of the equation

$$\nabla^2\, G(\mathbf{r},\, \mathbf{r}') = \delta^{(3)}(\mathbf{r} - \mathbf{r}') \tag{16.133}$$

that vanishes as $r \to \infty$ is given by

$$G(\mathbf{r},\, \mathbf{r}') = -\frac{1}{4\pi\, |\mathbf{r} - \mathbf{r}'|}\,. \tag{16.134}$$

This fundamental result will be derived formally in Chap. 29, Sect. 29.3.2.

Let **r** and **r**′ be the position vectors of any two points in three-dimensional space, and let γ be the angle between them. (See Fig. 16.2.)

Let r_s be the smaller of r and r', and r_g the larger of the two. That is,

$$r_\mathrm{s} = \min(r,\, r') \quad \text{and} \quad r_\mathrm{g} = \max(r,\, r'). \tag{16.135}$$

Then, if either $r > r'$, or $r' > r$, we have the following expansion in powers of $r_\mathrm{s}/r_\mathrm{g}$:

$$\boxed{\frac{1}{|\mathbf{r} - \mathbf{r}'|} = \frac{1}{(r^2 - 2rr' \cos\gamma + r'^2)^{1/2}} = \frac{1}{r_\mathrm{g}} \sum_{n=0}^{\infty} \left(\frac{r_\mathrm{s}}{r_\mathrm{g}}\right)^n P_n(\cos\gamma).} \tag{16.136}$$

The infinite series converges absolutely for $r_\mathrm{s}/r_\mathrm{g} < 1$. This expansion follows directly from the generating function for the Legendre polynomials, given by Eq. (16.96) of Sect. 16.4.1.

How do we write this expansion in terms of the respective angular coordinates $(\theta,\, \varphi)$ and $(\theta',\, \varphi')$ of $\mathbf{r}$ and $\mathbf{r}'$ in spherical polar coordinates? The answer is given by the generalization of a trigonometric identity, namely,

$$\boxed{\cos\gamma = \cos\theta\, \cos\theta' + \sin\theta\, \sin\theta'\, \cos(\varphi - \varphi').} \tag{16.137}$$

This identity is an elementary result in spherical trigonometry. It is known there as the **law of cosines** as applied to a triangle on the surface of a sphere, and is used extensively in observational astronomy. The generalization of Eq. (16.137) is the addition theorem for spherical harmonics, which reads

$$\boxed{P_n(\cos\gamma) = \frac{4\pi}{2n+1} \sum_{m=-n}^{n} Y^*_{nm}(\theta',\, \varphi')\, Y_{nm}(\theta,\, \varphi).} \tag{16.138}$$

The law of cosines is just the addition theorem in the case $n = 1$. Inserting the expansion (16.138) in Eq. (16.136), we arrive at an important formula: The Coulomb kernel "factorizes" in spherical polar coordinates according to

$$\boxed{\frac{1}{|\mathbf{r} - \mathbf{r}'|} = \frac{1}{r_\mathrm{g}} \sum_{n=0}^{\infty} \sum_{m=-n}^{n} \frac{4\pi}{2n+1} \left(\frac{r_\mathrm{s}}{r_\mathrm{g}}\right)^n Y^*_{nm}(\theta',\, \varphi')\, Y_{nm}(\theta,\, \varphi),} \tag{16.139}$$

where $r_\mathrm{s} = \min(r, r')$ and $r_\mathrm{g} = \max(r, r')$. In Chap. 29, Sect. 29.3.3, I will use this expansion to simplify the solution of Poisson's equation in the case of a spherically symmetric "source function".

The addition theorem of Eq. (16.138) gives an interesting identity when $\cos\gamma = 1$, i.e., when the vectors $\mathbf{r}$ and $\mathbf{r}'$ are collinear. Then $\theta = \theta'$ and $\varphi = \varphi'$. Since $P_n(1) = 1$ for all integers $n \geq 0$, we get a **sum rule** for spherical harmonics:

$$\frac{4\pi}{2n+1} \sum_{m=-n}^{n} |Y_{nm}(\theta,\, \varphi)|^2 = 1 \quad (n = 0,\, 1,\, 2,\, \ldots). \tag{16.140}$$

With the help of the symmetry property (16.129), Eq. (16.140) reduces to

$$\left[P_n(\cos\theta)\right]^2 + 2\sum_{m=1}^{n} \frac{(n-m)!}{(n+m)!}\left[P_n^m(\cos\theta)\right]^2 = 1 \quad (n = 0,\ 1,\ 2,\ \ldots). \quad (16.141)$$

★ **18.** As always, you will find it instructive to work out the steps to verify the results quoted.

(a) Verify that Eq. (16.140) reduces to Eq. (16.141).
(b) Verify Eq. (16.141) explicitly in the case $n = 1$.
(c) Similarly, set $n = 1$ in the addition theorem (16.138), and verify that the law of cosines, Eq. (16.137), is correctly recovered.

16.5 Solutions

1. Note that any polynomial of degree $m < n$, including the monomial x^m, can be expanded uniquely in terms of orthogonal polynomials of degree $\leq m$.

Remark Is the "complement" of Eq. (16.14) true, i.e., can we assert that $\int_a^b dx\, \rho(x)\, x^m\, \phi_n(x) = 0$ when m is *greater* than n? No. The expansion of x^m in orthogonal polynomials will, in general, involve a linear combination of all polynomials $\phi_j(x)$ where $0 \leq j \leq m$, including $j = n$. ▶

2. Square both sides of Eq. (16.15), multiply by $\rho(x)$ and integrate over x from a to b. Use the orthogonality relation (16.5). ▶

3. This can be done almost by inspection. Multiply both sides of Eq. (16.19) by $\rho(x)\, x^m$, and integrate over x from a to b. Observe that the integral of the determinant is the same determinant, but with each element x^j of the first row (where $0 \leq j \leq n$) replaced by its integral, which is M_{j+m}. (This works because the elements of all the other rows are just constants.) Hence some two rows of the determinant become identical for every $m < n$, and the determinant vanishes. ▶

5. (a) Observe that $F^{\text{her}}(-x, -t) = F^{\text{her}}(x, t)$. This yields $H_n(-x) = (-1)^n\, H_n(x)$, i.e., Hermite polynomials of even (respectively, odd) degree are even (respectively, odd) functions of x. Hence $H_{2n+1}(0) = 0$. Further, setting $x = 0$, we have

$$F^{\text{her}}(0, t) = e^{-t^2} = \sum_{n=0}^{\infty} \frac{H_{2n}(0)}{(2n)!}\, t^{2n}.$$

Expand e^{-t^2} and equate the coefficients of t^{2n}. The expression for $H_{2n}(0)$ follows.

(b) Regard Eq. (16.40) as the Taylor expansion of $F^{\text{her}}(x,t)$ about the point $t=0$. Then $H_n(x)$ is the nth derivative of $F^{\text{her}}(x,t)$ with respect to t, evaluated at $t=0$. Write the generating function in the form

$$F^{\text{her}}(0,t) = e^{x^2}\, e^{-(t-x)^2}.$$

Then

$$e^{-x^2} H_n(x) = \frac{d^n}{dt^n} e^{-(t-x)^2}\bigg|_{t=0} = (-1)^n \frac{d^n}{du^n} e^{-u^2}\bigg|_{u=x},$$

on setting $u = x - t$. The Rodrigues formula for $H_n(x)$ follows at once.

(c) Differentiate both sides of Eq. (16.40) with respect to x, and note that the left-hand side is $2t\, F^{\text{her}}(x,t)$. Expand $F^{\text{her}}(x,t)$ in powers of t, and equate the coefficients of t^n on both sides.

(d) From the Rodrigues formula,

$$H_{n+1}(x) = (-1)^{n+1}\, e^{x^2}\, \frac{d^{n+1}}{dx^{n+1}}\, e^{-x^2} = 2(-1)^n\, e^{x^2}\, \frac{d^n}{dx^n}\,(x\, e^{-x^2}).$$

Now use the chain rule for the nth derivative of a product of two functions of x, namely,

$$(u\, v)^{(n)} = u^{(n)}\, v + \binom{n}{1} u^{(n-1)}\, v^{(1)} + \cdots,$$

where the superscripts in parentheses denote derivatives with respect to x. Setting $u = e^{-x^2}$ and $v = x$, the recursion relation sought follows immediately.

(e) Since $H'_n = 2n\, H_n$, we have $H''_n = 4n^2\, H_{n-2}$. Hence

$$H''_n - 2x\, H'_n + 2n\, H_n = 2n\,(H_n - 2x\, H_{n-1} + 2n\, H_{n-2}) = 0,$$

using the recursion relation satisfied by H_n.

(f) Let $s = t/\sqrt{2},\ u = x\sqrt{2},\ v = y\sqrt{2}$. Then $2t(x+y) - t^2 = (2su - s^2) + (2sv - s^2)$. Exponentiate both sides and use the formula for the generating function of the Hermite polynomials. Hence

$$\sum_{n=0}^{\infty} \frac{H_n(x+y)}{n!}\, t^n = \sum_{l=0}^{\infty}\sum_{k=0}^{\infty} \frac{H_l(\sqrt{2}\,x)}{l!}\, \frac{H_k(\sqrt{2}\,y)}{k!}\, \frac{t^{l+k}}{2^{(l+k)/2}}.$$

Set $l + k = n$, and use the identity

$$\sum_{l=0}^{\infty}\sum_{k=0}^{\infty} f(l,k) = \sum_{n=0}^{\infty}\sum_{k=0}^{n} f(n-k,k).$$

Equating the coefficients of t^n on both sides in the result gives the addition theorem quoted. ▶

6. The derivation of (16.54) is simple. Equation (16.55) can be obtained without repeating the calculation. All you have to do is to use $\widetilde{\Phi}_n(p)$ in the place of $\Phi_n(x)$ in Eq. (16.51). Observe that the effect is simply to replace x by p, and z by $-iz$ (i.e., replace z_1 by z_2, and z_2 by $-z_1$). ▶

7. (a) Set $x = 0$ in Eq. (16.59) for the generating function $F^{\text{lag}}(x,t)$. It follows that $L_n^{\alpha}(0)$ is the coefficient of t^n in the binomial expansion

$$(1-t)^{-(\alpha+1)} = 1 + (\alpha+1)t + (\alpha+1)(\alpha+2)\frac{t^2}{2!} + \cdots .$$

Hence $L_n^{\alpha}(0) = \dfrac{(\alpha+1)\cdots(\alpha+n)}{n!} = \dfrac{\Gamma(n+\alpha+1)}{\Gamma(\alpha+1)\,n!}$.

(b) The polynomials concerned follow from $F^{\text{lag}}(x,t)$ and its first two derivatives with respect to t, evaluated at $t = 0$.

(c) Observe that the generating function for $L_n^{\alpha+1}(x)$ is equal to $(1-t)^{-1}$ times the generating function for $L_n^{\alpha}(x)$. Expanding $(1-t)^{-1}$ in its binomial series,

$$\sum_{n=0}^{\infty} L_n^{\alpha+1}(x)\,t^n = (1 + t + t^2 + \cdots)\sum_{k=0}^{\infty} L_k^{\alpha}(x)\,t^k.$$

Equating the coefficients of t^n on both sides, it follows that $L_n^{\alpha+1}(x) = \sum_{k=0}^{n} L_k^{\alpha}(x)$.

(d) Observe that the generating function for $L_n^{\alpha+\beta+1}(x+y)$ is just the product of the generating functions for $L_n^{\alpha}(x)$ and $L_n^{\beta}(y)$. Hence

$$\sum_{n=0}^{\infty} L_n^{\alpha+\beta+1}(x+y)\,t^n = \sum_{l=0}^{\infty}\sum_{k=0}^{\infty} L_l^{\alpha}(x)\,L_k^{\beta}(y)\,t^{l+k}.$$

Set $l + k = n$, and proceed exactly as in the case of the addition theorem for Hermite polynomials. The addition theorem for generalized Laguerre polynomials follows in a straightforward manner. Note that

$$\sum_{k=0}^{n} L_{n-k}^{\alpha}(x)\,L_k^{\beta}(y) = \sum_{k=0}^{n} L_k^{\alpha}(x)\,L_{n-k}^{\beta}(y),$$

because we can sum over $n - k$ instead of k without changing the value of the sum. ▶

8. These properties can be established using the generating function or the Rodrigues formula for $T_n(x)$. But it is simplest to use the trigonometric function representation of Eq. (16.84). ▶

9. As in the case of $T_n(x)$, these relations are most easily derived using the trigonometric representation (16.94). ▶

10. If m and n have the same parity, $P_m(x)\, P_n(x)$ is an even function of x. Its integral from 0 to 1 is equal to half its integral from -1 to 1. The result required then follows trivially from the orthonormality relation. ▶

11. Change independent variables from x to $\frac{1}{2}(1 - x)$. The relationship sought is given in Eq. (16.102). ▶

12. (a) Observe that $(1 - 2xt + t^2)\, \partial F^{\text{leg}}/\partial x = t\, F^{\text{leg}}$. Write out both sides as power series in t and equate coefficients of t^n.

(b) Use the fact that $t\, \partial F^{\text{leg}}/\partial t = (x - t)\, \partial F^{\text{leg}}/\partial x$, and proceed as before.

(c) Start from $t\, \partial(t\, F^{\text{leg}})/\partial t = t^2\, \partial F^{\text{leg}}/\partial t + t\, F^{\text{leg}}$. Substitute for $t\, \partial F^{\text{leg}}/\partial t$ and $t\, F^{\text{leg}}$ from the identities used in (a) and (b) above. The result is the identity

$$t\, \frac{\partial}{\partial t}(t\, F^{\text{leg}}) = (1 - xt)\, \frac{\partial F^{\text{leg}}}{\partial x}\, .$$

Expand both sides in powers of t and equate the coefficients of t^n.

(d) (i) Write the identity in (c) for $n + 1$ instead of n and add the result to the identity in (b).

(ii) Multiply the identity in (b) by x, and subtact the identity in (c).

(iii) Consider the identity in (a) for $n + 1$ instead of n, and write it as

$$(n + 1)\, P_{n+1} - (n + 1)x\, P_n = nx\, P_n - n P_{n-1}.$$

Substitute for the right-hand side from the identity in (ii).

(e) Differentiating the identity in (iii) above,

$$(x^2 - 1)\, P''_n + 2x\, P'_n = (n + 1)\, (P'_{n+1} - x\, P'_n) - (n + 1)\, P_n\, .$$

In the first term on the right-hand side, use the identity in (c) written for $n+1$ instead of n. Moving all the terms to the left-hand side, it follows that

$$(1-x^2)\,P''_n - 2x\,P'_n + n(n+1)\,P_n = 0.$$

(f) Setting $x = 1$ in the generating function gives

$$\frac{1}{1-t} = \sum_{n=0}^{\infty} t^n = \sum_{n=0}^{\infty} P_n(1)\,t^n \implies P_n(1) = 1.$$

Therefore $P_n(-1) = (-1)^n P_n(1) = (-1)^n$.

(g) Since $P_{2n+1}(x)$ is a bounded, continuous, odd function of x, it must vanish when $x = 0$, i.e., $P_{2n+1}(0) = 0$. Setting $x = 0$ in the generating function then gives

$$\frac{1}{(1+t^2)^{1/2}} = \sum_{n=0}^{\infty} P_n(0) t^n = \sum_{n=0}^{\infty} P_{2n}(0)\,t^{2n}.$$

The binomial expansion of the left-hand side gives $P_{2n}(0) = \dfrac{(-1)^n\,(2n)!}{2^{2n}\,(n!)^2}$. ▶

14. (a) The volume V is easily computed by integrating the volume element in spherical polar coordinates over the volume of the solid.

$$V = 2\pi \int_{-1}^{1} d(\cos\theta) \int_0^{r(\theta)} dr'\,r'^2 = \frac{2\pi a^3}{3} \int_{-1}^{1} d(\cos\theta)\,[1+\epsilon\,P_2(\cos\theta)]^3.$$

Integrate and simplify to arrive at the expression quoted.

(b) The surface area of the solid is given by $S = \int_0^{\pi} (2\pi r\,\sin\theta)(r\,d\theta)$, where $r = a\,[1+\epsilon\,P_2(\cos\theta)]$. Integration yields the result quoted.

Remark Such expressions are useful in many contexts, from geophysics to nuclear physics. In the latter, for instance, they are relevant in the description of the shapes of deformed nuclei. Note, incidentally, that neither V nor S has a term that is of first order in ϵ. ▶

16. Let m be an integer such that $0 \le m \le n$. Then $P_n^{-m}(x)$ is obtained by substituting $-m$ wherever m appears in Eq. (16.120). To show that this coincides with the right-hand side of Eq. (16.122), you have to show that

$$\frac{d^{n-m}}{dx^{n-m}}(1-x^2)^n = (-1)^m \frac{(n-m)!}{(n+m)!}(1-x^2)^m \frac{d^{n+m}}{dx^{n+m}}(1-x^2)^n.$$

You can do this by checking that the coefficient of each power of x is the same on both sides of this equation. ▶

18. (a) Use Eqs. (16.129) and (16.125).

(b) Note that $P_1^1(\cos\theta) = -\sin\theta$.

(c) First show that $Y_{11}(\theta,\varphi) = -Y^*_{1,-1}(\theta,\varphi) = -\sqrt{3/(8\pi)}\,(\sin\theta)\,e^{i\varphi}$. ▶

Chapter 17
Fourier Series

17.1 Series Expansion of Periodic Functions

Fourier series occupy a unique place in the history of mathematics. In a sense, one may regard them as having been present at the birth of mathematical physics: the latter event is often identified with the publication of Joseph Fourier's treatise, *Théorie Analytique de la Chaleur* (*Analytic Theory of Heat*) in 1822. Fourier series and transforms have been very extensively studied and vastly generalized since then. The part of modern mathematics concerned with this subject is called **harmonic analysis**.

The treatment here will be rather concise. It is convenient to begin with Fourier series. In Chap. 18, we will go on to Fourier integrals and transforms.[1]

17.1.1 Dirichlet Conditions

The basic facts about Fourier series are as follows. We consider the set of functions $f(x)$ $(-\infty < x < \infty)$, not necessarily real-valued, satisfying the following **Dirichlet conditions**:

(i) $f(x)$ is a **periodic function** of x, with a period L. The **fundamental interval** is taken to be (a, b), so that $L = (b - a)$, and $f(x) = f(x + L)$. Hence $f(x) = f(x + nL)$ where n is any integer.
(ii) In its fundamental interval, $f(x)$ is of bounded variation, and has at most a finite number of maxima, minima, and finite discontinuities (or jumps).

[1]A rather annoying aspect of Fourier series and integrals is the large number of different conventions for signs, numerical factors like 2π, and various other multiplicative constants. This makes it quite tedious to move from one source of information to another in a consistent manner. I have adopted a particular set of conventions in this regard, and will stick to it, for both Fourier series and Fourier integrals.

V. Balakrishnan, *Mathematical Physics*,
https://doi.org/10.1007/978-3-030-39680-0_17

A function $f(x)$ satisfying these conditions has a Fourier series expansion. It is important to note that these are *sufficiency* conditions, rather than *necessary* ones. One often encounters Fourier series (and Fourier integrals) representing functions that do not satisfy these conditions—in particular, with regard to generalized functions or distributions. Notable among these are the Dirac δ-function $\delta(x)$, the unit step function $\theta(x)$, etc. We will also consider some of these cases in the sequel.

17.1.2 Orthonormal Basis

The Fourier series expansion of a periodic function is based on the following fact: The set of functions

$$\phi_n(x) = e^{2\pi n i x/L} \quad (\text{where} \quad n \in \mathbb{Z}) \tag{17.1}$$

forms an orthonormal basis in the space of periodic functions of x with fundamental interval (a, b) and period $L = (b - a)$. This basis is sometimes called the "complex basis", as opposed to the "real basis" in terms of the trigonometric functions $\sin(n\pi x/L)$ and $\cos(n\pi x/L)$ (see below). Note that, in the complex basis, n runs over *all* integers from $-\infty$ to $+\infty$. The advantages of the complex basis are precisely those of the exponential function over the sine and cosine functions:

- Differentiation and integration are a little easier to handle.
- Phase angles are incorporated in the complex expansion coefficients.
- The superposition of harmonics is algebraically simpler.

It is trivially verified that the orthonormality condition is given by

$$\frac{1}{L}\int_a^b dx\, \phi_n^*(x)\, \phi_m(x) = \frac{1}{L}\int_a^b dx\, e^{2\pi i\,(m-n)x/L} = \delta_{nm} \quad (\text{orthonormality}). \tag{17.2}$$

This relation holds good no matter what the actual value of a is the start of the fundamental interval need not coincide with any special point such as a node or a maximum or a minimum in the sine or cosine function. The completeness relation is

$$\frac{1}{L}\sum_{n=-\infty}^{\infty} \phi_n^*(x)\, \phi_n(x\,') = \frac{1}{L}\sum_{n=-\infty}^{\infty} e^{-2\pi n i\,(x-x\,')/L} = \delta\,(x - x\,') \quad (\text{completeness}). \tag{17.3}$$

It is obvious that the infinite series of exponentials in (17.3) cannot converge absolutely, since each term has a magnitude equal to 1. Note the interesting fact that an infinite sum of equally spaced oscillating exponentials is a singular object, namely, a δ-function. We will recall this observation later on, when we deal with Fourier integrals.

17.1.3 Fourier Series Expansion and Inversion Formula

The expansion of a function $f(x)$ in Fourier series is given by

$$\boxed{f(x) = \frac{1}{L}\sum_{n=-\infty}^{\infty} f_n\, e^{2\pi nix/L} \qquad \text{(expansion formula).}} \tag{17.4}$$

The corresponding inversion formula for the Fourier coefficients is

$$\boxed{f_n = \int_a^b dx\, f(x)\, e^{-2\pi nix/L} \qquad \text{(inversion formula).}} \tag{17.5}$$

In the special case in which $f(x)$ is a real-valued function, it is easy to see that the Fourier coefficient f_n satisfies the symmetry property $f_n^* = f_{-n}$.

The real basis is obtained by rewriting the exponentials in terms of sines and cosines. Set

$$\frac{f_0}{L} = \frac{a_0}{2}, \quad \frac{f_n + f_{-n}}{L} = a_n, \quad \text{and} \quad \frac{i(f_n - f_{-n})}{L} = b_n \ (n \geq 1). \tag{17.6}$$

The Fourier series expansion of $f(x)$ can then be written in the alternative form

$$\boxed{f(x) = \tfrac{1}{2}\, a_0 + \sum_{n=1}^{\infty}\Big\{a_n\, \cos\,(2\pi nx/L) + b_n\, \sin\,(2\pi nx/L)\Big\}.} \tag{17.7}$$

The corresponding inversion formulas are

$$\boxed{a_n = (2/L)\int_a^b dx\, f(x)\, \cos\,(2\pi nx/L) \ \ (n \geq 0)} \tag{17.8}$$

and

$$\boxed{b_n = (2/L)\int_a^b dx\, f(x)\, \sin\,(2\pi nx/L) \ \ (n \geq 1).} \tag{17.9}$$

The quantity $\frac{1}{2}a_0 = (1/L)\int_a^b dx\, f(x)$ is just the average value of $f(x)$ over a complete period. You can go back to the complex basis by using the inverse of the relations (17.6), namely,

$$\frac{f_n}{L} = \frac{a_n - ib_n}{2}, \quad \frac{f_{-n}}{L} = \frac{a_n + ib_n}{2} \quad (n \geq 1). \tag{17.10}$$

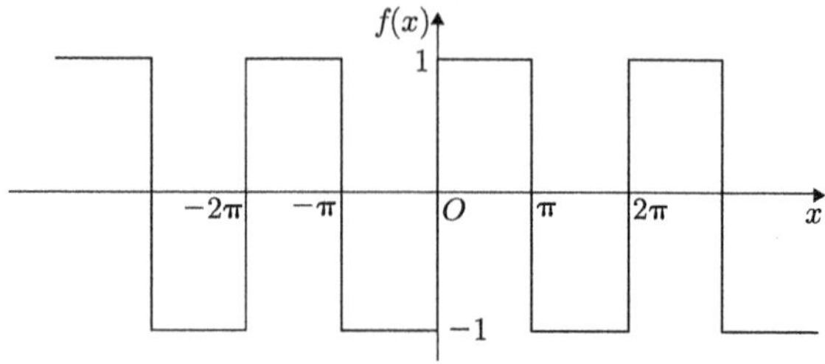

Fig. 17.1 The periodic function defined in Eq. (17.11)

The real basis above is particularly convenient when $f(x)$ has a specific parity, i.e., it is an even or odd function of x. If $f(x)$ is an even function, all the coefficients b_n vanish, and $f(x)$ is expandable in a **Fourier cosine series**. If $f(x)$ is an odd function, all the coefficients a_n vanish, and $f(x)$ is expanded in a **Fourier sine series**. Moreover, if $f(x)$ is a real function, it is evident that the coefficients a_n and b_n are also real numbers. Note that the index n in the real basis runs only over the nonnegative integers.

A simple example helps illustrate several features of the Fourier series representations of periodic functions. Consider the periodic function $f(x)$ defined in its fundamental interval $(0,\ 2\pi)$ by

$$f(x) = \begin{cases} +1 & \text{for } \ 0 < x < \pi \\ -1 & \text{for } \ \pi < x < 2\pi\,. \end{cases} \tag{17.11}$$

Figure 17.1 depicts this periodic function. $f(x)$ is a sequence of rectangular pulses. It is clearly an odd function of x, and so you may expect a_n to vanish for every $n \geq 0$. This is easily confirmed. The Fourier sine series for $f(x)$ works out to

$$f(x) = (4/\pi)\big(\sin x + \tfrac{1}{3}\sin 3x + \tfrac{1}{5}\sin 5x + \cdots\big) = (4/\pi)\sum_{n=0}^{\infty}\frac{\sin(2n+1)x}{(2n+1)}\,. \tag{17.12}$$

Setting $x = \frac{1}{2}\pi$ yields the famous Madhava–Leibniz formula[2]

$$1 - \tfrac{1}{3} + \tfrac{1}{5} - \cdots = \tfrac{1}{4}\pi. \tag{17.13}$$

Observe that $x = 0$ is a point of discontinuity for $f(x)$, which jumps from -1 to $+1$ at that point. The arithmetic average of these two values is 0. But this is precisely what the series yields when you set $x = 0$ in it. This is not an accident, as we will see in Sect. 17.2.1. As $n \to \infty$, the leading asymptotic behavior of the Fourier coefficient is given by $b_n \sim n^{-1}$ in this case. Again, this is related to the fact that $f(x)$ has finite discontinuities, as will be seen in Sect. 17.2.2.

[2]The value $\frac{1}{4}\pi$ for the sum of the infinite series on the left-hand side of Eq. (17.13) was first deduced by the Indian mathematician Madhava of Sangamagrama (1350–1425) from the more general series for $\tan^{-1} x$ discovered by him. Rediscovered by James Gregory in 1668, and again by Gottfried Leibniz some years later, it is now called the Madhava–Gregory series.

★ **1.** Sketch the periodic function $f(x)$ defined in its fundamental interval $(0,\, \pi)$ by

$$f(x) = \begin{cases} \sin x + \cos x & \text{for } 0 < x \le \frac{1}{2}\pi \\ \sin x - \cos x & \text{for } \frac{1}{2}\pi \le x < \pi . \end{cases}$$

Find its Fourier series expansion.

17.1.4 Parseval's Formula for Fourier Series

Recall that Parseval's Theorem has already been introduced in connection with the expansion of elements of an LVS in orthonormal basis sets, in Chap. 10, Sect. 10.2.3. Subsequently, we encountered the version relevant to the expansion of square-integrable functions in terms of families of orthogonal polynomials, in Chap. 16, Sects. 16.1.3 and 16.4.3. I reiterated there that the result was a general one, applicable to any $\mathcal{L}_2$ function. The counterpart of the formula in the case of Fourier series is as follows.

If $f(x)$ is also square-integrable in its fundamental interval, i.e., if $\int_a^b dx\, |f(x)|^2 < \infty$, then

$$\boxed{\int_a^b dx\, |f(x)|^2 = \frac{1}{L} \sum_{n=-\infty}^{\infty} |f_n|^2.} \tag{17.14}$$

Hence, if $f(x) \in \mathcal{L}_2[a,\, b]$, then its Fourier coefficients $f_n \in \ell_2$. In terms of the coefficients a_n and b_n, Parseval's formula reads

$$\boxed{(1/L) \int_a^b dx\, |f(x)|^2 = \tfrac{1}{4}|a_0|^2 + \tfrac{1}{2} \sum_{n=1}^{\infty} \left(|a_n|^2 + |b_n|^2\right).} \tag{17.15}$$

This is the form in which the formula is written most commonly. Once again, I repeat that Parseval's formula is simply the expression for the norm of a vector in an LVS in terms of the coefficients occurring in its expansion in a given basis. Equation (17.15) is a special case of this general result as applied to a periodic function expanded in the Fourier basis, i.e., in terms of sine and cosine functions.

A generalization of Parseval's formula: In the context of Fourier series, the term "Parseval's formula" is also used, on occasion, for the following somewhat more general result. Let $f(x)$ and $F(x)$ be two functions with the same fundamental interval $(a,\, b)$ and period L, with respective Fourier coefficients f_n and F_n in the complex basis, or $(a_n,\, b_n)$ and $(A_n,\, B_n)$ in the real basis. Then, if both functions are square-integrable in the fundamental interval,

$$\frac{1}{L}\int_a^b dx\, f(x)\, F(x) = \frac{1}{L^2}\sum_{n=-\infty}^{\infty} f_n\, F_{-n}$$
$$= \frac{1}{L^2}\, f_0\, F_0 + \frac{1}{L^2}\sum_{n=1}^{\infty}(f_n\, F_{-n} + f_{-n}\, F_n)$$
$$= \tfrac{1}{4}\, a_0\, A_0 + \tfrac{1}{2}\sum_{n=1}^{\infty}(a_n\, A_n + b_n\, B_n)\,. \tag{17.16}$$

Equation (17.15) is a special case of this result, corresponding to the choice $F(x) = f^*(x)$.

★ **2.** Derive Eq. (17.16).

There exists an analog of Eq. (17.16) in the case of Fourier integrals. This is given in Eq. (18.11) of Chap. 18, Sect. 18.1.6.

17.1.5 Simplified Formulas When $(a,\, b) = (-\pi,\, \pi)$

The formulas connected with Fourier series expansions look a bit simpler if one chooses the fundamental interval to be $(-\pi,\, \pi)$, so that $L = 2\pi$. In order to remind ourselves of this specific choice, I shall write it explicitly on the right of the expansion formula. The Fourier series for $f(x)$ is then given by

$$f(x) = \frac{1}{2\pi}\sum_{n=-\infty}^{\infty} f_n\, e^{nix}, \quad \text{when} \quad (a,\, b) = (-\pi,\, \pi). \tag{17.17}$$

The inversion formula is simply

$$f_n = \int_{-\pi}^{\pi} dx\, f(x)\, e^{-nix}, \quad n \in \mathbb{Z}. \tag{17.18}$$

In the real basis formed by sine and cosine functions,

$$f(x) = \tfrac{1}{2}\, a_0 + \sum_{n=1}^{\infty}\{a_n\, \cos(nx) + b_n\, \sin(nx)\}, \quad \text{when} \quad (a,\, b) = (-\pi,\, \pi). \tag{17.19}$$

The corresponding inversion formulas are

$$a_n = \frac{1}{\pi}\int_{-\pi}^{\pi} dx\, f(x)\, \cos(nx) \quad (n \ge 0) \tag{17.20}$$

and

$$b_n = \frac{1}{\pi}\int_{-\pi}^{\pi} dx\, f(x)\, \sin(nx) \quad (n \geq 1). \tag{17.21}$$

17.2 Asymptotic Behavior and Convergence

17.2.1 Uniform Convergence of Fourier Series

A basic question that arises in the context of Fourier series is the following: In what sense does the series expansion represent the actual function? That is, what sort of convergence property does the infinite series expansion have? Most important, what happens at points where the original periodic function $f(x)$ has a finite discontinuity or jump?

These questions are very important in any rigorous discussion of Fourier series. Our emphasis here, however, is on elementary applications of Fourier series. I therefore merely state the basic results.

(i) At any point where $f(x)$ is continuous, its Fourier series converges *uniformly* to the value of $f(x)$ at that point.

Uniform convergence means, in effect, that you can carry out certain operations on the series without affecting its convergence—for instance, you can differentiate it term by term.

(ii) At a point where $f(x)$ has a finite discontinuity or jump, the series converges to the *mean value* of $f(x)$ at that point.

Let $x = c$ be a point of discontinuity of $f(x)$, and let ϵ be a positive number. Suppose the values of $f(x)$ on approaching $x = c$ from the left and right, respectively, are $f^{\mp}(c)$. That is,

$$\lim_{\epsilon\to 0} f(c-\epsilon) = f^{-}(c) \text{ and } \lim_{\epsilon\to 0} f(c+\epsilon) = f^{+}(c), \tag{17.22}$$

where $0 < |f^{-}(c) - f^{+}(c)| < \infty$. Then, setting $x = c$ in the Fourier series expansion yields the mean value $\frac{1}{2}[f^{-}(c) + f^{+}(c)]$. This remains true even when c is an end-point (a or b) of the fundamental interval of the periodic function. You have already seen an example of this result in the case of the rectangular pulse sequence given by Eq. (17.11) and its Fourier series expansion (17.12). We will encounter further examples below.

(iii) However, in the neighborhood of a point of discontinuity, the partial sum of the Fourier series displays *oscillations* (overshoot as well as undershoot) that persist no matter how many terms are included in the partial sum.

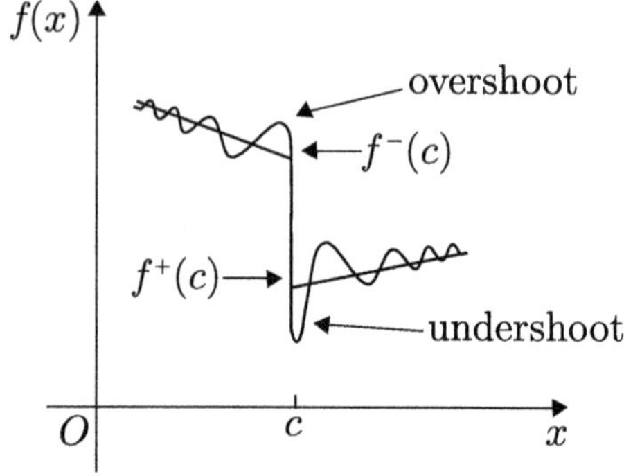

Fig. 17.2 Illustrating the Gibbs phenomenon. The overshoot and undershoot of the partial sum $S_N(x)$ persist no matter how large N becomes

Let $S_N(x)$ denote the partial sum up to $n = N$ in the Fourier series (17.7), i.e.,

$$S_N(x) = \tfrac{1}{2}\, a_0 + \sum_{n=1}^{N} \big\{a_n \cos\,(2\pi n x/L) + b_n \sin\,(2\pi n x/L)\big\}. \tag{17.23}$$

Let $x = c$ be the point of discontinuity of $f(x)$, with $f^-(c) > f^+(c)$, say. Then, the assertion is that the oscillatory *overshoot* of $S_N(x)$ over the actual value $f^-(c)$ persists as $x \to c$ from the left, no matter how large N gets. Similarly, the oscillatory *undershoot* of $S_N(x)$ below the actual value $f^+(c)$ persists as $x \to c$ from the right, no matter how large N gets. This is the famous **Gibbs phenomenon**, illustrated in Fig. 17.2. For a more detailed discussion of these assertions, you must refer to any standard text on the subject. The question of convergence is naturally related to the asymptotic $(n \to \infty)$ behavior of the Fourier coefficients f_n. Let us turn to this aspect next.

17.2.2 Large-n Behavior of Fourier Coefficients

The behavior of the Fourier coefficient f_n (or that of the coefficients a_n and b_n) as $n \to \infty$ gives us valuable information about the function $f(x)$. For instance, it is clear from Parseval's formula (17.14) that, if $|f_n|$ falls off faster than $1/\sqrt{n}$, then $f(x)$ is square-integrable in the fundamental interval. But considerably stronger results can be deduced. We can also say something about the *finite* discontinuities of $f(x)$ from the behavior of its Fourier coefficients. A very useful result is the following:

- Suppose the asymptotic behavior of the Fourier coefficient is a power-law decay of the form $f_n \sim 1/n^{k+1}$, where k is a nonnegative integer. Then the functions $f,\ df/dx,\ \ldots,\ d^{k-1} f/dx^{k-1}$ are continuous in the fundamental interval $(a,\, b)$; but the kth derivative $d^k f/dx^k$ has a finite number of finite discontinuities or jumps in that interval.
- In particular, if f_n decays as slowly as $1/n$, then $f(x)$ itself has a finite number of finite discontinuities in the interval.

We may also expect that if f_n decays *more slowly* than $1/n$ as $n \to \infty$, then $f(x)$ would have a more singular behavior than just finite jumps—for instance, it could become unbounded at one or more points. Such a function would not satisfy the Dirichlet conditions. But the latter are only sufficiency conditions, and not necessary ones, as I have stated already. We have, in fact, encountered an example of a slow decay of the coefficients already, namely, the case of the Dirac δ-function. It is an important case, and therefore worth a closer look.

17.2.3 Periodic Array of δ-Functions: The Dirac Comb

Recall the completeness relation in the complex basis, Eq. (17.3). Setting $x' = 0$ in that relation, we have

$$\frac{1}{L}\sum_{n=-\infty}^{\infty} e^{-2\pi n i\, x/L} = \frac{1}{L}\sum_{n=-\infty}^{\infty} e^{2\pi n i\, x/L} = \delta\,(x). \tag{17.24}$$

Suppose we regard this equation as a Fourier series expansion in the complex basis. We then have here an extreme situation in which $f_n = 1$ for each n, i.e., f_n does not decay at all as n increases! The outcome is a *singular* quantity—in this case, the Dirac δ-function, which we studied in Chap. 4, and whose Fourier integral representation we have already written down in Eq. (4.19), Sect. 4.2.4. We know that the δ-function is actually a distribution rather than a function in the conventional sense. But Fourier *series* (as opposed to Fourier integrals) are meant to represent *periodic* functions. Therefore, if Eq. (17.24) above is interpreted as a Fourier series expansion, it really represents a periodic, infinite *array* of δ-functions, with a fundamental interval $(-\frac{1}{2}L,\, \frac{1}{2}L)$. In explicit form, we may write it as $\sum_{n=-\infty}^{\infty} \delta(x - nL)$. Such an array is called a **Dirac comb** (see Fig. 17.3). The interesting fact is the following:

- An infinite periodic array of δ-functions is equal to the superposition of an infinite number of equally spaced oscillating exponentials.

Writing the exponentials in terms of sines and cosines, the odd part of the function vanishes owing to the summation over n. We thus get

$$\boxed{\sum_{n=-\infty}^{\infty} \delta(x - nL) = \frac{1}{L}\sum_{n=-\infty}^{\infty} e^{2\pi n i\, x/L} = \frac{1}{L} + \frac{2}{L}\sum_{n=1}^{\infty} \cos\left(\frac{2\pi n x}{L}\right).} \tag{17.25}$$

A "physical" explanation of this result is as follows. In effect, the interference between the different harmonics on the right-hand side causes their sum to vanish at all points except for sharp spikes at the set of points $x = nL$, where $n \in \mathbb{Z}$. I will return to this result when we discuss the Poisson summation formula in connection with Fourier integrals, in Chap. 18, Sect. 18.4.1. At present, we merely note that

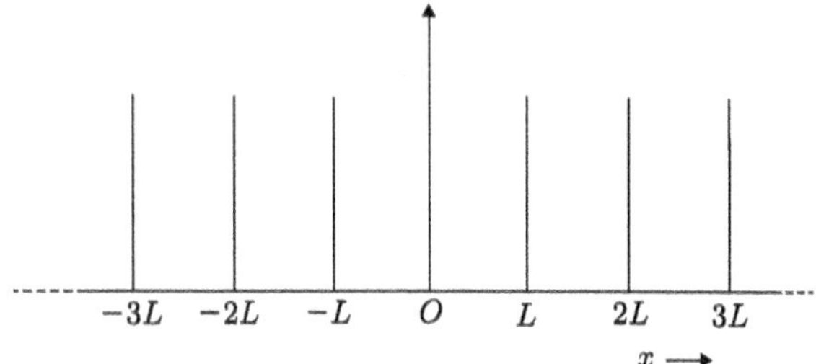

Fig. 17.3 The Dirac comb is an infinite array of regularly spaced δ-functions

- if $f_n \sim n^0$ as $n \to \infty$, the periodic function $f(x)$ may be expected to have infinite discontinuities of the δ-function kind.

17.3 Summation of Series

17.3.1 Some Examples

One of the by-products of Fourier series expansions is a method for the summation of certain infinite series. The values of some infinite sums can be determined by setting x equal to appropriate values in the Fourier series expansions of suitable periodic functions, as you have seen already in some of the examples above. Here are a few more illustrations of this technique.

★ **3.** Consider the periodic function $f(x) = |\sin x|$ with fundamental interval $(-\frac{1}{2}\pi, \frac{1}{2}\pi)$.

(a) Sketch this periodic function, and find its Fourier series expansion.
(b) Using this result, show that

$$\text{(i)} \sum_{k=1}^{\infty} \frac{1}{(2k-1)(2k+1)} = \frac{1}{2} \qquad \text{(ii)} \sum_{k=1}^{\infty} \frac{(-1)^k}{(2k-1)(2k+1)} = \frac{1}{2} - \frac{\pi}{4}.$$

★ **4.** Infinite series that are less trivial than the foregoing can also be summed by this method. Here is an example.

(a) Sketch the periodic function given by $f(x) = x^2$ with fundamental interval $(-1, 1)$, and find its Fourier series expansion.
(b) Use the result to show that

$$\frac{1}{1^2} + \frac{1}{2^2} + \frac{1}{3^2} + \cdots = \frac{\pi^2}{6} \quad \text{and} \quad \frac{1}{1^2} - \frac{1}{2^2} + \frac{1}{3^2} - \cdots = \frac{\pi^2}{12}.$$

17.3.2 The Riemann Zeta Function $\zeta(2k)$

The series $\sum_{n=1}^{\infty} 1/n^2$ is equal to $\zeta(2)$, which is the **Riemann zeta function** $\zeta(z)$ when its argument $z = 2$. What we have just found is that $\zeta(2) = \frac{1}{6}\pi^2$. The zeta function is of paramount importance in number theory. It is defined as

$$\boxed{\zeta(z) \stackrel{\text{def.}}{=} \sum_{n=1}^{\infty} \frac{1}{n^z} \quad \text{in the region } \operatorname{Re} z > 1} \tag{17.26}$$

of the complex z-plane. I will return to the zeta function more than once in the sequel,[3] and discuss its analytic properties, including its continuation to the whole of the complex z-plane, in a little more detail.

★ **5.** From the Fourier series expansion of the periodic function given by $f(x) = x^4$ in the fundamental interval $(-1,\ 1)$, deduce the value of $\zeta(4)$.

In principle, the Fourier cosine series for the function $f(x) = x^{2k}$ in the fundamental interval $(-1,\ 1)$ provides a way of determining the value of $\zeta(2k)$ (where k is a positive integer), recursively. Another way of deducing the values of the quantities $\zeta(2)$, $\zeta(4)$, etc., will be encountered in Chap. 18, Sect. 18.4.2. I will consider yet another method, involving contour integration, in Chap. 23, Sect. 23.4.

17.3.3 Fourier Series Expansions of $\cos \alpha x$ and $\sin \alpha x$

Any periodic function $f(x)$ that satisfies the Dirichlet conditions can be expanded in a Fourier series of sines and cosines. What happens when $f(x)$ is itself a single sine or cosine function?

Let us take the fundamental interval to be $(a,\ b) = (-\pi,\ \pi)$, for simplicity. Consider the periodic function $f(x) = \cos \alpha x$. It is straightforward to show that

$$\cos \alpha x = \frac{\sin \pi\alpha}{\pi\alpha}\left[1 + 2\alpha^2 \sum_{n=1}^{\infty} \frac{(-1)^n \cos nx}{\alpha^2 - n^2}\right], \quad \alpha \neq \text{integer}. \tag{17.27}$$

Similarly, in the case of the periodic function $f(x) = \sin \alpha x$, we get

$$\sin \alpha x = \frac{2 \sin \pi\alpha}{\pi} \sum_{n=1}^{\infty} \frac{(-1)^n\, n \sin nx}{\alpha^2 - n^2}, \quad \alpha \neq \text{integer}. \tag{17.28}$$

[3] In Chap. 18, Sect. 18.4.2; Chap. 23, Sect. 23.4; Chap. 25, Sect. 25.2.5; and Chap. 26, Sect. 26.2.3.

★ **6.** Consider the functions cos αx and sin αx in the fundamental interval $(-\pi,\ \pi)$, where α is a positive constant.

(a) Sketch these periodic functions and verify that $b_n = 0$ in the case of cos αx, while $a_n = 0$ in the case of sin αx.
(b) In the case $\alpha = r$ (a positive integer), check that a_n (respectively, b_n) becomes equal to δ_{nr}, as expected. Equivalently, check that the right-hand sides of Eqs. (17.27) and (17.28) reduce to cos rx and sin rx, respectively, in the limit $\alpha \to r$.
(c) Establish the Fourier series expansions (17.27) and (17.28) in the case when α is not equal to an integer.

As you know, trigonometric functions are related to hyperbolic functions according to

$$\cos(i\theta) \equiv \cosh\theta \quad \text{and} \quad \sin(i\theta) \equiv i\,\sinh\theta. \tag{17.29}$$

For *real* values of θ, of course, the functions cosh θ and sinh θ are not periodic functions.[4] But we can *define* periodic functions that are given, in the fundamental interval $(-\pi,\ \pi)$, by cosh αx and sinh αx, respectively, where α is a positive constant. Their Fourier series expansions are then

$$\cosh\alpha x = \frac{\sinh\pi\alpha}{\pi\alpha}\left[1 + 2\alpha^2\sum_{n=1}^{\infty}\frac{(-1)^n\cos nx}{\alpha^2+n^2}\right] \tag{17.30}$$

and

$$\sinh\alpha x = \frac{2\sinh\pi\alpha}{\pi}\sum_{n=1}^{\infty}\frac{(-1)^{n-1}\,n\sin nx}{\alpha^2+n^2}, \tag{17.31}$$

respectively.

★ **7.** Sketch the periodic functions given by cosh αx and sinh αx, respectively, in the fundamental interval $(-\pi,\ \pi)$. Obtain the Fourier series expansions in Eqs. (17.30) and (17.31).

★ **8.** Consider the Fourier series expansions (17.27)–(17.31) of the periodic functions that are equal to cos αx, sin αx, cosh αx, and sinh αx, respectively, in the fundamental interval $(-\pi,\ \pi)$.

(a) Explain why the nth Fourier coefficient falls off asymptotically like $1/n^2$ in the case of cos αx and cosh αx, but only like $1/n$ in the case of sin αx and sinh αx.
(b) In the series for sin αx and sinh αx, if we set $x = \pm\pi$, the right-hand sides vanish identically, although the left-hand sides do not. What is the explanation of this apparent inconsistency?

---

[4]They *are* periodic in the complex domain, their period being equal to $2\pi i$.

17.4 Solutions

1. $f(x)$ is a symmetric function. Hence $b_n = 0$. We find

$$a_n = -\frac{4\,[1 + (-1)^n]}{\pi\,(4n^2 - 1)}, \quad \text{so that } f(x) = \frac{4}{\pi} - \frac{8}{\pi}\sum_{k=1}^{\infty} \frac{\cos 2kx}{(16k^2 - 1)}.$$

Remark Setting $x = 0$ gives

$$\frac{1}{3 \times 5} + \frac{1}{7 \times 9} + \frac{1}{11 \times 13} + \cdots = \frac{1}{2} - \frac{1}{8}\pi.$$

Similarly, setting $x = \frac{1}{4}\pi$ gives

$$\frac{1}{3 \times 5} - \frac{1}{7 \times 9} + \frac{1}{11 \times 13} - \cdots = \frac{\pi}{4\sqrt{2}} - \frac{1}{2}.$$

The sum and difference of these two series further yield the values of the sums $\frac{1}{3\times 5} + \frac{1}{11\times 13} + \cdots$ and $\frac{1}{7\times 9} + \frac{1}{15\times 17} + \cdots$, and so on. ▶

2. Expand the functions $f(x)$ and $F(x)$ on the left-hand side in Fourier series (in the complex basis). The first equation follows from orthonormality relation (17.2). The final equation follows from the relations (17.10). What is the reason for requiring that $f(x)$ and $F(x)$ be square-integrable in the fundamental interval? ▶

3. (a) It is obvious that $f(x)$ is a symmetric function. We find

$$a_n = -\frac{4}{\pi\,(4n^2 - 1)}, \quad \text{so that } f(x) = \frac{2}{\pi} - \frac{4}{\pi}\sum_{n=1}^{\infty} \frac{\cos 2nx}{4n^2 - 1}.$$

(b) Setting $x = 0$ and $x = \frac{1}{2}\pi$, respectively, yields the values of the sums concerned.[5] ▶

4. (a) $f(x)$ is an even function. We find $a_0 = \frac{2}{3}$, while for $n \geq 1$,

$$a_n = \frac{4\,(-1)^n}{n^2\,\pi^2}, \quad \text{so that } x^2 = \frac{1}{3} + \frac{4}{\pi^2}\sum_{n=1}^{\infty} \frac{(-1)^n}{n^2} \cos n\pi x.$$

[5]Recall (from high school) that both the series given can actually be summed by an extremely elementary method: break up the summands into partial fractions, and watch successive terms cancel each other out!

(b) Setting $x = 1$ and 0, respectively, in the result above gives the sums sought. Further, adding the two sums gives $\sum_{n=1}^{\infty} 1/(2n-1)^2 = \frac{1}{8}\pi^2$. ▶

5. The coefficients of the Fourier cosine series expansion of $f(x)$ are found to be

$$a_0 = \frac{2}{5} \text{ and } a_n = 2\int_0^1 dx\, x^4 \cos n\pi x = \frac{4(-1)^n}{n^2\pi^2}\left(1 - \frac{6}{n^2\pi^2}\right), \quad n \geq 1.$$

Hence the Fourier expansion

$$x^4 = \frac{1}{5} + \frac{4}{\pi^2}\sum_{n=1}^{\infty} \frac{(-1)^n}{n^2}\left(1 - \frac{6}{n^2\pi^2}\right)\cos n\pi x.$$

Set $x = 1$, use the fact that $\sum_1^{\infty} 1/n^2 = \frac{1}{6}\pi^2$, and simplify. The result is $\sum_1^{\infty} 1/n^4 = \zeta(4) = \frac{1}{90}\pi^4$. ▶

7. Observe that the expansions (17.30) and (17.31) follow directly from the expansions (17.27) and (17.28), respectively, on replacing α by $i\alpha$. This can be justified rigorously. What is being done is an *analytic continuation* in the complex α-plane, from the positive real axis to the imaginary axis.[6] For a fixed value of x, the functions $\cos \alpha x$ and $\sin \alpha x$ are analytic functions of α that have no singularities for any finite value of $|\alpha|$. ▶

8. (a) Note how the periodic functions are constructed for a general value of α. In each case, the segment of the plot of the function in the fundamental interval $(-\pi, \pi)$ is repeated in the intervals $\ldots, (-3\pi, -\pi), (\pi, 3\pi), \ldots$. It is evident that the periodic functions in the cases $\cos \alpha x$ and $\cosh \alpha x$ are *continuous* everywhere, including the end-points of the intervals, while those corresponding to $\sin \alpha x$ and $\sinh \alpha x$ have *finite discontinuities* at these end-points. This implies that the corresponding Fourier coefficients fall-off asymptotically like $1/n^2$ in the former cases, and like $1/n$ in the latter cases, respectively.

(b) The periodic functions corresponding to $\cos \alpha x$ and $\cosh \alpha x$ have nonzero values at the end-points of the intervals, and are continuous at those points. The Fourier series will therefore converge to those values. On the other hand, the periodic functions corresponding to $\sin \alpha x$ and $\sinh \alpha x$ have finite discontinuities at the end-points. The values on either side of the discontinuity are nonzero, but equal in magnitude and opposite in sign. Hence the Fourier series will converge to the mean value at each point of discontinuity, namely, zero. ▶

[6] We will discuss analytic functions of a complex variable and analytic continuation in Chaps. 22–27.

Chapter 18
Fourier Integrals

18.1 Expansion of Nonperiodic Functions

18.1.1 Fourier Transform and Inverse Fourier Transform

We have seen that periodic functions can be expanded in Fourier series. Recall that, if the fundamental interval of the function $f(x)$ is (a, b) so that the period is $L = (b - a)$, the expansion and inversion formulas (Eqs. (17.4) and (17.5)) are

$$f(x) = (1/L) \sum_{n=-\infty}^{\infty} f_n \, e^{2\pi nix/L} \quad \text{and} \quad f_n = \int_a^b dx \, f(x) \, e^{-2\pi nix/L}. \tag{18.1}$$

What happens if the function $f(x)$ is not periodic?

By letting $a \to -\infty$ and $b \to \infty$, and hence passing to the limit $L \to \infty$, we can extend the idea of the expansion of an arbitrary periodic function in terms of elementary periodic functions (sines and cosines) to functions that need not be periodic. The number of "harmonics" required now becomes *uncountably* infinite. Therefore, instead of a summation over the integer index n, we require an *integration* over the continuous variable k to which $2\pi n/L$ tends in the limit. The general coefficient in the original expansion, f_n, becomes a function of the continuous variable k. In order to avoid confusion with $f(x)$, we may denote this function by $\widetilde{f}(k)$. The "dictionary" to go from Fourier series to Fourier integrals is

$$\frac{2\pi n}{L} \longrightarrow k, \quad \frac{1}{L} \sum_{n=-\infty}^{\infty} \longrightarrow \frac{1}{2\pi} \int_{-\infty}^{\infty} dk, \quad f_n \longrightarrow \widetilde{f}(k). \tag{18.2}$$

We then have the expansion formula

$$\boxed{f(x) = \frac{1}{2\pi} \int_{-\infty}^{\infty} dk \, e^{ikx} \, \widetilde{f}(k) \quad \text{(expansion formula).}} \tag{18.3}$$

V. Balakrishnan, *Mathematical Physics*,
https://doi.org/10.1007/978-3-030-39680-0_18

The inversion formula is

$$\boxed{\widetilde{f}(k) = \int_{-\infty}^{\infty} dx\, e^{-ikx}\, f(x) \quad \text{(inversion formula).}} \tag{18.4}$$

The function $\widetilde{f}(k)$ is obtained from the function $f(x)$ by acting upon the latter with an **integral operator** whose **kernel** is e^{-ikx}. Similarly, the function $f(x)$ is obtained from the function $\widetilde{f}(k)$ by acting upon the latter with an integral operator whose kernel is $(2\pi)^{-1}\, e^{ikx}$. The functions $f(x)$ and $\widetilde{f}(k)$ are **Fourier transforms** of each other. Equivalently, we could have called $\widetilde{f}(k)$ the Fourier transform of $f(x)$, and $f(x)$ the inverse Fourier transform of $\widetilde{f}(k)$. Which of the two we call the transform, and which we call the inverse transform, is a matter of convention. A Fourier transform is an example of an **integral transform**. A function $f(x)$ has a Fourier transform if it satisfies conditions analogous to the Dirichlet conditions for periodic functions. Broadly speaking, if

(i) $f(x)$ has at most a finite number of finite discontinuities or jumps, and
(ii) $f(x)$ is absolutely integrable in $(-\infty, \infty)$, i.e., $\int_{-\infty}^{\infty} dx\, |f(x)| < \infty$,

then its Fourier transform exists.

Note also the Fourier transform conventions used here (these follow from the convention we have already adopted in the case of Fourier series):

- when integrating over k, the kernel is $(2\pi)^{-1} e^{+ikx}$;
- when integrating over x, the kernel is e^{-ikx}.

I repeat that these are just matters of convention,[1] but it is quite important to choose some specific convention and stick to it consistently, in order to avoid errors.

18.1.2 Parseval's Formula for Fourier Transforms

If $f(x)$ is not only integrable but is also square-integrable, that is, if $f(x) \in \mathcal{L}_2(-\infty\,,\,\infty)$, it follows on applying Parseval's formula that its Fourier transform $\widetilde{f}(k)$ is *also* an element of $\mathcal{L}_2(-\infty\,,\,\infty)$, we have

$$\boxed{\int_{-\infty}^{\infty} dx\, |f(x)|^2 = \frac{1}{2\pi} \int_{-\infty}^{\infty} dk\, |\widetilde{f}(k)|^2.} \tag{18.5}$$

Equation (18.5) is just the continuum analog of the discrete version for Fourier series, Eq. (17.14). Recall that this connection between $\mathcal{L}_2$ functions and Fourier transforms has already been anticipated and discussed briefly in Chap. 13. (At this stage, you might find it

[1] For instance, as you will see in Eq. (18.25) below, the kernels $e^{ikx}/\sqrt{2\pi}$ and $e^{-ikx}/\sqrt{2\pi}$ are often used in the expansion formula and the inversion formula, instead of the kernels $e^{ikx}/(2\pi)$ and e^{-ikx} that we have used.

helpful to read through Sects. 13.2.4 and 13.2.5 of Chap. 13 once again.) To repeat what has been said earlier:

- The Fourier transform, as applied to elements of the function space $\mathcal{L}_2$, may be regarded as a change of basis.
- In the context of quantum mechanics, Parseval's formula (18.5) implies the following: If the position-space wave function of a particle is normalizable, then so is its momentum-space wave function, because these two wave functions form a Fourier transform pair. (This statement extends to one or more particles moving in any number of spatial dimensions.)

18.1.3 Fourier Transform of the δ-Function

The Dirichlet conditions and the conditions stated above are *sufficient* conditions for a function $f(x)$ to have a Fourier series expansion or a Fourier transform, as the case may be. They are by no means *necessary* conditions. For instance, numerous functions that are integrable, but not absolutely integrable (i.e., their absolute values are not integrable) have Fourier transforms. Functions that are more *singular* than what is permitted by the Dirichlet conditions may also have Fourier representations. Indeed, the theory of generalized functions or distributions is very closely linked with the Fourier transforms of these objects. As I have mentioned already, there is a highly developed area of mathematics called harmonic analysis that deals with these matters and their generalizations.

The Dirac δ-function, which is so useful in applications, also has a Fourier transform. Since we know (from Eq. (4.19) of Chap. 4, Sect. 4.2.4) that

$$\delta(x) = \frac{1}{2\pi}\int_{-\infty}^{\infty} dk\, e^{ikx}, \tag{18.6}$$

it follows that the Fourier transform of $\delta(x)$ is just unity, i.e., $\widetilde{\delta}(k) = 1$.

18.1.4 Examples of Fourier Transforms

★ **1.** Sketch the functions $f(x)$ listed below, and show that their Fourier transforms are as specified. (a and σ are positive constants, while b and μ are real constants.)

(a) $f(x) = e^{-a|x|}\sin bx \quad \Longrightarrow \quad \widetilde{f}(k) = \dfrac{-4iabk}{(a^2+b^2+k^2)^2 - 4b^2k^2}$.

(b) $f(x) = e^{-a|x|}\cos bx \quad \Longrightarrow \quad \widetilde{f}(k) = \dfrac{2a(a^2+b^2+k^2)}{(a^2+b^2+k^2)^2 - 4b^2k^2}$.

(c) $f(x) = \dfrac{[\theta(x+a) - \theta(x-a)]}{2a} \quad \Longrightarrow \quad \widetilde{f}(k) = \dfrac{\sin ka}{ka}$.

(d) $f(x) = \dfrac{\sin ax}{x} \quad \Longrightarrow \quad \widetilde{f}(k) = \pi\,[\theta\,(k+a) - \theta\,(k-a)].$

(e) $f(x) = \dfrac{1}{\sqrt{2\pi\,\sigma^2}}\, e^{-(x-\mu)^2/(2\sigma^2)} \quad \Longrightarrow \quad \widetilde{f}(k) = e^{-i\mu k - \frac{1}{2}k^2\sigma^2}.$

Remark The functions $f(x)$ in (c) and (e) above can be regarded as normalized probability density functions corresponding to a random variable x. The first corresponds to a *uniform distribution*, while the second corresponds, of course, to the normal or Gaussian distribution. The Fourier transform of the probability density of a random variable is called its *characteristic function*. Knowing the characteristic function of a random variable is equivalent to knowing all its moments and cumulants. Random variables and probability distributions will be discussed in Chaps. 19 and 20.

★ **2.** Sketch the functions $\widetilde{f}(k)$ listed below, and show that their inverse Fourier transforms $f(x)$ are as specified. (Here σ, k_0 and λ are positive constants.)

(a) $\widetilde{f}(k) = e^{-\frac{1}{2}(k-k_0)^2\sigma^2} \quad \Longrightarrow \quad f(x) = \dfrac{1}{\sqrt{2\pi\sigma^2}}\, e^{ik_0x}\, e^{-x^2/(2\sigma^2)}.$

(b) $\widetilde{f}(k) = \dfrac{k}{|k|}\,\theta\,(k_0 - |k|) \quad \Longrightarrow \quad f(x) = \dfrac{i(1-\cos k_0x)}{\pi x}.$

(c) $\widetilde{f}(k) = \theta\,(k_0 - |k|)\,(k_0 - |k|) \quad \Longrightarrow \quad f(x) = \dfrac{1-\cos(k_0x) + k_0x\,\sin(k_0x)}{2\pi x^2}.$

(d) $\widetilde{f}(k) = e^{-\lambda|k|} \quad \Longrightarrow \quad f(x) = \dfrac{\lambda}{\pi\,(x^2+\lambda^2)}.$

Remark (c) The function $\widetilde{f}(k)$ is a triangular pulse-shaped function in this case. Figure 18.1 shows $\widetilde{f}(k)$ and its inverse Fourier transform.

(d) Figure 18.2 depicts $\widetilde{f}(k)$ and $f(x)$ in this case. $f(x)$ is a "Lorentzian". It is the probability distribution function corresponding to the *Cauchy distribution*, which will be discussed in Chap. 20, Sect. 20.5.3.

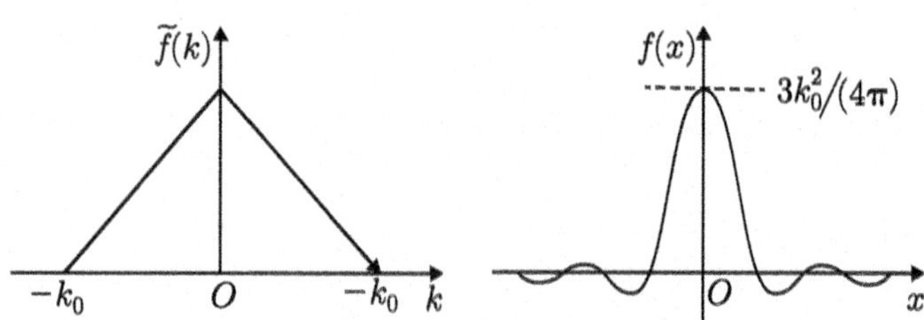

Fig. 18.1 A triangular pulse-shaped $\widetilde{f}(k)$ and its inverse Fourier transform

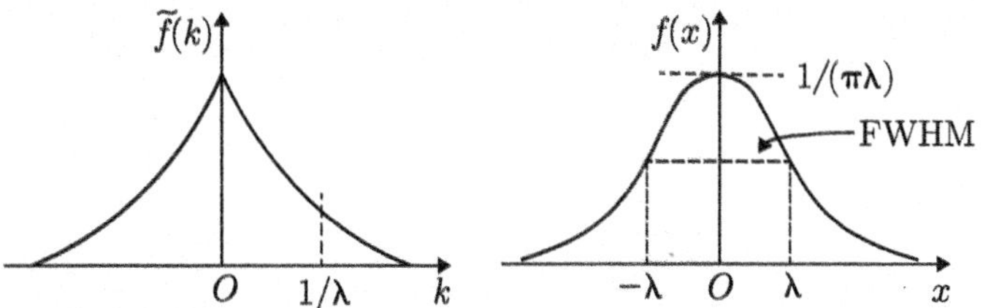

Fig. 18.2 The Fourier transform pair for a Lorentzian function

18.1.5 Relative "Spreads" of a Fourier Transform Pair

A general feature of great importance emerges from the examples above. Roughly speaking, when the function $f(x)$ is of *compact support*, i.e., it is nonzero in only a finite interval in x and is zero outside it, or is mostly concentrated in a finite interval and rapidly decreases to zero outside it, its transform $\widetilde{f}(k)$ is spread out in k. This is why the rectangular pulse considered earlier, which is strictly zero for $|x| > a$, has a Fourier transform that decays to zero relatively slowly, like $|k|^{-1}$. The same statement is applicable with the roles of $\widetilde{f}(k)$ and $f(x)$ interchanged. The compact pulses represented by $\widetilde{f}(k)$ in the examples above have transforms that decay like $|x|^{-1}$ for large $|x|$. When one member of the transform pair decays exponentially (which implies that it falls off faster than any inverse power of the argument), the other member decays like the inverse square of the argument. This is demonstrated by the case of the Lorentzian. Finally, when one of the pair is a Gaussian (with possible linear terms in the exponent), so is the other member of the pair. This result is so useful that it is worth writing it down again for ready reference:

$$\boxed{f(x) = \frac{1}{\sqrt{2\pi\sigma^2}}\, e^{-(x-\mu)^2/(2\sigma^2)} \quad \Longleftrightarrow \quad \widetilde{f}(k) = e^{-i\mu k - \frac{1}{2}k^2\sigma^2}.} \tag{18.7}$$

You have already seen an illustration of this property in the position-space and momentum-space wave functions corresponding to any coherent state of the quantum mechanical linear harmonic oscillator, Eqs. (16.54) and (16.55) of Chap. 16, Sect. 16.2.6. Recall that, when $z_1 = 0$ and $z_2 = 0$, these wave functions reduce to the Gaussian wave functions corresponding to the ground state of the harmonic oscillator.

The foregoing property of a Fourier transform pair has a far-reaching physical consequence.

- The "duality" described above—compactness in one variable, spreading in the "conjugate" variable—is at the very heart of the Heisenberg Uncertainty Principle.

The fact that $\hbar$ is not identically equal to zero necessitates a quantum mechanical description of systems, in terms of state vectors (or wave functions). In turn, the "fuzziness" implicit in such a description is inevitably subject to the duality mentioned above.

18.1.6 The Convolution Theorem

The convolution theorem for Fourier transforms is a basic and most useful relationship. Given two functions $f(x)$ and $g(x)$, the function

$$h(x) = \int_{-\infty}^{\infty} dx'\, f(x')\, g(x - x') = \int_{-\infty}^{\infty} dx'\, f(x - x')\, g(x') \tag{18.8}$$

is called the **convolution** of the functions $f(x)$ and $g(x)$. This relationship is sometimes written as $h = f * g$. (It should not be confused with ordinary multiplication of the functions concerned.) The **convolution theorem** for Fourier transforms is the following:

- If $f(x)$ and $g(x)$ have Fourier transforms $\widetilde{f}(k)$ and $\widetilde{g}(k)$, respectively, then the Fourier transform of their convolution $h(x)$ is given by

$$\boxed{\widetilde{h}(k) = \widetilde{f}(k)\, \widetilde{g}(k).} \tag{18.9}$$

- The Fourier transform operation therefore converts the convolution of functions to an ordinary multiplication of functions.

★ **3.** Establish the convolution theorem (18.9).

It is fairly obvious that this works in reverse, too. Consider the Fourier transform of the convolution of $\widetilde{f}$ and $\widetilde{g}$, namely, of the function of k given by

$$\int_{-\infty}^{\infty} dk'\, \widetilde{f}(k')\, \widetilde{g}(k - k') = \int_{-\infty}^{\infty} dk'\, \widetilde{f}(k - k')\, \widetilde{g}(k') = \widetilde{H}(k), \quad \text{say.} \tag{18.10}$$

Then the inverse Fourier transform of $\widetilde{H}(k)$ is simply the product $H(x) = f(x)\, g(x)$.

★ **4.** Check this out explicitly.

18.1.7 Generalized Parseval Formula

We have seen that Eq. (17.16) of Chap. 17, Sect. 17.1.4 is a generalization of Parseval's formula for Fourier series. As you might expect, this generalization has a counterpart in the case of Fourier transforms. Let $f(x)$ and $F(x)$ be two **"good" functions**[2] of x, i.e., they have derivatives of all orders for all x, and vanish (along with their derivatives of all orders) as $x \to \infty$. Then

$$\boxed{\int_{-\infty}^{\infty} dx\, f(x)\, F(x) = \frac{1}{2\pi} \int_{-\infty}^{\infty} dk\, \widetilde{f}(k)\, \widetilde{F}(-k).} \tag{18.11}$$

[2] I believe this simple but expressive terminology is due to M. J. Lighthill (see the Bibliography).

Note that neither side of this equation is in the form of a convolution of two functions. I will use Eq. (18.11) in Sect. 18.4.1, in the derivation of the Poisson summation formula.

★ **5.** Derive Eq. (18.11).

Finally, note that Parseval's formula for Fourier transforms, Eq. (18.5), follows as a special case of Eq. (18.11). All you have to do is to set $F(x)$ with $f^*(x)$, and observe that

$$F(x) = f^*(x) \quad \Longleftrightarrow \quad \widetilde{F}(-k) = [\widetilde{f}(k)]^*. \tag{18.12}$$

(Remember that, In general, the Fourier transform of the complex conjugate of a function is not the same as the complex conjugate of the Fourier transform of that function!)

18.2 The Fourier Transform Operator in $\mathcal{L}_2$

18.2.1 Iterates of the Fourier Transform Operator

You have seen that the Fourier transform $\widetilde{f}(k)$ of a function $f(x)$ may be regarded as the result of applying a certain integral operator, the Fourier transform operator $\mathcal{F}$, to the function. The kernel of this operator is just e^{-ikx}. In order to make the notion precise, we need to specify a function space, such that both the function $f(x)$ and its transform $\widetilde{f}(k)$ belong to the same space. As we know, the space $\mathcal{L}_2(-\infty\,,\,\infty)$ satisfies this requirement. Let us therefore restrict ourselves to this function space in what follows.

Given a function $f \in \mathcal{L}_2(-\infty\,,\,\infty)$, we have

$$[\mathcal{F}f](x) \equiv \widetilde{f}(x) = \int_{-\infty}^{\infty} dy\, e^{-ixy}\, f(y)\,. \tag{18.13}$$

(Observe the notation in this equation! I have written x for the argument of the *output* function *after* the Fourier transform operation is performed.) Something very interesting happens when the operator $\mathcal{F}$ is *iterated*, i.e., applied repeatedly to a function. We have

$$\begin{aligned}
[\mathcal{F}^2 f](x) &\equiv [\mathcal{F}\mathcal{F}f](x) = [\mathcal{F}\widetilde{f}\,](x) = \int_{-\infty}^{\infty} dz\, e^{-ixz}\, \widetilde{f}(z) \\
&= \int_{-\infty}^{\infty} dz\, e^{-ixz} \int_{-\infty}^{\infty} dy\, e^{-izy}\, f(y) = \int_{-\infty}^{\infty} dy\, f(y) \int_{-\infty}^{\infty} dz\, e^{-i(x+y)z} \\
&= \int_{-\infty}^{\infty} dy\, f(y)\, 2\pi\, \delta\,(x+y) = 2\pi\, f(-x)\,.
\end{aligned} \tag{18.14}$$

Now, we know that the **parity operator** $\mathcal{P}$ changes the sign of the argument when acting on any function of x. That is,

$$(\mathcal{P}f)(x) \equiv f(-x). \tag{18.15}$$

Since this is true for *any* $f(x) \in \mathcal{L}_2(-\infty\,,\,\infty)$, Eq. (18.14) implies that we have the *operator* relation

$$\boxed{\mathcal{F}^2 = 2\pi\,\mathcal{P}.} \tag{18.16}$$

- The square of the Fourier transform operator is just 2π times the parity operator.

Moreover, since the square of the parity operator is obviously just the unit operator I, we have the operator relationship

$$\boxed{\mathcal{F}^4 = (2\pi)^2\,\mathcal{P}^2 = (2\pi)^2\,I.} \tag{18.17}$$

- The Fourier transform operator is thus proportional to a "fourth root" of the unit operator!

These results for $\mathcal{P}$ and $\mathcal{F}$ have applications in **Fourier optics**. The factors of 2π in Eqs. (18.16) and (18.17) are a consequence of the particular Fourier transform convention I have used. It should be obvious that these factors can be made equal to unity by a suitable choice of convention.

18.2.2 Eigenvalues and Eigenfunctions of $\mathcal{F}$

The fact that $\mathcal{F}^4 = (2\pi)^2\,I$ seems to suggest that the eigenvalues of the Fourier transform operator in $\mathcal{L}_2$ are given by $(2\pi)^{1/2}$ times the fourth roots of unity, namely, the four numbers $\pm(2\pi)^{1/2}$ and $\pm i\,(2\pi)^{1/2}$. Can we find the eigenvalues and the corresponding eigenfunctions explicitly?

The answer to this question is that we have already done so! Recall that the Hamiltonian of the quantum mechanical linear harmonic oscillator has exactly the same form in the position basis and in the momentum basis, as in Eq. (14.40) of Chap. 14, Sect. 14.4.2. As a result, the position-space wave functions $\Phi_n(x)$ and the momentum-space wave functions $\widetilde{\Phi}_n(p)$ representing the eigenstates of this Hamiltonian are identical in form. These wave functions have already been written down in Eqs. (16.47) and (16.49) of Chap. 16, Sect. 16.2.4. But we also know that $\Phi_n(x)$ and $\widetilde{\Phi}_n(p)$ are elements of $\mathcal{L}_2(-\infty\,,\,\infty)$, and that they are Fourier transform pairs. It follows that these are precisely the eigenfunctions of $\mathcal{F}$ that we are looking for. It remains to prove this assertion directly, and also to find the eigenvalues of $\mathcal{F}$ explicitly. Let us now do so.

As we are concerned here with Fourier transforms rather than quantum mechanics itself, let us set $\hbar = 1$ and work with dimensionless variables x and k. We want to show that the functions

$$\Phi_n(x) = \frac{1}{\left(2^n n!\sqrt{\pi}\right)^{1/2}}\,e^{-x^2/2}\,H_n(x) \quad (n = 0, 1, \ldots), \tag{18.18}$$

where $H_n(x)$ is the Hermite polynomial of degree n, are eigenfunctions of $\mathcal{F}$. The n-dependent normalization constant is irrelevant in an eigenvalue equation, and so let us consider just the function $e^{-x^2/2}\,H_n(x)$. Recall Eq. 16.40 of Chap. 16, Sect. 16.2.4, for the generating function for the Hermite polynomials:

$$e^{2tx-t^2} = \sum_{n=0}^{\infty} H_n(x)\,\frac{t^n}{n!}\,. \tag{18.19}$$

As I have emphasized in Chap. 16, Sect. 16.2.3, an equation such as (18.19) must be regarded as an equation between two *analytic* functions of the complex variable t. A better understanding of what this statement implies will follow after we discuss complex analysis in Chaps. 22–27. But for the present, all we need to know is that the function on the left-hand side is such a well-behaved function of t that the power series on the right-hand side converges absolutely for all finite values of the modulus $|t|$. In other words, the equation is valid for all finite, complex values of t, including pure imaginary ones, as already mentioned.

With this information at hand, we start with Eq. (18.19), multiply both sides of the equation by $\exp\,(-ikx - \frac{1}{2}x^2)$, and integrate over x from $-\infty$ to ∞. The left-hand side becomes a Gaussian integral that can be evaluated. The result is

$$(2\pi)^{1/2}\,e^{-k^2/2}\,e^{-2ikt+t^2} = \sum_{n=0}^{\infty} \mathcal{F}\big[e^{-x^2/2}\,H_n(x)\big]\,\frac{t^n}{n!}\,. \tag{18.20}$$

But the factor $e^{-2ikt+t^2}$ on the left-hand side is again a generating function for Hermite polynomials. We have

$$e^{-2ikt+t^2} = e^{2k(-it)-(-it)^2} = \sum_{n=0}^{\infty} H_n(k)\,\frac{(-it)^n}{n!}\,. \tag{18.21}$$

Equating the coefficients of t^n of the two absolutely convergent power series in t, we finally get

$$\begin{aligned}\int_{-\infty}^{\infty} dx\,e^{-ikx}\,e^{-x^2/2}\,H_n(x) &= \mathcal{F}\big[e^{-x^2/2}\,H_n(x)\big]\\ &= (2\pi)^{1/2}\,(-i)^n\,\big(e^{-k^2/2}\,H_n(k)\big).\end{aligned} \tag{18.22}$$

We have therefore established the following results:

- The function $e^{-x^2/2}\,H_n(x)$, or this function multiplied by the normalization constant $(2^n\,n!\,\sqrt{\pi})^{-1/2}$, is an eigenfunction of the Fourier transform operator $\mathcal{F}$ in the space $\mathcal{L}_2(-\infty\,,\,\infty)$. Here n runs over the values $0,\ 1,\ 2,\ \ldots.$
- The corresponding eigenvalue of $\mathcal{F}$ is $(2\pi)^{1/2}\,(-i)^n$, or that of the operator $(2\pi)^{-1/2}\,\mathcal{F}$ is $(-i)^n$.

★ **6.** Starting from Eq. (18.19), work through the steps of this derivation to arrive at Eq. (18.22).

Since $(-i)^n$ is always equal to 1, $-i$, -1, or i, the operator $(2\pi)^{-1/2}\,\mathcal{F}$ has just these four distinct eigenvalues. Each of these is infinitely degenerate, i.e., there is a countable infinity of linearly independent eigenfunctions corresponding to each eigenvalue. These eigenfunctions are, respectively,

$$\Phi_{4n}(x),\ \Phi_{4n+1}(x),\ \Phi_{4n+2}(x) \text{ and } \Phi_{4n+3}(x)\quad (n = 0,\ 1,\ 2,\ \ldots), \tag{18.23}$$

where $\Phi_n(x)$ is given by Eq. (18.18).

As I have already mentioned, the factor of $(2\pi)^{1/2}$ in the eigenvalues of $\mathcal{F}$ arises from the Fourier transform convention we have adopted. Let me repeat this for clarity: we have defined

$$\left.\begin{aligned} [\mathcal{F}\, f](k) \quad &= \widetilde{f}(k) = \int_{-\infty}^{\infty} dx\, e^{-ikx}\, f(x), \\ [\mathcal{F}^{-1}\, \widetilde{f}\,](x) &= f(x) = \frac{1}{2\pi}\int_{-\infty}^{\infty} dk\, e^{ikx}\, \widetilde{f}(k). \end{aligned}\right\} \tag{18.24}$$

Had we used the alternative (and more symmetrical) Fourier transform convention

$$[\mathcal{F}\, f](k) = \frac{1}{\sqrt{2\pi}}\int_{-\infty}^{\infty} dx\, e^{-ikx}\, f(x),\quad [\mathcal{F}^{-1}\, \widetilde{f}](x) = \frac{1}{\sqrt{2\pi}}\int_{-\infty}^{\infty} dk\, e^{ikx}\, \widetilde{f}(k), \tag{18.25}$$

we would have found that $\mathcal{F}^4 = I$, and the eigenvalues of $\mathcal{F}$ would have been simply ± 1 and $\pm i$. Indeed, the customary normalization of quantum mechanical wave functions uses precisely this convention—see Eqs. (13.30) and (13.32) of Chap. 13, Sect. 13.2.5. But the convention we have chosen is used in many other applications of Fourier analysis, and so we shall stay with it.

18.2.3 The Adjoint of an Integral Operator

Taking the Fourier transform of a function involves applying an integral operator to the function. It is natural to ask: What is the *adjoint* of this operator? Before answering this question, we must digress briefly to consider the adjoint of a general integral operator.

Let $f(x)$ be the function representing the vector $|\, f\,\rangle$ in the given function space (here, $\mathcal{L}_2(-\infty\,,\ \infty)$). If K is the integral operator concerned, the vector $|\mathsf{K}f\rangle$ is represented by the function

$$\langle\, x\,|\,\mathsf{K}f\rangle \equiv \langle\, x\,|\mathsf{K}\,|\, f\,\rangle = \int_{-\infty}^{\infty} dy\, K(x,\, y)\, f(y), \tag{18.26}$$

where $K(x,\, y)$ is the kernel of the operator. The adjoint of K, denoted by $\mathsf{K}^\dagger$, is of course to be identified by applying Eq. (14.8) of Chap. 14, Sect. 14.2.2, namely, the condition $\langle g\,|\,\mathsf{K}f\rangle = \langle \mathsf{K}^\dagger g\,|\, f\,\rangle$ for every pair of elements $|\, f\,\rangle,\ |\, g\,\rangle$ in the function space. But

$$\begin{aligned}
\langle g \,|\, \mathsf{K} f\rangle &= \int_{-\infty}^{\infty} dx\, g^*(x) \int_{-\infty}^{\infty} dy\, K(x, y)\, f(y) \\
&= \int_{-\infty}^{\infty} dy \int_{-\infty}^{\infty} dx\, g^*(x)\, K(x, y)\, f(y)' \quad \text{(changing the order of integration)} \\
&= \int_{-\infty}^{\infty} dx \int_{-\infty}^{\infty} dy\, K(y, x)\, g^*(y)\, f(x) \quad \text{(re-labeling } x \leftrightarrow y) \\
&= \int_{-\infty}^{\infty} dx \int_{-\infty}^{\infty} dy\, [K^*(y, x)\, g(y)]^*\, f(x) = \langle \mathsf{K}^\dagger g \,|\, f\,\rangle. \qquad (18.27)
\end{aligned}$$

It follows that the kernel of the integral operator $\mathsf{K}^\dagger$ is $K^*(y, x)$. Note the close similarity between this result and that for the adjoint or hermitian conjugate of a matrix. If the (ij)th element of a matrix A is a_{ij}, the (ij)th element of its hermitian conjugate $A^\dagger$ is a^*_{ji}. The close analogy between integral operators and matrices (as operators in an LVS) plays a role in the theory of integral equations. The latter will be discussed in Chap. 32.

18.2.4 *Unitarity of the Fourier Transformation*

Returning to the case at hand, we see that $(2\pi)^{-1/2}\,\mathcal{F}$ has the kernel $K(k, x) = (2\pi)^{-1/2}\, e^{-ikx}$, while its inverse $[(2\pi)^{-1/2}\,\mathcal{F}]^{-1}$ has the kernel $(2\pi)^{-1/2}\, e^{ikx}$. But the latter quantity is precisely $K^*(x, k)$. We therefore have the operator identity

$$[(2\pi)^{-1/2}\,\mathcal{F}]^{-1} = [(2\pi)^{-1/2}\,\mathcal{F}]^\dagger. \qquad (18.28)$$

In other words:

- The operator $(2\pi)^{-1/2}\,\mathcal{F}$ is a *unitary* operator.

It is not surprising, then, that all its eigenvalues lie on the unit circle in the complex plane, just like those of a unitary matrix!

Hence the Fourier transform in $\mathcal{L}_2(-\infty\,,\,\infty)$ is not only a change of basis in the space, but also a unitary transformation. This fact has implications in quantum mechanics. For instance, it guarantees that the description of a particle (or a system of particles) in position space and in momentum space are *unitarily equivalent*—that is, you can use either description without altering the underlying physics.

18.3 Generalization to Several Dimensions

The Fourier transform and most of the results in the foregoing are generalized in a straightforward manner to functions of several real variables. Let $\mathbf{r} = (x_1, x_2, \ldots, x_d)$ be a d-dimensional vector, where $d = 2, 3, \ldots$. We then have

$$\boxed{f(\mathbf{r}) = \frac{1}{(2\pi)^d} \int d^d k \, e^{i\mathbf{k}\cdot\mathbf{r}} \, \widetilde{f}(\mathbf{k}) \quad \Longleftrightarrow \quad \widetilde{f}(\mathbf{k}) = \int d^d r \, e^{-i\mathbf{k}\cdot\mathbf{r}} \, f(\mathbf{r}).} \tag{18.29}$$

Here $d^d r = dx_1 \, dx_2 \cdots dx_d$ and $d^d k = dk_1 \, dk_2 \cdots dk_d$ denote the volume elements in d-dimensional $\mathbf{r}$-space and $\mathbf{k}$-space, respectively. The Fourier representation of the d-dimensional δ-function is

$$\boxed{\delta^{(d)}(\mathbf{r}) = \frac{1}{(2\pi)^d} \int d^d k \, e^{i\,\mathbf{k}\cdot\mathbf{r}} .} \tag{18.30}$$

This is just the d-dimensional counterpart of the formula written down for the case $d = 3$ in Eq. (4.34).

Fourier expansions of vector-valued functions can also be written down in an analogous manner. Thus, if $\mathbf{u}(\mathbf{r})$ is a vector field in $\mathbb{R}^d$, we have

$$\mathbf{u}(\mathbf{r}) = \frac{1}{(2\pi)^d} \int d^d k \, e^{i\mathbf{k}\cdot\mathbf{r}} \, \widetilde{\mathbf{u}}(\mathbf{k}) \quad \Longleftrightarrow \quad \widetilde{\mathbf{u}}(\mathbf{k}) = \int d^d r \, e^{-i\mathbf{k}\cdot\mathbf{r}} \, \mathbf{u}(\mathbf{r}) \, . \tag{18.31}$$

One of the great advantages of the Fourier expansion of functions now becomes evident. Consider the usual case, $d = 3$. A vector field is generally specified by differential equations for its divergence and curl, respectively.[3] The reasons for doing so have been discussed in Chap. 6, Sect. 6.2.7. Recalling Eq. (6.69), we have

$$\nabla \cdot \mathbf{u}(\mathbf{r}) = \frac{1}{(2\pi)^3} \int d^3 k \, e^{i\mathbf{k}\cdot\mathbf{r}} \, i\mathbf{k} \cdot \widetilde{\mathbf{u}}(\mathbf{k}) \tag{18.32}$$

and

$$\nabla \times \mathbf{u}(\mathbf{r}) = \frac{1}{(2\pi)^3} \int d^3 k \, e^{i\mathbf{k}\cdot\mathbf{r}} \, i\mathbf{k} \times \widetilde{\mathbf{u}}(\mathbf{k}). \tag{18.33}$$

Further, the set of functions $\left\{ e^{i\mathbf{k}\cdot\mathbf{r}} \,\middle|\, \mathbf{k} \in \mathbb{R}^3 \right\}$ forms an orthonormal basis for integrable functions of $\mathbf{r} \in \mathbb{R}^3$. Hence:

- *Partial differential equations* for the divergence and curl of a vector field $\mathbf{u}(\mathbf{r})$ reduce to *algebraic equations* for its Fourier transform $\widetilde{\mathbf{u}}(\mathbf{k})$.

The utility of this result should now be obvious to you.

18.4 The Poisson Summation Formula

The Poisson summation formula is a very useful result that has many applications. For instance, it can be used to sum many infinite series. It is also related to much deeper mathematical results (which I will not go into here).

---

[3] Maxwell's equations for electromagnetic fields provide a prominent example of this statement, as you have seen in Chap. 9.

18.4.1 Derivation of the Formula

Consider the infinite periodic array of δ-functions (a Dirac comb) given by

$$F(x) = \sum_{n=-\infty}^{\infty} \delta(x - nL), \tag{18.34}$$

where L is a positive constant. It is trivial to verify that $F(x) = F(-x)$. As you have seen in Chap. 17, Sect. 17.2.3, if we regard $F(x)$ as a periodic function of x with fundamental interval $(-\frac{1}{2}L, \frac{1}{2}L)$, we have $F(x) = \delta(x)$ in the fundamental interval. It can then be expanded in a Fourier series according to

$$F(x) = \frac{1}{L} \sum_{n=-\infty}^{\infty} F_n \, e^{2\pi nix/L}, \quad \text{where} \quad F_n = \int_{-L/2}^{L/2} dx \, e^{-2\pi nix/L} \, \delta(x) = 1. \tag{18.35}$$

Therefore, as we have already seen (in Eq. (17.25)), $F(x)$ can be written in two different ways—either as an infinite sum of exponentials, or as an infinite sum of δ-functions:

$$F(x) = \sum_{n=-\infty}^{\infty} \delta(x - nL) = \frac{1}{L} \sum_{n=-\infty}^{\infty} e^{2\pi nix/L}. \tag{18.36}$$

But

$$\delta(x - nL) = \delta(nL - x) = \frac{1}{L} \, \delta\left(n - \frac{x}{L}\right). \tag{18.37}$$

We thus have a useful identity relating a sum of equally spaced δ-functions to a sum of exponentials, namely,

$$\boxed{\sum_{n=-\infty}^{\infty} \delta\left(n - \frac{x}{L}\right) = \sum_{n=-\infty}^{\infty} e^{2\pi nix/L}.} \tag{18.38}$$

Pause once again for a moment to think about this most remarkable formula. Neither side of this equation is anything like an absolutely convergent series!

Next, consider the Fourier *transform* of $F(x)$. This is

$$\begin{aligned} \widetilde{F}(k) &= \int_{-\infty}^{\infty} dx \, e^{-ikx} \, F(x) \\ &= \sum_{n=-\infty}^{\infty} \int_{-\infty}^{\infty} dx \, e^{-ikx} \, \delta(x - nL) = \sum_{n=-\infty}^{\infty} e^{inkL}. \end{aligned} \tag{18.39}$$

But Eq. (18.38) says that an infinite sum of exponentials of this kind can be written as an infinite sum of δ-functions. Using this relationship, we get

$$\widetilde{F}(k) = \sum_{n=-\infty}^{\infty} \delta\left(n - \frac{kL}{2\pi}\right). \tag{18.40}$$

Once again, it is easy to see that $\widetilde{F}(k) = \widetilde{F}(-k)$. The final step is to insert the expressions of Eq. (18.36) for $F(x)$ and Eq. (18.40) for $\widetilde{F}(k) = \widetilde{F}(-k)$ in the identity given by Eq. (18.11), namely,

$$\int_{-\infty}^{\infty} dx\, f(x)\, F(x) = \frac{1}{2\pi} \int_{-\infty}^{\infty} dk\, \widetilde{f}(k)\, \widetilde{F}(-k). \tag{18.41}$$

I mentioned earlier that this identity was valid for *good* functions, but we are now using it for singular functions like the δ-function. I merely state here without proof that this step can be justified. We get

$$\int_{-\infty}^{\infty} dx\, f(x) \sum_{n=-\infty}^{\infty} \delta(x - nL) = \frac{1}{2\pi} \int_{-\infty}^{\infty} dk\, \widetilde{f}(k) \sum_{n=-\infty}^{\infty} \delta\left(n - \frac{kL}{2\pi}\right), \tag{18.42}$$

where $f(x)$ is any *arbitrary* function with a Fourier transform $\widetilde{f}(k)$. The integrals on the two sides of Eq. (18.42) are trivially done using the δ-functions. Finally, therefore,

$$\boxed{\sum_{n=-\infty}^{\infty} f(nL) = (1/L) \sum_{n=-\infty}^{\infty} \widetilde{f}(2\pi n/L).} \tag{18.43}$$

Here L is an arbitrary positive parameter. This is the famous **Poisson summation formula**.

- The formula is helpful in the summation of series in cases when the left-hand side is difficult to evaluate, but the right-hand side is more tractable; or *vice versa*.
- Note that the parameter L occurs in the *numerator* of the argument of the function on the left-hand side, but on the right-hand side it occurs in the *denominator*.
- Hence one of the two representations may be useful for *small* values of L, while the other is useful for *large* values of L, in physical applications.

18.4.2 Some Illustrative Examples

Here are some instances of the use of Poisson's summation formula to derive useful identities and to sum certain infinite series.

(i) If $f(x)$ is chosen to be a Gaussian function, the formula immediately gives

$$\boxed{\sum_{n=-\infty}^{\infty} e^{-\pi n^2 \lambda^2} = (1/\lambda) \sum_{n=-\infty}^{\infty} e^{-\pi n^2/\lambda^2}, \quad (\lambda > 0).} \tag{18.44}$$

★ **7.** Derive Eq. (18.44) from the Poisson summation formula.

The identity (18.44), due to Jacobi, is so useful and important that it is sometimes called the Poisson summation formula itself! A sum over the exponentials of the *squares* of the integers is called a **Gaussian sum**. A vast literature exists on this subject and its ramifications in several parts of mathematics and mathematical physics.

A physical application of Eq. (18.44) arises in the context of the phenomenon of diffusion. Here, the parameter λ^2 is proportional to Dt, where D is the diffusion coefficient and t is the time. The two sides of the identity (18.44) can then be used to obtain valuable insight into the behavior of the solution to the diffusion equation at long times and short times, respectively. I will return to this aspect in Chap. 30, Sect. 30.2.7.

(ii) Another useful case is that of the function

$$f(x) = \frac{\cos bx}{x^2 + a^2} \quad (a > 0,\ 0 \le b < 2\pi). \tag{18.45}$$

The Fourier transform of this function is easily found by contour integration, but we have not yet discussed that method. You can, however, write down $\widetilde{f}(k)$ using the information that is already available. Recall, from one of the exercises in the foregoing, that the inverse Fourier transform of $e^{-\lambda|k|}$ is a Lorentzian. That is,

$$f(x) = \frac{\lambda}{\pi(x^2 + \lambda^2)} \quad \Longleftrightarrow \quad \widetilde{f}(k) = e^{-\lambda\,|k|}. \tag{18.46}$$

Hence the Fourier transform of the function $f(x)$ in Eq. (18.45) is given by

$$\begin{aligned} \widetilde{f}(k) &= \int_{-\infty}^{\infty} dx\, \frac{\cos bx}{x^2 + a^2}\, e^{-ikx} = \frac{1}{2}\int_{-\infty}^{\infty} dx\, \frac{\left[e^{-i(k-b)x} + e^{-i(k+b)x}\right]}{x^2 + a^2} \\ &= \frac{\pi}{2a}\left[e^{-a\,|k-b|} + e^{-a\,|k+b|}\right]. \end{aligned} \tag{18.47}$$

Setting $L = 1$ in the Poisson summation formula (18.43), we therefore have

$$\sum_{n=-\infty}^{\infty} \frac{\cos(nb)}{n^2 + a^2} = \sum_{n=-\infty}^{\infty} \frac{\pi}{2a}\left[e^{-a\,|2\pi n - b|} + e^{-a\,|2\pi n + b|}\right]. \tag{18.48}$$

Some simplification then yields the identity

$$\sum_{n=1}^{\infty} \frac{\cos(nb)}{n^2 + a^2} = \frac{\pi}{2a}\left\{\frac{\cosh a(\pi - b)}{\sinh(\pi a)} - \frac{1}{\pi a}\right\}. \tag{18.49}$$

Setting $b = 0$ in this relation, we get an important and useful result for a certain sum, which I will denote by $S(a)$:

$$\boxed{S(a) \stackrel{\text{def.}}{=} \sum_{n=1}^{\infty} \frac{1}{n^2 + a^2} = \frac{\pi}{2a} \left\{ \coth \pi a - \frac{1}{\pi a} \right\}.} \tag{18.50}$$

In Chap. 23, Sect. 23.4, we shall see how this formula can be derived with the help of contour integration. As usual, a relation like Eq. (18.50) may be regarded as a relation between two analytic functions—in this case, as functions of the variable a. The quantity in curly brackets on the right-hand side occurs in physical applications—for instance, in the elementary theory of paramagnetism, where it is called the **Langevin function**.

A noteworthy by-product of Eq. (18.50) follows upon letting $a \to 0$. The outcome is the value of $\zeta(2)$, where $\zeta(z)$ is the Riemann zeta function defined in Eq. (17.26) of Chap. 17, Sect. 17.3. We find, once again,

$$\sum_{n=1}^{\infty} 1/n^2 \equiv \zeta(2) = \tfrac{1}{6}\,\pi^2 \ (\simeq 1.6449). \tag{18.51}$$

★ **8.** Consider Poisson's summation formula in the case of the function $f(x)$ given by Eq. (18.45).

(a) Work through the steps leading from Eq. (18.47) to Eq. (18.49).
(b) Take the limit $a \to 0$ in Eq. (18.50) to obtain Eq. (18.51).

The zeta function for even positive integers, $\zeta(2k)$**:** Differentiating $S(a)$ with respect to a leads to a number of related results. For instance,

$$\sum_{n=1}^{\infty} \frac{1}{(n^2 + a^2)^2} = -\frac{1}{2a}\frac{dS(a)}{da} = \frac{\pi}{4a^3} \left\{ \coth \pi a + \frac{\pi a}{\sinh^2 \pi a} - \frac{2}{\pi a} \right\}. \tag{18.52}$$

Once again, this result can also be derived using contour integration. As before, passing to the limit $a \to 0$ yields the sum

$$\sum_{n=1}^{\infty} 1/n^4 = \zeta(4) = \tfrac{1}{90}\,\pi^4 \ (\simeq 1.0823). \tag{18.53}$$

★ **9.** Verify Eqs. (18.52) and (18.53).

It should be obvious that we can continue this process recursively to find the sum $\sum_1^{\infty} (n^2 + a^2)^{-k}$ for every positive integer k. Taking the limit $a \to 0$ will then give us the value of $\zeta(2k)$. A slightly tedious but straightforward calculation leads to the sum

$$\sum_{n=1}^{\infty} 1/n^6 = \zeta(6) = \tfrac{1}{945}\,\pi^6 \;(\simeq 1.0173). \tag{18.54}$$

From the value of $\zeta(2) = \frac{1}{6}\pi^2$, we deduced that $\sum_{n=1}^{\infty}(-1)^{n-1}/n^2 = \frac{1}{12}\pi^2$. Similarly, from the values of $\zeta(4)$ and $\zeta(6)$, it follows that

$$\sum_{n=1}^{\infty}(-1)^{n-1}/n^4 = \tfrac{7}{720}\,\pi^4 \quad\text{and}\quad \sum_{n=1}^{\infty}(-1)^{n-1}/n^6 = \tfrac{31}{30240}\,\pi^6. \tag{18.55}$$

★ **10.** Verify Eqs. (18.54) and (18.55).

In Chap. 23, Sect. 23.4, you will see how the results in Eqs. (18.50)–(18.55) can be derived by an alternative method, namely, contour integration.

18.4.3 Generalization to Higher Dimensions

The Poisson summation formula is readily generalized to higher dimensions. If $\widetilde{f}(\mathbf{k})$ is the Fourier transform of $f(\mathbf{r})$ where $\mathbf{r} \in \mathbb{R}^d$, then

$$\sum_{\mathbf{n}\in\mathbb{Z}^d} f(\mathbf{n}L) = (1/L^d)\sum_{\mathbf{n}\in\mathbb{Z}^d} \widetilde{f}\,(2\pi\mathbf{n}/L)\,. \tag{18.56}$$

Here $\mathbf{n} = (n_1, n_2, \ldots, n_d)$ where each n_i is an integer, and $\mathbb{Z}^d$ denotes the set of such d-tuples of integers, in an obvious notation. Further generalizations are also possible, such as the counterpart of Eq. (18.56) for $f(\mathbf{n}L + \mathbf{r}_0)$, where $\mathbf{r}_0$ is any given vector in $\mathbb{R}^d$.

Finally, as I have stated earlier, the generalizations of the Poisson summation formula are related to deep results in advanced mathematics. I merely mention some of these topics in passing, to pique your curiosity: the asymptotic behavior of the heat kernel, the spectra of the Laplacian operator and general elliptic operators on manifolds, Eisenstein series, automorphic functions, the Selberg trace formula, and so on.

18.5 Solutions

1. (a) and (b) are straightforward: you need the integrals

$$\int_0^{\infty} dx\, e^{-ax}\cos cx = \frac{a}{a^2+c^2} \quad\text{and}\quad \int_0^{\infty} dx\, e^{-ax}\sin cx = \frac{c}{a^2+c^2} \quad (c \text{ real}).$$

(c) $f(x)$ is a rectangular pulse of height $1/(2a)$ from $x = -a$ to $x = a$, and zero elsewhere. Its Fourier transform $\widetilde{f}(k)$ is a "sinc" function. This is the name often used, especially in

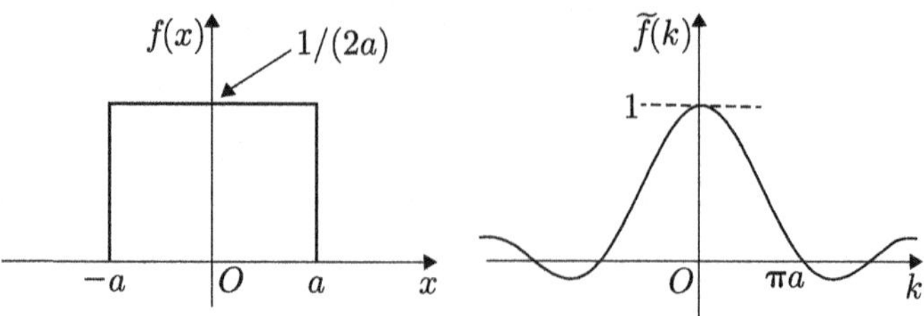

Fig. 18.3 A rectangular pulse function and its Fourier transform

signal processing, for the function $(\sin\,\theta)/\theta$. Figure 18.3 shows schematically the Fourier transform pair $f(x)$ and $\widetilde{f}(k)$ in this case.
(d) You might therefore expect $\widetilde{f}(k)$ to be a rectangular pulse when $f(x)$ is a sinc function. This is indeed so. We have

$$\widetilde{f}(k) = \int_{-\infty}^{\infty} \frac{dx}{x}\,\sin\,(ax)\,e^{-ikx} = 2\int_{0}^{\infty} \frac{dx}{x}\,\sin\,ax\,\cos\,kx,$$

because the imaginary part of the integral vanishes as the integrand is an odd function. Therefore

$$\widetilde{f}(k) = \int_{0}^{\infty} \frac{dx}{x}\,[\sin\,(k+a)x - \sin\,(k-a)x] = \tfrac{1}{2}\pi\,[\varepsilon(k+a) - \varepsilon(k-a)]$$

(where $\varepsilon(x)$ stands for the signum function), on using the Dirichlet integral in Eq. (2.18) of Chap. 2, Sect. 2.3. A moment's thought will show you that

$$\tfrac{1}{2}[\varepsilon(k+a) - \varepsilon(k-a)] = \theta\,(k+a) - \theta\,(k-a).$$

(e) You need to use the Gaussian integral given in Eq. (2.4) of Chap. 2, Sect. 2.1.1. Recall that this integral is valid for complex values of the parameter b, as I have already stated. ▶

3. Use Eq. (18.4) to define $\widetilde{h}(k)$. Substitute for $h(x)$ from Eq. (18.8). Use Eq. (18.3) to Fourier-expand $f(x')$ and $g(x-x')$. Using Eq. (18.6) twice to identify δ-functions, the result sought follows. ▶

7. Apply the formula to the transform pair in Eq. (18.7). Set $\mu = 0$ and $L/\sqrt{2\pi\sigma^2} = \lambda$. ▶

8. (a) In going from (18.48) to (18.49), pay attention to the modulus sign in the exponents.

(b) You need to use the fact that, as $a \to 0$, the function $\coth\,\pi a \simeq (\pi a)^{-1} + \frac{1}{3}\pi a + \mathcal{O}(a^3)$. ▶

Chapter 19
Discrete Probability Distributions

We now turn to a topic that is not only a very fundamental part of mathematical physics, but also impinges on every quantitative subject that exists random variables and the theory of probability. The subject is a vast one, and several courses are needed to do justice to it. But my objective here is far more modest: to familiarize you with the elementary aspects of the subject that are most relevant to physical applications.

There exists a rigorous mathematical theory of probability, developed over the course of the twentieth century.[1] It relies heavily on measure theory. The treatment in this book will not cover this formal theory of probability.

In this chapter I discuss random variables that can only take on a discrete set of values (usually a set of integers). In Chap. 20, we shall consider the case of continuous random variables. Finally, in Chap. 21, we will turn to random processes, i.e., random variables that change with time. I shall assume that you are already familiar with at least some of the elementary concepts in probability and statistics. As you know, **statistical mechanics** comprises one of the most important physical applications of the subject. I will therefore use the terminology of statistical mechanics whenever the opportunity presents itself, in order to reinforce this point.

19.1 Some Elementary Distributions

19.1.1 Mean and Variance

We begin with a quick recapitulation of some elementary properties of a random variable n that can take on values in the set of integers (or some subset of integers), called the **sample space** of n. Properly speaking, we must use distinct symbols for a random variable per se, and for a general *value* of this variable in its sample space.

---

[1]The foundations of the mathematical theory of probability owe much to the mathematicians A. A. Markov and A. N. Kolmogorov.

V. Balakrishnan, *Mathematical Physics*,
https://doi.org/10.1007/978-3-030-39680-0_19

But I will generally ignore this nicety in the interests of simplicity, except when it is likely to result in confusion. (See also the opening remarks in Chap. 20, Sect. 20.1.1.)

Let P_n be the normalized probability that the variable takes on the particular value n. Thus $P_n \geq 0$, and $\sum_n P_n = 1$. The **mean value** and mean squared value are given by

$$\boxed{\langle n \rangle \stackrel{\text{def.}}{=} \sum_n n\, P_n \quad \text{and} \quad \langle n^2 \rangle \stackrel{\text{def.}}{=} \sum_n n^2\, P_n\,,} \tag{19.1}$$

while the **variance** of n is given by

$$\boxed{\text{Var}\; n \stackrel{\text{def.}}{=} \langle n^2 \rangle - \langle n \rangle^2 = \langle \left(n - \langle n \rangle\right)^2 \rangle.} \tag{19.2}$$

The final equation in (19.2) shows that the variance is strictly positive for a random variable. It vanishes if and only if n takes on just one value with probability 1, i.e., if n is a *sure* or *deterministic* variable rather than a random variable. The **standard deviation** (often denoted by Δn) is the square root of the variance. It is the basic measure of the *scatter* of the random variable about its mean value, i.e., of the effect of statistical **fluctuations**. In dimensionless units, one often uses the ratio

$$\text{relative fluctuation} \;=\; \Delta n / \langle n \rangle \tag{19.3}$$

to measure this scatter about the mean. More information about the probability distribution P_n is contained in the higher moments $\langle n^k \rangle$, where $k \geq 3$.

★ **1.** A pair of (distinguishable) dice is tossed once. Each die can give a score of 1, 2, 3, 4, 5, or 6. Let s denote the total score of the pair of dice. It is obvious that the sample space of s is the set of integers from 2 to 12.

(a) Write down the set of probabilities $\{P_s\}$. What is the most probable value of s?
(b) Find the mean, standard deviation and relative fluctuation of s.

***Digression: Indistinguishability in the quantum mechanical sense*:** There is an important subtlety here that is worth noting. The score $s = 3$, for instance, is obtained by throwing either $(1, 2)$ or $(2, 1)$. Hence $P_3 = \frac{1}{36} + \frac{1}{36} = \frac{1}{18}$. *This remains true even if two dice look so alike that you cannot tell them apart*, i.e., even if they appear to be "indistinguishable". The dice are classical objects, and not quantum mechanical particles! In other words, no matter how "indistinguishable" the two dice appear to be, for all practical purposes they are not, and cannot be put, in an **entangled state**. Hence $(1, 2)$ and $(2, 1)$ are indeed distinct microstates of the system, and they must be counted separately. On the other hand, when one says that two identical particles are *indistinguishable in the quantum mechanical sense*, one means that they *can* be put in an entangled state, and that such a state must be counted as a single state.

- This is the fundamental reason why the counting rules in quantum statistics (both Fermi–Dirac statistics and Bose–Einstein statistics) are so drastically different from the counting rule in classical statistics.

In the next two problems, this point is illustrated by a quantum mechanical analog of a pair of dice.

★ **2.** Consider a system of two identical bosons. Let the set of energy eigenvalues of each particle be 1, 2, 3, 4, 5, and 6 in suitable units, with corresponding normalized eigenstates $|\, j\,\rangle$, $1 \leq j \leq 6$. The direct product state $|\, j\,\rangle \otimes |\, k\,\rangle \equiv |\, j\, k\,\rangle$ is a state in which particle 1 in the state $|\, j\,\rangle$ and particle 2 in the state $|\, k\,\rangle$, with a total energy $j + k$. The orthonormality condition is $\langle\, j\, k\,|\, j'\, k'\,\rangle = \delta_{jj'}\, \delta_{kk'}$.

(a) List the normalized eigenstate(s) of the system corresponding to each eigenvalue of the total energy ε.
(b) Given that each of these eigenstates is equally probable, what is the probability that $\varepsilon = 3$? This is clearly the analog of a total score $s = 3$ in the case of a pair of dice.
(c) What is the probability P_7 that the total energy has the value $\varepsilon = 7$?

★ **3.** Suppose the two particles are identical fermions rather than bosons, all the other conditions remaining the same as in the preceding exercise.

(a) What are the normalized energy eigenstates of the system?
(b) Given that each of these states is equally probable, what is the probability P_3 that the system has a total energy $\varepsilon = 3$?
(c) What is the probability P_7 that the total energy of the system is $\varepsilon = 7$?

19.1.2 Bernoulli Trials and the Binomial Distribution

Consider an event in which the outcome can be either success, with an a priori (or before-the-event) probability p; or failure, with an a priori probability $q \equiv 1 - p$. Here $0 < p < 1$ in general. Such an event is called a **Bernoulli trial**. A simple example is a coin toss, in which the outcome can be a head (success, say) or a tail (failure). If the coin is a *fair* one, then $p = q = \frac{1}{2}$. If the coin is *biased*, then $p \neq \frac{1}{2}$. We make N trials. The question is what is the probability P_n that the total number of successes is n? Here n is the random variable, and its sample space is the set of integers from 0 to N.

The trials are independent of each other. The outcome of any one trial cannot affect the outcome of any other. Hence the probability of any particular sequence of successes (heads) and failures (tails), such that the total number of successes is n, is just $p^n q^{N-n}$. Each such sequence contributes to P_n, and there are ${}^N C_n$ such sequences. Therefore[2]

$$\boxed{P_n = \binom{N}{n}\, p^n\, q^{N-n}, \quad 0 \leq n \leq N.} \tag{19.4}$$

[2]The preferred notation for the binomial coefficient ${}^N C_n$ is $\binom{N}{n}$.

This probability distribution is called the **binomial distribution**, for an obvious reason: each P_n is a term in the binomial expansion of $(p+q)^N$. The binomial distribution is a *two-parameter* distribution, the independent parameters being N and p.

We are interested in various moments of P_n. These are most easily obtained from the generating function of P_n. This function is given, for a given N, by

$$f(z) \stackrel{\text{def.}}{=} \sum_{n=0}^{N} P_n\, z^n = (pz+q)^N. \tag{19.5}$$

Recall our use of generating functions for different families of orthogonal polynomials. Generating functions are usually power series in some variable. As you know, a power series is absolutely convergent inside a circle of convergence in the complex plane. Hence it is appropriate to regard that variable as a complex variable. This is why I use z rather than x as the argument in all the generating functions to be encountered from now on. In the present instance, $f(z)$ is just a polynomial in z, so that the question of the convergence of the power series does not arise.

The normalization of the probability distribution is trivially verified, since $\sum_0^N P_n = f(1) = 1$. The kth **factorial moment** of n is given by

$$\langle n(n-1)\,\ldots\,(n-k+1)\rangle = \left.\frac{d^k f(z)}{dz^k}\right|_{z=1}. \tag{19.6}$$

The mean and variance of the binomial distribution are given by

$$\boxed{\langle n\rangle = N\,p \quad \text{and} \quad \text{Var}\; n = N\,p\,q,} \tag{19.7}$$

respectively. Hence the relative fluctuation for the binomial distribution is

$$\frac{\Delta n}{\langle n\rangle} = \sqrt{\frac{q}{Np}} = \sqrt{\frac{q}{\langle n\rangle}}\,. \tag{19.8}$$

The kth factorial moment of the binomial distribution is easily found to be

$$\langle n(n-1)\,\ldots\,(n-k+1)\rangle = \begin{cases} N(N-1)\,\ldots\,(N-k+1)\,p^k\,, & k \le N \\ 0\,, & k > N. \end{cases} \tag{19.9}$$

★ **4.** Verify Eqs. (19.7)–(19.9).

The vanishing of the kth factorial moment for every $k > N$ in the case of the binomial distribution (19.4) follows trivially from the fact that the generating function $f(z)$ in this case is a polynomial of order N. Hence all its derivatives higher than the Nth derivative vanish identically. This does not mean that the moments $\langle n^k\rangle$ themselves vanish for $k > N$. Rather, all these moments can be expressed in terms of the lower moments $\langle n\rangle, \langle n^2\rangle, \ldots, \langle n^N\rangle$.

***The most probable value in a binomial distribution*:** It is "intuitively obvious" that, if you toss a fair coin a N times, the most probable outcome will be $\frac{1}{2}N$ heads (if N is even) or $\frac{1}{2}(N \pm 1)$ (if N is odd). This is indeed borne out: the binomial distribution (19.4) reduces to $P_n = {}^N C_n \, 2^{-n}$ for $p = q = \frac{1}{2}$, and this is largest precisely at $n = \frac{1}{2}N$ when N is even, and at $n = \frac{1}{2}(N \pm 1)$ when N is odd.

For a general value of p, the most probable value $\overline{n}$ of the random variable n is easily determined. It is the integer lying in between $(N+1)\,p - 1$ and $(N+1)\,p$, i.e.,

$$(N+1)\,p - 1 < \overline{n} < (N+1)\,p. \tag{19.10}$$

In particular,

$$\left.\begin{aligned} \overline{n} &= 0 \quad \text{if } (N+1)\,p < 1, \\ \overline{n} &= N \quad \text{if } (N+1)\,p > N. \end{aligned}\right\} \tag{19.11}$$

★ **5.** Establish Eqs. (19.10) and (19.11).

19.1.3 Number Fluctuations in a Classical Ideal Gas

Note the $N^{-1/2}$ fall-off of the relative fluctuation with increasing N in Eq. (19.8). This feature is not restricted to the binomial distribution. It is characteristic of numerous physical situations in equilibrium statistical mechanics. In that context, N is usually proportional to the number of degrees of freedom of the system concerned. If N is of the order of Avogadro's number, for instance, then $N^{-1/2} \sim 10^{-12} \ll 1$.

- This is essentially why thermodynamics, which deals only with *mean* values of macroscopic observables, and neglects *fluctuations* in them, provides a satisfactory description of physical phenomena under the conditions in which it is applicable.

The following simple physical example illustrates the point.

Consider a (classical, ideal) gas of N molecules in thermal equilibrium in a container of volume V. The molecules are taken to be point masses that move about at random, suffering elastic collisions with each other. The average number density of the particles is $\rho = N/V$. Now consider a sub-volume v in the container (Fig. 19.1). It is obvious that the number of molecules that are present in the sub-volume is a rapidly fluctuating quantity. In principle, this number can be any integer from 0 to N at any given instant of time. Let n be this random number (at a given instant of time). We want the probability P_n that there are exactly n molecules in the sub-volume. It is assumed that the molecules are independent of each other, i.e., there is no correlation between the locations of different molecules. The a priori probability that a given molecule is in the sub-volume is just v/V, because it has an equal probability of being in any volume element in the container. Therefore the probability that it lies outside v is $(1 - v/V)$. Since the n molecules inside V can be chosen in ${}^N C_n$ ways, we have

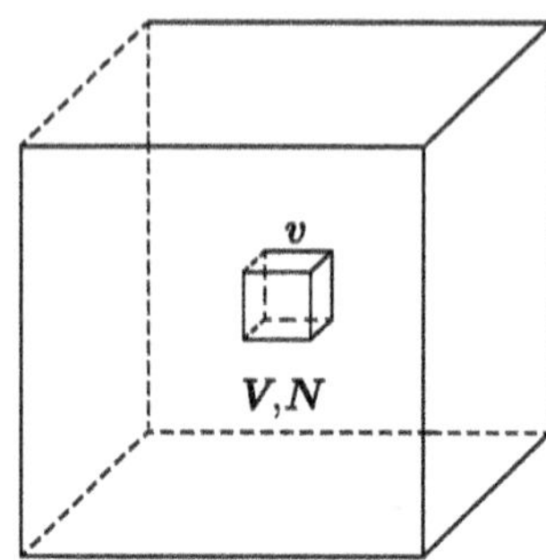

Fig. 19.1 Number fluctuations in a sub-volume of a gas

$$P_n = \binom{N}{n}\left(\frac{v}{V}\right)^n\left(1-\frac{v}{V}\right)^{N-n} = \binom{N}{n}\left(\frac{\rho v}{N}\right)^n\left(1-\frac{\rho v}{N}\right)^{N-n}. \tag{19.12}$$

Thus P_n is a binomial distribution, where the a priori probability p is identified with $v/V = \rho v/N$. The mean value is given by $\langle n\rangle = \rho v$, as expected. The relative fluctuation in n is

$$\frac{\Delta n}{\langle n\rangle} = \left[\frac{1}{\rho v}\left(1-\frac{\rho v}{N}\right)\right]^{1/2}. \tag{19.13}$$

The so-called **thermodynamic limit** corresponds, in this case, to letting N and V tend to infinity simultaneously, keeping their ratio $N/V = \rho$ fixed at a finite value. In this limit, the sample space of n becomes the set of all non-negative integers. Further, the binomial distribution (19.12) goes over into the Poisson distribution, as we shall see in Sect. 19.2.1. The mean value $\langle n\rangle$ remains equal to ρv, of course. The relative fluctuation becomes $1/\sqrt{(\rho v)}$. For air at standard temperature and pressure, it is easy to see that this quantity is of the order of 10^{-12} for a volume $v = 1\,\text{m}^3$.

19.1.4 The Geometric Distribution

Consider once again a coin that has a priori probabilities p for heads and $q \equiv 1-p$ for tails, when it is tossed. The experiment now consists of tossing the coin repeatedly till heads is obtained *for the first time*. What is the probability P_n that this happens on the $(n+1)$th toss?

It is clear that the random variable in this case is n. Its sample space is the set of nonnegative integers 0, 1, ... ad inf. Quite obviously,

$$P_n = q^n\, p. \tag{19.14}$$

Such a probability distribution is called a **geometric distribution**, because the successive terms in the sequence $P_0\,,\; P_1\,,\;\ldots$ form a geometric progression.

★ **6.** The properties of a geometric distribution are easily deduced.

(a) Show that the generating function for P_n is $f(z) = p/(1 - qz)$.
(b) Hence show that $\langle n \rangle = q/p$, $\text{Var}(n) = q/p^2$, and $\Delta n/\langle n \rangle = 1/\sqrt{q}$.

Denoting the mean $\langle n \rangle$ by μ, the geometric distribution (19.14) and its generating function are given by

$$\boxed{P_n = \frac{1}{\mu+1}\left(\frac{\mu}{\mu+1}\right)^n, \quad f(z) = \frac{1}{1+\mu-\mu z}.} \tag{19.15}$$

The geometric distribution is a single-parameter distribution, the parameter being μ. All its higher moments must therefore be functions of μ alone.

19.1.5 Photon Number Distribution in Blackbody Radiation

A physical situation in which the geometric distribution occurs is provided by blackbody radiation, which is just a gas of photons in thermal equilibrium at some absolute temperature T. I start with a brief recapitulation of some preliminaries, in order to make the account self-contained.

Recall that a photon is a particle with zero rest mass, satisfying the energy-momentum relation $\varepsilon = c\,p$, where $p = |\mathbf{p}|$ is the magnitude of its momentum. The connection with electromagnetic waves in free space is expressed via the relations

$$\varepsilon = h\nu, \quad \mathbf{p} = \hbar\mathbf{k}, \quad \text{and} \quad \lambda = 2\pi/k. \tag{19.16}$$

Here ν is the frequency, λ is the wavelength, $\mathbf{k}$ is the wave vector, and $k = |\mathbf{k}|$ is the wave number. Hence the "particle picture" relation $\varepsilon = c\,p$ is equivalent to the familiar "wave picture" relation $c = \nu\,\lambda$.

A photon is specified not only by its momentum $\mathbf{p}$ (or wave number $\mathbf{k}$), but also its spin $\mathbf{S}$. The spin quantum number of a photon is $S = 1$. Based on quantum mechanics, you might then expect the spin angular momentum along any specified direction to have only $(2S + 1) = 3$ possible values, namely, $-\hbar$, 0 and $+\hbar$. However, the photon has zero rest mass (i.e., it is "massless"), as already mentioned.

- A massless particle has only one possible speed in a vacuum, namely, the fundamental (or limiting) speed c.
- It is a consequence of relativistic quantum mechanics that a particle whose speed is always c has only *two* possible spin states, no matter what the quantum number S is.

The spin angular momentum of a photon is always directed either parallel or antiparallel to its momentum $\mathbf{p}$. That is, the quantity $(\mathbf{S} \cdot \mathbf{p})/(Sp)$ (called the **helicity** of the particle) can only take on the values $+1$ and -1, in units of $\hbar$. These correspond, in the wave language, to right circular polarization and left circular polarization, respectively. They are the only possible states of polarization of a single free photon.

Blackbody radiation is also called thermal radiation. It comprises photons of all wave vectors (and hence, all frequencies). It is also unpolarized, which means that it is an *incoherent superposition* of photons of both states of polarization for every wave vector. Since the spin quantum number of a photon is an integer ($S = 1$), photons obey Bose–Einstein statistics. Now consider all photons of a given wave vector $\mathbf{k}$ (and hence a given wave number k and frequency ν) and a given state of polarization (either left or right circular polarization) in blackbody radiation at a temperature T. We want to find the probability P_n that there are n such photons at any instant of time. (We should really write $n(\mathbf{k}, +1)$ or $n(\mathbf{k}, -1)$ to indicate the fact that these are photons of a given wave vector and polarization state, but let us keep the notation simple.)

There is a subtlety involved here that you must appreciate. A photon gas differs in two important ways from a gas of particles of nonzero rest mass in a container:

(i) In a gas of particles of nonzero rest mass, collisions between the particles help maintain the thermal equilibrium of the gas. In a photon gas, in contrast, the mutual interaction between photons themselves is quite negligible.[3] Hence the photons may be regarded as comprising an ideal gas of massless bosons.

(ii) In a gas of particles of nonzero rest mass, the number of massive particles in a closed container is constant. In stark contrast, the number of photons in a blackbody cavity fluctuates, because of their absorption and emission by the atoms in the walls of the radiation cavity.

In fact, it is precisely this atom–photon interaction that is responsible for maintaining the thermal equilibrium of the radiation. Although the photon number n fluctuates, all ensemble averages (including the mean photon number and all its higher moments) are guaranteed to remain constant in time, because the system is in thermal equilibrium. In other words, the probability distribution P_n must be independent of time.

Since the total energy of n photons of frequency ν is just $nh\nu$, P_n must be proportional to $e^{-\beta nh\nu}$, where $\beta = (k_B T)^{-1}$ and k_B is Boltzmann's constant. The constant of proportionality is easily determined by using the normalization condition $\sum_{n=0}^{\infty} P_n = 1$. We find

$$P_n = \left(1 - e^{-\beta h\nu}\right) e^{-n\beta h\nu}. \tag{19.17}$$

Hence, the mean number of photons of a given wave vector $\mathbf{k}$ and a given state of (circular) polarization in blackbody or thermal radiation at a temperature T is

$$\langle n \rangle = \sum_{n=0}^{\infty} n\, P_n = \frac{1}{e^{\beta h\nu} - 1}, \tag{19.18}$$

[3]There *does* exist a very small but nonzero cross-section for photon–photon scattering (called **Delbrück scattering**), induced by a quantum-electrodynamic process involving the creation and subsequent annihilation of a virtual electron-positron pair. The effect of this interaction is extremely small at normal intensities of the radiation field.

where $\nu = ck/(2\pi)$. You will readily recognize the right-hand side of Eq. (19.18) as the "Bose factor". The exponent in the denominator is $e^{\beta h\nu}$ rather than $e^{\beta(h\nu-\mu)}$, where μ is the chemical potential, because the chemical potential of a photon gas vanishes identically. This is a consequence of the fact that a photon has zero rest mass.

Question: Why is the expression in (19.18) the mean number of photons (of a given state of polarization and) of a given wave *vector* **k** and not wave *number* k?

Turning Eq. (19.18) around, we have

$$e^{-\beta h\nu} = \frac{\langle n\rangle}{\langle n\rangle + 1}. \tag{19.19}$$

Putting this into Eq. (19.17), P_n may be written in the alternative form

$$P_n = \frac{1}{\langle n\rangle + 1}\left(\frac{\langle n\rangle}{\langle n\rangle + 1}\right)^n \quad \text{where} \quad n = 0,\ 1,\ \ldots. \tag{19.20}$$

- The probability distribution of the number of photons of a given wave vector and polarization in blackbody radiation is therefore a geometric distribution.

Because of this connection with Bose–Einstein statistics, the geometric distribution itself is sometimes called the Bose–Einstein distribution, especially in the quantum optics literature concerned with photon-counting statistics. It follows from Eq. (19.20) that the variance and relative fluctuation of n are given, respectively, by

$$\mathrm{Var}\,(n) = \langle n\rangle\,\big(\langle n\rangle + 1\big) \quad \text{and} \quad \frac{\Delta n}{\langle n\rangle} = \left(1 + \frac{1}{\langle n\rangle}\right)^{1/2}. \tag{19.21}$$

The kth factorial moment of n is

$$\langle n(n-1)\ldots(n-k+1)\rangle = \langle n\rangle^k\, k! \tag{19.22}$$

★ **7.** Verify Eqs. (19.21) and (19.22).

I reiterate that the geometric distribution of the photon number in blackbody radiation is a direct consequence of the fact that photons obey Bose–Einstein statistics. Observe that the relative fluctuation in n is always greater than unity.[4] Thus, *in thermal radiation, the photon number has a large scatter about its mean value.* As you will see in Sect. 19.2.2, this feature is in stark contrast to what happens in the case of coherent radiation. The same remark applies to the factorial moments as well.

---

[4]This feature implies that the photon number distribution in blackbody radiation is "super-Poissonian". It is also related to the phenomenon of **photon bunching** in the photon-counting statistics of thermal radiation.

19.2 The Poisson Distribution

19.2.1 From the Binomial to the Poisson Distribution

The **Poisson distribution** is a discrete probability distribution that occurs in a very large number of physical situations. It can be understood as a limiting case of the binomial distribution when the number N of Bernoulli trials tends to infinity, while the probability p of success in a single trial simultaneously tends to zero, such that their product tends to a finite positive limit. That is,

$$\lim_{\substack{N\to\infty \\ p\to 0}} (N\,p) = \mu. \tag{19.23}$$

In this limit, the binomial distribution of Eq. (19.4) goes over into the Poisson distribution

$$\boxed{P_n = e^{-\mu}\,\frac{\mu^n}{n!}\,, \quad \text{where } n = 0,\, 1,\, \ldots.} \tag{19.24}$$

★ **8.** Start with the binomial distribution

$$P_n = \frac{N!}{(N-n)!\,n!}\;p^n\,(1-p)^{N-n}\,.$$

Eliminate p by setting it equal to μ/N. Use Stirling's formula for the factorial of a large number, and show that, in the limit $N \to \infty$ (n remaining finite), Eq. (19.24) is obtained.

The sample space of a Poisson-distributed random variable n is the infinite set of nonnegative integers $0,\, 1,\, \ldots$. The parameter μ is just the mean value of n, i.e., $\langle n\rangle = \mu$. The generating function of the Poisson distribution is

$$f(z) = \sum_{n=0}^{\infty} P_n\, z^n = e^{\mu(z-1)}\,. \tag{19.25}$$

The Poisson distribution is a single-parameter distribution—there is only one parameter, the mean value μ. Hence all the higher moments of the random variable n must be expressible in terms of μ. Indeed, the kth factorial moment in this case is just

$$\langle n(n-1)\,\ldots\,(n-k+1)\rangle = \left.\frac{d^k f(z)}{dz^k}\right|_{z=1} = \mu^k\,. \tag{19.26}$$

In particular, it is easy to see that

$$\boxed{\text{Var}\,(n) = \langle n\rangle \quad \text{for a Poisson distribution.}} \tag{19.27}$$

- The equality of the variance and the mean is a characteristic signature of the Poisson distribution.

This equality is a particular case of an even more special property of the Poisson distribution: all its *cumulants* are equal to μ. (Cumulants will be defined in Chap. 20, Sect. 20.1.3.)

Returning to the example of density fluctuations in an ideal gas (used earlier to illustrate the binomial distribution), the passage to the limiting Poisson distribution has a physical interpretation. The distribution is given by Eq. (19.12). Thus $p = \rho v/N$, and the limit in which Np tends to a finite limit (which is clearly ρv) is precisely the thermodynamic limit: the number of molecules $N \to \infty$ and the volume $V \to \infty$ such that the ratio N/V tends to a finite value, the mean number density ρ. In this limit, the relative fluctuations in thermodynamic quantities vanish, and thermodynamics (which involves only average values of macroscopic observables) becomes exact. In this example, the probability that there are n gas molecules in a volume v becomes a Poisson distribution,

$$P_n = e^{-\rho v}\,\frac{(\rho v)^n}{n!}\,, \tag{19.28}$$

with a mean value given by $\langle n\rangle = \rho v$, as expected.

19.2.2 Photon Number Distribution in Coherent Radiation

Another physical situation in which the Poisson distribution occurs is again provided by a collection of photons, but in a state very different from blackbody radiation.

In an ideal single-mode laser, the radiation is made up of photons of the same wave vector and polarization, and is said to be in a coherent state of the electromagnetic field. The properties of the quantum mechanical linear harmonic oscillator enable us to describe coherent radiation quantum mechanically. In a nutshell, what happens is as follows. When the electromagnetic field is *quantized*, each "mode" of the field—that is, each component of a given frequency and state of polarization—behaves like a quantum mechanical linear harmonic oscillator. The annihilation and creation operators of the quanta (or photons) of each given wave vector and state of polarization are exactly like the lowering and raising operators (or ladder operators) a and $a^\dagger$ of the oscillator. They satisfy the canonical commutation relation $[a\,,\,a^\dagger] = I$. The operator $N = a^\dagger a$ is now the photon number operator.[5] Its eigenvalues are, of course, the integers $n = 0,\ 1,\ \ldots,$ with corresponding eigenvectors $|\,0\,\rangle\,,\ |\,1\,\rangle\,,\ \ldots.$ As I have pointed out earlier, the energy eigenstate $|\,n\,\rangle$ of the oscillator is equivalent to the Fock state with exactly n photons of a given wave vector and polarization.

---

[5] It is clear that we should label a and $a^\dagger$ with the wave vector $\mathbf{k}$ and the state of polarization of the mode concerned. But as we are concerned here with just a single mode, let us keep the notation simple.

The coherent state $|\, z \,\rangle$ is an eigenstate of the lowering operator a with eigenvalue $z \in \mathbb{C}$. (Recall Eq. (14.51) of Chap. 14, Sect. 14.4.2.) It can be written as a superposition of the basis set $\{|\, n \,\rangle\}$, as in Eq. (15.64) of Chap. 15, Sect. 15.4.1. Repeating that expansion for ready reference, the normalized coherent state is given by

$$\boxed{|\, z \,\rangle = e^{-\frac{1}{2}|z|^2} \sum_{n=0}^{\infty} \frac{z^n}{\sqrt{n!}} \,|\, n \,\rangle.} \tag{19.29}$$

★ **9.** Derive Eq. (19.29) directly from the eigenvalue equation $a\,|\, z \,\rangle = z\,|\, z \,\rangle$.

I have already pointed out[6] the physical significance of the real and imaginary parts of the eigenvalue $z = z_1 + iz_2$. Further, the mean number of photons in the state $|\, z \,\rangle$ is given by $|z|^2$, as we have found in Eq. (15.72) of Chap. 15, Sect. 15.4.1. This result actually follows quite trivially: we have $a\,|\, z \,\rangle = z\,|\, z \,\rangle$ and its adjoint, $\langle\, z\,|\, a^\dagger = z^*\,\langle\, z\,|$, so that

$$\boxed{\langle\, z\,|N|\, z \,\rangle = \langle\, z\,|(a^\dagger\, a)|\, z \,\rangle = |z|^2 \langle z\,|\, z \,\rangle = |z|^2.} \tag{19.30}$$

- $|z|^2$ is the *average* number of photons in the coherent state $|\, z \,\rangle$.

But we can go on to find the probability distribution itself (of the number of photons in a coherent state). According to a basic rule of quantum mechanics, the **probability amplitude** that there are exactly n photons in the coherent state $|\, z \,\rangle$ is given by the inner product $\langle n|\, z \,\rangle$. The actual probability (that the number of photons in this state is n) is of course the square of the modulus of the probability amplitude. Using the orthonormality of the Fock states, we get

$$\langle n|\, z \,\rangle = \frac{e^{-\frac{1}{2}|z|^2}\, z^n}{\sqrt{n!}} \implies \boxed{P_n \equiv |\langle n|\, z \,\rangle|^2 = \frac{e^{-|z|^2}\, |z|^{2n}}{n!}.} \tag{19.31}$$

This is a Poisson distribution with mean value $|z|^2$.

- The photon number distribution in ideal, single-mode coherent radiation is a Poisson distribution.

The variance in the photon number is therefore equal to the mean value. The relative fluctuation in the photon number is $1/|z|$. It decreases as the mean number of photons in the coherent state increases, in marked contrast with the case of thermal radiation (where it never falls below unity).

In practice, radiation from a laser may contain more than one mode. But even in the case of a single mode, the coherent radiation is likely to be mixed with some thermal radiation as well. Photon-counting statistics enables us to analyze this admixture

---

[6]See Eqs. (15.70) of Chap. 15, Sect. 15.4.1, and the discussion following it; and Eqs. (16.54) and (16.55) of Chap. 16, Sect. 16.2.6.

by quantifying the deviation from Poisson statistics—for instance, by modeling the photon statistics of the admixture in terms of the negative binomial distribution, to be discussed in Sect. 19.3 below.

Following the discussion of coherent states in Chap. 15, Sect. 15.4.1, we considered the class of squeezed vacuum states in Sect. 15.4.2. In a similar vein, before continuing with the discussion of Poisson-distributed random variables, I digress briefly to examine the photon number distribution in a squeezed vacuum state.

19.2.3 Photon Number Distribution in the Squeezed Vacuum State

Recall, from Chap. 15, Sect. 15.4.2 that the squeezed vacuum state is given by

$$\begin{aligned} |\sigma(z)\rangle &= S(z)|\,0\,\rangle = e^{\frac{1}{2}(za^{\dagger 2}-z^{*}a^{2})}|\,0\,\rangle \\ &= \frac{1}{\sqrt{\cosh r}}\sum_{n=0}^{\infty}\frac{\sqrt{(2n)!}}{2^{n}\,n!}\left(e^{i\theta}\tanh r\right)^{n}|\,2n\,\rangle, \end{aligned} \tag{19.32}$$

repeating Eqs. (15.79), (15.77) and (15.88) for ready reference. The state is labeled by the complex parameter $z = r\,e^{i\theta}$. It follows at once that the probability distribution of the photon number in the state $|\sigma(z)\rangle$ is given by

$$P_k = |\langle\,k\,|\sigma(z)\rangle|^2 = \begin{cases} \dfrac{(2n)!}{2^{2n}\,(n!)^2}\,\dfrac{(\tanh r)^{2n}}{\cosh r} & \text{when } k = 2n, \\ 0 & \text{when } k = 2n+1. \end{cases} \tag{19.33}$$

Note that the distribution is independent of the phase θ of the parameter z, and depends only on its magnitude r. The zero-photon probability P_0 is just sech r. The generating function corresponding to the distribution (19.33) is given by[7]

$$f(u) = \sum_{k=0}^{\infty} P_k\,u^k = \sum_{n=0}^{\infty}\frac{(2n)!}{2^{2n}\,(n!)^2}\,\frac{(u\tanh r)^{2n}}{\cosh r}. \tag{19.34}$$

But the series on the right-hand side of Eq. (19.34) is just a binomial series! It can be summed to give

$$f(u) = (\cosh^2 r - u^2\sinh^2 r)^{-1/2}. \tag{19.35}$$

Since $\sum_{k=0}^{\infty} P_k = f(1) = 1$, the probability distribution P_k is correctly normalized. The moments of the photon number k (i.e., the expectation values of powers of the

---

[7] I use u instead of z as the argument of the generating function because z has already been used to denote the parameter in the state $|\sigma(z)\rangle$.

number operator $N = a^\dagger a$) in a squeezed vacuum state are now computed quite easily.

★ **10.** Starting from Eq. (19.32), work out the steps to arrive at Eq. (19.35). Then verify, using the generating function $f(u)$, that the mean and variance of the photon number are given by

$$\langle N \rangle = \sinh^2 r \quad \text{and} \quad \text{Var}\,(N) = \tfrac{1}{2}\,\sinh^2(2r).$$

These expressions have been derived earlier (recall Eqs. (15.83) and (15.87) of Chap. 15, Sect. 15.4.2).

19.2.4 The Sum of Poisson-Distributed Random Variables

Often, one has to deal with a random variable that is the sum of two or more independent random variables, each of which is Poisson-distributed. What is the probability distribution of this sum? Let us first consider the case of two random variables. The more general case will turn out to be a simple extension of this case, owing to a fundamental property of the Poisson distribution.

Let m and n be two independent random variables, each of which has a Poisson distribution, given by

$$P_1(m) = e^{-\mu}\,\frac{\mu^m}{m!} \quad \text{and} \quad P_2(n) = e^{-\nu}\,\frac{\nu^n}{n!}\,, \tag{19.36}$$

respectively.[8] Let $s = m + n$ denote the sum of the two random variables. It is obvious that the sample space of s is again the set of nonnegative integers. We want to find the probability distribution $P_{\text{sum}}(s)$ of the sum s. This is obtained as follows.

(a) Multiply together the two individual probability distributions to get the joint probability distribution of m and n, since the variables are independent of each other.
(b) Sum over all the possibilities for the individual random variables, *subject to the constraint* that the sum $m + n$ be equal to s.

Thus,

$$P_{\text{sum}}(s) = \sum_{m=0}^{\infty}\sum_{n=0}^{\infty} P_1(m)\,P_2(n)\,\delta_{m+n\,,\,s}\,. \tag{19.37}$$

[8] A trivial point of notation: I use two different symbols P_1 and P_2 for the two distributions in order to distinguish between them—they involve different parameters μ and ν.

Carrying out the summations, the final result is

$$\boxed{P_{\text{sum}}(s) = e^{-(\mu+\nu)}\,\frac{(\mu+\nu)^s}{s!}\,, \quad s = 0,\ 1,\ \ldots\,.} \tag{19.38}$$

★ **11.** Start with the formal expression in Eq. (19.37) and derive the result in Eq. (19.38).

Equation (19.38) shows that the sum of the two Poisson-distributed random variables m and n is also Poisson-distributed, with a mean value that is just the sum of the mean values of m and n:

$$\langle s \rangle = \mu + \nu = \langle m \rangle + \langle n \rangle. \tag{19.39}$$

It follows that

$$\text{Var}\,(s) = \text{Var}\,(m) + \text{Var}\,(n) = \mu + \nu. \tag{19.40}$$

Further, the sum of any number of Poisson-distributed, independent random variables is also Poisson-distributed, with mean values and variances simply adding up. In this sense the Poisson distribution is a kind of discrete, integer-valued analog of the so-called **stable distributions**. I will discuss the latter in Chap. 20, Sect. 20.5.

19.2.5 The Difference of Two Poisson-Distributed Random Variables

Consider again the pair of Poisson-distributed random variables m and n of the preceding section. Let $r = m - n$ be the *difference* of the two random variables. The task is to find the probability distribution of r.

It is obvious that r cannot have a Poisson distribution, because it can take on all possible integer values, including *negative* ones—its sample space is $\mathbb{Z}$. Analogous to Eq. (19.37), the probability distribution of r is given by

$$P_{\text{diff}}(r) = \sum_{m=0}^{\infty}\sum_{n=0}^{\infty} P_1(m)\, P_2(n)\, \delta_{m-n\,,\,r}\,. \tag{19.41}$$

Once again, the sum over n can be eliminated using the Kronecker delta to replace n by $m - r$. As before, we must be careful about the limits of the subsequent summation over m. On the $(m\,,n)$ plane, mark the lattice points $m \geq 0\,,\ n \geq 0$. Note the set of points satisfying the constraint $m - n = r$. It is immediately clear that two cases arise, corresponding to $r \geq 0$ and $r < 0$, respectively, as shown in Fig. 19.2. Consider the case $r \geq 0$ first. The summation over m must obviously run from r upward, because n does not take on negative values. Hence the expression for $P_{\text{diff}}(r)$ reduces to

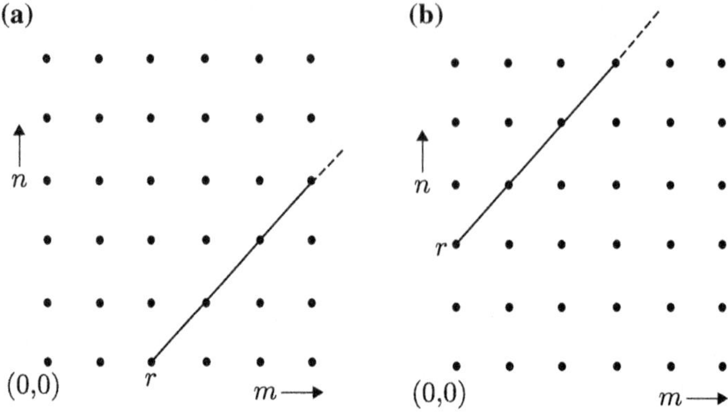

Fig. 19.2 The evaluation of the double sum in Eq. (19.41), in the cases **a** $r \geq 0$ and **b** $r < 0$. The diagonal line in each case runs through the (infinite set of) values of (m, n) selected by the Kronecker delta for an illustrative value of $r (= 2$ in case **a**, and -2 in case **b**)

$$P_{\text{diff}}(r) = \sum_{m=r}^{\infty} P_1(m)\, P_2(m-r) \quad (r \geq 0). \tag{19.42}$$

Note the lower limit of the summation. Now consider the case $r < 0$. It should be clear from the figure that all values of m from 0 upward contribute to the sum in this case. Hence

$$P_{\text{diff}}(r) = \sum_{m=0}^{\infty} P_1(m)\, P_2(m+|r|) \quad (r < 0). \tag{19.43}$$

It remains to substitute the given Poisson distributions for P_1 and P_2 in the sums above, and to carry out the summations.

It turns out that the answer can be expressed in terms of the **modified Bessel function** of the first kind and of order l, denoted by $I_l(z)$. This function is defined, when l is not a negative integer, by the power series[9]

$$\boxed{I_l(z) = \sum_{m=0}^{\infty} \frac{1}{m!\,\Gamma(m+l+1)} \left(\tfrac{1}{2}z\right)^{l+2m}.} \tag{19.44}$$

When $l = 0, 1, 2, \ldots$, the gamma function reduces to a factorial, and we have

$$I_l(z) = \sum_{m=0}^{\infty} \frac{1}{m!\,(m+l)!} \left(\tfrac{1}{2}z\right)^{l+2m}. \tag{19.45}$$

[9]More will be said about the modified Bessel function after we discuss analytic functions of a complex variable, in Chap. 22, Sect. 22.6.2, and again when we consider Laplace transforms, in Chap. 28, Sect. 28.3.

This series (as also the series in Eq. (19.44)) converges absolutely for all finite values of $|z|$. (I will comment further on this aspect at the end of Sect. 22.6.2, Chap. 22.) For negative integer values of l, the modified Bessel function is *defined* by the symmetry property

$$\boxed{I_{-l}(z) \stackrel{\text{def.}}{=} I_l(z), \quad l = \text{integer.}} \tag{19.46}$$

Using these properties, we get an explicit formula for the probability distribution of r. For any integer value of r, we have

$$\boxed{P_{\text{diff}}(r) = e^{-(\mu+\nu)}\,(\mu/\nu)^{r/2}\,I_r(2\sqrt{\mu\nu}), \quad r \in \mathbb{Z}.} \tag{19.47}$$

This distribution is also known as the **Skellam distribution**. Note that (i) it is a distribution on the *full* set of integers, and (ii) it is characterized by *two* parameters, μ and ν.

Now, the generating function for the modified Bessel function of the first kind is given by

$$\boxed{\sum_{r=-\infty}^{\infty} I_r(t)\,z^r = \exp\left\{\tfrac{1}{2}t\left(z + z^{-1}\right)\right\}.} \tag{19.48}$$

Note that the summation on the left-hand side is over *all* integers r, rather than just the nonnegative integers. The series converges absolutely for all z in the region $0 < |z| < \infty$.[10] Using the identity (19.48), it follows that the generating function of the Skellam distribution is given by

$$f(z) = \sum_{r=-\infty}^{\infty} P_{\text{diff}}(r) z^r = e^{-(\mu+\nu)}\,e^{\mu z + \nu z^{-1}}. \tag{19.49}$$

★ **12.** The formula (19.47) for $P_{\text{diff}}(r)$ must be verified.

(a) Substitute for P_1 and P_2 in Eqs. (19.42) and (19.43) from Eqs. (19.36). Carry out the summations using the definition of the modified Bessel function given above, to obtain Eq. (19.47).

(b) Use the generating function for the modified Bessel function to check that the normalization condition $\sum_{r=-\infty}^{\infty} P_{\text{diff}}(r) = 1$ is satisfied.

★ **13.** As in the case of the sum $s = m + n$ of two Poisson-distributed random variables, the distribution of the difference $r = m - n$ is also derived most easily

[10] Series of this kind, that involve both positive and negative powers of z, are called **Laurent series**. We will discuss such series in Chap. 23, Sect. 23.2.4. Characteristically, the region of convergence of such a series is an *annular* region. In the present instance, $0 < |z| < \infty$ is an annular region because it is the complex plane with a hole in it (the origin is excluded).

by using the corresponding generating functions. Multiply both sides of the defining relation, Eq. (19.41), by z^r and sum over all integer values of r. Use the Kronecker delta on the right-hand side to eliminate the sum over r, obtaining the relation

$$f_{\text{diff}}(z) = f_1(z)\, f_2(1/z) = e^{-(\mu+\nu)}\, e^{\mu z + \nu z^{-1}}.$$

Now use the formula (19.48) for the generating function of the modified Bessel function to read off the result (19.47) for $P_{\text{diff}}(r)$.

It is obvious that the mean value of $r = m - n$ is given by

$$\langle r \rangle = \langle m \rangle - \langle n \rangle = \mu - \nu. \tag{19.50}$$

Since r is not Poisson-distributed, however, we cannot conclude at once that its variance is equal to its mean value, namely, $(\mu - \nu)$. Besides, the mean value $(\mu - \nu)$ can have either sign, whereas the variance must be positive definite. But it is easy to show that

$$\text{Var}\,(r) = \text{Var}\,(m - n) = \mu + \nu. \tag{19.51}$$

Thus, although the difference $r = m - n$ is itself not Poisson distributed, its variance is just the *sum* of the variances of the Poisson-distributed variables m and n. This interesting result is a special case of the following simple general result: If a and b are arbitrary real constants, and μ and ν are the mean values of Poisson-distributed random variables m and n, then

$$\text{Var}\,(am + bn) = a^2\,\mu + b^2\,\nu. \tag{19.52}$$

★ **14.** Establish Eqs. (19.51) and (19.52).

★ **15.** The Skellam distribution is, in a sense, a generalization (to the full set of integers) of the Poisson distribution. You can see this by letting $\nu \to 0$ in Eq. (19.47) while retaining μ as a positive constant. Show that the distribution $P_{\text{diff}}(r)$ tends, in this limit, to the Poisson distribution $e^{-\mu}\,\mu^r/r!$, where $r = 0, 1, 2, \ldots$.

In Chap. 21, Sect. 21.5.3, you will see how the Skellam distribution reappears as the solution to the problem of a biased random walk on a linear lattice in continuous time.

19.3 The Negative Binomial Distribution

We have seen that the number distribution of photons of a given wave vector and state of polarization in coherent radiation is a Poisson distribution, while that in thermal radiation is a geometric distribution. The natural question to ask is does there exist a *family* of distributions, of which both the distributions above are members?

The **negative binomial distribution** provides an answer to this question. This is a discrete distribution in which the random variable n takes the integer values 0, 1, ... ad infinitum, with probability

$$\boxed{P_n = \binom{N+n-1}{n} p^N q^n,} \tag{19.53}$$

where $0 < p < 1$, $q = 1 - p$, and the parameter N is a fixed positive integer. This distribution is therefore characterized by two independent parameters, p and N. It is immediately evident that P_n reduces to the geometric distribution, Eq. (19.14), in the case $N = 1$. The generating function of the negative binomial distribution is

$$f(z) = \sum_{n=0}^{\infty} P_n z^n = p^N \sum_{n=0}^{\infty} \frac{(N+n-1)!}{(N-1)!\, n!} (qz)^n. \tag{19.54}$$

The series above can be summed by elementary methods, once it is recognized as just a binomial series! The result is

$$\boxed{f(z) = \frac{p^N}{(1-qz)^N}.} \tag{19.55}$$

This expression tells you why the distribution is called the negative binomial distribution. From the generating function, it follows easily that the mean, variance, and relative fluctuation of n are given, respectively, by

$$\langle n \rangle = \frac{Nq}{p}, \quad \text{Var}\,(n) = \frac{Nq}{p^2}, \quad \frac{\Delta n}{\langle n \rangle} = \frac{1}{\sqrt{(Nq)}}. \tag{19.56}$$

★ **16.** Verify that the infinite series on the right-hand side of Eq. (19.54) is precisely the binomial expansion of the function $(1 - qz)^{-N}$, so that Eq. (19.55) follows. Use the latter to check out the expressions given in Eqs. (19.56).

***Generalization of the negative binomial distribution*:** There exists a more general form of the negative binomial distribution. Here the parameter N is allowed to be *any* positive number, not necessarily an integer. The binomial coefficient in the definition of the distribution must then be expressed in terms of gamma functions, i.e.,

$$P_n = \binom{N+n-1}{n} p^N q^n = \frac{\Gamma(N+n)}{\Gamma(N)\, n!} p^N q^n. \tag{19.57}$$

This form enables us to see how the Poisson distribution arises as a limiting case of the negative binomial distribution, as follows.

★ **17.** Denote the mean value $\langle n \rangle$ by μ, so that $\mu = N(1-p)/p$. Hence $p = N/(N+\mu)$ and $q = \mu/(N+\mu)$. Substitute these expressions in Eq. (19.57), and pass to the limit $N \to \infty$. Use Stirling's approximation for the gamma functions, to arrive at the result

$$\lim_{N\to\infty} P_n = e^{-\mu} \frac{\mu^n}{n!},$$

the Poisson distribution with mean value μ.

- Thus, the negative binomial distribution with parameter N reduces to
  - the geometric distribution for $N = 1$;
  - the Poisson distribution as $N \to \infty$ keeping $\langle n \rangle$ $(= \mu)$ finite.

As I have already mentioned, the negative binomial distribution is used, among other applications, in photon-counting statistics to characterize admixtures of thermal and coherent radiation. Recall that, for a given mode, the photon number distribution is geometric in thermal radiation and Poisson in coherent radiation. The family of negative binomial distributions *interpolates* between these extreme cases.

19.4 The Simple Random Walk

Random walks of various kinds are paradigms for a huge number of physical processes and phenomena in just about every subject one can think of. It is therefore important for you to be familiar with this topic, at least at a rudimentary level. You have already come across some specific examples random walks and Markov chains (to which random walks in discrete time are closely related), in Chap. 12, Sects. 12.3.9, 12.3.10, and 12.5.2. We will now discuss the simplest random walk, and return to the topic several times in the sequel: in Chap. 20, Sect. 20.4.1; Chap. 21, Sect. 21.5.2; and Chap. 30, Sect. 30.2.1.

19.4.1 Random Walk on a Linear Lattice

Consider an infinite one-dimensional lattice of sites, labeled by the integer $j \in \mathbb{Z}$. Any of the sites can be taken to be the origin $(j = 0)$, because the lattice is of infinite extent both to the left and to the right. The random walker starts at time 0 from $j = 0$. Time is measured in integer multiples of a time step τ. After each time step, the walker takes a single-step to the neighboring site to her right or left, with respective probabilities α and $\beta = 1 - \alpha$ (see Fig. 19.3). If $\alpha = \beta = \frac{1}{2}$, the random walk is *unbiased*; if $\alpha \neq \beta$, it is a *biased* random walk. The primary question is What is the probability $P(j, n)$ that the walker will be at any arbitrary site j at the end of n time steps, i.e., at time $n\tau$?

Note that it is the site index j that is the random variable here, for a given number n of time steps. What is the sample space of j? It is obvious that $P(j, n)$ must vanish

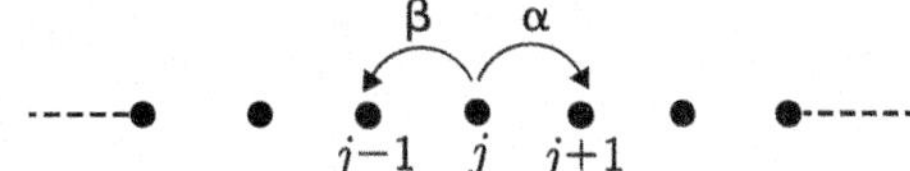

Fig. 19.3 A biased random walk on a linear lattice

identically for $|j| > n$, because the walker cannot reach these points in n time steps. A moment's thought shows that j can only be an even integer if n is even, and it must be an odd integer if n is odd. Therefore the sample space is the set of $(n+1)$ integers $\{-n, \ -(n-2), \ \ldots, \ (n-2), \ n\}$. Moreover, in order to reach the site j at time $n\tau$, the walker must be either at the site $(j-1)$ or at the site $(j+1)$ at time $(n-1)\tau$, and then jump to j with probabilities α and β, respectively, at the final time step. As *successive steps are independent of each other*, we have the following evolution equation for the time-dependent probability $P(j, n)$:

$$P(j, n) = \alpha \, P(j-1, n-1) + \beta \, P(j+1, n-1). \tag{19.58}$$

Since time is a discrete variable in this problem, we have a **difference equation**, rather than the more common differential equations that one usually encounters in dynamics. Moreover, it is a difference equation in *two* independent variables, the spatial site variable j and the discrete time variable n. We seek the solution of Eq. (19.58) subject to the initial condition $P(j, 0) = \delta_{j,0}$.

Observe that the probability distribution at time $n\tau$ is only dependent on the distribution at the *immediately previous* instant of time $(n-1)\tau$, and not on the distribution at any *earlier* instants of time. This is very important. It implies that the discrete time random walk we are studying here is an example of a **Markov chain**. We shall discuss continuous-time Markov processes in Chap. 21, and return to the continuous-time version of the random walk in Sect. 21.5.2. When *both* the spatial variable and the time variable are continuous, the biased random walk under discussion goes over into the process of diffusion in the presence of a drift. This process will be studied in Chap. 30, Sect. 30.2.1.

Returning to Eq. (19.58), the most convenient way to solve such difference equations is via the use of a generating function. In the present instance, however, we can write down the solution with the help of a simple argument. As already noted, n and j must have the same parity, i.e., $(n \pm j)$ must be an even integer. Now suppose j is positive, so that it lies to the right of the origin. To reach j, the walker must take at least j steps to the right. The remaining steps, $(n-j)$ of them, must be steps to the right and left that cancel each other out exactly. Hence the total number of steps to the right must be $j + \frac{1}{2}(n-j) = \frac{1}{2}(n+j)$, while the total number of steps to the left must be $\frac{1}{2}(n-j)$. The probability of any particular contributing sequence of steps is therefore $\alpha^{(n+j)/2} \, \beta^{(n-j)/2}$. But the *order* in which the steps are taken does not matter. Hence the probability above must be multiplied by the number of ways in which we can choose $\frac{1}{2}(n+j)$ steps to the right (or $\frac{1}{2}(n-j)$ steps to the left), out of the total of n steps. Putting all this together, the final result is

$$P(j,n) = \begin{cases} \binom{n}{\frac{1}{2}(n+j)} \alpha^{(n+j)/2}\,\beta^{(n-j)/2}, & \text{if } |j| \le n \text{ and } (n-j) \text{ is even} \\ 0 & \text{otherwise.} \end{cases} \tag{19.59}$$

It is left to the reader to check that exactly the same answer is obtained if j is negative.

The expression in (19.59) is the binomial distribution once again, in a (very slightly) disguised form. Let us set $\frac{1}{2}(n+j) = k$ for a moment. Then, summing j over the set of values $\{-n, -(n-2), \ldots, (n-2), n\}$ is precisely the same as summing k over the set of values $\{0, 1, \ldots, (n-1), n\}$. The distribution over j in Eq. (19.59) becomes, in terms of k, the distribution ${}^nC_k\, \alpha^k\, \beta^{n-k}$, which is precisely the binomial distribution.

The generating function for $P(j,n)$ is given by

$$f(z) = \sum_{\substack{j=-n \\ (n-j)\text{ even}}}^{n} P(j,n)\, z^j = \sum_{\substack{j=-n \\ (n-j)\text{ even}}}^{n} \binom{n}{\frac{1}{2}(n+j)} (\alpha z)^{(n+j)/2}\,(\beta z^{-1})^{(n-j)/2}$$
$$= \left(\alpha z + \beta z^{-1}\right)^n. \tag{19.60}$$

It is now trivial to check that the probability $P(j,n)$ is correctly normalized to unity, because $f(1) = 1$. The mean displacement of the walker in an n-step random walk is given by

$$\langle j(n)\rangle = n(\alpha - \beta). \tag{19.61}$$

This means that a biased ($\alpha \neq \beta$) random walk leads to a uniform **drift velocity** that is just $(\alpha - \beta)$ in the units we have chosen. The variance of the displacement is given by

$$\langle j^2(n)\rangle - \langle j(n)\rangle^2 = 4n\alpha\beta. \tag{19.62}$$

The most important feature of the random walk now emerges:

- The linear ($\sim n^1$) growth in time of the variance of the displacement is the defining property of the phenomenon of **diffusion**, as you will see in the sequel.
- In the case of an unbiased random walk ($\alpha = \beta = \frac{1}{2}$), the mean displacement vanishes, as expected, while its variance becomes exactly equal to n.

★ **18.** Use the generating function in (19.60) to derive Eqs. (19.61) and (19.62).

As I have mentioned earlier, the continuum limit of the random walk will be considered in Chap. 30, Sect. 30.2.1. We shall see how, when the simultaneous limits $a \to 0$, $\tau \to 0$ and $(\alpha - \beta) \to 0$ are taken appropriately, the difference Eq. (19.58) for the probability $P(j,n)$ goes over into the diffusion equation for the positional probability density function of a particle diffusing on a line in the presence of a constant applied force.

19.4.2 Some Generalizations of the Simple Random Walk

This is a good place to mention a few of the generalizations of the simple random walk we have just discussed, in view of their importance.

***Lattices in higher dimensions*:** The generalization of the random walk on a linear lattice to regular lattices in higher dimensions $d \geq 2$ is straightforward. The linear growth with time of the variance of the displacement remains a characteristic feature, *independent of the dimensionality* of the lattice.

***Continuous time*:** We have considered a random walk in discrete time. The time variable can be made continuous by letting the time step $\tau \to 0$ and $n \to \infty$, such that $\lim\, n\tau = t$. The difference Eq. (19.58) will then become a differential equation in time. I will consider this case in Chap. 21, Sect. 21.5.2.

***Continuous medium*:** When the random walk in continuous time occurs in a continuous medium rather than a lattice, the spatial variable also becomes continuous. The random walk goes over into the diffusion process in the limit of zero step size. In the absence of a bias, the positional probability density function becomes a Gaussian in the spatial coordinates. We will see how this happens in Chap. 20, Sect. 20.4.1.

***The remarkable generality of the basic result*:** In all these cases, there is a powerful result from statistics at work, namely, the **Central Limit Theorem**. We will discuss this theorem briefly in Chap. 20, Sect. 20.3.2. It ensures that

- the linear increase with time of the variance of the displacement persists under very general circumstances.

These circumstances include

- discrete or continuous time;
- discrete or continuous spatial variables;
- any spatial dimensionality;
- arbitrary probability distributions (in space as well as in time) of the individual steps of the random walk or flight, as long as these distributions have finite mean values and variances.

19.5 Solutions

1. (a) $P_s = (s-1)/36$ for $2 \leq s \leq 7$, and $P_s = (13-s)/36$ for $8 \leq s \leq 12$. It is obvious that $\sum_s P_s = 1$. The most probable value of s is 7, with a probability $P_7 = \frac{1}{6}$.
(b) $\langle s \rangle = 7,\ \ \Delta s = \sqrt{35/6} \simeq 2.415,\ \ \Delta s/\langle s \rangle \simeq 0.345$.

Remark It is instructive to recall the language of statistical mechanics. Each possible configuration of the two dice is an **accessible microstate** of the system comprising the two dice. Each die can yield a score of 1, 2, 3, 4, 5, or 6. Hence there are $6 \times 6 = 36$

possible microstates, each with an a priori probability of $\frac{1}{36}$. This should remind you of the postulate of equal a priori probabilities in the microcanonical ensemble! The total score s may be taken to label the possible **macrostates** of the system. There are 11 possible macrostates, corresponding to $s = 2, 3, \ldots, 12$. The most probable macrostate corresponds to $s = 7$, *because it has the largest number of accessible microstates* (= 6). This is why $P_7 = 6 \times \frac{1}{36} = \frac{1}{6}$. ▶

2. (a) As the particles are identical bosons, any state of the system must be *symmetric* under the exchange of the two particles. The state $|\, j\, j\, \rangle$ is symmetric under this exchange, but the state $|\, j\, k\, \rangle$ is not, when $j \neq k$. The symmetrized and normalized version of this state is the entangled state $(|\, j\, k\, \rangle + |\, k\, j\, \rangle)/\sqrt{2}$. Hence the energy eigenvalues and normalized eigenstates of the system are as follows:

$$
\begin{aligned}
\varepsilon &= 2 : |\,1\,1\,\rangle \\
\varepsilon &= 3 : (|\,1\,2\,\rangle + |\,2\,1\,\rangle)/\sqrt{2} \\
\varepsilon &= 4 : (|\,1\,3\,\rangle + |\,3\,1\,\rangle)/\sqrt{2}, \ |\,2\,2\,\rangle \\
\varepsilon &= 5 : (|\,1\,4\,\rangle + |\,4\,1\,\rangle)/\sqrt{2}, \ (|\,2\,3\,\rangle + |\,3\,2\,\rangle)/\sqrt{2} \\
\varepsilon &= 6 : (|\,1\,5\,\rangle + |\,5\,1\,\rangle)/\sqrt{2}, \ (|\,2\,4\,\rangle + |\,4\,2\,\rangle)/\sqrt{2}, \ |\,3\,3\,\rangle
\end{aligned}
$$

$$
\begin{aligned}
\varepsilon &= 7 : \ (|\,1\,6\,\rangle + |\,6\,1\,\rangle)/\sqrt{2}, \ (|\,2\,5\,\rangle + |\,5\,2\,\rangle)/\sqrt{2}, \ (|\,3\,4\,\rangle + |\,4\,3\,\rangle)/\sqrt{2} \\
\varepsilon &= 8 : \ (|\,2\,6\,\rangle + |\,6\,2\,\rangle)/\sqrt{2}, \ (|\,3\,5\,\rangle + |\,5\,3\,\rangle)/\sqrt{2}, \ |\,4\,4\,\rangle \\
\varepsilon &= 9 : \ (|\,3\,6\,\rangle + |\,6\,3\,\rangle)/\sqrt{2}, \ (|\,4\,5\,\rangle + |\,5\,4\,\rangle)/\sqrt{2} \\
\varepsilon &= 10 : (|\,4\,6\,\rangle + |\,6\,4\,\rangle)/\sqrt{2}, \ |\,5\,5\,\rangle \\
\varepsilon &= 11 : (|\,5\,6\,\rangle + |\,6\,5\,\rangle)/\sqrt{2} \\
\varepsilon &= 12 : |\,6\,6\,\rangle
\end{aligned}
$$

(b) There are 21 possible energy eigenstates. Since we are given that each of these is equally probable, and there is only one state corresponding to $\varepsilon = 3$, it follows that $P_3 = \frac{1}{21}$. Compare this with the probability $P_3 = \frac{1}{18}$ for a pair of dice.
(c) There are 3 energy eigenstates corresponding to the eigenvalue $\varepsilon = 7$. Hence the probability that the energy of the system has this value is $P_7 = \frac{3}{21} = \frac{1}{7}$. Note, however, that 7 is not the *unique* most probable value of ε in this case, since the probabilities P_6 and P_8 are also equal to $\frac{1}{7}$. ▶

3. (a) As the particles are identical fermions, any state of the system must be *antisymmetric* under the exchange of the two particles. The state $|\, j\, j\, \rangle$ is not possible, as cannot be made antisymmetric. (This is the **Pauli Exclusion Principle**, which says that two identical fermions cannot be in the same state!) The antisymmetrized form of the state $|\, j\, k\, \rangle$ (where $j \neq k$) is the entangled state $(|\, j\, k\, \rangle - |\, k\, j\, \rangle)/\sqrt{2}$. Hence the energy eigenvalues and normalized eigenstates of the system are as follows:

$$\varepsilon = 3: \ (|1\,2\rangle - |2\,1\rangle)/\sqrt{2}$$
$$\varepsilon = 4: \ (|1\,3\rangle - |3\,1\rangle)/\sqrt{2}$$
$$\varepsilon = 5: \ (|1\,4\rangle - |4\,1\rangle)/\sqrt{2}, \ (|2\,3\rangle - |3\,2\rangle)/\sqrt{2}$$
$$\varepsilon = 6: \ (|1\,5\rangle - |5\,1\rangle)/\sqrt{2}, \ (|2\,4\rangle - |4\,2\rangle)/\sqrt{2}$$
$$\varepsilon = 7: \ (|1\,6\rangle - |6\,1\rangle)/\sqrt{2}, \ (|2\,5\rangle - |5\,2\rangle)/\sqrt{2}, \ (|3\,4\rangle - |4\,3\rangle)/\sqrt{2}$$
$$\varepsilon = 8: \ (|2\,6\rangle - |6\,2\rangle)/\sqrt{2}, \ (|3\,5\rangle - |5\,3\rangle)/\sqrt{2}$$
$$\varepsilon = 9: \ (|3\,6\rangle - |6\,3\rangle)/\sqrt{2}, \ (|4\,5\rangle - |5\,4\rangle)/\sqrt{2}$$
$$\varepsilon = 10: (|4\,6\rangle - |6\,4\rangle)/\sqrt{2}$$
$$\varepsilon = 11: (|5\,6\rangle - |6\,5\rangle)/\sqrt{2}$$

Note that the eigenvalues $\varepsilon = 2$ and $\varepsilon = 12$ are no longer allowed, as there are no antisymmetrized states corresponding to these.
(b) There are now 15 possible energy eigenstates. As they are all equally probable, we have $P_3 = \frac{1}{15}$.
(c) As in the classical case, we see that 7 is the unique most probable value of ε. It now occurs with a probability $P_7 = \frac{3}{15} = \frac{1}{5}$.

Remark You will find it instructive to tabulate the probability distribution P_s as a function of s, and P_ε as a function of ε in the bosonic and fermionic cases, and compare the results. In particular, observe that $P_7(\text{bosonic}) = \frac{1}{7}$, $P_7(\text{classical}) = \frac{1}{6}$, $P_7(\text{fermionic}) = \frac{1}{5}$. ▶

5. To find $\overline{n}$, the most probable value of n, impose the requirement that that $P_{\overline{n}}$ be larger than both $P_{\overline{n}-1}$ and $P_{\overline{n}+1}$. ▶

9. Let $|\,z\,\rangle = \sum_{n=0}^{\infty} c_n\,|\,n\,\rangle$. The eigenvalue equation $a\,|\,z\,\rangle = z\,|\,z\,\rangle$ and the relations $a\,|\,0\,\rangle = 0, \ a\,|\,n\,\rangle = \sqrt{n}\,|\,n-1\,\rangle$ give

$$\sqrt{1}\,c_1\,|0\rangle + \sqrt{2}\,c_2\,|1\rangle + \sqrt{3}\,c_3\,|2\rangle + \cdots = z\,c_0\,|0\rangle + z\,c_1\,|1\rangle + z\,c_2\,|2\rangle + \cdots.$$

Equating the respective coefficients of $|\,0\,\rangle, \ |\,1\rangle, \ldots$ (because they form a complete orthonormal set of states) yields the constants c_n recursively, in terms of c_0. The normalization condition $\langle\,z|\,z\,\rangle = 1$ determines $|c_0|$. Taking the phase of c_0 to be zero gives $c_0 = e^{-|z|^2/2}$. ▶

10. To find the moments $\langle N\rangle$ and $\langle N^2\rangle$, use the fact that $f\,'(1) = \langle k\rangle$ and $f\,''(1) = \langle k(k-1)\rangle$. ▶

11. This problem is clearly the discrete analog of a shift and interchange of the order of integration in a double integral. Since $m + n$ must always be equal to s, you can eliminate the sum over n by replacing n by $s - m$. But you must be careful about the limits of the subsequent summation over m. Figure 19.4 shows the lattice points corresponding to non-negative integer values of m and n in the $(m\,,\,n)$ plane. For

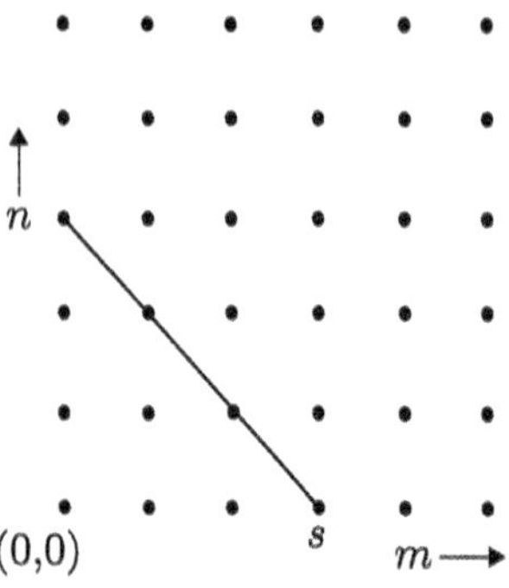

Fig. 19.4 The evaluation of the double sum in Eq. (19.37). The diagonal line runs through the finite set of vales of (m, n) selected by the Kronecker delta for each given value of $s (= 4$ in this figure)

any fixed integer value of s, note the set of points corresponding to the constraint $m + n = s$. Only these points must be summed over in each case. Since n only runs over the integers from 0 upward, the sum over m must be restricted to values $\leq s$, for a given s. Thus

$$P_{\text{sum}}(s) = \sum_{m=0}^{\infty} \sum_{n=0}^{\infty} P_1(m)\, P_2(n)\, \delta_{m+n\,,\,s} = \sum_{m=0}^{s} P_1(m)\, P_2(s - m).$$

Insert the expressions in Eqs. (19.36) for P_1 and P_2 and simplify, to get

$$P_{\text{sum}}(s) = e^{-(\mu+\nu)}\, \frac{\nu^s}{s!} \left(1 + \frac{\mu}{\nu}\right)^s = e^{-(\mu+\nu)}\, \frac{(\mu + \nu)^s}{s!}\,.$$

This is again a Poisson distribution.

Remark As you might have guessed, this result is most easily deduced with the help of the generating function. Let $f_1(z)$, $f_2(z)$, and $f_{\text{sum}}(z)$ be the generating functions corresponding to the random variables m, n, and $s = m + n$, respectively. Multiply both sides of Eq. (19.37) by z^s and sum over all nonnegative integer values of s. Use the Kronecker delta $\delta_{m+n\,,\,s}$ to eliminate the sum over s on the right-hand side, replacing the factor z^s by $z^m\, z^n$. This immediately yields the relation

$$f_{\text{sum}}(z) = f_1(z)\, f_2(z) = e^{(\mu+\nu)(z-1)}.$$

Expanding the exponential and picking out the coefficient of z^s yields $P_{\text{sum}}(s)$ at once. ▶

14. Equation (19.52) can be proved directly from the definition of the variance! Note that, since m and n are *independent* random variables, the average of the product $\langle mn \rangle$ reduces to the product of the individual averages. That is, $\langle mn \rangle = \langle m \rangle\, \langle n \rangle = \mu\, \nu$. Equation (19.51) is just a special case of (19.52), with $a = 1$ and $b = -1$. ▶

15. As $\nu \to 0$, the argument of the modified Bessel function in Eq. (19.47) tends to zero. The leading behavior of $I_l(z)$ as $z \to 0$, for a nonnegative integer value of l, is easily found from Eq. (19.45). It is just the $m = 0$ term in the series on the right-hand side. Use this in Eq. (19.47). ▶

18. Recall that $[df/dz]_{z=1} = \langle j \rangle$ and $[d^2 f/dz^2]_{z=1} = \langle j(j-1) \rangle$. ▶

Chapter 20
Continuous Probability Distributions

20.1 Continuous Random Variables

20.1.1 Probability Density and Cumulative Distribution

We now consider (real) random variables whose sample space is a continuum. The probability that a continuous random variable takes any particular, precisely specified, value in a continuous set is actually zero, in general.[1] The reason is that a single point, specified to infinite precision, is a set of measure zero in a continuum. For *continuous* random variables, therefore, we must speak of a **probability density function** (which I will frequently abbreviate as PDF). It is useful to distinguish between a continuous random variable per se and the *values* it can take, by using different (but related) symbols for the two—e.g., upper case and lower case letters such as X and x, respectively. Wherever it is preferable to do so, I will make this distinction. Suppose the sample space of a random variable X is $(-\infty, \infty)$, and $p(x)$ is its PDF. Then:

(i) $p(x)\,dx$ is the probability that the random variable X has a value in an infinitesimal interval dx at the value x, i.e., in the range $(x,\; x+dx)$.
(ii) $p(x) \geq 0$ for all x.
(iii) $\int_{-\infty}^{\infty} dx\; p(x) = 1$.

Note that $p(x)$ itself does not have to be less than unity. In fact, it can even become unbounded at a point or points lying in the domain of x. But it must be integrable, because of condition (iii) above.

The **cumulative distribution function** (CDF) $F(x)$, also called simply the distribution function, is the total probability that the random variable X is less than, or equal to, any given value x. Thus, in the case of a sample space $(-\infty, \infty)$,

[1] This is true unless there is a "δ-function contribution" at that point. See the comments below.

V. Balakrishnan, *Mathematical Physics*,
https://doi.org/10.1007/978-3-030-39680-0_20

$$\boxed{F(x) \stackrel{\text{def.}}{=} \Pr(X \le x) = \int_{-\infty}^{x} dx'\, p(x').} \tag{20.1}$$

$F(x)$ is a *non-decreasing*, nonnegative function of x that satisfies

$$F(-\infty) = 0, \quad F(\infty) = 1. \tag{20.2}$$

It is obvious that

$$\boxed{\frac{dF}{dx} \equiv p(x).} \tag{20.3}$$

There is a technical point worth noting here. The CDF is a more general concept than the PDF. All probability distributions need not necessarily have well-behaved functions as corresponding probability densities. For instance, there may be some specific value of the random variable, say x_0, with a nonzero probability α, say, of its occurrence. In the heuristic approach we have taken, we would simply say that the PDF $p(x)$ has a "spike" of the form $\alpha\,\delta(x - x_0)$. In a more rigorous treatment, care must be exercised in handling such singularities. A δ-function spike in $p(x)$ leads to a *finite* jump or step in $F(x)$. This is one reason why it is found to be more convenient, mathematically, to deal with the CDF rather than the PDF.

As mentioned above, the CDF is frequently called the distribution function. In physical applications, one often goes a step further and refers to the PDF itself as the "probability distribution" or just the "distribution". While this terminology is loose, it is very common in the physics literature. (I shall also use it quite frequently.) No confusion should arise, as matters will be quite clear from the context. Note that

- $p(x)$ has the physical dimensions of x^{-1}, while $F(x)$ is dimensionless.
- $F(x)$ can never exceed unity. (But there is no such restriction on $p(x)$, of course.)

20.1.2 The Moment-Generating Function

What follows below is applicable both to continuous random variables as well as discrete ones, with integration replaced by summation wherever appropriate. You will find it instructive to supply the appropriate modifications to the formulas that follow. The important point is that

- probability distributions themselves are not measurable or "physical" quantities. All that we can measure directly, or find, are various averages.

For a single random variable, this is the set of its **moments** $\langle X^r \rangle$, $r = 1, 2, \ldots$. It is very useful, in this context, to introduce the **moment-generating function** $M(u)$, defined as[2]

$$\boxed{M(u) \stackrel{\text{def.}}{=} \langle e^{uX} \rangle = 1 + \sum_{r=1}^{\infty} \frac{\langle X^r \rangle}{r!}\, u^r.} \tag{20.4}$$

But the series above has precisely the form of a Taylor series. It follows that $\langle X^r \rangle$ is just the rth derivative of $M(u)$ with respect to u, evaluated at $u = 0$:

$$\boxed{\langle X^r \rangle = \left[\frac{d^r M(u)}{du^r} \right]_{u=0}.} \tag{20.5}$$

In particular, the first moment or mean value $\langle X \rangle \equiv \mu$ is the coefficient of u in the power series expansion of $M(u)$ in powers of u.

More useful than the moments are the **central moments** of the random variable. These are the moments of the *deviation* of X from its mean value, namely, the moments of $\delta X \equiv X - \mu$:

$$\langle (\delta X)^r \rangle = \langle (X - \mu)^r \rangle. \tag{20.6}$$

The first of these is zero, by definition. The second is the variance,

$$\sigma^2 \equiv \text{Var}\,(X) \stackrel{\text{def.}}{=} \langle (X - \mu)^2 \rangle = \langle X^2 \rangle - \mu^2. \tag{20.7}$$

As you know already, the variance of a random variable is always positive, because it is the average value of a squared quantity. (Recall that the variance vanishes if and only if the variable is deterministic.) It provides the leading measure of the *scatter* of a random variable about its mean value.

In Chap. 19, we defined and used the generating function $f(z)$ of a distribution. In the present notation, this is just

$$f(z) = \int dx\, p(x)\, z^x = \langle z^X \rangle. \tag{20.8}$$

Hence the generating function f and the moment-generating function M are related to each other according to

$$\boxed{M(u) = f(e^u),} \tag{20.9}$$

a very useful relation.

---

[2]A minor point: I will use the symbol r for a general integer when discussing the moments of random variables, rather than n, simply because I have already used n to denote an integer-valued random variable. This is to help avoid any confusion.

20.1.3 The Cumulant-Generating Function

The idea behind the variance of a random variable is extended to its higher moments by means of the **cumulants** κ_r of a probability distribution. These quantities arise in a natural manner as follows. The moment-generating function $M(u)$ is the average value of the exponential e^{uX}. The idea is to write this average value as the exponential of a function of u. The latter is, in general, a power series in u, whose coefficients are proportional to the cumulants of the random variable X. We have

$$\boxed{M(u) = \langle e^{uX}\rangle \equiv e^{K(u)}, \quad \text{where } K(u) \stackrel{\text{def.}}{=} \sum_{r=1}^{\infty} \frac{\kappa_r}{r!}\, u^r.} \tag{20.10}$$

The quantities κ_r are called the cumulants of X, and $K(u)$ is the **cumulant-generating function**. Since $M(0) = 1$ for a normalized probability distribution, $K(0) = 0$. Therefore $K(u)$ starts with the first-order term in u. The cumulant-generating function is related to the moment-generating function by

$$\boxed{K(u) = \ln M(u).} \tag{20.11}$$

The rth cumulant is then found from the inversion formula

$$\kappa_r = \left[\frac{d^r K(u)}{du^r}\right]_{u=0}. \tag{20.12}$$

The cumulants can therefore be expressed in terms of the moments, and *vice versa*. The leading term in κ_r is $\langle X^r\rangle$, followed by terms involving the lower moments of X. The first cumulant is just the mean. The second and third cumulants are the corresponding central moments. Thus

$$\left.\begin{aligned} \kappa_1 &= \mu = \langle X\rangle, \\ \kappa_2 &= \sigma^2 = \langle(\delta X)^2\rangle = \langle X^2\rangle - \langle X\rangle^2, \\ \kappa_3 &= \langle(\delta X)^3\rangle = \langle X^3\rangle - 3\langle X^2\rangle\,\langle X\rangle + 2\langle X\rangle^3\,. \end{aligned}\right\} \tag{20.13}$$

The fourth cumulant is given by

$$\left.\begin{aligned} \kappa_4 &= \langle(\delta X)^4\rangle - 3\langle(\delta X)^2\rangle^2 \\ &= \langle X^4\rangle - 4\langle X^3\rangle\,\langle X\rangle - 3\langle X^2\rangle^2 + 12\langle X^2\rangle\,\langle X\rangle^2 - 6\langle X\rangle^4\,. \end{aligned}\right\} \tag{20.14}$$

A central problem in mathematical statistics is the so-called **problem of moments**: Given all the moments (and hence all the cumulants) of a distribution, can the distribution be reconstructed uniquely? I will not digress into this formal question here. In practice, however, the first four cumulants of a distribution provide a reasonable approximate description of the salient properties of the probability distribution of the random variable concerned.

The cumulant κ_r is a **homogeneous function** of order r in the following sense: if the random variable X is multiplied by a constant c, the rth cumulant of cX is c^r times the corresponding cumulant of X. Observe that each κ_r starts with the rth moment $\langle X^r \rangle$, but involves subtracting away contributions from the products of lower order moments. This should remind you of the way in which we obtained spherical tensors from Cartesian tensors, to get irreducible sets of quantities under rotations, in Sect. 5.3.1 of Chap. 5. Something similar is involved here, too:

- For $r \geq 2$, the cumulants κ_r are invariant under *translations* of the random variable, i.e., under a change of variables $X \mapsto X + a$ where a is a constant.

More than the moments themselves, or even the central moments, the different cumulants help characterize *distinct* properties of a distribution. In general terms:

- The second cumulant κ_2 (i.e., the variance) is a measure of the spread or dispersion of the random variable about its mean value.
- The third cumulant κ_3 measures the asymmetry or **skewness** of the distribution about the mean value.
- The fourth cumulant κ_4 characterizes, for a symmetric distribution, the extent to which the distribution deviates from the normal or Gaussian distribution. We will return to this aspect in Sect. 20.2.2.

20.1.4 Application to the Discrete Distributions

Let us briefly re-visit the discrete distributions considered in Chap. 19, and consider their cumulants. Let $n,\ P_n$, and $f(z)$ denote the (discrete) random variable, its probability distribution, and its generating function, as before. The moment-generating and cumulant-generating functions are related to $f(z)$ according to

$$M(u) = \langle e^{un} \rangle = \sum_n P_n\, e^{un} = f(e^u), \quad K(u) = \ln\, f(e^u). \tag{20.15}$$

Consider the binomial, geometric, Poisson, Skellam, and negative binomial distributions (Eqs. (19.4), (19.15), (19.24), (19.47), and (19.53), respectively). Eliminate parameters wherever applicable in favor of μ, which represents the mean value (except in the case of the Skellam distribution). The cumulant-generating functions are then given by

$$K(u) = \begin{cases} (\mu/p)\ \ln\,(1 - p + p\, e^u) & \text{(binomial)} \\ -\ \ln\,(1 + \mu - \mu e^u) & \text{(geometric)} \\ \mu\,(e^u - 1) & \text{(Poisson)} \\ \mu\,(e^u - 1) - \nu(1 - e^{-u}) & \text{(Skellam)} \\ N\ \ln\, N - N\ \ln\,(N + \mu - \mu\, e^u) & \text{(negative binomial).} \end{cases} \tag{20.16}$$

★ **1.** The expressions given above for $K(u)$ are very useful in practice.

(a) Verify that the cumulant-generating functions in the five cases are as given in (20.16).
(b) Find the first four cumulants in each of the five cases.

Note, in particular, that the Poisson distribution enjoys the property $\kappa_r = \mu$ for all $r \geq 1$. Hence:

- *All* the cumulants of the Poisson distribution are equal to the mean value μ itself.

This is the generalization of the property already found, namely, that the variance is equal to the mean for Poisson distribution.

There is again a minor generalization of this property in the case of the Skellam distribution, as you might expect. In this case we have

$$\kappa_r = \mu + (-1)^r \, \nu. \tag{20.17}$$

Hence

- All the *odd* cumulants of the Skellam distribution are equal to the mean value $\kappa_1 = \mu - \nu$, while all the *even* ones are equal to the variance $\kappa_2 = \mu + \nu$.

★ **2.** Here is a simple example that shows how the cumulants κ_r $(r \geq 2)$ remain unchanged under a shift of the random variable by a constant. Let n be a Poisson-distributed random variable with mean value μ. Consider the random variable $X = an + b$, where a and b are arbitrary positive constants (not necessarily integers).

(a) Find the probability distribution (or the PDF) of X, and its moment-generating function.
(b) Hence find the cumulant-generating function of X, and show that its cumulants are given by $\kappa_1 = a\mu + b$ and $\kappa_r = a^r \, \mu$ for $r \geq 2$.

20.1.5 The Characteristic Function

We return to the case of a continuous random variable X with PDF $p(x)$. A most useful quantity associated with $p(x)$ is its Fourier transform $\widetilde{p}(k)$, defined as

$$\boxed{\widetilde{p}(k) \stackrel{\text{def.}}{=} \int_{-\infty}^{\infty} dx \, e^{-ikx} \, p(x).} \tag{20.18}$$

This definition is consistent with the Fourier transform convention we have chosen (see Eq. (18.4) of Chap. 18, Sect. 18.1.1). $\widetilde{p}(k)$ is called the **characteristic function** of the distribution. It is closely related to the moment-generating function of the random variable: $\widetilde{p}(k)$ can also be regarded as the expectation value of $\exp(-ikX)$, so that

$$\boxed{\widetilde{p}(k) = \langle e^{-ikX} \rangle = M(-ik).} \tag{20.19}$$

Thus $\langle X^r \rangle / r!$ is the coefficient of $(-ik)^r$ in the power series expansion of $\widetilde{p}(k)$.

- The characteristic function carries the same information about the random variable as does its PDF.

The significance and utility of the characteristic function will become clear shortly.

A function of $k \in (-\infty, \infty)$ has to have some special features in order to qualify as the characteristic function $\widetilde{p}(k)$ of a probability distribution:

(i) $\widetilde{p}(0)$ must be equal to unity, to ensure that the random variable X is a proper random variable, with a normalized probability distribution.
(ii) The inverse Fourier transform of $\widetilde{p}(k)$ must be a real, nonnegative function of x, in order to be an acceptable PDF $p(x)$.

The second of these requirements, in particular, places a very nontrivial restriction on the function $\widetilde{p}(k)$.

20.1.6 The Additivity of Cumulants

A crucial property of cumulants emerges when we consider *sums* of independent random variables. Let X_1 and X_2 be independent random variables, each with a sample space comprising all real numbers, and with PDFs $p_1(x_1)$ and $p_2(x_2)$, respectively. Let $X = X_1 + X_2$ be their sum. The PDF of X is given by the continuum analog of the formula we have already used in the discrete case, namely, Eq. (19.37) of Chap. 19, Sect. 19.2.4. In the present instance, because both x_1 and x_2 run over *all* real values from $-\infty$ to ∞, you can use the δ-function right away to carry out one of the integrations. Thus

$$\begin{aligned} p(x) &= \int_{-\infty}^{\infty} dx_1 \int_{-\infty}^{\infty} dx_2\, p_1(x_1)\, p_2(x_2)\, \delta\big(x - (x_1 + x_2)\big) \\ &= \int_{-\infty}^{\infty} dx_1\, p_1(x_1)\, p_2(x - x_1). \end{aligned} \tag{20.20}$$

But this is just a convolution of the PDFs of x_1 and x_2. Hence, by the convolution theorem for Fourier transforms, the Fourier transform of the function p is just the product of the Fourier transforms of the functions p_1 and p_2. In other words, the corresponding characteristic functions are related according to

$$\widetilde{p}(k) = \widetilde{p}_1(k)\, \widetilde{p}_2(k), \quad \text{or} \quad M(-ik) = M_1(-ik)\, M_2(-ik), \tag{20.21}$$

where M_1 and M_2 are the moment-generating functions of X_1 and X_2. Taking logarithms, it follows at once that the corresponding cumulant-generating functions simply add up:

$$K(-ik) = K_1(-ik) + K_2(-ik). \tag{20.22}$$

Consequently, we have the following:

- The rth cumulant of $X_1 + X_2$ is the sum of the rth cumulants of X_1 and X_2, for every positive integer r.
- It is obvious that the results above can be extended immediately to the sum (more generally, to any linear combination) of any number of independent random variables.
- Unlike the cumulants, none of the moments higher than the first moment (or mean value) has this property of additivity. *This is a major advantage that the cumulants of random variables enjoy over the corresponding moments.*

20.2 The Gaussian Distribution

20.2.1 The Normal Density and Distribution

The **Gaussian distribution**, also called the **normal distribution**, plays a central role in statistics. In applications, too, it appears everywhere. The distribution of a continuous random variable X that takes values in $(-\infty, \infty)$ is Gaussian if its normalized PDF is given by

$$\boxed{p(x) = \frac{1}{\sqrt{(2\pi\sigma^2)}} \exp\left\{-\frac{(x-\mu)^2}{2\sigma^2}\right\},} \tag{20.23}$$

where μ is any real number and σ is a positive number. The parameter μ is the mean value of X, σ^2 is its variance, and σ is its standard deviation:

$$\langle X \rangle = \mu, \quad \langle (X-\mu)^2 \rangle = \sigma^2. \tag{20.24}$$

In statistics, the notation $\mathcal{N}(\mu, \sigma^2)$ is often used to denote the normal distribution. $\mathcal{N}(0, 1)$ is the **standard normal distribution**, with zero mean and unit variance.

The PDF (20.23) is an example of a *symmetric, unimodal* density: it has a single peak or **mode**, which also happens to be located at $x = \mu$, and the plot of $p(x)$ is symmetric about this point. The variance σ^2 is a direct measure of the width of the Gaussian. The full-width-at-half-maximum (FWHM) is given by

$$\text{FWHM}\Big|_{\text{Gaussian}} = 2\,(2\ \ln\ 2)^{1/2}\sigma \simeq 2.355\,\sigma. \tag{20.25}$$

Figure 20.1 shows these features schematically.

***A physical example*:** Here is a physical instance in which the Gaussian distribution appears. Recall the number fluctuations in a gas, considered in Chap. 19, Sect. 19.1.3.

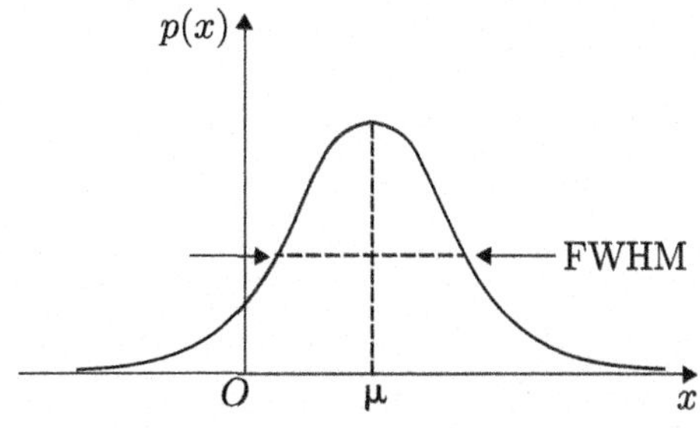

Fig. 20.1 The probability density function of a Gaussian distribution

We have seen in Sect. 19.2.1 how, in the thermodynamic limit, the binomial distribution goes over into the Poisson distribution. To recall Eq. (19.28), the probability that a volume v of the gas has n molecules is given by $P_n = e^{-\langle n \rangle}\, \langle n \rangle^n / n!$, where $\langle n \rangle = \rho v$ and ρ is the mean number density of the gas. At normal temperature and pressure, the mean number of molecules in a cubic meter of air is of the order of 10^{22}. This is so large compared to unity that we may regard the *deviation* $X = n - \langle n \rangle$ in the number of molecules around its mean value to be an essentially continuous variable. The distribution of X is then a Gaussian with zero mean, with a variance that remains equal to $\langle n \rangle$.

★ **3.** Establish the result just stated.

The cumulative distribution function of a Gaussian random variable X is given by

$$\Pr(X \le x) = F(x) = \frac{1}{\sqrt{2\pi\sigma^2}} \int_{-\infty}^{x} dx' \, \exp\left\{-\frac{(x' - \mu)^2}{2\sigma^2}\right\}. \tag{20.26}$$

We have $F(-\infty) = 0$ and $F(\infty) = 1$, as required. Further, $F(\mu) = \frac{1}{2}$. The CDF above can be expressed in terms of an error function. Recall the definition of this function from Eq. (3.4) of Chap. 3, Sect. 3.1.2. Changing the variable of integration in Eq. (20.26) from x' to $x' - \mu$, we get

$$\boxed{F(x) = \frac{1}{2}\left\{1 + \text{erf}\left(\frac{x - \mu}{\sqrt{2\sigma^2}}\right)\right\}.} \tag{20.27}$$

20.2.2 *Moments and Cumulants of a Gaussian Distribution*

Owing to the symmetry of the Gaussian PDF about the mean value, all the *odd* moments of the deviation from the mean value (that is, all the odd central moments) vanish identically:

$$\langle (X - \mu)^{2l+1} \rangle = \int_{-\infty}^{\infty} dx \, (x - \mu)^{2l+1} \, p(x) = 0, \quad l = 0, 1, \ldots . \tag{20.28}$$

On the other hand, all the *even* central moments of a Gaussian random variable are determined completely in terms of σ^2, i.e., in terms of the variance of the distribution. We find, for $l = 0, 1, \ldots,$

$$\langle (X-\mu)^{2l} \rangle = \int_{-\infty}^{\infty} dx\, (x-\mu)^{2l}\, p(x) = \frac{(2\sigma^2)^l}{\sqrt{\pi}}\, \Gamma(l+\tfrac{1}{2}). \tag{20.29}$$

Simplifying the gamma function, we get

$$\boxed{\langle (X-\mu)^{2l} \rangle = \frac{(2l)\,!}{2^l\, l!}\, \sigma^{2l} \quad (l = 0, 1, \ldots).} \tag{20.30}$$

Note, in particular, that

$$\langle (\delta X)^4 \rangle = 3\,\sigma^4 = 3\,\langle (\delta X)^2 \rangle^2. \tag{20.31}$$

Hence the fourth cumulant κ_4, defined in Eq. (20.14), is exactly zero for the Gaussian distribution. I will return to this point shortly.

★ **4.** Use the Gaussian integral in Eq. (3.16) of Chap. 3, Sect. 3.1.5 to verify the result (20.29). Simplify the gamma function to obtain Eq. (20.30).

***A combinatorial sidelight*:** The expression in Eq. (20.30) for the even moments of a Gaussian random variable can also be written (for $l = 1, 2, \ldots$) as

$$\boxed{\langle (X-\mu)^{2l} \rangle = \sigma^{2l}\,(2l-1)(2l-3)\cdots(3)(1) \equiv \sigma^{2l}\,(2l-1)!!} \tag{20.32}$$

This result has a combinatorial interpretation. The double factorial $(2l-1)!!$ is the number of distinct ways in which the $2l$ individual factors of the integrand (other than the Gaussian factor), $(x-\mu)^{2l} = (x-\mu)(x-\mu)\cdots(x-\mu)$, can be paired off, i.e., the number of *distinct* pairs that can be formed from them. This fact has far-reaching connections—for instance, in the so-called contractions involved in *Wick's Theorem* in quantum field theory.

Returning to the general form of the Gaussian pdf, the moment-generating function for the Gaussian distribution is

$$M(u) = \langle e^{u\,X} \rangle = \frac{1}{\sqrt{2\pi\sigma^2}} \int_{-\infty}^{\infty} dx\, \exp\left\{-\frac{(x-\mu)^2}{2\sigma^2} + u\,x\right\}. \tag{20.33}$$

Evaluating the integral, we get

$$M(u) = \exp\left(\mu\,u + \tfrac{1}{2}\sigma^2\,u^2\right). \tag{20.34}$$

★ **5.** Obtain Eq. (20.34) for $M(u)$ from Eq. (20.33), using the formula (2.4) of Chap. 2, Sect. 2.1 for a shifted Gaussian integral.

The characteristic function of the Gaussian distribution is therefore

$$\boxed{\widetilde{p}(k) = M(-ik) = \exp\left(-i\mu k - \tfrac{1}{2}\sigma^2 k^2\right).} \tag{20.35}$$

The corresponding cumulant-generating function is just

$$\boxed{K(u) = \ln\, M(u) = \mu u + \tfrac{1}{2}\sigma^2 u^2,} \tag{20.36}$$

a quadratic function of u. A fundamental property follows at once:

- *All the cumulants of a Gaussian distribution higher than the second cumulant vanish identically.*

In particular, the fourth cumulant κ_4, defined in Eq. (20.14), is identically equal to zero for a Gaussian, as shown explicitly in Eq.(20.31). This brings us to the significance of κ_4 for a general distribution that is symmetric about its mean value. The dimensionless ratio

$$\frac{\kappa_4}{\kappa_2^2} = \frac{\langle(\delta x)^4\rangle - 3\,\langle(\delta x)^2\rangle^2}{\langle(\delta x)^2\rangle^2} \tag{20.37}$$

is called the **excess of kurtosis**. This quantity is exactly zero for a Gaussian distribution. A positive excess of kurtosis (a **leptokurtic distribution**) implies that the fourth central moment dominates over the square of the second central moment. Hence larger values of $|x - \mu|$ contribute more significantly than they do in a Gaussian. In qualitative terms, this means that the PDF has a *leaner* peak than a Gaussian, and fatter "tails" on either side of the peak. On the other hand, a negative excess of kurtosis (a **platykurtic distribution**) implies that the weight of large values of $|x - \mu|$ is less than what it would be in the case of a Gaussian. Hence the peak is *broader* than that of a Gaussian, while the tails on either side of the peak are thinner.

***An interesting question*:** $K(u)$ is a quadratic function of u for a Gaussian distribution. Hence κ_r vanishes identically for all $r \geq 3$ in this case. It is natural to ask: Are there continuous distributions for which $K(u)$ is a cubic, or quartic, or some polynomial of *finite* degree $l \geq 3$, so that all cumulants with $r \geq l + 1$ would vanish identically? Interestingly, the answer is that *there are no such distributions.*

20.2.3 Simple Functions of a Gaussian Random Variable

An arbitrary *function* of a Gaussian random variable will, in general, have a PDF that looks quite different from a Gaussian. Here is a simple example. Let the random variable X have a distribution $\mathcal{N}(0, \sigma^2)$, so that its normalized PDF is

$$p(x) = \frac{1}{\sqrt{(2\pi\sigma^2)}}\, e^{-x^2/(2\sigma^2)}. \tag{20.38}$$

What is the normalized PDF $\rho(\xi)$ of the random variable X^2?

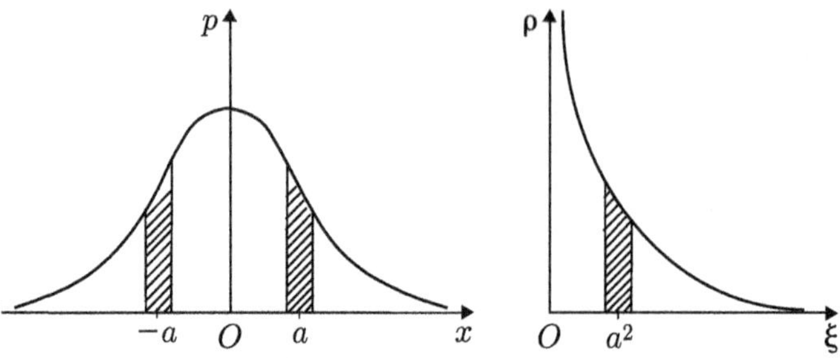

Fig. 20.2 The twofold contribution of the PDF $p(x)$ to the PDF $\rho(\xi)$

The sample space of X^2 is, of course, just the half-line $0 \le \xi < \infty$. If the mapping $x \mapsto \xi = x^2$ had been one-to-one, then we could have used the formal identity $p(x)\,|dx| = \rho(\xi)\,|d\xi|$ to find $\rho(\xi)$. This identity is simply the following statement: if x lies in the infinitesimal interval dx about any value a, then ξ lies in the infinitesimal interval $d\xi$ about the value a^2, and the corresponding probabilities are obviously equal to each other. But both the values $-a$ and $+a$ of x correspond to the same value a^2 of ξ. Hence there is a *two*-fold contribution to $\rho(\xi)$, as you can see from Fig. 20.2.

This gives an extra factor of 2, and we have

$$\rho(\xi) = 2\,p(\sqrt{\xi}\,)\left|\frac{dx}{d\xi}\right| = \frac{1}{\sqrt{(2\pi\sigma^2\,\xi)}}\,e^{-\xi/(2\sigma^2)}. \tag{20.39}$$

Alternatively, you could also use the formal expression

$$\rho(\xi) = \int_{-\infty}^{\infty} dx\, p(x)\,\delta(x^2 - \xi) = \int_{-\infty}^{\infty} dx\, p(x)\left\{\frac{\delta(x + \sqrt{\xi}) + \delta(x - \sqrt{\xi})}{2\sqrt{\xi}}\right\}, \tag{20.40}$$

to arrive at the same result as in (20.39). Note that $\rho(\xi)$ is properly normalized, since $\int_0^\infty d\xi\,\rho(\xi) = 1$.

***A physical example*:** The most well-known example of a Gaussian distribution in physics is, of course, the Maxwellian distribution of velocities in a classical ideal gas in thermal equilibrium at an absolute temperature T. (Recall Eq. (13.19) of Chap. 13, Sect. 13.2.3.) Each Cartesian component of the velocity of a molecule of mass m has a Gaussian PDF, with zero mean and a variance equal to $k_B T/m$. Thus

$$p^{\text{eq}}(v_x) = \left(\frac{m}{2\pi k_B T}\right)^{1/2} \exp\left\{-\frac{m v_x^2}{2k_B T}\right\}, \quad (-\infty < v_x < \infty) \tag{20.41}$$

and similarly for v_y and v_z. The PDF of the velocity $\mathbf{v} = (v_x,\ v_y\ v_z)$ itself is just the product of these PDFs. In a slight abuse of notation, let us continue to use the same symbol (p) for this joint PDF. Then

$$p^{\text{eq}}(\mathbf{v}) = \left(\frac{m}{2\pi k_B T}\right)^{3/2} \exp\left\{-\frac{m\mathbf{v}^2}{2k_B T}\right\}. \tag{20.42}$$

Correspondingly, the PDF $\rho^{\text{eq}}(v)$ of the speed v of a molecule is given by

$$\rho^{\text{eq}}(v) = \left(\frac{m}{2\pi k_B T}\right)^{3/2} 4\pi v^2 \exp\left\{-\frac{mv^2}{2k_B T}\right\}, \quad (0 \le v < \infty), \tag{20.43}$$

as stated in Eq. (13.19). It then follows that the PDF $\phi^{\text{eq}}(\varepsilon)$ of the (kinetic) energy ε of a molecule is given by

$$\phi^{\text{eq}}(\varepsilon) = \frac{2}{\sqrt{\pi}\,(k_B T)^{3/2}}\, \varepsilon^{1/2}\, e^{-\varepsilon/(k_B T)}, \quad (0 \le \varepsilon < \infty). \tag{20.44}$$

★ **6.** Starting with Eq. (20.41) for each Cartesian component of the velocity of a molecule, derive Eqs. (20.43) and (20.44). Sketch the functions $\rho^{\text{eq}}(v)$ and $\phi^{\text{eq}}(\varepsilon)$ schematically.

The PDF in Eq. (20.44) is an example of a **gamma distribution**. This is a two-parameter distribution of a random variable taking values in $[0, \infty)$, with a PDF proportional to $x^{a-1}\, e^{-x/b}$. The positive constants a and b are called the shape factor and the scale parameter, respectively. In the present example, $a = \frac{3}{2}$ and $b = k_B T$.

20.2.4 Mean Collision Rate in a Dilute Gas

This is a convenient place to digress for a moment into an interesting physical application of Gaussian integrals. In the kinetic theory of gases, one needs to calculate the mean rate of collisions between two molecules ("binary" collisions) in a dilute gas with a uniform mean number density (i.e., number of molecules per unit volume) n, in thermal equilibrium at temperature T. If the effective range of inter-molecular interaction is r_0, we have an effective collision cross-section $\sigma = \pi r_0^2$. In a time interval δt, the mean number of collisions of molecules with respective velocities $\mathbf{v}_1$ and $\mathbf{v}_2$ is given by n times the volume of a "collision cylinder" of basal area σ and height $|\mathbf{v}_2 - \mathbf{v}_1|\,\delta t$. Averaging $\mathbf{v}_1$ and $\mathbf{v}_2$ over the Maxwellian distribution, the mean collision rate we seek is given by

$$\nu = n\sigma \int d^3v_1 \int d^3v_2\, p^{\text{eq}}(\mathbf{v}_1)\, p^{\text{eq}}(\mathbf{v}_2)\, |\mathbf{v}_2 - \mathbf{v}_1|. \tag{20.45}$$

Inserting the expression in Eq. (20.42) for the PDFs in Eq. (20.45) and carrying out the integrations, we get

$$\nu = 4n\sigma\left(\frac{k_B T}{m\pi}\right)^{1/2}. \tag{20.46}$$

★ **7.** Derive Eq. (20.46) from Eq. (20.45).

20.3 The Gaussian as a Limit Law

20.3.1 Linear Combinations of Gaussian Random Variables

The statistical properties of *sums* (more generally, of linear combinations) of independent Gaussian random variables are of considerable interest. Consider the simplest case first. Here is a fundamental result:

(i) Let $X_1, X_2, \ldots, X_n$ be n independent, identically distributed random variables,[3] each with a distribution $\mathcal{N}(\mu, \sigma^2)$. Then the random variable

$$Z_n = \frac{X_1 + \ldots + X_n - n\mu}{\sqrt{n\sigma^2}} \quad \text{has the distribution } \mathcal{N}(0, 1). \tag{20.47}$$

That is, Z_n is is also a Gaussian random variable, with zero mean and unit variance.

★ **8.** Establish this result.

(ii) A more general version of the result just established is as follows. Let $X_1, X_2, \ldots, X_n$ be independent Gaussian random variables, and let the mean and variance of X_i be μ_i and σ_i^2, respectively. Consider the linear combination

$$\xi_n = \sum_{i=1}^{n} a_i X_i, \tag{20.48}$$

where the constants a_i are real numbers. It follows from the additivity of the cumulants κ_1 and κ_2 that the mean and variance of ξ_n are given, respectively, by

$$\langle \xi_n \rangle = \sum_{i=1}^{n} a_i \mu_i \quad \text{and} \quad \langle (\xi_n - \langle \xi_n \rangle)^2 \rangle = \langle (\delta\xi_n)^2 \rangle = \sum_{i=1}^{n} a_i^2 \sigma_i^2. \tag{20.49}$$

Then, the random variable

$$\chi_n = \frac{(\xi_n - \langle \xi_n \rangle)}{\langle (\delta\xi_n)^2 \rangle^{1/2}} \quad \text{has the distribution } \mathcal{N}(0, 1). \tag{20.50}$$

★ **9.** Work through the steps to verify the statement in (20.50).

20.3.2 The Central Limit Theorem

One of the most important theorems of statistics is the **Central Limit Theorem**. Widely regarded as the "crown jewel" of the subject of probability and statistics, the

---

[3] The abbreviation *iidrv* is often used for "independent, identically distributed random variables".

theorem is actually a generic name for a class of convergence theorems in statistics. What we are concerned with here is the most common of these results. Let $X_1, X_2, \ldots, X_n$ be *iidrv* with mean μ and variance σ^2. They need not be Gaussian random variables! Then:

- In the limit $n \to \infty$, the probability distribution of the random variable $Z_n = (X_1 + \ldots + X_n - n\mu)/(\sigma\sqrt{n})$ tends to a Gaussian distribution with zero mean and unit variance (the standard normal distribution).

A Gaussian is therefore a **limit law** or **limit distribution** in this sense. (We will encounter other examples in Sect. 20.5.) Several of the conditions stated above can be relaxed without affecting the validity of the theorem. For instance, under certain conditions, the random variables need not be identically distributed. The crucial requirement, however, is *the finiteness of the mean and variance of each of the random variables* making up the sum. The Central Limit Theorem helps us understand why the Gaussian distribution occurs so frequently in all physical applications. In very broad, qualitative terms:

- When a large number of independent and uncorrelated effects contribute to a cause, one may expect the resultant, suitably shifted and rescaled, to have a normal distribution.

This is essentially how the Maxwellian distribution of velocities arises in a gas in thermal equilibrium.

★ **10.** A random variable X is *uniformly distributed* in the unit interval $[0, 1]$. That is, its PDF is given by

$$p(x) = \begin{cases} 1 \text{ for } & 0 \le x \le 1 \\ 0 \text{ otherwise.} \end{cases}$$

Another, independent, random variable Y is also uniformly distributed in $[0, 1]$. Show that the PDF $\rho_2(z)$ of their sum $Z = X + Y$ is given by

$$\rho_2(z) = \begin{cases} z & \text{for } \; 0 \le z \le 1 \\ 2 - z & \text{for } \; 1 < z \le 2. \end{cases}$$

20.3.3 An Explicit Illustration of the Central Limit Theorem

The example just considered shows how the uniform distribution in x and y leads to a tent-shaped PDF for their sum. As one keeps adding more such variables, the PDF of the resultant acquires more and more bends, and approximates a Gaussian ever more closely. Here is how the Gaussian emerges as a limiting distribution in this instance.

Let $X_1, \ldots, X_n$ be *iidrv*, each distributed uniformly in [0, 1]. The mean value of each variable is obviously $\mu = \frac{1}{2}$, while its variance is $\sigma^2 = \langle x_i^2 \rangle - \mu^2 = \frac{1}{3} - \frac{1}{4} = \frac{1}{12}$. Now consider the random variable

$$Z_n = \frac{X_1 + \ldots + X_n - \frac{1}{2}n}{\sqrt{n/12}}\,. \tag{20.51}$$

We are going to show that, as $n \to \infty$, the distribution of Z_n tends to the standard normal distribution $\mathcal{N}(0, 1)$. This will be done in two slightly different (but equivalent) ways, as an exercise.

(i) Since the PDF of each X_i is just unity, the PDF of Z_n is

$$\rho_n(z) = \int_0^1 dx_1 \ldots \int_0^1 dx_n\, \delta(z - z_n), \tag{20.52}$$

where $z_n = \sum_1^n (x_i - \frac{1}{2})/\sqrt{n/12}$. Using the Fourier representation

$$\delta(z - z_n) = \frac{1}{2\pi}\int_{-\infty}^{\infty} dk\, e^{ik(z - z_n)}, \tag{20.53}$$

the multiple integral in Eq. (20.52) is easily evaluated. We get

$$\rho_n(z) = \frac{1}{2\pi}\int_{-\infty}^{\infty} dk\, e^{ikz}\, e^{ik\sqrt{3n}}\, I^n, \tag{20.54}$$

where the integral

$$I = \int_0^1 dx\, e^{-ikx\sqrt{12/n}}. \tag{20.55}$$

Simplifying, we have

$$\rho_n(z) = \frac{1}{2\pi}\int_{-\infty}^{\infty} dk\, e^{ikz} \left\{\frac{\sin\,(k\sqrt{3/n}\,)}{k\sqrt{3/n}}\right\}^n. \tag{20.56}$$

It is now easy to pass to the limit $n \to \infty$. Expand the sine in its power series, and note that

$$\lim_{n\to\infty}\left\{\frac{\sin\,(k\sqrt{3/n})}{k\sqrt{3/n}}\right\}^n = \lim_{n\to\infty}\left(1 - \frac{k^2}{2n}\right)^n = e^{-\frac{1}{2}k^2}. \tag{20.57}$$

Equation (20.57) is the crucial step. The resulting Gaussian integral can be written down using the standard formula in Eq. (2.4) of Chap. 2, Sect. 2.1. The final answer is

$$\lim_{n\to\infty} \rho_n(z) = \rho(z) = \frac{1}{\sqrt{2\pi}}\, e^{-\frac{1}{2}z^2}, \tag{20.58}$$

a Gaussian PDF with zero mean and unit variance, as asserted.

★ **11.** Starting with Eq. (20.52), work out the steps outlined above to derive the final result in Eq. (20.58).

(ii) Alternatively, we can arrive at the characteristic function of Z_n with a little less work, by the following argument. Since each X_i is uniformly distributed in [0, 1] with a PDF equal to unity, the random variable $Y_i = \sqrt{12}\,(X_i - \frac{1}{2})/\sqrt{n}$ is uniformly distributed in $[-\sqrt{3/n},\ \sqrt{3/n}]$, with a constant PDF equal to $\sqrt{n/12}$. Hence the moment-generating function of any Y_i is given by

$$M_Y(u) = \sqrt{(n/12)} \int_{-\sqrt{3/n}}^{\sqrt{3/n}} dy\, e^{uy} = \frac{\sinh\,(u\sqrt{3/n})}{u\sqrt{3/n}}. \tag{20.59}$$

The cumulant-generating function of any Y_i is given by $\ln M_Y(u)$. Therefore the cumulant-generating function of $Z_n = \sum_1^n Y_i$, which is a sum of n independent random variables, is just $n \ln M_Y(u) = \ln\,[M_Y(u)]^n$. Exponentiating this quantity, the moment-generating function of Z_n is

$$M(u) = \left\{\frac{\sinh\, u\sqrt{3/n}}{u\sqrt{3/n}}\right\}^n. \tag{20.60}$$

The characteristic function of Z_n is therefore

$$M(-ik) = \widetilde{p}_n(k) = \left\{\frac{\sin\,(k\sqrt{3/n})}{k\sqrt{3/n}}\right\}^n. \tag{20.61}$$

Taking the inverse Fourier transform gives us Eq. (20.56). The rest of the derivation proceeds as before.

★ **12.** Work out the foregoing steps explicitly to derive Eq. (20.61), and then proceed to arrive at Eq. (20.58).

20.4 Random Flights

20.4.1 From Random Flights to Diffusion

Recall that we discussed a simple random walk in discrete time and discrete space (a one-dimensional lattice) in Chap. 19, Sect. 19.4.1. Subsequently, in Sect. 19.4.2, I mentioned that the random walk goes over into the diffusion process in the limit in which the step sizes in both time and space tend to zero. Seeing how this comes about serves also as a beautiful illustration of the Central Limit Theorem.

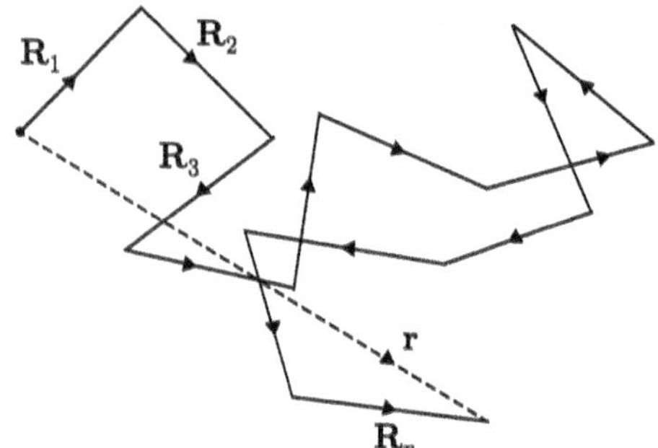

Fig. 20.3 An n-step random walk: each step is of a fixed length l, but is in a random direction

In order to be specific, let us consider random flights of the following kind, in three-dimensional space of infinite extent.[4] The walker starts at an arbitrary origin, and at the end of each time step τ, takes a step of fixed length l in an *arbitrary* direction in space. This assumption can be relaxed to include a distribution of step lengths, but I will not do so here. Each successive step is taken independently, and is uncorrelated with the steps preceding it. We ask for the normalized probability density function $p\,(\mathbf{r}, n\tau)$ of the position vector $\mathbf{r}$ of the walker at the end of n time steps, i.e., at time $n\tau$. I have written $p\,(\mathbf{r}, n\tau)$ rather than $p\,(\mathbf{r}, n)$ in order to make the time-dependence explicit. The objective:

- To calculate this quantity, and to show that it reduces to the fundamental Gaussian solution of the diffusion equation when the limits $n \to \infty$, $l \to 0$ and $\tau \to 0$ are taken in a specific manner.

Let the successive steps of the walker be given by the (randomly directed) vectors $\mathbf{R}_1, \mathbf{R}_2, \ldots, \mathbf{R}_n$, so that

$$\mathbf{r} = \sum_{j=1}^{n} \mathbf{R}_j, \tag{20.62}$$

as shown in Fig. 20.3. As usual, $\langle \cdots \rangle$ denotes the statistical average. In the present instance, this is an average over all realizations of the random flight, i.e., *over all possible configurations of the sequence of steps*. In effect, it is as if we have a chain with n straight, rigid segments, with loose-jointed hinges connecting every pair of adjacent segments. As each step is equally likely to be in any direction, it is obvious that the average value of the *vector* representing any step vanishes, i.e., $\langle \mathbf{R}_j \rangle = 0$. Hence the mean displacement vector $\langle \mathbf{r} \rangle = 0$, as you would expect intuitively. Further, using the fact that the magnitude of every step is equal to l, we have

$$\langle \mathbf{r} \cdot \mathbf{r} \rangle = \langle r^2 \rangle = nl^2 + l^2 \sum_{\substack{i,j=1 \\ i \neq j}}^{n} \langle \cos\,\theta_{ij} \rangle, \tag{20.63}$$

[4]For convenience, I will use the term random *walk* for random motion on regular lattices, and the term random *flight* for random motion with finite-sized steps in a continuous space.

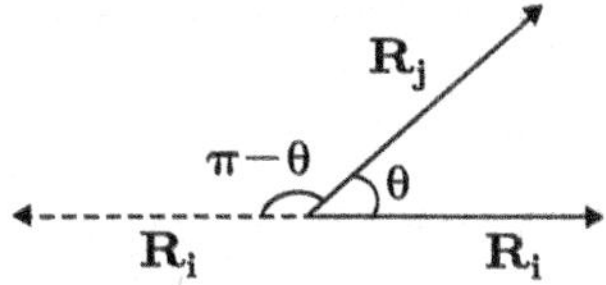

Fig. 20.4 For each given $\mathbf{R}_j$ and θ, the vector $\mathbf{R}_i$ is equally likely to have either of the two directions shown. Hence the average value of $\cos\,\theta_{ij}$ is zero

where θ_{ij} is the angle between the vectors $\mathbf{R}_i$ and $\mathbf{R}_j$. But $\langle\cos\,\theta_{ij}\rangle = 0$, because the angle between the two vectors is as likely to have a value θ as it is to have a value $(\pi - \theta)$, and $\cos\,(\pi - \theta) = -\cos\,\theta$. The contributions from the two possibilities add up to zero (see Fig. 20.4). It follows that

$$\boxed{\langle r^2\rangle = nl^2.} \tag{20.64}$$

Since $\langle\mathbf{r}\rangle = 0$, this is actually the variance of the *displacement* in an n-step random flight.[5] As n is proportional to the duration t of the walk, you can already see the emergence of the *linear growth in time* of the variance of the displacement. This behavior is characteristic of diffusive motion, as I have already emphasized in Chap. 19, Sect. 19.4.2. Note, in particular, that this result for the variance is actually *independent* of the number of dimensions of the space in which the random flight occurs!

Let us return to the calculation of the PDF $p\,(\mathbf{r}, n\tau)$. It turns out to be easier to compute its Fourier transform $\widetilde{p}\,(\mathbf{k}, n\tau)$ first. The latter is just the characteristic function of the random variable $\mathbf{r}$. Recall the multiplicative property of characteristic functions established in Eq. (20.21). Since $\mathbf{r}$ is the sum of the independent random variables $\mathbf{R}_j$ $(1 \le j \le n)$, it follows that $\widetilde{p}\,(\mathbf{k}, n\tau)$ must be the product of the characteristic functions of the individual steps. It is instructive to see this explicitly, as follows. Consistent with the Fourier transform conventions that we have used throughout, we have

$$\widetilde{p}(\mathbf{k}, n\tau) = \int d^3r\; e^{-i\mathbf{k}\cdot\mathbf{r}}\; p(\mathbf{r}, n\tau) \Longleftrightarrow p(\mathbf{r}, n\tau) = \frac{1}{(2\pi)^3}\int d^3k\; e^{i\mathbf{k}\cdot\mathbf{r}}\; \widetilde{p}(\mathbf{k}, n\tau). \tag{20.65}$$

But $\widetilde{p}(\mathbf{k}, n\tau)$ can also be interpreted as the mean value of the quantity $\exp\,(-i\mathbf{k}\cdot\mathbf{r})$ with respect to the PDF $p(\mathbf{r}, n\tau)$, i.e., an average of this quantity over all possible n-step random walks. Therefore we have

[5]It is not the variance of the end-to-end *distance*, because the mean distance $\langle r\rangle \neq 0$. We will consider $\langle r\rangle$ and $\mathrm{Var}\,(r)$ in Sect. 20.4.2.

$$\widetilde{p}(\mathbf{k}, n\tau) = \left\langle e^{-i\mathbf{k}\cdot\mathbf{r}} \right\rangle = \left\langle e^{-i\sum_{j=1}^{n}\mathbf{k}\cdot\mathbf{R}_j} \right\rangle = \left\langle e^{-i\mathbf{k}\cdot\mathbf{R}_1} \ldots e^{-i\mathbf{k}\cdot\mathbf{R}_n} \right\rangle. \tag{20.66}$$

A great simplification occurs now, *because the individual steps are completely independent of each other*: The average value of the product in the final step above becomes equal to the product of average values. As a result, the final expression in (20.66) reduces to

$$\widetilde{p}(\mathbf{k}, n\tau) = \left\langle e^{-i\mathbf{k}\cdot\mathbf{R}_1} \right\rangle \left\langle e^{-i\mathbf{k}\cdot\mathbf{R}_2} \right\rangle \cdots \left\langle e^{-i\mathbf{k}\cdot\mathbf{R}_n} \right\rangle = \prod_{j=1}^{n} \left\langle e^{-i\mathbf{k}\cdot\mathbf{R}_j} \right\rangle. \tag{20.67}$$

Now let $p(\mathbf{r}, \tau) \equiv p_1(\mathbf{r})$ denote the PDF of a *single* step. Since this PDF is the *same* for each of the steps, we have, for every j,

$$\left\langle e^{-i\mathbf{k}\cdot\mathbf{R}_j} \right\rangle = \int d^3\mathbf{R}_j \, e^{-i\mathbf{k}\cdot\mathbf{R}_j} \, p_1(\mathbf{R}_j) = \widetilde{p}_1(\mathbf{k}), \tag{20.68}$$

the Fourier transform of the single-step PDF. Thus

$$\widetilde{p}(\mathbf{k}, n\tau) = [\widetilde{p}_1(\mathbf{k})]^n \,. \tag{20.69}$$

Next, we must find the single-step PDF $p_1(\mathbf{r})$ and calculate its Fourier transform $\widetilde{p}_1(\mathbf{k})$. The only condition imposed on a step is that its magnitude be equal to l. Therefore $p_1(\mathbf{r})$ must be proportional to $\delta(r - l)$. But normalization implies that $\int d^3r \, p_1(\mathbf{r})$ must be equal to unity. This condition fixes the constant of proportionality. The normalized PDF is

$$p_1(\mathbf{r}) = \frac{1}{4\pi l^2} \, \delta(r - l). \tag{20.70}$$

Therefore

$$\widetilde{p}_1(\mathbf{k}) = \int d^3r \, e^{-i\mathbf{k}\cdot\mathbf{r}} \, p_1(\mathbf{r}) = \frac{1}{4\pi l^2} \int d^3r \, e^{-i\mathbf{k}\cdot\mathbf{r}} \, \delta(r - l). \tag{20.71}$$

Carrying out the integration, we get

$$\widetilde{p}_1(\mathbf{k}) = \frac{\sin kl}{kl}, \quad \text{where } k = |\mathbf{k}|. \tag{20.72}$$

★ **13.** Set $p_1(\mathbf{r}) = c\,\delta(r - l)$, where c is a constant.

(a) Use the normalization condition $\int d^3r \, p_1(\mathbf{r}) = 1$ to obtain $c = 1/(4\pi l^2)$.
(b) Next, carry out the integration in (20.71) to obtain the expression quoted in Eq. (20.72) for $\widetilde{p}_1(\mathbf{k})$.

From Eqs. (20.69) and (20.72), the characteristic function of the position vector $\mathbf{r}$ for an n-step random flight is given by

$$\widetilde{p}(\mathbf{k}, n\tau) = \left(\frac{\sin kl}{kl}\right)^n. \tag{20.73}$$

Using this in the second equation in (20.65), we get

$$\boxed{p(\mathbf{r}, n\tau) = \frac{1}{(2\pi)^3}\int d^3k\, e^{i\mathbf{k}\cdot\mathbf{r}} \left(\frac{\sin kl}{kl}\right)^n.} \tag{20.74}$$

This is an *exact* formula for the PDF of the end-to-end displacement vector $\mathbf{r}$ in a random flight of n steps, each of length l, in three-dimensional space. Observe its striking similarity to the expression found in Eq. (20.56) for the PDF of a sum of n random variables uniformly distributed in [0, 1]—a problem that is seemingly quite different from the one we are discussing now!

As both the integrand and the region of integration in Eq. (20.74) are rotationally invariant, the integral itself must be a function of the radial coordinate r alone, rather than the vector $\mathbf{r}$ itself. The integration over all the directions of $\mathbf{k}$ is easily carried out if we choose spherical polar coordinates with the polar axis along the direction of $\mathbf{r}$, exploiting rotational invariance. The result is

$$p(\mathbf{r}, n\tau) = \frac{1}{2\pi^2 r}\int_0^\infty dk\, k\, \sin kr \left(\frac{\sin kl}{kl}\right)^n. \tag{20.75}$$

The evaluation of the integral in Eq. (20.75) for successive values of n is an interesting exercise in its own right. For sufficiently small values of n $(= 2, 3, \ldots)$, the integration over k can be done explicitly without much difficulty. In Sect. 20.4.2, you will see how the PDF $p(\mathbf{r}, n\tau)$, which starts out for $n = 1$ as a singular density proportional to $\delta(r - l)$, becomes "smoother" as n increases.

Our immediate purpose, however, is to examine $p(\mathbf{r}, n\tau)$ for very *large* n—in fact, in the limit $n \to \infty$ and $\tau \to 0$ such that $\lim n\tau = t$. Now, $|(\sin kl)/(kl)| < 1$ for all $kl \neq 0$. Hence the factor $[(\sin kl)/(kl)]^n$ in the integrand causes the integral to vanish as $n \to \infty$ as long as l remains finite and nonzero. But if l tends to zero simultaneously, we have

$$\left(\frac{\sin kl}{kl}\right)^n \simeq \left(1 - \frac{k^2\, l^2}{6}\right)^n. \tag{20.76}$$

It is then clear that if $l^2 \to 0$ like n^{-1}, the right-hand side of Eq. (20.76) has a finite limit. This is the *only* possible way in which a nontrivial limiting PDF can arise. But $n^{-1} \sim \tau$, so that we must let l^2 tend to zero like τ. Therefore, let $\tau \to 0$ and $l \to 0$, such that

$$\boxed{\lim_{l,\,\tau\to 0} l^2/(6\tau) = D,} \tag{20.77}$$

where D is a finite positive quantity called the **diffusion coefficient**.[6] This limit of a random flight is called the **diffusion limit**. Then

$$\left(1 - \frac{k^2 l^2}{6}\right)^n = \left(1 - \frac{Dk^2 t}{n}\right)^n \xrightarrow[n\to\infty]{} e^{-Dk^2 t}. \tag{20.78}$$

We thus obtain

$$p(\mathbf{r}, n\tau) \to p(\mathbf{r}, t) = \frac{1}{(2\pi)^3} \int d^3k \; e^{-Dk^2 t + i\mathbf{k}\cdot\mathbf{r}}. \tag{20.79}$$

This last integral is most easily evaluated in *Cartesian* coordinates—it then factors into the product of three Gaussian integrals. Put $\mathbf{r} = (x, y, z)$, $\mathbf{k} = (k_1, k_2, k_3)$, and use the familiar formula of Eq. (2.4) for the shifted Gaussian integral. Re-combine $x^2 + y^2 + z^2$ into r^2, to arrive finally at

$$\boxed{p(\mathbf{r}, t) = \frac{1}{(4\pi Dt)^{3/2}} \, e^{-r^2/(4Dt)}.} \tag{20.80}$$

This is precisely the normalized fundamental Gaussian solution to the three-dimensional diffusion equation, as we shall see in Chap. 30, Sect. 30.1.3.

- Thus, in the diffusion limit, the simple random flight goes over into the diffusion process.

It is easily checked that the mean squared displacement is

$$\langle r^2 \rangle = \int d^3r \, r^2 \, p(\mathbf{r}, t) = \frac{4\pi}{(4\pi Dt)^{3/2}} \int_0^\infty dr \, r^4 \, e^{-r^2/(4Dt)} = 6Dt, \tag{20.81}$$

on using the Gaussian integral in Eq. (3.16) of Chap. 3, Sect. 3.1.5. The linear growth of the variance of the displacement with time is thus retained in the diffusion limit.

20.4.2 *The Probability Density for Short Random Flights*

Let us go back to the solution for $p(\mathbf{r}, n\tau)$ in Eq. (20.75), and consider *small* values of n $(= 2, 3, 4)$. The integration can then be carried out explicitly. But you must be careful in doing so, because the integrals concerned are not absolutely convergent. Interestingly, they involve precisely the sequence of apparently singular integrals that we derived from the Dirichlet integral in Chap. 2, Sect. 2.3.

---

[6] The choice of the precise numerical factor $(\frac{1}{6})$ in the above is not important for the argument to go through. It has been made so as to obtain the same numerical factors in the limiting expression for $p(\mathbf{r}, n\tau)$ as the fundamental Gaussian solution of the diffusion equation in three-dimensional space, to be derived in Chap. 30, Sect. 30.1.3.

It is obvious that $p(\mathbf{r}, n\tau)$ must vanish identically for $r > nl$. For $n = 2$, the result is

$$p(\mathbf{r}, 2\tau) = \frac{1}{8\pi l^2 r}\,\theta(2l - r), \tag{20.82}$$

where $\theta(x)$ is the unit step function. For $n = 3$, we find

$$\begin{aligned} p(\mathbf{r}, 3\tau) &= \frac{1}{32\pi l^3 r}\big(2r - 3|r - l| + |r - 3l|\big) \\ &= \begin{cases} 1/(8\pi l^3), & 0 \le r \le l \\ (3l - r)/(16\pi l^3 r), & l \le r \le 3l \\ 0, & r > 3l. \end{cases} \end{aligned} \tag{20.83}$$

For $n = 4$, we get

$$\begin{aligned} p(\mathbf{r}, 4\tau) &= \frac{1}{128\pi l^4 r}\big[8rl - 3r^2 + 4(r - 2l)\,|r - 2l| - (r - 4l)\,|r - 4l|\big] \\ &= \begin{cases} (8l - 3r)/(64\pi l^4), & 0 \le r \le 2l \\ (4l - r)^2/(64\pi l^4 r), & 2l \le r \le 4l \\ 0, & r > 4l. \end{cases} \end{aligned} \tag{20.84}$$

★ **14.** Derive the results in Eqs. (20.82)–(20.84), and verify the normalization $\int d^3r\, p(\mathbf{r}, n\tau) = 1$ in each case.

With increasing values of n, the calculation rapidly becomes quite tedious. The expressions for $p(\mathbf{r}, n\tau)$ up to $n = 6$ were first deduced by Lord Rayleigh, way back in 1919. Today, however, computational software packages enable us to write down closed-form expressions for $p(\mathbf{r}, n\tau)$ for fairly large values of n, although the expressions become rather lengthy.

To summarize what we have learnt so far regarding this simple model of random flights:

(i) The variance of the displacement vector $\mathbf{r}$ in a random flight of n steps is equal to nl^2 for all $n \ge 1$ (Eq. (20.64)).
(ii) The characteristic function of $\mathbf{r}$ is $[(\sin kl)/(kl)]^n$ (Eq. (20.73)), providing a formal solution for the PDF $p(\mathbf{r}, n\tau)$ of $\mathbf{r}$ (Eq. (20.74)).
(iii) For small values of n, explicit expressions may be written down for $p(\mathbf{r}, n\tau)$, as in Eqs. (20.82)–(20.84) in the cases $n = 2$, 3 and 4, respectively.
(iv) In the diffusion limit, $p(\mathbf{r}, n\tau)$ tends to the fundamental Gaussian solution to the diffusion equation (Eq. (20.80)).

***End-to-end distance distribution*:** $p(\mathbf{r}, n\tau)\, d^3r$ is, of course, the probability that the end-point of the random flight of n time steps is in a volume element d^3r located at the point $\mathbf{r}$. It is natural to ask: What is the PDF of the end-to-end *distance* r in such a random flight?

Let us denote the PDF of r by $\rho(r, n\tau)$. To find it, we must obviously integrate $p(\mathbf{r}, n\tau)$ over all possible *directions* of $\mathbf{r}$. Since $p(\mathbf{r}, n\tau)$ has turned out to be a

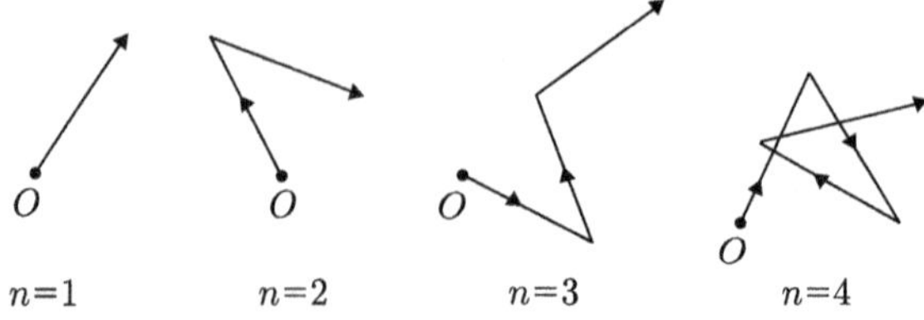

Fig. 20.5 Short random flights of 1, 2, 3 and 4 steps

function of r alone, and has no dependence on the direction of $\mathbf{r}$, we get

$$\rho(r, n\tau) = 4\pi r^2\, p(\mathbf{r}, n\tau) = \frac{2r}{\pi}\int_0^\infty dk\, k\, \sin kr \left(\frac{\sin kl}{kl}\right)^n. \tag{20.85}$$

It is important to remember the presence of the extra factor of r^2 in this PDF.[7] In the cases $n = 1$–4 we thus have, using Eqs. (20.70) and (20.82)–(20.84), the following results:

$$\left.\begin{aligned}
\rho(r, \tau) &= \delta(r - l),\\
\rho(r, 2\tau) &= \frac{r}{2l^2}\,\theta(2l - r),\\
\rho(r, 3\tau) &= \frac{r}{8l^3}\,\big(2r - 3|r - l| + |r - 3l|\big),\\
\rho(r, 4\tau) &= \frac{r}{32l^4}\,\big[8rl - 3r^2 + 4(r - 2l)\,|r - 2l| - (r - 4l)\,|r - 4l|\big].
\end{aligned}\right\} \tag{20.86}$$

Figure 20.5 shows a configuration of the random flights in the cases $n = 1$–4. Figure 20.6 shows the corresponding PDFs of the end-to-end distance r. Observe the neat progression by which the singularities in the PDF $\rho(r, n\tau)$ get smoothed out as n increases:

- $\rho(r, \tau)$ has a δ-function spike (at $r = l$).
- $\rho(r, 2\tau)$ has a finite discontinuity (at $r = 2l$).
- $\rho(r, 3\tau)$ is continuous, and only has a point of non-differentiability (at $r = l$) inside its support $0 < r < 3l$.
- $\rho(r, 4\tau)$ is continuous *and* smooth (i.e., its first derivative is also continuous) everywhere inside its support $0 < r < 4l$. This property persists for all higher values of n.
- As n increases, the singularities become "milder", i.e., get shifted to higher and higher derivatives of the PDF. In the limit $n \to \infty$, the PDF $\rho(r, n\tau)$ tends to a smooth, infinitely differentiable function that is proportional to $r^2\, e^{-r^2/(4Dt)}$ for all $r > 0$.

[7]To give an analogy: the PDF of the radial distance r in the problem of a quantum mechanical particle in a central potential is $4\pi r^2\, |\psi(\mathbf{r}, t)|^2$, and not just $|\psi(\mathbf{r}, t)|^2$. If you do not take this into account, you end up with the erroneous conclusion that, in the ground state of a hydrogen atom, the most probable distance of the electron from the nucleus is 0 rather than the Bohr radius a_0!.

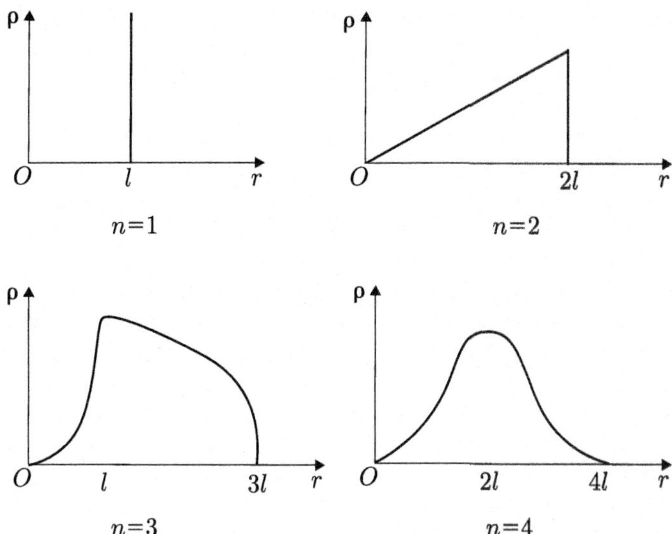

Fig. 20.6 The PDF $\rho(r, n\tau)$ of the end-to-end distance in short random flights (schematic)

★ **15.** Find the mean, variance and relative fluctuation of r in each case. Compare the last of these with the relative fluctuation in the diffusion limit of the random flight. $\langle r \rangle$ can be found directly in the case $n = 2$. In this case $\mathbf{r} = \mathbf{R}_1 + \mathbf{R}_2$, so that $r^2 = 2l^2(1 + \cos\,\theta)$, where θ is the angle between the vectors $\mathbf{R}_1$ and $\mathbf{R}_2$. Therefore $\langle r \rangle = l\sqrt{2}\langle(1 + \cos\,\theta)^{1/2}\rangle$. Take $\mathbf{R}_1$ to lie along the polar axis, and average over all directions of $\mathbf{R}_2$. This yields

$$\langle r \rangle = l\sqrt{2}/(4\pi) \int d\Omega\,(1 + \cos\,\theta)^{1/2} = \tfrac{4}{3}l,$$

corroborating what we know already. A calculation along these lines seems to be very difficult, if not impossible, for $n \geq 3$.

On the other hand, we can turn the matter around and use the known values of $\langle r \rangle$ to write down the values of certain otherwise formidable integrals! Here is what happens for $n = 3$. Let us set $l = 1$ (as it is just a scale factor here). Take $\mathbf{R}_1$ to be directed along the polar axis, and let $(\theta,\, \varphi)$ and $(\theta',\, \varphi')$ be the polar and azimuthal angles of the vectors $\mathbf{R}_2$ and $\mathbf{R}_3$, respectively. The angle γ between $\mathbf{R}_2$ and $\mathbf{R}_3$ is then given by the law of cosines, Eq. (16.137) of Chap. 16, Sect. 16.4.8. Repeating this formula for ready reference, we have

$$\cos\,\gamma = \cos\,\theta\,\cos\,\theta' + \sin\,\theta\,\sin\,\theta'\,\cos\,(\varphi - \varphi').$$

The result found above for $\langle r \rangle$ in the case $n = 3$ now enables us to assert that

$$\int d\Omega \int d\Omega' \left[3 + 2(\cos\,\theta + \cos\,\theta' + \cos\,\gamma)\right]^{1/2} = 26\,\pi^2.$$

Clearly, this is a far-from-obvious result!

20.5 The Family of Stable Distributions

20.5.1 *What Is a Stable Distribution?*

As I have mentioned at the beginning of Sect. 20.3.2, a topic of great importance in statistics is the probability distribution of the sums of independent, identically distributed random variables (*iidrv*). In particular, a crucial question is the existence of a limit law when the number of random variables becomes infinite. We have already seen an example of such a limit law, namely, the Gaussian distribution. The Gaussian is one member (arguably the most *important* member, as far as physical applications are concerned) of a whole family of distributions intimately connected with such limit laws.

Suppose we have n *iidrv* $X_1, X_2, \ldots, X_n$, each with a cumulative distribution function (CDF) F. We ask, Are there forms of F such that the sum $\sum_{i=1}^{n} X_i$, possibly shifted by an n-dependent constant and rescaled by another n-dependent constant, also has the *same* CDF? The complete answer to this question is one of the key results in statistics. There is a whole family of distributions with the property required, called the **Lévy skew alpha-stable distributions**, or **stable distributions** for short. There are several alternative (and equivalent) ways of stating the defining property of these probability distributions.

Definition 20.1 The CDF F is a stable distribution if and only if,[8] for *every* positive integer $n \geq 2$, it is possible to find a positive constant a_n and a real constant b_n such that the probability distribution of the quantity

$$Z_n = \frac{(X_1 + X_2 + \cdots + X_n) - b_n}{a_n} \tag{20.87}$$

is also given by F itself. If this condition can be satisfied with $b_n = 0$ for all n, the distribution F is said to be **strictly stable**. The latter constitute a subset of the class of stable distributions.

Definition 20.2 Let F be the CDF of two *iidrv* X_1 and X_2. Then F is a stable distribution if and only if, given any two arbitrary positive numbers a_1 and a_2, a

[8] Remember that "if and only if" means that the conditions are both necessary and sufficient.

positive number a and a real number b can be found such that $(a_1 X_1 + a_2 X_2 - b)/a$ also has the distribution F. Again, if this can be done without a shift constant b, the distribution is strictly stable.

This formulation of the defining property can be re-expressed as a property of the distribution function itself, as follows.

Definition 20.3 F is a stable distribution if and only if, given any two positive numbers a_1 and a_2, we can find a positive number a and a real number b such that F satisfies

$$F(x/a_1) * F(x/a_2) = F\left((x-b)/a\right), \tag{20.88}$$

where the symbol $*$ denotes the convolution of the two distributions. If this relation is satisfied with $b = 0$ in all cases, F is strictly stable.

20.5.2 *The Characteristic Function of Stable Distributions*

Not surprisingly, the most explicit way of specifying the stable distributions is in terms of their characteristic functions. The definition (20.88), involving the convolution of distributions, suggests that the characteristic function of a stable distribution might satisfy some sort of "multiplication property", and be related to exponential functions. This intuitive guess is indeed borne out.

The family of stable distributions is characterized by four different parameters, but I shall not go into these details here. The important points are as follows.

- It turns out that the scaling constant a_n in the definition of Z_n in Eq. (20.87) must have the power-law form $a_n = n^{1/\alpha}$, where $0 < \alpha \leq 2$.
- The exponent α is the primary characterizer of the members of the family of stable distributions.
- The stable distributions have continuous PDFs that are unimodal (i.e., have single peaks).

Let X denote a random variable with a stable distribution, and $p(x)$ its PDF. *Except* for certain special values of α (to be discussed in Sect. 20.5.3), $p(x)$ cannot be written down in explicit form in terms of elementary functions. In fact, it cannot be written down even in terms of the so-called "higher transcendental functions" for general α, although complicated expressions in terms of hypergeometric functions are possible in various special cases for rational values of α. The characteristic function $\widetilde{p}(k)$, however, *can* be expressed in terms of elementary functions for all $0 < \alpha \leq 2$.

- In essence, the modulus of the characteristic function $\widetilde{p}(k)$ for a stable distribution with exponent α behaves like $\exp\left(-|k|^{\alpha}\right)$.

The reasons for restricting the exponent α to the range $0 < \alpha \leq 2$ are as follows: (i) If $\alpha \leq 0$, the function $\widetilde{p}(k) \to 1$ as $|k| \to \infty$. Therefore $\int_{-\infty}^{\infty} dk\, e^{ikx}\, \widetilde{p}(k)$ diverges, and hence the PDF $p(x)$ does not exist.

(ii) At the other end, if $\alpha > 2$, the inverse Fourier transform of $\widetilde{p}(k)$ is no longer guaranteed to be a real, nonnegative function of x, as it must be in order to represent a PDF. (This is much harder to prove. I will not attempt to do so here.)

20.5.3 Three Important Cases: Gaussian, Cauchy, and Lévy

There are three notable and important cases in which the formula for $\widetilde{p}(k)$ can be inverted to yield explicit expressions for the PDF $p(x)$ in terms of *elementary* functions. They are also the cases that appear most often in physical applications.

(i) ***The Gaussian distribution*****:** For $\alpha = 2$, we have the Gaussian distribution parametrized by the mean μ and variance σ^2. Repeating Eqs. (20.23) and (20.35) for ready reference, the PDF and the characteristic function are

$$\boxed{p(x) = \frac{1}{(2\pi\sigma^2)^{1/2}}\, e^{-(x-\mu)^2/(2\sigma^2)}, \quad x \in (-\infty,\, \infty)} \tag{20.89}$$

and

$$\widetilde{p}(k) = e^{-i\mu k - \sigma^2 k^2/2}, \tag{20.90}$$

respectively. Recall that the PDF has already been sketched schematically, in Fig. 20.1.

(ii) ***The Cauchy distribution*****:** For $\alpha = 1$, we have the Cauchy distribution. The PDF has the familiar Lorentzian shape. It is symmetric about its center μ, and is given by

$$\boxed{p(x) = \frac{\lambda}{\pi[(x-\mu)^2 + \lambda^2]}, \quad x \in (-\infty,\, \infty)} \tag{20.91}$$

where λ is a positive constant. The characteristic function in this case can be shown to be

$$\widetilde{p}(k) = e^{-i\mu k - \lambda |k|}. \tag{20.92}$$

(iii) ***The Lévy distribution*****:** For $\alpha = \frac{1}{2}$, we have the Lévy distribution,[9] given by the PDF

$$\boxed{p(x) = \left(\frac{c}{2\pi x^3}\right)^{1/2} e^{-c/(2x)}, \quad (0 \le x < \infty)} \tag{20.93}$$

where c is a positive constant. Figure 20.7 depicts the PDF schematically. The characteristic function in this case can be shown to be

[9] This label can be a little confusing: as I have mentioned, the complete name for the whole *family* of stable distributions is "Lévy skew alpha-stable distributions".

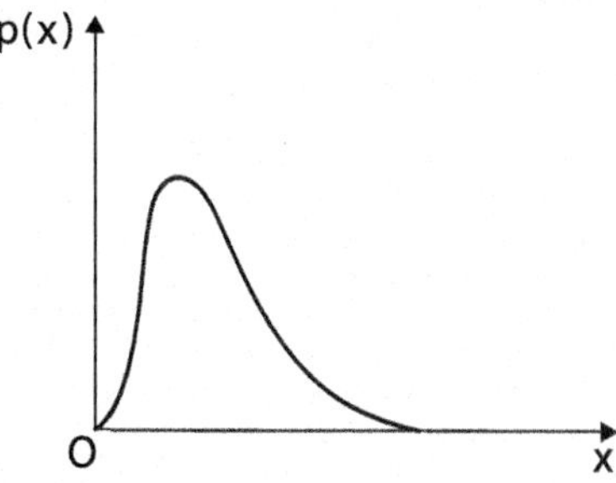

Fig. 20.7 The probability density function of a Lévy distribution

$$\widetilde{p}(k) = e^{-c|k|^{1/2}\,(1+i\,\varepsilon(k))}, \qquad (20.94)$$

where $\varepsilon(k)$ is the signum function ($= +1$ for $k > 0$ and -1 for $k < 0$).
The evaluation of the Fourier transform of the PDF (20.91) to obtain the characteristic function in Eq. (20.92) is most easily done using contour integration, as we will see in Chap. 23, Sect. 23.3.2. The derivation of Eq. (20.94) for the Lévy distribution requires the *Mellin inversion formula* for a Laplace transform, to be discussed in Chap. 28, Sect. 28.2.1. In the interests of accuracy, it must be stated that the PDFs in Eqs. (20.91) and (20.93) actually correspond to the most important special cases of more general stable distributions with $\alpha = 1$ and $\alpha = \frac{1}{2}$, respectively, for certain special values of some other parameters.

***The special nature of the Gaussian distribution*:** The Gaussian distribution, corresponding to $\alpha = 2$, differs markedly from the rest of the family of stable distributions (corresponding to $0 < \alpha < 2$).

- The stable distributions with $\alpha < 2$ are **heavy-tailed distributions**, in the sense that the PDF $p(x)$ has the following leading power-law asymptotic behavior:

$$p(x) \sim 1/|x|^{\alpha+1} \text{ as } |x| \to \infty. \qquad (20.95)$$

It then follows that the CDF $F(x)$ has the following leading asymptotic behavior as $x \to \pm\infty$:

$$\left.\begin{aligned} F(x) &\sim 1/|x|^{\alpha} \quad \text{as } x \to -\infty, \\ 1 - F(x) &\sim 1/x^{\alpha} \quad\ \ \text{as } x \to \infty. \end{aligned}\right\} \qquad (20.96)$$

- In sharp contrast, the Gaussian PDF decays to zero much more rapidly than any power law as $x \to \pm\infty$.
- As a consequence, *of all the stable distributions, only the Gaussian has a finite variance*. All the other stable distributions have infinite variance.
- For $\alpha \le 1$, even the first moment or mean value does not exist. Once again, this happens because the PDF $p(x)$ decays to zero too slowly as $|x| \to \infty$.

Finally, I mention that there is a generalization of the Central Limit Theorem that is applicable to the family of stable distributions with $\alpha < 2$.

20.5.4 Some Connections Between the Three Cases

There are close connections between the three special distributions listed above. Here are a couple of these relationships.

(i) *The reciprocal of the square of a Gaussian random variable* has a Lévy distribution. In Eq. (20.44) of Sect. 20.2.3, you have seen that the *square* of a Gaussian random variable (e.g., the kinetic energy of a molecule in a classical ideal gas) has a gamma distribution. Now consider the *reciprocal* of the square. Let X be a Gaussian random variable with zero mean and variance σ^2. Then the random variable $\xi = 1/X^2$ has a Lévy distribution with $c = 1/\sigma^2$. Its PDF is given by

$$\rho(\xi) = (2\pi\sigma^2\xi^3)^{-1/2}\, e^{-1/(2\sigma^2\xi)} \quad (0 \le \xi < \infty). \tag{20.97}$$

★ **16.** Derive Eq. (20.97).

An amusing consequence of the result just established is as follows. Given a set of *iiddv* with a specified distribution, it is quite difficult, in general, to deduce the probability distribution of some intricate function of these *iiddv*. Occasionally, however, there may exist a trick to do so. Let $X_1, \ldots, X_n$ be *iiddv*, each with a Gaussian distribution $\mathcal{N}(0, \sigma^2)$. Then the random variable

$$Z_n = \frac{X_1 X_2 \ldots X_n}{\left[(X_2^2 X_3^2 \ldots X_n^2) + (X_1^2 X_3^2 \ldots X_n^2) + \cdots + (X_1^2 X_2^2 \ldots X_{n-1}^2)\right]^{1/2}} \tag{20.98}$$

is also a Gaussian random variable, with the distribution $\mathcal{N}(0, \sigma^2/n)$.

★ **17.** Establish this result.

***An interesting duality*:** The foregoing relationship between the Gaussian and Léve distributions is a special case of a **duality** that exists between different stable distributions:

- A stable distribution with exponent α (where $1 \le \alpha \le 2$) for the random variable X is essentially equivalent to a stable distribution with exponent α^{-1} (so that $\frac{1}{2} \le \alpha^{-1} \le 1$) for the random variable $X^{-\alpha}$.

(ii) *The ratio of two Gaussian random variables*, each with a mean equal to zero, is Cauchy-distributed. Let us see how this happens, using as an example the physical context of two independent random walks on a line, in the diffusion limit. Consider two particles, each starting from the origin at $t = 0$, diffusing on the x-axis. The particles are assumed to have no interaction with each other (and to be able to "pass through" each other). Let X_1 and X_2 be the positions of the particles, and let D_1 and D_2 be their respective diffusion coefficients. The respective PDFs $p_1(x_1, t)$ and $p_2(x_2, t)$ of X_1 and X_2 are given by the fundamental Gaussian solution to the one-dimensional diffusion equation (to be discussed in Chap. 30, Sect. 30.1.3). This solution, a Gaussian with zero mean, is the one-dimensional version of the formula (20.80) derived in Sect. 20.4.1 for $p(\mathbf{r}, t)$. Accordingly, we have

$$p_i(x_i, t) = \frac{1}{(4\pi D_i t)^{1/2}}\, e^{-x_i^2/(4D_i t)}, \quad \text{where } i = 1, 2. \tag{20.99}$$

Let $\xi = X_1/X_2$ be the ratio of the positions of the two particles. It is clear that the sample space of ξ is also $(-\infty, \infty)$. The PDF of ξ is given by

$$\rho(\xi, t) = \int_{-\infty}^{\infty} dx_1 \int_{-\infty}^{\infty} dx_2\; p_1(x_1, t)\, p_2(x_2, t)\, \delta\Big(\xi - \frac{x_1}{x_2}\Big). \tag{20.100}$$

Substituting for $p_i(x_i, t)$ from Eqs. (20.99) and carrying out the integrations, we obtain

$$\rho(\xi, t) = \frac{\lambda}{\pi}\, \frac{1}{(\xi^2 + \lambda^2)}, \quad \text{where } \lambda = \sqrt{D_1/D_2}. \tag{20.101}$$

There is no time-dependence on the right-hand side! Remarkably enough, the ratio $\xi = X_1/X_2$ has precisely the same Cauchy distribution for *all* $t > 0$.

★ **18.** Derive Eq. (20.101).

20.6 Infinitely Divisible Distributions

20.6.1 Divisibility of a Random Variable

To recapitulate: Let $X_1, \ldots, X_n$ be a set of *iidrv* with a stable probability distribution. Then, there exist constants a_n and b_n such that the (shifted and rescaled) sum

$$Z_n = \frac{\left(\sum_{i=1}^{n} X_i\right) - b_n}{a_n} \tag{20.102}$$

also has the same stable distribution, for *every* positive integer n.

Going the other way, it is natural to ask the opposite question: *given* a random variable X with a specified probability distribution, when can we write it as the sum of n *iidrv* $X_1, \ldots, X_n$ for *every* positive integer value of n? Whenever this can be done, X is said to be an *infinitely divisible* random variable. Its probability distribution is an **infinitely divisible distribution**. It should come as no surprise that stable distributions are intimately related to infinitely divisible distributions.

It is convenient to call the individual random variables X_i (into which X is decomposed) the "components" of X. The following points will help you understand the notion of the divisibility of a random variable:

(i) The components X_i *need not* have the same distribution as their sum X itself! All that is required is that, for each value of n, all the components X_i (where $i \le 1 \le n$) have the same distribution.

(ii) All stable distributions are infinitely divisible. *The converse is not necessarily true.* Stable distributions comprise a subset of infinitely divisible distributions.

(iii) The special feature of a stable distribution is that, in this case, each component X_i also has a distribution of the *same* functional form as the distribution of the sum X, for every n.
(iv) A random variable may be "n-divisible", but not "m-divisible", where n and m are different positive integers.
Here is a very simple example to illustrate the last point above. Suppose the sample space of a random variable X is the set of integers $\{0, 1, 2, 3\}$. It can then be decomposed into 3 components X_1, X_2, and X_3, where each X_i has the sample space $\{0, 1\}$. Thus X has the possibility of being 3-divisible (see below). But it can never be 2-divisible, because there is no way you can find two *iidrv* X_1 and X_2 that sum up to X with the sample space specified above, namely, $\{0, 1, 2, 3\}$. It is not hard to see why. If X_1 and X_2 have the sample space $\{0, 1\}$, then their sum can never reach the value 3. If the sample space of X_1 and X_2 is $\{0, 1, 2\}$, then their sum can have the value 4, which is not in the sample space of X. If $\frac{3}{2}$ is in the sample space of X_1 and X_2, then, since 0 must also be in this sample space, $\frac{3}{2}$ must also be in the sample of space of X, which is not the case. And so on.

Continuing with this example, it must also be understood that not *every* random variable with the sample space $\{0, 1, 2, 3\}$ is even 3-divisible, because the distribution of X actually gets fixed by that requirement. For, if each component X_i has the sample space $\{0, 1\}$ with probabilities p and $q = 1 - p$ for the two values (and this is the only possibility), then the distribution of X must necessarily be a binomial distribution with exponent 3. That is, the respective probabilities of the values $0, 1, 2$, and 3 must be p^3, $3p^2q$, $3pq^2$ and q^3, in order that X be 3-divisible. It should now be obvious that a random variable X with a binomial distribution with parameter N (as in Eq. (19.4) of Chap. 19, Sect. 19.1.2) is N-divisible: each of its N components is a Bernoulli trial, i.e., it can take the values 0 and 1, with respective probabilities p and q.
(v) A moment's thought shows us that, if X is an *infinitely* divisible random variable, then its sample space *must* be unbounded in at least one direction—from below or from above, or both.

The definitive property of an infinitely divisible distribution is found by looking at its characteristic function. We know that the characteristic function of a sum of *iidrv* is just the product of the characteristic functions of its components. Hence:
(vi) X can be an infinitely divisible random variable only if its characteristic function $\widetilde{p}(k)$ can be written as the nth power of a characteristic function (of any of its components X_i) for every n. That is, for *every* positive integer n, we must be able to find a characteristic function $\widetilde{p}_n(k)$ such that

$$\widetilde{p}(k) = \left[\, \widetilde{p}_n(k) \,\right]^n. \tag{20.103}$$

Recall, now, the features required of a characteristic function, listed in Sect. 20.1.5. The characteristic functions of the all the Lévy skew alpha-stable distributions can be shown to satisfy the property (20.103). Hence all stable distributions are also infinitely divisible distributions, as already stated. It is easy to see how this happens in the special cases of the Gaussian, Cauchy and Lévy distributions. Their characteristic functions have been written down in Eqs. (20.90), (20.92) and (20.94), respectively. It is almost trivial to observe that the characteristic functions of the components in these cases are given by

$$\widetilde{p}_n(k) = \begin{cases} \exp\left\{-i(\mu/n)\,k - \frac{1}{2}(\sigma/\sqrt{n}\,)^2\,k^2\right\} & \text{(Gaussian)} \\ \exp\left\{-i(\mu/n)\,k - (\lambda/n)\,|k|\right\} & \text{(Cauchy)} \\ \exp\left\{-(c/n)\,|k|^{1/2}\left(1 + i\,\varepsilon(k)\right)\right\} & \text{(Lévy).} \end{cases} \tag{20.104}$$

That is, all that we need to be do is to *scale down the parameters* of the distributions by n (and by $\sqrt{n}$ in the case of the standard deviation of the Gaussian), to read off $\widetilde{p}_n(k)$ from $\widetilde{p}(k)$ itself. It is obvious that the *exponential form of the characteristic function* makes this special property possible.

A question now arises naturally. Let $\widetilde{p}(k)$ be the characteristic function of an *arbitrary* probability distribution. But $\widetilde{p}(k) \equiv \left[\left(\widetilde{p}(k)\right)^{1/n}\right]^n$, where n can be any positive integer. The condition (20.103) is then satisfied, with $\widetilde{p}_n(k) = \left(\widetilde{p}(k)\right)^{1/n}$. Can we conclude, then, that any arbitrary probability distribution is infinitely divisible? Certainly not! The given characteristic function $\widetilde{p}(k)$ is guaranteed, by definition, to have an inverse Fourier transform that is real and non-negative, and hence is a valid PDF. But there is no guarantee that $\left(\widetilde{p}(k)\right)^{1/n}$ shares this property for any $n \geq 2$. In fact, in general it does *not* do so. You need special forms of $\widetilde{p}_n(k)$ for this to happen, such as those corresponding to the stable distributions, the Poisson and Skellam distributions, etc.

20.6.2 *Infinite Divisibility Does Not Imply Stability*

It remains to give instances of distributions that are infinitely divisible, but are *not* stable distributions. A prominent example is the Poisson distribution, which is not a member of the family of Lévy skew alpha-stable distributions. The moment-generating function of a Poisson distribution with mean value μ is

$$M(u) = \exp\{\mu(e^u - 1)\} \quad \text{(Poisson).} \tag{20.105}$$

The characteristic function is therefore $\widetilde{p}(k) = M(-ik) = \exp\{\mu\,(e^{-ik} - 1)\}$. But this expression can be written as

$$\widetilde{p}(k) = \left[\exp\left\{(\mu/n)\,(e^{-ik} - 1)\right\}\right]^n \tag{20.106}$$

for every positive integer n, entailing a simple rescaling of the mean value. It is therefore obvious that the Poisson distribution is infinitely divisible. Moreover, for each n, the components of a Poisson-distributed random variable are themselves Poisson-distributed.

Similarly, the moment-generating function of a Skellam distribution with mean value $(\mu - \nu)$ and variance $(\mu + \nu)$ is given by

$$M(u) = \exp\{\mu(e^u - 1) - \nu(1 - e^{-u})\} \quad \text{(Skellam)}. \tag{20.107}$$

Hence

$$\widetilde{p}(k) = \left[\exp\left\{(\mu/n)\,(e^{-ik} - 1) - (\nu/n)\,(1 - e^{ik})\right\}\right]^n. \tag{20.108}$$

Once again, it is evident that the distribution is infinitely divisible.

19. The specific divisibility properties of the binomial and negative binomial distributions are also deduced easily.

(a) We have already seen that a random variable n, with a binomial distribution (Eq. (19.4) of Chap. 19, Sect. 19.1.2) given by

$$P_n = \binom{N}{n} p^n\,(1-p)^{N-n} \quad (0 \le n \le N),$$

is N-divisible into N Bernoulli trials (recall that $\mu = Np$ in this case). Establish this result formally, using the criterion in Eq. (20.103).

(b) Show that the negative binomial distribution with parameter N (Eq. (19.53) of Chap. 19, Sect. 19.3), given by

$$P_n = \binom{N+n-1}{n} p^N\,(1-p)^n \quad (n = 0, 1, \ldots \textit{ ad inf.}),$$

is N-divisible into N geometric distributions.

20.7 Solutions

2. (a) Although the sample space of X is obviously the discrete set of values $\{an + b\}$ where $n = 0,\ 1,\ \ldots,$ it is convenient to treat it formally as a continuous random variable with a PDF $p(x)$ that has δ-function contributions at these discrete values. Clearly, this PDF is given by

$$p(x) = \sum_{n=0}^{\infty} e^{-\mu}\,\frac{\mu^n}{n!}\,\delta(x - an - b).$$

It follows that the moment-generating function of X is

$$M(u) = \int_{-\infty}^{\infty} dx\, e^{ux}\, p(x) = e^{-\mu} \sum_{n=0}^{\infty} \frac{\mu^n}{n!}\, e^{u(an+b)} = \exp\left[ub + \mu(e^{au} - 1)\right].$$

Note that $M(0) = 1$, showing that $p(x)$ is a properly normalized PDF.
(b) Hence the cumulant-generating function is

$$K(u) = \ln\, M(u) = ub + \mu(e^{au} - 1).$$

It follows that $\kappa_1 = [dK/du]_{u=0} = a\mu + b$, while $\kappa_r = [d^r K/du^r]_{u=0} = a^r\,\mu$ for every integer $r \geq 2$.

Remark In particular, if $a = 1$, then X is just a shifted version of the Poisson-distributed random variable n. In consequence, κ_1 is shifted from μ to $\mu + b$, as required, while the higher cumulants remain unaltered, all of them being equal to μ as before. ▶

3. Consider the logarithm of P_n. Use Stirling's formula for $\ln\, n!$, replace n by $\langle n\rangle + x$, and P_n by the PDF $p(x)$. Use the fact that

$$\ln\,\big(\langle n\rangle + x\big) = \ln\,\langle n\rangle + \ln\,\left(1 + \frac{x}{\langle n\rangle}\right) \simeq \ln\,\langle n\rangle + \frac{x}{\langle n\rangle} - \frac{x^2}{2\langle n\rangle^2}\,.$$

After the cancelation of various terms, you will find that $\ln\, p(x)$ is proportional to $-x^2/(2\langle n\rangle)$, to leading order. We therefore conclude that X has the normal distribution $\mathcal{N}(0, \langle n\rangle)$. ▶

6. It is obvious that you must change to spherical polar coordinates in velocity space, and integrate over all the directions of the velocity, to get the PDF of the speed. For the PDF of the energy, use the relation $\phi^{\rm eq}(\varepsilon)\,d\varepsilon = \rho^{\rm eq}(v)\,dv$, where $\varepsilon = \frac{1}{2}mv^2$. ▶

7. The combination $(k_BT/m)^{1/2}$ has the physical dimensions of a velocity. (Recall that the rms velocity, the mean speed, etc., are all proportional to this combination.) Since $(n\sigma)$ has the physical dimensions of a length, it is obvious that the rate ν must be equal to $n\sigma(k_BT/m)^{1/2}$ times a numerical factor. Evaluating the multiple integral in Eq. (20.45) determines this numerical factor.

The mean collision rate is given by

$$\nu = n\sigma\Big(\frac{m}{2\pi k_BT}\Big)^3 \int d^3v_1 \int d^3v_2\, |\mathbf{v}_2 - \mathbf{v}_1|\, \exp\left\{-\,\frac{m\,(v_1^2 + v_2^2)}{2k_BT}\right\}.$$

The temperature-dependence of ν is immediately extracted by scaling the velocities appropriately: set

$$\mathbf{v}_1 = (k_BT/m)^{1/2}\,\mathbf{u}_1 \quad \text{and} \quad \mathbf{v}_2 = (k_BT/m)^{1/2}\,\mathbf{u}_2,$$

so that

$$d^3v_1 = (k_BT/m)^{3/2}\,d^3u_1 \quad \text{and} \quad d^3v_2 = (k_BT/m)^{3/2}\,d^3u_2.$$

Then

$$\nu = \frac{n\sigma}{(2\pi)^3}\Big(\frac{k_BT}{m}\Big)^{1/2} I,$$

where

$$I = \int d^3u_1 \int d^3u_2\, |\mathbf{u}_2 - \mathbf{u}_1|\, e^{-\frac{1}{2}(u_1^2+u_2^2)}.$$

Note that $\mathbf{u}_1$ and $\mathbf{u}_2$ are dimensionless, so that I is a pure number. The obvious next step is to change variables of integration from $\mathbf{u}_2$ to $\mathbf{u} = \mathbf{u}_2 - \mathbf{u}_1$. Doing so yields

$$I = \int d^3u_1 \int d^3u\, u\, e^{-u_1^2 - \frac{1}{2}u^2 - \mathbf{u}_1\cdot\mathbf{u}}.$$

It is most tempting to try to carry out $\int d^3u\,(\cdots)$ by changing to spherical polar coordinates, and choosing the polar axis along the direction of $\mathbf{u}_1$. Resist this temptation(!), because it leads to integrals of error functions in the subsequent integration over $\mathbf{u}_1$. Instead, interchange the order of integration and carry out $\int d^3u_1(\cdots)$ first, *in Cartesian coordinates*. The integral over $\mathbf{u}_1$ then *factorizes* into three shifted Gaussian integrals. If $\mathbf{u} = (x, y, z)$ and $\mathbf{u}_1 = (\xi, \eta, \zeta)$, say, then

$$\begin{aligned} I &= \int d^3u\, u\, e^{-\frac{1}{2}u^2} \int_{-\infty}^{\infty} d\xi\, e^{-\xi^2-\xi x} \int_{-\infty}^{\infty} d\eta\, e^{-\eta^2-\eta y} \int_{-\infty}^{\infty} d\zeta\, e^{-\zeta^2-\zeta z} \\ &= \pi^{3/2} \int d^3u\, u\, e^{-\frac{1}{4}u^2} = 4\,\pi^{5/2} \int_0^{\infty} du\, u^3\, e^{-\frac{1}{4}u^2} = 32\,\pi^{5/2}. \end{aligned}$$

Hence the mean collision rate $\nu = 4n\sigma[k_BT/(m\pi)]^{1/2}$. ▶

8. The "brute force" way would be to compute the PDF of Z_n as follows: Take the product of the n individual Gaussian PDFs $p(x_i)$ of the random variables X_i, multiply this by $\delta(z - z_n)$ where $z_n = \sum_{i=1}^{n}(x_i - \mu)/(\sigma\sqrt{n})$ (to take care of the definition of Z_n), and integrate the result over all the n variables x_i. It would be foolish to use the δ-function to carry out one of the integrations, because you would still be left with an $(n-1)$-fold integral. A better way is to write the Fourier representation for the δ-function. This would immediately factorize the argument of the δ-function into a product, and each of the resulting Gaussian integrals can be evaluated. The result is the PDF $(2\pi)^{-1/2}\, e^{-z^2/2}$, which corresponds to a standard normal distribution.

But there is an even simpler way to write down the answer! Recall the additivity property of the cumulants (or their generating function) for independent random variables, and the fact that they are invariant under a shift of the random variable by a constant. The cumulant-generating function of each $(X_i - \mu)$ is $\frac{1}{2}\sigma^2u^2$. It follows that the cumulant-generating function of Z_n is given by $\frac{1}{2}n\sigma^2u^2/(n\sigma^2) = \frac{1}{2}u^2$. Comparing this with the general expression in Eq. (20.36), we may conclude at once that Z_n has a Gaussian distribution with zero mean and unit variance. ▶

10. The sample space of Z is $[0, 2]$. Its PDF is of course given by

$$\rho_2(z) = \int_0^1 dx \int_0^1 dy\, \delta(x + y - z),$$

since the PDFs of X and Y are just unity in their sample spaces. Draw the unit square in the xy-plane, and note how the line $x + y = z$ intersects it when $0 < z < 1$ and $1 < z < 2$, respectively. Carrying out the integration with respect to y using the δ-function thus *restricts the range of integration* over x to the intervals $[0, z]$ and $[z - 1, 1]$, respectively, in the two cases. The result quoted follows immediately. ▶

13. (b) Choose spherical polar coordinates to evaluate the integral in (20.71). Without loss of generality, the polar axis may be chosen to be directed along the vector $\mathbf{k}$, greatly simplifying the evaluation of the integral. ▶

14. In the case $n = 2$, first write the product of sines as

$$\sin(kr)\,\sin^2(kl) = \tfrac{1}{2}\sin(kr) - \tfrac{1}{4}\big[\sin k(r - 2l) + \sin k(r + 2l)\big].$$

Now use the Dirichlet integral (Eq. (2.18) of Chap. 2, Sect. 2.3),

$$\int_0^\infty du\,\frac{(\sin bu)}{u} = \tfrac{1}{2}\pi\,\varepsilon(b).$$

Keep careful track of the various signum functions that occur.

In the case $n = 3$, write the product of sines in the form

$$\sin(kr)\,\sin^3(kl) = \tfrac{3}{8}\big[\cos k(r - l) - \cos k(r + l)\big] + \tfrac{1}{8}\big[\cos k(r + 3l) - \cos k(r - 3l)\big].$$

Add and subtract unity appropriately, and use the integral in Eq. (2.20), namely,

$$\int_0^\infty du\,\frac{(1 - \cos cu)}{u^2} = \tfrac{1}{2}\pi|c|.$$

When $n = 4$, express the product of sines in the form

$$\sin(kr)\,\sin^4(kl) = \tfrac{3}{8}\sin(kr) - \tfrac{1}{4}\big[\sin k(r - 2l) + \sin k(r + 2l)\big] + \tfrac{1}{16}\big[\sin k(r - 4l) + \sin k(r + 4l)\big].$$

The integral in Eq. (2.22), namely,

$$\int_0^\infty du\,\frac{(\sin cu - cu)}{u^3} = \tfrac{1}{4}\pi c\,|c|$$

will now lead you to the expression sought. ▶

15. The case $n = 1$ is trivial, since r takes on just one value, l. The mean squared value of r for a flight of n steps is, of course, nl^2. You have seen how this conclusion (Eq. (20.64)) emerges quite trivially from the simple geometrical argument given in Sect. 20.4.1. The mean value $\langle r\rangle$, however, is not found via such a simple argument,

except in the case $n = 2$. But it is quite straightforward to find $\langle r \rangle = \int_0^{nl} dr\, r\, \rho(r, n\tau)$ from the explicit solutions found above for the PDF $\rho(r, n\tau)$. A little calculation gives the following values:
(i) $n = 2$: $\langle r \rangle = \frac{4}{3}\, l$, $\text{Var}\,(r) = \frac{2}{9} l^2$, $\Delta r/\langle r \rangle = \frac{1}{2\sqrt{2}} \simeq 0.3536$.
(ii) $n = 3$: $\langle r \rangle = \frac{13}{8}\, l$, $\text{Var}\,(r) = \frac{23}{64} l^2$, $\Delta r/\langle r \rangle = \frac{\sqrt{23}}{13} \simeq 0.3689$.
(iii) $n = 4$: $\langle r \rangle = \frac{28}{15}\, l$, $\text{Var}\,(r) = \frac{116}{225} l^2$, $\Delta r/\langle r \rangle = \frac{\sqrt{29}}{14} \simeq 0.3847$.
It is evident that the relative fluctuation increases slowly with increasing n. In the diffusion limit, the solution (20.80) for $p(\mathbf{r}, t)$ gives

$$\begin{aligned}\langle r \rangle &= (4\pi D t)^{-3/2} \int d^3 r\, r\, e^{-r^2/(4Dt)} \\ &= (4\pi)(4\pi D t)^{-3/2} \int_0^\infty dr\, r^3\, e^{-r^2/(4Dt)} = 4(Dt/\pi)^{1/2}.\end{aligned}$$

Since $\langle r^2 \rangle = 6Dt$, as already found in (20.81), the variance is $(6\pi - 16)\, Dt/\pi$. Hence the relative fluctuation is

$$\Delta r/\langle r \rangle = \left(\tfrac{3}{8}\pi - 1\right)^{1/2} \simeq 0.4220 \text{ in the diffusion limit.}$$

This is the saturation value of the relative fluctuation in the end-to-end distance. ▶

16. The sample space of ξ is obviously $[0, \infty)$. We have

$$\begin{aligned}\rho(\xi) &= \frac{1}{(2\pi\sigma^2)^{1/2}} \int_{-\infty}^{\infty} dx\, e^{-x^2/(2\sigma^2)}\, \delta(\xi - x^{-2}) \\ &= \frac{1}{(2\pi\sigma^2)^{1/2}} \int_{-\infty}^{\infty} dx\, e^{-x^2/(2\sigma^2)} \left\{ \frac{\delta(x + \xi^{-1/2}) + \delta(x - \xi^{-1/2})}{2\xi^{3/2}} \right\}\end{aligned}$$

on writing $\delta(\xi - x^{-2})$ in terms of δ-functions at $x = \pm 1/\sqrt{\xi}$, using the formula in Eq. (4.23) of Chap. 4, Sect. 4.2.5. The integration over x is now trivially done. ▶

17. The random variable $\xi_j = 1/X_j^2$ has a Lévy distribution, with parameter $c = 1/\sigma^2$. Since $X_1, \ldots, X_n$ are *iiddv*, so are $\xi_1, \ldots, \xi_n$. Let $\zeta_n = 1/Z_n^2$. Now observe that

$$\frac{1}{Z_n^2} = \frac{1}{X_1^2} + \cdots + \frac{1}{X_n^2}, \quad \text{so that} \quad \zeta_n = \xi_1 + \cdots + \xi_n\,.$$

The characteristic function of each ξ_j is given by Eq. (20.94), and is

$$e^{-c|k|^{1/2}\,(1+i\,\varepsilon(k))}.$$

The characteristic function of ζ_n is just the nth power of this quantity, and is therefore

$$e^{-nc|k|^{1/2}\,(1+i\,\varepsilon(k))}.$$

That is, $\zeta_n = 1/Z_n^2$ is also Lévy-distributed, with parameter nc. Hence Z_n has a Gaussian distribution with zero mean and variance $1/nc = \sigma^2/n$, i.e., its distribution is $\mathcal{N}(0, \sigma^2/n)$.

Remark A similar result holds good when $X_1, \ldots, X_n$ are independent, but not *identically* distributed, Gaussian random variables. Let X_j have the distribution $\mathcal{N}(0, \sigma_j^2)$. Then Z_n is a Gaussian random variable with zero mean and variance $\big[\sum_1^n (1/\sigma_j^2)\big]^{-1}$. ▶

18. Write the δ-function in Eq. (20.100) as $|x_2|\,\delta(x_1 - \xi\, x_2)$ and eliminate the integration over x_1. What remains is a simple integral over x_2. ▶

19. (a) The characteristic function of the binomial distribution with parameter N is $\widetilde{p}(k) = (q + p\, e^{-ik})^N$. This is a valid characteristic function for every positive integer N, and hence for $N = 1$ as well. Therefore the distribution is N-divisible, and each component has the characteristic function $q + p\, e^{-ik}$.

Now consider a Bernoulli trial whose sample space is $\{0, 1\}$, with respective probabilities $P_0 = q$ and $P_1 = p$, or, equivalently, a PDF $p(x) = q\delta(x) + p\delta(x - 1)$. Its characteristic function is then given by

$$\widetilde{p}(k) = \int_{-\infty}^{\infty} dx\, e^{-ikx}\, p(x) = q + p\, e^{-ik}.$$

Hence the binomial distribution with parameter N is N-divisible, each component being a Bernoulli trial.
(b) The characteristic function of the negative binomial distribution with parameter N and mean value μ is given by

$$\widetilde{p}(k) = \left(\frac{N}{N + \mu - \mu\, e^{-ik}}\right)^N.$$

Setting $N = 1$ in this expression gives us the characteristic function of the geometric distribution with mean value μ. Now, we can write $\widetilde{p}(k)$ as

$$\widetilde{p}(k) = \left(\frac{1}{1 + (\mu/N) - (\mu/N)\, e^{-ik}}\right)^N.$$

But $[1 + (\mu/N) - (\mu/N)\, e^{-ik}]^{-1}$ is precisely the characteristic function of a geometric distribution with mean value μ/N. The corresponding probability distribution is given by

$$P_n = \frac{N}{\mu + N}\left(\frac{\mu}{\mu + N}\right)^n, \quad n = 0, 1, \ldots.$$

Hence a random variable specified by a negative binomial distribution with an exponent N and a mean value μ is N-divisible into N geometrically distributed *iidrv*, each with a mean value μ/N. ▶

Chapter 21
Stochastic Processes

In this third and final chapter on random variables, I give a very brief introduction to **stochastic processes**, also known as random processes: that is, random variables which take on different values from their sample spaces as time elapses. Stochastic processes comprise a rather extensive subject by themselves. I will focus here on an important subclass of these processes, namely, Markov processes. These are the processes that are used most often in applications.

It is to be understood that, as usual, we deal with probabilities and probability densities, respectively, for discrete and continuous sample spaces. As before, I shall use the symbols P and p, respectively, for these quantities. For notational simplicity, I shall deal with the discrete case for the most part. But what follows is adaptable, with obvious modifications, to the continuous case. We will discuss this case briefly in Sect. 21.6.

21.1 Multiple-Time Joint Probabilities

Consider a random variable X whose sample space consists of a discrete set of N values, $\{x_1, \ldots, x_N\}$. Instead of referring to the values themselves, it is notationally convenient to specify them as "states" labeled from 1 to N. The indices $j, k, l, \ldots$ will be used to denote these state labels. Alternatively, I will use $j_1, j_2, \ldots$ for the state labels, when the need arises to introduce an arbitrarily large number of labels. The formalism below is also applicable when N is infinite, provided certain convergence conditions are satisfied. But we will not get into these technicalities here. As the random process evolves in time, the state of the variable changes randomly.

Let $t_1 < t_2 < t_3 < \ldots$ be an arbitrary sequence of instants of time. The statistical properties of the random process are specified completely by an infinite set of multiple-time **joint probabilities**

V. Balakrishnan, *Mathematical Physics*,
https://doi.org/10.1007/978-3-030-39680-0_21

$$P_1(j,\, t_1),\; P_2(k,\, t_2\,;\; j,\, t_1),\; P_3(l,\, t_3\,;\; k,\, t_2\,;\; j,\, t_1),\; \dots \quad \textit{ad infinitum}. \tag{21.1}$$

The n-time joint probability $P_n(j_n,\, t_n\,;\; j_{n-1},\, t_{n-1}\,;\; \dots\,;\; j_1,\, t_1)$ is the probability that the random variable has the values corresponding to the state j_1 at time t_1, the state j_2 at time t_2, and so on, and the state j_n at time t_n. This n-time joint probability is expressible as the product of an n-time **conditional probability** and an $(n-1)$-time joint probability, according to the **chain rule**

$$\begin{aligned} P_n(j_n,\, t_n\,;\; j_{n-1},\, t_{n-1}\,;\dots;\; j_1,\, t_1) = \;& P_n(j_n,\, t_n \mid j_{n-1},\, t_{n-1}\,;\; \dots\,;\; j_1,\, t_1) \times \\ & \times P_{n-1}(j_{n-1},\, t_{n-1}\,;\; j_{n-2},\, t_{n-2}\,;\; \cdots\,;\; j_1,\, t_1). \end{aligned} \tag{21.2}$$

The first factor on the right-hand side is a conditional probability: it is the probability that the state j_n occurs at time t_n, *given* that the events to the right of the vertical bar have occurred at the successively earlier instants $t_{n-1},\, t_{n-2},\, \dots,\, t_1$, respectively. In turn, the joint probability P_{n-1} on the right-hand side can be written as the product of an $(n-1)$-time conditional probability and a joint probability P_{n-2}, and so on. Thus,

- a stochastic process is specified by an infinite hierarchy of multiple-time conditional probabilities, and a single-time probability $P_1(j, t)$.

This is convenient: in all physical applications of probability and random processes, we can only write equations for, or model, *conditional* probabilities or probability densities—rather than the corresponding absolute probabilities themselves.

21.2 Discrete Markov Processes

21.2.1 The Two-Time Conditional Probability

In practice, it is essentially impossible to determine or specify all the members of the infinite set of conditional probabilities. We must make further simplifying assumptions. Fortunately for us, most physical stochastic processes are also amenable to modeling in terms of such simplified processes. The most notable of these is the class of **Markov processes**.

Markov processes can have discrete or continuous sample spaces. As already stated, the latter case will be considered in Sect. 21.6. Discrete Markov processes may also be defined in discrete *time*, in which case they are called **Markov chains**. There is a vast literature on this subject. But we shall restrict ourselves here to the case of continuous time. Arguably, this is the case that occurs most commonly in physical applications such as nonequilibrium statistical mechanics. The results to be discussed below are appropriate extensions of the corresponding results for Markov chains.

A Markov process is one with a "memory" that is restricted, at any instant of time, to the *immediately preceding time argument alone*. That is

$$\boxed{P_n(j_n, t_n \mid j_{n-1}, t_{n-1}; \ldots; j_1, t_1) = P_2(j_n, t_n \mid j_{n-1}, t_{n-1}) \quad \text{for all} \quad n \geq 2.} \tag{21.3}$$

- The single-time-step memory characterizing a Markov process is equivalent to saying that the *future* state of the process is only dependent on its *present* state, and not on the history of *how* the process reached the present state.

Although this appears to be a mild technical assumption, it leads to a great deal of simplification. It has an immediate implication:

- *All* the joint probabilities of a Markov process are expressible as products of just two independent probabilities: (i) the *single-time* probability $P_1(j, t)$, and (ii) the *two-time* conditional probability $P_2(k, t \mid j, t')$, where $t' < t$.

Repeated application of Eq. (21.3) to Eq. (21.2) shows that, for a Markov process,

$$\begin{aligned} P_n(j_n, t_n; j_{n-1}, t_{n-1}; \ldots; j_1, t_1) = \; & P_2(j_n, t_n \mid j_{n-1}, t_{n-1}) \times \\ & P_2(j_{n-1}, t_{n-1} \mid j_{n-2}, t_{n-2}) \cdots \times \\ & P_2(j_2, t_2 \mid j_1, t_1)\, P_1(j_1, t_1). \end{aligned} \tag{21.4}$$

Further simplification occurs in the case of a **stationary random process**.

- A stationary random process is one whose *statistical* properties do not depend on the choice of the origin of time.

You will recognize that stationarity is the probabilistic equivalent of the assumption of **time-translation invariance** in deterministic dynamics (classical or quantum). In the mathematics literature, a stationary random process is often called a *homogeneous* random process, because stationarity is the same thing as homogeneity in time. For a stationary random process, the single-time probability $P_1(j, t)$ is actually *independent* of t:

$$P_1(j, t) = P_1(j). \tag{21.5}$$

Moreover, the two-time conditional probability

$$P_2(k, t \mid j, t') = P_2(k, t - t' \mid j, 0). \tag{21.6}$$

That is, it is a function of the time *difference* $(t - t')$ alone. It is convenient to use the notation

$$P_2(k, t - t' \mid j) \stackrel{\text{def.}}{=} P_2(k, t - t' \mid j, 0). \tag{21.7}$$

Further, let us drop the subscripts 1 and 2 in P_1 and P_2, and write the two different functions as simply $P(j)$ and $P(k, t - t' \mid j)$, for notational simplicity. No confusion should arise, although the same symbol P is used for two different functions: The first probability is time-independent, while the second is not. Whenever necessary, I shall refer to $P(j)$ as the *stationary probability*, and to $P(k,\, t - t' \mid j)$ or $P(k,\, t \mid j)$ as the *conditional probability*. We therefore have, for a stationary Markov process,

$$\boxed{P_n(j_n,\, t_n\,;\; j_{n-1},\, t_{n-1}\,;\; \dots\,;\; j_1,\, t_1) = \left\{\prod_{r=1}^{n-1} P(j_{r+1},\, t_{r+1} - t_r \mid j_r)\right\} P(j_1),} \tag{21.8}$$

for every $n \geq 2$. From this point onward, I consider stationary processes, unless otherwise specified. This is entirely for convenience—the notation simplifies somewhat in the case of stationary processes.

Next, we need inputs for the stationary probability $P(j)$ and the conditional probability $P(k,\, t \mid j)$. These are, a priori, independent quantities. Now, in physical situations modeled by Markov processes, the dynamics underlying the random process generally enjoys a sufficient degree of **mixing**. This is a technical term in dynamical systems theory. It refers to a kind of "scrambling up" that occurs in phase space under time evolution for most real-life dynamical systems. (I do not digress into further detail here.) The mixing property ensures that the following important relationship holds good:

$$\boxed{\lim_{t\to\infty} P(k,\, t \mid j) = P(k).} \tag{21.9}$$

In other words, the memory of the initial state j is "lost" as $t \to \infty$, and the conditional probability simply tends to the stationary probability $P(k)$ corresponding to the final state k. As a consequence,

- the single conditional probability distribution $P(k,\, t \mid j)$ completely determines *all* the statistical properties of such a stationary Markov process.

The property implied by Eq. (21.9) appears to be quite plausible on physical grounds. As I have indicated, however, there are technical issues involved here. One of them, which is pertinent to current work in nonequilibrium statistical physics, is whether there exists a unique stationary distribution, as opposed to a whole set of such distributions. The latter possibility is indeed realized in many physical situations. Once again, we cannot go into these nontrivial and interesting questions here as they would take us too far afield.

21.2.2 The Master Equation

The next step is to find an equation satisfied by the conditional probability $P(k,\, t \mid j)$ of a Markov process. A starting point for the derivation of this equation is the **Chapman–Kolmogorov equation** satisfied by Markov processes:

$$P(k, t \mid j) = \sum_{l=1}^{N} P(k, t - t' \mid l)\; P(l, t' \mid j) \quad \text{for any } t' \in (0, t). \tag{21.10}$$

In effect, this equation says

- The probability of "propagating" from an initial state j to a final state k is the product of the probabilities of propagating from j to any *intermediate* state l, and subsequently from l to k, summed over all possible intermediate states.

Equation (21.10) is a necessary *but not sufficient* condition for a random process to be Markovian. Contrary to a common impression, it is not unique to Markov processes. It is an identity (a so-called **renewal equation**) that is satisfied by a more general family of stochastic processes. But here we are only concerned with Markov processes. The Chapman–Kolmogorov equation is applicable to continuous Markov processes as well. The corresponding version of the equation will be written down in Eq. (21.69) of Sect. 21.6.1.

Observe that the Chapman–Kolmogorov equation is a *nonlinear* equation for the conditional probability, since the right-hand side is quadratic in P. Nonlinear equations are generally much harder to handle than linear ones. Equation (21.10) can, however, be reduced to a linear equation by defining a transition probability per unit time, or **transition rate** $w(k \mid j)$ between any two distinct states j and k, as follows. Recall that, for a stationary process, $P(k, t + \delta t \mid j, t) = P(k, \delta t \mid j)$. We now assume that, in any infinitesimal time interval δt, the probability $P(k, \delta t \mid j)$ is proportional to δt, with a coefficient of proportionality that depends on j and k, i.e.,

$$P(k, \delta t \mid j) = w(k \mid j)\, \delta t \quad (j \neq k). \tag{21.11}$$

Such a relation is not always guaranteed for all Markov processes. When it does hold good, it *defines* the transition rate $w(k \mid j)$. What follows below is therefore valid for those Markov processes for which finite transition rates exist between the states. The quantity $w(k \mid j)$ is the probability per unit time that the random variable jumps from (its value in) the state j to (its value in) the state k. It can then be shown that Eq. (21.10) leads to the following differential equation for the set of conditional probabilities $\{P(k, t \mid j)\}$:

$$\boxed{\frac{d}{dt} P(k, t \mid j) = \sum_{\substack{l=1 \\ l \neq k}}^{N} \Big\{ w(k \mid l)\, P(l, t \mid j) - w(l \mid k)\, P(k, t \mid j) \Big\}.} \tag{21.12}$$

For each initial state j, this equation is satisfied for every k from 1 to N. The initial condition is of course

$$P(k, 0 \mid j) = \delta_{jk}\,. \tag{21.13}$$

Equation (21.12) is called the **master equation** (for a discrete Markov process). It has the form of a **rate equation** of the sort that is familiar, for instance, in chemical physics. Viewed this way, it is clear that the first term on the right in (21.12) is a "gain" term, while the second is a "loss" term.

★ **1.** Derive the master equation (21.12) from the Chapman–Kolmogorov equation (21.10).

Why are Markov processes so important? Observe that the master equation (21.12) for the conditional probability distribution is a *first-order* differential equation in the time variable. *This is a direct consequence of the Markovian nature of the random process.* The solution of the equation requires a single piece of information: namely, the specification of the initial distribution $P(k, 0 \mid j)$. No *earlier* history is necessary, i.e., no information is required about *how* that initial distribution was attained. But this is precisely what holds good in the case of *deterministic* dynamics as well, both classical and quantum! In classical dynamics, Hamilton's equations for the variables in phase space are first-order differential equations in t. In quantum mechanics, the Schrödinger equation for the state vector, or the Liouville equation for the density matrix, are first-order differential equations in t.

- In this sense, the master equation for a Markov process is the stochastic analog of the fundamental equations of deterministic dynamics.

Recall that classical dynamics is "first-order" dynamics, provided we identify and work in terms of the right number of dependent variables, i.e., the generalized coordinates *as well as* the generalized momenta pertaining to a system.[1] Similarly, a given physical situation may appear to require modeling by a stochastic process that is more history-dependent than a Markov process. But it often turns out that this is in fact a Markov process in a coupled *set* of random variables, i.e., by a *multi-component* Markov process. This possibility is the reason why Markov processes make their appearance everywhere.

21.2.3 Formal Solution of the Master Equation

Next, let us turn to the formal solution of the master equation. Let $\mathsf{P}(t)$ denote the column vector whose kth row is $P(k, t \mid j)$. Although the initial state j is not explicit in this notation, we will keep it in mind. (The master equation is a time evolution equation for *conditional* probabilities.) Equation (21.12) can then be written as the matrix equation

$$\frac{d\mathsf{P}}{dt} = W\,\mathsf{P}, \tag{21.14}$$

where W is called the **relaxation matrix** in the physics literature. The reason will become clear shortly. This $(N \times N)$ matrix has off-diagonal elements

[1] A simple example: as you know, the familiar second-order equation $m\ddot{x} = F(x)$ actually arises from the pair of coupled *first*-order equations $\dot{x} = p/m$ and $\dot{p} = F(x)$.

$$W_{kj} \stackrel{\text{def.}}{=} w(k \mid j) \quad (k \neq j) \tag{21.15}$$

and diagonal elements

$$W_{kk} = -\sum_{\substack{l=1 \\ l \neq k}}^{N} w(l \mid k). \tag{21.16}$$

W is *not* a symmetric matrix, in general. You must note that, in the literature, the relaxation matrix is often defined such that its (kj)th element is equal to $w(j \mid k)$ (rather than $w(k \mid j)$). But I will stick to the notation in Eq. (21.15).

Since the elements of W do not depend on time (we are considering a stationary process), the formal solution to the matrix differential equation (21.14) is simply

$$\mathsf{P}(t) = e^{Wt}\, \mathsf{P}(0), \tag{21.17}$$

where $\mathsf{P}(0)$ is a column vector whose jth row is 1, and all other elements are 0. (This is where the initial state j makes its appearance.) It follows from Eq. (21.17) that the time-dependence of the probability distribution is essentially determined by the eigenvalue spectrum of the matrix W.

Finding the exponential of the matrix Wt in explicit form is generally not a simple task. Recall that in Chap. 11, Sect. 11.2, we have seen how to exponentiate an arbitrary (2×2) matrix. But it was also pointed out that there is no simple general formula of this sort for matrices of higher order. A more convenient way of dealing with the problem is to work with Laplace transforms. We will discuss these transforms in Chap. 28, but I mention the relevant result here. Let $\widetilde{\mathsf{P}}(s)$ denote the Laplace transform of $\mathsf{P}(t)$. Then Eq. (21.17) is equivalent to

$$\widetilde{\mathsf{P}}(s) = \big(sI - W\big)^{-1}\, \mathsf{P}(0), \tag{21.18}$$

where I is the $(N \times N)$ unit matrix. The matrix $(sI - W)^{-1}$ is, of course, the resolvent of the transition matrix W. (Recall the definition of the resolvent of a matrix given in Eq. (11.54) of Chap. 11, Sect. 11.4.4.) The problem of exponentiating the matrix W is replaced by the relatively simpler one of finding the *inverse* of a related matrix. We are still left with the task of inverting the Laplace transform, of course.

The known properties of the relaxation matrix W make it possible to say something about its eigenvalue spectrum. Let us restrict ourselves here to the most general statements, without going into the details of the necessary and sufficient conditions for their validity, exceptional cases, and so on. Note that the sum of the elements of each column of W is equal to zero. It follows at once that $\det W = 0$. Hence 0 is an eigenvalue of this matrix. (The physical implication of this fact will become clear shortly.) The elements of W are real. Hence its eigenvalues occur in complex conjugate pairs. Now apply Gershgorin's Circle Theorem (Chap. 11, Sect. 11.4.2) to W. The centers of the Gershgorin discs are located on the negative real axis of the complex plane, at the points $W_{kk} = -|W_{kk}|$, where k runs from 1 to N. The

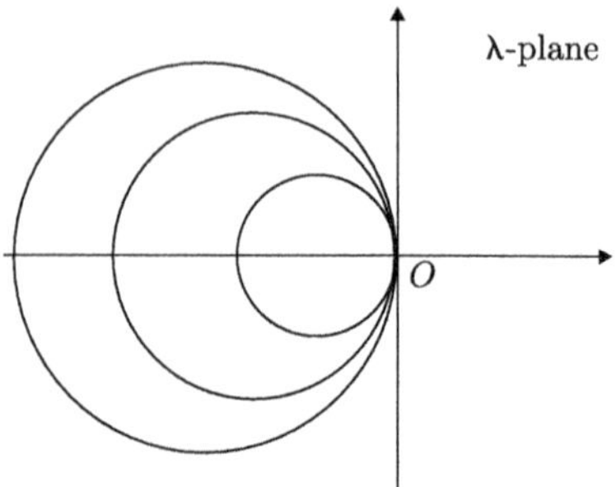

Fig. 21.1 Gershgorin discs for a relaxation matrix W. All the eigenvalues of W must lie in the union of these discs

radii of the circles are given by $|W_{kk}|$. The right-most points of all the Gershgorin discs are therefore at the origin, as shown in Fig. 21.1. Hence the real parts of all the eigenvalues are negative, *except for the eigenvalue* 0. It follows that the time evolution of the probabilities, which is governed by e^{Wt}, is given by *decaying* exponentials in t, possibly multiplied by factors involving sinusoidal functions of t. This property justifies the term "relaxation matrix" for W.

21.2.4 The Stationary Distribution

To what distribution, if any, do the conditional probabilities $P(k, t \mid j)$ "relax" as $t \to \infty$? An important aspect of the master equation concerns the conditions under which there exists a unique time-independent distribution P^{st} whose kth row is the stationary probability $P(k)$, as in Eq. (21.9). This question becomes even more nontrivial for discrete Markov processes with an an infinite-dimensional state space ($N \to \infty$), and for continuous Markov processes. Here I shall only consider the case when N is finite and, further, a unique stationary distribution P^{st} exists. This distribution is given by the solution of

$$\frac{d\mathsf{P}^{\text{st}}}{dt} = W\,\mathsf{P}^{\text{st}} = 0. \tag{21.19}$$

- The stationary distribution P^{st} is, therefore, given by the elements of the appropriately normalized *right* eigenvector of the matrix W corresponding to the eigenvalue 0.

As each column of W adds up to zero, it is evident that the uniform *row* vector $\big(1\ 1\ \cdots\ 1\big)$ is the (unnormalized) *left* eigenvector of W with eigenvalue zero. Since W is not symmetric in general, the right and left eigenvectors are not adjoints of each other. It is also necessary to show that the elements of the right eigenvector are nonnegative numbers that add up to unity, after normalization. I do not go into the proof of this assertion here, but merely mention that it is based on the **Frobenius–Perron Theorem** for real nonnegative matrices.

From the explicit form (21.12) of the master equation in terms of the transition rates, it is clear that the stationary distribution is given by the solution of the set of N homogeneous simultaneous equations

$$\sum_{\substack{l=1 \\ l \neq k}}^{N} \Big\{ w(k\,|\,l)\, P(l) - w(l\,|\,k)\, P(k) \Big\} = 0, \quad (k = 1, 2, \ldots, N). \tag{21.20}$$

Since $\det W = 0$, we are guaranteed that this set of homogeneous simultaneous equations has unique nontrivial solutions for the *ratios* of the unknowns, say $P(2)/P(1),\; P(3)/P(1), \ldots, P(N)/P(1)$. We need one more equation, an *inhomogeneous* one, to fix the actual values of the probabilities. As you would have guessed, this is provided by the normalization condition $\sum_{k=1}^{N} P(k) = 1$. All the probabilities $\{P(k)\}$ are now fully determined.

- The stationary distribution P^{st} is thus expressible in terms of the set of transition rates of the Markov process.

When $N = 2$, we have a 2-state Markov process, also called a **dichotomous Markov process** (DMP). I will consider the DMP in greater detail in Sect. 21.4. The stationary distribution is given in this case by

$$P(1) = \frac{W_{12}}{W_{12} + W_{21}} \quad \text{and} \quad P(2) = \frac{W_{21}}{W_{12} + W_{21}}. \tag{21.21}$$

The stationary distribution for a general 3-state Markov process is already considerably more complicated. It is given by

$$P(k) = \mathcal{N}_k / \mathcal{D}_k, \tag{21.22}$$

where

$$\begin{aligned} \mathcal{N}_1 &= W_{12} W_{13} + W_{13} W_{32} + W_{12} W_{23}, \\ \mathcal{D}_1 &= \mathcal{N}_1 + W_{21} W_{13} + W_{31} W_{12} + W_{32} W_{21} + W_{23} W_{31} + W_{31} W_{32} + W_{21} W_{23}, \end{aligned} \tag{21.23}$$

and $\mathcal{N}_2,\ \mathcal{D}_2,\ \mathcal{N}_3,\ \mathcal{D}_3$ are given by cyclic permutations of the expressions above.

★ **2.** Verify the solutions in Eqs. (21.21) and (21.22)–(21.23) for 2-state and 3-state Markov processes, respectively.

21.2.5 Detailed Balance

It is clear that, even for the stationary distribution, the expressions rapidly get more and more complicated as N increases. But there is a very important special case in which the solution is simplified considerably. This is the situation in which the **detailed balance** condition applies: namely, when *each* term in the summand in Eq. (21.20) vanishes by itself. We then have

$$\boxed{w(k\,|\,l)\,P(l) = w(l\,|\,k)\,P(k)} \quad \text{(detailed balance)} \tag{21.24}$$

for *every pair* of distinct states k and l. The detailed balance condition has its origin in **time reversal invariance**. Its importance arises from the fact that *it is applicable to systems in the state of thermal equilibrium*, under normal circumstances. The corresponding stationary distribution may then be termed the **equilibrium distribution**, with probabilities denoted by $P^{\text{eq}}(k)$. These probabilities can be found easily in terms of the transition rates. The result is the compact formula

$$\boxed{P^{\text{eq}}(k) = \left\{1 + \sum_{\substack{l=1\\ l\neq k}}^{N} \frac{w(l\,|\,k)}{w(k\,|\,l)}\right\}^{-1}.} \tag{21.25}$$

★ **3.** Use detailed balance and the normalization condition $\sum_{k=1}^{N} P^{\text{eq}}(k) = 1$ to derive Eq. (21.25).

The case of a symmetric relaxation matrix W**:** In the special case of a symmetric relaxation matrix W, i.e., when $w(l\,|\,k) = w(k\,|\,l)$, more can be said on general grounds. As W is a real symmetric matrix, it is diagonalizable by an orthogonal transformation. All its eigenvalues are real. Taken together with the conclusions we drew in Sect. 21.2.3 based on the Gershgorin Circle Theorem, this implies that the eigenvalues are negative, except for the eigenvalue 0. The corresponding normalized right eigenvector is the equilibrium distribution, which reduces in this case to the *uniform distribution* $P^{\text{eq}}(k) = 1/N$ for every k. This should remind you of the postulate of equal a priori probabilities for the accessible microstates of an isolated macroscopic system in thermal equilibrium (the fundamental postulate of equilibrium statistical mechanics).

21.3 The Autocorrelation Function

One of the most important quantities associated with any stochastic process is its **autocorrelation function**. This object is a generalization of the variance of a random variable. In essence, it is the primary measure of the degree of "memory" possessed by the random variable as it evolves in time. What follows below pertains to the case of a discrete-valued random process $X(t)$. The extension of the formulas to a continuous random process is straightforward, and is given in Sect. 21.6.3. Bear in mind that the formulas of the present section are valid for all stationary discrete stochastic processes, and not just Markov processes. In the case of a Markov process, however, a knowledge of the autocorrelation function is of particular significance. This is because the autocorrelation function depends on the conditional probability; and the latter essentially determines *all* the joint probabilities associated with a Markov process, as we know from Eqs. (21.4) and (21.8) above.

Consider the average value of the product of the random variable at time t_1 with itself at time t_2. We have, by definition,

$$\begin{aligned}\langle X(t_1)\,X(t_2)\rangle &\stackrel{\text{def.}}{=} \sum_j\sum_k x_j\,x_k\;P_2(k,t_2\,;\;j,\,t_1)\\ &= \sum_j\sum_k x_j\,x_k\;P_2(k,t_2\,|\,j,\,t_1)\,P_1(j,\,t_1).\end{aligned} \tag{21.26}$$

For a stationary random process this becomes, on writing $t_2 - t_1$ as simply t,

$$\boxed{C_X(t) \stackrel{\text{def.}}{=} \langle X(0)\,X(t)\rangle = \sum_j\sum_k x_j\,x_k\;P(k,t\,|\,j)\,P(j),} \tag{21.27}$$

in terms of the stationary probability and the conditional probability. This is the autocorrelation function $C_X(t)$ of the stationary, discrete random process $X(t)$, in the case when the mean value is zero at all times.

When the mean value of a stationary random process is nonzero, its autocorrelation function is defined in terms of the *deviation* from the mean value, namely, $\delta X = X - \langle X\rangle$. We have in this case

$$C_X(t) \stackrel{\text{def.}}{=} \langle \delta X(0)\,\delta X(t)\rangle. \tag{21.28}$$

We then find

$$C_X(t) = \Big\{\sum_j\sum_k x_j\,x_k\;P(k,t\,|\,j)\,P(j)\Big\} - \langle X\rangle^2, \tag{21.29}$$

where, of course,

$$\langle X\rangle = \sum_j x_j\;P(j). \tag{21.30}$$

★ **4.** Verify Eq. (21.29).

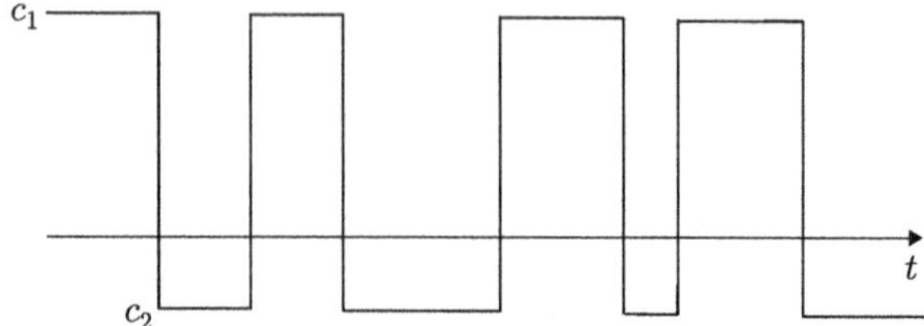

Fig. 21.2 Schematic of a realization of a dichotomous Markov process

21.4 The Dichotomous Markov Process

The dichotomous Markov process (or DMP) is a very basic random process that appears in various forms in numerous applications in science and engineering. It is a stationary Markov, discrete (or jump) process $X(t)$ in which the sample space consists of just two values, c_1 and c_2. The random variable X flips back and forth between these values (or between the states 1 and 2) at random instants of time. Figure 21.2 shows schematically (a part of) a realization of such a process. Let λ_1 be the mean transition rate from c_1 to c_2, and λ_2 the mean transition rate from c_2 to c_1. Successive transitions are supposed to be completely uncorrelated with each other. The transition matrix for this 2-state process is therefore

$$W = \begin{pmatrix} -\lambda_1 & \lambda_2 \\ \lambda_1 & -\lambda_2 \end{pmatrix}. \tag{21.31}$$

It is natural to expect the DMP to play an important role in the description of the stochastic dynamics of two-level systems, in general.

21.4.1 The Stationary Distribution

Before we go on to the solution of the master equation in this case, let us understand the physical significance of the parameters λ_1 and λ_2. Suppose the system is in the state 1 at any instant of time. The probability that it will make a transition to the state 2 in the next infinitesimal time interval δt is just $\lambda_1\,\delta t$. Now, δt is taken to be so small that the probability of two or more transitions occurring within this time interval is negligible. Hence the probability that it will *not* make a transition in the interval δt is $(1 - \lambda_1\,\delta t)$. From this fact, it is easy to calculate the probability that, if the system has just arrived at the state 1 at some instant of time t_0, it remains in that state at time $t\ (\geq t_0)$ without having made any transitions in between. All we have to do is to divide the interval $(t - t_0)$ into n subintervals, each of duration δt. Since the process is Markovian, there is no history-dependence. The no-transition or zero-transition probability sought is therefore $(1 - \lambda_1\,\delta t)^n$, in the limit $n \to \infty$ and $\delta t \to 0$ such that the product $n\,\delta t = (t - t_0)$. Hence

$$\Pr\left\{\text{stay in state 1 during an interval } (t-t_0)\right\} = \lim_{\substack{n\to\infty \\ \delta t\to 0}} (1-\lambda_1\,\delta t)^n$$
$$= e^{-\lambda_1 (t-t_0)}. \tag{21.32}$$

Exactly the same argument applies to the state 2, with λ_1 replaced by λ_2. Thus

$$\Pr\left\{\text{stay in state 2 during an interval } (t-t_0)\right\} = e^{-\lambda_2 (t-t_0)}. \tag{21.33}$$

It follows that the **mean residence time** in the state 1 between two successive transitions is $\tau_1 = \lambda_1^{-1}$. Similarly, the mean residence time in state 2, between successive transitions, is $\tau_2 = \lambda_2^{-1}$. We may therefore expect that, over a very long interval of time, the fraction of the total time that the system spends in states 1 and 2 are, on the average,

$$\frac{\tau_1}{\tau_1+\tau_2} = \frac{\lambda_2}{\lambda_1+\lambda_2} \quad\text{and}\quad \frac{\tau_2}{\tau_1+\tau_2} = \frac{\lambda_1}{\lambda_1+\lambda_2}, \quad\text{respectively.} \tag{21.34}$$

This would imply that the a priori probabilities $P(1)$ and $P(2)$ of finding the system in states 1 and 2 are, respectively, precisely the fractions written down in (21.34). But you have already seen that this is indeed so—recall Eq. (21.21) of Sect. 21.2.4 for the stationary distribution of a 2-state stationary Markov process.

Given these expressions for the stationary probabilities $P(1)$ and $P(2)$, it is easy to write down the mean and variance of X. We find

$$\langle X\rangle = \sum_{j=1}^{2} c_j\, P(j) = \frac{\tau_1\, c_1 + \tau_2\, c_2}{\tau_1+\tau_2} = \frac{c_1\,\lambda_2 + c_2\,\lambda_1}{\lambda_1+\lambda_2}, \tag{21.35}$$

and similarly

$$\langle (X-\langle X\rangle)^2\rangle = \frac{\tau_1\,\tau_2\,(c_1-c_2)^2}{(\tau_1+\tau_2)^2} = \frac{\lambda_1\,\lambda_2\,(c_1-c_2)^2}{(\lambda_1+\lambda_2)^2}. \tag{21.36}$$

★ **5.** Verify Eqs. (21.35) and (21.36).

21.4.2 Solution of the Master Equation

As already defined for an N-state Markov process, let $\mathsf{P}(t)$ denote the column vector whose kth row is $P(k, t\,|\,j)$. For a given initial state j (which can be either 1 or 2), k now runs over the values 1 and 2. The master equation is $d\mathsf{P}/dt = W\,\mathsf{P}$, where W is the (2×2) matrix given in Eq. (21.31). The solution is $\mathsf{P}(t) = e^{Wt}\,\mathsf{P}(0)$, where $\mathsf{P}(0) = \binom{1}{0}$ or $\binom{0}{1}$, depending on whether $j = 1$ or 2. The problem therefore reduces to finding the exponential of the matrix Wt. Define the *mean transition rate* for the DMP as

$$\lambda = \tfrac{1}{2}(\lambda_1 + \lambda_2). \tag{21.37}$$

Then

$$e^{Wt} = \frac{1}{2\lambda}\begin{pmatrix} \lambda_2 + \lambda_1\, e^{-2\lambda t} & \lambda_2\,(1 - e^{-2\lambda t}) \\ \lambda_1\,(1 - e^{-2\lambda t}) & \lambda_1 + \lambda_2\, e^{-2\lambda t} \end{pmatrix}. \tag{21.38}$$

The four normalized conditional probabilities characterizing the DMP are given by

$$\left.\begin{aligned} P(c_1, t\,|\,c_1) &= \frac{\lambda_2 + \lambda_1\, e^{-2\lambda t}}{\lambda_1 + \lambda_2}, & P(c_2, t\,|\,c_1) &= \frac{\lambda_1\,(1 - e^{-2\lambda t})}{\lambda_1 + \lambda_2} \\ P(c_1, t\,|\,c_2) &= \frac{\lambda_2\,(1 - e^{-2\lambda t})}{\lambda_1 + \lambda_2}, & P(c_2, t\,|\,c_2) &= \frac{\lambda_1 + \lambda_2\, e^{-2\lambda t}}{\lambda_1 + \lambda_2}. \end{aligned}\right\} \tag{21.39}$$

★ **6.** Find the matrix e^{Wt}, to arrive at Eq. (21.38). Hence obtain the expressions given in Eq. (21.39) for the conditional probabilities of the DMP.

Observe that

$$\lim_{t\to\infty} e^{Wt} = \frac{1}{2\lambda}\begin{pmatrix} \lambda_2 & \lambda_2 \\ \lambda_1 & \lambda_1 \end{pmatrix}. \tag{21.40}$$

Hence the stationary probability distribution of the DMP is simply

$$\mathsf{P}^{\text{st}} = \begin{pmatrix} P(1) \\ P(2) \end{pmatrix} = \begin{pmatrix} \lambda_2/(\lambda_1 + \lambda_2) \\ \lambda_1/(\lambda_1 + \lambda_2), \end{pmatrix}, \tag{21.41}$$

as already written down in Eq. (21.21).

***The DMP is exponentially correlated*:** The autocorrelation function of a dichotomous Markov process is easily calculated. The result is

$$\langle \delta X(0)\, \delta X(t)\rangle = \frac{\lambda_1\, \lambda_2\, (c_1 - c_2)^2}{(\lambda_1 + \lambda_2)^2}\, e^{-2\lambda t}, \tag{21.42}$$

where $\delta X(t) \equiv X(t) - \langle X\rangle$.

★ **7.** Use Eqs. (21.39) and (21.41) in the formula (21.28) for the autocorrelation function, to derive the result in Eq. (21.42).

Thus, the DMP is "exponentially correlated", i.e., the autocorrelation function is a single decaying exponential in t. The **correlation time** is

$$\tau_{\text{corr}} = \frac{1}{2\lambda} = \frac{1}{\lambda_1 + \lambda_2} = \frac{\tau_1\, \tau_2}{\tau_1 + \tau_2}, \tag{21.43}$$

rather than just λ^{-1}, as one might guess at first sight. In other words

- The correlation time of a dichotomous Markov process is the harmonic mean of the mean residence times in the two states of the process.

The symmetric DMP: When $c_1 = -c_2 = c$ (say), and $\lambda_1 = \lambda_2 = \lambda$, we have a *symmetric* DMP, and all the foregoing expressions simplify considerably. We then have

$$\langle X \rangle = 0, \quad \langle X^2 \rangle = c^2, \quad \langle X(0)\, X(t) \rangle = c^2\, e^{-2\lambda t}. \tag{21.44}$$

Further,

$$P(\pm c,\, t \mid \pm c) = e^{-\lambda t} \cosh \lambda t, \quad P(\pm c,\, t \mid \mp c) = e^{-\lambda t} \sinh \lambda t. \tag{21.45}$$

21.5 Birth-and-Death Processes

A birth-and-death process is a Markov process in which the random variable takes on integer values, and transitions occur between neighboring values (or states) with prescribed mean rates. Common examples include the Poisson pulse process and a simple random walk on a linear lattice.

21.5.1 The Poisson Pulse Process and Radioactive Decay

In the symmetric DMP discussed in Sect. 21.4.2, suppose we focus on the instants of *time* (sometimes called the *epochs*) at which the DMP flips from one of its two values to the other. These instants themselves constitute a random sequence called a **Poisson pulse process** with a mean pulse rate[2] λ: that is, the probability that exactly r epochs occur in a given time interval t is Poisson-distributed, being given by $(\lambda t)^r\, e^{-\lambda t}/r!$. Let us now see how this comes about.

It is instructive to work in the context of a physical example, such as the decay events in a radioactive isotope containing a sufficiently large number of nuclei. The latter condition ensures that, during the time interval in which we observe the sample, the events can be regarded as an on-going *stationary* random process. Starting at an arbitrary instant of time, the sequence of epochs at which a decay takes place comprise a Poisson pulse process. These epochs are marked by crosses in Fig. 21.3a. The number of nuclei that have decayed till time t is an integer-valued random variable $n(t)$ that starts at the value zero, and increases in steps of unity at random instants of time. A typical realization of $n(t)$ is shown in Fig. 21.3b.

Consider a sufficiently large sample of a radioactive isotope with decay constant λ. At the level of the individual nucleus, λ is the *mean rate* of decay. The basic assumptions are as follows:

---

[2]This rate is called the *intensity* of the Poisson process in the statistics literature.

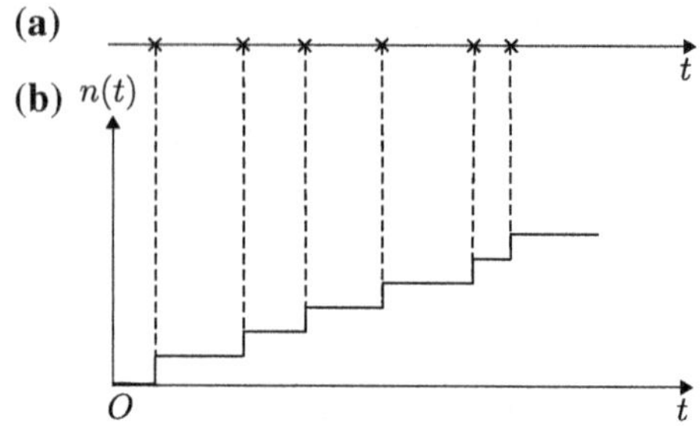

Fig. 21.3 **a** Schematic of a Poisson pulse process: the crosses denote the epochs. They are uncorrelated with each other. **b** The corresponding birth process $n(t)$. The height of each step is 1

(i) In any sufficiently small time interval δt, only one of two things can happen: either a single decay occurs, with probability $\lambda\,\delta t$, or there is no decay at all, with probability $(1 - \lambda\,\delta t)$.

If δt is not sufficiently small, then of course multiple decays can occur in that interval. The idea is that, for any given finite value of λ, the time increment δt can be chosen to be small enough for the assumption (of either no decay or just one decay) to be valid.

(ii) Successive decays are statistically independent of each other.

That is, the locations of the crosses in Fig. 21.3a are completely uncorrelated with each other. In technical terms, this means that the number of decays is a Markov process. In the present instance, since the number n of decay events cannot decrease as t increases, we have a Markovian **birth process**.

Let $P(r,\ t + \delta t)$ denote the probability that exactly r nuclei have decayed in a time interval $t + \delta t$. This probability is given by the sum of the contributions from two possibilities:

- *Either* $(r - 1)$ decays have occurred till time t, for which the probability is $P(r - 1,\ t)$, and one more decay occurs in the incremental interval δt, with probability $\lambda\,\delta t$;
- *or*, all r decays have already occurred in time t, for which the probability is $P(r,\ t)$, and no further decay occurs in the incremental interval δt, with probability $1 - \lambda\,\delta t$.

As these two possibilities are *mutually exclusive events*, their respective probabilities add up. We therefore have

$$P(r, t + \delta t) = (\lambda\,\delta t)\,P(r - 1, t) + (1 - \lambda\,\delta t)\,P(r, t), \quad \text{where} \quad r \geq 1. \quad (21.46)$$

Move the term $P(r, t)$ to the left-hand side, divide both sides by δt and let $\delta t \to 0$. The outcome is the differential equation

$$\frac{dP(r, t)}{dt} = \lambda\big\{P(r - 1, t) - P(r, t)\big\}, \quad r \geq 1. \quad (21.47)$$

This is the master equation for the probability distribution. Note that it is actually a set of coupled differential equations for the infinite set of probabilities $\{P(r,t)\}$, where $r = 1, 2, \ldots$. We need an initial condition for each $P(r,t)$, which is provided by the obvious condition $P(r,0) = \delta_{r,0}$. Hence $P(r,0) = 0$ for every $r \geq 1$, while $P(0,0) = 1$.

The zero-decay probability $P(0,t)$ itself obeys an even simpler equation. The counterpart of Eq. (21.46) in this case is simply

$$P(0, t + \delta t) = (1 - \lambda\, \delta t)\, P(0,t). \tag{21.48}$$

The corresponding differential equation is therefore

$$\frac{dP(0,t)}{dt} = -\lambda\, P(0,t). \tag{21.49}$$

The solution to this equation, with the initial condition $P(0,0) = 1$, is just

$$P(0,t) = e^{-\lambda t}. \tag{21.50}$$

In other words: The probability that no decay event occurs in a time interval t decreases exponentially with increasing t.

The coupled set of equations in (21.47) can be solved in several ways. As you would expect by now, a convenient way is to use a generating function. Accordingly, let us define

$$f(z,t) = \sum_{r=0}^{\infty} P(r,t)\, z^r, \tag{21.51}$$

where z is a complex variable. Multiply both sides of Eq. (21.47) by z^r, sum over r, and add on Eq. (21.49). The result is the differential equation

$$\frac{\partial f}{\partial t} = \lambda(z-1)\, f. \tag{21.52}$$

Since $P(r,0) = \delta_{r,0}$, the initial condition on $f(z,t)$ is $f(z,0) = 1$. The solution to Eq. (21.52) is

$$f(z,t) = e^{\lambda(z-1)t}. \tag{21.53}$$

It is now trivial to pick out the coefficient of z^r in $f(z,t)$ to get the probability distribution

$$P(r,t) = \frac{e^{-\lambda t}\, (\lambda t)^r}{r!}, \quad r \geq 0. \tag{21.54}$$

Therefore $P(r,t)$ is a Poisson distribution with a mean value λt: that is, the mean number of decays that take place in a time interval t is just λt. This is in complete accord with the identification of λ as the mean rate of decay of a nucleus. The half-life of the isotope concerned is $(\ln 2)/\lambda$.

21.5.2 Biased Random Walk on a Linear Lattice

After the example above of a pure birth process, let us turn to a prototype of a **birth-and-death process**. In Chap. 19, Sect. 19.4.1, we have worked out the problem of a biased random walk on a linear lattice in *discrete* time. Consider, now, the continuous-time version of this random walk.

Recall the notation used earlier in the random walk problem. The sites of an infinite one-dimensional lattice are labeled by the integer j. The random walker jumps from any site j to the site $(j+1)$ with probability α, or to the site $(j-1)$ with probability $\beta = (1-\alpha)$. (See Fig. 19.3 of Chap. 19, Sect. 19.4.1.) If $\alpha \neq \beta$, the random walk is biased. A bias models the effect of a constant force acting on the random walker, causing a systematic drift toward the left (if $\alpha < \beta$) or the right (if $\alpha > \beta$). The difference between the discrete-time random walk considered earlier and the continuous-time version is as follows: we now assume that the successive steps of the random walk are taken at *random* instants of time, with a mean jump rate λ. The random variable of current interest, however, is the *position* (or site label j) of the random walker at any time t. On an infinite linear lattice, we may take the random walker to start from the site $j = 0$ at $t = 0$, without loss of generality. What is the probability $P(j, t)$ that she is at the site j at time t? The crucial technical assumption is that

- the jump events themselves constitute a Poisson birth process—that is, the number of steps taken in any given time interval t is Poisson-distributed with a mean value λt.

This makes $j(t)$ a Markov process. Moreover, since j can both increase as well as decrease with time as the walker moves back and forth, $j(t)$ is an example of a birth-and-death process.

As before, we write down an equation for $P(j, t + \delta t)$ based on the assumption that successive steps are independent of each other, and that δt is small enough to ensure that only three things can possibly happen in that incremental time interval: namely, the walker (i) either takes a step to the right, (ii) or takes a step to the left, (iii) or does not take a step at all. These three possibilities are mutually exclusive events. Hence their contributions to $P(j, t + \delta t)$ add up, and we get

$$P(j, t + \delta t) = (\lambda\, \alpha\, \delta t)\, P(j-1, t) + (\lambda\, \beta\, \delta t)\, P(j+1, t) + (1 - \lambda\, \delta t)\, P(j, t). \tag{21.55}$$

In the limit $\delta t \to 0$, this yields the master equation

$$\boxed{\frac{dP(j,t)}{dt} = \lambda \big\{\alpha\, P(j-1, t) - P(j, t) + \beta\, P(j+1, t)\big\}, \quad j \in \mathbb{Z}.} \tag{21.56}$$

The initial condition is of course $P(j, 0) = \delta_{j,0}$. Once again, it is convenient to solve the master equation (21.56) using the generating function of $P(j, t)$. Let

$$f(z, t) = \sum_{j=-\infty}^{\infty} P(j, t)\, z^j . \tag{21.57}$$

Note that the summation over j runs over *all* integers (and not just the nonnegative ones). It then follows from Eq. (21.56) that $f(z, t)$ satisfies the first-order differential equation

$$\frac{\partial f}{\partial t} = \lambda\,(pz - 1 + qz^{-1})\, f, \tag{21.58}$$

with the initial condition $f(z, 0) = 1$. The solution is easy to write down. It is

$$f(z, t) = \exp\left\{\lambda t\,(\alpha z - 1 + \beta z^{-1})\right\} = e^{-\lambda t}\, \exp\left\{\lambda t\,(\alpha z + \beta z^{-1})\right\}. \tag{21.59}$$

We need the coefficient of z^j in the expansion of $f(z, t)$ in integer powers of z. But recall, now, the expression given in Eq. (19.48) of Chap. 19, Sect. 19.2.5, for the generating function of the modified Bessel function of the first kind, $I_j(u)$. For ready reference, here it is:

$$\sum_{j=-\infty}^{\infty} I_j(u)\, z^j = \exp\left\{\tfrac{1}{2}u\,(z + z^{-1})\right\}. \tag{21.60}$$

It is easy to see, now, that the probability distribution we seek is given by

$$\boxed{P(j, t) = e^{-\lambda t}\,(\alpha/\beta)^{j/2}\, I_j\big(2\lambda t\sqrt{\alpha\beta}\big).} \tag{21.61}$$

This solution is valid for all integer values of j—positive, negative, and zero. Since $I_j(u) \equiv I_{-j}(u)$ when j is an integer, the effect of the bias in the random walk is essentially carried by the factor $(\alpha/\beta)^{j/2}$ in the solution above. It is obvious that, when $\alpha > \beta$, the probability is larger for positive values of j than it is for negative values. The situation is reversed when $\beta > \alpha$, as expected.

★ **8.** Start with the master equation (21.56) and work through the steps given above to arrive at the solution in Eq. (21.61).

★ **9.** ***Inclusion of a sojourn probability*:** Here is a slightly more generalized version of the biased random walk discussed above. Suppose the walker jumps from any site j to neighboring sites $j + 1$ and $j - 1$ with respective probabilities α and β, as before, and *stays* at the site j with a probability γ, where $\alpha + \beta + \gamma = 1$. See Fig. 21.4. What is $P(j, t)$ in this case?

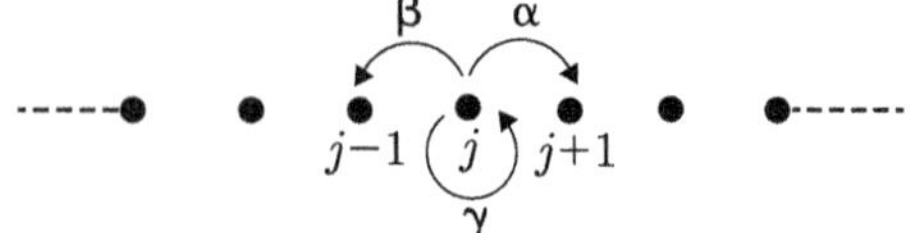

Fig. 21.4 A biased random walk on a linear lattice with a stay or sojourn probability γ

21.5.3 Connection with the Skellam Distribution

The expression in Eq. (21.61) should look familiar to you. Recall that, in Chap. 19, Sect. 19.2.5, we have found the probability distribution of the *difference* of two independent random variables, each of which has a Poisson distribution. Given two Poisson-distributed random variables with mean values μ and ν, respectively, the probability distribution of their difference r is given by the Skellam distribution, Eq. (19.47). Repeating this equation for ready reference,

$$P_{\text{diff}}(r) = e^{-(\mu+\nu)}\,(\mu/\nu)^{r/2}\,I_r(2\sqrt{\mu\nu}) \quad \text{for all } r \in \mathbb{Z}. \tag{21.62}$$

But Eq. (21.61) is precisely of this form, with

$$\mu = \alpha\lambda t \quad \text{and} \quad \nu = \beta\lambda t, \quad \text{so that} \quad \mu + \nu = \lambda t. \tag{21.63}$$

In other words:

- The position j of the random walker in a biased random walk on a linear lattice can be interpreted physically as the difference of two Poisson processes.
- The respective mean rates of these processes are $\alpha\lambda$ and $\beta\lambda$, so that the mean values of the individual processes in a time interval t are $\alpha\lambda t$ and $\beta\lambda t$.
- It is the Markov property of the random walk that leads to such a simple interpretation of the process.

Finally, from the known values of the mean and variance of the Skellam distribution, we can write down the mean and variance of the displacement of the random walker at any time t. These are given by

$$\boxed{\langle j(t)\rangle = (\alpha-\beta)\lambda t, \quad \text{Var}\, j(t) \stackrel{\text{def.}}{=} \langle j^2(t)\rangle - \langle j(t)\rangle^2 = \lambda t.} \tag{21.64}$$

The first of the equations above shows how a bias in the random walk leads to a drift, as measured by the mean displacement, that increases linearly with time. The second shows how the expected *diffusive behavior* of the random walk emerges, once the effect of the drift is subtracted out from the mean squared displacement. This is the most important feature of the random walk.

21.5.4 Asymptotic Behavior of the Probability

The fact that the variance of $j(t)$ becomes unbounded as $t \to \infty$ suggests that the probability $P(j, t)$ does not, in fact, tend to any stationary distribution in that limit. In order to see this explicitly, consider the asymptotic ($t \to \infty$) behavior of the exact solution in Eq. (21.61). What you need for this purpose is the leading asymptotic behavior of the modified Bessel function when its argument tends to infinity. This is given by

$$I_j(z) \sim \frac{e^z}{\sqrt{2\pi z}}, \quad |z| \to \infty. \tag{21.65}$$

It follows that the leading asymptotic behavior of $P(j, t)$ as $t \to \infty$ is given by

$$P(j, t) \sim \frac{(\alpha/\beta)^{j/2}\, e^{-\lambda t(1-2\sqrt{\alpha\beta})}}{\left(4\pi\lambda t\sqrt{\alpha\beta}\right)^{1/2}}. \tag{21.66}$$

But $1 - 2\sqrt{\alpha\beta} = 1 - 2\sqrt{\alpha(1-\alpha)} \geq 0$. The equality sign is attained only when $\alpha = \beta = \frac{1}{2}$. We conclude that

- $P(j, t)$ decays *exponentially* to zero as $t \to \infty$ for all $\alpha \neq \frac{1}{2}$, i.e., when the random walk is biased (either to the right or to the left).
- When the random walk is unbiased ($\alpha = \beta = \frac{1}{2}$), this asymptotic behavior is drastically modified. $P(j, t)$ now decays to zero as $t \to \infty$ like an inverse *power* of t, namely, like $1/\sqrt{t}$.
- In either case, $P(j, t)$ vanishes in the limit $t \to \infty$, for all j. Hence the random walk process has *no* stationary distribution.

21.6 Continuous Markov Processes

21.6.1 Master Equation for the Conditional density

I turn now to a brief account of some aspects of a continuous Markov process $X(t)$. The sample space of the random variable is some continuous set of values. The discussion of the preceding sections can be extended to this case with obvious modifications, such as the replacement of probabilities by the corresponding probability density functions. Thus, the counterpart of Eq. (21.8) for a stationary, continuous Markov process is

$$p_n(x_n, t_n\,;\, x_{n-1}, t_{n-1}\,;\, \ldots\,;\, x_1, t_1) = \left\{\prod_{r=1}^{n-1} p(x_{r+1}, t_{r+1} - t_r \,|\, x_r)\right\} p(x_1), \tag{21.67}$$

for every $n \geq 2$. (Here, the values $\{x_k\}$ belong to the sample space of $X(t)$.) Hence the fundamental quantity characterizing a stationary continuous Markov process is the conditional density $p(x, t \,|\, x_0)$. The stationary PDF $p(x)$ is expected to be related to this quantity according to

$$\lim_{t\to\infty} p(x, t \,|\, x_0) = p(x), \tag{21.68}$$

independent of the initial value x_0. This relation is the continuum analog of Eq. (21.9). As in the discrete case, the conditional density satisfies the Chapman–Kolmogorov equation. This equation now reads

$$p(x, t \,|\, x_0) = \int dx'\, p(x, t - t' \,|\, x')\; p(x', t' \,|\, x_0) \quad (0 \leq t' \leq t). \tag{21.69}$$

The integration runs over the range of values assumed by the random variable.

As before, this nonlinear equation may be converted to a linear one. Let $w(x \,|\, x')\, dx$ be the probability per unit time of a transition from a given value x' to the range of values $(x, x + dx)$ of the random variable. Therefore $w(x \,|\, x')$ is the **transition probability density** per unit time. The chain equation can then be reduced to the master equation

$$\boxed{\frac{\partial}{\partial t}\, p(x, t \,|\, x_0) = \int dx' \Big\{ p(x', t \,|\, x_0)\, w(x \,|\, x') - p(x, t \,|\, x_0)\, w(x' \,|\, x) \Big\}.} \tag{21.70}$$

The initial condition on the conditional density is obviously $p(x, 0 \,|\, x_0) = \delta(x - x_0)$.

★ **10.** Starting from Eq. (21.69), obtain Eq. (21.70).

21.6.2 *The Fokker–Planck Equation*

Even though the master equation (21.70) is a linear equation for the conditional PDF $p(x, t \,|\, x_0)$, it is an *integro-differential* equation. This is the price we pay for the reduction of the nonlinear chain equation to a linear equation! Solving it exactly is far from an easy task. One approach is to convert it to a partial differential equation. The problem is that, in general, the latter is of *infinite* order in the partial derivative $\partial/\partial x$. This is called the **Kramers–Moyal expansion** of the master equation. I will not go into this here, as it would take us too far afield.

In many physical applications, however, the master equation can either be reduced to, or be well-approximated by, a *second-order* partial differential equation of the form

$$\boxed{\frac{\partial}{\partial t}\, p(x,\,t\,|\,x_0) = -\frac{\partial}{\partial x}\,[A_1(x)\, p(x,\,t\,|\,x_0)] + \frac{1}{2}\frac{\partial^2}{\partial x^2}\,[A_2(x)\, p(x,\,t\,|\,x_0)].} \tag{21.71}$$

The functions $A_1(x)$ and $A_2(x)$ are defined in Eq. (21.72) below. Equation (21.71) is known as the **forward Kolmogorov equation** or, more commonly in the physics literature, the **Fokker–Planck equation**. It is the most frequently used form of the master equation for the conditional PDF in physical applications of continuous Markov processes. There is an interesting exact result in this regard:

- The master equation (21.70) reduces to a partial differential equation that is either of *infinite* order in $\partial/\partial x$, or of *second* order—nothing in between!

The diffusion equation to be studied in Chap. 30 is a special case of the Fokker–Planck equation above. We shall also consider there the diffusion equation in the presence of a constant external force. That equation, called the **Smoluchowski equation**, is also a special case of the Fokker–Planck equation. You will encounter another example of a Smoluchowski equation in this chapter itself, in Sect. 21.7.5 below, in the context of the diffusion equation for a linear harmonic oscillator.

In the mathematical literature on stochastic processes, all continuous Markov processes whose conditional densities satisfy the Fokker–Planck equation (21.71) are called **diffusion processes**. The functions $A_1(x)$ and $A_2(x)$ are referred to as the **drift coefficient** and **diffusion coefficient**, respectively. They are essentially the first two moments of the transition rate. Their precise definitions are

$$A_1(x) = \int dx'\, x'\, w(x + x'\,|\,x), \quad A_2(x) = \int dx'\, x'^{\,2}\, w(x + x'\,|\,x). \tag{21.72}$$

The solution of the Fokker–Planck equation itself is again a nontrivial task. Part of the technical problem is that the differential operator on the right-hand side of Eq. (21.71) is not a self-adjoint operator. There exists a considerable body of literature on the analysis of the Fokker–Planck equation and its solution in various cases.

Some features of the equation, however, can be deduced quite easily. Equation (21.71) can be written in the form of an *equation of continuity*, according to

$$\frac{\partial}{\partial t}\, p(x,\,t\,|\,x_0) + \frac{\partial}{\partial x}\, j(x,\,t\,|\,x_0) = 0, \tag{21.73}$$

where the *probability current density* j is given by

$$j(x,\,t\,|\,x_0) = -\frac{\partial}{\partial x}\left\{\tfrac{1}{2}A_2(x)\, p(x,\,t\,|\,x_0)\right\} + A_1(x)\, p(x,\,t\,|\,x_0). \tag{21.74}$$

Consider, now, what happens as $t \to \infty$. From Eq. (21.68), we have $p(x,\,t\,|\,x_0) \to p(x)$. Correspondingly, $j(x,t)$ tends to the stationary current

$$j^{\text{st}}(x) = -\frac{d}{dx}\left\{\tfrac{1}{2}A_2(x)\, p(x)\right\} + A_1(x)\, p(x). \tag{21.75}$$

Obviously, $\partial p(x)/\partial t = 0$. Hence Eq. (21.73) reduces to $dj^{\text{st}}/dx = 0$, so that $j^{\text{st}}(x)$ is actually a constant (independent of x). This means that the stationary density itself can be found by solving the *first-order, ordinary* differential equation

$$\frac{d}{dx}\left\{\tfrac{1}{2}A_2(x)\,p(x)\right\} - A_1(x)\,p(x) = \text{constant}, \tag{21.76}$$

where the constant is obtained from the boundary conditions in any given instance.

I shall not enter here into a discussion of the general Fokker–Planck equation. But there is an important special case that will be considered shortly. Before that, let us write down the general expression for the autocorrelation function in the case of a stationary, continuous Markov process.

21.6.3 The Autocorrelation Function for a Continuous Process

The formulas that follow are straightforward extensions of those in Sect. 21.3 for a stationary discrete stochastic process. Writing out the analogs of Eqs. (21.26)–(21.30), for ready reference, we have

$$\begin{aligned}\langle X(t')\,X(t)\rangle &= \int dx'\int dx\; x\,x'\;p_2(x,t\,;\,x',t')\\ &= \int dx'\int dx\; x\,x'\;p_2(x,t\,|\,x',t')\,p_1(x',t').\end{aligned} \tag{21.77}$$

For a stationary process this becomes, on setting $t' = 0$,

$$\langle X(0)\,X(t)\rangle = \int dx'\int dx\; x\,x'\;p(x,t\,|\,x')\,p(x'), \tag{21.78}$$

in terms of the stationary and conditional PDFs. This is the autocorrelation function $C_X(t)$ when the mean value of $X(t)$ is zero.

When the mean value of the random process is nonzero, the autocorrelation function is defined in terms of the deviation $\delta X = X - \langle X\rangle$ from the mean value. We then have

$$C_X(t) \stackrel{\text{def.}}{=} \langle \delta X(0)\,\delta X(t)\rangle = \int dx'\int dx\; x\,x'\;p(x,t\,|\,x')\,p(x') - \langle X\rangle^2, \tag{21.79}$$

where

$$\langle X\rangle = \int dx\, x\, p(x). \tag{21.80}$$

21.7 The Stationary Gaussian Markov Process

21.7.1 The Ornstein–Uhlenbeck Process

Among continuous Markov processes, there is a very unique one:

- There is only one continuous random process that is stationary, Markov, as well as Gaussian. This is the **Ornstein–Uhlenbeck process.**

By a "Gaussian process" we mean that all the joint probability densities of the random variable are Gaussians in functional form. It is understood that the range of the random variable is $(-\infty, \infty)$.

The Ornstein–Uhlenbeck process corresponds to the case when the drift coefficient is proportional to x, and the diffusion coefficient is a constant:

$$A_1(x) = -a_1 x \quad \text{and} \quad A_2(x) = a_2, \quad \text{where} \quad a_1, a_2 = \text{positive constants.} \tag{21.81}$$

In Sects. 21.7.3 and 21.7.5, I will discuss two physical examples of this random process. The Fokker–Planck equation in this case is

$$\boxed{\frac{\partial}{\partial t}\, p(x, t \,|\, x_0) = a_1 \frac{\partial}{\partial x}\, [x\, p(x, t \,|\, x_0)] + \frac{a_2}{2} \frac{\partial^2}{\partial x^2}\, p(x, t \,|\, x_0).} \tag{21.82}$$

The physical dimensions of a_1 and a_2 are, respectively, 1/[time] and $[x^2]$/[time]. The initial condition on $p(x, t \,|\, x_0)$ is of course given by

$$p(x, 0 \,|\, x_0) = \delta(x - x_0). \tag{21.83}$$

The simplest case corresponds to the natural boundary conditions

$$p(x, t \,|\, x_0) = 0 \quad \text{as} \quad x \to \pm\infty. \tag{21.84}$$

The solution of Eq. (21.82) that satisfies these conditions will be written down in Sect. 21.7.2 (see Eq. (21.90)). The main physical features of this solution are as follows:

(i) The conditional density starts at $t = 0$ as a δ-function peak at x_0, and widens as t increases. It is a Gaussian for all $t > 0$, and attains a limiting Gaussian form as $t \to \infty$.
(ii) As t increases, the mean value of $X(t)$ (and hence the peak of the Gaussian) drifts monotonically to zero like a decaying exponential function of t.
(iii) Simultaneously, the variance of $X(t)$ increases monotonically from zero, and tends to a saturation value as $t \to \infty$.
(iv) The autocorrelation function of $X(t)$ decays to zero exponentially in time, with a characteristic time constant given by $1/a_1$.

Some of these features can be deduced even without solving the Fokker–Planck equation explicitly. Consider, for instance, the mean value $\langle X(t)\rangle$ and the second moment $\langle X^2(t)\rangle$. These are defined as

$$\mu_X(t) \equiv \langle X(t)\rangle \stackrel{\text{def.}}{=} \int_{-\infty}^{\infty} dx\, x\; p(x,\, t\,|\,x_0) \;\text{ and }\; \langle X^2(t)\rangle \stackrel{\text{def.}}{=} \int_{-\infty}^{\infty} dx\, x^2\; p(x,\, t\,|\,x_0), \tag{21.85}$$

respectively. These moments satisfy ordinary first-order differential equations in time that are easily found. Multiply both sides of the Fokker–Planck equation by x and x^2, respectively, and integrate over x. After some simplification, one gets

$$\frac{d}{dt}\langle X(t)\rangle + a_1\langle X(t)\rangle = 0 \;\text{ and }\; \frac{d}{dt}\langle X^2(t)\rangle + 2a_1\langle X^2(t)\rangle - a_2 = 0. \tag{21.86}$$

It is important to remember that the moments under discussion are *conditional* averages: that is, they are averages over all realizations of the random process $X(t)$, *given* the specific initial value x_0. Therefore the initial conditions on the moments are simply

$$\langle X(0)\rangle = x_0 \;\text{ and }\; \langle X^2(0)\rangle = x_0^2\,. \tag{21.87}$$

The solutions of Eq. (21.86) are then given by

$$\langle X(t)\rangle \equiv \mu_X(t) = x_0\, e^{-a_1 t} \;\text{ and }\; \langle X^2(t)\rangle = \frac{a_2}{2a_1} + \Big(x_0^2 - \frac{a_2}{2a_1}\Big)e^{-2a_1 t}, \tag{21.88}$$

for all $t \geq 0$.

★ **11.** From the Fokker–Planck equation (21.82), derive the expressions in Eq. (21.88), following the procedure outlined above.

From the expressions in Eq. (21.88), we find that the time-dependent variance of $X(t)$ is

$$\text{Var}\, X(t) \equiv \sigma_X^2(t) = \langle X^2(t)\rangle - \mu_X^2(t) = \frac{a_2}{2a_1}\big(1 - e^{-2a_1 t}\big). \tag{21.89}$$

The variance thus increases monotonically from its initial value of zero to the limiting value $a_2/(2a_1)$. Observe that it is actually independent of the initial value x_0 for all t.

21.7.2 The Ornstein–Uhlenbeck Distribution

Let us now consider the exact solution of the Fokker–Planck equation (21.82) with the initial condition $p(x,\, 0\,|\,x_0) = \delta(x - x_0)$ and natural boundary conditions $p(x,\, t\,|\,x_0) = 0$ as $x \to \pm\infty$. There are several ways of solving the equation. I will not discuss these here, but will merely write down the solution, based on the assertion that the PDF is a Gaussian at all times. A knowledge of the mean and variance

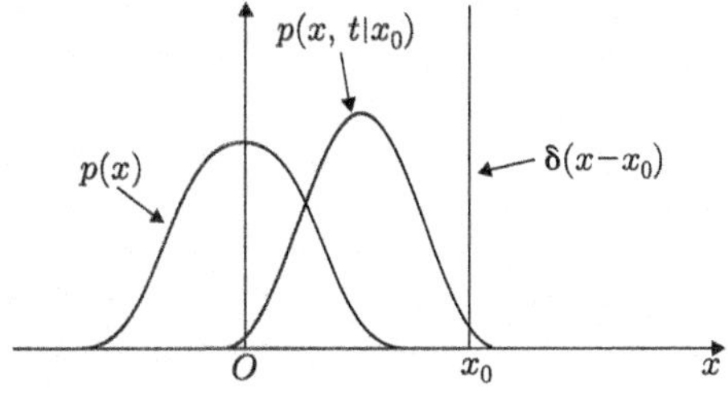

Fig. 21.5 The time evolution of the PDF corresponding to the Ornstein–Uhlenbeck distribution

therefore suffices for our purposes, because a Gaussian is fully determined by its mean and variance. The normalized conditional probability density is given by

$$\boxed{p(x,\,t\,|\,x_0) = \frac{1}{\sqrt{2\pi\sigma_X^2(t)}}\,\exp\left\{-\frac{\big(x-\mu_X(t)\big)^2}{2\sigma_X^2(t)}\right\},} \tag{21.90}$$

where the mean $\mu_X(t)$ and variance $\sigma_X^2(t)$ are give by Eqs. (21.88) and (21.89), respectively. This is the **Ornstein–Uhlenbeck distribution** (or rather, the corresponding PDF). The claim is that it is the PDF of the only continuous stochastic process that is stationary, Markov, as well as Gaussian. Figure 21.5 shows schematically how the PDF evolves in time from an initial δ-function to a Gaussian centered at the origin. Observe that, as $t \to \infty$, it tends to the stationary PDF given by

$$\boxed{p(x) = \sqrt{a_1/(\pi a_2)}\; e^{-a_1 x^2/a_2}.} \tag{21.91}$$

★ **12.** Derive the expression for the stationary PDF in Eq. (21.91) directly from Eq. (21.76), by setting $A_1(x) = -a_1 x$ and $A_2 = a_2$.

The autocorrelation function of the Ornstein–Uhlenbeck process has already been stated to be a decaying exponential function of t. Let us see how this comes about.

The first step is to recognize that the mean value of the process, as opposed to the *conditional* mean $\mu_X(t) = \langle X(t)\rangle = x_0\, e^{-a_1 t}$ found above, is actually zero. This is easily done, because all that is needed is a *further* averaging of $\langle X(t)\rangle$ over all initial conditions weighted with the stationary density $p(x_0)$. But the latter is a Gaussian centered at $x_0 = 0$, and hence a symmetric function of x_0. Therefore the full, or unconditional, average of X is zero.[3] The autocorrelation function is then given by Eq. (21.78). In the present instance, we have

[3]One must really use a better notation to distinguish between conditional and full averages. I have not done so here merely because we are not going to elaborate on this aspect much farther.

$$C_X(t) = \langle X(0)\, X(t) \rangle = \int_{-\infty}^{\infty} dx_0 \int_{-\infty}^{\infty} dx\; x\, x_0\; p(x, t \,|\, x_0)\; p(x_0). \tag{21.92}$$

$p(x,\ t \,|\, x_0)$ is given by Eq. (21.90), while $p(x_0)$ can be read off from Eq. (21.91). The result of carrying out the integrations is

$$C_X(t) = \frac{a_2}{2a_1}\, e^{-a_1 t}. \tag{21.93}$$

As claimed earlier, this is a decaying exponential function of time, with a correlation time given by $1/a_1$.

★ **13.** It is an instructive exercise to carry out the integrations above and derive the result in Eq. (21.93). Do so.

Next, I turn to a brief discussion of two physical examples of the Ornstein–Uhlenbeck process and distribution.

21.7.3 Velocity Distribution in a Classical Ideal Gas

The Fokker–Planck equation arose originally in the context of the velocity distribution of the molecules of a classical ideal gas in thermal equilibrium at a temperature T. From elementary statistical physics, you know that the normalized stationary PDF of each Cartesian component of the velocity of a molecule is a Gaussian. For notational simplicity, let us denote any one of the Cartesian components of the velocity of a molecule by U (or u, depending on whether we are talking about the random variable or its value).[4] The stationary or equilibrium PDF of U is given by the Maxwellian distribution

$$p(u) = \left(\frac{m}{2\pi k_B T}\right)^{1/2} \exp\left\{-\frac{mu^2}{2k_B T}\right\}, \tag{21.94}$$

where m is the mass of a molecule and k_B is Boltzmann's constant. It is quite natural to ask a related question: Suppose we focus on any one particular molecule (the "tagged" particle), and find that its instantaneous velocity component at $t = 0$ is equal to u_0. How does its velocity distribution change with time, and attain the Maxwellian form as $t \to \infty$? This simple-looking question already takes us beyond the purview of *equilibrium* statistical mechanics, as it involves the *time-dependent* conditional probability density $p(u,\ t \,|\, u_0)$.

The simplest model that describes the physical situation is based on a random or *stochastic differential equation* for the velocity of the tagged particle, called the **Langevin equation**. It then turns out that the velocity component $U(t)$ of the tagged particle is a stationary, Gaussian, Markov process. (But see the comments at the end of this section.) Its conditional probability density satisfies the Fokker–Planck

---

[4] I use u because the standard symbol v has already been used to denote the *speed* of a molecule, in Chap. 13, Sect. 13.2.3.

equation

$$\boxed{\frac{\partial}{\partial t}\, p(u,\, t\,|\,u_0) = \gamma\, \frac{\partial}{\partial u}\, [u\, p(u,\, t\,|\,u_0)] + \frac{\gamma k_B T}{m}\, \frac{\partial^2}{\partial u^2}\, p(u,\, t\,|\,u_0).} \tag{21.95}$$

Here γ is a positive constant with the physical dimensions of $[\text{time}]^{-1}$. It is directly proportional to the viscosity of the fluid. It is evident that $U(t)$ is an Ornstein–Uhlenbeck process, with

$$a_1 = \gamma \quad \text{and} \quad a_2 = 2\gamma k_B T/m\,. \tag{21.96}$$

From Eqs. (21.88) and (21.89), it follows that the conditional mean and variance of the velocity are given by

$$\mu_U(t) = u_0\, e^{-\gamma t} \quad \text{and} \quad \sigma_U^2(t) = \frac{k_B T}{m}\, \big(1 - e^{-2\gamma t}\big). \tag{21.97}$$

We can now write down the normalized fundamental solution to the Fokker–Planck equation (21.95) with the initial condition

$$p(u,\, 0\,|\,u_0) = \delta(u - u_0) \tag{21.98}$$

and natural boundary conditions

$$p(u,\, t\,|\,u_0) \to 0 \quad \text{as} \quad u \to \pm\infty. \tag{21.99}$$

It is the Ornstein–Uhlenbeck distribution

$$\boxed{p(u,\, t\,|\,u_0) = \left[\frac{m}{2\pi k_B T\big(1 - e^{-2\gamma t}\big)}\right]^{1/2} \exp\left\{-\frac{m\big(u - u_0\, e^{-\gamma t}\big)^2}{2k_B T\big(1 - e^{-2\gamma t}\big)}\right\}.} \tag{21.100}$$

★ **14.** Verify that the expression in Eq. (21.100) satisfies the Fokker–Planck equation (21.95).

The autocorrelation function of the velocity component is, as expected, a decaying exponential in t. Setting $a_1 = \gamma$ and $a_2 = 2\gamma k_B T/m$ in Eq. (21.93), we have

$$C_U(t) = \frac{k_B T}{m}\, e^{-\gamma t} \quad (t \geq 0). \tag{21.101}$$

It is immediately evident that $1/\gamma$ is the correlation time of the velocity process. It is the characteristic relaxation time (or **equilibration time**) over which the velocity "thermalizes", starting from any specified initial value u_0.

Finally, in the interests of technical accuracy I must mention an important fact. The foregoing model is too simplistic, as it stands, to be applicable directly to an individual molecule as the tagged particle. The Langevin equation and the associated Fokker–Planck equation (21.95) are actually more appropriate for describing the random motion of a much more massive tagged particle moving in a fluid of much less massive molecules.

21.7.4 Solution for an Arbitrary Initial Velocity Distribution

The fact that the solution (21.100) corresponds to the sharp initial condition $p(u, 0 \mid u_0) = \delta(u - u_0)$ implies that we can use it to go further: The solution of the Fokker–Planck equation for an *arbitrary* initial distribution of velocities can be written down at once! The Fokker–Planck equation is a *linear* equation, and the fundamental solution in (21.100) is essentially the **Green function** for the differential operator in that equation.[5] Let the initial distribution be given by the normalized PDF $p_{\text{init}}(u_0)$. Then the PDF of the velocity component at any time $t > 0$ is given by the expression

$$p(u, t) = \left[\frac{m}{2\pi k_B T\left(1 - e^{-2\gamma t}\right)}\right]^{1/2} \int_{-\infty}^{\infty} du_0 \, \exp\left\{-\frac{m\left(u - u_0\, e^{-\gamma t}\right)^2}{2k_B T\left(1 - e^{-2\gamma t}\right)}\right\} p_{\text{init}}(u_0). \tag{21.102}$$

This solution exhibits two interesting properties:

(i) Regardless of the initial density $p_{\text{init}}(u_0)$, the solution tends to the stationary Gaussian PDF $p(u)$ as $t \to \infty$.

(ii) Suppose the initial probability density is the Maxwellian distribution itself, i.e., $p_{\text{init}}(u_0) = p(u_0)$. Then $p(u, t)$ remains equal to $p(u)$ at *all* times.

★ **15.** Verify the statements (i) and (ii) above.

Ponder over these facts. They show how robust and stable the state of thermal equilibrium is.

21.7.5 Diffusion of a Harmonically Bound Particle

Our second physical example of the Ornstein–Uhlenbeck distribution arises from the diffusion equation[6] for a harmonically bound particle. The particle undergoes random motion on a line (the x-axis, say), while it is under the influence of a harmonic oscillator potential $\frac{1}{2}m\omega^2 x^2$. The particle is also subject to a frictional force

---

[5] We will discuss Green functions for some basic partial differential operators in Chaps. 29–31.

[6] As I have already mentioned, we will study the phenomenon of diffusion and the diffusion equation in greater detail in Chap. 30.

$-m\gamma\dot{x}(t)$ arising from the medium in which it moves (a fluid in thermal equilibrium at temperature T), where γ is a positive constant with the physical dimensions of $[\text{time}]^{-1}$. We are interested in the conditional probability density $p(x, t \mid x_0)$ of the position $X(t)$ of the particle, given that it starts at $t = 0$ from some point x_0.

Recall from elementary physics that a linear harmonic oscillator is underdamped when $\gamma < 2\omega$, and overdamped when $\gamma > 2\omega$. It turns out that, in the *over*damped case, and for $\gamma t \gg 1$, $p(x, t \mid x_0)$ satisfies the following Fokker–Planck equation:

$$\frac{\partial}{\partial t}\, p(x, t \mid x_0) = \frac{\omega^2}{\gamma}\,\frac{\partial}{\partial x}\big[x\, p(x, t \mid x_0)\big] + \frac{k_B T}{m\gamma}\,\frac{\partial^2}{\partial x^2}\, p(x, t \mid x_0). \tag{21.103}$$

This is an example of a diffusion equation for the positional PDF, in the presence of an applied force or in an external potential. As already mentioned, such an equation is called a Smoluchowski equation. Equation (21.103) is of the same form as Eq. (21.82), with

$$a_1 = \omega^2/\gamma \quad \text{and} \quad a_2 = 2k_B T/(m\gamma)\,. \tag{21.104}$$

Thus $X(t)$ is an Ornstein–Uhlenbeck process, under the conditions mentioned above. For this reason, the Ornstein–Uhlenbeck process itself is sometimes called the **oscillator process**.

The fundamental solution of Eq. (21.103) is the Gaussian given in Eq. (21.90), with the conditional mean and variance given by

$$\mu_X(t) = x_0\, e^{-\omega^2 t/\gamma} \quad \text{and} \quad \sigma_X^2(t) = \frac{k_B T}{m\omega^2}\left(1 - e^{-2\omega^2 t/\gamma}\right). \tag{21.105}$$

Letting $t \to \infty$ in this solution, we get

$$\lim_{t\to\infty} p\,(x, t\,|x_0) = p(x) = \left(\frac{m\omega^2}{2\pi k_B\, T}\right)^{1/2} \exp\left\{-\,\frac{m\omega^2 x^2}{2k_B T}\right\}. \tag{21.106}$$

This is just the normalized stationary or equilibrium PDF that we would write down in the canonical ensemble in equilibrium statistical mechanics.

I reiterate that Eq. (21.103) and its Gaussian solution are not exact relations for the positional probability density $p(x, t \mid x_0)$ of a harmonically bound diffusing particle. They are approximations that are only valid in the highly overdamped case, and at times $t \gg \gamma^{-1}$. Only under these circumstances does the position variable $X(t)$ become a stationary Markov process, with an autocorrelation function given by the decaying exponential

$$C_X(t) = \frac{k_B T}{m\omega^2}\, e^{-\omega^2 t/\gamma}. \tag{21.107}$$

Note, too, that the existence of a nonzero stationary PDF $p(x)$ as given by (21.106) has another implication. It means that, in stark contrast to a free particle, a harmonically bound particle in thermal equilibrium with a heat bath does not undergo any long-range diffusion at all! The variance of its displacement does not increase linearly with t—in fact, it does not diverge like any power of t. Instead, it saturates (as $t \to \infty$) to the value $k_B T/m\omega^2$. This expression is precisely what you would write down based on an elementary application of the equipartition theorem, according to which $\frac{1}{2}m\omega^2 \langle x^2 \rangle = \frac{1}{2}k_B T$.

21.8 Solutions

1. Subtract $P(k, t - t' \,|\, j)$ from both sides of Eq. (21.10), and set $t' = t - \delta t$. Use the fact that, for *any* given initial state j, the sum over the final states of the conditional probability must be equal to unity, i.e.,

$$\sum_{k=1}^{N} P(k, t \,|\, j) = 1 \quad \text{for each } j.$$

(Obviously, this condition is just the statement of the conservation of probability.) You will need to make use of this relation in the form

$$1 - P(k, \delta t \,|\, k) = \sum_{\substack{l=1 \\ l \neq k}}^{N} P(l, \delta t \,|\, k) = \sum_{\substack{l=1 \\ l \neq k}}^{N} w(l \,|\, k)\, \delta t.$$

A bit of algebra now leads to Eq. (21.12). ▶

6. A simple way to compute e^{Wt}, where W is the matrix in Eq. (21.31), is to note that $W^2 = -2\lambda\, W$. ▶

9. A nonzero stay or *sojourn* probability implies that a "jump" $j \to j$ is made with an average rate $\lambda\gamma$. Hence Eq. (21.55) now has an additional term on the right-hand side, namely, $(\lambda\,\gamma\,\delta t)\, P(j, t)$. The master equation (21.56) becomes

$$\frac{dP(j,t)}{dt} = \lambda \left\{ \alpha\, P(j-1, t) - (1-\gamma) P(j, t) + \beta\, P(j+1, t) \right\}, \quad j \in \mathbb{Z}.$$

Define the generating function for $P(j, t)$ and proceed as before, to get

$$P(j, t) = e^{-\lambda(1-\gamma)t}\, (\alpha/\beta)^{j/2}\, I_j\big(2\lambda t \sqrt{\alpha\beta}\big).$$

Remark The maximum value of $\sqrt{\alpha\beta}$ corresponds to the unbiased walk, with $\alpha = \beta = \frac{1}{2}(1-\gamma)$. Then $P(j, t) = e^{-\lambda(1-\gamma)t}\, I_j\big(\lambda(1-\gamma)t\big)$. The effect of the sojourn probability is merely to reduce the transition rate from λ to $\lambda(1-\gamma)$. ▶

10. Follow the same procedure as the one already used to go from Eq. (21.10) to Eq. (21.12), in the case of a discrete Markov process. ▶

11. Note that

$$\frac{d\langle X(t)\rangle}{dt} = \int_{-\infty}^{\infty} dx\, x\, \frac{\partial p}{\partial t} \quad \text{and} \quad \frac{d\langle X^2(t)\rangle}{dt} = \int_{-\infty}^{\infty} dx\, x^2\, \frac{\partial p}{\partial t}\,.$$

Use the Fokker–Planck equation for $\partial p/\partial t$, and integrate by parts. All the boundary terms at $x = \pm\infty$ may be set equal to zero: you may assume that $p(x,\, t \,|\, x_0)$ and its derivatives with respect to x tend to zero faster than any inverse power of x as $|x| \to \infty$. (This will be justified *post facto* by the solution for $p(x,\, t \,|\, x_0)$ to be derived below.) Once you arrive at Eq. (21.86), it is straightforward to solve them with the initial conditions (21.87). ▶

12. Equation (21.76) becomes, in this case,

$$\frac{a_2}{2}\frac{dp(x)}{dx} + a_1\, x\, p(x) = \text{constant.}$$

The constant on the right-hand side is the negative of the stationary current density. It must be zero in this case because of the boundary conditions, by the following argument: The stationary current density j^{st} must vanish as $|x| \to \infty$. But this current must also be independent of x. Hence it must vanish for all x. The normalized solution to the differential equation above is then trivially found. It is precisely the Gaussian given by Eq. (21.91). ▶

13. The calculation becomes much easier if you carry out the integration over x first! Shift the variable of integration from x to $x - \mu_X(t)$. ▶

Chapter 22
Analytic Functions of a Complex Variable

The theory of functions of a complex variable (and its generalization, the theory of functions of *several* complex variables), or complex analysis, is one of the richest and most beautiful branches of mathematics, with deep results and far-reaching implications and applications. It is also a vast subject. In this chapter and the next four chapters, a brief account will be given of some aspects of this subject, at the general level of the rest of this book. In keeping with the spirit and purpose of the book, I skip formal proofs of theorems, and focus on techniques for solving a variety of problems. Some standard topics in the subject will necessarily have to be omitted, owing to limitations of space. On the other hand, several topics that are usually not covered at this level have been included, because they are (in my opinion) both interesting and important for mathematical physics.

I shall assume that you have a knowledge of the basic properties of complex numbers, and also some familiarity with analytic functions at an elementary level. The treatment of these aspects of the topic will therefore be cursory, and in the nature of a recapitulation.

22.1 Some Preliminaries

22.1.1 Complex Numbers

Given a complex number $z = x + iy = re^{i\theta}$, its real and imaginary parts are x and y, respectively; its modulus and argument are r and θ, respectively. Its complex conjugate is $z^* = x - iy = re^{-i\theta}$. It is important to remember that the specification of a complex number implies the specification of two *independent* pieces of information, namely, x and y; or r and θ; or, equivalently, you could take this pair to be z and z^* themselves! Once you bear in mind that z and z^* *are linearly independent*, it becomes much easier to understand the concept of analytic functions of a complex variable z. Recall also the basic relations

V. Balakrishnan, *Mathematical Physics*,
https://doi.org/10.1007/978-3-030-39680-0_22

$$x = r\cos\theta = \frac{(z+z^*)}{2}, \quad y = r\sin\theta = \frac{(z-z^*)}{2i}, \tag{22.1}$$

as well as

$$r = (x^2+y^2)^{1/2} = zz^* = |z|^2, \quad \theta = \tan^{-1}\left(\frac{y}{x}\right) = \frac{1}{2i}\ln\left(\frac{z}{z^*}\right). \tag{22.2}$$

★ **1.** Show that the real and imaginary parts of the complex numbers listed below are as indicated. Remember the standard phase convention, according to which $i = e^{i\pi/2}$ and $-i = e^{-i\pi/2} = e^{3\pi i/2}$.

(a) $(i)^{i^i} = \cos\left(\frac{1}{2}\pi e^{-\pi/2}\right) + i\sin\left(\frac{1}{2}\pi e^{-\pi/2}\right)$.

(b) $\displaystyle\sum_{n=0}^{\infty}\frac{(i\pi)^{2n+1}}{(2n+1)!} = 0$.

(c) $\displaystyle\sum_{n=0}^{\infty}\frac{e^{i\pi n/4}}{(2n)!} = \cosh\left(\cos\tfrac{1}{8}\pi\right)\cos\left(\sin\tfrac{1}{8}\pi\right) + i\sinh\left(\cos\tfrac{1}{8}\pi\right)\sinh\left(\sin\tfrac{1}{8}\pi\right)$.

(d) The infinitely iterated square root

$$\left(i + \left(i + \left(i + (\ldots)^{1/2}\right)^{1/2}\right)^{1/2}\right)^{1/2} = \frac{1}{2} + \frac{\sqrt{\sqrt{17}+1}}{2\sqrt{2}} + i\,\frac{\sqrt{\sqrt{17}-1}}{2\sqrt{2}}$$
$$\simeq 1.3002 + 0.6248\,i.$$

(e) The infinite continued fraction

$$\frac{i}{1+}\,\frac{i}{1+}\,\frac{i}{1+}\cdots = -\frac{1}{2} + \frac{\sqrt{\sqrt{17}+1}}{2\sqrt{2}} + i\,\frac{\sqrt{\sqrt{17}-1}}{2\sqrt{2}}$$
$$\simeq 0.3002 + 0.6248\,i.$$

(f) The infinite continued fraction

$$\frac{1}{i+}\,\frac{1}{i+}\,\frac{1}{i+}\cdots = e^{-i\pi/6} = \tfrac{1}{2}(\sqrt{3}-i).$$

In (d), (e) and (f), calling the left-hand side z, you have to solve the respective equations $z = (i+z)^{1/2}$, $z = i/(1+z)$ and $z = 1/(i+z)$. Each of these reduces to a quadratic equation, and you must take care to identify the correct root in each case.

22.1.2 *Equations to Curves in the Plane in Terms of z*

As you know, the equation to a curve in the xy-plane has the general form $\phi(x, y) = 0$. In elementary coordinate geometry, we study the equations to basic curves such as conic sections.

The equations specifying curves are often very conveniently expressed in terms of z and z^*. For instance, the equation of a circle of radius a with center at (x_0, y_0) is just $|z - z_0| = a$, where $z_0 = x_0 + i\, y_0$.

★ **2.** Here are some examples of familiar curves, expressed in terms of z and z^*:

(a) Let a and c be positive constants, where $a > c$. Show that the equation $|z + c| + |z - c| = 2a$ specifies an ellipse whose foci are located at $\pm c$ on the x-axis, and whose semi-major axis is equal to a.
(b) Show that the equations (i) $z^2 + z^{*2} = C$ and (ii) $z^2 - z^{*2} = iC$, where C is a real constant, specify rectangular hyperbolas.
(c) Show that the locus of points satisfying the equation $\big|(z - 1)/(z - 3)\big| = 1$ lie on the straight line $x = 2$.
(d) Show that the equation $|z - 1| - |z + 1| = \pm 1$ represent, respectively, the left and right branches of the hyperbola $12x^2 - 4y^2 = 3$, whose asymptotes are the straight lines $y = \pm\sqrt{3}\, x$.

In case (a), recall that an ellipse is the locus of a point moving such that the *sum* of its distances from two fixed points is a (positive) constant. Similarly, in case (d), recall that a branch of hyperbola is the locus of a point moving such that the *difference* of its distances from two fixed points is a constant.

22.2 The Riemann Sphere

22.2.1 *Stereographic Projection*

We know that the number line extends from $-\infty$ on the left to $+\infty$ on the right. In the complex plane, z may tend to infinity along any one of an infinite number of directions. It is convenient, for mathematical purposes, to *compactify* the plane by bringing together all these points at infinity and "gluing" them together into a single point. This can be done by a *mapping* between the complex plane and the surface of a sphere, called a **stereographic projection**.

Consider a sphere of unit radius with its center at O, the origin of coordinates, as shown in Fig. 22.1. The coordinates of any point on the surface are given by (ξ_1, ξ_2, ξ_3), where

$$\xi_1^2 + \xi_2^2 + \xi_3^2 = 1. \tag{22.3}$$

In terms of spherical polar coordinates (θ, φ) on the unit sphere, we have of course

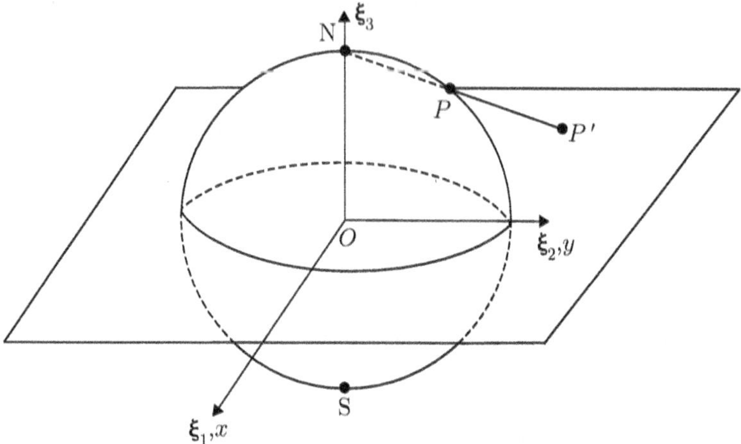

Fig. 22.1 The Riemann sphere $\mathcal{S}$ and stereographic projection to the complex plane $\mathbb{C}$. The north pole of the sphere, N, has been chosen as the point of projection. The point P on $\mathcal{S}$ is projected to the point P′ on $\mathbb{C}$

$$\xi_1 = \sin\theta\,\cos\varphi,\ \xi_2 = \sin\theta\,\sin\varphi,\ \xi_3 = \cos\theta\,. \tag{22.4}$$

The coordinates of the north pole N are $(0, 0, 1)$, while those of the south pole S are $(0, 0, -1)$.

Now consider the plane in which the equator of the sphere lies. Let the Cartesian coordinates on this plane be (x, y), with respect to the same origin O, and with the x and y axes running along the ξ_1 and ξ_2 axes, respectively. The stereographic projection of any point P on the sphere onto the equatorial plane is obtained by joining N and P by a straight line, and extending it (if necessary) till it meets that plane. The projection of P is the point P′ where the line cuts the equatorial plane. The set of all points on this plane with finite coordinates comprise the **complex plane**, denoted by $\mathbb{C}$. The sphere is called the **Riemann sphere**, and will be denoted by $\mathcal{S}$. The coordinates (x, y) of P′ in the complex plane $(z = x + iy)$ are related to the coordinates (ξ_1, ξ_2, ξ_3) of P on the Riemann sphere by

$$\boxed{x = \frac{\xi_1}{1-\xi_3} = \cot\tfrac{1}{2}\theta\,\cos\varphi \quad\text{and}\quad y = \frac{\xi_2}{1-\xi_3} = \cot\tfrac{1}{2}\theta\,\sin\varphi.} \tag{22.5}$$

Hence

$$\boxed{z = \frac{\xi_1 + i\xi_2}{1-\xi_3} = e^{i\varphi}\cot\tfrac{1}{2}\theta,\quad z^* = \frac{\xi_1 - i\xi_2}{1-\xi_3} = e^{-i\varphi}\cot\tfrac{1}{2}\theta.} \tag{22.6}$$

Referring to Fig. 22.1, we note the following:

(i) The south pole of $\mathcal{S}$ (the point S) is mapped to the origin in the complex plane.
(ii) Points on the equator of $\mathcal{S}$ are mapped to themselves as points on the **unit circle** $|z| = 1$ in the complex plane.
(iii) Points of the *northern* hemisphere of $\mathcal{S}$ are mapped to points *outside* the unit circle in the complex plane.
(iv) Points in the *southern* hemisphere of $\mathcal{S}$ are mapped to points *inside* the unit circle in the complex plane.

As P gets closer to the point of projection N, it is clear that the point P′ moves farther and farther away from the origin in the complex plane. This happens regardless of the direction along which P approaches N on the sphere.

- The **point at infinity** in the complex plane is *defined* as the image of the north pole itself under the projection, and is denoted by ∞.
- The finite part of the complex plane, *together with* the point at infinity, i.e., $\mathbb{C} \cup \{\infty\}$, is called the **extended complex plane**. It will be denoted by $\hat{\mathbb{C}}$.

In what follows, by the term "complex plane" I shall generally mean the extended complex plane. It is obvious that the point at infinity corresponds to the complex number $z = re^{i\theta}$ with $r \to \infty$, for all values of θ.

- It is very advantageous to have "infinity" identified with a single point in this fashion, and to be able to specify ∞ as a specific value for the complex number z.

The stereographic projection described above, i.e., the mapping from the Riemann sphere $\mathcal{S}$ to the complex plane $\hat{\mathbb{C}}$, is *invertible*. That is, *given* the coordinates x and y (or, equivalently, z and z^*), each of the quantities ξ_1, ξ_2, and ξ_3 can be uniquely determined. Thus, the inverse relations corresponding to (22.5) are

$$\boxed{\xi_1 = \frac{2x}{x^2 + y^2 + 1}, \quad \xi_2 = \frac{2y}{x^2 + y^2 + 1}, \quad \xi_3 = \frac{x^2 + y^2 - 1}{x^2 + y^2 + 1}.} \tag{22.7}$$

Equivalently, the inverse relations corresponding to (22.6) are

$$\boxed{\xi_1 = \frac{z + z^*}{|z|^2 + 1}, \quad \xi_2 = \frac{z - z^*}{i(|z|^2 + 1)}, \quad \xi_3 = \frac{|z|^2 - 1}{|z|^2 + 1}.} \tag{22.8}$$

It is evident that $\xi_1^2 + \xi_2^2 + \xi_3^2 \equiv 1$, as required.

★ **3.** Establish the relations (22.5) and the inverse relations (22.7).

We have used the north pole of the Riemann sphere, N, as the point of projection. In fact, *any* point on the sphere can be chosen as the point of projection.

- The great advantage of using the Riemann sphere as a representation of the extended complex plane is that the point at infinity has the same status as any other point.

22.2.2 Maps of Circles on the Riemann Sphere

Under the stereographic projection, circles of latitude on $\mathcal{S}$ are mapped to concentric circles (with their center at the origin) in the complex plane. On a circle of latitude, the polar angle θ (which is just the co-latitude) remains constant. Such a circle is mapped to the circle in the complex plane given by

$$x^2 + y^2 = k^2, \quad \text{where} \quad k^2 = \frac{1 + \cos\,\theta}{1 - \cos\,\theta}. \tag{22.9}$$

It is obvious that the radius of this circle is greater than unity for $0 < \theta < \frac{1}{2}\pi$, and less than unity for $\frac{1}{2}\pi < \theta < \pi$. Similarly, the azimuthal angle φ is constant on a meridian of longitude. This is mapped in $\hat{\mathbb{C}}$ to a straight line starting from the origin, namely,

$$y = x\,\tan\,\varphi. \tag{22.10}$$

★ **4.** Establish Eqs. (22.9) and (22.10).

These results are special cases of a very important property of the stereographic projection:

- *Any circle on the Riemann sphere is mapped into either a circle or a straight line in the complex plane by stereographic projection.*

***Proof*:** Any circle on the Riemann sphere is the intersection of $\mathcal{S}$ with a plane whose (perpendicular) distance s from the center of the sphere is less than unity. Hence an arbitrary circle on $\mathcal{S}$ is given by the equation

$$\mathbf{n} \cdot \boldsymbol{\xi} = s, \quad \text{where} \quad \boldsymbol{\xi} = (\xi_1,\,\xi_2,\,\xi_3) \quad \text{and} \quad \xi_1^2 + \xi_2^2 + \xi_3^2 = 1. \tag{22.11}$$

Here $\mathbf{n} = (n_1,\,n_2,\,n_3)$ is the unit normal to the plane, and s (where $0 < s < 1$) is the distance of the plane from the center of the Riemann sphere. Substituting the expressions in Eq. (22.7) for the ξ_i in terms of x and y, we get

$$(s - n_3)(x^2 + y^2) - 2n_1\,x - 2n_2\,y + n_3 + s = 0. \tag{22.12}$$

You will recall from elementary coordinate geometry that this is the equation to a circle in the complex plane, provided $s \neq n_3$. Completing the squares in x and y, we find that the center of the circle is at $\big(n_1/(s - n_3),\,n_2/(s - n_3)\big)$, and that its radius is equal to $\sqrt{1 - s^2}/|s - n_3|$.

The case $n_3 = s$ is precisely that of a circle on $\mathcal{S}$ that passes through N, the point of projection: since the coordinates of N are given by $\xi_1 = 0,\ \xi_2 = 0,\ \xi_3 = 1$, the condition $\mathbf{n} \cdot \boldsymbol{\xi} = s$ immediately implies that $n_3 = s$. Equation (22.12) then reduces to

$$n_1\,x + n_2\,y = s. \tag{22.13}$$

This is the equation of a straight line in the complex plane. In the special case $s = 0$, the straight line in the complex plane passes through the origin. This line is obviously the map of a full circle of longitude on $\mathcal{S}$, i.e., of a meridian ($\varphi =$ constant) and its antipodal meridian ($\varphi + \pi =$ constant).

22.2.3 A Metric on the Extended Complex Plane

The usual notion of a distance in the complex plane is what you would expect from elementary Euclidean geometry in a plane: given two points $z_1 = x_1 + iy_1$ and $z_2 = x_2 + iy_2$, the distance between them is $|z_1 - z_2| = [(x_1 - x_2)^2 + (y_1 - y_2)^2]^{1/2}$. It is zero if and only if z_1 and z_2 coincide. Moreover, this distance function satisfies the triangle inequality

$$|z_1 - z_3| \leq |z_1 - z_2| + |z_2 - z_3| \tag{22.14}$$

for any three points z_1, z_2, and z_3. (The equality sign obviously applies when z_2 lies on the line segment between z_1 and z_3.) The problem with this definition of the distance is that it is not very useful when we have to deal with the point at infinity.

As you might expect, this difficulty is removed when we use the Riemann sphere as the representation of the extended complex plane. Let the points z_1 and z_2 in the complex plane correspond to the points P and Q on the Riemann sphere, with coordinates $\boldsymbol{\xi} = (\xi_1, \xi_2, \xi_3)$ and $\boldsymbol{\zeta} = (\zeta_1, \zeta_2, \zeta_3)$, respectively. See Fig. 22.2. A natural definition of the distance between these two points on $\mathcal{S}$ would be the **geodesic distance** between them—namely, the length of the (shorter) arc of the great circle on $\mathcal{S}$ that passes through P and Q. This is the shortest path between P and Q that lies entirely on the sphere. Since the radius of the sphere is unity, this arc length is numerically equal to the angle (γ, say) between the two vectors. But $\cos\gamma = \boldsymbol{\xi} \cdot \boldsymbol{\zeta}$, because both $\boldsymbol{\xi}$ and $\boldsymbol{\zeta}$ are unit vectors. The geodesic distance $\gamma = \cos^{-1}(\boldsymbol{\xi} \cdot \boldsymbol{\zeta})$ can

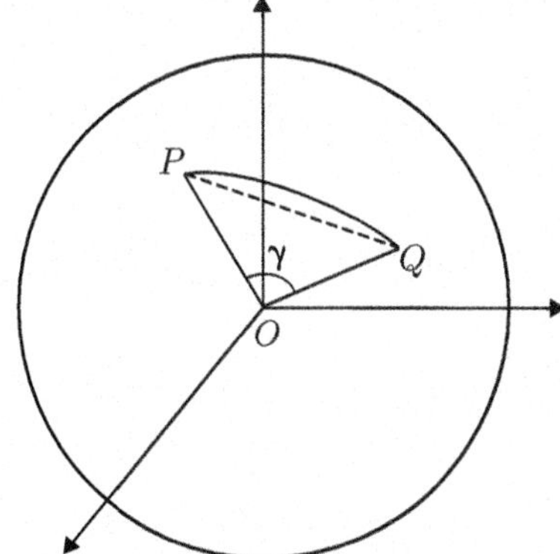

Fig. 22.2 Distance between two points P and Q on the Riemann sphere. The curved line shows the *geodesic* or great circle distance, which is equal to the dihedral angle γ between the position vectors of P and Q with respect to the origin O. The dashed line shows the chordal distance between P and Q

then be expressed in terms of z_1, z_2 and their complex conjugates.[1] The result is

$$\gamma(z_1, z_2) = \cos^{-1}\left(1 - \frac{2|z_1 - z_2|^2}{(|z_1|^2 + 1)(|z_2|^2 + 1)}\right). \tag{22.15}$$

It follows that the distance of any point z from the point at infinity is just

$$\gamma(z, \infty) = \cos^{-1}\left(\frac{|z|^2 - 1}{|z|^2 + 1}\right) = \theta, \tag{22.16}$$

where θ is the polar angle of the point on the Riemann sphere that is mapped to z in the complex plane.

★ **5.** Derive Eqs. (22.15) and (22.16).

As a measure of the distance, the arc length γ is somewhat cumbersome. A more convenient measure is the *chordal distance* between P and Q. This is the length of a chord drawn from P to Q (treating the sphere as a hollow surface in three-dimensional Euclidean space), rather than that of the great circle arc between these points. This quantity, expressed in terms of z_1, z_2 and their complex conjugates, is

$$d(z_1, z_2) = \frac{2\,|z_1 - z_2|}{\sqrt{(|z_1|^2 + 1)(|z_2|^2 + 1)}}. \tag{22.17}$$

★ **6.** Establish Eq. (22.17).

Note that a compact formula for the distance is

$$d(z_1, z_2) = 2\,|\sin \tfrac{1}{2}\gamma|. \tag{22.18}$$

The quantity $d(z_1, z_2)$ satisfies the properties expected of a distance function: $d(z_1, z_2) = d(z_2, z_1)$, and $d(z_1, z_2) = 0$ if and only if $z_1 = z_2$. Further, for any three points z_1, z_2, and z_3, we have

$$d(z_1, z_3) \leq d(z_1, z_2) + d(z_2, z_3). \tag{22.19}$$

★ **7.** Verify the triangle inequality (22.19) for the distance function defined in Eq. (22.17).

With this definition of the distance, the exceptional role played by the point at infinity causes no trouble, as it does in the case of the ordinary Euclidean definition $|z_1 - z_2|$. We have

$$d(z, \infty) = \frac{2}{\sqrt{|z|^2 + 1}}. \tag{22.20}$$

[1] Recall, incidentally, that $\cos\gamma$ is expressed in terms of the polar and azimuthal angles of the two vectors $\boldsymbol{\xi}$ and $\boldsymbol{\zeta}$ by the law of cosines, given by Eq. (16.137) of Chap. 16, Sect. 16.4.8.

Clearly, $d(z_1, z_2) \leq 2$ for any two points z_1 and z_2 in the extended complex plane.

Setting $z_1 = z$ and $z_2 = z + dz$ in Eq. (22.17), we have the following expression for the square of the distance between two infinitesimally separated points:

$$(ds)^2 = \frac{4[(dx)^2 + (dy)^2]}{(x^2 + y^2 + 1)^2} = \frac{4|dz|^2}{(|z|^2 + 1)^2}. \tag{22.21}$$

The interval thus depends on the location in the complex plane. That is, the **metric** on the extended complex plane is not a "flat" or Euclidean metric. I do not go into these aspects any further here.

22.3 Analytic Functions of z

22.3.1 The Cauchy–Riemann Conditions

In the analysis of real functions $f(x)$ of a real variable x, special interest is attached to those functions with a specified degree of smooth or regular behavior—such as continuous functions, functions that are differentiable a given number of times, functions that are infinitely differentiable (sometimes termed "real analytic" functions), and so on. In the case of complex-valued functions $f(z)$ of a complex variable z, the functions that are of special interest are the so-called **analytic functions**.[2] These functions have an enormous number of deep and important properties. You will get a glimpse of some of these as we go along. For the present, we restrict our attention to **univalent** or **single-valued** functions: that is, functions such that $f(z)$ has a unique value for each given z. In Chap. 26, we will consider multivalued (or multivalent) functions.

Consider a function $f(z)$ of the complex variable z. Let $u(x, y)$ and $v(x, y)$ denote its real and imaginary parts, respectively.

- $f(z)$ is said to be *analytic* (more precisely, **holomorphic**) in some region of the z-plane if its real and imaginary parts satisfy the **Cauchy–Riemann conditions**

$$\boxed{\frac{\partial u}{\partial x} = \frac{\partial v}{\partial y} \quad \text{and} \quad \frac{\partial u}{\partial y} = -\frac{\partial v}{\partial x}} \tag{22.22}$$

  at every point in this region.

Strictly speaking, a function $f(z)$ is (complex) analytic in a neighborhood of a point z_0 if it can be represented as an absolutely convergent power series in nonnegative powers of $(z - z_0)$, i.e., a Taylor series in $(z - z_0)$, in that neighborhood. It turns out that *complex analytic* and *holomorphic* are equivalent properties. I will, therefore,

[2]The term "analytic" is a convenient abbreviation for "complex analytic".

not distinguish between "complex analytic" and "holomorphic", and will generally use the term "analytic" to indicate this property.

There are several ways of understanding the meaning and significance of the Cauchy–Riemann conditions. We shall return to some of these later on. The region in which these conditions are satisfied is the region of analyticity, or **domain of holomorphy**, of the function concerned. If a function $f(z)$ is holomorphic in $\mathbb{C}$, the whole of the *finite* part of the complex plane (that is, for all $|z| < \infty$), then it is called an **entire function**. Any constant is trivially an entire function. The simplest example of a nontrivial entire function is $f(z) = z$ itself. More generally, all polynomials in z are entire functions. So is the exponential function e^z, and so are linear combinations of exponentials such as $\cos z$, $\sin z$, $\cosh z$, and $\sinh z$. (More will be said on entire functions in Sect. 22.6.) The question that arises naturally is: Can a function be holomorphic everywhere in the *extended* complex plane $\hat{\mathbb{C}}$?

- **Liouville's Theorem** asserts that a function that is holomorphic at every point in $\hat{\mathbb{C}}$ must be a constant.
- This means that an entire function (that is not trivially a constant) *cannot* be analytic at the point at infinity. In other words, it must be *singular* at $z = \infty$. We will discuss singularities subsequently.

Here is one way to understand what is meant by an analytic function of z. Recall that, just as x and y are linearly independent coordinates on the plane, so are the linear combinations $z = x + iy$ and $z^* = x - iy$. Hence any function of x and y can equally well be written as a function of z and z^*.

- An analytic function of z is a sufficiently well-behaved function of x and y that depends on the combination $x + iy$ alone, and that does *not* involve the other combination, namely, $x - iy$.
- Here, the phrase "sufficiently well-behaved" means that the real and imaginary parts of the function have continuous partial derivatives with respect to x and y. That is, all four first derivatives $\partial u/\partial x$, $\partial u/\partial y$, $\partial v/\partial x$, and $\partial v/\partial y$ are continuous functions of x and y.

Thus, f is an analytic function of z if $\partial f/\partial z^* \equiv 0$. Since $z^* = x - iy$ we have, by the chain rule of differentiation,

$$\frac{\partial f}{\partial z^*} = 0 \implies \frac{\partial f}{\partial x} + i\frac{\partial f}{\partial y} = 0. \tag{22.23}$$

The real and imaginary parts of the last equation yield precisely the Cauchy–Riemann conditions (22.22).

- The Cauchy–Riemann conditions may, therefore, be written compactly as

$$\boxed{\frac{\partial f}{\partial z^*} \equiv 0 \quad \text{(Cauchy–Riemann conditions).}} \tag{22.24}$$

Writing any function of x and y in terms of z and z^* enables you to quickly identify functions that *cannot* be analytic functions of z, merely by checking whether z^* makes its appearance in the function. If it does so, the function cannot be an analytic function of z.

★ **8.** In the list of functions given below, identify the functions that are analytic functions of z in some region, and find their regions of analyticity.

(a) x	(b) $ix - y$	(c) r
(d) $e^{i\theta}$	(e) $x^2 - iy^2$	(f) $x - iy$
(g) $x^2 - y^2 + 2ixy$	(h) $x^2 - y^2 - 2ixy$	(i) $i\,\tan^{-1}(y/x)$
(j) $(iy + x^2 + y^2)$	(k) $[(x+i)^2 - y^2]^{1/2}$	(l) $x^4 + 2ix^2y^2 - y^4$
(m) $i\,e^x\,\sin\,y$	(n) $x^2 + x + 1 - y^2 + iy\,(2x+1)$	(o) $(x - iy)/(x^2 + y^2)$.

***Cauchy–Riemann conditions in polar form*:** Instead of the "Cartesian" representation $z = x + iy$ and $f(z) = u + iv$, suppose we consider the "polar" representations $z = r\,e^{i\theta}$ and $f(z) = R\,e^{i\psi}$. Thus $R = |f(z)|$, while $\psi = \arg f(z)$. The Cauchy–Riemann conditions now read

$$\boxed{\frac{\partial \ln R}{\partial \ln r} = \frac{\partial \psi}{\partial \theta}, \quad \frac{\partial \ln R}{\partial \theta} = -\frac{\partial \psi}{\partial \ln r}.} \tag{22.25}$$

★ **9.** Derive Eq. (22.25) from the Cauchy–Riemann conditions (22.22).

22.3.2 The Real and Imaginary Parts of an Analytic Function

I have stated above that the real and imaginary parts of an analytic function have continuous first derivatives with respect to x and y. It turns out that the condition of analyticity actually implies that u and v have smooth derivatives of *all* orders with respect to x and y.

Anticipating this result, let us differentiate both sides of the Cauchy–Riemann conditions in Eq. (22.22) with respect to x and y. It follows at once that both u and v individually satisfy Laplace's equation in two dimensions, i.e.,

$$\nabla^2 u = 0 \quad \text{and} \quad \nabla^2 v = 0. \tag{22.26}$$

- Hence the real and imaginary parts of an analytic function are *harmonic functions*. (Recall the brief discussion of harmonic functions in Chap. 8, Sect. 8.2.)
- The regions in which $u(x, y)$ and $v(x, y)$ are harmonic may, in general, be different. The combination $u + iv$ then constitutes an analytic function of z in the *intersection* of the regions in which they are individually harmonic.

- It is also clear from the Cauchy–Riemann conditions that an analytic function cannot be *identically* equal to a purely real or purely imaginary function, except in the trivial case when the function is just a constant.

An analytic function that is real when its argument is real, i.e., a function such that $v(x, 0) \equiv 0$, so that $f(x) = u(x, 0)$, is called a **real analytic** function. Examples of real analytic functions are easy to find: z, e^z, $\cos z$, $\sin z$, and so on.

Another property that follows from the Cauchy–Riemann conditions is the relation

$$\nabla u \cdot \nabla v = 0. \tag{22.27}$$

- Therefore, for *any* analytic function, the curves $u =$ constant and $v =$ constant constitute two mutually orthogonal families of curves in the complex plane.

This fact has an important implication for *conformal mapping*, which will be touched upon in a later chapter.

★ **10.** Given u (or v) as indicated below, find the corresponding v (respectively, u) by integrating the Cauchy–Riemann conditions, and hence write down the analytic function $f(z)$ (up to an additive constant). Indicate also the region of analyticity of $f(z)$ in each case.

(a) $u = x^2 - y^2$ (b) $u = e^x \cos y$ (c) $u = \ln (x^2 + y^2)^{1/2}$

(d) $u = \cos x \cosh y$ (e) $v = e^{2x} \sin (2y)$ (f) $v = 2xy$

(g) $v = \cos x \sinh y$ (h) $v = 2 \tan^{-1}(y/x)$ (i) $v = e^{x^2 - y^2} \sin (2xy)$.

22.4 The Derivative of an Analytic Function

The condition of analyticity is so strong that an analytic function is guaranteed to have a derivative that is, moreover, itself an analytic function. The formal proof of this statement is not difficult, but I will not give it here. From the various representations for analytic functions that will be encountered as we go along, the validity of this statement will be more or less obvious. It then follows at once that an analytic function has derivatives of arbitrarily high order that are also analytic functions. This is in marked contrast to the case of functions of a real variable, where a function maybe once differentiable but not twice differentiable; or, in general, differentiable some r times, but not $(r + 1)$ times.

- Any analytic function of a complex variable is infinitely differentiable.

Defining the derivative of an analytic function helps us understand the Cauchy–Riemann conditions and the meaning of analyticity in yet another (albeit related) way. In analogy with the definition of the derivative of a function of a real variable, it is natural to define the derivative of $f(z)$ with respect to z as

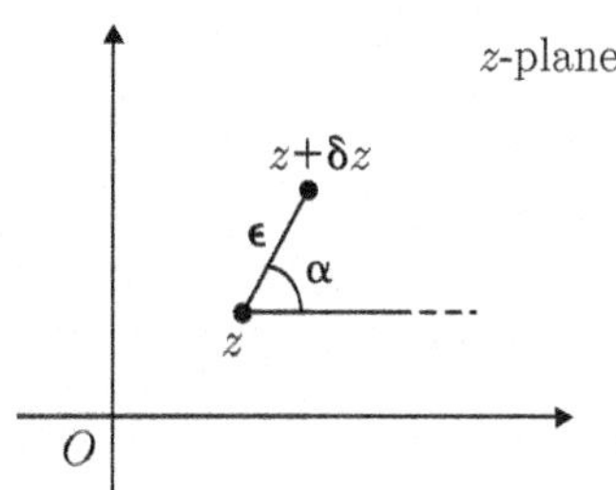

Fig. 22.3 The derivative of an analytic function: its lack of dependence on the phase angle α is equivalent to the Cauchy–Riemann conditions

$$\frac{df(z)}{dz} = \lim_{\delta z \to 0} \frac{f(z + \delta z) - f(z)}{\delta z}, \tag{22.28}$$

where δz is a complex number of infinitesimal magnitude. But the question that arises at once is: in what *direction* in the complex plane should the point $z + \delta z$ lie, relative to the point z? That is, what should the argument (or phase angle) of the increment δz be? Suppose the complex number δz has a magnitude ϵ and an argument α, i.e., $\delta z = \epsilon\, e^{i\alpha}$, as in Fig. 22.3. Then, provided the real and imaginary parts of $f(z)$ have continuous partial derivatives, the definition (22.28) yields, in the limit $\epsilon \to 0$,

$$\frac{df(z)}{dz} = e^{-i\alpha} \left\{ \left(\frac{\partial u}{\partial x} \cos \alpha + i \frac{\partial v}{\partial y} \sin \alpha \right) + i \left(\frac{\partial v}{\partial x} \cos \alpha - i \frac{\partial u}{\partial y} \sin \alpha \right) \right\}. \tag{22.29}$$

(Check this out explicitly.) The remarkable fact is that *this expression becomes completely independent of* α if and only if $\partial u/\partial x = \partial v/\partial y$ and $\partial u/\partial y = -\partial v/\partial x$, which are precisely the Cauchy–Riemann conditions.

- The analyticity of a function $f(z)$ may therefore be understood as equivalent to the requirement that the limit defining the derivative of $f(z)$ be *unique*, *independent* of the direction along which $z + \delta z$ approaches z in the complex plane.

And this is such a strong condition that it automatically implies the existence of derivatives of *all* orders that are themselves analytic functions!

★ **11.** In its region of analyticity, an analytic function $f(z) = u + iv$ may also be regarded as a **map** from (a region of) the complex plane to (a region of) the complex plane (Fig. 22.4).

(a) Show that the Jacobian determinant of the transformation $(x, y) \mapsto (u, v)$ is just $|f'(z)|^2$, where $f'(z)$ denotes the derivative of $f(z)$.
(b) Show that $\nabla^2(|f(z)|^2) = 4\,|f'(z)|^2$.

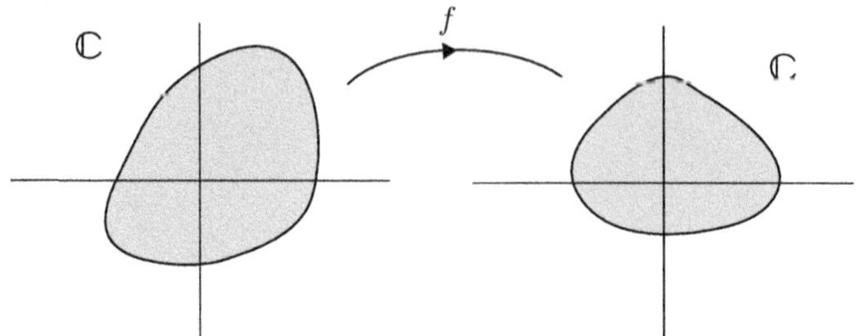

Fig. 22.4 An analytic function is a map of some region of the complex plane to some region of the complex plane

22.5 Power Series as Analytic Functions

22.5.1 *Radius and Circle of Convergence*

I turn now to the most important way of representing an analytic function in its domain of holomorphy: namely, as a **power series**. As pointed out in following Eq. (22.22), the existence of a valid power series representation may in fact be taken to be the *definition* of the analyticity of a function.

Let $f(z)$ be analytic inside a region $\mathcal{R}$. If z_0 is a point in the region, then there generally exists a neighborhood of z_0 such that $f(z)$ can be expanded in an absolutely convergent series in nonnegative powers of $(z - z_0)$, called a **Taylor series**:

$$\boxed{f(z) = \sum_{n=0}^{\infty} a_n\,(z - z_0)^n\,.} \tag{22.30}$$

As you know, the **ratio test**[3] tells us that such an infinite series is absolutely convergent if

$$\lim_{n\to\infty}\left|\frac{a_{n+1}\,(z - z_0)^{n+1}}{a_n\,(z - z_0)^n}\right| = \lim_{n\to\infty}\left|\frac{a_{n+1}}{a_n}\right|\,|z - z_0| < 1. \tag{22.31}$$

It follows that the neighborhood concerned is the interior of a circle centered at z_0, i.e., it is given by $|z - z_0| < R$, where R is called the **radius of convergence** of the

[3]As indicated earlier, I assume that you are familiar with the basic ideas related to infinite series and their convergence, including the ratio test for absolute convergence. Recall, in particular, the following: The harmonic series $1 + \frac{1}{2} + \frac{1}{3} + \frac{1}{4} + \cdots$ is divergent, while the alternating series $1 - \frac{1}{2} + \frac{1}{3} - \frac{1}{4} + \cdots$ is conditionally convergent. Rearranging the terms of a conditionally convergent series *may* change its value or even cause it to diverge. I will not go into details such as the concept of uniform convergence, which permits us to differentiate an infinite series term by term, or rearrange the terms of the series, and so on. Further remarks on the behavior of a power series *on* its circle of convergence will be made in Sect. 22.5.3.

series. The circle $|z - z_0| = R$ is called the **circle of convergence** of the series. It lies within $\mathcal{R}$, the region of analyticity of $f(z)$. (Its boundary may coincide with that of $\mathcal{R}$ at one or more points.) **Absolute convergence** means that the sum of the *magnitudes* of the terms of the series is also finite: that is, $\sum_{n=0}^{\infty} |a_n(z - z_0)^n| < \infty$. Such convergence permits us to manipulate the series and perform operations such as differentiation or integration on it term by term, as long as z remains inside the circle of convergence.

If $f(z)$ and z_0 are specified, the coefficients a_n in the Taylor series above are uniquely determined. We have

$$\boxed{a_n = \frac{1}{n!}\left[\frac{d^n f(z)}{dz^n}\right]_{z=z_0}.} \tag{22.32}$$

The analyticity of $f(z)$ guarantees the existence of the derivatives of $f(z)$ of all orders. The radius of convergence itself is determined by the set of coefficients $\{a_n\}$, or, more precisely, by the *asymptotic* behavior of a_n as $n \to \infty$. The radius of convergence is given by

$$\boxed{R = \lim_{n\to\infty}\left|\frac{a_n}{a_{n+1}}\right|,} \tag{22.33}$$

provided the limit concerned exists. More generally,

$$\boxed{R^{-1} = \limsup_{n\to\infty} |a_n|^{1/n},} \tag{22.34}$$

where "lim sup" (or supremum) stands for the *least upper bound* of the quantity concerned. It is important to keep the following in mind:

- At all points *inside* its circle of convergence, a power series in z converges absolutely, and is a *representation* of an analytic function of z.
- The function itself may have many other representations, each with its own region of validity.
- A power series in z with an infinite radius of convergence represents an entire function of z.

22.5.2 An Instructive Example

A very familiar power series provides a most instructive example of a general aspect of analytic functions. Consider the binomial series (which also happens to be a geometric progression in this case)

$$\frac{1}{(1-z)} = \sum_{n=0}^{\infty} z^n, \quad |z| < 1. \tag{22.35}$$

The ratio test shows that the radius of convergence of this power series is unity, i.e., $R = 1$.

- The series on the right-hand side of Eq. (22.35) is a *representation* of the function on the left-hand side that is valid in the interior of the unit circle in the z-plane.

Other power series representing the *same* function are easily deduced. Let $a \neq 1$ be any complex number of finite modulus. Note the simple identity

$$(1 - z) = (1 - a) - (z - a) = (1 - a)\left\{1 - \frac{z - a}{1 - a}\right\}. \tag{22.36}$$

It follows that

$$\frac{1}{(1 - z)} = \frac{1}{(1 - a)} \sum_{n=0}^{\infty} \left(\frac{z - a}{1 - a}\right)^n, \quad \text{provided} \quad |z - a| < |1 - a|. \tag{22.37}$$

The single function $(1 - z)^{-1}$ thus has an infinite number of representations in the form of power series, one for each value of a. For instance, setting $a = 0, -\frac{1}{2}, i$ and 2 in succession, we get the four corresponding power series

$$\frac{1}{(1 - z)} = \begin{cases} \sum_{n=0}^{\infty} z^n, & |z| < 1 \\ \frac{2}{3}\sum_{n=0}^{\infty} \left[\frac{2}{3}(z + \frac{1}{2})\right]^n, & |z + \frac{1}{2}| < \frac{3}{2} \\ \frac{1}{2}(1 + i)\sum_{n=0}^{\infty} [(z - i)/(1 - i)]^n, & |z - i| < \sqrt{2} \\ -\sum_{n=0}^{\infty} (2 - z)^n, & |z - 2| < 1. \end{cases} \tag{22.38}$$

Every one of the power series on the right-hand side is a representation of the *same* function, $1/(1 - z)$, valid in the region indicated in each case. That is, the value of the series coincides pointwise with the value of $(1 - z)^{-1}$, for every point in the region in which the series converges. For instance, setting $z = 0$ on the left-hand side yields the value 1. It is easily checked that the first three series on the right-hand side also yield the value 1 when z is set equal to 0, because $z = 0$ is inside the circle of convergence in each case. On the other hand, the fourth series diverges (becomes infinite) if you set $z = 0$ in it. But then the point $z = 0$ is outside the circle of convergence of this series, and it makes no sense to use the series to find the value of the function at such a point.

It is clear that the function in this example, and similarly every other analytic function, can be written as a power series in an infinite number of ways. Each representation is an **analytic continuation** of the function to the region in which the corresponding power series converges.

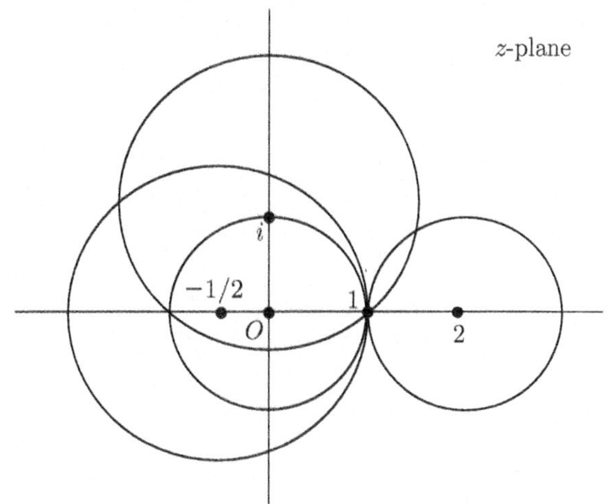

Fig. 22.5 Circles of convergence for four different power series representations (Eq. (22.38)) of the same function, $f(z) = (1 - z)^{-1}$. The centers of the various circles are located at 0, $-\frac{1}{2}$, i, and 2, respectively. The point $z = 1$ is a singularity of the function

The circles of convergence for the series representations in Eq. (22.38) are shown in Fig. 22.5. The functional form $(1 - z)^{-1}$ may be regarded as a *master representation* of this function, because it is evidently valid at all points in the z-plane. At $z = 1$, the function has a *singularity*. Note, too, that the boundary of the region of convergence for each of the series above passes through the point $z = 1$. This is not just a coincidence. I will return to analytic continuation and singularities in greater detail in the chapters that follow, but the example just discussed should help give you a rough idea of what these concepts entail.

★ **12.** The ratio test gives us a quick way of deducing the region of convergence of a large number of series. Find the region of absolute convergence of each of the following power series:

$$\text{(a)} \sum_{n=0}^{\infty} \frac{z^n}{(n+1)^3} \quad \text{(b)} \sum_{n=1}^{\infty} \frac{z^n}{\ln(n+1)} \quad \text{(c)} \sum_{n=1}^{\infty} \frac{z^n}{\ln\ln(n+1)}$$

$$\text{(d)} \sum_{n=0}^{\infty} n^7 z^n \quad \text{(e)} \sum_{n=1}^{\infty} \frac{(\ln n)\, z^n}{n} \quad \text{(f)} \sum_{n=2}^{\infty} \frac{(\ln\ln n)\, z^n}{\ln n}$$

$$\text{(g)} \sum_{n=1}^{\infty} \frac{(z/2)^n}{n(n+1)} \quad \text{(h)} \sum_{n=0}^{\infty} \frac{(2z)^n}{n^2+1} \quad \text{(i)} \sum_{n=0}^{\infty} \frac{z^n}{e^{2n}+1}$$

$$\text{(j)} \sum_{n=0}^{\infty} \frac{z^n}{(n+1)^n} \quad \text{(k)} \sum_{n=0}^{\infty} \frac{(z-2)^{2n}}{(2n)!} \quad \text{(l)} \sum_{n=0}^{\infty} \frac{(z-3)^n}{\sqrt{n!}}$$

$$\text{(m)} \sum_{n=0}^{\infty} \frac{z^n}{(n!)^{1/6}} \quad \text{(n)} \sum_{n=0}^{\infty} \left(\frac{1-z}{1+z}\right)^n \quad \text{(o)} \sum_{n=0}^{\infty} \left(\frac{z^2-1}{z^2+1}\right)^n.$$

22.5.3 Behavior on the Circle of Convergence

The question of what happens to a power series at points *on* its circle of convergence is both important and interesting.

A power series *diverges* (i.e., becomes infinite) outside its circle of convergence. *On* the circle of convergence itself, its behavior can be quite complicated, and may change from one point on the circle to another. The series may diverge at one or more points on the circle; or it may converge at some points, diverge at others, and oscillate at others.

Even when the series does not converge, but rather oscillates at a point on its circle of convergence, it can provide useful information, as you can see from the following example. Consider, once again, the geometric series in Eq. (22.35), whose circle of convergence is the unit circle $|z| = 1$. Setting $z = -1$ in this series produces the oscillating series $1 - 1 + 1 - 1 + \cdots$. The two distinct partial sums of this series are 1 and 0. The arithmetic average of the two sums, called the **Cesàro mean**, is $\frac{1}{2}$. But this is precisely the value of the "master function" $(1 - z)^{-1}$ at $z = -1$. Consider another value on the unit circle, say $z = i$. The series becomes $1 + i - 1 - i + 1 + \cdots$. There are now four distinct partial sums, namely, $1,\ 1 + i,\ i$, and 0. These yield the Cesàro mean $\frac{1}{2}(1 + i)$. Once again, this is precisely the value of $(1 - z)^{-1}$ at $z = i$.

A power series may be convergent—or even *absolutely* convergent!—at *all* points on its circle of convergence. An example of this possibility is provided by the series $\sum_{n=1}^{\infty} z^n/n^2$, whose circle of convergence is the unit circle $|z| = 1$. The series of course converges absolutely at all points inside the unit circle. Since the coefficients $a_n\ (= 1/n^2)$ are all positive numbers in this case, it is intuitively clear that the magnitude of the infinite sum attains its largest value on the unit circle at the point $z = 1$. (There is a theorem to this effect.) At this point, the series becomes equal to $\zeta(2)$, which has the finite value $\frac{1}{6}\pi^2$, as we have already seen.[4] This is a finite value. We are therefore guaranteed that the series also converges to a finite value at every other point on the unit circle. In fact, this is an example of a power series that converges absolutely at every point *on* its circle of convergence, over and above the interior of this circle.

No matter how a power series behaves on its circle of convergence, however, the following assertion holds good:

- The analytic *function* that a power series represents inside its circle of convergence has at least one singularity *on* that circle.

Recall the example of the function $(1 - z)^{-1}$, which has a singularity at $z = 1$. As you have already seen, every power series representation of this function (Eqs. (22.37) and (22.38)) has a circle of convergence that passes through the point $z = 1$. Even in the case of the series $\sum_{n=1}^{\infty} z^n/n^2$, which is absolutely convergent even at $z = 1$, the analytic *function* represented by the series does have a singularity at $z = 1$, although this is not obvious by looking at its *value* at $z = 1$ alone. It turns out that the

[4]For instance, in Eq. (18.51) of Chap. 18, Sect. 18.4.2.

function has what is known as a *logarithmic singularity* at $z = 1$. This *singular* part is proportional to $(1 - z)\ \ln(1 - z)$, which happens to vanish in the limit $z \to 1$.

22.5.4 Lacunary Series

At the other extreme, we have cases in which the function represented by the series has a dense set of singularities *everywhere* on the circle of convergence. A well-known example of such a series is given by

$$\sum_{n=0}^{\infty} (z)^{2^n} = z + z^2 + z^4 + \cdots . \tag{22.39}$$

The radius of convergence of the series is 1. Setting $z = 1$, it is clear that the series diverges at that point. Let $f(z)$ denote the function represented by the series. Then $f(z)$ is singular at $z = 1$. But $f(z)$ satisfies the obvious identity $f(z) = z + f(z^2)$. Since z itself is not singular in the finite part of the complex plane, any singularity of $f(z)$ must come from $f(z^2)$. But $f(z^2)$ is singular wherever its argument is equal to unity, i.e., at both $z = 1$ as well as $z = -1$. Hence $f(z)$ must be singular at $z = \pm 1$. Continuing this argument, we have $f(z) = z + z^2 + f(z^4)$. The last term on the right-hand side is singular at $z = \pm 1$ and at $z = \pm i$, and hence so is $f(z)$. In this way we find that $f(z)$ is singular at all the $(2^n)^{\text{th}}$ roots of unity for every positive integer value of n. These points are dense on the unit circle. A more delicate argument now leads to the conclusion that the barrier of singularities of $f(z)$ on the unit circle prevents the function from having an analytic continuation outside the region $|z| < 1$. The unit circle forms what is known as a **natural boundary** for this function.

Series of this kind are called **lacunary series**, or lacunary functions. The name comes from the fact that the successive powers of z in such series have large gaps, or lacunas. In the example above, the sequence of powers is 1, 2, 4, 8, 16, ..., with ever-increasing gaps. A theorem due to Hadamard tells us when we can expect a lacunary series to represent a function with a dense set of singularities on its circle of convergence, so that the function has no analytic continuation beyond the circle: Asymptotically, the sequence of powers must increase at least geometrically, i.e., like $(1 + \epsilon)^n$, where ϵ is a positive number. That is, the ratio of successive *powers* must be greater than $(1 + \epsilon)$ in the limit $n \to \infty$. Thus the series

$$\sum_{n=0}^{\infty} z^{n!} = 1 + z^2 + z^6 + z^{24} + \ldots \tag{22.40}$$

is a lacunary series. But series like

$$\sum_{n=0}^{\infty}(z)^{n^2} = 1 + z + z^4 + z^9 + \ldots, \quad \sum_{n=0}^{\infty}(z)^{n^3} = 1 + z + z^8 + z^{27} + \ldots, \tag{22.41}$$

etc., are not. Hence these series do not represent lacunary functions.

22.6 Entire Functions

22.6.1 *Representation of Entire Functions*

As we have just seen, lacunary series cannot be continued analytically beyond their circles of convergence. In contrast, we have entire functions—as mentioned already, these are analytic functions whose domain of holomorphy is $\mathbb{C}$, the whole of the finite part of the complex z-plane. Remember that

- an entire function (unless it is just a constant) must necessarily be singular at $z = \infty$ (the point at infinity).

Since the exponential function e^z is an entire function of z, so are its even and odd parts. These are, respectively,

$$\left.\begin{aligned} \cosh z &= \frac{(e^z + e^{-z})}{2} = \sum_{n=0}^{\infty} \frac{z^{2n}}{(2n)!}, \\ \sinh z &= \frac{(e^z - e^{-z})}{2} = \sum_{n=0}^{\infty} \frac{z^{2n+1}}{(2n+1)!}. \end{aligned}\right\} \tag{22.42}$$

Obviously, e^{iz} is also an entire function. Hence so are the functions

$$\left.\begin{aligned} \cos z &= \frac{(e^{iz} + e^{-iz})}{2} = \sum_{n=0}^{\infty} \frac{(-1)^n z^{2n}}{(2n)!}, \\ \sin z &= \frac{(e^{iz} - e^{-iz})}{2} = \sum_{n=0}^{\infty} \frac{(-1)^n z^{2n+1}}{(2n+1)!}. \end{aligned}\right\} \tag{22.43}$$

In Chap. 23, Sect. 23.2.5, we will consider the nature of the singularity of these functions at $z = \infty$.

The zeroes of an entire function, i.e., the points where $f(z) = 0$, are of particular interest. We know that a polynomial of order N can be written uniquely as a product of factors corresponding to its roots $\alpha_1, \ldots, \alpha_N$, according to

$$\sum_{n=0}^{N} a_n z^n = a_N \prod_{k=1}^{N} (z - \alpha_k). \tag{22.44}$$

This means that

- any polynomial is determined, up to an overall multiplicative constant, by its zeroes.

The natural question is whether some such result is valid for *all* entire functions. Note that there are entire functions like e^z that have no zeroes at all, as well as entire functions like $\sin z$ that have an infinite number of zeroes. In the latter case, the zeroes cannot accumulate at any finite point in the complex plane, because such a point would be a singularity, and that is not allowed for an entire function. Moreover, when you have an *infinite* product of factors $(z - \alpha_k)$, other questions arise: Does the infinite product converge? Can the order of the factors be changed without affecting the convergence? These are nontrivial issues, and I will not go into them here. The main result in this regard is called **Hadamard's Factorization Theorem**. It involves what is known as the **Weierstrass canonical product**. Broadly speaking, the result is that, given a sequence of points with a limit point at infinity, you can construct an entire function that vanishes at each of those points, and nowhere else. Such a function is clearly not unique. Rather, it is determined only up to an overall factor that is of the form $e^{g(z)}$, where $g(z)$ is itself an entire function.

In Eq. (25.25) of Chap. 25, Sect. 25.2.4, we will encounter an example of a Weierstrass canonical product: an infinite product representation of an entire function, namely, the reciprocal of the gamma function. This entire function has a simple zero at each nonpositive integer.

22.6.2 The Order of an Entire Function

Although polynomials in z and exponential functions such as e^z, e^{z^2}, and e^{e^z} are all entire functions, they become unbounded as $|z| = r \to \infty$ in very different ways. The asymptotic behavior (i.e., the behavior as $r \to \infty$) of entire functions provides a quantitative way to classify them. I will only mention the basic idea here, namely, that of the **order** of an entire function. Roughly speaking, if an entire function $f(z) \sim \exp(z^\rho)$ as $r \to \infty$, then $f(z)$ is of order ρ. More precisely,

$$\boxed{\rho = \limsup_{r\to\infty} \frac{\ln \ln |f(z)|}{\ln r}.} \tag{22.45}$$

By this definition:

(i) $\rho = 0$ for any polynomial in z;
(ii) $\rho = 1$ for functions like e^z, $\cos z$, $\sin z$, $\cosh z$, $\sinh z$, etc.;

(iii) $\rho = 2$ for the function e^{z^2};
(iv) the function e^{e^z} is of infinite order.

When $f(z)$ is an entire function of finite order, the exponent $g(z)$ in its canonical product representation must be a polynomial in z.

If $f(z)$ is an entire function, its power series representation $f(z) = \sum_0^\infty a_n z^n$ must have an infinite radius of convergence. Hence the coefficients $\{a_n\}$ must decay very rapidly to zero as $n \to \infty$. The decay must certainly be faster than any negative power of n, and even faster than e^{-cn}, where c is any positive constant. The reciprocal of the factorial, $1/n!$, satisfies this requirement. Recall from Stirling's formula that the leading behavior of $(n!)^{-1}$ is $\sim n^{-n} = e^{-n \ln n}$. This is why e^z is an entire function. There is a very useful formula that tells us the order of an entire function, given the asymptotic behavior of the coefficient a_n. It is

$$\boxed{\rho = \limsup_{n\to\infty} \frac{n \ln n}{\ln (1/|a_n|)}} \tag{22.46}$$

Using this formula, it is easy to see that the order of the entire function e^{az}, where $a \in \mathbb{C}$ is a constant, is given by $\rho = 1$.

★ **13.** Use the formula of Eq. (22.46), and Stirling's formula wherever needed, to verify that the order ρ of each of the entire functions listed below is as stated.

(a) $f(z) = e^{z^2}$ is an entire function of order $\rho = 2$.

(b) The series $\displaystyle\sum_{n=0}^{\infty} \frac{z^n}{(n!)^a}$, where $a > 0$, is an entire function of order $\rho = 1/a$.

It follows from the last result above that

$$\sum_{n=0}^{\infty} \frac{z^n}{(n!)^2} \quad \text{is of order } \tfrac{1}{2}, \quad \sum_{n=0}^{\infty} \frac{z^n}{(n!)^{1/2}} \quad \text{is of order } 2,$$

and so on. But you must be careful! The next example will show you why.

★ **14.** ***ρ for the modified Bessel function of the first kind*:**

(a) The modified Bessel function $I_0(z)$ of the first kind and of order 0 (this "order" is not to be confused with the index ρ) is given by the power series

$$I_0(z) = \sum_{n=0}^{\infty} \frac{1}{(n!)^2} \left(\tfrac{1}{2}z\right)^{2n}.$$

It is obvious that the series converges absolutely for all $|z| < \infty$. It therefore represents an entire function. Show that the order of this entire function is $\rho = 1$, and *not* $\frac{1}{2}$ as one might guess at first sight (because of the coefficient $1/(n!)^2$ in the power series).

(b) Now consider the modified Bessel function $I_l(z)$ of the first kind and of order l, where l is a positive integer. Repeating Eq. (19.45) of Chap. 19, Sect. 19.2.5 for ready reference, we have

$$I_l(z) = \sum_{n=0}^{\infty} \frac{1}{n!\,(n+l)!}\,\left(\tfrac{1}{2}z\right)^{l+2n}.$$

Show that $I_l(z)$ is an entire function of order $\rho = 1$ for any positive integer value of l.

Finally, here is the comment promised after Eq. (19.45) in Chap. 19, Sect. 19.2.5. The summand in the power series for the modified Bessel function $I_l(z)$ has *two* factorials in the denominator, and so you should expect it to be absolutely convergent for all finite values of z. (As you know, even a single factorial in the denominator suffices for this purpose.) Comparing the power series for $I_l(z)$ with the series for the general hypergeometric function ${}_pF_q$ given in Eq. (16.34) of Chap. 16, Sect. 16.2.2, it is readily seen that

$$I_l(z) = \frac{1}{l!}\left(\tfrac{1}{2}z\right)^l\,{}_0F_1(l+1\,;\,\tfrac{1}{4}z^2). \tag{22.47}$$

As I have mentioned earlier, ${}_pF_q$ is an entire function of its argument when $p \le q$.

22.7 Solutions

3. Consider Fig. 22.6, which is a projection of Fig. 22.1 in the (ξ_1, ξ_3) plane. It follows from elementary geometry (similarity of triangles) that $\xi_1/x = (1-\xi_3)/1$, so that $x = \xi_1/(1-\xi_3)$. A similar projection of Fig. 22.1 in the (ξ_2, ξ_3) plane yields the relation $y = \xi_2/(1-\xi_3)$. It follows that $x^2+y^2 = (\xi_1^2+\xi_2^2)/(1-\xi_3)^2 = (1+\xi_3)/(1-\xi_3)$. Hence $(1-\xi_3) = 2/(1+x^2+y^2)$. The inverse relations (22.7) follow readily. ▶

5. $\gamma(z_1, z_2) = \cos^{-1}(\boldsymbol{\xi}\cdot\boldsymbol{\zeta})$. Use Eq. (22.8) to express the components of $\boldsymbol{\xi}$ and $\boldsymbol{\zeta}$ in terms of z_1, z_2 and their complex conjugates. Simplify to obtain Eq. (22.15). Now set $z_1 = z$ and let $z_2 \to \infty$, to arrive at (22.16). ▶

6. The chordal distance between P and Q is

$$[(\xi_1-\zeta_1)^2+(\xi_2-\zeta_2)^2+(\xi_3-\zeta_3)^2]^{1/2} = [2-2(\boldsymbol{\xi}\cdot\boldsymbol{\zeta})]^{1/2} = [2(1-\cos\gamma)]^{1/2}.$$

But $\cos\gamma$ can be read off from Eq. (22.15). ▶

7. The proposition is obvious, because the chordal lines between three points on the Riemann sphere form a *planar* triangle. ▶

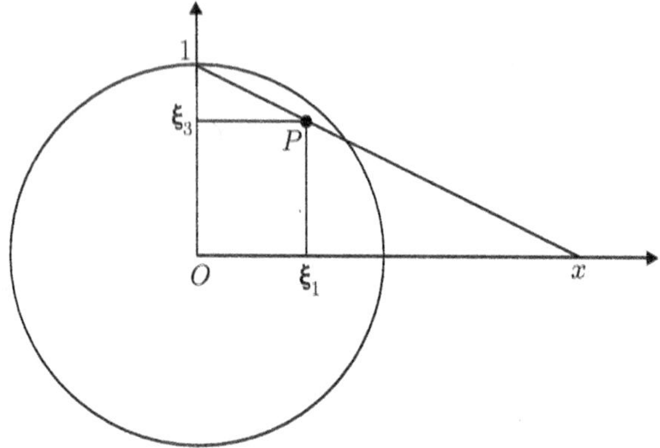

Fig. 22.6 Geometry for stereographic projection

8. You could, of course, check in each case whether the Cauchy–Riemann conditions are satisfied. But it is simpler to write each function in terms of z and z^*, eliminating x and y. Except for the functions (b) iz, (g) z^2, (n) $z^2 + z + 1$, and (o) z^{-1}, all the others involve z^*, and so cannot be analytic functions of z. The functions in (b), (g), and (n) are polynomials in z, and hence the domain of holomorphy is $\mathbb{C}$, the finite complex plane. The function in (o) has a singularity at $z = 0$. Its domain of holomorphy is the extended complex plane $\hat{\mathbb{C}}$ with the origin removed.[5] ▶

10. Given the real part $u(x, y)$ (respectively, the imaginary part $v(x, y)$), Integration of the Cauchy–Riemann equations will obviously determine $v(x, y)$ (respectively, $u(x, y)$) up to an additive constant (set equal to zero in the solutions given below). The analytic functions $f(z)$ are as follows.

$$\text{(a) } z^2 \quad \text{(b) } e^z \quad \text{(c) } \ln z \quad \text{(d) } \cos z \quad \text{(e) } e^{2z}$$
$$\text{(f) } z^2 \quad \text{(g) } \sin z \quad \text{(h) } \ln z^2 \quad \text{(i) } e^{z^2}.$$

In all these cases, the form of the function v (or u), and hence the analytic function $f(z)$, can be guessed by mere inspection. ▶

11. (a) Let us use the convenient notation $\partial u/\partial x = u_x$, $\partial u/\partial y = u_y$, etc. Then the determinant of the Jacobian of the transformation is

$$\left|\frac{\partial(u, v)}{\partial(x, y)}\right| = \begin{vmatrix} u_x & u_y \\ v_x & v_y \end{vmatrix} = u_x\, v_y - u_y\, v_x = u_x^2 + v_x^2,$$

on using the Cauchy–Riemann conditions. But

$$f'(z) = \frac{df}{dz} = \frac{\partial f}{\partial x}\frac{\partial x}{\partial z} + \frac{\partial f}{\partial y}\frac{\partial y}{\partial z} = \frac{1}{2}\Big(\frac{\partial f}{\partial x} - i\frac{\partial f}{\partial y}\Big),$$

[5]In mathematical notation, $\hat{\mathbb{C}} \setminus \{0\}$.

using the fact that $x = (z + z^*)/2$ and $y = (z - z^*)/(2i)$. Since $f = u + iv$, this gives

$$f'(z) = u_x + i v_x$$

on using the Cauchy–Riemann conditions. Hence

$$\left| \frac{\partial(u, v)}{\partial(x, y)} \right| = u_x^2 + v_x^2 = |f'(z)|^2.$$

(b) We have

$$\begin{aligned} \nabla^2 |f|^2 &= \nabla^2 (u^2 + v^2) \\ &= 2\big(u_x^2 + u\, u_{xx} + v_x^2 + v\, v_{xx} + u_y^2 + u\, u_{yy} + v_y^2 + v\, v_{yy}\big) \\ &= 4(u_x^2 + v_x^2) = 4|f'(z))|^2, \end{aligned}$$

on using the relations $u_{xx} + u_{yy} = v_{xx} + v_{yy} = 0$.

Remark Note that the Jacobian can be written in the alternative forms $|f'(z)|^2 = u_x^2 + v_x^2 = u_x^2 + u_y^2 = u_y^2 + v_y^2 = v_x^2 + v_y^2$. ▶

12. When the coefficient a_n is a polynomial in n, or more generally any rational function of n, it is easy to see that the ratio $|a_n/a_{n+1}| \to 1$ as $n \to \infty$. The same thing happens when a_n involves any logarithmic function of n. Hence the radius of convergence of the power series $\sum a_n z^n$ is just unity in these cases. This is what happens in cases (a) to (f), so that the region of convergence is $|z| < 1$ in all these cases. Similarly, for (g), (h), and (i) we have, respectively, the circles of convergence $|z| = 2$, $|z| = \frac{1}{2}$ and $|z| = e^2$.

When a_n decays like n^{-n} or like any negative power of $n!$, it is easy to show that the ratio $|a_n/a_{n+1}| \to \infty$ as $n \to \infty$. This implies that the radius of convergence of the series $\sum a_n z^n$ is infinite, i.e., the power series represents an entire function of z. This is what happens in cases (j) to (m).

Cases (n) and (o) are geometric series, but in the variables $(1 - z)/(1 + z)$ and $(z^2 - 1)/(z^2 + 1)$, respectively. They converge absolutely in the region in which the modulus of each of these variables is less than unity. The condition $|1 - z| < |1 + z|$ implies that the distance between z and 1 must be less than that between z and -1. This is true for all points in the right-half plane, $x = \operatorname{Re} z > 0$. Similarly, the condition $|z^2 - 1| < |z^2 + 1|$ is easily shown to imply that $|x| > |y|$. For positive x, this is the conical region lying between the lines $y = -x$ and $y = x$; for negative x, this is the conical region lying between the lines $y = x$ and $y = -x$. ▶

14. (a) The denominator in the coefficient of the general term in the power series for $I_0(z)$ involves the *square* of $n!$. One might, therefore, jump to the conclusion that the order of the entire function is $\frac{1}{2}$. Note, however, that the power series is in the variable z^2 rather than z itself. Roughly speaking, therefore, we then have an entire function of order $\frac{1}{2}$ in z^2, or an entire function of order 1 in z. You can establish this formally quite easily. Set $2n = k$, so that

$$I_0(z) = \sum_{\substack{k=0 \\ k \text{ even}}}^{\infty} \frac{1}{\left[\left(\frac{1}{2}k\right)!\right]^2} \left(\tfrac{1}{2}z\right)^k.$$

The order of $I_0(z)$ is given by

$$\rho = \overline{\lim_{k\to\infty}} \frac{k \ln k}{\ln(1/a_k)}, \quad \text{where} \quad \frac{1}{a_k} = 2^k \left[\left(\tfrac{1}{2}k\right)!\right]^2.$$

Use Stirling's formula in the form $\ln N! \simeq N \ln N - N$ $(N \gg 1)$. After a bit of simplification, we get

$$\rho = \lim_{k\to\infty} \frac{k \ln k}{k \ln k - k} = 1.$$

(b) A similar procedure yields $\rho = 1$ for the order of $I_l(z)$ when l is any positive integer. Note that the factor z^l multiplying z^{2n} in the power series for $I_l(z)$ does not affect the order of this entire function. Moreover, since $I_{-l}(z) \equiv I_l(z)$ for all integer values of l, $I_l(z)$ is an entire function of order 1 for every integer l. ▶

Chapter 23
More on Analytic Functions

23.1 Cauchy's Integral Theorem

We know that the property of analyticity implies that

- an analytic function $f(z)$ is infinitely differentiable, and
- its derivative of every order is also an analytic function,

in the domain of holomorphy of the function. We now ask: what does analyticity imply for the *integral* of a function?

Once again, the Cauchy–Riemann conditions can be used to arrive at the following conclusion. Let $f(z)$ be holomorphic in some region $\mathcal{R}$.

- The **line integral**
$$\int_{z_1}^{z_2} dz\, f(z) \tag{23.1}$$
is *independent of the actual path* between z_1 and z_2 , provided the path and its end points lie entirely in $\mathcal{R}$.

The value of the line integral is dependent only on the end points z_1 and z_2.[1] As a consequence, the path connecting these end points can be distorted like a flexible and stretchable piece of elastic thread (while staying within $\mathcal{R}$), without changing the value of the integral, as shown in Fig. 23.1a. **Cauchy's Integral Theorem** follows at once: Let C be an *oriented, closed,* contour lying entirely in $\mathcal{R}$, as in Fig. 23.1b. Then

[1]This property should remind you at once of the corresponding property of a conservative vector field $\mathbf{u}(\mathbf{r})$, i.e., a vector field that can be expressed as the gradient of a scalar field $\nabla\,\phi(\mathbf{r})$. In that case, too, the value of the line integral of the vector field from a point $\mathbf{r}_1$ to another point $\mathbf{r}_2$ is independent of the actual path between the two points, and is given by

$$\int_{\mathbf{r}_1}^{\mathbf{r}_2} \mathbf{u}\cdot d\mathbf{r} = \int_{\mathbf{r}_1}^{\mathbf{r}_2} \nabla\,\phi\cdot d\mathbf{r} = \phi(\mathbf{r}_2) - \phi(\mathbf{r}_1).$$

V. Balakrishnan, *Mathematical Physics*,
https://doi.org/10.1007/978-3-030-39680-0_23

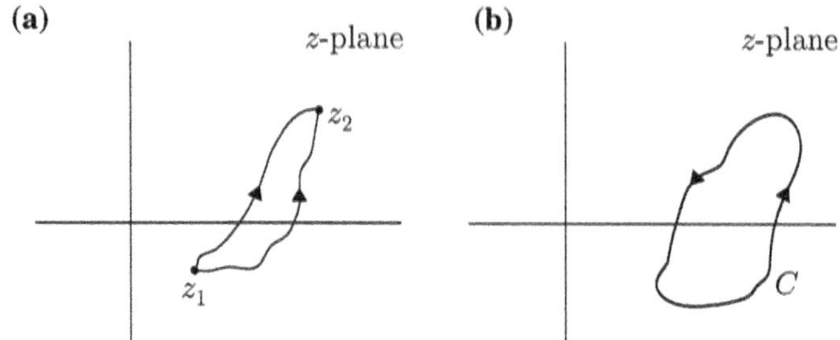

Fig. 23.1 **a** The line integral (23.1) is independent of the path taken between the end points z_1 and z_2. **b** The integral of $f(z)$ over a closed contour is zero

$$\boxed{\oint_C dz\, f(z) = 0.} \tag{23.2}$$

It is evident that the contour C may be distorted like a rubber band (to any other contour like C') without changing the property above, as long as it does not leave $\mathcal{R}$. These properties are responsible for much of the power of contour integration in evaluating definite integrals, and hence in solving a variety of problems that can be reduced to the evaluation of such integrals.

23.2 Singularities

We now come to the **singularities** of an analytic function. These are the most interesting features of analytic functions of one or more complex variables. Even in the case of a function of a single complex variable, singularities come in a remarkable number of varieties. But we shall consider here only the simplest (and the most frequently-encountered kinds) of singularities. As I have stated in Chap. 22, Sect. 22.3, we are dealing here with univalent functions. Multivalent functions have another kind of singularity called a *branch point*, which is not included in the list that follows below. We will discuss branch points in Chap. 26.

23.2.1 Simple Pole; Residue at a Pole

Let us dispose of the case of a **removable singularity** first, with the help of an example. The function

$$f(z) = \frac{\sin z}{z} \tag{23.3}$$

is not defined at $z = 0$, and appears to have a singularity at that point because its denominator vanishes there. However, $\lim_{z\to 0} f(z) = 1$, and we may *set* $f(0) = 1$ by definition. This prescription gets rid of the "removable" singularity and ensures the continuity of $f(z)$ at $z = 0$. In all that follows, I shall assume that all removable singularities have been taken care of in precisely the same manner as in the example just considered.

The simplest nontrivial singularity is a **simple pole**:

- $f(z)$ has a simple pole at the point $z = a$ if, in a (sufficiently small) neighborhood of that point, it can be expressed in the form

$$f(z) = \underbrace{\frac{c_{-1}}{(z-a)}}_{\text{singular part}} + \underbrace{\sum_{n=0}^{\infty} c_n\,(z-a)^n}_{\text{regular part}}, \tag{23.4}$$

  where the coefficients c_n $(n \geq -1)$ are constants, and the second term on the right-hand side is a convergent powers series.

The first term on the right-hand side is called the **singular part** of $f(z)$ at $z = a$. It becomes infinite at $z = a$. The singular part is sometimes called the 'principal part', but I shall stick to the term "singular part" because it is more suggestive and less confusing.

- The coefficient c_{-1} is a nonzero, finite constant (in general, a complex number) that is called the **residue** of $f(z)$ at $z = a$.

The notation c_{-1} is used to remind us that it is the coefficient of $(z-a)^{-1}$ in the expansion of $f(z)$ in powers of $(z-a)$. The **regular part** denotes a function that is analytic in the neighborhood concerned. Hence it is expandable there in a convergent power series in nonnegative powers of $(z-a)$. In some cases this series may, of course, terminate after a finite number of terms, or even be absent altogether.

The crucial point is that, at a simple pole, the singular part comprises just the single term $c_{-1}\,(z-a)^{-1}$. But this term may or may not be *explicit*, as you will see from the elementary examples below.

Example 1 Consider the function $(\sin z)/z^2$. Using the power series expansion of $\sin z$, we have

$$\frac{\sin z}{z^2} = \underbrace{\frac{1}{z}}_{\text{singular part}} + \underbrace{\left(-\frac{z^2}{3!} + \frac{z^4}{5!} - \cdots\right)}_{\text{regular part}}. \tag{23.5}$$

This function therefore has a simple pole at $z = 0$, with residue equal to 1. Incidentally, it has no other singularities in the finite part of the complex plane (remember that $\sin z$ itself is an entire function).

Example 2 Consider the function cosec $z = 1/(\sin z)$. Since $\sin z$ has simple zeroes at all integer multiples of π, i.e., at $z = n\pi$, where $n \in \mathbb{Z}$, it follows that cosec z has simple poles at those points. Using the fact that $\sin n\pi = 0$, the Taylor expansion of $\sin z$ in the neighborhood of $z = n\pi$ is given by

$$\begin{aligned} \sin z &= (z - n\pi)\cos n\pi - \tfrac{1}{6}(z - n\pi)^3 \cos n\pi + \tfrac{1}{120}(z - n\pi)^5 \cos n\pi - \cdots \\ &= (-1)^n (z - n\pi)\Big\{1 - \tfrac{1}{6}(z - n\pi)^2 + \tfrac{1}{120}(z - n\pi)^4 - \cdots\Big\}. \end{aligned} \tag{23.6}$$

Hence

$$\begin{aligned} \text{cosec}\, z &= \frac{(-1)^n}{(z - n\pi)}\Big\{1 - \tfrac{1}{6}(z - n\pi)^2 + \tfrac{1}{120}(z - n\pi)^4 - \cdots\Big\}^{-1} \\ &= \underbrace{\frac{(-1)^n}{(z - n\pi)}}_{\text{singular part}} + \underbrace{\tfrac{1}{6}(-1)^n\Big\{(z - n\pi) + \tfrac{7}{60}(z - n\pi)^3 + \cdots\Big\}}_{\text{regular part}}. \end{aligned} \tag{23.7}$$

Therefore cosec z has a simple pole at $z = n\pi$ for every integer n, with residue equal to $(-1)^n$.

***Convenient formulas for the residue at a simple pole*:** In general, if $f(z)$ has a simple pole at $z = a$, then its residue at the pole is given by

$$\boxed{\operatorname*{Res}_{z=a} f(z) \stackrel{\text{def.}}{=} c_{-1} = \lim_{z \to a} [(z - a)\, f(z)].} \tag{23.8}$$

A commonly occurring situation is when $f(z)$ is of the form $g(z)/h(z)$, where $g(z)$ and $h(z)$ are analytic at $z = a$, and $h(z)$ has a simple zero at $z = a$ while $g(a) \neq 0$. In such a case, $f(z)$ has a simple pole at $z = a$. Its residue at the pole is then given by

$$\boxed{\operatorname*{Res}_{z=a} \frac{g(z)}{h(z)} = \frac{g(a)}{h'(a)},} \tag{23.9}$$

where the prime denotes the derivative of the function concerned. This result is useful, for instance, in identifying the poles of commonly occurring trigonometric and hyperbolic functions.

★ **1.** Verify that, for every $n \in \mathbb{Z}$,

(a) cosech z has a simple pole at $z = in\pi$, with residue $(-1)^n$.
(b) $\sec z$ has a simple pole at $z = (n + \frac{1}{2})\pi$, with residue $(-1)^{n+1}$.
(c) sech z has a simple pole at $z = i(n + \frac{1}{2})\pi$, with residue $i\,(-1)^{n+1}$.
(d) $\tan z$ has a simple pole at $z = (n + \frac{1}{2})\pi$, with residue -1.
(e) $\tanh z$ has a simple pole at $z = i(n + \frac{1}{2})\pi$, with residue $+1$.
(f) $\cot z$ has a simple pole at $z = n\pi$, with residue $+1$.
(g) $\coth z$ has a simple pole at $z = in\pi$, with residue $+1$.

Note how the residue becomes independent of n in the last four cases. This property is of use in the summation of certain infinite series, as you will see in Sect. 23.4 below.

23.2.2 Multiple pole

The next case is that of a **multiple pole**, or a pole of higher order. Let m be a positive integer > 1. The function $f(z)$ has a pole of order m at $z = a$ if, in the neighborhood of that point, it can be expressed in the form

$$f(z) = \underbrace{\frac{c_{-m}}{(z-a)^m} + \cdots + \frac{c_{-1}}{(z-a)}}_{\text{singular part}} + \underbrace{\sum_{n=0}^{\infty} c_n (z-a)^n}_{\text{regular part}}, \tag{23.10}$$

where all the coefficients c_j are constants. It is very important to note that

- the residue of $f(z)$ at the multiple pole is still given by the coefficient c_{-1} of $(z-a)^{-1}$.

The significance of the residue will become clear shortly, when we consider Cauchy's integral formula. In order to "extract" the coefficient c_{-1} from the expansion (23.10), we multiply both sides of the equation by $(z-a)^m$, differentiate the resulting expression $(m-1)$ times, and then set $z = a$. Thus, the residue at a multiple pole of order m is given by

$$\boxed{c_{-1} = \operatorname*{Res}_{z=a} f(z) = \lim_{z\to a} \frac{1}{(m-1)!} \frac{d^{m-1}}{dz^{m-1}} [(z-a)^m f(z)].} \tag{23.11}$$

This formula is a generalization of Eq. (23.8) for the residue at a simple pole.

23.2.3 Essential Singularity

Going a step further, the singular part may involve an unbounded number of negative integral powers of $(z-a)$. The function $f(z)$ has an **isolated essential singularity** at $z = a$ if, in the neighborhood of that point, it can be expressed in the form

$$f(z) = \underbrace{\sum_{n=1}^{\infty} \frac{c_{-n}}{(z-a)^n}}_{\text{singular part}} + \underbrace{\sum_{n=0}^{\infty} c_n (z-a)^n}_{\text{regular part}}. \tag{23.12}$$

Once again, the coefficient c_{-1} is the residue of $f(z)$ at the singularity.

The standard example of an isolated essential singularity is provided by the exponential function

$$e^{1/z} = \sum_{n=0}^{\infty} \frac{1}{n!\, z^n} = \sum_{n=1}^{\infty} \frac{1}{n!\, z^n} + 1. \tag{23.13}$$

The series on the right-hand side converges absolutely for all $z \neq 0$. This function has an essential singularity at the origin $z = 0$, with a residue equal to 1. Its regular part is just a constant, also equal to 1.

A function that has an isolated essential singularity is guaranteed to be unbounded *at* that point. In the neighborhood of the singularity, however, its behavior is quite complicated. Again, the function $e^{1/z}$ offers a simple illustration. If $z \to 0$ along the positive real axis, $e^{1/x} \to \infty$ monotonically. If $z \to 0$ along the negative real axis, $e^{1/x}$ tends to zero monotonically. On the other hand, $e^{1/z}$ oscillates with increasing amplitude as $z \to 0$ in the right-half-plane (except along the real axis), while it oscillates with decreasing amplitude as $z \to 0$ in the left-half plane (again, except along the real axis). If $z \to 0$ along the imaginary axis, the function $e^{i/y}$ is oscillatory, has a modulus equal to unity, and does not approach any limit as $y \to 0$.

The general question of the behavior of a function in the neighborhood of an essential singularity is an important and nontrivial topic. There exist several deep theorems in this regard that are beyond the scope of the treatment given here. I mention, however, a fundamental result, called **Picard's great theorem**:

- Suppose the analytic function $f(z)$ has an isolated essential singularity at $z = a$. Then, in every neighborhood of $z = a$, $f(z)$ equals every complex number an infinite number of times, with the possible exception of at most one value.

This theorem shows how "wild" the behavior of an analytic function is, in the vicinity of an essential singularity. In the case of the function $e^{1/z}$, which has an isolated essential singularity at $z = 0$, the exceptional value is 0. This is the *only* value the function never takes on. It is trivial to shift this exceptional value to any given complex number c: the function $e^{1/z} + c$ never has the value c.

23.2.4 Laurent Series

Representations like those in Eqs. (23.4), (23.10), and (23.12), which involve positive as well as negative powers of $(z - a)$, are called **Laurent series**. What is the region in which such a representation is valid?

The regular part, as we know, is convergent inside some circle of convergence centered at $z = a$, and of radius r_1, say. If the regular part happens to be a polynomial in z, or, more generally, an entire function of z, then r_1 is infinite. In order to examine the singular part, we set $w = (z - a)^{-1}$. The singular part is then either a polynomial in w (if $z = a$ is a pole), or an infinite series in positive integer powers of w (if $z = a$ is

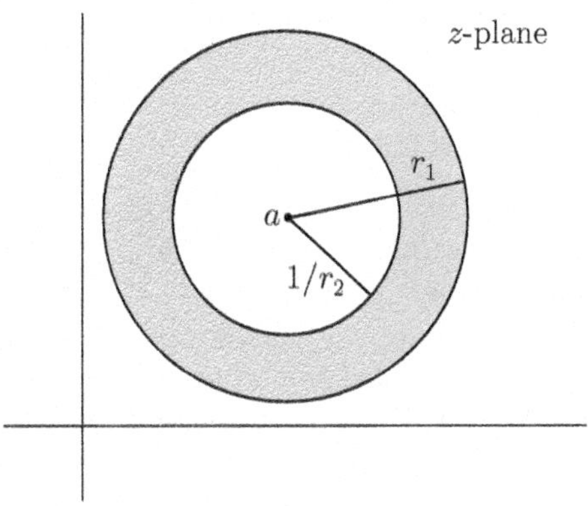

Fig. 23.2 The annular region in which a Laurent series is typically valid

an essential singularity). Therefore, it is convergent inside some circle of convergence of radius r_2, say, centered at the origin in the complex w-plane. Once again, r_2 is infinite when the singular part is either a polynomial in w, or more generally, an entire function of w. Reverting to the z-plane, the singular part is convergent *outside* a circle of radius $1/r_2$, centered at $z = a$. Therefore

- If $1/r_2 < r_1$, i.e., if $r_1 r_2 > 1$, there is *an annular overlap region* centered at $z = a$, of inner radius $1/r_2$ and outer radius r_1, in which the two infinite series representing the regular part and the singular part are both absolutely convergent.
- This **annulus** is the region in which the Laurent series is convergent, as illustrated in Fig. 23.2.

As I have already mentioned, it may so happen that r_2 is infinite, i.e., $1/r_2 = 0$. Then the Laurent series is convergent in the *punctured disc* of radius r_1 centered at $z = a$, with only the point $z = a$ left out. For example, in the case of the function $e^{1/z}$ already referred to, $r_1 = \infty$ while $1/r_2 = 0$. Hence the Laurent series on the right-hand side of Eq. (23.13) is convergent for all $z \neq 0$, *including* the point at infinity in the z-plane. That is, it is valid in the region $|z| > 0$. In general

- A Laurent series is convergent (i.e., provides a valid representation of an analytic function) in some *annular* region.
- In some cases, the inner radius of the annulus may shrink to zero; similarly, in some cases the outer radius of the annulus may extend to infinity.

23.2.5 Singularity at Infinity

What about possible singularities at the point at infinity, i.e., at $z = \infty$ in the extended complex plane? The nature of such a singularity, if any, can be deduced by first changing variables to $w = 1/z$. This maps the point $z = \infty$ in the extended z-plane to the point $w = 0$ in the w-plane. The singularity of the corresponding transformed function $f(1/w) \equiv \phi(w)$ at the origin $w = 0$ then determines the nature of the singularity of $f(z)$ at $z = \infty$. For example, if $f(z) = z$ itself, then $\phi(w) = 1/w$, which has a simple pole at $w = 0$. Hence $f(z) = z$ has a simple pole at $z = \infty$. Likewise, it follows that

- A polynomial of order n in z has a pole of order n at $z = \infty$.
- The function e^{cz} (where c is any nonzero complex constant) has an essential singularity at $z = \infty$.
- Hence so do the functions $\cosh z$, $\sinh z$, $\cos z$ and $\sin z$, each of these being a linear combination of two exponential functions. As you have seen already in Chap. 22, Sect. 22.6.1, each of these is an entire function of z. Their power series expansions, given in Eqs. (22.42) and (22.43), may be regarded as Laurent series valid in the region $0 \leq |z| < \infty$, i.e., in $\mathbb{C}$, the *finite* part of the complex plane.

Example The function

$$e^{1/z} + e^{z} = \sum_{n=1}^{\infty} \frac{1}{n!\, z^n} + 2 + \sum_{n=1}^{\infty} \frac{z^n}{n!} \tag{23.14}$$

has essential singularities at $z = 0$ as well as $z = \infty$. The Laurent series on the right-hand side of Eq. (23.14) is valid in the annular region $0 < |z| < \infty$.

The *residue at infinity*, however, is a little different from the usual idea of the residue at a singularity at any point in the finite part of the complex plane. This notion will be discussed in Sect. 23.3.5.

23.2.6 Accumulation Points

The singularities we have discussed so far are *isolated* singularities—that is, each singularity has a neighborhood that is free of any other singularity. But analytic functions can have *non-isolated* singularities as well. You have already encountered an instance of such a singularity: the circle of convergence of a lacunary series (Chap. 22, Sect. 22.5.4) is a natural boundary. This is a non-isolated singularity, as the singularities of the function concerned are dense on the circle.

Another kind of non-isolated singularity, called an **accumulation point**, is illustrated by the function

$$f(z) = \operatorname{cosec} \pi z. \tag{23.15}$$

This function has simple poles at all the integers, at the set of points $z = n$ where $n = 0, \pm 1, \pm 2, \ldots$. All these points lie on the x-axis. What is the nature of the singularity, if any, of $\operatorname{cosec} \pi z$ at $z = \infty$? The answer is provided by considering the Riemann sphere. Recall that the mapping from the complex plane $\hat{\mathbb{C}}$ to the Riemann sphere S is given by Eq. (22.8): The coordinates on S are given by

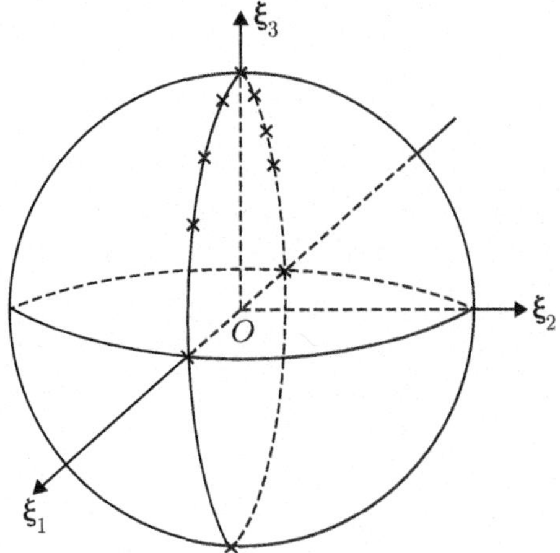

Fig. 23.3 The poles of cosec πz (marked by crosses) on the Riemann sphere. The point at infinity (the north pole) is an accumulation point of poles

$$\xi_1 = \frac{z+z^*}{|z|^2+1}, \quad \xi_2 = \frac{z-z^*}{i(|z|^2+1)}, \quad \xi_3 = \frac{|z|^2-1}{|z|^2+1}. \tag{23.16}$$

The positive and negative segments of the x-axis on the complex plane map onto the meridians of longitude 0° and 180°, respectively, on the Riemann sphere. Hence the poles of cosec z lie on these meridians, at the coordinates

$$\xi_1 = \frac{2n}{n^2+1}, \quad \xi_2 = 0, \quad \xi_3 = \frac{n^2-1}{n^2+1}, \quad \text{where } n \in \mathbb{Z}. \tag{23.17}$$

Except for the south pole (corresponding to $n = 0$), and the points $(\pm 1, 0, 0)$ on the equator (corresponding to $n = \pm 1$), all the poles lie in the northern hemisphere. It is evident that they get more and more crowded together as $n \to \pm\infty$, and approach the point $(0, 0, 1)$, i.e., the north pole, as shown in Fig. 23.3. They become dense on the two longitude lines in an infinitesimal neighborhood of the north pole. On the extended complex plane, $z = \infty$ is, therefore, an accumulation point of poles. The singularity is no longer an *isolated* singularity.

23.2.7 Meromorphic Functions

An analytic function whose only singularities (if any) in the *finite* part of the complex plane are isolated removable singularities or poles is called a **meromorphic function** on $\mathbb{C}$.

- Entire functions are also meromorphic, as they have no singularities at all in $\mathbb{C}$.
- Any rational function of z, i.e., a ratio of two polynomials in z, is a meromorphic function.

- So are the functions cosec z, sec z, cosech z, sech z, tan z, tanh z, cot z, and coth z.

Is cosec $(1/z)$ *not* a meromorphic function? No, because it has an accumulation point of poles at $z = 0$.

★ **2.** Obtain the Laurent series expansions of the following functions, valid in the region indicated in each case:

(a) $(z + z^{-1})^{100}$ $(0 < |z| < \infty)$ (b) $z^2\, e^{1/z}$ $(0 < |z| < \infty)$
(c) $z^{-1}\, e^{-1/z^2}$ $(|z| > 0)$ (d) $e^{1/z}\,(1 - z)^{-1}$ $(0 < |z| < 1)$
(e) $z^2\, e^{1/(z-2)}$ $(0 < |z - 2| < \infty)$ (f) $(z - 1)^{-1}(z - 2)^{-2}$ $(1 < |z| < 2)$.

★ **3.** Find the location, order, and residue of each pole or essential singularity of the functions listed below, and the nature of the singularity (if any) at the point at infinity.[2]

(a) e^{-z^2} (b) $(1 - z^{100})/(1 - z)$ (c) $(2^z - 1)/z$
(d) $\coth z - z^{-1}$ (e) $(e)^{e^z}$ (f) $(1 - \cos z)/z^2$
(g) $1/(e^z + 1)$ (h) $z \operatorname{cosec} z$ (i) $z \operatorname{cosec}^2 z$
(j) $(z \cos z - \sin z)/z^4$ (k) $1/(2^z - 1)$ (l) $\operatorname{cosec}(z^2)$
(m) $1/(\sin z - \cos z)$ (n) $1/(\cosh z - \cos z)$ (o) $1/(3^z - 2^z)$
(p) $1/(e^{z^2} - 1)$ (q) $1/(e^{2z} - e^z)$ (r) $\sum_{n=1}^{\infty} z^n/n^n$
(s) $\sum_{n=0}^{\infty} z^n/(n!)^2$ (t) $\sum_{n=0}^{\infty}[(1 - z)/(1 + z)]^n$ (u) $\exp[2/(z^2 - 1)]$.

Many of the functions given above have removable singularities. It is understood that they are defined to have their limiting values at the points concerned. You must also remember that the equation $e^z - 1 = 0$ has an infinite number of solutions in the complex plane—namely, $z = 2\pi n i$, where $n \in \mathbb{Z}$. Wherever it is obvious that a pole arises from a zero of the denominator, expand the denominator in a Taylor series. This makes it easier to find the residue at the pole.

Meromorphic functions on $\hat{\mathbb{C}}$ or S: In the foregoing, we have considered meromorphic functions on $\mathbb{C}$. We could also define meromorphic functions on the extended complex plane $\hat{\mathbb{C}}$, or, equivalently, on the Riemann sphere S. These are functions that have only removable singularities or poles, and no other singularities, anywhere in $\hat{\mathbb{C}}$ or S. Examples of such functions include all rational functions of z, all polynomials in z, etc. On the other hand, functions like e^z, trigonometric or hyperbolic functions of z, and so on, are not meromorphic functions in this more restricted sense, since they have an essential singularity at the point at infinity.

---

[2] As I have stated already, the *residue* at infinity is a different matter that will be discussed in Sect. 23.3.5.

23.3 Contour Integration

We now come to a powerful mathematical technique, contour integration. It has a large number of applications, of which I will spell out a few. The focus will be on diverse specific cases, rather than an exhaustive account of the subject.

23.3.1 A Basic Formula

Consider the contour integral $\oint_C z^n\, dz$, where $n = 0, 1, \ldots,$ and C is an arbitrary simple closed contour that encircles the origin once in the positive sense, as in Fig. 23.4. ("Simple" means that it does not intersect itself.) As z^n is an entire function, we can use Cauchy's Integral Theorem (Eq. (23.2)) and distort the contour without changing the value of the integral. Hence C can be replaced by a circle of arbitrary radius r with its center at the origin. Since $z = r\, e^{i\theta}$ on this circle, we have $dz = r\, e^{i\theta}\, i\, d\theta$. Hence

$$\oint_C z^n\, dz = i\, r^n \int_0^{2\pi} d\theta\, e^{(n+1)i\theta}$$
$$= i\, r^n \int_0^{2\pi} d\theta\, [\cos{(n+1)i\theta} + i\, \sin{(n+1)i\theta}] = 0. \qquad (23.18)$$

The vanishing of the integral is quite obvious, even without writing it out as above: the integrand z^n is holomorphic everywhere inside and on the contour C. Hence C can, in fact, be distorted till it shrinks to a point, and the integral is then equal to zero.

Now consider the contour integral $\oint_C dz/z^{n+1}$, where $n = 0, 1, \ldots,$ and C is again a simple closed contour that encircles the origin once in the positive sense. The integrand z^{-n-1} has a pole of order $(n+1)$ at the origin, and no other singularities. Therefore C can again be deformed to a circle of some radius r centered at the origin. But now it cannot be shrunk to a point, because of the singularity of the integrand at

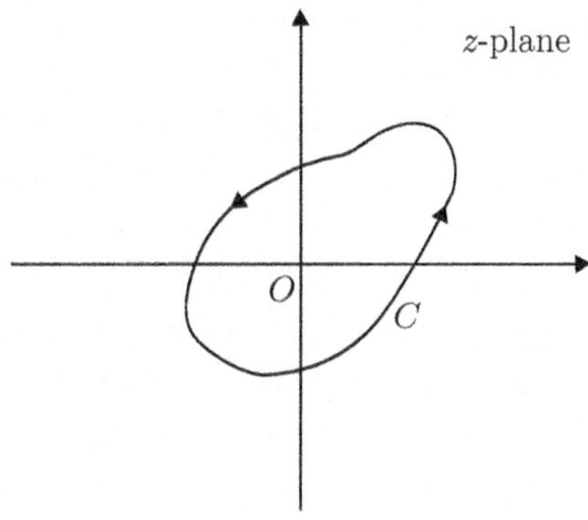

Fig. 23.4 A closed contour C encircling the origin once in the positive sense

the origin. On the other hand, we may write

$$\oint_C \frac{dz}{z^{n+1}} = \frac{i}{r^n} \int_0^{2\pi} d\theta \, e^{-ni\theta}. \tag{23.19}$$

But this integral, too, vanishes identically in all cases *except* one: namely, when $n = 0$. In that case, and in that case alone, we get

$$\boxed{\oint_C \frac{dz}{z} = 2\pi i.} \tag{23.20}$$

Alternatively, observe that we can now *expand* the circle and let its radius tend to infinity without affecting the value of the integral, because $z^{-(n+1)}$ is analytic at ∞. This causes the integral to vanish, *except* in the case $n = 0$. In that case the integral is precisely equal to $2\pi i$, independent of the value of the radius r.

Collecting the results in Eqs. (23.18) and (23.20), we have

$$\frac{1}{2\pi i} \oint_C \frac{dz}{z^{n+1}} = \delta_{n,0} \quad \text{where } n \text{ is any integer.} \tag{23.21}$$

($\delta_{n,0}$ is the Kronecker delta.) A trivial generalization of this relation is obtained by shifting from z to $z - a$, where a is any point in the complex plane. If the contour C encircles the point a once in the positive sense, then

$$\boxed{\frac{1}{2\pi i} \oint_C \frac{dz}{(z-a)^{n+1}} = \delta_{n,0} \quad \text{where} \quad n \in \mathbb{Z}.} \tag{23.22}$$

This is a most useful representation of the Kronecker delta. The whole of the **calculus of residues** is essentially based on it!

23.3.2 *Cauchy's Residue Theorem*

Equation (23.22) and Cauchy's Integral Theorem lead immediately to a very important result. In its simplest form, it reads as follows:

***Cauchy's Residue Theorem*:** Let C be a closed contour lying entirely in a region in which the function $f(z)$ is analytic, except for isolated poles and essential singularities at the set of points $\{z_k, k = 1, 2, \ldots\}$. No singularities must lie on C itself. Then, if C encircles the singularities once in the *positive* or anticlockwise sense (as in Fig. 23.5),

$$\boxed{\oint_C dz \, f(z) = (2\pi i) \sum_k \operatorname*{Res}_{z=z_k} f(z).} \tag{23.23}$$

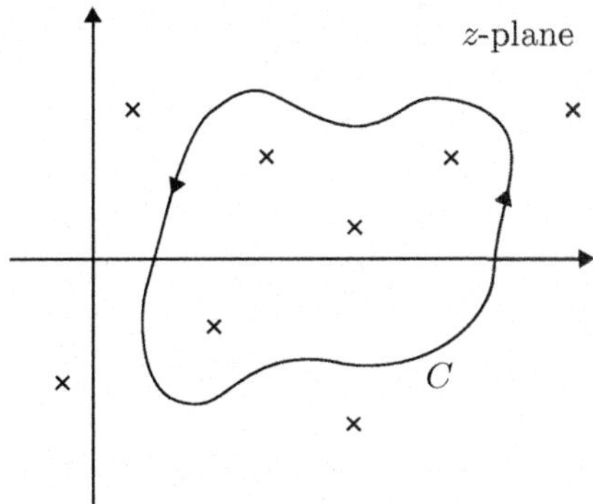

Fig. 23.5 Closed contour C for Cauchy's Residue Theorem. The small crosses denote possible poles or isolated essential singularities of $f(z)$

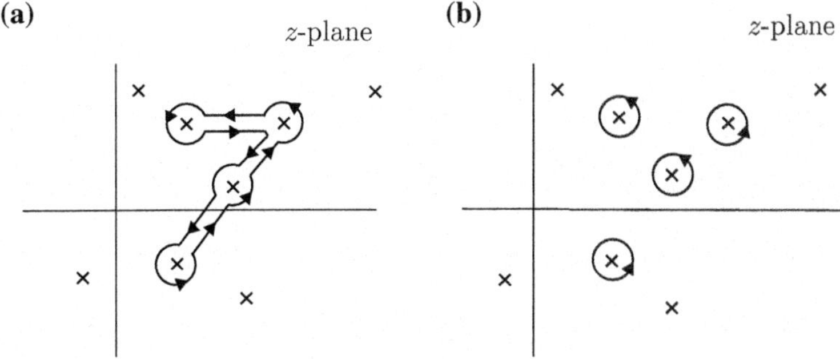

Fig. 23.6 Contour deformation leading to the residue theorem

The right-hand side is multiplied by -1 if C encircles the singularities in the negative or clockwise sense. If a singularity is encircled r times, the contribution to the integral is $\pm r$ times the residue at that singularity, the negative sign being applicable when the contour winds around the singularity in the negative sense. r is called the **winding number** of the contour at the singularity.

The proof of the residue theorem will not be given here. But you can understand how the result arises by noting that the value of the integral in (23.23) is unchanged under an arbitrary deformation of the contour C, as long as two conditions are satisfied:

(i) No part of C lies outside the region of analyticity of $f(z)$.
(ii) No singularity of $f(z)$ is crossed by the contour during the deformation (which also implies that the winding number of C at each singularity is not altered).

Figure 23.6a and b show how the residue theorem comes about: The contour C in Fig. 23.5 is first deformed to that in Fig. 23.6a. This does not change the value of the contour integral. The contributions of the sides of the "channels" in this figure cancel out as they are brought closer together, till the contour is pinched off into separate small circles around each of the poles originally enclosed by C. This is shown in Fig. 23.6b. The theorem now follows, upon applying Eq. (23.22) at each singularity.

Note that the residue theorem applies to a *closed* contour. When you use it to evaluate a definite integral of the form $\int_a^b dx\, f(x)$, you must first relate the latter to the integral of an analytic function $f(z)$ over a closed contour C. There are standard tricks to do this, and the exercises that follow illustrate some of these. In many cases, the piece that is adjoined to the original line integral in order to close the contour contributes zero to the total integral. In this sense, you may say that the trick is to add a well-chosen zero to arrive at the result sought!

★ **4.** Let a and b denote positive constants, and let $n = 0, 1, \ldots$. Use contour integration and the residue theorem to show that

(a) $$\int_{-\infty}^{\infty} \frac{dx}{(x^2+a^2)(x^2+b^2)} = \frac{\pi}{ab(a+b)}$$

(b) $$\int_0^{\infty} \frac{dx\, x^2}{(x^2+a^2)^3} = \frac{\pi}{16a^3}$$

(c) $$\int_{-\infty}^{\infty} \frac{dx\, x\, \sin x}{(x^2+a^2)} = \pi\, e^{-a}$$

(d) $$\int_{-\infty}^{\infty} \frac{dx\, (x^2-x+2)}{x^4+10x^2+9} = \frac{5\pi}{12}$$

(e) $$\int_0^{2\pi} \frac{d\theta}{a-b\cos\theta} = \frac{2\pi}{\sqrt{a^2-b^2}} \quad (a > b \geq 0)$$

(f) $$\int_0^{2\pi} d\theta\, e^{a\cos\theta} \cos(a\sin\theta - n\theta) = \frac{2\pi a^n}{n!}$$

(g) $$\int_0^{2\pi} d\theta\, e^{a\cos\theta} \sin(a\sin\theta - n\theta) = 0$$

(h) $$\int_0^{2\pi} d\theta\, e^{a\sin\theta} \cos(a\cos\theta + n\theta) = \frac{2\pi a^n}{n!} \cos\frac{n\pi}{2}$$

(i) $$\int_0^{2\pi} d\theta\, e^{a\sin\theta} \sin(a\cos\theta + n\theta) = \frac{2\pi a^n}{n!} \sin\frac{n\pi}{2}.$$

23.3.3 *The Dirichlet Integral; Cauchy Principal Value*

Recall the *Dirichlet integral*

$$\int_0^{\infty} dx\, \frac{\sin bx}{x} = \frac{\pi}{2} \quad (\text{for any } b > 0) \tag{23.24}$$

that was evaluated way back in Eq. (2.18) of Chap. 2, Sect. 2.3. The method used there was an elementary one. We are now ready derive the same result by a somewhat more sophisticated method. Interestingly, several small subtleties are involved in this simple example.

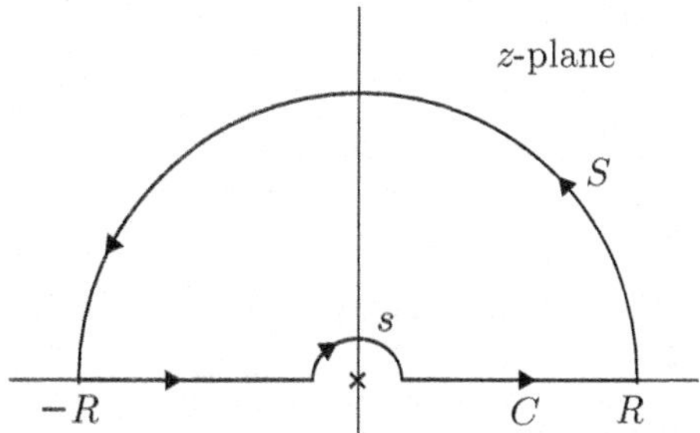

Fig. 23.7 Contour for the Dirichlet integral

First write the integral $\int_0^\infty dx\, \sin(bx)/x$ as $\frac{1}{2}$ times the integral from $-\infty$ to ∞, and *then* write $\sin bx = \text{Im}\, e^{ibx}$. Now consider the integrand e^{ibz}/z over a closed contour C running from $-R$ to $+R$ on the x-axis, together with a semicircle of radius R in the upper half-plane (because that is where the contribution from the semicircle will vanish as $R \to \infty$). But there is now an additional complication, because the integrand has a pole at the origin, unlike the original integrand $(\sin bx)/x$. You can avoid this pole by *indenting* the contour, i.e., let the path of integration takes a detour into the upper half-plane, in the form of a small semicircle of radius ϵ around $z = 0$. The closed contour is now as shown in Fig. 23.7. Call the small and large semicircles s and S, respectively. Now, the function e^{ibz}/z is holomorphic everywhere inside and on C. Cauchy's Integral Theorem therefore gives

$$0 = \oint_C dz\, \frac{e^{ibz}}{z} = \left\{ \int_{-R}^{-\epsilon} dx\, \frac{e^{ibx}}{x} + \int_{\epsilon}^{R} dx\, \frac{e^{ibx}}{x} \right\} + \int_s dz\, \frac{e^{ibz}}{z} + \int_S dz\, \frac{e^{ibz}}{z}. \tag{23.25}$$

The idea is to pass to the limits $R \to \infty$ and $\epsilon \to 0$ of the integrals on the right-hand side.

***Cauchy principal value*:** The quantity inside the curly brackets on the right-hand side of Eq. (23.25) is an integral along with the two segments of the real axis *omitting a symmetric infinitesimal neighborhood of the singularity* at the origin. In the limit $\epsilon = 0$, it is called the **Cauchy principal value** integral. It is written as

$$\lim_{\epsilon \to 0} \left\{ \int_{-R}^{-\epsilon} dx\, \frac{e^{ibx}}{x} + \int_{\epsilon}^{R} dx\, \frac{e^{ibx}}{x} \right\} \stackrel{\text{def.}}{=} \text{P} \int_{-R}^{R} dx\, \frac{e^{ibx}}{x}. \tag{23.26}$$

In the limit $R \to \infty$, we have

$$\lim_{R \to \infty} \text{P} \int_{-R}^{R} dx\, \frac{e^{ibx}}{x} = \text{P} \int_{-\infty}^{\infty} dx\, \frac{e^{ibx}}{x}. \tag{23.27}$$

I will return to the Cauchy principal value of a singular integral in Sect. 23.3.4, and again in Chap. 24, Sect. 24.2.1, when discussing dispersion relations.

The contribution from the integral over s in Eq. (23.25) is evaluated easily. Since $z = \epsilon\, e^{i\theta}$ on this piece of the contour, we have $dz = i\epsilon\, e^{i\theta}\, d\theta$, so that $dz/z = i\, d\theta$. Hence

$$\lim_{\epsilon\to 0}\int_s dz\, \frac{e^{ibz}}{z} = \lim_{\epsilon\to 0} i \int_\pi^0 d\theta\, \exp\,(ib\,\epsilon\, e^{i\theta}) = -i\pi. \tag{23.28}$$

(Note how the integration over θ runs from π to 0, rather than the other way about.) Similarly, we have $z = R\, e^{i\theta}$ on S. But the factor e^{ibz} in the integrand vanishes exponentially as $R \to \infty$ when z is in the upper half-plane, where $\text{Im}\, z > 0$. The contribution from S is therefore zero when we pass to the limit $R \to \infty$. Equation (23.25) then reduces to

$$\text{P}\int_{-\infty}^{\infty} dx\, \frac{e^{ibx}}{x} = i\pi. \tag{23.29}$$

Equating the respective imaginary parts of the two sides of this equation, we get

$$\text{P}\int_{-\infty}^{\infty} dx\, \frac{\sin\, bx}{x} = \pi. \tag{23.30}$$

But the principal value restriction can now be dropped, because $(\sin\, x)/x$ does not actually have a singularity at $x = 0$, in contrast to e^{ibx}/x, which does. Therefore

$$\int_{-\infty}^{\infty} dx\, \frac{\sin\, bx}{x} = 2\int_0^{\infty} dx\, \frac{\sin\, bx}{x} = \pi. \tag{23.31}$$

This is how the Dirichlet integral is evaluated by contour integration.

★ **5.** You will find it a simple but instructive exercise to evaluate the Dirichlet integral using the following alternative closed contours for C:

(a) The large semicircle S in the upper half-plane, and the small semicircle s in the lower half-plane.
(b) Both S and s in the lower half-plane.
(c) S in the upper half-plane, s in the lower half-plane.

23.3.4 The "$i\epsilon$-Prescription" for a Singular Integral

The example above illustrates a useful device for handling integrals in which the integrand has a specific kind of singularity on the path of integration.

Consider an integral of the form

$$f(x_0) = \int_a^b dx\, \frac{\phi(x)}{x - x_0}, \tag{23.32}$$

where $a < x_0 < b$ are real numbers, $\phi(x)$ is a sufficiently smooth function, and $\phi(x_0) \neq 0$. The integrand obviously diverges at $x = x_0$ because of the factor $(x - x_0)$ in the denominator. As it stands, the Riemann integral in Eq. (23.32) does not exist, because of this nonintegrable singularity of the integrand. Suppose, however, we move the singularity away from the path of integration by giving it either a positive imaginary part $+i\epsilon$ (where $\epsilon > 0$), or a negative imaginary part $-i\epsilon$. This step, called an *$i\epsilon$-prescription*, makes the original integral well-defined. The integrals in the two cases are given, respectively, by

$$f(x_0 \pm i\epsilon) = \int_a^b dx\, \frac{\phi(x)}{x - (x_0 \pm i\epsilon)} = \int_a^b dx\, \frac{\phi(x)}{x - x_0 \mp i\epsilon}. \tag{23.33}$$

The question is: what happens as $\epsilon \to 0$? We can continue to keep the integral well-defined by distorting the path of integration away from the approaching singularity, to form a small semicircle of radius ϵ. This semicircle lies in the *lower* half-plane in the case of $f(x_0 + i\epsilon)$, and in the *upper* half-plane in the case of $f(x_0 - i\epsilon)$. Figure 23.8a and b depict the two cases. On the small semicircles, the variable of integration is $z = x_0 + \epsilon\, e^{i\theta}$, so that $dz = \epsilon\, e^{i\theta} i d\theta$. The argument θ runs from π to 2π in the case of $f(x + i\epsilon)$, and from π to 0 in the case of $f(x - i\epsilon)$. Taking the limit $\epsilon \to 0$ then yields the Cauchy principal value integral from a to b, plus the contribution from the semicircle:

$$\boxed{\lim_{\epsilon \to 0} \int_a^b dx\, \frac{\phi(x)}{x - x_0 \mp i\epsilon} = \mathrm{P} \int_a^b dx\, \frac{\phi(x)}{x - x_0} \pm i\pi\phi(x_0).} \tag{23.34}$$

This is a very useful result. It is often written in the form of the "formula"

$$\boxed{\text{“}\lim_{\epsilon \to 0} \frac{1}{x - x_0 \mp i\epsilon} = \mathrm{P}\, \frac{1}{x - x_0} \pm i\pi\, \delta(x - x_0).\text{”}} \tag{23.35}$$

The quotation marks are meant to remind you that (23.35) is just a mnemonic. It is to be understood in the following sense: Multiply both sides of the equation by any suitable function $\phi(x)$, and integrate over an interval on the x-axis that includes the point x_0.

Note, in particular, that

$$\lim_{\epsilon \to 0} \big[f(x_0 + i\epsilon) - f(x_0 - i\epsilon)\big] = 2\pi i\, \phi(x_0). \tag{23.36}$$

The function f thus has a **discontinuity** for all values of its argument lying in the interval (a, b). This result has an immediate implication for the singularity structure of the function $f(z)$ of the complex variable z defined by the integral

$$f(z) = \int_a^b dt\, \frac{\phi(t)}{t - z}. \tag{23.37}$$

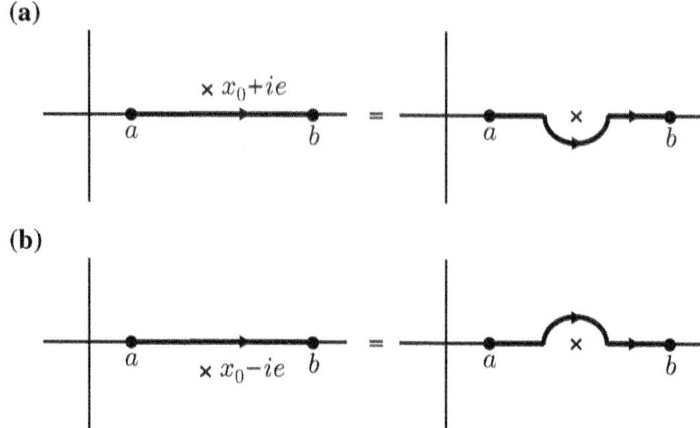

Fig. 23.8 The "$i\epsilon$ prescription" to handle a simple pole of the integrand at the point x_0 on the path of integration, by displacing the pole into (**a**) the upper half-plane, (**b**) the lower half-plane

I will return to this point in Chap. 26, Sect. 26.3, when we discuss the singularities of functions defined by integrals.

The $i\epsilon$-prescription for "taming" singular integrals has an important practical application. Solving differential equations using Green functions often involves singular integrals similar to that in Eq. (23.32). A suitable $i\epsilon$-prescription then not only makes such an integral well-defined, but also enables some specific boundary condition to be incorporated in the solution to the differential equation. In Chap. 31, Sect. 31.1.2, and again in Chap. 32, Sect. 32.2.3, we will see how appropriate $i\epsilon$-prescriptions can be used to obtain the physically relevant Green functions for the wave operator and the Helmholtz operator, respectively.

23.3.5 Residue at Infinity

Suppose an analytic function $f(z)$ has an isolated singularity at some *finite* point $z = a$. Recall that the residue of $f(z)$ at $z = a$ is the coefficient of $(z - a)^{-1}$ in the Laurent series expansion of $f(z)$ in powers of $(z - a)$. It is given by $(2\pi i)^{-1}$ times the integral of $f(z)$ over a simple closed contour enclosing no singularities other than the one at $z = a$, traversed once in the positive sense.

- If $f(z)$ is regular at $z = a$, the residue at that point vanishes.
- But the converse is not true: the residue may vanish simply because the term proportional to $(z - a)^{-1}$ is missing in the singular part of $f(z)$. (For example, the residue of the function $f(z) = 1/z^2$ at $z = 0$ is zero.)

The **residue at infinity** is *defined* as follows:

- If a single-valued function $f(z)$ has an isolated singularity at ∞, *or even if it has no singularity at that point*, its residue at infinity is given by

$$\boxed{\operatorname*{Res}_{z=\infty} f(z) \stackrel{\text{def.}}{=} -\frac{1}{2\pi i}\oint_C dz\, f(z),} \tag{23.38}$$

where C is a circle traversed in the anticlockwise or positive sense, with a sufficiently *large* radius R, so as to enclose *all* the singularities of $f(z)$ in the finite part of the complex plane.[3] Note, in particular, the minus sign in the definition above.

Clearly, the contour integral in Eq. (23.38) is equal to $(-2\pi i)$ times the sum of the residues of $f(z)$ in the finite part of the complex plane.

- It follows that

$$\boxed{\operatorname*{Res}_{z=\infty} f(z) = -\sum_j \operatorname*{Res}_{z=a_j} f(z),} \tag{23.39}$$

where the sum runs over all the isolated singularities (located at the points a_j) in the finite part of the complex plane.
- Hence, if the right-hand side of Eq. (23.39) does not vanish, $f(z)$ has a residue at infinity even if it is *not* singular at $z = \infty$.
- On the other hand, a function may be *singular* at $z = \infty$, and yet have *no* residue at infinity. This is what happens when $f(z)$ is a (non-constant) entire function.

Example 1 The function $f(z) = (z-1)^{-1} + (z-2)^{-1}$ has simple poles at $z = 1$ and $z = 2$, and is regular at $z = \infty$. But its residue at infinity is nonzero. It is equal to (-1) times the sum of its residues at $z = 1$ and $z = 2$, namely, -2.

Example 2 In contrast, the function $f(z) = e^z$ is an entire function with an isolated essential singularity at ∞. But its residue at infinity is zero, because the contour C in Eq. (23.38) can be shrunk to a point in the finite part of $\hat{\mathbb{C}}$, making the integral vanish. The same thing is true of any polynomial in z: any polynomial of degree n is an entire function that has a pole of order n at $z = \infty$, but its residue at infinity vanishes identically.

There is an equivalent way of finding the residue at infinity. Change variables from z to $w = 1/z$. Equation (23.38) becomes

$$\boxed{\operatorname*{Res}_{z=\infty} f(z) = -\frac{1}{2\pi i}\oint_c dw\, \frac{f(1/w)}{w^2},} \tag{23.40}$$

where c is a circle of infinitesimal radius $1/R$ encircling the origin in the w-plane, traversed in the positive sense. It follows that

[3]It is helpful to think about what C looks like when mapped onto the Riemann sphere $\mathcal{S}$: it is a *small* circle encircling the point at infinity.

- the residue at infinity of $f(z)$ is the coefficient of w^{-1} in the Laurent expansion of $-(1/w^2)\,f(1/w)$ in powers of w.

It is now trivial to check that the residue at infinity of $(z-1)^{-1}+(z-2)^{-1}$ is indeed equal to -2, as found above. Similarly, it is easily checked that neither e^z nor any polynomial in z has a residue at infinity.

***Application to the evaluation of certain contour integrals*:** A practical use of the concept of the residue at infinity is in the evaluation of certain contour integrals involving rational functions. Here is a simple example.

★ **6.** Let

$$p_n(z) = a_n\, z^n + a_{n-1}\, z^{n-1} + \cdots + a_0$$

and

$$q_{n+1}(z) = b_{n+1}\, z^{n+1} + b_n\, z^n + \cdots + b_0$$

be polynomials of order n and $n+1$, respectively. If C is a simple closed contour that encloses all the roots of $q_{n+1}(z) = 0$ once, in the positive sense, show that

$$\oint_C dz\, \frac{p_n(z)}{q_{n+1}(z)} = \frac{2\pi i a_n}{b_{n+1}}\,.$$

23.4 Summation of Series Using Contour Integration

In Sect. 23.2.1, we have seen that the function $f(z) = \cot z$ has a simple pole at $z = n\pi$ (where n is any integer), with residue equal to 1. Similarly, $\operatorname{cosec} z$ has a simple pole at $z = n\pi$ (where $n \in \mathbb{Z}$), with residue equal to $(-1)^n$. We may rewrite these statements as follows:

- The function $\pi \cot(\pi z)$ has a simple pole at every integer $z = n$, with residue equal to $+1$.
- The function $\pi \operatorname{cosec}(\pi z)$ has a simple pole at every integer $z = n$, with residue equal to $(-1)^n$.

These facts can be used to sum certain infinite series with the help of contour integration. The method works when the summand is an even function of n, so that the series can be written as a sum over both positive and negative integers.

In order to illustrate the general method, I consider a series that has already been summed, namely, Eq. (18.50) of Chap. 18, Sect. 18.4.2:

$$S(a) \equiv \sum_{n=1}^{\infty} \frac{1}{n^2 + a^2} = \frac{\pi}{2a}\left\{\coth \pi a - \frac{1}{\pi a}\right\}. \tag{23.41}$$

Here is how this result may be derived using contour integration. First write $S(a)$ as $\frac{1}{2}$ times the sum over positive as well as negative integers n. Each term in the sum is just the residue of the analytic function

$$f(z) = \frac{\pi \cot \pi z}{z^2 + a^2} \tag{23.42}$$

at its simple pole at $z = n$, where $n = \pm 1, \pm 2, \ldots$. Hence it is $(2\pi i)^{-1}$ times the contour integral of $f(z)$ over a small circle c_n encircling the pole at $z = n$ in the positive sense. Thus

$$S(a) = \frac{1}{2} \sum_{\substack{n=-\infty \\ n \neq 0}}^{\infty} \frac{1}{n^2 + a^2} = \frac{1}{4\pi i} \sum_{\substack{n=-\infty \\ n \neq 0}}^{\infty} \oint_{c_n} dz\, f(z). \tag{23.43}$$

Figure 23.9 shows the first few of the contours $\{c_n\}$ over which the integral of $f(z)$ is to be evaluated. Observe that the pole of the factor $\cot(\pi z)$ at $z = 0$ is *not* included in this set. The poles of the factor $(z^2 + a^2)^{-1}$ at $z = \pm ia$ are also not enclosed by any contour.

And now the power of the freedom to distort the contour without altering the value of the integral becomes evident! The small circles c_n can expand till they merge, yielding the two hairpin contours straddling the x-axis from $-\infty$ to -1 and from 1 to ∞, respectively—as depicted by the solid lines in Fig. 23.10. These two disjoint contours, in turn, can be made part of a single closed contour by attaching large semicircles in the upper and lower half-planes, as shown by the dotted lines in Fig. 23.10. The crucial point is that the contributions from both these semicircles vanish when their radius $R \to \infty$, because $\cot(\pi z)$ remains bounded as $|z| \to \infty$, while $dz/(z^2 + a^2) \to 0$ like R^{-1} on the semicircles. So the attachment of the two large semicircles is, in this sense, the clever addition of a well-chosen zero!

But this closed contour now encircles just three poles of $f(z)$, namely, the simple poles at $z = 0,\ z = ia$, and $z = -ia$. These poles are encircled in the *clockwise* sense. You can now *shrink* the contour to three small circles (traversed in the negative sense) around the three poles. The contour integral is then just $(-2\pi i)$ times the sum of the corresponding residues. Simplification yields the answer on the right-hand side of Eq. (23.41), if you use the fact that $\cot(i\pi a) = -i \coth(\pi a)$, etc.

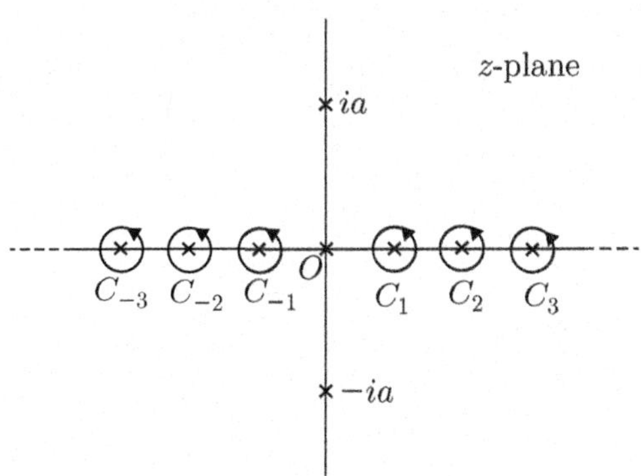

Fig. 23.9 Some of the contours $\{c_n\}$ in the sum of integrals in Eq. (23.43)

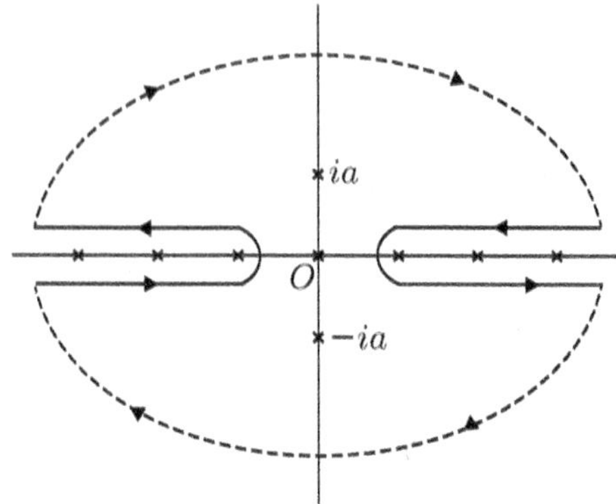

Fig. 23.10 The set of contours $\{c_n\}$ in Fig. 23.9 can be deformed to the two hairpin contours straddling the real axis as shown (solid lines). Attaching two large semicircles (dotted lines) to these hairpin contours yields a single closed contour

★ **7.** Work out the steps just described to evaluate $S(a)$, to arrive at the answer quoted in Eq. (23.41).

★ **8.** As you know, taking the limit $a = 0$ in Eq. (23.41) for $S(a)$ yields the value $\frac{1}{6}\pi^2$ for the Riemann zeta function $\zeta(2) = \sum_1^\infty 1/n^2$. You can, of course, use contour integration to establish this result directly. Use the function

$$f(z) = \frac{\pi \cot \pi z}{z^2}$$

and repeat the procedure, to show that $\sum_1^\infty 1/n^2 = \frac{1}{6}\pi^2$. Note that the singularity of $f(z)$ at $z = 0$ is now a pole of order 3.

★ **9.** Use the function

$$f(z) = \frac{\pi \operatorname{cosec} \pi z}{z^2 + a^2}$$

to show that

$$\sum_{n=1}^{\infty} \frac{(-1)^{n-1}}{n^2 + a^2} = \frac{\pi}{2a} \left\{ \frac{1}{\pi a} - \frac{1}{\sinh \pi a} \right\}.$$

Remark Here is a simple way to verify this result. Observe that the sum required is given by $S(a) - \frac{1}{2}\, S(\frac{1}{2}a)$.

★ **10.** Use the function

$$f(z) = \frac{\pi \cot \pi z}{(z^2 + a^2)^2}$$

to show that

$$\sum_{n=1}^{\infty} \frac{1}{(n^2 + a^2)^2} = \frac{\pi}{4a^3} \left\{ \coth \pi a + \frac{\pi a}{\sinh^2 \pi a} - \frac{2}{\pi a} \right\}.$$

Remark Recall that this result has already been written down in Eq. (18.52) of Chap. 18, Sect. 18.4.2. There, we used the fact that the sum required is just $-(2a)^{-1} dS(a)/da$.

***The zeta function for positive even integers*:** A useful by-product of the last formula above is the value of the zeta function $\zeta(4)$. Passing to the limit $a \to 0$ in this result gives

$$\sum_{n=1}^{\infty} \frac{1}{n^4} = \zeta(4) = \tfrac{1}{90}\pi^4. \tag{23.44}$$

(Check this out.) It is evident that repeated differentiation of $S(a)$ in Eq. (23.41) with respect to the parameter a will yield the sum $\sum_{n=1}^{\infty} 1/(n^2 + a^2)^r$, where $r = 2, 3, \ldots$. Taking the limit $a \to 0$ in the result will then give us the value of $\zeta(2r)$. I will return to some other properties of the zeta function in Chap. 25, Sect. 25.2.5 and in Chap. 26, Sects. 26.2.3 and 26.2.4.

★ **11.** Use the function

$$f(z) = \frac{\pi \cot \pi z}{(z^2 + a^2)(z^2 + b^2)}$$

to show that

$$\sum_{n=1}^{\infty} \frac{1}{(n^2 + a^2)(n^2 + b^2)} = \frac{\pi(b \coth \pi a - a \coth \pi b)}{2ab(b^2 - a^2)} - \frac{1}{2a^2 b^2}.$$

Remark Once again, you can verify this result by observing that the sum required is just $[S(a) - S(b)]/(b^2 - a^2)$.

23.5 Linear Recursion Relations with Constant Coefficients

23.5.1 The Generating Function

At several places in the preceding chapters, you have seen how useful generating functions can be. They also provide a convenient way to solve difference equations or recursion relations satisfied by sequences $c_0, c_1, c_2, \ldots$. Given the recursion relation, the problem is to find c_n explicitly as a function of n.

The basic idea is to set up and solve the corresponding equation for the generating function of the sequence, defined as

$$f(z) = \sum_{n=0}^{\infty} c_n z^n. \tag{23.45}$$

In a sufficiently small neighborhood of the origin in the z-plane, Eq. (23.45) is a Taylor series representation of the analytic function $f(z)$. On the other hand, the difference equation satisfied by the set $\{c_n\}$ leads to an equation for $f(z)$. The solution of this equation usually yields a closed form for this function. In particular, this is so when the difference equation has constant (i.e., n-independent) coefficients, which is the only case I will consider here. This form is the analytic continuation of the Taylor series to a bigger region than the circle of convergence of the Taylor series. The solution to the original difference equation (i.e., the coefficient c_n) is then found by the inversion of Eq. (23.45). The inversion formula is of course

$$c_n = \frac{1}{n!}\left[\frac{d^n f}{dz^n}\right]_{z=0}. \tag{23.46}$$

While this is a formal solution, it is not very convenient in practice: explicitly finding the nth derivative of even a relatively simple-looking function (such as a rational function, for instance) is quite tedious.

We therefore use the more convenient formula

$$c_n = \frac{1}{2\pi i}\oint_C \frac{f(z)\,dz}{z^{n+1}}, \tag{23.47}$$

where C is a simple closed contour enclosing the origin (and *no other singularities* of $f(z)$) once, in the positive sense. The integrand in Eq. (23.47) has a pole of order $(n+1)$ at the origin. Evaluating its residue at this pole will, of course, just take us back to Eq. (23.46). But the idea is to avoid having to compute the nth derivative of $f(z)$. We can do this by *expanding* the contour out to infinity in all directions, and picking up the residues at the poles of $f(z)$ (with a minus sign, because the poles would obviously be encircled in the negative sense!). This procedure will usually yield c_n much more easily, provided of course that $f(z)$ has only isolated singularities. This is certainly the case when the difference equation for the sequence $\{c_n\}$ is a *linear* one, because $f(z)$ then turns out to be a *rational* function of z. Let us now see how the method works in practice.

★ **12.** This is a very simple exercise. Consider a linear *one*-step difference equation, i.e., a recursion relation of the form

$$c_{n+1} = a c_n + b$$

where a, b, and c_0 are given constants. Such a recursion relation can be solved directly by repeated iteration. Show that the solution is

$$c_n = a^n c_0 + \frac{(a^n - 1)}{(a-1)}\,b.$$

Now obtain this solution once again, using the generating function method.

23.5.2 Hemachandra-Fibonacci Numbers

Two-step linear recursion relations, i.e., those that involve a linear relation between the three successive coefficients c_n, c_{n+1}, and c_{n+2}, are a little more involved. Let us consider a very famous example.

The **Hemachandra-Fibonacci sequence** $\{h_0, h_1, \ldots\}$ is defined, as you are undoubtedly aware, by the rule that each number in the sequence is the sum of the preceding two numbers. Thus

$$\boxed{h_{n+2} = h_{n+1} + h_n\,, \quad n = 0,\ 1,\ \ldots\,.} \tag{23.48}$$

Specifying the first two numbers h_0 and h_1 then determines the entire sequence. (This is true of all two-step difference equations. Such equations are the discrete analogs of second-order differential equations.) Multiply both sides of Eq. (23.48) by z^n, and sum over n. After some simplification, the generating function $f(z) = \sum_0^\infty h_n z^n$ is found to be given by the rational function

$$f(z) = \frac{(h_0 - h_1)z - h_0}{z^2 + z - 1}\,. \tag{23.49}$$

In order to be specific, consider the case $h_0 = 0$ and $h_1 = 1$. This choice corresponds to the most familiar form[4] of the Hemachandra-Fibonacci sequence, namely,

$$0,\ 1,\ 1,\ 2,\ 3,\ 5,\ 8,\ 13,\ 21,\ 34,\ 55,\ 89,\ 144,\ \ldots\,. \tag{23.50}$$

Then h_n is given by

$$h_n = \frac{1}{2\pi i}\oint_C \frac{(-z)\,dz}{z^{n+1}(z^2 + z - 1)}\,. \tag{23.51}$$

The contour C encircles the multiple pole at the origin once, in the *negative* sense. We wish to avoid evaluating the residue of the integrand at this multiple pole. The other poles of the integrand are *simple* poles, located at the roots of the quadratic equation $z^2 + z - 1 = 0$, i.e., at $z = \frac{1}{2}(\sqrt{5} - 1)$ and $z = -\frac{1}{2}(\sqrt{5} + 1)$. The contour C can be expanded out to infinity in all directions, leaving behind two small closed contours that encircle these two poles in the *clockwise* sense. The contributions from these poles are easily written down. The contribution from the large circle at infinity vanishes, since the integrand vanishes faster than $|z|^{-1}$ as $|z| \to \infty$. After some simplification, the outcome is the following explicit expression for the Hemachandra-Fibonacci numbers:

[4] Presented by the Indian scholar and polymath Acharya Hemachandra *ca.* the year 1150, and a few decades later by the Italian mathematician Fibonacci in the year 1202. The sequence was known even earlier, and it appears (for instance) *ca.* 1135 in the work of the Indian mathematician Gopala.

$$\boxed{h_n = \frac{1}{2^n\sqrt{5}}\left[(\sqrt{5}+1)^n - (-1)^n(\sqrt{5}-1)^n\right], \qquad n = 0,\,1,\,\ldots.} \tag{23.52}$$

In spite of the appearance of the irrational number $\sqrt{5}$ in this formula, every h_n (for $n \geq 1$) is a positive integer.

★ **13.** Work out the steps described above to obtain Eq. (23.52).
As you are no doubt aware, the irrational number $\tau = \frac{1}{2}(\sqrt{5}+1)$ is called the **golden mean**. It has a large number of interesting properties. So do the Hemachandra-Fibonacci numbers, since $h_n = [\tau^n - (-\tau)^{-n}]/\sqrt{5}$. Equation (23.52) shows that the asymptotic (large-n) growth of the Hemachandra-Fibonacci sequence (23.50) is exponential, i.e., h_n has a leading large-n behavior $\sim e^{\lambda n}$, where $\lambda = \ln\tau \approx 0.48$.

★ **14.** Use the method described above to solve the recursion relations listed below. In each case, n runs over all the nonnegative integers.

(a) $c_{n+2} - 2c_{n+1} + c_n = 0, \quad c_0 = 1,\ c_1 = 1.$
(b) $c_{n+2} - c_{n+1} - 2c_n = 0, \quad c_0 = 1,\ c_1 = 2.$
(c) $c_{n+2} - 2c_{n+1} - c_n = 0, \quad c_0 = 0,\ c_1 = 1.$

23.5.3 Catalan Numbers

The recursion relations we have considered above are linear, second-order difference equations with constant coefficients. Occasionally, one encounters a *nonlinear* recursion relation that can be solved by elementary methods. An interesting and rather important instance is the following n-step recursion relation.

Let c_0 be a given number. Consider the set of nonlinear difference equations

$$c_1 = c_0\,c_0,\ c_2 = c_0c_1 + c_1c_0,\ c_3 = c_0c_2 + c_1c_1 + c_2c_0,\ \ldots. \tag{23.53}$$

That is,

$$c_n = c_0c_{n-1} + c_1c_{n-2} + c_2c_{n-3} + \cdots + c_{n-1}c_0 = \sum_{j=0}^{n-1} c_j\,c_{n-1-j}\,, \quad n \geq 1. \tag{23.54}$$

Remarkably enough, this recursion relation can be solved quite easily. Define the generating function $f(z) = \sum_0^\infty c_n\,z^n$, as usual. Squaring it and collecting powers of z, we see that $f(z)$ satisfies the equation

$$f^2(z) = z^{-1}[f(z) - c_0], \quad \text{or} \quad zf^2(z) - f(z) + c_0 = 0. \tag{23.55}$$

The correct root of this quadratic equation is identified by the requirement that $f(0)$ must be equal to c_0. The solution is

$$f(z) = \frac{1 - \sqrt{1 - 4c_0 z}}{2z}. \tag{23.56}$$

(Check that $f(z) \to c_0$ as $z \to 0$.) c_n can now be read off by using the binomial expansion of $(1 - 4c_0 z)^{1/2}$. After some simplification, we get

$$c_n = \frac{(2n)!\, c_0^{n+1}}{n!\,(n+1)!} = \frac{c_0^{n+1}}{(n+1)} \binom{2n}{n}. \tag{23.57}$$

★ **15.** Starting from the recursion relation (23.54), work through the steps to arrive at the solution (23.57).

For the special choice $c_0 = 1$, the numbers c_n are called **Catalan numbers**. In order to keep track of this choice, let us use the notation C_n (instead of c_n) for the numbers in this case. Thus,

$$\boxed{C_n \stackrel{\text{def.}}{=} \frac{(2n)!}{n!\,(n+1)!} = \frac{1}{(n+1)} \binom{2n}{n}.} \tag{23.58}$$

It is not immediately obvious from this definition that C_n is always an integer. In order to see this, you have merely to note that C_n is the difference of two binomial coefficients (which are, of course, integers) according to the identity

$$\binom{2n}{n} - \binom{2n}{n+1} = \frac{1}{n+1}\binom{2n}{n} = C_n\,. \tag{23.59}$$

The first few Catalan numbers are

$$C_0 = 1,\;\; C_1 = 1,\;\; C_2 = 2,\;\; C_3 = 5,\;\; C_4 = 14,\;\; C_6 = 42,\; \ldots\,. \tag{23.60}$$

***Catalan numbers and random walks*:** The Catalan numbers occur in an astonishingly large number of combinatorial problems. Here is a basic one, couched in the language of random walks. Consider a random walk on an $(n \times n)$ square lattice, with the sites labeled by the pair of integers (j, k), where $0 \le j \le n$ and $0 \le k \le n$. It is specified that the walker starts at the origin $(0, 0)$ and ends at the point (n, n), and takes exactly $2n$ steps. Obviously, this means that the walker can only take steps to the right or upward, but not to the left or downward. Moreover, n of these steps must be to the right, and n steps must be upward. It is obvious that the number of distinct random walks of this kind is $\binom{2n}{n} = (2n)!/(n!)^2$. Figure 23.11 shows two such walks. Now consider the subset of these walks in which the walker does not *cross* the diagonal line between the origin and the end point. The number of such random walks is precisely the Catalan number C_n. There are several ways of proving this assertion, but I will not give a proof here.

***Asymptotic (large-n) behavior*:** An application of Stirling's formula for the factorial of a large integer (Eq. (2.11) of Chap. 2, Sect. 2.2) shows that the leading asymptotic

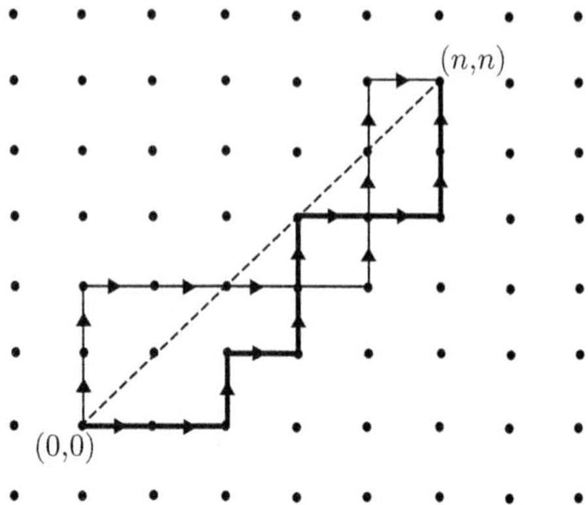

Fig. 23.11 Two $2n$-step random walks from $(0, 0)$ to (n, n) on a square lattice. Thin line: an arbitrary random walk; thick line: a walk that does not cross the diagonal (dashed line)

behavior of the Catalan numbers for large n is given by

$$C_n \sim \frac{2^{2n}}{\pi^{1/2}\, n^{3/2}} \,. \tag{23.61}$$

Reverting to the case of a general c_0, the leading asymptotic behavior of c_n is given by

$$c_n \sim \frac{(4c_0)^n}{\pi^{1/2}\, n^{3/2}}\, c_0 \,. \tag{23.62}$$

It follows that, as $n \to \infty$, the sequence $\{c_n\}$

- diverges exponentially when $c_0 > \frac{1}{4}$;
- converges exponentially to zero when $c_0 < \frac{1}{4}$;
- converges to zero according to a power law, $\sim n^{-3/2}$, when $c_0 = \frac{1}{4}$.

★ **16.** Establish (23.62).

While the two-step linear recursion relation (23.54) has constant coefficients, it is easy to check that c_n also satisfies the *one*-step recursion relation

$$c_n = \frac{2(2n-1)}{(n+1)}\, c_{n-1}, \quad n \geq 1. \tag{23.63}$$

But the coefficient multiplying c_{n-1} is now a function of n. This is what makes the solution of the recursion relation nontrivial.

23.5.4 Connection with Wigner's Semicircular Distribution

Here is another instance in which the Catalan numbers occur naturally. In Chap. 20, we have studied some of the properties of the probability distributions of a continuous random variable. A distribution that appears in several physical applications, most notably in the context of the eigenvalue distribution of *random matrices*, is **Wigner's semicircular distribution**. (The original physical context was that of the statistical distribution of the energy levels of complex atomic nuclei.) It pertains to a random variable X that takes values in the range $[-a, a]$ (where a is a positive constant), with a normalized PDF given by

$$p(x) = \frac{2}{\pi a^2}\sqrt{(a^2 - x^2)}\,. \tag{23.64}$$

The name of the distribution arises from the fact that, in terms of dimensionless variables, the plot of $\pi a p(x)/2$ versus x/a is a semicircle of unit radius, centered at the origin.

Since $p(x)$ is an even function of x, the odd moments of X vanish identically. The even moments are found easily with the help of the integral written down in Eq. (3.21) of Chap. 3, Sect. 3.1.5, namely,

$$\int_0^{\pi/2} d\theta\, \sin^l\theta = \frac{1}{2}\int_0^{\pi} d\theta\, \sin^l\theta = \frac{\sqrt{\pi}\,\Gamma\left(\frac{1}{2}(l+1)\right)}{2\Gamma\left(1+\frac{1}{2}l\right)} \quad (l = 0,\, 1,\, 2,\, \ldots). \tag{23.65}$$

We find

$$\langle X^{2n}\rangle = \int_{-a}^{a} dx\, x^{2n}\, p(x) = \left(\frac{a}{2}\right)^{2n} \frac{(2n)!}{n!\,(n+1)!} = \left(\frac{a}{2}\right)^{2n} C_n\,. \tag{23.66}$$

In particular, in the case $a = 2$ (the "standard" form of Wigner's semicircular distribution), $\langle X^{2n}\rangle$ is precisely the Catalan number C_n.

The moment-generating function corresponding to Wigner's semicircular distribution is also of interest. Recall, from Eq. (20.4) of Chap. 20, Sect. 20.1.2, that the moment-generating function of a distribution is defined as $M(u) = \langle e^{uX}\rangle$. In the present instance, we find

$$M(u) = \int_{-a}^{a} dx\, e^{ux}\, p(x) = \frac{2}{au}\, I_1(au), \tag{23.67}$$

where I_1 is the modified Bessel function of the first kind, of order 1.

★ **17.** Verify the result in (23.67).

23.6 Mittag-Leffler Expansion of Meromorphic Functions

We have seen that analytic functions can be represented by Taylor series in their domains of holomorphy. In the neighborhood of poles and isolated essential singularities, Taylor series are replaced by Laurent series.

A further generalization is possible for meromorphic functions. Recall that a meromorphic function in the complex plane $\mathbb{C}$ is one whose only singularities in the (finite) complex plane, if any, are poles. Suppose we are given the locations of all the poles of such a function, and the singular parts at these poles. Can we find a representation for the function that involves a "sum over poles"? The **Mittag-Leffler expansion** of a meromorphic function provides such a representation.

In general, suppose the meromorphic function $f(z)$ has poles at the points $a_j \ (j = 1, 2, \ldots)$, with singular parts $s_j(z)$. For example, if a_j is a pole of order m, then the singular part corresponding to this pole is

$$s_j(z) = \frac{c_{-m}^{(j)}}{(z - a_j)^m} + \cdots + \frac{c_{-1}^{(j)}}{(z - a_j)}. \tag{23.68}$$

The Mittag-Leffler expansion (or representation) of $f(z)$ is then of the form

$$f(z) = \sum_j s_j(z) + g(z), \tag{23.69}$$

where $g(z)$ is an *entire* function. When the number of poles is finite, the sum over j presents no problem. This is the case when $f(z)$ is a *rational function*: that is, $f(z)$ is the ratio of two polynomials.

- In such cases (i.e., when the number of poles of $f(z)$ is finite), the Mittag-Leffler expansion is nothing but the resolution of the function into partial fractions.

The function $f(z) = (z - a)/(z - b)^2$, for instance, has the Mittag-Leffler representation

$$\frac{z - a}{(z - b)^2} = \frac{b - a}{(z - b)^2} + \frac{1}{z - b}. \tag{23.70}$$

***An instructive example*:** When $\sum_j s_j(z)$ is an *infinite* sum, we must worry about its convergence. The matter then becomes nontrivial. Consider, for instance, the meromorphic function $\pi \cot(\pi z)$, which has (as you have seen already) a simple pole at every integer value of z, with a residue equal to unity. In order to find its Mittag-Leffler representation, we start with Eq. (23.41) for the sum $S(a)$, repeated once more for ready reference:

$$S(a) \equiv \sum_{n=1}^{\infty} \frac{1}{n^2 + a^2} = \frac{\pi}{2a}\left\{\coth(\pi a) - \frac{1}{\pi a}\right\}. \tag{23.71}$$

Although the infinite series on the left-hand side was summed originally for real positive values of the parameter a, Eq. (23.71) is really a relation between two analytic functions of a, by virtue of *the principle of analytic continuation*. Set $a = iz$ in it, and use the relation $\coth(i\pi z) = -i \cot(\pi z)$. Equation (23.71) can then be written as

$$\pi \cot(\pi z) = \frac{1}{z} + 2\sum_{n=1}^{\infty} \frac{z}{z^2 - n^2} = \sum_{n=-\infty}^{\infty} \frac{z}{z^2 - n^2}. \tag{23.72}$$

We are almost there, because this expansion seems to express the meromorphic function $\pi \cot(\pi z)$ as a sum over the singular parts at its poles (which are located at all integer values of z). But the poles of $\cot \pi z$ are *simple* poles, whereas the summand in the last equation involves $z^2 - n^2$, a *quadratic* function of z, in the denominator. It is a simple matter to write

$$\sum_{n=1}^{\infty} \frac{2z}{z^2 - n^2} = \sum_{n=1}^{\infty} \left\{ \frac{1}{(z-n)} + \frac{1}{(z+n)} \right\}. \tag{23.73}$$

But we *cannot* go on to write the right-hand side of Eq. (23.73) as

$$\sum_{n=1}^{\infty} \frac{1}{(z-n)} + \sum_{n=1}^{\infty} \frac{1}{(z+n)},$$

because each of the two individual infinite series in the last line above *diverges* logarithmically. In order to find the correct Mittag-Leffler representation, let us go back to

$$\pi \cot(\pi z) = \sum_{n=-\infty}^{\infty} \frac{z}{z^2 - n^2} = \frac{1}{z} + \sum_{\substack{n=-\infty \\ n \neq 0}}^{\infty} \frac{z}{z^2 - n^2}. \tag{23.74}$$

But

$$\begin{aligned}
\sum_{\substack{n=-\infty \\ n \neq 0}}^{\infty} \frac{z}{z^2 - n^2} &= \tfrac{1}{2} \sum_{\substack{n=-\infty \\ n \neq 0}}^{\infty} \frac{2z}{z^2 - n^2} = \tfrac{1}{2} \sum_{\substack{n=-\infty \\ n \neq 0}}^{\infty} \left\{ \frac{1}{(z-n)} + \frac{1}{(z+n)} \right\} \\
&= \tfrac{1}{2} \sum_{\substack{n=-\infty \\ n \neq 0}}^{\infty} \left[\left\{ \frac{1}{(z-n)} + \frac{1}{n} \right\} + \left\{ \frac{1}{(z+n)} - \frac{1}{n} \right\} \right] \\
&= \sum_{\substack{n=-\infty \\ n \neq 0}}^{\infty} \left\{ \frac{1}{(z-n)} + \frac{1}{n} \right\}.
\end{aligned} \tag{23.75}$$

The last line above is obtained by using the fact that the sum over all nonzero integers remains unchanged if $n \to -n$. Note that the term $1/n$ inside the curly brackets in the summand on the right-hand side is necessary. Without it, the sum (over n) of $1/(z - n)$ would diverge. With the $1/n$ term present, the summand is sufficiently well-behaved to guarantee convergence of the series. We thus obtain

$$\boxed{\pi \cot(\pi z) = \frac{1}{z} + \sum_{\substack{n=-\infty \\ n\neq 0}}^{\infty} \left\{ \frac{1}{(z-n)} + \frac{1}{n} \right\} = \frac{1}{z} + \sum_{\substack{n=-\infty \\ n\neq 0}}^{\infty} \frac{z}{n(z-n)} .} \tag{23.76}$$

This is the correct Mittag-Leffler representation of the meromorphic function $\pi \cot(\pi z)$.

The series on the right-hand side in Eq. (23.76) is sufficiently well-behaved to be differentiated term by term with respect to z. The result is the Mittag-Leffler representation of the function $\operatorname{cosec}^2(\pi z)$:

$$\operatorname{cosec}^2(\pi z) = \frac{1}{(\pi z)^2} + \frac{1}{\pi^2} \sum_{\substack{n=-\infty \\ n\neq 0}}^{\infty} \frac{1}{(z-n)^2} . \tag{23.77}$$

★ **18.** Use Eq. (23.77) to check, one more time, that $\zeta(2) = \frac{1}{6}\pi^2$.
Finally, rewrite Eq. (23.77) as

$$\boxed{\sum_{n=-\infty}^{\infty} \frac{1}{(z-n)^2} = \frac{\pi^2}{\sin^2(\pi z)} .} \tag{23.78}$$

Step back, now, and admire this remarkably beautiful formula!

23.7 Solutions

2. Recall that, in general, a Laurent series is convergent in an annular region. A region of the form $0 < |z-a| < \infty$ is a punctured plane: the points $z = a$ and $z = \infty$ are excluded, and the Laurent series is in (positive as well as negative) integer powers of $(z-a)$.

(a) In the region $0 < |z| < \infty$,

$$\begin{aligned} \left(z + z^{-1}\right)^{100} &= \sum_{n=0}^{100} \binom{100}{n} z^{2n-100} \\ &= \frac{1}{z^{100}} + \frac{100}{z^{98}} + \cdots + \frac{100!}{(50)!)^2} + \cdots + 100\, z^{98} + z^{100} . \end{aligned}$$

(b) In the region $0 < |z| < \infty$, $z^2 e^{1/z} = \sum_{n=1}^{\infty} \frac{1}{(n+2)!\, z^n} + \frac{1}{2} + z + z^2$.

(c) In the region $|z| > 0$, $z^{-1} e^{-1/z^2} = \sum_{n=0}^{\infty} \frac{(-1)^n}{n!\, z^{2n+1}}$.

(d) The exponential factor can be expanded in a power series in z^{-1} for all $|z| > 0$. The factor $(1-z)^{-1}$ can be expanded in a binomial series for $|z| < 1$. Therefore both series are valid in the overlap region $0 < |z| < 1$. Multiply them together and collect terms corresponding to each power of z. The result is

$$e^{1/z}(1-z)^{-1} = e\sum_{n=0}^{\infty} z^n + \sum_{n=1}^{\infty} \frac{c_n}{z^n}, \quad \text{where } c_n = \sum_{k=n}^{\infty} \frac{1}{k!}.$$

(e) Since the Laurent series is in powers of $(z-2)$, write $z^2 = (z-2)^2 + 4(z-2) + 4$, expand the exponential factor in a power series in $(z-2)^{-1}$, and collect terms. The result is

$$z^2 e^{1/(z-2)} = \sum_{n=1}^{\infty} \frac{(4n^2+16n+17)}{(n+2)!\,(z-2)^n} + \frac{17}{2} + 5(z-2) + (z-2)^2.$$

(f) The annular region $1 < |z| < 2$ is centered at $z = 0$. Hence the Laurent series is in powers of z. Resolving the given function into partial fractions,

$$\frac{1}{(z-1)(z-2)^2} = \frac{1}{(z-1)} - \frac{1}{(z-2)} + \frac{1}{(z-2)^2}$$
$$= \frac{1}{z}\left(1-\frac{1}{z}\right)^{-1} + \frac{1}{2}\left(1-\frac{z}{2}\right)^{-1} + \frac{1}{4}\left(1-\frac{z}{2}\right)^{-2}.$$

Note that the first term on the right-hand side can now be expanded in a binomial series for $|z| > 1$, while the other two can be expanded in binomial series for $|z| < 2$. Writing out the corresponding binomial series and collecting terms, we get, for $1 < |z| < 2$,

$$\frac{1}{(z-1)(z-2)^2} = \sum_{n=1}^{\infty} \frac{(-1)^{n-1}}{z^n} + \frac{1}{4}\sum_{n=0}^{\infty} \frac{(n+3)}{2^n} z^n.$$

This Laurent series involves all negative integral powers of z, but you *cannot* conclude from this that the function represented by the series has an essential singularity at $z = 0$. Remember that the series is only valid in $1 < |z| < 2$, and *not* in a neighborhood of $z = 0$. The function $(z-1)^{-1}(z-2)^{-2}$ has poles at $z = 1$ and $z = 2$, which lie on the boundaries of the annular region in which the Laurent series in powers of z converges. ▶

3. (a) Entire function; essential singularity at ∞.
(b) $z = 1$ is a removable singularity. The function given is just the polynomial $1 + z + \cdots + z^{99}$. It is an entire function, with a pole of order 99 at ∞.
(c) Write 2^z as $e^{z\ln 2}$. There is a removable singularity at 0, and an essential singularity at ∞.

(d) It is easy to show that $\coth z = z^{-1} + \frac{1}{3}z + \mathcal{O}(z^3)$ in the neighborhood of the origin. The singularity at $z = 0$ is canceled out when z^{-1} is subtracted from $\coth z$. The function therefore has simple poles at all the points $z = n\pi i$, where $n = \pm 1, \pm 2, \ldots$. The residue is 1 at each of these poles. There is an accumulation point of poles at ∞.
(e) Entire function; essential singularity at ∞.
(f) Removable singularity at 0, essential singularity at ∞.
(g) Simple poles at $z = (2n+1)i\pi$, where $n \in \mathbb{Z}$, i.e., at the points $\pm i\pi, \pm 3i\pi, \ldots$. The residue at each pole is -1. Accumulation point of poles at ∞.
(h) Removable singularity at 0. Simple poles at $z = n\pi$ where $n = \pm 1, \pm 2, \ldots$, with residue $(-1)^n\, n\pi$. Accumulation point of poles at ∞.
(i) Simple pole at the origin, with residue 1. In the vicinity of $z = n\pi$ (where $n = \pm 1, \pm 2, \ldots$), we have $z \operatorname{cosec}^2 z \simeq n\pi(z - n\pi)^{-2} + (z - n\pi)^{-1} + \frac{1}{3}n\pi + \mathcal{O}(z - n\pi)$. Therefore the function has double poles at these points. The residue at each of these poles is 1. The poles accumulate at ∞.
(j) Note that the numerator has a leading behavior $\sim \frac{1}{3}z^3$ as $z \to 0$. Therefore the function given has a simple pole at $z = 0$, with residue equal to $\frac{1}{3}$. It also has an essential singularity at ∞.
(k) Simple poles at $z = 2\pi n i/(\ln 2)$, where $n \in \mathbb{Z}$, with residue $1/(\ln 2)$ at each pole. Accumulation point of poles at ∞.
(l) In the vicinity of the origin in the z-plane, $\operatorname{cosec}(z^2) \simeq z^{-2} + \frac{1}{6}z^2 + \ldots$. Hence the function has a double pole at $z = 0$, with a residue equal to 0. The function has simple poles at all the points $z^2 = \pm\pi, \pm 2\pi, \ldots$. Let n label the positive integers. The residue is $\pm(-1)^n/[2\sqrt{n\pi}]$ at the poles at $z = \pm\sqrt{n\pi}$; and it is $\pm(-1)^n/[2i\sqrt{n\pi}]$ at the poles at $z = \pm i\sqrt{n\pi}$. There is an accumulation point of poles at ∞.
(m) Simple poles at the points where $\tan z = 1$, i.e., $z = \left(n + \frac{1}{4}\right)\pi,\ n \in \mathbb{Z}$, with residue $(-1)^n/\sqrt{2}$. Accumulation point of poles at ∞.
(n) The singularities of the function occur at the roots of the equation $\cosh z = \cos z$. Observe that, if α is a root of this equation, so are $-\alpha$, $i\alpha$, and $-i\alpha$. As you know, $\cosh x \geq 1$ while $\cos x \leq 1$ for all real values of the argument x. It is therefore clear that 0 is the only real root of the equation. It follows that there are no purely imaginary roots, either. A little more work is needed to show that there are no roots other than 0. Since $(\cosh z - \cos z)^{-1} \simeq z^{-2} + \frac{1}{360}z^2 + \ldots$ near $z = 0$, it follows that 0 is a double pole of the function, with residue equal to 0. There is, of course, an essential singularity at ∞.
(o) Simple poles at the points $z = 2\pi n i/\ln(3/2)$, where $n \in \mathbb{Z}$. Residue given by $\left(\ln \frac{3}{2}\right)^{-1} \exp\left\{-2\pi n i\,(\ln 2)/\left(\ln \frac{3}{2}\right)\right\}$. Accumulation point of poles at ∞.
(p) In the vicinity of the origin, we have $(e^{z^2} - 1)^{-1} \simeq z^{-2} - \frac{1}{2} + \mathcal{O}(z^2)$. Hence the function has a double pole at $z = 0$ with residue equal to 0. It also has simple poles at all the roots of $z^2 = 2\pi n i$, where $n = \pm 1, \pm 2, \ldots$. That is, at the points

$$(1+i)\sqrt{\pi n},\ -(1+i)\sqrt{\pi n},\ (1-i)\sqrt{\pi n} \text{ and } -(1-i)\sqrt{\pi n}, \text{ where } n = 1, 2, \ldots.$$

The corresponding residues are given by

$$\frac{(1-i)}{4\sqrt{\pi n}}, \quad -\frac{(1-i)}{4\sqrt{\pi n}}, \quad \frac{(1+i)}{4\sqrt{\pi n}}, \quad \text{and} \quad -\frac{(1-i)}{4\sqrt{\pi n}}.$$

There is an accumulation point of poles at ∞.
(q) Simple poles at $z = 2\pi n i$ where $n \in \mathbb{Z}$, with residue equal to 1 at each pole. Accumulation point of poles at ∞.
(r) The ratio test shows that the power series converges absolutely for all finite values of $|z|$. Hence the function represented by the series is an entire function. The order is easily found to be $\rho = 1$, using Eq. (22.46) of Chap. 22, Sect. 22.6. The function has an essential singularity at ∞.
(s) Again, as we have seen in Chap. 22, Sect. 22.6, this is an entire function of order $\rho = 2$. It has an essential singularity at ∞.
(t) As we saw in Chap. 22, Sect. 22.5.1, this is a geometric series that converges absolutely in the region $|1 - z| < |1 + z|$, i.e., in the right-half-plane $x > 0$. There can be no singularity in this region. Staying within this region, the series is easily summed to give the closed-form expression $(1 + z)/(2z)$. In this form, the function may be continued analytically to the rest of the complex plane. It is evident that the function has a simple pole at $z = 0$, with a residue equal to $\frac{1}{2}$.
(u) Write the function as

$$f(z) = \exp\left(\frac{2}{z^2-1}\right) = \exp\left(\frac{1}{z-1} - \frac{1}{z+1}\right) = \exp\left(\frac{1}{z-1}\right)\exp\left(\frac{-1}{z+1}\right).$$

The function therefore has two essential singularities, and $z = -1$ and $z = 1$, respectively. In finding the corresponding residues, you have to be a little careful. Consider the singularity at $z = 1$. The first factor has the Laurent series expansion

$$\exp\left(\frac{1}{z-1}\right) = 1 + \sum_{n=1}^{\infty} \frac{1}{n!\,(z-1)^n}, \quad |z-1| > 0.$$

The second factor is analytic at $z = 1$, and therefore can be expanded in a Taylor series about that point. The residue of $f(z)$ at $z = 1$ is the coefficient of $(z - 1)^{-1}$ in the product of the two factors. It is clear that there is a contribution from every term of the Taylor series. We find

$$\operatorname*{Res}_{z=1} f(z) = \sum_{n=1}^{\infty} \frac{1}{n!\,(n-1)!}\left[\frac{d^{n-1}}{dz^{n-1}}\exp\left(\frac{-1}{z+1}\right)\right]_{z=1}.$$

Similarly, we get

$$\operatorname*{Res}_{z=-1} f(z) = \sum_{n=1}^{\infty} \frac{(-1)^n}{n!\,(n-1)!}\left[\frac{d^{n-1}}{dz^{n-1}}\exp\left(\frac{1}{z-1}\right)\right]_{z=-1}$$

for the residue of $f(z)$ at $z = -1$. ▶

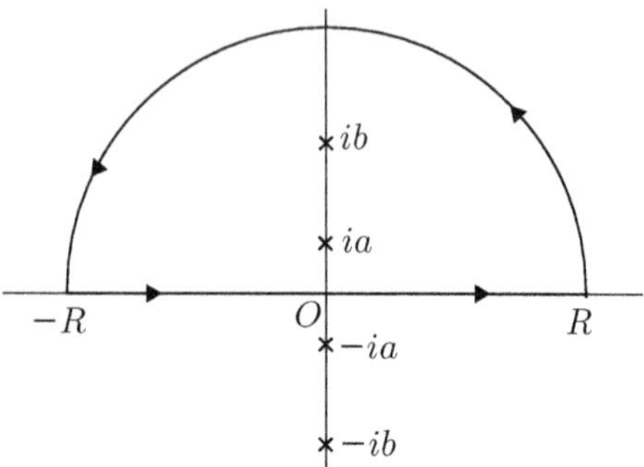

Fig. 23.12 Contour for Problem 4(a)

4. (a) Consider the integral from $-R$ to R on the x-axis, and attach a semicircle of radius R in the upper half-plane, to close the contour. Now consider the integral of the analytic function $[(z^2 + a^2)(z^2 + b^2)]^{-1}$ over the closed contour C (see Fig. 23.12). The function has simple poles at $z = ia$ and $z = ib$ in the upper half-plane, and at $z = -ia$ and $z = -ib$ in the lower half-plane. By Cauchy's theorem, this integral is $2\pi i$ times the sum of the residues of the function at the poles enclosed by C. Let $R \to \infty$. The crucial point is that, in this limit, *the contribution from the semicircle vanishes, while the integral along the x-axis becomes the integral required to be evaluated.*

Remark Check out that the same answer is obtained if you close the contour with a semicircle in the *lower* half-plane. The contour integral is then equal to *minus* $2\pi i$ times the sum of the residues at the poles at $z = -ia$ and $z = -ib$.

(b) Write the integral as $\frac{1}{2}$ times the integral from $-\infty$ to ∞ along the x-axis, and proceed as in the preceding example. Remember that the pole enclosed by C is a pole of order 3. You must use the formula (23.11) of Sect. 23.2.2 to find the residue at the pole.

(c) In this case, if we consider the integrand as it stands, the contour cannot be closed with a semicircle in either the upper or the lower half-plane, because $\sin z$ is a linear combination of e^{iz} and e^{-iz}. If we let $|z| \to \infty$ in the lower half-plane, the term e^{iz} diverges; on the other hand, the term e^{-iz} diverges if $|z| \to \infty$ in the upper half-plane. You can easily get around this difficulty by writing $\sin z = (e^{iz} - e^{-iz})/(2i)$ and splitting the integral into two integrals. Close the contour in the upper half-plane in the case of e^{iz} and in the lower half-plane in the case of e^{-iz}, and pick up the residue at the enclosed pole in each case.

Alternatively, you can first write $\sin x$ as the imaginary part of e^{ix}, and then consider the integral $\int_{-\infty}^{\infty} dx\, x\, e^{ix}/(x^2 + a^2)$. The contour can now be closed in the upper half-plane (and not in the lower half-plane). After you evaluate the integral, take its imaginary part to get the value of the original integral.

(d) Consider the rational function $f(z) = (z^2 - z + 2)/(z^4 + 10z^2 + 9)$. The poles of the denominator are at $z = \pm i$ and $z = \pm 3i$. Further, $f(z) \to 0$ as $|z| \to \infty$ like $1/z^2$, i.e., more rapidly than $1/z$. You can therefore close the contour with a large

semicircle in either the upper or the lower half-plane, and pick up the contributions from the poles enclosed.
(e) Let $z = e^{i\theta}$. Then $\cos\theta = \frac{1}{2}(z + z^{-1})$. It is obvious that the integration over θ from 0 to 2π now becomes an integral over z around the unit circle in the complex plane, traversed once in the positive sense. Since $dz = e^{i\theta}\, i\, d\theta$, or $d\theta = dz/(iz)$, we get

$$\int_0^{2\pi} \frac{d\theta}{a - b\cos\theta} = \frac{2i}{b} \oint_{|z|=1} \frac{dz}{(z-\alpha)(z-\beta)},$$

where α and β are the roots of $z^2 - (2a/b)z + 1 = 0$, i.e.,

$$\alpha,\ \beta = (a \pm \sqrt{a^2 - b^2})/b.$$

It is easily checked that the pole at β lies inside the unit circle, while the pole at α lies outside it. Using the residue theorem, the value of the integral is $(2i/b)(2\pi i)/(\beta - \alpha) = 2\pi/\sqrt{a^2 - b^2}$.
(f), (g) The integrals are the real and imaginary parts, respectively, of the integral

$$\int_0^{2\pi} d\theta\, e^{a\cos\theta}\, e^{i(a\sin\theta - n\theta)} = \int_0^{2\pi} d\theta\, e^{a\, e^{i\theta}}\, e^{-ni\theta} = -i \oint \frac{dz\, e^{az}}{z^{n+1}},$$

on setting $e^{i\theta} = z$ to convert the integral to a contour integral around the unit circle in the z-plane. Expand the factor e^{az} in powers of z, and use Eq. (23.21). Since the integral turns out to be a real quantity, its imaginary part vanishes identically.
(h), (i) Proceeding exactly as in (f) and (g) above, the integrals are the real and imaginary parts, respectively, of

$$-i \oint \frac{dz\, e^{ia/z}}{z^{1-n}}.$$

Expand $e^{ia/z}$ and use Eq. (23.21). The integral becomes equal to $2\pi(ia)^n/n!$. Since

$$i^n = e^{i\pi n/2} = \cos\tfrac{1}{2}n\pi + i\sin\tfrac{1}{2}n\pi,$$

the results quoted follow at once. ▶

5. (a) The contribution from s is now $i\int_\pi^{2\pi} d\theta = i\pi$. But C now encircles the simple pole of the integrand e^{ibz}/z at $z = 0$. The residue at the pole is unity, so that $\int_C dz\, e^{ibz}/z = 2\pi i$.
(b) Since the large semicircle now lies in the *lower* half-plane, you must consider the integrand e^{-ibz}/z, so that the contribution from S vanishes in the limit $R \to \infty$. The contour integral over C is zero, since no singularity of the integrand is enclosed by C. On s, the angular variable θ runs from π to 2π.

(c) Once again, the integrand e^{-ibz}/z must be considered. The pole at the origin is now encircled once in the *negative* sense by C. Hence $\int_C dz\, e^{-ibz}/z = -2\pi i$. On s, θ runs from π to 0. ▶

6. The integral required is just $(-2\pi i)\ \operatorname*{Res}_{z=\infty}\, [p_n(z)/q_{n+1}(z)]$, and is equal to

$$\oint_c \frac{dw}{w}\,\frac{(a_n + a_{n-1}w + \cdots + a_0 w^n)}{(b_{n+1} + b_n w + \cdots + b_0 w^{n+1})},$$

where c encircles only the singularity at the origin, in the positive sense. ▶

12. Multiply both sides of the recursion relation $c_{n+1} = ac_n + b$ by z^n, and sum over n from zero to infinity. A little simplification gives

$$f(z) = \frac{c_0 + (b - c_0)z}{(1 - az)(1 - z)}.$$

Resolve the denominator into partial fractions, and use the binomial theorem to expand $(1 - az)^{-1}$ and $(1 - z)^{-1}$ in powers of z. Pick out the coefficient of z^n in the power series expansion of $f(z)$ to obtain the result quoted for c_n. ▶

14. In all three cases, the generating function turns out to be so simple that a binomial expansion suffices to identify c_n.
(a) The generating function, for arbitrary values of c_0 and c_1, is easily found to be

$$f(z) = \frac{c_0 + (c_1 - 2c_0)z}{(1 - z)^2}.$$

From the binomial expansion $(1 - z)^{-2} = 1 + 2z + 3z^2 + \ldots$, it follows that $c_n = c_0 + n(c_1 - c_0)$. For the choice $c_0 = 1,\ c_1 = 1$, we have $c_n = 1$. If $c_0 = 0$ and $c_1 = 1$, then $c_n = n$; and so on.

Remark We could have guessed that c_n would turn out to be a linear function of n in this case, because each coefficient is just the arithmetic average of its neighbors on either side: the recursion relation is $c_{n+1} = \frac{1}{2}(c_{n+2} + c_n)$.

(b) The generating function in this case is just $f(z) = (1 - 2z)^{-1}$. Hence a binomial expansion immediately yields $c_n = 2^n$.
(c) Unlike case (a) above, in the present instance c_n is equal to one-half times the *difference* of the numbers on either side of it, i.e., $c_{n+1} = \frac{1}{2}(c_{n+2} - c_n)$. The generating function is easily found to be $f(z) = z/(1 - 2z - z^2)$. Resolving it into partial fractions and using the binomial theorem, we find the following solution to the recursion relation:

$$c_n = \frac{1}{2\sqrt{2}}\left[(\sqrt{2} + 1)^n - (-1)^n(\sqrt{2} - 1)^n\right], \qquad n = 0,\ 1,\ \ldots.$$

Note the similarity of the structure of this expression with that for the Hemachandra-Fibonacci sequence, Eq. (23.52). The sequence of numbers in the present instance is $0, 1, 2, 5, 12, 29, 70, \ldots$. This is the sequence of **Pell numbers**, which also has numerous interesting and important mathematical properties. ▶

17. Expand e^{ux} in powers of ux and integrate term by term. Use the result in Eq. (23.66). Identify the sum obtained with a Bessel function, using the definition of the latter given in Eq. (19.45) of Chap. 19, Sect. 19.2.5. ▶

18. Move the term $1/(\pi z)^2$ to the left-hand side, and let $z \to 0$. ▶

Chapter 24
Linear Response and Analyticity

In very broad terms, all our knowledge of physical systems comes, ultimately, from an analysis of how these systems respond to various kinds of stimuli. This response is often characterized by so-called response functions and their associated generalized susceptibilities. A notable physical application of the analyticity properties of functions of complex variables is the derivation of dispersion relations for these generalized susceptibilities. This is the theme of this chapter.

24.1 The Dynamic Susceptibility

24.1.1 Linear, Causal, Retarded Response

Under very general conditions, the physical response of a system to an applied time-dependent stimulus displays certain basic features. These are not strict requirements or rigorously provable properties. Rather, they are assumptions that are quite plausible on physical grounds, and are valid in a wide variety of cases. Let us denote the stimulus (or "force") by $F(t)$, and the response by $R(t)$. Some examples of such stimulus-and-response pairs are: mechanical force and displacement; electric field and polarization; magnetic field and magnetization; stress and strain; electromotive force and current; and so on. For notational simplicity, I have suppressed the appropriate indices in the stimulus and the response when these are vectors, tensors, etc. The basic properties assumed about $R(t)$ are then as follows:

(i) The response is *linear* in the applied force. This implies at once that the **superposition principle** is applicable.
(ii) The response at any instant of time t depends on the force history at all *earlier* instants of time, but *not* later ones. This is the **principle of causality**, which says that *an effect cannot precede its cause*.

V. Balakrishnan, *Mathematical Physics*,
https://doi.org/10.1007/978-3-030-39680-0_24

(iii) The effect of a force $F(t')$ applied at the instant t' on the response $R(t)$ at a later instant t depends only on the *elapsed* time interval $(t - t')$, rather than on both t and t' individually. This is called **retarded response**.

Retarded response implies that the choice of the actual origin of the time coordinate, i.e., the particular instant at which we set $t = 0$, does not matter.

We are therefore dealing with the case of *linear, causal, retarded response*. Let us take the force to be applied from $t = -\infty$ onward: this makes it easy to specialize to all other starting times. The most general linear functional of $F(t)$ subject to the requirements (i)–(iii) listed above is given by

$$R(t) = \int_{-\infty}^{t} dt'\, \phi(t - t')\, F(t'). \tag{24.1}$$

Note that t is the upper limit of the integration over t', in accordance with the requirement of causality. The quantity $\phi(t - t')$ is called the **response function**. The fact that it is a function of the time difference $(t - t')$ indicates that we are dealing with a retarded response. $\phi(t - t')$ represents the weight with which a force applied at time t' contributes to the response $R(t)$ at any later time t. We may expect it to be a decreasing (or at least a nonincreasing) function of its argument, such as a decaying exponential. But it could oscillate as a function of $(t - t')$, although we would expect it to do so with decreasing amplitude, in most instances. I reiterate that these are not strict requirements, but features that are plausible on physical grounds.

24.1.2 Frequency-Dependent Response

It is natural to decompose general time-dependent functions such as $F(t)$ and $R(t)$ into their Fourier (or frequency) components. Thus

$$F(t) = \frac{1}{2\pi}\int_{-\infty}^{\infty} d\omega\, e^{-i\omega t}\, \widetilde{F}(\omega) \quad \text{and} \quad R(t) = \frac{1}{2\pi}\int_{-\infty}^{\infty} d\omega\, e^{-i\omega t}\, \widetilde{R}(\omega). \tag{24.2}$$

The inverse transforms are then given by

$$\widetilde{F}(\omega) = \int_{-\infty}^{\infty} dt\, e^{i\omega t}\, F(t) \quad \text{and} \quad \widetilde{R}(\omega) = \int_{-\infty}^{\infty} dt\, e^{i\omega t}\, R(t). \tag{24.3}$$

***Digression*:** A remark is in order on the Fourier transform convention adopted here. In Chap. 18, Sect. 18.1.1, a Fourier transform pair was defined according to

$$f(x) = \frac{1}{2\pi}\int_{-\infty}^{\infty} dk\, e^{+ikx}\, \widetilde{f}(k) \quad \text{and} \quad \widetilde{f}(k) = \int_{-\infty}^{\infty} dx\, e^{-ikx}\, f(x). \tag{24.4}$$

In the present instance, t replaces x, while ω replaces k. The sign convention in Eqs. (24.2) and (24.3) is opposite to that adopted for $f(x)$ and $\widetilde{f}(k)$: there is a difference of sign in the exponents between the kernels $e^{-i\omega t}$ and e^{+ikx}. This deliberate choice is a matter of convenience, and is not serious. But there is a reason for it. We shall usually associate x with a spatial variable, and t with the time. When it is exponentiated, the phase factor $(kx - \omega t)$ (respectively, $(\mathbf{k} \cdot \mathbf{r} - \omega t)$ in more than one spatial dimension) represents a plane wave propagating along the positive x-axis (respectively, along the wave vector $\mathbf{k}$). I shall retain this form for the basis functions when defining Fourier transforms of functions that depend on the spatial coordinates as well as time—for instance, when discussing the solution of the wave equation in Chap. 31, Sect. 31.1.1.

Let us return to Eq. (24.1) Substituting Eq. (24.2) in it and simplifying, we find

$$\widetilde{R}(\omega) = \chi(\omega)\, \widetilde{F}(\omega), \tag{24.5}$$

where

$$\boxed{\chi(\omega) \stackrel{\text{def.}}{=} \int_0^\infty dt\, e^{i\omega t}\, \phi(t).} \tag{24.6}$$

★ **1.** Establish Eqs. (24.5) and (24.6). ▶

The function $\chi(\omega)$ is called the **dynamic susceptibility** (or the frequency-dependent susceptibility, or the **generalized susceptibility**) corresponding to this particular stimulus-and-response pair. It measures the response of the system to a sinusoidally varying "force" of unit amplitude and frequency ω. Note that $\chi(\omega)$ is a complex-valued function (even when its argument ω is real). It will become clear, from the examples to be considered below, that both its real and imaginary parts are of physical significance.

Observe that $\chi(\omega)$ is *not* the Fourier transform of $\phi(t)$. The lower limit of integration in its definition (Eq. (24.6)) is $t = 0$ rather than $-\infty$. Going back a step, we see that this lower limit is a direct consequence of the fact that t is the upper limit of integration in the basic expression for $R(t)$. In turn, this follows directly from causality.

- The dynamic susceptibility is the *one-sided* Fourier transform (sometimes also called the **Fourier-Laplace transform**) of the response function $\phi(t)$, because of causality.

Physical examples of the dynamic susceptibility include the frequency-dependent (or AC) susceptibility of a magnetic substance, the electric polarizability of a dielectric material, the dynamic mobility of a fluid, the elastic compliances (inverses of the elastic moduli) of a solid, and so on.

24.1.3 Symmetry Properties of the Dynamic Susceptibility

An important property of the dynamic susceptibility can be extracted from its definition, Eq. (24.6). Since $\phi(t)$ is a real quantity, we find by inspection that

$$\chi(-\omega) = \chi^*(\omega) \quad \text{for all } \textit{real} \text{values of } \omega, \tag{24.7}$$

where the asterisk denotes the complex conjugate as usual. I have added the phrase "for all *real* values of ω" for the following reason. As you will see shortly, it makes sense to consider $\chi(\omega)$ as an analytic function of the *complex* variable ω, although the physically accessible values of the frequency are of course real and nonnegative. It follows at once from Eq. (24.7) that

$$\boxed{\text{Re } \chi(-\omega) = \text{Re } \chi(\omega) \quad \text{and} \quad \text{Im } \chi(-\omega) = -\text{Im } \chi(\omega) \quad (\omega \text{ real}).} \tag{24.8}$$

That is, the *real* part of the dynamic susceptibility is an *even* function of the frequency, while the *imaginary* part is an *odd* function of the frequency. You can, of course, infer these properties directly by writing $e^{i\omega t} = (\cos \omega t + i \sin \omega t)$ in the definition of $\chi(\omega)$.

Now consider the formula (24.6) for $\chi(\omega)$ as the one-sided Fourier transform of $\phi(t)$. We have assumed tacitly that the integral exists for real values of the frequency ω. It is obvious that it will continue to do so even if ω is a complex variable, provided $\text{Im } \omega$ is *positive*. This is because the factor $e^{i\omega t}$ then provides an extra damping factor $e^{-(\text{Im } \omega) t}$ in the integrand. Such a factor can only *improve* the convergence of the integral, since the integration over t is restricted to nonnegative values. It follows that:

(i) The dynamic susceptibility $\chi(\omega)$ can be *continued analytically* to the upper half-plane in ω, and it is holomorphic in that region.
(ii) The defining expression (24.6) itself provides a representation of $\chi(\omega)$ for all $\text{Im } \omega \geq 0$.

More generally,

- causality ensures that the Fourier-Laplace transform of a causal response function is holomorphic in a certain half-plane in ω.

Whether this is the upper or lower half-plane depends on the Fourier transform convention chosen. With our particular convention in which a function $f(t)$ of the time is expanded as in (24.2), it is the *upper* half of the ω-plane in which the susceptibility is analytic. Had we chosen the opposite sign convention and used the kernel $e^{+i\omega t}$ in the Fourier expansion of $f(t)$, the region of analyticity in ω would have been the *lower* half-plane. The convention I have adopted is the more commonly used one, at least in physics.

Given that the dynamic susceptibility is analytic in the upper half-plane in ω, we can extend the relation $\chi(-\omega) = \chi^*(\omega)$ (that holds good for real ω) to complex values of ω. By inspection, we find that

$$\boxed{\chi(-\omega^*) = \chi^*(\omega), \quad \text{Im } \omega \geq 0.} \tag{24.9}$$

Note that, if ω lies in the upper half of the complex plane, so does $-\omega^*$. Hence the arguments of the functions on both sides of Eq. (24.9) do lie in the region in which we are guaranteed that the dynamic susceptibility is holomorphic. On general grounds, and without further input, we really cannot say very much about the possible behavior of $\chi(\omega)$ in the lower half of the complex ω-plane. But we know that an analytic function of a complex variable cannot be holomorphic at *all* points of the extended complex plane $\hat{\mathbb{C}}$, unless it is (trivially) a constant—recall Liouville's Theorem (Chap. 22, Sect. 22.3.1). In general, therefore,

- $\chi(\omega)$ will have singularities in the *lower* half-plane in ω.

24.2 Dispersion Relations

24.2.1 Derivation of the Relations

The analyticity of $\chi(\omega)$ in the upper half-plane in ω enables us to derive certain important relations between its real and imaginary parts called **dispersion relations.** (The origin of the term will be explained below.)

Let us assume that $|\chi(\omega)| \to 0$ as $\omega \to \infty$ along any direction in the upper half-plane. This assumption is physically plausible in most circumstances. The reason is as follows. Even for *real* ω, we expect the susceptibility to vanish as the frequency becomes very large because the *inertia* present in any system will not permit it to respond significantly to a sinusoidal applied force oscillating at a frequency much higher than all the natural frequencies present in the system. And when ω is a complex number with a positive imaginary part, the factor $e^{i\omega t}$ in the formula for $\chi(\omega)$ lends an *extra* damping factor $e^{-(\text{Im }\omega)\,t}$. This would make $\chi(\omega)$ decay to zero even more rapidly as $|\omega| \to \infty$. The assumption is thus a reasonable one to make. But it is not absolutely essential. It can be relaxed, as we shall see in Sect. 24.2.3.

Let ω be a fixed, real, positive frequency. Consider the quantity

$$f(\omega') = \frac{\chi(\omega')}{\omega' - \omega} \tag{24.10}$$

as a function of the complex variable ω'. This function is analytic everywhere in the upper half-plane in ω', as well on the real axis in that variable, except for a simple pole at $\omega' = \omega$ located on the real axis. By Cauchy's integral theorem, its integral over any closed contour C' lying entirely in the upper half-plane is identically equal to zero. The contour C' can be expanded to the contour C (see Fig. 24.1) without changing the value of the integral, namely, zero. Thus

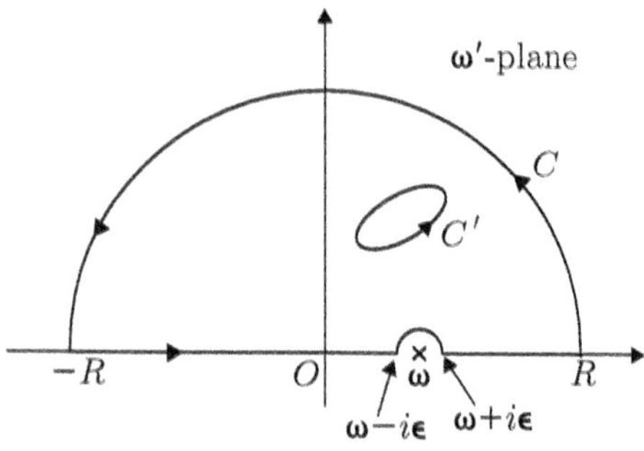

Fig. 24.1 The contours C' and C

$$\oint_{C'} d\omega' \, f(\omega') = \oint_C d\omega' \, \frac{\chi(\omega')}{\omega' - \omega} = 0. \tag{24.11}$$

The closed contour C is made up of the following:

(i) a large semicircle of radius R in the upper half-plane;
(ii) a line integral on the real axis running from $-R$ to $\omega - \epsilon$;
(iii) a small semicircle, from $\omega - \epsilon$ to $\omega + \epsilon$, lying in the upper half-plane so as to avoid the simple pole of the integrand; and, finally,
(iv) a line integral on the real axis from $\omega + \epsilon$ to R.

Now, since $\chi(\omega') \to 0$ as $|\omega'| \to \infty$ by assumption, the function $f(\omega') \to 0$ more rapidly than $1/\omega'$ as $\omega' \to \infty$ along all directions in the upper half-plane. The contribution from the large semicircle, therefore, vanishes in the limit $R \to \infty$. On the small semicircle, we have $\omega' = \omega + \epsilon \, e^{i\theta}$, where θ runs from π to 0. In the limit $\epsilon = 0$, the quantity $\chi(\omega')$ in the integrand can be replaced by $\chi(\omega)$. Hence the contribution from the small semicircle is $-i\pi\chi(\omega)$. We thus get

$$\lim_{\epsilon \to 0} \left\{ \int_{-\infty}^{\omega-\epsilon} d\omega' \, \frac{\chi(\omega')}{\omega' - \omega} + \int_{\omega+\epsilon}^{\infty} d\omega' \, \frac{\chi(\omega')}{\omega' - \omega} \right\} - i\pi\chi(\omega) = 0. \tag{24.12}$$

But the $\epsilon = 0$ limit of the expression in curly brackets on the left-hand side of this equation is precisely the *Cauchy principal value integral* of the integrand $\chi(\omega')/(\omega' - \omega)$. This concept has already been encountered in Chap. 23, Sect. 23.3.3, when we evaluated the Dirichlet integral. Recall that the Cauchy principal value is a specific prescription for avoiding the divergence or infinity that would otherwise arise—in the present instance, owing to the singularity (a simple pole) of the integrand at $\omega' = \omega$. It amounts to omitting, from the range of integration, an infinitesimal interval that is *symmetrically* located on either side of the singularity. Denoting this principal value integral by $\mathrm{P} \int (\cdots)$ as before, Eq. (24.12) reduces to

$$\chi(\omega) = -\frac{i}{\pi} \, \mathrm{P} \int_{-\infty}^{\infty} d\omega' \, \frac{\chi(\omega')}{\omega' - \omega}. \tag{24.13}$$

Note that this equation expresses the susceptibility at any *real* frequency ω as a certain weighted sum of the susceptibility over all other *real* frequencies ω'. No complex frequencies appear anywhere in this formula. We made an excursion into the (upper half of the) complex frequency plane. This was made possible by the analyticity properties of the susceptibility. But we have returned to the real axis, bringing Eq. (24.13) back. Equating the respective real and imaginary parts of the two sides of this equation, we get

$$\left.\begin{aligned} \text{Re}\ \chi(\omega) &= \frac{1}{\pi}\,\text{P}\int_{-\infty}^{\infty} d\omega'\,\frac{\text{Im}\ \chi(\omega')}{\omega'-\omega}, \\ \text{Im}\ \chi(\omega) &= -\frac{1}{\pi}\,\text{P}\int_{-\infty}^{\infty} d\omega'\,\frac{\text{Re}\ \chi(\omega')}{\omega'-\omega}. \end{aligned}\right\} \tag{24.14}$$

These formulas are the dispersion relations we seek. They imply that the two real functions of a real argument, Re $\chi(\omega)$ and Im $\chi(\omega)$, form what is known as a **Hilbert transform pair**. I will return to this aspect in Sect. 24.2.4.

Recall (from Chap. 22, Sect. 22.3.2) that a *real analytic function* of the complex variable $z = x + iy$ is an analytic function that is real whenever z itself is real, i.e., for all $y = 0$. It should now be clear that the dynamic susceptibility $\chi(\omega)$ *cannot*, in general, be a real analytic function of ω. For, if that were so, we must have Im $\chi(\omega) \equiv 0$ for all real ω. Hence, from the first of Eq. (24.14), Re $\chi(\omega) \equiv 0$. Thus $\chi(\omega)$ itself would vanish identically.

The dispersions relations (24.14) still involve integrals over negative as well as positive frequencies, whereas physically accessible frequencies are nonnegative. But the symmetry properties of Re $\chi(\omega)$ and Im $\chi(\omega)$ in Eq. (24.8) can be used to restrict the range of integration to physically accessible frequencies, namely, $0 \le \omega' < \infty$. We then find that

$$\left.\begin{aligned} \text{Re}\ \chi(\omega) &= \frac{2}{\pi}\,\text{P}\int_{0}^{\infty} d\omega'\,\frac{\omega'\,\text{Im}\ \chi(\omega')}{\omega'^2-\omega^2}, \\ \text{Im}\ \chi(\omega) &= -\frac{2\omega}{\pi}\,\text{P}\int_{0}^{\infty} d\omega'\,\frac{\text{Re}\ \chi(\omega')}{\omega'^2-\omega^2}. \end{aligned}\right\} \tag{24.15}$$

★ **2.** Use Eqs. (24.8) to reduce the dispersions relations (24.14) to the form given in Eqs. (24.15).

The physical significance of **Re $\chi(\omega)$ and Im $\chi(\omega)$:** The term "dispersion relation" originates from the fact that these relations were first derived in the context of optics. The real and imaginary parts of the frequency-dependent refractive index of an optical medium are, respectively, measures of the **dispersion** and **absorption** of radiation by the medium. These are not independent quantities, but are related to each other via dispersion relations, which are also called **Kramers–Kronig relations** in physics. I reiterate that they are a direct consequence of the principle of causality as applied to response functions.

More generally, the real and imaginary parts of any dynamic susceptibility represent the *dispersive* or *reactive* part of the response, and the *absorptive* or *dissipative* part of the response. Although in many cases (including the optical one just mentioned) Re $\chi(\omega)$ represents the reactive part and Im $\chi(\omega)$ the dissipative part, the reverse could also be true in some instances. Which alternative is chosen turns out to depend on the time-reversal properties of the particular response function involved. In the example that follows below, it is Re $\chi(\omega)$ that represents the dissipative part, while Im $\chi(\omega)$ represents the reactive part.

24.2.2 Complex Admittance of an LCR Circuit

Here is an elementary illustration of a dynamic susceptibility and its properties. Consider an LCR series circuit (Fig. 24.2). If the applied EMF is $V(t)$, the instantaneous current $I(t)$ in the circuit is given by the equation

$$L\frac{dI}{dt} + RI + \frac{1}{C}\int_{-\infty}^{t} I(t')\,dt' = V(t). \tag{24.16}$$

Now insert the Fourier expansions

$$V(t) = \frac{1}{2\pi}\int_{-\infty}^{\infty} d\omega\, e^{-i\omega t}\,\widetilde{V}(\omega) \quad\text{and}\quad I(t) = \frac{1}{2\pi}\int_{-\infty}^{\infty} d\omega\, e^{-i\omega t}\,\widetilde{I}(\omega). \tag{24.17}$$

in Eq. (24.16). The last term on the left-hand side becomes (after a change of integration variable from t' to $\tau = t - t'$)

$$\frac{1}{2\pi C}\int_{-\infty}^{\infty} d\omega\, e^{-i\omega t}\,\widetilde{I}(\omega)\int_{0}^{\infty} d\tau\, e^{i\omega\tau}. \tag{24.18}$$

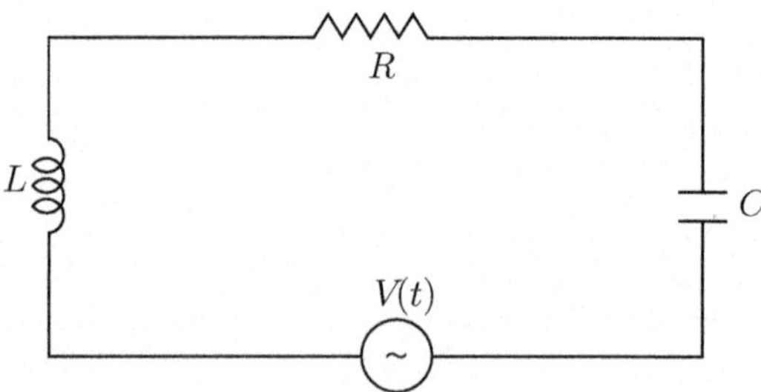

Fig. 24.2 A series LCR circuit

The problem is that the integral over τ diverges, i.e., it is formally infinite. We should really have used Laplace transforms[1] instead of Fourier transforms to tackle this problem! Had we done so, the integral $\int_0^\infty d\tau\, e^{i\omega\tau}$ would have been replaced by $\int_0^\infty d\tau\, e^{-s\tau}$, with the variable s lying in a region such that the convergence of the integral is guaranteed. In this case this region is simply $\text{Re}\, s > 0$, and the value of the integral is just $1/s$. Let us suppose that this has been done, so that we may interpret $\int_0^\infty d\tau\, e^{i\omega\tau}$ as standing for the *analytic continuation* to $s = -i\omega$ of the corresponding Laplace transform. This gives the value $1/(-i\omega) = i/\omega$ for the apparently divergent integral. With this small technicality taken care of, the Fourier transforms of the current and the voltage are related to each other according to

$$\Big(- i\omega L + R + \frac{i}{\omega C}\Big)\widetilde{I}(\omega) = \widetilde{V}(\omega). \tag{24.19}$$

The quantity in the brackets on the left-hand side is of course the **complex impedance** of the circuit, usually denoted by $Z(\omega)$. The corresponding dynamic susceptibility is the reciprocal of the impedance, namely, the **complex admittance**, customarily denoted by $Y(\omega)$. We have

$$\widetilde{I}(\omega) = Y(\omega)\, \widetilde{V}(\omega), \tag{24.20}$$

where

$$Y(\omega) = \Big(- i\omega L + R + \frac{i}{\omega C}\Big)^{-1} = \frac{i\omega}{L\,(\omega^2 + i\gamma\,\omega - \omega_0^2)}. \tag{24.21}$$

Here

$$\gamma = R/L \quad \text{and} \quad \omega_0 = 1/\sqrt{LC}. \tag{24.22}$$

You will recognize $\gamma^{-1} = L/R$ as the time constant of an LR circuit. This is the time scale characterizing the dissipation in the system, arising from the resistance in the circuit. ω_0 is, of course, the natural frequency of an undamped LC circuit (the reactive part of the circuit). Observe that $Y(\omega)$ is analytic in the upper half-plane in ω, as required of a dynamic susceptibility. In this elementary example we know its form explicitly. It has two simple poles in the lower half-plane, at

$$\omega_\pm = -\tfrac{1}{2}i\gamma \pm \omega_{\text{u}}\,, \quad \text{where} \quad \omega_{\text{u}} = \sqrt{\omega_0^2 - \tfrac{1}{4}\gamma^2}\,. \tag{24.23}$$

(The subscript in ω_{u} is meant to remind you that ω_{u} is a real frequency in the **u**nderdamped case $\omega_0 > \frac{1}{2}\gamma$.) Further, $|Y(\omega)|$ vanishes like $1/|\omega|$ as $|\omega| \to \infty$ in the upper half-plane. The real and imaginary parts of the admittance are given by

[1]Laplace transforms will be discussed in Chap. 28.

$$\left.\begin{aligned} \text{Re } Y(\omega) &= \frac{\gamma\omega^2}{L[(\omega^2-\omega_0^2)^2+\gamma^2\omega^2]}, \\ \text{Im } Y(\omega) &= \frac{\omega(\omega^2-\omega_0^2)}{L[(\omega^2-\omega_0^2)^2+\gamma^2\omega^2]}. \end{aligned}\right\} \tag{24.24}$$

These quantities must satisfy the dispersion relations given in Eq. (24.15).

★ **3.** Here is a simple but instructive exercise. Let us fix ω_0 at some (positive) value. In the complex ω-plane, indicate the locations of the poles of $Y(\omega)$ at ω_+ and ω_-, starting with the underdamped case in which $\omega_0 > \frac{1}{2}\gamma$. Track how these poles move as you increase γ through the critically damped case ($\omega_0 = \frac{1}{2}\gamma$) to the overdamped situation ($\omega_0 < \frac{1}{2}\gamma$). What happens as $\gamma \to \infty$?

★ **4.** Verify that the functions Re $Y(\omega)$ and Im $Y(\omega)$ given in Eq. (24.24) satisfy the dispersion relations

$$\text{Re } Y(\omega) = \frac{2}{\pi}\,\text{P}\int_0^\infty d\omega'\,\frac{\omega'\,\text{Im } Y(\omega')}{\omega'^2-\omega^2},$$

$$\text{Im } Y(\omega) = -\frac{2\omega}{\pi}\,\text{P}\int_0^\infty d\omega'\,\frac{\text{Re } Y(\omega')}{\omega'^2-\omega^2}.$$

In the present example, it is the *real* part of the susceptibility (the admittance) that measures the effect of dissipation, γ being proportional to the resistance R. The imaginary part represents the reactive or non-dissipative aspect of the response. The situation would have been reversed if we had chosen to consider the instantaneous *charge* $q(t)$ on the capacitor, instead of the *current* $I(t) = \dot{q}(t)$ in the circuit, as the response of the system. This conclusion follows easily from the fact that the Fourier transforms of $q(t)$ and $I(t)$ are related according to $\widetilde{q}(\omega) = (i/\omega)\widetilde{I}(\omega)$. Hence the corresponding dynamic susceptibility is now given by $(i/\omega)Y(\omega)$.

A further remark is in order. The charge $q(t)$ satisfies the familiar second-order differential equation $L\ddot{q} + R\dot{q} + q/C = V(t)$. As you know, this is of precisely the same form as the equation satisfied by the displacement of a damped linear harmonic oscillator. This is the basis of the well-known **electrical ↔ mechanical analogy** between the two systems. We will return to the equation for $q(t)$ under a sinusoidal applied voltage, and its solution using Laplace transforms, in Chap. 28, Sect. 28.2.2.

24.2.3 Subtracted Dispersion Relations

It may so happen that a dynamic susceptibility $\chi(\omega)$ does not vanish as $|\omega| \to \infty$ in the upper half-plane or along the real axis. (The latter is a more likely possibility.) Instead, it may tend to a constant as $|\omega| \to \infty$ along some direction or directions in the region Im $\omega \geq 0$. In this case, when we try to derive dispersion relations for the real and imaginary parts of $\chi(\omega)$, we find that the contribution from the large semicircle of radius R no longer vanishes as $R \to \infty$. If $\chi(\omega)$ tends *uniformly* to a constant ($= \chi_\infty$, say) as $|\omega| \to \infty$ along the real axis and along all directions in the upper half-plane, we could go ahead by including $i\pi\chi_\infty$ as the extra contribution from the large semicircle. But there is no guarantee that this will be the case. In fact, it does not happen, in general.

In order to get over the problem, we assume that the value of $\chi(\omega)$ at some particular real value of the frequency, say ω_1, is known. Then, instead of the function $\chi(\omega')/(\omega' - \omega)$ introduced in Eq. (24.10), consider the function

$$f(\omega') = \frac{\chi(\omega') - \chi(\omega_1)}{(\omega' - \omega_1)(\omega' - \omega)}. \tag{24.25}$$

Here ω is any real frequency other than ω_1. It is obvious that this new function vanishes faster than $1/\omega'$ as $|\omega'| \to \infty$, owing to the extra factor $1/(\omega' - \omega_1)$. Moreover, it is analytic everywhere in the upper half ω'-plane and on the real axis, except for a simple pole at $\omega' = \omega$, as before. The apparent singularity of $f(\omega')$ at $\omega' = \omega_1$ is a removable singularity. This is the reason for subtracting $\chi(\omega_1)$ from $\chi(\omega')$ in the numerator, in the definition of $f(\omega')$. Since $\chi(\omega')$ is analytic on the real axis in the ω'-plane, $f(\omega_1)$ has the finite value $\chi'(\omega_1)/(\omega_1 - \omega)$, where χ' is the derivative of χ.

Repeating the earlier derivation with this new definition of $f(\omega')$, we now get

$$\chi(\omega) = \chi(\omega_1) - \frac{i}{\pi}(\omega - \omega_1)\,\mathrm{P}\int_{-\infty}^{\infty} d\omega'\,\frac{\chi(\omega') - \chi(\omega_1)}{(\omega' - \omega_1)(\omega' - \omega)} \tag{24.26}$$

as the counterpart of Eq. (24.13). Equating the respective real and imaginary parts on both sides of Eq. (24.26) yields the dispersion relations satisfied by $\chi(\omega)$ in this case. These are called **once-subtracted dispersion relations,** ω_1 being the "point of subtraction".

★ **5.** Derive Eq. (24.26), and hence the dispersion relations satisfied by the real and imaginary parts of $\chi(\omega)$.

From the mathematical point of view, it is clear that the procedure given above can be extended to cover situations when the analytic function $\chi(\omega)$ actually diverges as $|\omega| \to \infty$ in the upper half-plane. For instance, if $\chi(\omega) \sim \omega^{n-1}$ asymptotically, where n is a positive integer, we can write down an n-fold subtracted dispersion relation for it. The latter will require n constants as inputs. These could be, for instance, the values of $\chi(\omega)$ at n different frequencies or points of subtraction.

24.2.4 Hilbert Transform Pairs

I have mentioned already that the dispersion relations (24.14) imply that, for real values of ω, the two functions Re $\chi(\omega)$ and Im $\chi(\omega)$ comprise an integral transform and its inverse, a Hilbert transform pair. The transform and its inverse are defined as follows.

Given a (possibly complex-valued) function $h(x)$ of the real variable x, its Hilbert transform (when it exists) is defined as

$$\mathcal{H}\big[h(x)\big] \equiv H(x) \stackrel{\text{def.}}{=} \frac{1}{\pi}\,\mathrm{P}\int_{-\infty}^{\infty} dx'\,\frac{h(x')}{x'-x}. \tag{24.27}$$

It can then be shown that the inverse transform is given by

$$\mathcal{H}^{-1}\big[H(x)\big] = h(x) = -\frac{1}{\pi}\,\mathrm{P}\int_{-\infty}^{\infty} dx'\,\frac{H(x')}{x'-x}. \tag{24.28}$$

In other words, the inverse of the Hilbert transform operator $\mathcal{H}$ is given by $\mathcal{H}^{-1} = -\mathcal{H}$. (In an alternative convention, the definitions of $\mathcal{H}$ and $\mathcal{H}^{-1}$ given above are interchanged.) The Hilbert transform is closely related to the Fourier transform, but I shall not go into this aspect here.

Eliminating the function $H(x)$ between Eqs. (24.27) and (24.28), we have

$$h(x) = -\frac{1}{\pi^2}\mathrm{P}\int_{-\infty}^{\infty} dx'\;\mathrm{P}\int_{-\infty}^{\infty} dx''\frac{h(x'')}{(x'-x)(x''-x')}. \tag{24.29}$$

This equation may be rewritten as

$$h(x) = \int_{-\infty}^{\infty} dx''\left\{\frac{1}{\pi^2}\mathrm{P}\int_{-\infty}^{\infty}\frac{dx'}{(x'-x)(x'-x'')}\right\} h(x''). \tag{24.30}$$

Since $h(x)$ is arbitrary, the quantity in the curly brackets in Eq. (24.30) must be the unit operator in function space. In other words, we must have

$$\boxed{\frac{1}{\pi^2}\mathrm{P}\int_{-\infty}^{\infty}\frac{dx'}{(x'-x)(x'-x'')} = \delta(x-x'').} \tag{24.31}$$

We thus have yet another representation of the Dirac δ-function.

★ **6.** Verify the identity (24.31) explicitly.

The iterate of a Hilbert transform is easily found, using Eq. (24.31). We have

$$\mathcal{H}^2\big[h(x)\big] = -h(x). \tag{24.32}$$

***Parseval's formula for the Hilbert transform*:** In Chap. 18, Sect. 18.1.2, you have seen that if $f(x)$ is square-integrable, so is its Fourier transform $\widetilde{f}(k)$. Further, the Fourier transform preserves the $\mathcal{L}_2$-norm: restating Parseval's formula in this case (Eq. (18.5)),

$$\int_{-\infty}^{\infty} dx\, |f(x)|^2 = \frac{1}{2\pi}\int_{-\infty}^{\infty} dk\, |\widetilde{f}(k)|^2. \tag{24.33}$$

(The factor $(2\pi)^{-1}$ is an artifact of the Fourier transform convention we have chosen to work with.) Similarly, if one member of a Hilbert transform pair is square-integrable, so is the other, and Parseval's Theorem holds good. Once again, Eq. (24.31) helps us establish that

$$\int_{-\infty}^{\infty} dx\, |h(x)|^2 = \int_{-\infty}^{\infty} dx\, |H(x)|^2. \tag{24.34}$$

How does one actually compute the Hilbert transform of a given function? That is, how does one evaluate the principal value integral that is involved? The simplest case is that in which we can recognize $h(x)$ as the real part of an analytic function $f(z)$, when its argument is real (i.e., $z = x$). The corresponding Hilbert transform $H(x)$ is then the imaginary part of this analytic function. Thus, if $h(x) = \cos x$, you might guess that $H(x) = \sin x$, and if $h(x) = \sin x$, then $H(x) = -\cos x$ (the corresponding analytic functions being e^{iz} and $-ie^{iz}$, respectively). Somewhat more generally, the method that suggests itself is as follows. Reverse the steps that led to the derivation of dispersion relations for functions that are analytic in the upper (or lower) half-plane and are sufficiently well-behaved as $z \to \infty$. In effect, this involves starting with the formula (23.34) in Chap. 23, Sect. 23.3.4, and rewriting it for our present purposes in the form

$$\mathrm{P}\int_{-\infty}^{\infty} dx'\, \frac{\phi(x')}{x' - x} = \lim_{\epsilon\to 0}\int_{-\infty}^{\infty} dx'\, \frac{\phi(x')}{x' - (x \pm i\epsilon)} \mp i\pi\phi(x). \tag{24.35}$$

The task reduces to evaluating the definite integral on the right-hand side. Note that the pole of the integrand at $x' = x \pm i\epsilon$ does not lie on the path of integration. In some (but not all) cases, it may be possible to use contour integration to compute the integral.

★ **7.** Find the Hilbert transform of the function $h(x) = 1/(1 + x^2)$, and verify that Parseval's formula (24.34) is satisfied.

Clearly, there exist functions $h(x)$ for which this straightforward method will not work. More sophisticated techniques are then required to find the corresponding Hilbert transforms.

24.2.5 Discrete and Continuous Relaxation Spectra

Recall that the dynamic susceptibility $\chi(\omega)$ is the Fourier-Laplace transform of the response function $\phi(t)$, as in Eq. (24.6). What can be said about the response function itself?

On physical grounds, we would expect $\phi(t)$ to decay with increasing t, for $t \geq 0$. The simplest physical model of a response function is an exponentially decaying function of the form $\phi(t) = \phi_0 \, e^{-t/\tau}$, where ϕ_0 and τ (> 0) are constants. This form corresponds to what is known as **Debye relaxation**, and is typical of (dielectric, magnetic, mechanical, $\cdots$) relaxation processes governed by a single characteristic relaxation time τ. It then follows that

$$\chi(\omega) = \frac{\phi_0 \tau}{1 - i\omega\tau}. \tag{24.36}$$

Thus $\chi(\omega)$ has a simple pole at $\omega = -i/\tau$ on the negative imaginary axis in the ω-plane. A commonly used phenomenological representation for the dynamic susceptibility is the **Cole–Cole plot**: this is just the Argand diagram of $\chi(\omega)$, i.e., the plot of Im $\chi(\omega)$ versus Re $\chi(\omega)$, as ω runs from zero to infinity.

★ **8.** In the case of Debye relaxation, let us scale out the constants by defining

$$u = \text{Re}\,\chi(\omega)/(\phi_0\tau) \quad \text{and} \quad v = \text{Im}\,\chi(\omega)/(\phi_0\tau).$$

Show that the Cole–Cole plot in the (u, v) plane is given by $\left(u - \frac{1}{2}\right)^2 + v^2 = \frac{1}{4}$.

- The Cole–Cole plot for Debye relaxation is just a semicircle in the first quadrant of the (u, v) plane, with center at $(\frac{1}{2}, 0)$ and radius equal to $\frac{1}{2}$. The plot starts at $(1, 0)$ when $\omega = 0$, and ends at $(0, 0)$ as $\omega \to \infty$.
- Any departure (of the observed data) from this shape implies that the relaxation is no longer of the pure single-relaxation-time form.

Several *empirical* modifications of Eq. (24.36) are in common use in the analysis of experimental data on relaxation phenomena. At a more basic level, the relaxation is often governed by a *spectrum* of characteristic times τ_j ($j = 1, 2, \ldots, n$), with corresponding weight factors σ_j.

★ **9.** Given that the real part of the dynamic susceptibility is given by

$$\text{Re}\,\chi(\omega) = \sum_{j=1}^{n} \frac{\sigma_j}{1 + \omega^2 \tau_j^2},$$

where σ_j and τ_j are positive constants, show that its imaginary part is

$$\text{Im}\,\chi(\omega) = \sum_{j=1}^{n} \frac{\omega\,\sigma_j\,\tau_j}{1 + \omega^2 \tau_j^2}.$$

The foregoing expressions correspond to the form

$$\chi(\omega) = \sum_{j=1}^{n} \frac{\sigma_j}{1 - i\omega\,\tau_j} \tag{24.37}$$

for the dynamic susceptibility. In this case, $\chi(\omega)$ has a set of simple poles on the negative imaginary axis in the ω-plane, at $\omega = -i/\tau_j\,,\ 1 \le j \le n$.

***Continuous relaxation spectrum*:** When the relaxation spectrum, i.e., the number of contributing relaxation modes is *infinite*, or if there is a *continuum* of such modes, much more complex dynamical behavior may result.

★ **10.** Consider a situation in which there is a continuous set (or spectrum) of relaxation times that lie in the interval $[\tau_{\min}\,,\ \tau_{\max}]$. Given that

$$\mathrm{Re}\,\chi(\omega) = \int_{\tau_{\min}}^{\tau_{\max}} \frac{\sigma(\tau)\,d\tau}{1+\omega^2\,\tau^2},$$

where the *spectral weight function* $\sigma(\tau)$ is a well-behaved, nonnegative function of τ, show that

$$\mathrm{Im}\,\chi(\omega) = \int_{\tau_{\min}}^{\tau_{\max}} \frac{\omega\tau\sigma(\tau)\,d\tau}{1+\omega^2\,\tau^2}\,.$$

The foregoing expressions correspond to a dynamic susceptibility of the form

$$\chi(\omega) = \int_{\tau_{\min}}^{\tau_{\max}} \frac{\sigma(\tau)\,d\tau}{1-i\omega\tau}\,. \tag{24.38}$$

The singularity structure of $\chi(\omega)$ in the lower half-plane in ω may now be quite intricate. (It continues to remain analytic in the upper half-plane, of course.) In general, the functional form given by Eq. (24.38) will have *logarithmic branch points*[2] at $\omega = -i/\tau_{\max}$ and $\omega = -i/\tau_{\min}$. Even more intricate possibilities arise when $(\tau_{\min}\,,\ \tau_{\max}) = (0\,,\ \infty)$. These, in turn, are associated with physically interesting kinds of non-Debye relaxation, including *slow* or *glassy dynamics*, *stretched exponential decay*, etc., in a variety of condensed matter systems with different kinds of *quenched disorder*—for instance, in amorphous materials like oxide glasses and metallic glasses, in spin glasses, and so on. As the name "glassy dynamics" itself suggests, a physical example is provided by the time-dependent elastic moduli of ordinary glasses, as deduced, for instance, from the phenomenon of *stress relaxation* in these materials. "Stretched exponential decay" means that the relaxation proceeds like $\exp[-(t/\tau)^{\alpha}]$, where $0 < \alpha < 1$. This is a faster fall-off than any inverse power of t, but slower than a decaying exponential like $\exp[-(t/\tau)]$.

24.3 Solutions

4. First write the dispersions relations back in their original form (24.14), with the integrals over ω' running from $-\infty$ to ∞. Insert the expressions in (24.24) for

---

[2]Multivalued functions and branch points will be discussed in Chap. 26.

Re Y and Im Y in the integrands, and work backwards through the steps in the derivation of the dispersion relations. (That is, add the small and large semicircles to the line integrals, close the contours and evaluate the corresponding integrals using the residue theorem.) ▶

6. First consider $x \neq x''$, and show that the integral on the left-hand side vanishes identically. Next, for any sufficiently well-behaved function $\phi(x)$, show that

$$\int_{-\infty}^{\infty} dx'' \left\{ \frac{1}{\pi^2} \mathrm{P} \int_{-\infty}^{\infty} \frac{dx'}{(x'-x)(x'-x'')} \right\} \phi(x'') = \phi(x).$$

Hence Eq. (24.31) is indeed a representation of the δ-function. ▶

7. Equation (24.35) yields, in this case,

$$H(x) = \lim_{\epsilon \to 0} \frac{1}{\pi} \int_{-\infty}^{\infty} \frac{dx'}{(x'^2+1)[x' - (x \pm i\epsilon)]} \mp \frac{i}{x^2+1}.$$

Choose the + sign, say, in $x \pm i\epsilon$. Let the integration run from $-R$ to R (where R is a sufficiently large number), and attach a semicircle of radius R to close the contour in either the upper or the lower half-plane. The integrand has simple poles at $x' = i$ and $x' = x + i\epsilon$ in the upper half-plane, and at $x' = -i$ in the lower half-plane. It is therefore (marginally) simpler to close the contour in the lower half-plane. Picking up $-2\pi i$ times the residue at $x' = -i$ and simplifying, we get

$$H(x) = -\frac{x}{x^2+1}.$$

The integrals $\int_{-\infty}^{\infty} dx/(x^2+1)^2$ and $\int_{-\infty}^{\infty} dx\, x^2/(x^2+1)^2$ are elementary (set $x = \tan\,\theta$). We find

$$\int_{-\infty}^{\infty} dx\, |h(x)|^2 = \int_{-\infty}^{\infty} dx\, |H(x)|^2 = \tfrac{1}{2}\pi,$$

in accord with Parseval's Theorem.

Remark Verify that the same result is obtained for $H(x)$ if (a) the contour is closed in the upper half-plane; (b) the $-$ sign is chosen in $x \pm i\epsilon$. ▶

9. Use the dispersion relation for Im $\chi(\omega)$. ▶

10. Once again, use the dispersion relation for Im $\chi(\omega)$. ▶

Chapter 25
Analytic Continuation and the Gamma Function

25.1 Analytic Continuation

25.1.1 What Is Analytic Continuation?

A fundamental objective in the study of analytic functions of a complex variable is the following: Suppose we know that a given function is holomorphic (i.e., satisfies the Cauchy–Riemann conditions) in some regions of the complex plane. How can we extend this domain of holomorphy, i.e., find an **analytic continuation** of the function to regions outside the original one?

In Eqs. (22.37) and (22.38) of Chap. 22, Sect. 22.5.2, you have seen how different power series could represent the same "master" function $1/(1-z)$ in different regions. The idea underlying analytic continuation is as follows:

- Suppose two *apparently* different functions, say $f_1(z)$ and $f_2(z)$, are analytic in the respective regions $\mathcal{R}_1$ and $\mathcal{R}_2$ (including their boundaries).
- Suppose, further, that these regions have an *overlap*, and that the numerical values of $f_1(z)$ and $f_2(z)$ are equal, point by point, in this overlap region.
- Then $f_1(z)$ and $f_2(z)$ are analytic continuations of each other.
- Equivalently, they represent the same analytic function, now defined in the union of $\mathcal{R}_1$ and $\mathcal{R}_2$.

Technically, the overlap must occur on a dense set. It could be an area in the complex plane, as in Fig. 25.1a, or even a common continuous line segment on the boundaries of $\mathcal{R}_1$ and $\mathcal{R}_2$, as in Fig. 25.1b.

In practice, there are numerous methods of analytically continuing a function from a region in which a local representation of the function is available, to a larger region. Some of these are the following:

(i) Chains of power series with overlapping circles of convergence, as shown schematically in Fig. 25.2. This was essentially the method used in the example considered in Eqs. (22.37) and (22.38).

V. Balakrishnan, *Mathematical Physics*,
https://doi.org/10.1007/978-3-030-39680-0_25

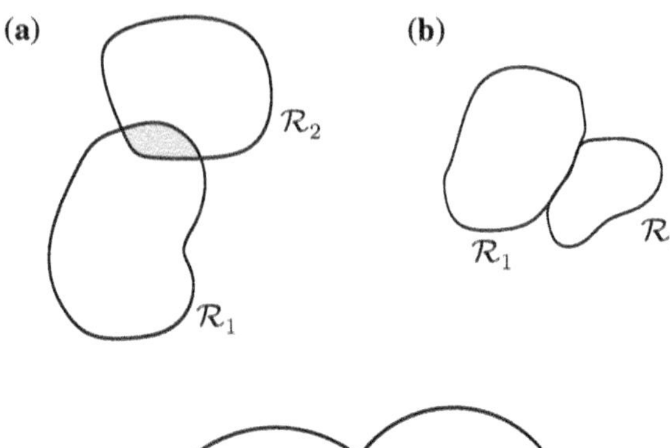

Fig. 25.1 Analytic continuation of a function. **a** Overlapping regions; the overlap is the shaded part. **b** Regions with a common boundary segment

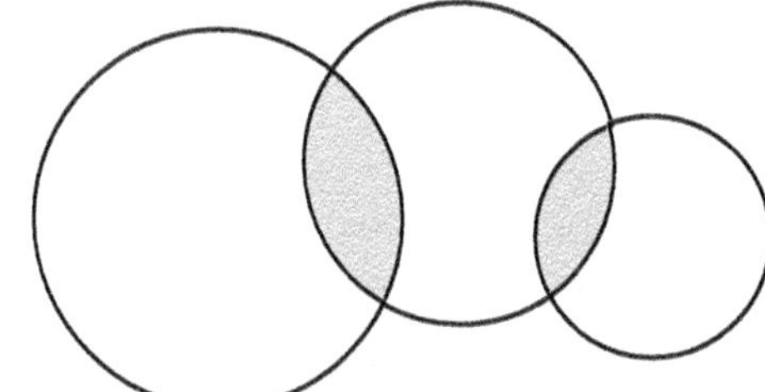

Fig. 25.2 Analytic continuation using a chain of power series with overlapping circles of convergence

(ii) The Schwarz reflection principle, applicable to real analytic functions (see below).
(iii) The use of the "permanence" of functional equations.
(iv) Summability methods for power series.
(v) Integral representations for functions in terms of contour integrals, using the distortability of the contours.

I will touch upon some of these by means of specific examples. Before doing so, I comment briefly on the Schwarz reflection principle, since we shall not return to it in the sequel.

The Schwarz reflection principle provides a convenient and powerful way of finding the analytic continuation of a *real analytic* function—that is, an analytic function that is real when its argument is real. Let $f(z)$ be analytic in a region $\mathcal{R}$ of the upper half-plane in z, such that a segment (or all) of the real axis is a boundary of $\mathcal{R}$. Further, let $f(x)$ be continuous on that segment. Then the Schwarz reflection principle is the symmetry property

$$f(z^*) = f^*(z). \tag{25.1}$$

Note that if z is located in the upper half-plane, then z^* is in the lower half-plane, at a location obtained by reflecting the position of z about the x-axis. Let $\mathcal{R}'$ be the mirror image of $\mathcal{R}$, reflected about the real axis, as shown in Fig. 25.3. Then, all you have to do to find the value of $f(z)$ at any point in $\mathcal{R}'$ is to take the complex conjugate of its value at the corresponding image point in $\mathcal{R}$. Equation (25.1) therefore provides an analytic continuation of $f(z)$ to the region $\mathcal{R}'$. The Schwarz reflection principle has many practical uses—in particular, in connection with the technique of *conformal*

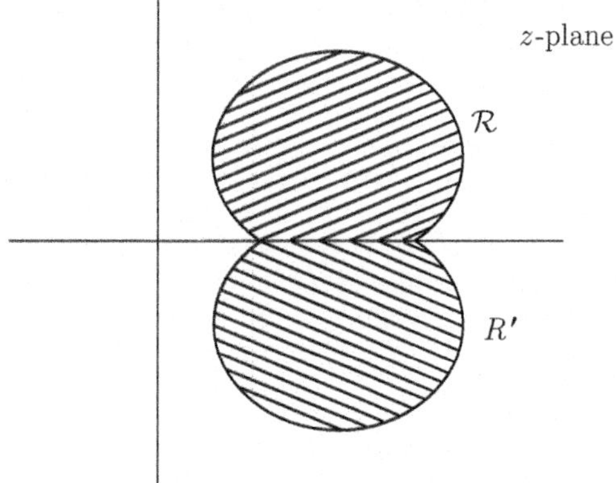

Fig. 25.3 Illustrating the Schwarz reflection principle. $\mathcal{R}'$ is the reflection of the region $\mathcal{R}$ about the real axis

mapping. Although conformal mapping *per se* will not be discussed in this book, owing to limitations of space, we will study an important class of conformal maps at some length, in Chap. 27.

The uniqueness of an analytic continuation is what makes the concept significant. When we analytically continue a function outside its original region of analyticity, we are guaranteed that there is a *unique* function that is being so continued. *Representations* of a function may be many in number, of course. But the underlying *function* is unique. If this were not so, the whole exercise would be arbitrary and pointless.

25.1.2 The Permanence of Functional Relations

The permanence of functional relations is connected to what has just been stated above. It is a feature that appears to be so obvious that one might take it for granted. But it is nontrivial, and only valid under specific conditions—namely, *the analyticity of the two sides* of the relation concerned! The point is as follows:

- Suppose two analytic functions $f(z)$ and $g(z)$ satisfy the relation $f(z) = g(z)$ in some region. Then this relation continues to be valid for the analytic continuations of $f(z)$ and $g(z)$.
- That is, if the function $f(z) - g(z)$ is identically zero in some region, its analytic continuation must be zero as well.

This is where you must be careful to distinguish between a region and a set of points. At the risk of sounding trivial, here is an example. The equation $\sin \pi z = 0$ is only valid at the points $z = n$, where n is an integer. It is obvious that $\sin \pi z$ is not *identically* equal to zero. On the other hand, the equation $\sin^2 z + \cos^2 z = 1$ is a relation between analytic functions of z, and is "permanent", in the sense that it is valid for all values of z.

It is instructive to elaborate a bit on this aspect, with the help of a less trivial example. Consider the relation between Legendre polynomials in Eq. (16.104) of Chap. 16, Sect. 16.4.1, namely,

$$P_n(x) = P_{-n-1}(x), \quad \text{where} \quad -1 \le x \le 1 \text{ and } n = 0, 1, \ldots . \tag{25.2}$$

The interval $[-1, 1]$ is a dense set of points. $P_n(x)$ can be analytically continued to all complex values of the argument. Hence we have the relation

$$P_n(z) = P_{-n-1}(z) \quad (z \in \hat{\mathbb{C}},\ n \in \mathbb{Z}) \tag{25.3}$$

between two analytic functions of z. In this sense, the "permanence" of the equation helps us write down the analytic continuation of the two sides of the equation.

In *this* particular instance we can, in fact, go even further. As I have mentioned briefly in Chap. 16, Sect. 16.4.5, the Legendre function $P_\nu(z)$ is an analytic function of both the argument z as well as the index ν, regarded as complex variables. The general relation, of which Eq. (25.2) is a very special case, is actually

$$\boxed{P_\nu(z) = P_{-\nu-1}(z), \quad (z \in \hat{\mathbb{C}},\ \nu \in \hat{\mathbb{C}})} \tag{25.4}$$

valid for complex values of both z and ν. You must appreciate the fact that this further generalization is *not* obvious. The set of integers $\{n\}$ is not a dense set, and we cannot automatically "analytically continue" a relation that is originally valid only on this set. (For example, an extra term $\sin \pi\nu$ could have been present on one side of Eq. (25.4), without affecting the validity of Eq. (25.3).) Further conditions have to be imposed to ensure the uniqueness of such an extrapolation. These conditions happen to be satisfied in the present instance, so that Eq. (25.4) is indeed valid as it stands. It is an example of a **reflection formula**, because it relates the Legendre function of the first kind at any value ν of the order to the same function at the value $-\nu - 1$ of the order, and $-\nu - 1$ is the image of ν reflected in a mirror on the line $\text{Re}\, \nu = -\frac{1}{2}$ in the complex ν-plane.

Here is a related example to prove the point just made, involving the Legendre function of the *second* kind. In Chap. 16, Sect. 16.4.5, you have encountered the Legendre function of the second kind, $Q_n(x)$. This function happens to satisfy a relation that is analogous to Eq. (25.2), but for *half*-odd-integer order:

$$Q_{n+\frac{1}{2}}(x) = Q_{-n-\frac{3}{2}}(x), \quad \text{where} \quad -1 \le x \le 1 \text{ and } n = 0, 1, \ldots . \tag{25.5}$$

As in the case of $P_n(x)$, the function $Q_n(x)$ can also be continued analytically to the function $Q_n(z)$, with a complex argument z. Since the interval $[-1, 1]$ is a dense set of points, the *relation* (25.5) itself can be continued analytically to

$$Q_{n+\frac{1}{2}}(z) = Q_{-n-\frac{3}{2}}(z), \quad \text{where } z \in \hat{\mathbb{C}} \text{ and } n = 0, 1, \ldots . \tag{25.6}$$

But $Q_n(z)$ is the value, at $\nu = n$, of the analytic function $Q_\nu(z)$ of the complex variable ν. The generalization of Eq. (25.6), however, is *not* $Q_\nu(z) = Q_{-\nu-1}(z)$, but rather

$$\boxed{Q_\nu(z) - Q_{-\nu-1}(z) = (\pi \cot \pi\nu) P_\nu(z) \quad (z \in \hat{\mathbb{C}},\ \nu \in \hat{\mathbb{C}}).} \tag{25.7}$$

The relation (25.7) is the reflection formula for $Q_\nu(z)$. The factor $\cot(\pi\nu)$ vanishes, of course, when $\nu = n + \frac{1}{2}$. The relation (25.6) is then recovered.

An interesting property of $Q_\nu(z)$ follows from the identity (25.7). It turns out that, as a function of the order ν, this function has a singularity at every negative integer value of ν. Moreover, $P_\nu(z)$ has no singularities as a function of ν for any finite value of ν.

★ **1.** Given these facts, use the identity (25.7) to show that $Q_\nu(z)$ has a simple pole at $\nu = -n - 1$, where $n = 0, 1, \ldots$, with a residue equal to $P_n(z)$.

Contour integral representations for the Legendre functions $P_\nu(z)$ and $Q_\nu(z)$, valid for general complex values of the argument z and the order ν, will be given in Chap. 26, Sect. 26.2.5. More will be said there about the analytic properties of the Legendre functions of both kinds.

25.2 The Gamma Function for Complex Argument

25.2.1 Stripwise Analytic Continuation of $\Gamma(z)$

The gamma function that was introduced in Chap. 3, Sect. 3.1.4, provides an excellent illustration of how analytic continuation works. I will, therefore, devote the rest of this chapter to a discussion of various properties of this specific function.

Recall the definition of $\Gamma(x)$ for positive values of the argument x, given in Eq. (3.13) of Chap. 3, Sect. 3.1.4:

$$\Gamma(x) \stackrel{\text{def.}}{=} \int_0^\infty dt\, t^{x-1}\, e^{-t}, \quad x > 0. \tag{25.8}$$

As pointed out there, the condition $x > 0$ is necessary in order to ensure the convergence of the integral, because of the behavior of the factor t^{x-1} at the lower limit of integration, $t = 0$. Suppose, now, that the real variable x is replaced by the complex variable $z = x + iy$. The convergence properties of the integral are not affected at all by the *imaginary* part of z, because $t^{iy} = e^{iy \ln t}$ is just a phase factor whose modulus is equal to unity. Therefore the integral defines, *as it stands*, an analytic function of z in the region $\text{Re}\, z > 0$. That is

$$\boxed{\Gamma(z) \stackrel{\text{def.}}{=} \int_0^\infty dt\, t^{z-1}\, e^{-t}, \quad \text{Re}\, z > 0.} \tag{25.9}$$

This function is the analytic continuation of the gamma function from the positive real axis to the whole of the right-half-plane in z.

The interesting question that arises now is: how do we analytically continue the gamma function to the *left* half-plane, i.e., to the region $\text{Re}\, z \le 0$ in the complex z-plane? In order to do so, we need to improve the behavior of the factor t^{z-1} at the lower limit of integration, $t = 0$. Obviously, integration by parts will do just this. *Keeping z in the region* $\text{Re}\, z > 0$, we integrate by parts to get

$$\begin{aligned}\Gamma(z) &= \frac{t^z\, e^{-t}}{z}\bigg|_{t=0}^{\infty} + \frac{1}{z}\int_0^\infty dt\, t^z\, e^{-t} \\ &= \underbrace{\frac{1}{z}\int_0^\infty dt\, t^z\, e^{-t}}_{\text{conv. in Re}\, z > -1}. \end{aligned} \tag{25.10}$$

The second line follows from the first precisely because z is *restricted* to the region $\text{Re}\, z > 0$, so that the factor t^z vanishes at $t = 0$. But now, *having obtained the last expression*, we see that the integral in it actually converges in the *extended* region $\text{Re}\, z > -1$, while the *explicit* factor $1/z$ outside the integral suggests that the gamma function itself has a simple pole at $z = 0$. And indeed it does, because the integral on the right-hand side does not vanish at $z = 0$. It is equal to unity when $z = 0$. We may, therefore, conclude that $\Gamma(z)$ has a simple pole at $z = 0$ with residue equal to unity. Thus, the region in which we have an explicit representation of the gamma function has been *extended* to the extra strip $-1 < \text{Re}\, z \le 0$. We now have the representation

$$\Gamma(z) = \frac{1}{z}\int_0^\infty dt\, t^z\, e^{-t}, \quad \text{Re}\, z > -1. \tag{25.11}$$

It is clear that we may repeat this procedure of integrating by parts as many times as we please. The result is the following representation, valid for any nonnegative integer n:

$$\boxed{\Gamma(z) = \frac{1}{z(z+1)\cdots(z+n)}\int_0^\infty dt\, t^{z+n}\, e^{-t}, \quad \text{Re}\, z > -n-1.} \tag{25.12}$$

We have thus achieved an analytic continuation of the gamma function, *strip by strip*, to the region $\text{Re}\, z > -n-1$, starting with a representation for the function that was only valid in the region $\text{Re}\, z > 0$ (see Fig. 25.4). In principle, this can be done for an arbitrarily large value of n.

The underlying mechanism is summarized in the *functional equation* satisfied by the gamma function. This equation follows from Eq. (25.11), on noting that the integral on the right-hand side is just $\Gamma(z+1)$. Hence

$$\Gamma(z) = \frac{1}{z}\,\Gamma(z+1), \quad \text{or} \quad \boxed{\Gamma(z+1) = z\,\Gamma(z).} \tag{25.13}$$

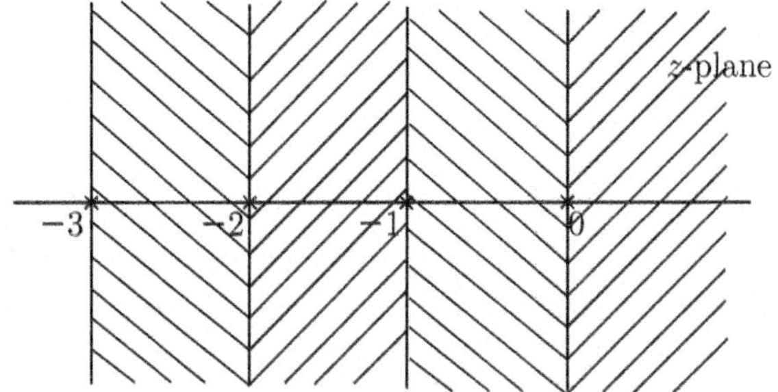

Fig. 25.4 Stripwise analytic continuation of the gamma function by integration by parts. The original region of analyticity, Re $z > 0$, is extended successively to Re $z > -1$, Re $z > -2$, and so on. $\Gamma(z)$ is found to have simple poles at $z = 0, -1, -2, \ldots$, indicated by small crosses

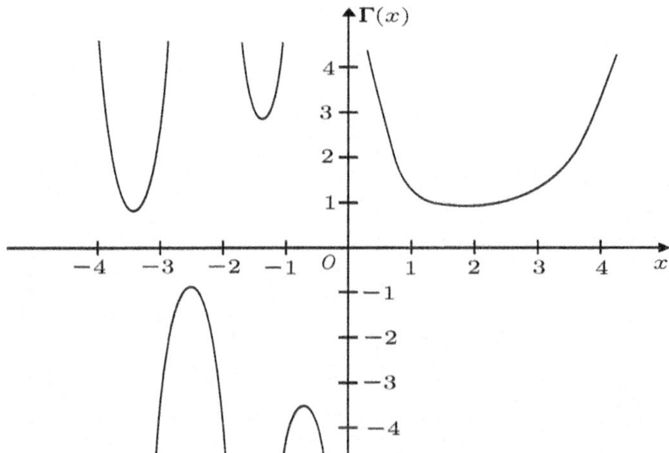

Fig. 25.5 $\Gamma(x)$ versus x

The permanence of this equation (i.e., its validity for all z) can be used to analytically continue the gamma function arbitrarily far to the left of the imaginary axis in the complex z-plane. It follows immediately that

- $\Gamma(z)$ has a simple pole at each of the nonpositive integers $z = -n$, where $n = 0, 1, 2, \ldots$, with a residue equal to $(-1)^n/n!$.

★ **2.** Establish this assertion.

Recall Fig. 3.1 of Chap. 3, Sect. 3.1.4, depicting $\Gamma(x)$ for positive values of x. Figure 25.5 now shows the function for negative values of x as well.

25.2.2 Mittag-Leffler Expansion of $\Gamma(z)$

Other than a simple pole at every nonpositive integer, $\Gamma(z)$ has no other singularities in the finite part of the complex plane. It is, therefore, a meromorphic function in $\mathbb{C}$. It is then natural to ask for the Mittag-Leffler expansion of $\Gamma(z)$—namely, a representation that explicitly separates out the contribution from all the poles of the function from the part that is an entire function, as explained in Chap. 23, Sect. 23.6. It is straightforward to show that the expansion required is

$$\Gamma(z) = \underbrace{\sum_{n=0}^{\infty} \frac{(-1)^n}{n!\,(z+n)}}_{\text{sum over poles}} + \underbrace{\int_1^{\infty} dt\, t^{z-1}\, e^{-t}}_{\text{entire function}}. \tag{25.14}$$

The first term on the right-hand side is a "sum over poles", while the second term is an entire function.

★ **3.** Establish the Mittag-Leffler expansion (25.14) of $\Gamma(z)$.

25.2.3 Logarithmic Derivative of $\Gamma(z)$

Since $\Gamma(z)$ has a simple pole at each nonpositive integer, its Laurent expansion in the neighborhood of the pole at $z = -n$ is of the form

$$\Gamma(z) = \frac{(-1)^n}{n!\,(z+n)} + \text{regular part}, \quad \text{where } n = 0, 1, \ldots. \tag{25.15}$$

The region of validity of the expansion (25.15) should be obvious to you, by now. It is clearly the annular region $0 < |z+n| < 1$. The "regular part" in Eq. (25.15) is an infinite series in powers of $(z+n)$ that must necessarily diverge at $z+n = -1$, because you know that $\Gamma(z)$ also has a pole at $z = -n-1$.

It follows from Eq. (25.15) that the *derivative* of the gamma function, $\Gamma'(z)$, has a *double* pole at $z = -n$. The corresponding Laurent expansion is

$$\Gamma'(z) = -\frac{(-1)^n}{n!\,(z+n)^2} + \text{regular part}, \quad \text{where } n = 0, 1, \ldots. \tag{25.16}$$

Note, however, that the residue of $\Gamma'(z)$ at each of its poles is identically equal to zero. (There is no term proportional to $(z+n)^{-1}$ in its Laurent expansion about the point $z = -n$.) The ratio

$$\frac{\Gamma'(z)}{\Gamma(z)} = \frac{d}{dz} \ln \Gamma(z) \stackrel{\text{def.}}{=} \psi(z) \tag{25.17}$$

is the **logarithmic derivative** of the gamma function. It is also called the **digamma function**. The analytic properties of $\psi(z)$ follow directly from those of $\Gamma(z)$. It is obvious that $\psi(z)$ also has a simple pole at each nonpositive integer. The functional equation (25.13) satisfied by $\Gamma(z)$ leads to the functional equation

$$\boxed{\psi(z+1) - \psi(z) = \frac{1}{z}.} \tag{25.18}$$

★ **4.** Establish the following:

(a) $\psi(z)$ has simple poles at $z = 0, -1, -2, \ldots$, with a residue equal to -1 at each pole.

(b) $\psi(z)$ satisfies the difference Eq. (25.18).

25.2.4 *Infinite Product Representation of* $\boldsymbol{\Gamma(z)}$

The Euler–Mascheroni constant: As you know, the harmonic series

$$\sum_{n=1}^{N} \frac{1}{n} = 1 + \frac{1}{2} + \frac{1}{3} + \cdots + \frac{1}{N} \tag{25.19}$$

diverges as $N \to \infty$. The divergence is logarithmic, i.e., the sum has a leading asymptotic behavior $\sim \ln N$ for large values of N. The following question then arises naturally: does the *difference* $\left(\sum_1^N 1/n\right) - \ln N$ have a finite nonzero limit as $N \to \infty$? The answer is yes. It turns out that

$$\boxed{\lim_{N\to\infty} \left\{\left(\sum_{n=1}^{N} \frac{1}{n}\right) - \ln N\right\} = \gamma,} \tag{25.20}$$

where γ is a number called the **Euler–Mascheroni constant**. Its numerical value is $0.5772\cdots$. In fact, γ is very likely to be an irrational, in fact transcendental number, like e or π.[1]

The constant γ has many close links with the gamma function. For instance, we know that $\Gamma(z)$ has a simple pole at $z = 0$, with residue equal to 1. If we subtract out this singular part, and then pass to the limit $z \to 0$, we get

$$\lim_{z\to 0} \left\{\Gamma(z) - \frac{1}{z}\right\} = -\gamma. \tag{25.21}$$

[1]Recall that a transcendental number is a number that is not the root of a polynomial equation of finite degree with rational coefficients. Let me mention, in passing, that rigorously *proving* the transcendental nature of a specific irrational number is quite a nontrivial matter, in general.

In other words, the behavior of the gamma function near in the neighborhood of $z = 0$ is given by

$$\Gamma(z) = \frac{1}{z} - \gamma + \mathcal{O}(z). \tag{25.22}$$

Another relationship is

$$\psi(1) = \Gamma'(1) = -\gamma. \tag{25.23}$$

The constant γ occurs explicitly in the following representation of the gamma function. As already stated, $\Gamma(z)$ is a meromorphic function of z, with a simple pole at every nonpositive integer. Moreover, it turns out to have no zeroes. Hence its reciprocal is an entire function of z, with a simple zero at every nonpositive integer. There is a representation of $\Gamma(z)$ in which these properties are explicit. This is the **Weierstrass infinite product representation** given by

$$\Gamma(z) = \frac{e^{-\gamma z}}{z} \prod_{n=1}^{\infty} \left(\frac{n}{z+n}\right) e^{z/n}. \tag{25.24}$$

Hence

$$\boxed{\frac{1}{\Gamma(z)} = z\, e^{\gamma z} \prod_{n=1}^{\infty} \left(1 + \frac{z}{n}\right) e^{-z/n}.} \tag{25.25}$$

★ **5.** Take logarithms in Eq. (25.24) and differentiate with respect to z to show that

$$\psi(z) = -\frac{1}{z} - \gamma + \sum_{n=1}^{\infty} \frac{z}{n(n+z)}.$$

Set $z = 1$ to verify that $\psi(1) = -\gamma$.

25.2.5 *Connection with the Riemann Zeta Function*

The Riemann zeta function has already been mentioned more than once so far.[2] As stated in Eq. (17.26) of Chap. 17, Sect. 17.3, it is defined as

$$\zeta(z) = \sum_{n=1}^{\infty} \frac{1}{n^z}, \quad \mathrm{Re}\, z > 1. \tag{25.26}$$

The half-plane $\mathrm{Re}\, z > 1$ is the region in which the series above converges absolutely. It is possible to continue $\zeta(z)$ analytically to the left of this region, as you will see

[2] In Chap. 17, Sect. 17.3; Chap. 18, Sect. 18.4.2; and Chap. 23, Sect. 23.4.

in Chap. 26, Sect. 26.2.3. To help you get a better picture, let me anticipate matters a bit, and state an important result:

- $\zeta(z)$ is an analytic function of z that has just one singularity in the finite part of the complex plane: namely, a simple pole at $z = 1$, with residue equal to unity.

Interestingly enough, the regular part of $\zeta(z)$ at $z = 1$ is also equal to γ. Analogous to Eq. (25.21), it turns out that

$$\lim_{z=1} \left\{ \zeta(z) - \frac{1}{z-1} \right\} = \gamma. \tag{25.27}$$

Getting back to Eq. (25.26), this is a good place to mention a few properties of the zeta function based on this representation. The values of the zeta function $\zeta(r)$ for integer values $r = 2, 3, 4, \ldots$ of its argument are of particular interest, because the infinite sums $\sum_{n=1}^{\infty} 1/n^r$ occur often in physical calculations. In Chap. 23, Sect. 23.4, you have seen how $\zeta(2), \zeta(4), \ldots$ can be evaluated by contour integration. $\zeta(2n)$, where n is a positive integer, is π^{2n} times a rational number. Intriguingly, there is no such simple method to find $\zeta(3), \zeta(5), \ldots$ in closed form, but their values can be numerically computed, of course. Some specific values are

$$\zeta(2) = \tfrac{1}{6}\pi^2 \simeq 1.645, \;\; \zeta(3) \simeq 1.202, \;\; \zeta(4) = \tfrac{1}{90}\pi^4 \simeq 1.082, \tag{25.28}$$

and so on. It is obvious that, as $z \to \infty$ along any ray in the right-half-plane, so that $\text{Re}\, z = x \to +\infty$, the function $\zeta(z) \to 1$.

- In the range $1 < x < \infty$, $\zeta(x)$ decreases monotonically with increasing x, and decays exponentially rapidly to its asymptotic value of unity.

The difference between $\zeta(r)$ and 1, therefore, tends to zero as r increases. Interestingly enough, there is a very simple answer for the sum, from $r = 2$ upwards, of the difference $[\zeta(r) - 1]$:

$$\sum_{r=2}^{\infty} [\zeta(r) - 1] = 1. \tag{25.29}$$

★ **6.** Establish Eq. (25.29).

The zeta function has many close connections with the Euler–Mascheroni constant γ and the function $\psi(z)$. Here are a couple of easy examples. It follows from Eq. (25.24) that

$$\sum_{r=2}^{\infty} \frac{(-1)^r \zeta(r)}{r} = \gamma. \tag{25.30}$$

Similarly, it follows that

$$\psi(z+1) = -\gamma - \sum_{r=1}^{\infty} \zeta(r+1)\,(-z)^r. \tag{25.31}$$

★ **7.** From the infinite product formula (25.24) for $\Gamma(z)$, establish (a) Eq. (25.30), and (b) Eq. (25.31).

25.2.6 The Beta Function

Recall that the gamma function arose from the so-called Euler integral of the second kind, namely, $\int_0^\infty dt\, t^{n-1}\, e^{-t} = (n-1)!$ for positive integer values of n. Likewise, the **Euler integral of the first kind** is defined as

$$B(m, n) = \int_0^1 dt\, t^{m-1}\, (1-t)^{n-1}, \quad \text{where } m, n = \text{positive integers.} \tag{25.32}$$

In a manner that should be familiar to you by now, we may replace the integers m and n with complex variables z and w without affecting the convergence of the integral, provided the *real parts* of z and w are positive. This yields the **beta function**, defined (to start with) as

$$\boxed{B(z, w) \stackrel{\text{def.}}{=} \int_0^1 dt\, t^{z-1}\, (1-t)^{w-1}, \quad \text{where } \operatorname{Re} z > 0,\ \operatorname{Re} w > 0.} \tag{25.33}$$

$B(z, w)$ is an analytic function of its arguments in the region indicated. A change of the variable of integration from t to $1-t$ shows that the beta function enjoys the symmetry property

$$\boxed{B(z, w) = B(w, z).} \tag{25.34}$$

As in the case of the gamma function, we can try to extend the region of analyticity to the left in the z-plane by integrating by parts with respect to t. But a problem arises now. Although integrating the factor t^{z-1} improves matters in the z-plane, it also worsens the situation in the w-plane, because the factor $(1-t)^{w-1}$ gets differentiated, and leads to the factor $(1-t)^{w-2}$. The resulting integral converges in the region $\operatorname{Re} z > -1,\ \operatorname{Re} w > 1$. The opposite happens if we integrate the factor $(1-t)^{w-1}$ and differentiate the factor t^{z-1}. Integration by parts cannot extend the region of analyticity in both z and w *simultaneously*. A more sophisticated trick (to be described in Chap. 26, Sect. 26.2.2) is needed to achieve this.

Relation between the beta and gamma functions: The actual structure of the beta function is clarified by relating it to the gamma function as follows. Consider the product of two Gaussian integrals given by

$$I(z, w) = \int_0^\infty du\, u^{2z-1}\, e^{-u^2} \int_0^\infty dv\, v^{2w-1}\, e^{-v^2}. \tag{25.35}$$

The integrals converge in the region Re $z > 0$, Re $w > 0$. Next, we need the Gaussian integral given in Eq. (3.16) of Chap. 3, Sect. 3.1.4 namely,

$$\int_0^\infty dx\, x^n\, e^{-ax^2} = \tfrac{1}{2}\Gamma\left(\tfrac{1}{2}(n+1)\right) a^{-(n+1)/2}, \quad a > 0,\ n > -1. \tag{25.36}$$

This formula will clearly continue to be valid if we replace n by any complex variable α, as long as Re $\alpha > -1$. Doing so, and setting $a = 1$, we have

$$\int_0^\infty dx\, x^\alpha\, e^{-x^2} = \tfrac{1}{2}\Gamma\left(\tfrac{1}{2}(\alpha+1)\right) \quad \text{Re}\ \alpha > -1. \tag{25.37}$$

Applying this formula to the integrals in Eq. (25.35), we get

$$I(z, w) = \tfrac{1}{4}\,\Gamma(z)\,\Gamma(w). \tag{25.38}$$

But we can also regard $I(z, w)$ as the double integral

$$I(z, w) = \int_0^\infty du \int_0^\infty dv\, e^{-(u^2+v^2)}\, u^{2z-1}\, v^{2w-1}, \tag{25.39}$$

the region of integration being the first quadrant of the uv-plane. Let us change variables to plane polar coordinates according to $u = r \cos\theta$ and $v = r \sin\theta$. The integration over r is again a Gaussian integral. Carrying it out yields

$$I(z, w) = \tfrac{1}{2}\Gamma(z+w) \int_0^{\pi/2} d\theta\, (\cos\theta)^{2z-1}\, (\sin\theta)^{2w-1}. \tag{25.40}$$

Changing variables of integration to $t = \cos^2\theta$, we get

$$I(z, w) = \tfrac{1}{4}\Gamma(z+w) \int_0^1 dt\, t^{z-1}\, (1-t)^{w-1} = \tfrac{1}{4}\Gamma(z+w)\, B(z, w). \tag{25.41}$$

Equating the two expressions for $I(z, w)$ found in Eqs. (25.38) and (25.41), we arrive at the basic relationship

$$\boxed{B(z, w) = \frac{\Gamma(z)\,\Gamma(w)}{\Gamma(z+w)}.} \tag{25.42}$$

As always, Eq. (25.42) must be regarded as an equation between two analytic functions, valid for all values of the arguments z and w. It leads directly to a number of important identities satisfied by the gamma function. I shall give two examples below: the reflection formula and the doubling formula for the gamma function.

Before turning to these, let us record the following result for a trigonometric integral, from which a number of other useful integrals follow. (Some of these have been listed in Chap. 3, Sect. 3.1.5.) As you have just seen, for any complex z and w

such that Re $z > 0$, Re $w > 0$,

$$\boxed{\int_0^{\pi/2} d\theta\,(\cos\theta)^{2z-1}\,(\sin\theta)^{2w-1} = \tfrac{1}{2}\,B(z, w) = \frac{\Gamma(z)\,\Gamma(w)}{2\,\Gamma(z+w)}.} \tag{25.43}$$

In particular, if n and m are positive integers,

$$\int_0^{\pi/2} d\theta\,(\cos\theta)^{n-1}\,(\sin\theta)^{m-1} = \tfrac{1}{2}\,B\big(\tfrac{1}{2}n, \tfrac{1}{2}m\big) = \frac{\Gamma\big(\tfrac{1}{2}n\big)\,\Gamma\big(\tfrac{1}{2}m\big)}{2\,\Gamma\big(\tfrac{1}{2}(n+m)\big)}. \tag{25.44}$$

This is the formula quoted in Eq. (3.20) of Chap. 3, Sect. 3.1.5. As stated there, it is a special case of the more general formula (25.43).

25.2.7 *Reflection Formula for* $\Gamma(z)$

Set $w = 1 - z$ in Eq. (25.42), to get

$$\Gamma(z)\,\Gamma(1-z) = B(z, 1-z). \tag{25.45}$$

Now use the defining representation of Eq. (25.33) for the beta function. Then, *provided z is restricted to the strip* $0 < \text{Re}\, z < 1$,

$$\Gamma(z)\,\Gamma(1-z) = \int_0^1 \frac{t^{z-1}\,dt}{(1-t)^z} \quad (0 < \text{Re}\, z < 1). \tag{25.46}$$

The restriction on z ensures the convergence of the integral on the right-hand side. Changing variables to $u = t/(1-t)$, Eq. (25.46) becomes

$$\Gamma(z)\,\Gamma(1-z) = \int_0^\infty \frac{u^{z-1}\,du}{u+1} \quad (0 < \text{Re}\, z < 1). \tag{25.47}$$

The integral on the right-hand side can be evaluated by contour integration, as you will see in Chap. 26, Sect. 26.2.1. The result is the reflection formula for the gamma function,

$$\boxed{\Gamma(z)\,\Gamma(1-z) = \pi \operatorname{cosec} \pi z.} \tag{25.48}$$

Note that z is no longer restricted to any region of the complex plane. Having obtained the value of the integral in (25.47) as $\pi \operatorname{cosec} \pi z$, we can remove the restriction on z by the principle of analytic continuation. Equation (25.48) it is a relation between analytic functions; by the permanence of functional equations, it is valid for *all* z. It is called the (Euler) reflection formula because the point $1 - z$ is just the reflection of the point z about the line $\text{Re}\, z = \frac{1}{2}$, and the left-hand side of the equation is the product of the gamma functions at these symmetrically situated points. As you know, cosec πz has a simple pole at every integer value of z. On the left-hand side, $\Gamma(z)$ supplies the poles at zero and the negative integers, while $\Gamma(1-z)$ supplies the poles at the positive integers.

★ **8.** From the identity (25.48), show that $\left|\Gamma\left(\frac{1}{2}+iy\right)\right|^2 = \pi\,\text{sech}\,(\pi y)$.

The reflection formula for $\Gamma(z)$ leads immediately to a difference equation for the digamma function $\psi(z)$. Equating the logarithmic derivatives of the two sides of Eq. (25.48), we get another beautiful identity, the reflection formula for $\psi(z)$:

$$\boxed{\psi(1-z) - \psi(z) = \pi\,\cot\,\pi z.} \tag{25.49}$$

25.2.8 *Legendre's Doubling Formula*

Another very useful identity that can be derived from the connection between the beta and gamma functions is **Legendre's doubling formula** (also called the **duplication formula**) for the gamma function. This relation expresses $\Gamma(2z)$ as a product of $\Gamma(z)$ and $\Gamma\left(z+\frac{1}{2}\right)$. We start with

$$B(z,z) = \int_0^1 dt\,[t\,(1-t)]^{z-1}, \quad \text{where} \quad \text{Re}\,z > 0. \tag{25.50}$$

But the integrand on the right-hand side is symmetric about the mid-point of integration, $t = \frac{1}{2}$. Hence

$$B(z,z) = 2\int_0^{1/2} dt\,[t\,(1-t)]^{z-1}. \tag{25.51}$$

Change the variable of integration to $u = 4t(1-t)$, to get

$$B(z,z) = 2^{1-2z}\,B\left(z,\tfrac{1}{2}\right). \tag{25.52}$$

Now use the relation (25.42) between the beta and gamma functions to obtain Legendre's doubling formula, namely,

$$\boxed{\Gamma(2z) = \frac{2^{2z-1}}{\sqrt{\pi}}\,\Gamma(z)\,\Gamma\left(z+\tfrac{1}{2}\right).} \tag{25.53}$$

★ **9.** Work through the steps given above to derive the doubling formula (25.53).

Taking logarithmic derivatives on both sides in the doubling formula, we get

$$\psi(2z) = \ln\,2 + \tfrac{1}{2}\psi(z) + \tfrac{1}{2}\psi\left(z+\tfrac{1}{2}\right). \tag{25.54}$$

★ **10.** Show that, for any positive integer n,

(a) $\psi(n+1) = -\gamma + \sum_{j=1}^{n} 1/j$

(b) $\psi(z+n) = \psi(z) + \sum_{j=1}^{n} 1/(z+j-1)$

(c) $\psi\left(\frac{1}{2}\right) = -\gamma - 2\ln 2$

(d) $\psi\left(n+\frac{1}{2}\right) = -\gamma - 2\ln 2 + 2\sum_{j=1}^{n} 1/(2j-1)$.

Multiplication theorem for the gamma function: The doubling formula (25.53) is actually a special case of a multiplication theorem for $\Gamma(z)$, which reads

$$\Gamma(nz) = (2\pi)^{(1-n)/2}\, n^{nz-\frac{1}{2}} \prod_{j=0}^{n-1} \Gamma\left(z+\frac{j}{n}\right), \quad n = 1, 2, \ldots . \tag{25.55}$$

The corresponding theorem for the digamma function reads

$$\psi(nz) = \ln n + \frac{1}{n}\sum_{j=0}^{n-1} \psi\left(z+\frac{j}{n}\right), \quad n = 1, 2, \ldots . \tag{25.56}$$

★ **11.** Equate logarithmic derivatives in Eq. (25.55) to derive Eq. (25.56).

25.3 Solutions

1. Recall that the function $\pi \cot \pi\nu$ has a simple pole at *every* integer value of ν. On the left-hand side of the identity (25.7), these poles are matched by the poles of $Q_\nu(z)$ at $\nu = -1, -2, \ldots$, and by the poles of $Q_{-\nu-1}(z)$ at $\nu = 0, 1, \ldots$. ▶

2. The result follows at once from Eq. (25.12) on using the fact that $\int_0^\infty dt\, e^{-t} = 1$. Alternatively, iterate (25.13) to get

$$\Gamma(z) = \frac{\Gamma(z+n+1)}{z(z+1)\cdots(z+n)},$$

and set $z = -n$ in all factors except $(z+n)^{-1}$ to read off the residue. ▶

3. It should be clear from the foregoing that the poles of $\Gamma(z)$ arise from the behavior of the factor t^{z-1} (or, more generally, the factor t^{z+n-1}) at $t = 0$ in its integral representation. You must, therefore, split this integral according to

$$\Gamma(z) = \int_0^\infty dt\, t^{z-1}\, e^{-t} = \int_0^1 dt\, t^{z-1}\, e^{-t} + \int_1^\infty dt\, t^{z-1}\, e^{-t}.$$

The second integral on the right-hand side does not have any singularities in the region $|z| < \infty$, and is therefore an entire function of z. In the integral from $t = 0$ to $t = 1$, expand e^{-t} in powers of t and integrate term by term, to obtain Eq. (25.14).

Remark Let a be any arbitrary positive number $\neq 1$. You could have chosen to split the integral in the form

$$\int_0^\infty dt\, t^{z-1}\, e^{-t} = \int_0^a dt\, t^{z-1}\, e^{-t} + \int_a^\infty dt\, t^{z-1}\, e^{-t},$$

and proceeded as before. You must explain why this will *not* yield the Mittag-Leffler expansion of $\Gamma(z)$. ▶

6. The left-hand side of Eq. (25.29) is $\sum_{r=2}^\infty \sum_{n=2}^\infty 1/n^r$. Interchange the order of summation in the double sum, and the result follows very easily! ▶

7. (a) Set $z = 1$ in Eq. (25.24), and take logarithms. Use the power series expansion of $\ln(1 + n^{-1})$ *inside* the summation over n.
(b) Take logarithms in Eq. (25.24). Use the power series for $\ln(1 + z/n)$ inside the summation over n. Differentiate both sides with respect to z. Use the functional equation (25.18). ▶

8. Set $z = \frac{1}{2} + iy$ in Eq. (25.48). Use the fact that $\Gamma(z)$ is a real analytic function, which implies the property $\Gamma(z^*) = \Gamma^*(z)$. ▶

9. In particular, check out how to go from Eq. (25.51) to Eq. (25.52): inverting the relation $u = 4t(1-t)$ gives $t = \frac{1}{2}[1 \pm \sqrt{1-u}]$. You must choose the correct root. It helps to draw the graph of u versus t. ▶

10. The results quoted follow quite easily from the difference equation (Eq. (25.18)) and the doubling formula (Eq. (25.54)) for $\psi(z)$. You must also use the fact that $\psi(1) = -\gamma$. ▶

Chapter 26
Multivalued Functions and Integral Representations

26.1 Multivalued Functions

26.1.1 Branch Points and Branch Cuts

An analytic function $f(z) = w$ should be regarded as a map $f : z \mapsto w$ of some region $\mathcal{R}$ of $\hat{\mathbb{C}}$ to some region $\mathcal{R}'$ of $\hat{\mathbb{C}}$. If $f(z)$ is single valued (univalent), this map yields a unique value of w for every value of z. All the functions we have considered so far satisfy this property. In general, however, we have to deal with **multivalued functions** (or multivalent functions).

A simple example is provided by the function $f(z) = z^{1/2}$. It is obvious that the whole of the complex z-plane is mapped to just the upper half-plane in w, because $z = r\, e^{i\theta}$ maps to $w = r^{1/2}\, e^{i\theta/2}$ (see Fig. 26.1). In order to cover the whole of the w-plane, the z-plane has to be covered *twice*: the argument θ of z has to run from 0 to 4π, rather than just 2π. As θ increases from 0 to 2π, we obtain the branch $+\sqrt{z}$ of the function $f(z) = z^{1/2}$. As θ increases further from 2π to 4π, we obtain the branch $e^{i\pi}\, z^{1/2} = -\sqrt{z}$ of the square root function.

In order to keep track of the correct branch, we therefore need *two* copies of the z-plane. The two copies may be imagined to be two *sheets*, one lying below the other, such that we descend to the second sheet by starting just above the positive real axis on the first sheet and traversing a path that encircles the origin once in the positive sense. On encircling 0 once again, this time on the second sheet, we ascend back to the first sheet.

The two sheets (labeled I and II) are called **Riemann sheets**, and they are supposed to be connected to each other along a slit in each sheet running from 0 to ∞ on the positive real axis. The top sheet, on which the phase of z runs from 0 to 2π, is called the **principal sheet** of the function concerned. The points where the two branches of the two-valued function $z^{1/2}$ coincide in value, namely, $z = 0$ and $z = \infty$ (recall that there is only one point at infinity in $\hat{\mathbb{C}}$), are called **branch points**. They are connected by a **branch cut**. The two sheets pasted together as described above form the **Riemann surface** of the function $z^{1/2}$. Figure 26.2a shows this schematically.

V. Balakrishnan, *Mathematical Physics*,
https://doi.org/10.1007/978-3-030-39680-0_26

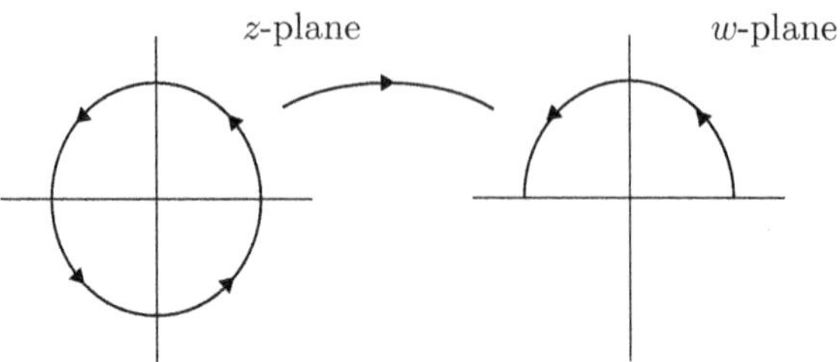

Fig. 26.1 The square root map $w = z^{1/2}$. The whole of the unit circle in the z-plane, for instance, is mapped to just the upper half of the unit circle in the w-plane

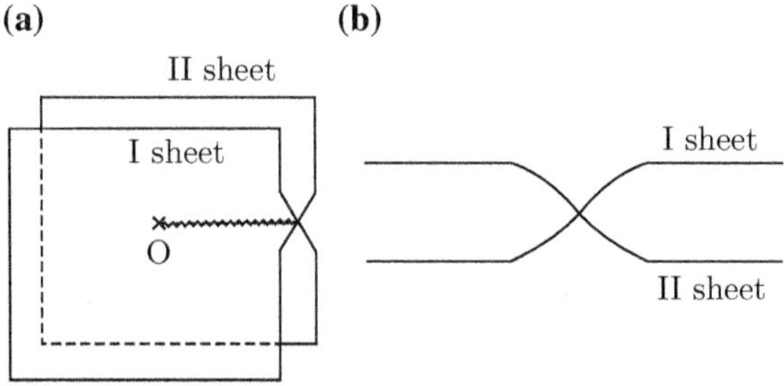

Fig. 26.2 **a** Schematic of the two-sheeted Riemann surface of $z^{1/2}$. **b** An end-view of the two sheets, looking along the positive real axis from infinity toward the origin. In both figures, the two sheets have been drawn with a nonzero separation for clarity of illustration

An "end-view" looking from ∞ toward 0 along the positive real axis is shown in Fig. 26.2b.

- The crucial point is that, though the function $f(z) = z^{1/2}$ is multivalued (in this case, double-valued) on the complex z-plane, it becomes single valued on the two-sheeted Riemann surface.
- This is the fundamental idea behind the introduction of Riemann sheets in the case of multivalued functions.

The branch cut joining the branch points at $z = 0$ and $z = \infty$ may be taken to run along any curve running between these two points. It is most convenient to take it to run along a straight line. We have chosen the positive real axis in the foregoing. But this is by no means the only possible choice. All that is needed is a specification of the phase (or argument) of the function concerned just above and just below the branch cut, so that we can calculate the discontinuity (or the jump in the value) of the function across the branch cut. Figure 26.3 illustrates this in the example under consideration. It shows the first Riemann sheet of the function $z^{1/2}$, and the phases of the function on the positive real axis, just above and just below the cut. For reference, I have also indicated the phase of the function on the negative real axis. Label the function on the sheets I and II as $f_{\mathrm{I}}(z)$ and $f_{\mathrm{II}}(z)$, respectively. By continuity, the value of the function on sheet I, as you approach the positive real axis from below, is

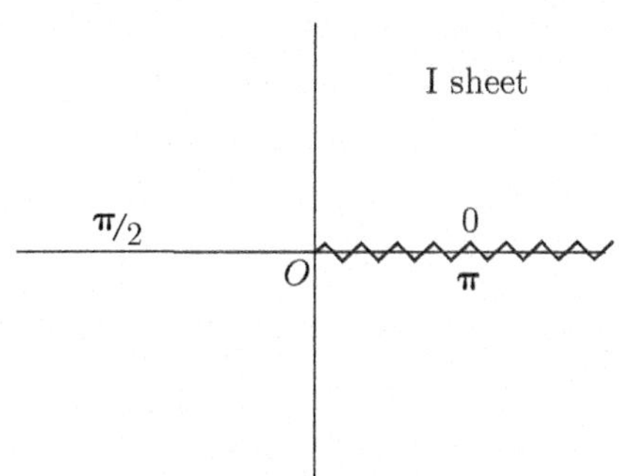

Fig. 26.3 The function $z^{1/2}$ on its first Riemann sheet, showing its phases on the positive real axis above and below the branch cut on the positive real axis, and on the negative real axis

the same as the value of the function on sheet II, as the real axis is approached from above. That is,

$$\lim_{\epsilon\to 0} f_{\rm I}(x - i\epsilon) = \lim_{\epsilon\to 0} f_{\rm II}(x + i\epsilon), \quad x > 0. \tag{26.1}$$

The discontinuity across the cut is then easily determined. In the present instance, it is given by

$$\text{disc } f(z)\Big|_{x>0} \stackrel{\text{def.}}{=} \lim_{\epsilon\to 0} [f_{\rm I}(x + i\epsilon) - f_{\rm I}(x - i\epsilon)] \tag{26.2}$$

$$= \lim_{\epsilon\to 0} [f_{\rm I}(x + i\epsilon) - f_{\rm II}(x + i\epsilon)]$$

$$= \sqrt{x} - (-\sqrt{x}) = 2\sqrt{x}. \tag{26.3}$$

26.1.2 Types of Branch Points

***Algebraic branch point*:** The branch points of $f(z) = z^{1/2}$ at $z = 0$ and ∞ are **algebraic branch points**. So are those of the function $z^{1/3}$, for instance. In this case the Riemann surface has three sheets. The end-view of these, the analog of Fig. 26.2b, is shown in Fig. 26.4. More generally, the function $f(z) = z^{p/q}$, where p is a nonzero integer, q is a positive integer, and p and q have no common factors, has algebraic branch points at $z = 0$ and $z = \infty$. Its Riemann surface comprises q sheets. Each sheet descends smoothly to the one below it as we cross the branch cut; crossing the cut on the lowest sheet brings us back to topmost sheet.

***Winding point*:** The function $f(z) = z^{\alpha}$, where α is not a rational real number, also has branch points at $z = 0$ and $z = \infty$. These are called **winding points**. The Riemann surface of this function has an infinite number of sheets, because $e^{2\pi n i\alpha}$ is never equal to unity for any nonzero integer value of n, positive or negative. These sheets are labelled by an integer $n \in \mathbb{Z}$. On the principal sheet, the phase of z runs from 0 to 2π, as usual, and $n = 0$.

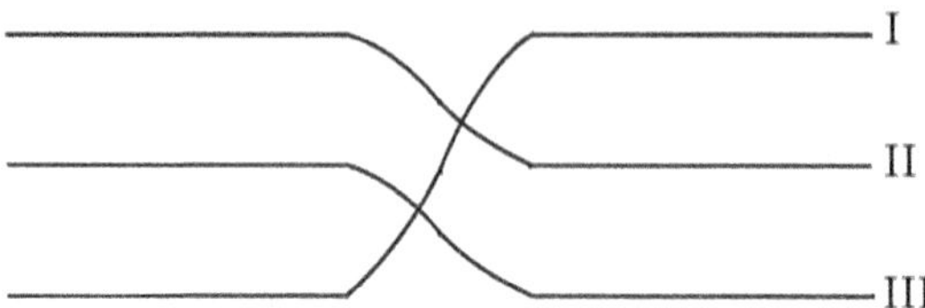

Fig. 26.4 End-view of the three sheets I, II and III of the Riemann surface of $z^{1/3}$, looking along the positive real axis from infinity toward the origin. The sheets are "glued together" as shown, along the positive real axis. As before, the sheets have been depicted with a nonzero separation and the positive real axis has been given a finite width in this sketch, for clarity of illustration

***Logarithmic branch point*:** The function $f(z) = \ln z$ has **logarithmic branch points** at $z = 0$ and $z = \infty$. The Riemann surface is again infinite-sheeted, the sheets being numbered by the full set of integers. On the principal sheet, $\ln z = \ln r + i\theta$, where $0 \leq \theta < 2\pi$. On the nth sheet,

$$\ln z = \ln r + i\theta + 2\pi n i, \quad \text{where } n \in \mathbb{Z} \text{ and } 0 \leq \theta < 2\pi. \tag{26.4}$$

Remember, in particular, that $\ln 1 = 0$ only on the principal sheet. On the nth sheet, $\ln 1 = 2\pi n i$.

Branch cuts run from one branch point to another. No function can have just one branch point; the smallest number of branch points that a function can have is two. If a closed path is traversed in the complex plane such that a branch point of a multivalued function is encircled once (and no other branch point is encircled), the function does *not* return to its original value when we return to the starting point of the path. This becomes obvious if we consider the path on the corresponding Riemann surface of the function, because a single circuit around the branch point will take us to a different Riemann sheet, and the path is therefore no longer a closed path. In order to obtain a closed contour that encircles a branch point, the contour must encircle the branch point (or other branch points) as many times as is necessary to return the function to its original value at the starting point (or, equivalently, to return to the starting point on the original Riemann sheet).

Consider the functions $(z-a)^{1/2}/(z-b)^{1/2}$ and $(z-a)^{1/2}\,(z-b)^{1/2}$, where a and b are any two finite complex numbers. They have algebraic (square root) branch points at $z = a$ and $z = b$, but their behavior at $z = \infty$ is regular. Therefore their branch cuts can be chosen to run over the *finite* segment from a to b. This is shown in Fig. 26.5 for the case in which a and b are real numbers.

For any arbitrary non-integer value of α, including complex values, the cuts of the function $(z-a)^{\alpha}/(z-b)^{\alpha}$ can also be chosen to run over the finite segment from $z = a$ to $z = b$ alone. But this is not possible for the product function $(z-a)^{\alpha}\,(z-b)^{\alpha}$, unless α happens to be a half-odd-integer. This function has branch points at $z = a$ and $z = b$, *as well as* $z = \infty$, for every value of α that is not an integer or half-odd-integer. The branch cut structure of this function must necessarily run up to ∞. Figure 26.6 illustrates these cases.

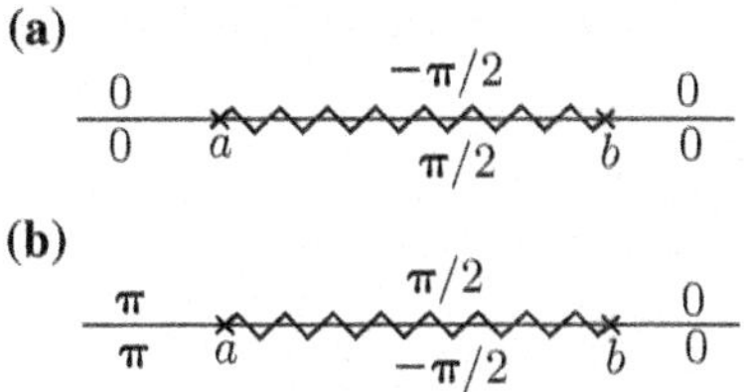

Fig. 26.5 Convenient branch cuts for the functions **a** $(z-a)^{1/2}/(z-b)^{1/2}$ and **b** $(z-a)^{1/2}\,(z-b)^{1/2}$. The phases of the functions in different parts of the real axis on the principal sheet are also indicated. Phase angles of 2π and 0 are equivalent, as are phase angles of $\frac{3}{2}\pi$ and $-\frac{1}{2}\pi$

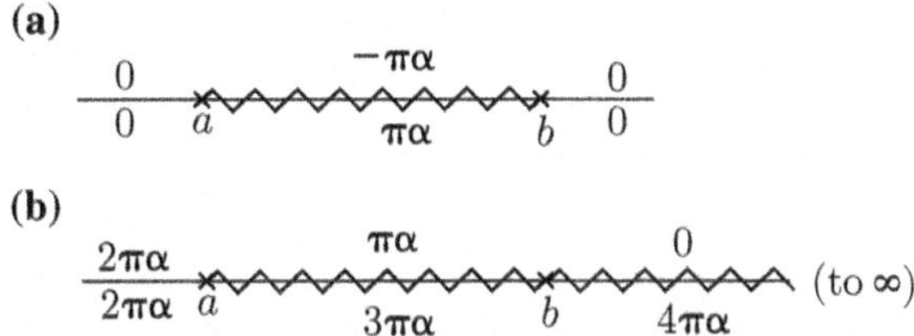

Fig. 26.6 Branch cuts **a** for the function $(z-a)^{\alpha}/(z-b)^{\alpha}$ when α is not an integer, and **b** for the function $(z-a)^{\alpha}\,(z-b)^{\alpha}$ when α is not an integer or half-odd integer. The phases of the functions in different parts of the real axis are also indicated

The singularity structure of multivalued functions can be quite intricate and interesting. Here is a simple example. The function

$$f(z) = z^{-1}\ln(1-z) \tag{26.5}$$

has logarithmic branch points at $z=1$ and $z=\infty$. On the *principal* sheet of the logarithm, we have $\ln 1 = 0$; hence $\ln(1-z) \simeq -z$ in the infinitesimal neighborhood of the origin on this sheet. The simple pole at $z=0$ of the factor z^{-1} is, therefore, "cancelled" by the simple zero of the logarithm, and $f(z)$ only has a removable singularity at $z=0$. There is no pole at $z=0$. On *every other* sheet of the logarithm, however, $\ln 1 = 2\pi n i \neq 0$. Hence $f(z)$ does have a simple pole at $z=0$ on each of these other sheets!

26.1.3 Contour Integrals in the Presence of Branch Points

As you have seen, the evaluation of integrals via contour integration relies, ultimately, on integrating analytic functions over *closed* contours. To repeat what has already been said: In the case of multivalued functions, however, if the contour starts at some point z and encircles a branch point before returning to the starting point, the function does not return to its original value. Hence the contour is not really a closed one—the final point is actually on another Riemann sheet of the function. In order to apply the theorems pertaining to integrals over closed contours, you must ensure that the function has returned to its starting value (or, equivalently, that z has returned to the starting point on the Riemann surface of the function). This might involve encircling more than one branch point, or the same branch point more than once, and so on, as will become clear from the examples that follow.

★ **1.** Let a and b be arbitrary real numbers, where $a < b$. Use contour integration to show that the integral

$$I = \int_a^b \frac{dx}{\sqrt{(b-x)(x-a)}} = 2\pi,$$

independent of a and b.

★ **2.** Use a similar method to show that

$$\int_C \frac{dz}{\sqrt{1+z+z^2}} = 2\pi i,$$

where C is the circle $|z| = 2$ traversed once in the positive sense.

An integral involving a class of rational functions: Consider the integral of a rational function, of the form

$$I = \int_0^\infty dx\, \frac{p(x)}{q(x)}, \tag{26.6}$$

where $p(x)$ and $q(x)$ are polynomials in x that satisfy the following two conditions: (i) The degree of $q(x)$ exceeds that of $p(x)$ by at least 2. Hence, the integrand decays at least as rapidly as $1/x^2$ as $x \to \infty$, and the convergence of the integral is guaranteed. (ii) $q(x)$ does not have any zeroes for $x \geq 0$. Hence there is no non-integrable singularity on the path of integration in Eq. (26.6), and the integral exists. As you know, such an integral can be evaluated by elementary means—for instance, by resolving the integrand into partial fractions.[1] But there is an easier way to evaluate I by contour integration, using a simple trick.

Consider, instead of I, the contour integral

$$I' = -\frac{1}{2\pi i}\int_C dz\, \frac{p(z)\ln z}{q(z)}, \tag{26.7}$$

where C is the familiar hairpin contour that comes in from infinity just below the positive real axis, encircles zero from the left, and runs just above the positive real

[1] If this method is used, care must be taken to avoid the appearance of any spurious logarithmic singularity owing to the upper limit of integration. Evaluate the integral up to some upper limit L, and take the limit $L \to \infty$ carefully, after the different terms are recombined properly.

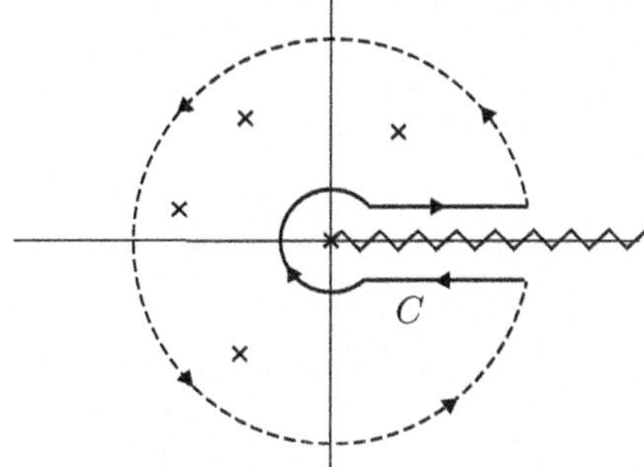

Fig. 26.7 Hairpin contour C (solid line) for the integral in Eq. (26.7). The dotted line shows the large circle to be attached to C to make it a closed contour. The crosses show the possible poles of the integrand (located at the zeroes of $q(z)$)

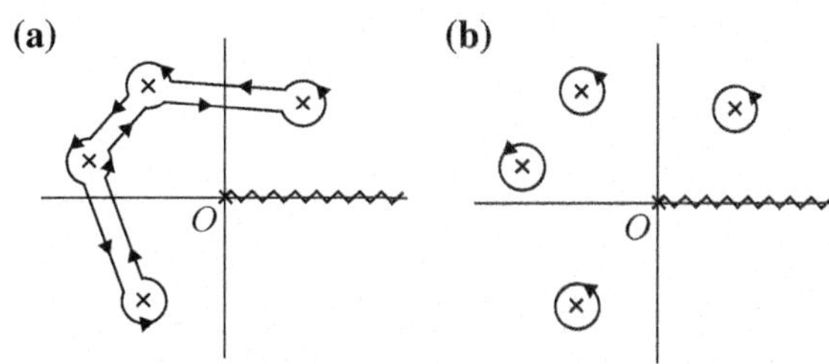

Fig. 26.8 Distortion of the closed contour in Fig. 26.7 to the contour in (**a**), and subsequently to that in (**b**)

axis. The branch cut of $\ln\, z$ is taken to run from 0 to ∞ along the positive real axis in the z-plane. Figure 26.7 shows the branch cut and the contour C. Since $\ln\, z = \ln\, x$ just above the cut, and $\ln\, z = \ln\, x + 2\pi i$ just below the cut, we have

$$I' = -\frac{1}{2\pi i}\int_{\infty}^{0} dx\, \frac{p(x)\,(\ln\, x + 2\pi i)}{q(x)} - \frac{1}{2\pi i}\int_{0}^{\infty} dx\, \frac{p(x)\,\ln\, x}{q(x)} = I. \qquad (26.8)$$

On the other hand, we can evaluate I' by completing the contour C by attaching to it a very large circle (that does not cross the cut), as shown by the dotted lines in Fig. 26.7. This circle is so large in size that all the poles of the integrand, located at the zeroes of $q(z)$, lie inside it. The contribution of the large circle vanishes as its radius R of the circle tends to infinity, because the integrand vanishes at least as fast as $(\ln\, R)/R$ on this circle. Adding the circle to the original contour C, therefore, does not change the value of the integral.

But the closed contour can now be *shrunk* to the contour in Fig. 26.8a, encircling each of the poles of the integrand (located at the zeroes of $q(z)$), and with "channels" connecting the small circles. As the contour shrinks further, the contributions from these channels cancel out, as the two sides of each channel are traversed in opposite directions. We are left with the disconnected small circles encircling each of the poles, as shown in Fig. 26.8b. The residue theorem may then be used to write down the contributions from these contours. The integral I is thus evaluated quite easily.

A couple of standard integrals whose values can be determined using the method described above are

$$\int_0^\infty \frac{dx}{x^n+1} = \frac{\pi}{n} \operatorname{cosec} \frac{\pi}{n} \quad (n = 2, 3, \ldots) \tag{26.9}$$

and

$$\int_0^\infty \frac{dx}{x^{n-1}+x^{n-2}+\cdots+x+1} = \frac{\pi}{n} \operatorname{cosec} \frac{2\pi}{n} \quad (n = 3, 4, \ldots). \tag{26.10}$$

★ **3.** Derive the results quoted in (a) Eq. (26.9) and (b) Eq. (26.10).

26.2 Contour Integral Representations

Contour integrals often provide "master representations" or analytic continuations of various special functions that are valid for all complex values of the arguments of these functions. Some examples follow.

26.2.1 The Gamma Function

In Chap. 25, we discussed analytic continuation with particular reference to the representative (as well as instructive) example of the gamma function. We found that integration by parts provided a way of continuing the function analytically into the left half-plane, *strip by strip*. The same thing could also be achieved with the help of the functional equation $\Gamma(z+1) = z\,\Gamma(z)$ satisfied by the gamma function. But we are now in a position to write down a single integral representation for the gamma function that is valid *throughout* the complex plane.

Recall the original, defining representation of the function given by Eq. (25.9) of Chap. 25, Sect. 25.2.1:

$$\Gamma(z) = \int_0^\infty dt\, t^{z-1}\, e^{-t}, \quad \operatorname{Re} z > 0. \tag{26.11}$$

Now consider the integrand as an analytic function of t, for any given complex value of z. The factor t^{z-1} has branch points at $t = 0$ and $t = \infty$, with a cut running between them. Take the branch cut to run along the positive t-axis. The phase of t^{z-1} just above the cut is 0, while it is $2\pi z$ just below the cut (the factor $e^{-2\pi i}$ is just 1). Consider the integral

$$I(z) = \int_C dt\, t^{z-1}\, e^{-t}, \tag{26.12}$$

where the contour C comes in from ∞ to ϵ just below the cut, encircles the origin in the negative sense in an arc of a circle of radius ϵ, and runs just above the cut from

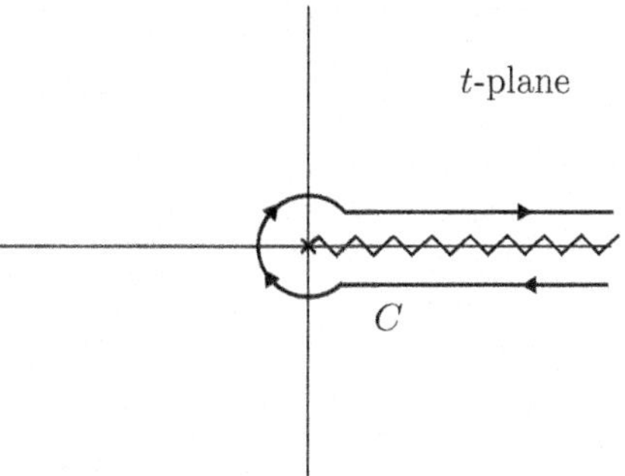

Fig. 26.9 Contour C for the integral in Eq. (26.12)

ϵ to ∞, as shown in Fig. 26.9. *As long as* $\text{Re}\, z > 0$, the contribution from the arc of the small circle vanishes as $\epsilon \to 0$. Moreover, the contributions from the two line segments equal $-e^{2\pi i z}\,\Gamma(z)$ and $\Gamma(z)$, respectively. Hence

$$I(z) = (1 - e^{2\pi i z})\,\Gamma(z) \quad \text{in the region} \quad \text{Re}\, z > 0. \tag{26.13}$$

But C does not pass through the point $t = 0$, and may be deformed to stay clear of the origin, in the form of a hairpin contour straddling the branch cut of t^{z-1}. The contour integral *is thus defined for all finite values of* z. On the other hand, the function $(1 - e^{2\pi i z})^{-1}$ has simple poles at all integer values of z. The product of these two factors is a meromorphic function of z, and represents $\Gamma(z)$ for all z. That is,

$$\boxed{\Gamma(z) = \frac{1}{(1 - e^{2\pi i z})} \int_C dt\, t^{z-1}\, e^{-t} \quad \text{for all} \quad z.} \tag{26.14}$$

Note that the hairpin contour C can be partially "straightened out", but we must always ensure that $\text{Re}\, t \to +\infty$ asymptotically at both ends of the open contour, so that the damping factor e^{-t} ensures the convergence of the integral. In particular, there is in this case no question of "closing the contour" by attaching a large circle to the hairpin contour as in Fig. 26.7.

★ **4.** It can be checked that the known analytic properties of $\Gamma(z)$ follow from the integral representation (26.14).

(a) Owing to the factor $1/(1 - e^{2\pi i z})$, it might appear that $\Gamma(z)$ has a simple pole at *every* integer value of z. But we know that $\Gamma(n)$ is finite (and equal to $(n-1)!$) when n is a positive integer. What happens is that the contour integral in (26.14) *also* vanishes when $z = 1, 2, \ldots$ (the integrand becomes single valued because t^{n-1} does not have any branch points). Verify that, as z tends to any positive integer n, the limiting value of the right-hand side of Eq. (26.14) is precisely $(n-1)!$.

(b) On the other hand, when $z = -n$ where $n = 0, -1, -2, \ldots$, the branch cut from $t = 0$ to $t = \infty$ disappears (the discontinuity of the factor t^{z-1} across the positive real axis vanishes), but now there is a *pole* of order $(n + 1)$ at $t = 0$ in the t-plane. The contour encloses this pole once, in the negative sense. The contour integral is then evaluated easily. (Essentially, you have to pick out the coefficient of $1/t$ in the integrand.) Verify that $\Gamma(z)$ has a simple pole at $z = -n$ with residue equal to $(-1)^n/n!$.

***Another useful integral*:** The argument used above for converting the line integral from 0 to ∞ to a hairpin contour that straddles the real positive axis, avoiding the origin, is a very helpful one. Recall the integral that arose in Eq. (25.47) of Chap. 25, Sect. 25.2.7, in the derivation of the reflection formula for $\Gamma(z)$. We can now show that, for $0 < \text{Re}\, z < 1$,

$$\int_0^\infty \frac{u^{z-1}\, du}{u+1} = \pi \operatorname{cosec} \pi z. \tag{26.15}$$

★ **5.** Derive Eq. (26.15).

26.2.2 The Beta Function

Recall the original definition of the beta function in Eq. (25.33) of Chap. 25, Sect. 25.2.6:

$$B(z, w) = \int_0^1 dt\, t^{z-1}\, (1-t)^{w-1}, \quad \text{where } \text{Re}\, z > 0,\ \text{Re}\, w > 0. \tag{26.16}$$

You have also seen that, in this case, it is not possible to extend the region of analyticity in z and w *simultaneously* by using integration by parts.

Once again, we can continue $B(z, w)$ analytically to all values of z and w by suitably exploiting the discontinuity of the integrand $t^{z-1}\,(1-t)^{w-1}$ across the branch cut running between the branch points at $t = 0$ and $t = 1$. For general values of z and w, there is a branch point at $t = \infty$ as well. The cut structure of the integrand is therefore a little more involved in this case, because the cut actually runs all the way to infinity. But all that we need to keep track of is the following fact: starting on the principal sheet, the integrand acquires a factor $e^{2\pi i z}$ (respectively, $e^{-2\pi i z}$) whenever the contour encircles the branch point at $t = 0$ in the positive (respectively negative) sense. Similarly, a factor $e^{2\pi i w}$ or $e^{-2\pi i w}$ results when the branch point at $t = 1$ is encircled.

Clearly, the least that is required in order to return to the original value of the function, and hence to close the contour, is a *double* encircling of each of the branch points $t = 0$ and $t = 1$, once in the positive sense and once in the negative sense. The outcome is a **Pochhammer contour** that is written symbolically

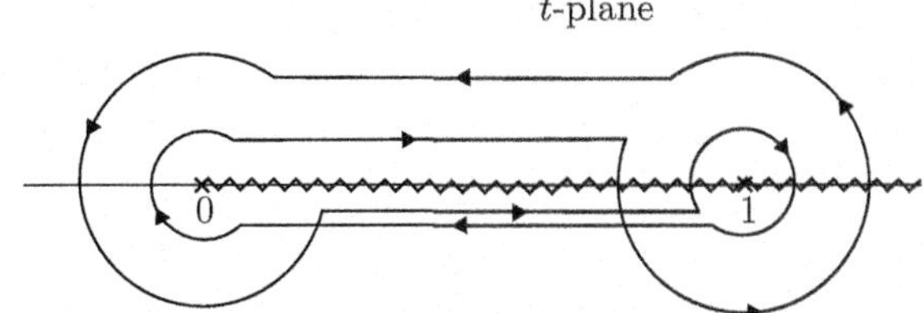

Fig. 26.10 The Pochhammer contour $C = (1-, 0-, 1+, 0+)$ for the integral representation of the beta function

as $C = (1-, 0-, 1+, 0+)$.[2] It is shown in Fig. 26.10. The contributions from the infinitesimal circles around 0 and 1 vanish as long as $\mathrm{Re}\, z > 0$ and $\mathrm{Re}\, w > 0$. Then the right-hand side of Eq. (26.16) can be written in the alternative form

$$\boxed{B(z, w) = \frac{1}{(1 - e^{-2\pi i z})(1 - e^{-2\pi i w})} \int_C dt\, t^{z-1} (1 - t)^{w-1}.} \tag{26.17}$$

But the contour of integration no longer passes through the singularities at $t = 0$ and $t = 1$. The contour can be distorted away from these points without changing the value of the integral. The integral representation of Eq. (26.17) thus provides an analytic continuation of the beta function for *all* finite values of z and w.

26.2.3 The Riemann Zeta Function

The Riemann zeta function $\zeta(z)$, defined as

$$\zeta(z) = \sum_{n=1}^{\infty} \frac{1}{n^z} \quad (\mathrm{Re}\, z > 1), \tag{26.18}$$

was last discussed in Chap. 25, Sect. 25.2.5. It was stated there that an analytic continuation of $\zeta(z)$ to the whole of the z-plane can be found. Once again, a contour integral representation serves the purpose. The infinite sum in Eq. (26.18) converges absolutely for $\mathrm{Re}\, z > 1$. Based on our experience so far, we may expect the analytic continuation of $\zeta(z)$ to have one or more singularities on the *boundary* of the region of convergence, i.e., on the line $\mathrm{Re}\, z = 1$.

In order to obtain a contour integral representation that is valid for all z, we start with z in the region $\mathrm{Re}\, z > 1$. The trick is to consider the product

[2] The term "Pochhammer contour" is used for the class of such contours that encircle several branch points in a specified order, each encirclement being traversed in a specific (either positive or negative) sense.

$$\zeta(z)\,\Gamma(z) = \sum_{n=1}^{\infty} \frac{1}{n^z} \int_0^{\infty} du\, u^{z-1}\, e^{-u}, \quad \text{where } \operatorname{Re} z > 1. \tag{26.19}$$

Change variables of integration by setting $u = nt$. Then

$$\begin{aligned}\zeta(z)\,\Gamma(z) &= \sum_{n=1}^{\infty} \int_0^{\infty} dt\, t^{z-1}\, e^{-nt} = \int_0^{\infty} dt\, t^{z-1} \sum_{n=1}^{\infty} e^{-nt} \\ &= \int_0^{\infty} dt\, \frac{t^{z-1}}{(e^t - 1)}, \quad \text{where } \operatorname{Re} z > 1.\end{aligned} \tag{26.20}$$

We see that the the divergence of the *sum* $\sum_{n=1}^{\infty} n^{-z}$ that occurs when $\operatorname{Re} z = 1$ now becomes a divergence of the *integral* in Eq. (26.20), as follows. Near $t = 0$, the integrand behaves like t^{z-2}, because an extra factor of t^{-1} comes from the denominator $(e^t - 1)^{-1}$. Hence the integral converges when $\operatorname{Re}(z - 2) > -1$, i.e., when $\operatorname{Re} z > 1$.

It is clear that we cannot achieve convergence of the integral to the left of $\operatorname{Re} z = 1$ by resorting to integration by parts, in this case. While $t^{z-1} \to t^z$ upon integration, $(e^t - 1)^{-1} \to (e^t - 1)^{-2}$ upon differentiation. Therefore the behavior of the integrand near $t = 0$ does not improve. But we may exploit (once again) the branch cut of t^{z-1} running from $t = 0$ to $t = \infty$ to convert the integral to a contour integral over a hairpin contour C straddling the branch cut, encircling the origin in the negative sense. The contour is the same as the one used for the gamma function, shown in Fig. 26.9. This gives

$$\boxed{\zeta(z) = \frac{1}{\Gamma(z)\,(1 - e^{2\pi i z})} \int_C dt\, \frac{t^{z-1}}{(e^t - 1)}.} \tag{26.21}$$

The path of integration no longer passes through $t = 0$. The integral representation of the zeta function given by Eq. (26.21) is valid for all z. It represents the analytic continuation of $\zeta(z)$ to the whole of the complex plane.

Now use the reflection formula for the gamma function, $\Gamma(z)\,\Gamma(1 - z) = \pi \operatorname{cosec}(\pi z)$ (Eq. (25.48) of Chap. 25, Sect. 25.2.7). Equation (26.21) can then be rewritten as

$$\boxed{\zeta(z) = \frac{i}{2\pi}\, e^{-i\pi z}\, \Gamma(1 - z) \int_C dt\, \frac{t^{z-1}}{(e^t - 1)}.} \tag{26.22}$$

The integral $\int_C dt\, t^{z-1}/(e^t - 1)$ on the right-hand side of this equation is an entire function of z. This fact enables us to read off a number of interesting properties of the zeta function.

★ **6.** From the formula (26.22), it follows that the only *possible* singularities of $\zeta(z)$ are those coming from the poles of the factor $\Gamma(1 - z)$ at $z = 1, 2, 3, \ldots$.

(a) Show that $\zeta(z)$ has a simple pole at $z = 1$, with residue equal to 1.
(b) Show that $\zeta(z)$ has *no* singularities at $z = n$, where $n = 2, 3, \ldots$, and that $\zeta(n)$ is given by

$$\zeta(n) = \frac{1}{(n-1)!}\int_0^\infty dt\, \frac{t^{n-1}}{(e^t - 1)}$$

in accord with Eq. (26.20).

Hence

- The Riemann zeta function $\zeta(z)$ is a meromorphic function of z in $\mathbb{C}$, whose only singularity is a simple pole at $z = 1$. The residue at this pole is equal to 1.

26.2.4 Connection with Bernoulli Numbers

We found in Chap. 25, Sect. 25.2.5 that $\zeta(2n)$, where n is a positive integer, is equal to π^{2n} multiplied by a rational number less than unity. I also mentioned there that no such simple closed-form expression exists for $\zeta(2n+1)$. In contrast, the value of $\zeta(z)$ when z is a nonnegative integer can be determined quite easily, as follows.

The function $1/(e^t - 1)$ has a simple pole at $t = 0$, with residue equal to unity. The function $t/(e^t - 1)$, therefore, has a removable singularity at the origin, and tends to 1 as $t \to 0$. We take 1 to be the value of the function at $t = 0$, as usual. The Taylor expansion of the function about $t = 0$ is then given by

$$\boxed{\frac{t}{(e^t - 1)} = \sum_{n=0}^{\infty} B_n \frac{t^n}{n!}}, \tag{26.23}$$

where the constants B_n are certain rational numbers, called the Bernoulli numbers. These are *defined* by the expansion above. Thus $t/(e^t - 1)$ is the generating function for the Bernoulli numbers. The first few numbers are found to be

$$B_0 = 1, \ B_1 = -\tfrac{1}{2}, \ B_2 = \tfrac{1}{6}, \ B_4 = -\tfrac{1}{30}, \ B_6 = \tfrac{1}{42}, \tag{26.24}$$

and so on. Interestingly enough, *all the odd Bernoulli numbers B_{2n+1} are equal to zero, except for B_1.*

★ **7.** What is the radius of convergence of the series on the right-hand side of Eq. (26.23)?

★ **8.** Use the expansion (26.23) in the formula (26.22) to establish the following results:

(a) $\zeta(0) = -\frac{1}{2}$.

(b) $\zeta(-2n) = 0, \quad \text{where} \quad n = 1, 2, \ldots.$
(c) $\zeta(1-2n) = -B_{2n}/(2n), \quad \text{where} \quad n = 1, 2, \ldots.$

As I have emphasized repeatedly, the the original series representation of the zeta function, Eq. (26.18), is only valid in the region $\text{Re}\, z > 1$. If we ignore this fact and simply set $z = 0$ in the series, we get the divergent series $1 + 1 + 1 + \cdots$. Similarly, if we set $z = -1$ and -2, respectively, in the series, we get the divergent series $1 + 2 + 3 + \cdots$ and $1^2 + 2^2 + 3^2 + \cdots$. The proper analytic continuation of $\zeta(z)$, however, yields the values $\zeta(0) = -\frac{1}{2}$, $\zeta(-1) = -\frac{1}{12}$ and $\zeta(-2) = 0$. We could (playfully!) write the "equations"[3]

$$\left.\begin{aligned} \text{“}1 + 1 + 1 + \cdots &= -\tfrac{1}{2}, \\ 1 + 2 + 3 + \cdots &= -\tfrac{1}{12}, \\ 1^2 + 2^2 + 3^2 + \cdots &= 0.\text{”} \end{aligned}\right\} \tag{26.25}$$

Although these "equations" are not correct, and appear to be meaningless, they are not without some import. In quantum field theory, one is often faced with formally divergent quantities when calculating physical quantities that are supposed to be finite. The technique of extracting finite values from these divergent expressions is called **regularization**. Several techniques have been evolved for this purpose. A powerful method among these is called **zeta-function regularization**, which involves the analytic continuation of $\zeta(z)$ and certain generalizations of this function. I do not go into any further details here, as it will take us too far afield. Another important regularization technique that is also based on analytic continuation is called *dimensional regularization.* I will use the latter technique in Chap. 29, Sects. 29.4.3 and 29.5, in connection with the problem of finding the inverse of the Laplacian operator.

The Riemann Hypothesis: You have seen that the Riemann zeta function has zeroes at the even negative integers, $-2, -4, \ldots$. These are called the **trivial zeroes** of the zeta function. $\zeta(z)$ also has an infinite number of other zeroes in the strip $0 < \text{Re}\, z < 1$. The famous **Riemann hypothesis** asserts that

- *all* of these nontrivial zeroes lie on the so-called critical line $\text{Re}\, z = \frac{1}{2}$.

It is known that an infinite number of zeroes do lie on the critical line. It is also known that all nontrivial zeroes lie on that line "almost surely" (in the sense of probability theory, i.e., with probability equal to 1). The first 13 billion or so zeroes have indeed been verified explicitly to lie on the critical line. $\zeta(z)$ can be shown to be a real

[3]In his theory of divergent series, Srinivasa Ramanujan arrived at precisely these values for the divergent series concerned. In his second letter to G. H. Hardy in 1913, he wrote, "... the sum of an infinite number of terms of the series $1 + 2 + 3 + 4 + \cdots = -\frac{1}{12}$ under my theory. If I tell you this you will at once point out to me the lunatic asylum as my goal." [*Ramanujan: Letters and Commentary*, Bruce C. Berndt and Robert A. Rankin, Affiliated East West Press, New Delhi, 1995, p. 53.] Ramanujan did not arrive at these results by analytic continuation of the zeta function. He did so by means of a "theory" of divergent series "based on tenuous and imprecise principles and ideas" [Berndt and Rankin, *op. cit.*, p. 33].

analytic function. Therefore if $\zeta(\frac{1}{2} + iy) = 0$ for any real value of y, it follows that $\zeta(\frac{1}{2} - iy) = 0$ as well.

- The Riemann hypothesis is generally regarded as the most important unsolved problem in mathematics today.[4]

It has resisted a rigorous and complete proof for over a century and a half. (You might want to give it a try!) A very large number of other results in mathematics rest on the validity of the hypothesis. There are also several intriguing connections between the distribution of the zeros of $\zeta(z)$ on the critical line, on the one hand, and physical problems, on the other—for instance, the level-spacing of the energy eigenvalues of quantum mechanical systems whose classical counterparts are chaotic, of the eigenvalues of certain classes of random matrices, and of the energy eigenvalues of complex nuclei. It is clear that the zeta function and its counterparts and analogs in classical and quantum dynamical systems hide many secrets yet to be discovered.

26.2.5 *The Legendre Functions $P_\nu(z)$ and $Q_\nu(z)$*

As our last example of integral representations of special functions, let us consider the Legendre functions of the first and second kinds, $P_\nu(z)$ and $Q_\nu(z)$, for complex z and ν. These functions appear in the solution of many physical problems—for instance, in the quantum theory of scattering. I have mentioned $P_\nu(z)$ and $Q_\nu(z)$ briefly in Chap. 16, Sect. 16.4.5, and discussed a few of their properties in Chap. 25, Sect. 25.1.2. They turn out to have the contour integral representations given in Eqs. (26.27) and (26.28) below, for complex values of both the argument z and the order ν.

Consider the function

$$f(t; z, \nu) = \left[\frac{(t^2 - 1)}{2(t - z)}\right]^\nu \tag{26.26}$$

of the complex variable t, for given (in general, complex) values of z and ν. This function has branch points (winding points, in general) at $t = 1$, $t = -1$ and $t = z$. Suppose we start at some point on the principal sheet of $f(t; z, \nu)$, and move in a path that encircles one or more of the branch points. In how many different ways can this be done, such that the function returns to its original value when we return to the starting point? In each case, the path then becomes a closed contour over which $f(t; z, \nu)$ can be integrated.

If either $t = 1$ or $t = -1$ is encircled once in the positive sense, the function acquires a phase factor $e^{2\pi i\nu}$. If $t = z$ is encircled once in the positive sense, it acquires a factor $e^{-2\pi i\nu}$. This shows that there are essentially three such independent

[4]The popular and semipopular literature on the Riemann hypothesis and related matters is fairly extensive. For a rather concise but eminently readable account, see B. Mazur and W. Stein, *Prime Numbers and the Riemann Hypothesis*, Cambridge University Press, 2016.

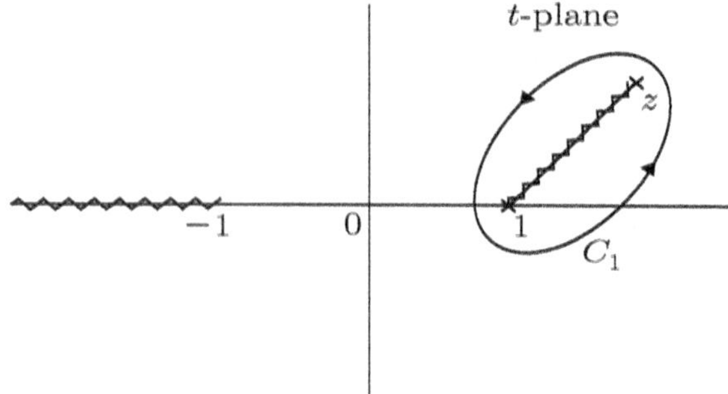

Fig. 26.11 Contour C_1 and the branch cuts of the integrand in the representation of $P_\nu(z)$

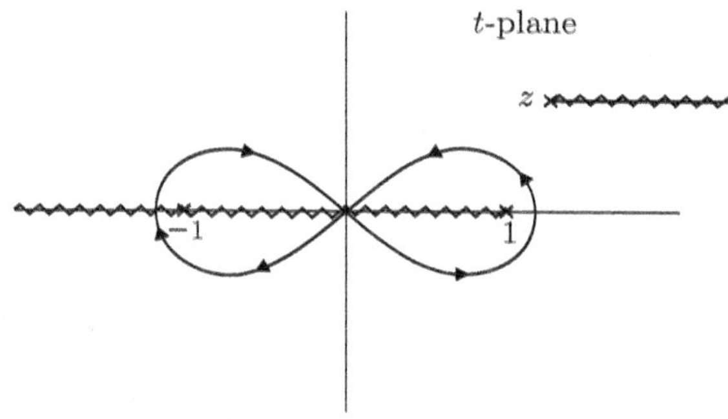

Fig. 26.12 Contour C_2 and the branch cuts of the integrand in the representation of $Q_\nu(z)$

paths. Using the Pochhammer notation, the contours are as follows:

$$\text{(i) } C_1 = (1+, z+), \text{ (ii) } C_2 = (1+, -1-), \text{ and (iii) } C_3 = (-1+, z+).$$

There exist, of course, the reverses of these paths (i.e., paths traversed in the opposite sense), but these will only lead to an overall minus sign in a contour integral. More complicated paths can be decomposed into linear combinations of these basic paths. (Convince yourself that this is so.)

The branch cut structure of $f(t; z, \nu)$ in each of the three cases is implicit in the statement that C_i is a closed contour for the function. In case (i), there is a cut of finite length running between $t = 1$ and $t = z$, and a cut from -1 to $-\infty$ along the negative real axis on the t-plane (say). This is shown, along with the contour C_1, in Fig. 26.11. In case (ii), there is a cut running from each of the three branch points to infinity, for instance as shown in Fig. 26.12. The closed contour C_2 is also shown. Case (iii) is similar to case (i), with the roles of the branch points at 1 and -1 interchanged.

The Legendre functions of the first and second kinds are then given by the formulas

$$\boxed{P_\nu(z) = \frac{1}{2\pi i}\oint_{C_1} \frac{dt}{(t-z)}\left[\frac{(t^2-1)}{2(t-z)}\right]^\nu} \tag{26.27}$$

and

$$Q_\nu(z) = \frac{1}{\pi(e^{2\pi\nu i} - 1)} \oint_{C_2} \frac{dt}{(t - z)} \left[\frac{(t^2 - 1)}{2(t - z)}\right]^\nu, \tag{26.28}$$

respectively. The analytic properties of $P_\nu(z)$ and $Q_\nu(z)$ can be deduced from these representations. Of the extensive set of these properties, I mention the following important ones:

(i) $P_\nu(z)$ ***for non-integer order*:** For general non-integer values of the order ν, the function $P_\nu(z)$ *is no longer a polynomial in* z. It has branch points at $z = -1$ and $z = \infty$. It is conventional to choose the branch cut to run from -1 to $-\infty$ on the real axis (the x-axis) in the z-plane. The discontinuity across the cut at any point x is again proportional to $P_\nu(|x|)$ itself.

(ii) $Q_\nu(z)$ ***for non-integer order*:** Similarly, for general non-integer values of the order ν, the function $Q_\nu(z)$ has branch points at $z = 1$, $z = -1$ and $z = \infty$. It is conventional to choose the branch cut to run from 1 to $-\infty$ on the x-axis. The discontinuity of $Q_\nu(x)$ for $x \in (-1, 1)$ is proportional to $P_\nu(x)$, while the discontinuity at $x < -1$ is proportional to $Q_\nu(|x|)$.

(iii) $P_\nu(z)$ ***for nonnegative integer order*:** Consider $P_\nu(z)$ when $\nu = n$, where $n = 0, 1, \ldots$. The contour C_1 now encloses no branch points, but only a pole of order $(n + 1)$ at $t = z$. The integral is then evaluated easily, and the Rodrigues formula for the polynomial $P_n(z)$ is recovered:

$$P_n(z) = \frac{1}{2^n\, n!} \frac{d^n}{dt^n}(t^2 - 1)^n \bigg|_{t=z} \equiv \frac{1}{2^n\, n!} \frac{d^n}{dz^n}(z^2 - 1)^n. \tag{26.29}$$

This formula has already been encountered in Eq. (16.98) for $P_n(x)$, where $x \in [-1, 1]$, in Chap. 16, Sect. 16.4.1. We see now that it is actually valid for complex values of the argument of $P_n(z)$. Note that $P_n(z)$ continues to remain a polynomial of order n in the complex variable z, as you would expect.

(iv) $P_\nu(z)$ ***for negative integer order*:** When ν is a negative integer $-(n + 1)$, the contour C_1 encloses a pole of order $(n + 1)$ at $t = 1$. Once again, the integral can be evaluated. The symmetry property $P_{-n-1}(z) = P_n(z)$ can be deduced. Recall that this is just a special case of the more general reflection symmetry $P_{-\nu-1}(z) = P_\nu(z)$ that has already been quoted in Eq. (25.4) of Chap. 25, Sect. 25.1.2.

(v) $Q_\nu(z)$ ***for nonnegative integer order*:** Turning to $Q_\nu(z)$, when $\nu = n\, (= 0, 1, \ldots)$, the contour C_2 encloses no singularity at all. Hence the contour integral vanishes. But so does the factor $(e^{2\pi\nu i} - 1)$ in the denominator of the formula (26.28). Their ratio has a finite limit as $\nu \to n$. The outcome is the function $Q_n(z)$, which turns out to have logarithmic branch points at $z = \pm 1$. Recall Eq. (16.117) of

Chap. 16, Sect. 16.4.5 for $Q_n(x)$, in which this logarithmic dependence is explicit. There is a simple, important, and very useful formula that connects $Q_n(z)$, for any arbitrary complex value of the argument z, to the Legendre polynomial of order n. It is

$$\boxed{Q_n(z) = \frac{1}{2}\int_{-1}^{1} dt\, \frac{P_n(t)}{(z-t)}, \quad n = 0, 1, \ldots .} \tag{26.30}$$

Setting z equal to any real value x lying between -1 and 1 appears to lead to a divergence of the integral on the right-hand side of this formula. But you must remember that $Q_n(z)$ has a branch cut running from $z = -1$ to $z = 1$ on the real axis. The *boundary values* of $Q_n(z)$ as z approaches the real axis from above the cut and from below the cut are well-defined. They are easily written down with the help of the formula (23.35) of Chap. 23, Sect. 23.3.4. Repeating this formula for ready reference, we have, after making the obvious replacements $x \to t,\ x_0 \to x$,

$$\lim_{\epsilon \downarrow 0} \frac{1}{t - x \mp i\epsilon} = \mathrm{P}\,\frac{1}{t-x} \pm i\pi\,\delta(t-x), \tag{26.31}$$

where P stands for the principal value integral Using this formula in Eq. (26.30) we get, for $x \in (-1, 1)$,

$$\left.\begin{aligned} \lim_{\epsilon \downarrow 0} Q(x+i\epsilon) &= \frac{1}{2}\,\mathrm{P}\int_{-1}^{1} dt\,\frac{P_n(t)}{x-t} - \frac{i\pi}{2}\,P_n(x), \\ \lim_{\epsilon \downarrow 0} Q(x-i\epsilon) &= \frac{1}{2}\,\mathrm{P}\int_{-1}^{1} dt\,\frac{P_n(t)}{x-t} + \frac{i\pi}{2}\,P_n(x). \end{aligned}\right\} \tag{26.32}$$

Since the integrals in Eq. (26.32) are principal value integrals, they are finite even though x lies on the path of integration. I will return to Eqs. (26.32) in Sect. 26.3.2 below.

(vi) $Q_\nu(z)$ ***for negative integer order*****:** When ν is a negative integer $-(n+1)$, the contour C_2 encloses poles of order $(n+1)$ at $z = -1$ and $z = 1$. The contour integral makes a finite, nonzero contribution. The factor $(e^{2\pi\nu i} - 1)$ in the denominator then leads to a *simple pole* of $Q_\nu(z)$ at $\nu = -(n+1)$. Hence $Q_\nu(z)$ is infinite when ν is a negative integer. The residue at the pole turns out to be $P_n(z)$, as I have stated already in Chap. 25, Sect. 25.1.2.

Based on the foregoing properties, it is possible to write down *dispersion relations* for the functions $P_\nu(z)$ and $Q_\nu(z)$. Equation (26.30) is a spacial case of such a relation in the case when ν is a nonnegative integer. The dispersion relations for a general value of ν are quoted below, in Eqs. (26.49) and (26.50), respectively.

Contour integral representations exist for all other special functions as well. In most cases, they serve as analytic continuations of the corresponding functions to the largest possible domain of their arguments as well as parameters such as the order, degree, and so on.

26.3 Singularities of Functions Defined by Integrals

Integral representations of functions are most useful for exhibiting their analytic properties. The question that arises naturally is: How does a function defined by an integral become singular, and what are the possible singularities? As you might expect, the answer to such a general question involves whole branches of mathematics (such as *homology theory*). Here, I shall narrow the question down very considerably, and examine the simplest possibilities. The treatment below is elementary, heuristic, and essentially based on a few simple examples.

26.3.1 End Point and Pinch Singularities

Consider functions of the form

$$f(z) = \int_a^b dt\, \phi(t, z), \tag{26.33}$$

where the path of integration is an open contour running from some point a to some other point b in the complex t-plane. The integrand $\phi(t, z)$ is assumed to be analytic in t and z, with some singularities. In general, when the integral exists, it defines an analytic function of z. Let us start with a value of z in the region in which $f(z)$ is holomorphic, and examine how the function could develop a singularity as z moves out of this region. This is determined by the behavior of the singularities in the t-plane of the integrand $\phi(t, z)$.

These singularities are of two kinds: they could be *fixed*, i.e., z-independent, such as a pole at some point $t = c$ (not lying on the original path of integration); or they could be *movable*, i.e., z-dependent, such as a pole at $t = z$. As z changes, one or more of the latter could approach the path of integration in the t-plane. If the contour can be distorted away so as to avoid the singularity (keeping it pinned down at the end points a and b, of course), we have an analytic continuation of $f(z)$, defined by the integral over the distorted contour. This is schematically shown in Fig. 26.13. There are two cases in which this simple device will not work. Each of them leads, in general, to a singularity of $f(z)$.

***End point singularity*:** If a moving singularity approaches one of the end points of integration, a or b, the contour cannot be moved away to avoid it. An **end point singularity** of $f(z)$ ensues. Consider the extremely elementary example

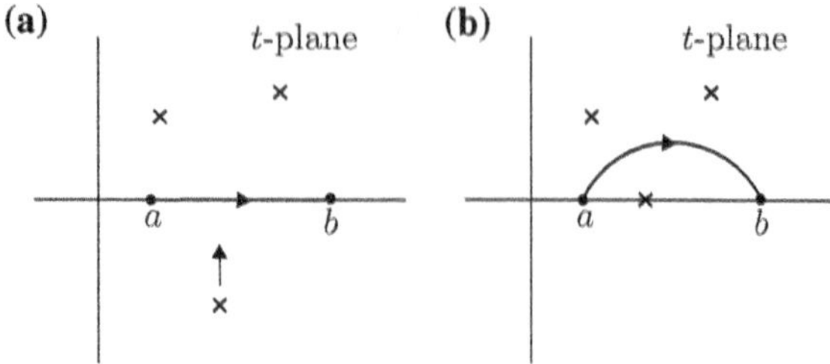

Fig. 26.13 Analytic continuation of $f(z)$ by distorting the contour of integration in the t-plane to avoid a singularity. The small crosses indicate the locations of the singularities of the integrand. **a** The original contour runs from a to b. One of the singularities moves toward the contour as z is varied. **b** The new contour from a to b, distorted so as to avoid a singularity of the integrand that has moved to lie on the original contour

$$f(z) = \int_0^1 \frac{dt}{t - z}. \tag{26.34}$$

The integrand $1/(t - z)$ has a moving pole at $t = z$. (That is, the location of the pole in the t-plane changes as we move from point to point in the z-plane.) As it stands, the integral exists for all $z \notin [0, 1]$. Suppose z approaches a point in the interval $(0, 1)$ of the real axis either from above the real axis or from below it. The pole at $t = z$ in the t-plane moves in the same fashion. The contour of integration can be moved away ahead of the pole, and the integral will continue to exist in either case. As you might guess, however, the analytic continuations of $f(z)$ will differ in the two cases, suggesting already that we are dealing with different branches of $f(z)$. As $z \to 0$ or 1 (in the z-plane), the pole in the t-plane approaches one of the end points of the contour, and a singularity of $f(z)$ occurs. This is corroborated by carrying out the trivial integral (26.34) to obtain the explicit form

$$f(z) = \ln\left(\frac{z - 1}{z}\right). \tag{26.35}$$

It is now obvious that $f(z)$ has logarithmic branch points at $z = 0$ and $z = 1$, confirming our expectation.

A little more generally, consider the function

$$f(z) = \int_0^1 dt\, \frac{\phi(t)}{(t - z)}, \tag{26.36}$$

where $\phi(t)$ is a well-behaved function (for instance, a polynomial in t) such that the integral exists as long as $z \notin [0, 1]$. Once again, $f(z)$ has end point singularities at $z = 0$ and $z = 1$, and these are again logarithmic branch points. Note the important point that

- it is *not necessary* to be able to carry out explicitly the integral defining $f(z)$, in order to reach this conclusion.

The *discontinuity* of $f(z)$ across the branch cut running from $z = 0$ to $z = 1$ on the real axis in the z-plane is also computed easily. Using the formula (26.31) in Eq. (26.36), we get

$$\text{disc } f(z)\Big|_{z=x\in(-1,1)} \stackrel{\text{def.}}{=} \lim_{\epsilon\downarrow 0} \big[f(x+i\epsilon) - f(x-i\epsilon)\big] = 2\pi i\,\phi(x). \tag{26.37}$$

Let us apply the foregoing to Eq. (26.30), an integral representation for the Legendre function $Q_n(z)$, $n = 0, 1, \ldots$. It follows that $Q_n(z)$ has logarithmic branch points at $z = \pm 1$, as we already know. The discontinuity across the branch cut running from -1 to 1 on the real axis is given by

$$\text{disc } Q_n(x)\Big|_{-1<x<1} = \lim_{\epsilon\downarrow 0} \big[Q_n(x+i\epsilon) - Q_n(x-i\epsilon)\big] = -i\pi\, P_n(x). \tag{26.38}$$

Recall that the values of the individual boundary values $Q_n(x + i\epsilon)$ and $Q_n(x - i\epsilon)$ have been written down in Eqs. (26.32).

***Pinch singularity*:** If two moving singularities approach the same point on the path of integration, but from *opposite* sides of the path, the contour gets pinched between them, as shown in Fig. 26.14. A **pinch singularity** of $f(z)$ may then occur. The same thing happens if a moving singularity traps the contour between itself and a fixed singularity of the integrand lying on the other side of the contour.

As an elementary example, consider the function

$$f(z) = \int_{-1}^{1} \frac{dt}{(z^2 - t^2)}\,. \tag{26.39}$$

The integral exists and defines an analytic function of z, as long as $z \notin [-1, 1]$. The integrand has moving poles at $t = \pm z$. We may, therefore, expect end point singularities of $f(z)$ at $z = \pm 1$. Further, the poles at $t = z$ and $t = -z$ pinch the

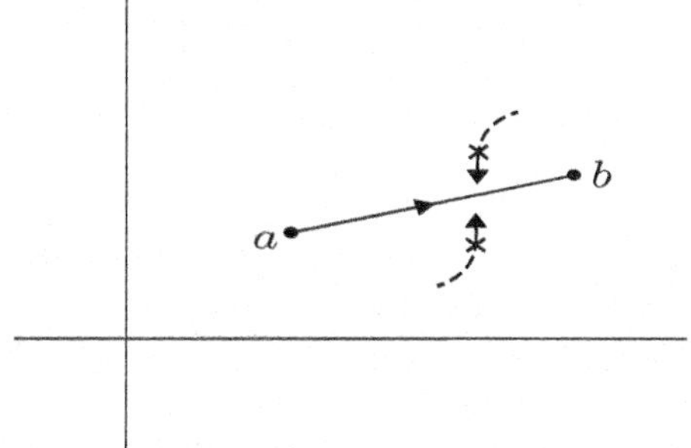

Fig. 26.14 "Pinching" of a contour by two singularities of the integrand approaching a point on the contour from opposite sides

contour of integration from opposite sides as $z \to 0$, and so a pinch singularity may be expected at $z = 0$. Explicit evaluation of the integral gives

$$f(z) = \frac{1}{z} \ln \left(\frac{z+1}{z-1}\right). \tag{26.40}$$

Thus, $f(z)$ does have singularities (logarithmic branch points) at $z = \pm 1$. It also has a pole at $z = 0$, on every sheet of the logarithm *except* the principal sheet, on which $\ln 1 = 0$.[5] As before, a similar analysis is applicable to the more general integral

$$f(z) = \int_{-1}^{1} dt \, \frac{\phi(t)}{z^2 - t^2}, \tag{26.41}$$

where $\phi(t)$ is a well-behaved function such as a polynomial.

The *nature* of an end point or pinch singularity of a function defined by an integral depends also on the kind of moving singularities (of the integrand) that are involved in producing the singularity. In the preceding examples, the singularities of the integrand were simple poles. As an example of what can happen when two branch-points pinch the contour of integration, consider the integal

$$f(z) = \int_{-1}^{1} \frac{dt}{\sqrt{z^2 - t^2}}. \tag{26.42}$$

Again, the integral is very easily evaluated explicitly. But let us first list the possible singularities of the integral as it stands. The integrand has square root branch points at $t = z$ and $t = -z$. End point singularities may be expected at $z = \pm 1$. These should be "mild" singularities, since $(t \pm z)^{-1/2}$ is an integrable singularity. (It is trivially seen that $f(z)$ does not diverge at $z = \pm 1$; instead, it has the finite value π.) Further, as $z \to 0$, the contour of integration is pinched between these branch points. Hence $z = 0$ must also be a singularity of $f(z)$. Note that if we simply *set* $z = 0$ in Eq. (26.42), the integral diverges.

In order to find $f(z)$ explicitly, let us start with z on the positive real axis, such that $z = x > 1$. A simple change of the variable of integration then yields $f(x) = 2 \sin^{-1}(1/x)$. But this is a multivalued function, and you must be careful about its branch structure when writing down its analytic continuation to the rest of the z-plane. It is convenient to rewrite the arcsine function as a logarithmic function, using the identity

$$\sin^{-1} u = -i \ln \left[iu \pm (1 - u^2)^{1/2}\right]. \tag{26.43}$$

In order to choose the right sign before the radical, note the following: When $1 \le x < +\infty$, the integral representing $f(x)$ is real positive; and it decreases monotonically from π to 0 as $x \to \infty$. It follows that

[5]Recall that we have already encountered an example of this feature in Eq. (26.5), Sect. 26.1.2.

$$f(x) = 2i \ln\left(\frac{x}{i + \sqrt{x^2 - 1}}\right), \quad 1 \le x < \infty. \tag{26.44}$$

In this form, the function is (trivially) analytically continued to

$$f(z) = 2i \ln\left(\frac{z}{i + \sqrt{z^2 - 1}}\right). \tag{26.45}$$

Thus, $f(z)$ has singularities at $z = 0$ and at $z = \pm 1$, corroborating our earlier conclusions. The singularity at $z = 0$ is a logarithmic branch point, while those at $z = \pm 1$ are square root branch points.

26.3.2 Singularities of the Legendre Functions

Finally, let us apply the foregoing to the integral representations for the Legendre functions $P_\nu(z)$ and $Q_\nu(z)$ given by Eqs. (26.27) and (26.28), respectively. Here is what happens for a general value of the index ν:

- As $z \to -1$, the contour C_1 gets pinched between the moving singularity of the integrand at $t = z$ and the fixed singularity at $t = -1$. Hence $P_\nu(z)$ has a singularity at $z = -1$. The branch cut of $P_\nu(z)$ is customarily taken to run from -1 to $-\infty$ along the negative real axis in the z-plane.
- As $z \to \pm 1$, the contour C_2 gets pinched between the singularities of the integrand at $t = z$ and at $t = \pm 1$, respectively. Hence $Q_\nu(z)$ has a singularities at $z = 1$ and $z = -1$. The branch cut of $Q_\nu(z)$ is customarily taken to run from 1 through -1 to $-\infty$ along the real axis in the z-plane.

With a little effort, you can show that the discontinuities of $P_\nu(z)$ and $Q_\nu(z)$ across their branch cuts are as follows:

$$\text{disc } P_\nu(x)\Big|_{-\infty < x < -1} = 2i \sin(\pi\nu)\, P_\nu(-x), \tag{26.46}$$

$$\text{disc } Q_\nu(x)\Big|_{-\infty < x < -1} = 2i \sin(\pi\nu)\, Q_\nu(-x), \tag{26.47}$$

$$\text{disc } Q_\nu(x)\Big|_{-1 < x < 1} = -i\pi\, P_\nu(x). \tag{26.48}$$

★ **9.** Establish Eqs. (26.46)–(26.48).

***Dispersion relations for the Legendre functions*:** Finally, based on Eqs. (26.46)–(26.48), *dispersions relations* can be derived for the Legendre functions of both kinds. As you would expect, the asymptotic ($|z| \to \infty$) behaviors of $P_\nu(z)$ and $Q_\nu(z)$ are also required to arrive at these formulas. I shall not go into this aspect here. To merely quote the dispersion relations obtained

$$P_\nu(z) = \frac{\sin \pi\nu}{\pi} \int_1^\infty dt\, \frac{P_\nu(t)}{(z+t)} \tag{26.49}$$

and

$$Q_\nu(z) = \frac{1}{2} \int_{-1}^{1} dt\, \frac{P_\nu(t)}{(z-t)} + \frac{\sin \pi\nu}{\pi} \int_1^\infty dt\, \frac{Q_\nu(t)}{(z+t)}. \tag{26.50}$$

Since $\sin \pi n = 0$ and $Q_n(t)$ is finite for any nonnegative integer n, it is evident that Eq. (26.30) is a special case Eq. (26.50).

26.4 Solutions

1. Consider the function $f(z) = (z-a)^{-1/2}\,(z-b)^{-1/2}$. Note that the behavior of the function at ∞ is regular. Choose the branch cut to run from a to b, and write down the phases of the function above and below the cut. Relate the line integral I to a contour integral surrounding the branch cut. Now open out the contour to make it a large circle whose radius tends to infinity, and pick up the (nonvanishing!) contribution from the circle to arrive at the result.

Remark Note that this method makes it obvious why the value of the integral is independent of the actual values of a and b. ▶

2. The branch points of the integrand are at the two cube roots of unity other than 1, namely, at $z = \omega = e^{2\pi i/3}$ and $z = \omega^2 = e^{4\pi i/3}$. The branch cut of the integrand may be taken to run between these points. As $|z| \to \infty$ along any direction, the integrand tends to $1/z$. Expand the contour to a large circle and exploit this fact. ▶

3. (a) The poles of the integrand that are enclosed by C are at the roots of the equation $z^n + 1 = 0$, or $z^n = -1 = e^{i\pi}$. These are the points $z = e^{i\pi/n}\,\omega^r$, where $\omega = e^{2\pi i/n}$ and $r = 0, 1, \ldots, n-1$.

(b) The poles of the integrand that are enclosed by C are at the roots of unity other than 1 itself, i.e., at $z = \omega^r$, where $r = 1, \ldots, n-1$.

Remark There is a simpler (and standard) way of evaluating the integral in Eq. (26.9). The original integral runs from 0 to ∞ along the positive real axis. Consider the integral of the function $1/(z^n + 1)$ around the closed contour shown in Fig. 26.15. The contribution from the arc of the large circle vanishes as its radius $R \to \infty$. On the slanted line, $z = r\, e^{2\pi i/n}$, so that $dz = dr\, e^{2\pi i/n}$ and $z^n = r^n$. The contour encloses a single pole of the integrand, at $z = e^{i\pi/n}$. The integral is therefore computed very easily, as you should verify. ▶

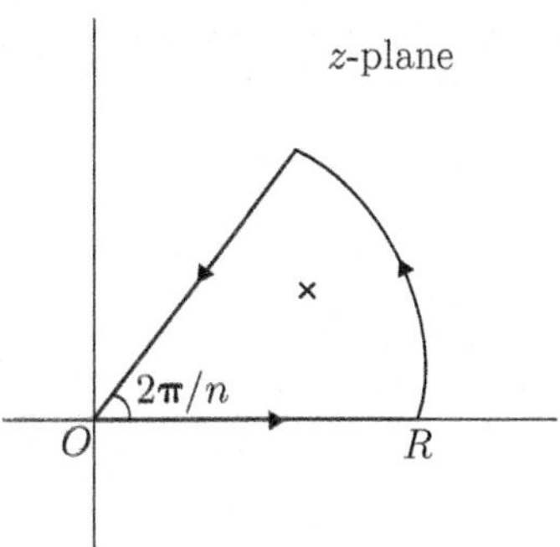

Fig. 26.15 Contour for the evaluation of the integral in Eq. (26.9). The small cross marks the simple pole of the integrand located at the point $e^{i\pi/n}$ on the unit circle

5. Let I denote the line interval to be evaluated. Consider the integral

$$\int_C du\, u^{z-1}/(u+1),$$

where C is the closed contour shown in Fig. 26.16. Choose the branch cut of the function u^z in the u-plane to run from 0 to ∞ along the positive real axis. The phase of u^{z-1} is 0 just above the cut, and $2\pi(z-1)$ just below it. The contribution from the small circle of radius ϵ vanishes (like ϵ^z) as $\epsilon \to 0$ as long as $0 < \text{Re}\, z < 1$. Similarly, the contribution from the large circle of radius R vanishes (like $1/R^{1-z}$) as $R \to \infty$, again as long as $0 < \text{Re}\, z < 1$. It follows that the integral over the closed contour C is equal to $(1 - e^{2\pi iz})\, I$. But now the contour C can be shrunk till it encircles the simple pole of the integrand at $z = -1 = e^{i\pi}$ once, in the positive sense. Hence

$$\int_C du\, \frac{u^{z-1}}{u+1} = 2\pi i\, e^{i\pi(z-1)} = -2\pi i\, e^{i\pi z}.$$

The quoted result for I follows immediately. ▶

6. (a) When $z = 1$, the branch point of t^z at $t = 0$ disappears. It is replaced by a simple pole that comes from the factor $(e^t - 1)^{-1}$ in the integrand, because

$$\frac{1}{e^t - 1} = \frac{1}{t} + \text{regular part}$$

in the neighborhood of $t = 0$. The discontinuity across the branch cut vanishes, and the contour reduces to a closed path encircling the pole at the origin once, in the negative sense. Hence the integral reduces to $\oint dt/t = -2\pi i$. The singular part of $\zeta(z)$ at $z = 1$ is therefore given by

$$\frac{i}{2\pi}\, e^{-i\pi}\, \frac{1}{(1-z)}\, (-2\pi i) = \frac{1}{z-1},$$

i.e., a simple pole at $z = 1$, with residue 1.

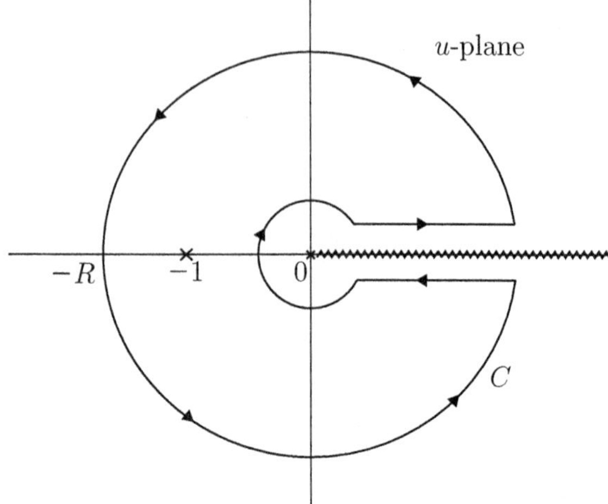

Fig. 26.16 Closed contour C for the evaluation of the line integral in Eq. (26.15)

(b) Again, when $z = n$ (where $n = 2, 3, \ldots$), the branch points of t^z at $t = 0$ and $t = \infty$ disappear, and the contour C encloses no singularities of the integrand. Hence the integral in Eq. (26.22) vanishes when $z = n$. On the other hand, the factor $\Gamma(1 - z)$ has a pole, and therefore diverges, at $z = n$. The product of the two factors yields to a finite, nonzero value for $\zeta(n)$ in the limit $z \to n$.

To find this limiting value, first write $\Gamma(1 - z)$ as $\pi/(\Gamma(z) \sin \pi z)$. The leading behavior of $e^{-i\pi z}\Gamma(1 - z)$ as $z \to n$ is then $1/[(n - 1)!\,(z - n)]$. Now consider the integral in Eq. (26.22). Regard the factor t^{z-1} as a function of z, and expand it in a Taylor series about the point $z = n$:

$$t^{z-1} = t^{n-1} + (z - n)\, t^{n-1} \ln t + \cdots .$$

The contribution from the first term (t^{n-1}) vanishes, as already explained. Hence the limiting value of $\zeta(z)$ as $z \to n$ is

$$\zeta(n) = \frac{i}{2\pi(n - 1)!} \int_C dt\, \frac{t^{n-1} \ln t}{(e^t - 1)} \quad (n = 2, 3, \ldots).$$

But the contour integral can now be written as

$$\int_C dt\, \frac{t^{n-1} \ln t}{(e^t - 1)} = \int_0^\infty dt\, \frac{t^{n-1} \ln t}{(e^t - 1)} + \int_\infty^0 dt\, \frac{t^{n-1} (\ln t + 2\pi i)}{(e^t - 1)}$$
$$= -\, 2\pi i \int_0^\infty dt\, \frac{t^{n-1}}{(e^t - 1)} .$$

Simplifying, we have

$$\zeta(n) = \frac{1}{(n-1)!} \int_0^\infty dt \, \frac{t^{n-1}}{e^t - 1}, \quad (n = 2, 3, \ldots)$$

in agreement with the expression that we get upon setting $z = n$ in Eq. (26.20). ▶

7. $|B_2|, |B_4|, |B_6|, \ldots$ *appears* to be a decreasing sequence. You might, therefore, guess that the series in Eq. (26.23) ought to converge at least as well as the series for the exponential e^t. The radius of convergence would then be infinite, and the series would represent an entire function. But the left-hand side of Eq. (26.23) shows that the function represented by the series has poles at $t = 2\pi n i$, where $n = \pm 1, \pm 2, \ldots$. Hence the radius of convergence of the series must be the distance from the origin to the nearest of these singularities, i.e., 2π. In fact, the magnitude $|B_{2n}|$ of the Bernoulli numbers actually *increases* quite rapidly with increasing n. (The sign of B_{2n} oscillates.) ▶

8. When z is zero or a negative integer, the factor t^{z-1} in the integrand in Eq. (26.22) does not have any branch points. Instead, there is a pole at $t = 0$. The contour C then collapses to a small circle encircling the pole once in the clockwise sense. The integral is easily evaluated by the residue theorem. All you have to do is to apply the basic result in Eq. (23.21) of Chap. 23, Sect. 23.3.1: for any integer r, the integral $\oint dt/t^{r+1} = -2\pi i\, \delta_{r,0}$ (the minus sign coming from the negative sense in which the pole is encircled here). Using this fact, we find

$$\text{(a)}\, \zeta(0) = \frac{i}{2\pi} \oint \frac{dt}{t^2} \Big(B_0 + B_1\, t + \frac{B_2 t^2}{2!} + \frac{B_4 t^4}{4!} + \cdots \Big) = B_1 = -\frac{1}{2}.$$

$$\text{(b)}\ \zeta(-2n) = \frac{i\, \Gamma(2n+1)}{2\pi} \oint \frac{dt}{t^{2n+2}} \Big(B_0 + B_1\, t + B_2 \frac{t^2}{2!} + B_4 \frac{t^4}{4!} + \cdots \Big) = 0.$$

$$\text{(c)}\ \zeta(1-2n) = \frac{-i\, \Gamma(2n)}{2\pi} \oint \frac{dt}{t^{2n+1}} \Big(B_0 + B_1\, t + \frac{B_2 t^2}{2!} + \frac{B_4 t^4}{4!} + \cdots \Big) = -\frac{B_{2n}}{2n}. \ ▶$$

9. In order to find the discontinuity of $P_\nu(z)$ across the cut from -1 to $-\infty$, consider its integral representation (26.27) in the respective cases when $z = x + i\epsilon$ and $z = x - i\epsilon$, where $x < -1$. The contour C_1 in the two cases is shown in Fig. 26.17a and (b), respectively, along with the branch cuts of the integrand. Write down the contour integrals segment by segment, keeping careful track of the phases of the various factors in the integrand. Compare the result with the contour integral for $P_\nu(-x)$ (where $-x > 1$), to arrive at Eq. (26.46). A similar procedure will yield Eqs. (26.47) and (26.48).

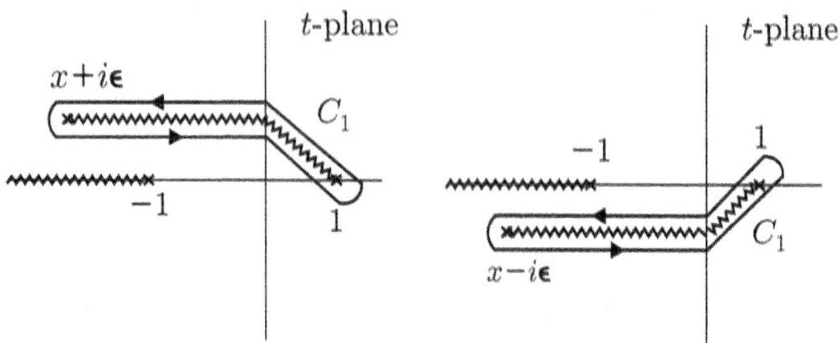

Fig. 26.17 The contour C_1 in the t-plane for $P_\nu(z)$ when $z = x + i\epsilon$ and $z = x - i\epsilon$, respectively, where $x < -1$

Remark Consider what happens when $\nu = n$, a nonnegative integer. As expected, the discontinuity of $P_n(z)$ vanishes: $P_n(z)$ is a polynomial in z, and it has no singularity at $z = -1$ or at any other point in $\mathbb{C}$, the finite part of the complex z-plane. Similarly, $Q_n(z)$ has no branch cut running from -1 to $-\infty$, since the discontinuity across this cut vanishes identically. The discontinuity of $Q_n(z)$ between -1 and 1 is nonzero, as you have seen already in Eq. (26.38). Equation (26.48) is a generalization of this result to arbitrary values of the order ν. ▶

Chapter 27
Möbius Transformations

27.1 Conformal Mapping

As stated at the beginning of Chap. 26, every analytic function $f(z)$ is a map of some region $\mathcal{R}$ of the complex plane to some region $\mathcal{R}'$ of the complex plane. In other words, given a complex number $z \in \mathcal{R}$, the map $f : z \mapsto w$ produces another complex number $w \in \mathcal{R}'$. Under such a mapping, points in $\mathcal{R}$ map to points in $\mathcal{R}'$, curves to curves, and areas to areas, in general.

- The special feature of an *analytic* function $f(z)$ is that this map is *conformal*. That is, it preserves the angles between curves.

Conformal mapping is a standard and very well-studied topic in complex analysis. It is also one that has a number of practical uses—for instance, in fluid dynamics and aerodynamics. I shall not go into conformal mapping *per se* here. Instead, in this chapter, the focus is on one specific kind of conformal map, because it is a very special one in a sense that will become clear shortly. It is also of fundamental importance in many areas of mathematical physics.

A natural question that arises is whether the mapping is one-to-one: namely, is there a *unique* $w = f(z)$ for every given z, and *vice versa*? If the function $f(z)$ is single-valued, then by definition there is a unique w for every z. The converse is not necessarily true, of course. It is clear that a *nonlinear* function such as $f(z) = z^2$ does not satisfy this requirement. While there is a unique $w = z^2$ for every z, we cannot find z uniquely if we are given w. We can only do so up to an overall sign. Or else, as you know, we agree to map the complex w-plane onto a *two-sheeted Riemann surface*, on which $z = w^{1/2}$ is single-valued. On the other hand, it may well be possible to map some of the specific, restricted part $\mathcal{R}$ of the complex plane to another such part $\mathcal{R}'$, in a one-to-one manner. Indeed, in most of the standard applications of conformal mapping, this is all that is sought.

V. Balakrishnan, *Mathematical Physics*,
https://doi.org/10.1007/978-3-030-39680-0_27

The more general question of whether there is a one-to-one conformal mapping $\mathbb{C} \mapsto \mathbb{C}$ has a simple and rather trivial answer:

- The most general one-to-one mapping that takes the whole of the *finite* part of the complex plane to itself is the linear map

$$f(z) = az + b, \tag{27.1}$$

  where a and b are complex constants.
- It is obvious that the point at infinity is mapped to itself under the map (27.1).

In terms of the real and imaginary parts of $z = x + iy$, this map amounts to the linear transformations

$$x \mapsto a_1\, x - a_2\, y + b_1\,, \quad y \mapsto a_2\, x + a_1\, y + b_2\,, \tag{27.2}$$

where $a = a_1 + ia_2$ and $b = b_1 + ib_2$. In geometrical terms, this map is made up (in general) of a rotation, dilation, and shear in the xy-plane, followed by a shift of the origin. Interesting as it is, the map is quite simple. The matter appears to end here. But this is not quite so, as you will see below.

27.2 Möbius (or Fractional Linear) Transformations

27.2.1 *Definition*

Things become much more interesting when we consider the *extended* complex plane $\hat{\mathbb{C}}$ (that is, we include the point at infinity), *and permit the point at infinity to be mapped to some other point.* Remember that $\hat{\mathbb{C}}$ is essentially equivalent to the Riemann sphere $\mathcal{S}$ introduced in Chap. 22, Sect. 22.2. A nontrivial result emerges now.

- The most general one-to-one, conformal map $\hat{\mathbb{C}} \mapsto \hat{\mathbb{C}}$ of the extended complex plane to itself (or of the Riemann sphere to itself, $\mathcal{S} \rightarrow \mathcal{S}$) is of the form

$$\boxed{w = f(z) = \frac{az + b}{cz + d}\,, \quad (ad - bc \neq 0)} \tag{27.3}$$

  where a, b, c, d are arbitrary finite, complex constants. This is called a **fractional linear transformation**, or a **Möbius transformation**.

The latter name is more commonly used when the transformation is regarded as a mapping of the Riemann sphere to itself. I will use this term for the most part. There are several other names for this transformation—*linear fractional transformation*, *homographic transformation*, *projective transformation*, and *bilinear transformation*.

Note that the condition $ad - bc \neq 0$ is required. If $ad - bc = 0$, then $w = f(z)$ trivially reduces to a constant for every z, which is not very interesting. Further, it is obvious that we can divide each of the four constants a, b, c, and d by any nonzero number without affecting the ratio $(az + b)/(cz + d)$ that defines the transformation. In particular, we can divide through by $(ad - bc)^{1/2}$. In effect, therefore,

- $(ad - bc)$ can always be taken to be equal to 1, without loss of generality.

I will assume henceforth that this has been done, unless specified otherwise.

The **inverse** of a Möbius transformation is also a Möbius transformation. We have

$$\boxed{z = f^{-1}(w) = \frac{dw - b}{-cw + a}.} \tag{27.4}$$

The same determinant condition, $ad - bc = 1$, remains valid for this transformation too.

Every Möbius transformation maps a certain point in the z-plane to ∞. If $c \neq 0$, this point is $z = -d/c$. Similarly, the point at infinity maps to a/c. That is,

$$z = -\frac{d}{c} \mapsto w = \infty \quad \text{and} \quad z = \infty \mapsto w = \frac{a}{c}. \tag{27.5}$$

If $c = 0$, we have just a linear transformation, and $z = \infty$ maps to $w = \infty$.

27.2.2 Fixed Points

The **fixed points** of a Möbius transformation, i.e., points that map to themselves under the transformation, are of fundamental importance. They help classify the transformations into distinct types, as you will see later.

The fixed points are the roots of the equation $f(z) = z$, or

$$cz^2 + (d - a)z - b = 0. \tag{27.6}$$

It is immediately clear that a Möbius transformation can have at most *two* fixed points. Let us denote these by ξ_1 and ξ_2. (The only exception is the identity transformation, which of course, maps every point to itself.) Two cases arise, each with a sub-case, so that there are four distinct possibilities:

(i) ***Finite, distinct fixed points*****:** In the general case, when $c \neq 0$ and $a + d \neq \pm 2$, there are two distinct, finite, fixed points. Remembering that $ad - bc = 1$, we have

$$\boxed{\xi_{1,2} = \frac{a - d \pm \sqrt{(a+d)^2 - 4}}{2c}, \quad \text{when } c \neq 0 \text{ and } a + d \neq \pm 2.} \tag{27.7}$$

(ii) ***Finite, coincident fixed points*:** If $c \neq 0$ but $a + d = \pm 2$, the two fixed points coincide. We have

$$\boxed{\xi_1 = \xi_2 = \frac{a - d}{2c} = \frac{a \mp 1}{c}, \quad \text{when } c \neq 0 \text{ and } a + d = \pm 2.} \tag{27.8}$$

(iii) ***One fixed point at*** ∞**:** If $c = 0$, the Möbius transformation reduces to the *linear* transformation

$$w = \frac{az + b}{d} = a^2 z + ab, \tag{27.9}$$

since $ad = 1$ in this case. The fixed points are

$$\boxed{\xi_1 = \frac{b}{d - a} = \frac{ab}{1 - a^2} \quad \text{and} \quad \xi_2 = \infty, \quad \text{when } c = 0.} \tag{27.10}$$

(iv) ***Both fixed points at*** ∞**:** If $c = 0$ and further $a = d$, then the linear transformation is merely a shift of the origin,

$$w = z + \frac{b}{a} = z \pm b, \tag{27.11}$$

since $a^2 = 1$ in this case. The fixed points then coincide at infinity. That is,

$$\boxed{\xi_1 = \xi_2 = \infty, \quad \text{when } c = 0 \text{ and } a = d.} \tag{27.12}$$

I will return shortly to the role played by these fixed points.

27.2.3 The Cross-Ratio and Its Invariance

The **cross-ratio** of four points is a basic concept in geometry. It plays a fundamental role in the subject of **projective geometry**. It is extremely useful in the context of Möbius transformations, as many results can be established quite easily with the help of its properties.

Let (z_1, z_2, z_3, z_4) be an ordered set of four distinct points in the complex plane. The cross-ratio of these points is defined as

$$\boxed{[z_1, z_2; z_3, z_4] \stackrel{\text{def.}}{=} \frac{(z_1 - z_3)(z_2 - z_4)}{(z_1 - z_4)(z_2 - z_3)}.} \tag{27.13}$$

Figure 27.1 shows the distances involved in the cross-ratio, for a particular configuration of points. It is obvious that the cross-ratio of four points depends on the order in which they are specified. Hence there are $4! = 24$ possible cross-ratios that can be associated with an unordered set of four points. However, there are several obvious symmetries among these, and the 24 cross-ratios are related to each other. Let us write $[z_i, z_j; z_k, z_l]$ as $[ij; kl]$, for brevity. It is easy to show that

$$[ij; kl] = \frac{1}{[ij; lk]} = \frac{1}{[ji; kl]} = [ji; lk] = [kl; ij]. \tag{27.14}$$

The identities

$$[ik; jl] = 1 - [ij; kl] \quad \text{and} \quad [il; jk] = 1 - \frac{1}{[ij; kl]} \tag{27.15}$$

are also easily established. Using these, it is possible to express all 24 cross-ratios among any set of 4 points in terms of any one cross-ratio (see below).

If one of the four points in a cross-ratio is the point at infinity, the cross-ratio is defined by a limiting process. For instance, if $z_1 = \infty$, we have

$$[\infty, z_2; z_3, z_4] \stackrel{\text{def.}}{=} \lim_{z_1 \to \infty} [z_1, z_2; z_3, z_4] = \frac{(z_2 - z_4)}{(z_2 - z_3)}. \tag{27.16}$$

Numerous results can be derived from the following fundamental property:

- A Möbius transformation leaves the cross-ratio of any four arbitrary points unaltered in value.

Let $z_i \mapsto w_i$ $(i = 1, 2, 3, 4)$ under the transformation (27.3). Then

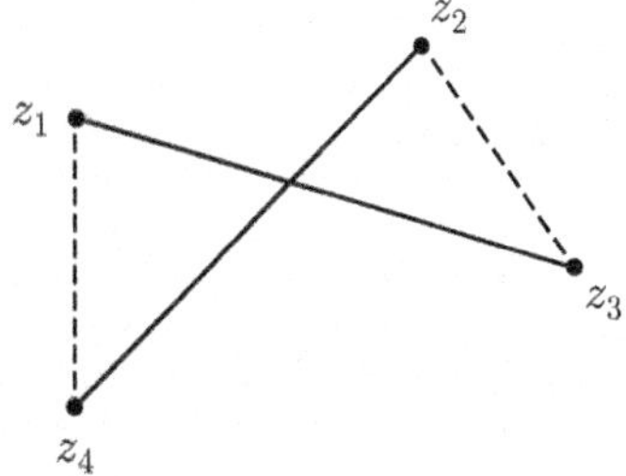

Fig. 27.1 The magnitude of the cross-ratio of the ordered set of four points z_1, z_2, z_3 and z_4 is the product of the differences indicated by full lines divided by the product of the differences indicated by dashed lines

$$\boxed{[z_1, z_2; z_3, z_4] = [w_1, w_2; w_3, w_4].} \tag{27.17}$$

★ **1.** The properties of the cross-ratio that have been stated above are easily established.

(a) Find the relations between the 24 cross-ratios associated with any four distinct points $(z_1, z_2, z_3, z_4) \in \mathbb{C}$.
(b) Verify Eq. (27.17), where $w_i = (az_i + b)/(cz_i + d), \; i = 1, 2, 3, 4$.

From the invariance of the cross-ratio under a Möbius transformation, the following property can be established:

- Given two sets of three distinct points (z_1, z_2, z_3) and (w_1, w_2, w_3) on the Riemann sphere, there exists a unique Möbius transformation that maps each z_i to the corresponding w_i for $i = 1, 2$ and 3.

There are several ways to prove this assertion. I shall not go into the complete proof here, but it is possible to *construct* the Möbius transformation concerned quite easily:

★ **2.** Assume that such a Möbius transformation exists. Construct it explicitly.

***A crucial geometrical property of Möbius transformations*:** Since a circle can be drawn through any three distinct noncollinear points, the result stated above has the following implication:

- A Möbius transformation maps circles to circles on the Riemann sphere.
- In the complex plane, a Möbius transformation maps circles and straight lines to circles and straight lines.

Remember that a straight line in the complex plane is a circle passing through the point of projection on the Riemann sphere. Figure 27.2a and b show examples of the mapping of the unit circle in the z-plane to a circle and a straight line, respectively. In case (b), note that $|z| = 1 \Rightarrow z = e^{i\theta}$, so that $w = -i(z-1)/(z+1) = \tan \frac{1}{2}\theta$.

★ **3.** Let z_1, z_2, z_3 be three arbitrary finite points in the complex plane. There is a unique circle [respectively, straight line] passing through these points if they are non-collinear [resp., collinear]. Show that the Möbius transformation that maps this circle [resp., straight line] to the real axis such that $z_1 \mapsto 0, \; z_2 \mapsto 1, \; z_3 \mapsto \infty$ is given by

$$w = \frac{(z - z_1)(z_3 - z_2)}{(z - z_3)(z_1 - z_2)}.$$

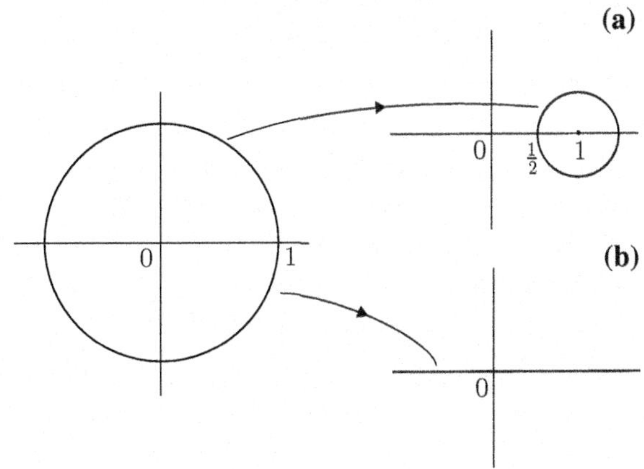

Fig. 27.2 The unit circle $|z| = 1$ is mapped **a** to the circle $|w - 1| = \frac{1}{2}$ by the map $w = \frac{1}{2}z + 1$, and **b** to the real axis by the map $w = -i(z - 1)/(z + 1)$

27.3 Normal Form of a Möbius Transformation

27.3.1 Normal Forms in Different Cases

The original form of a Möbius transformation, $z \mapsto w = (az + b)/(cz + d)$ with $ad - bc = 1$, is not the most convenient way to express the transformation for some purposes. The so-called **normal form** provides a better representation. As this involves the fixed points of the transformation, there is a normal form for each of the four cases listed in Sect. 27.2.2.

(i) ***Finite, distinct fixed points*****:** The two fixed points ξ_1 and ξ_2 are given by Eq. (27.7). Now consider the four points z, ∞, ξ_1 and ξ_2, where z is variable. Under the Möbius transformation, these points are mapped according to

$$z \mapsto w, \quad \infty \mapsto \frac{a}{c}, \quad \xi_1 \mapsto \xi_1\,, \quad \xi_2 \mapsto \xi_2\,. \tag{27.18}$$

Since the cross-ratio does not change under the mapping, we must have

$$[z, \infty; \xi_1, \xi_2] = [w, a/c; \xi_1, \xi_2]. \tag{27.19}$$

Define the constant

$$\boxed{K = \frac{a - c\,\xi_1}{a - c\,\xi_2} = \frac{a + d - \sqrt{(a+d)^2 - 4}}{a + d + \sqrt{(a+d)^2 - 4}}\,.} \tag{27.20}$$

Equation (27.19) then implies that the transformation can be expressed in the form

$$\boxed{\frac{w-\xi_1}{w-\xi_2} = K\left(\frac{z-\xi_1}{z-\xi_2}\right).} \tag{27.21}$$

Equation (27.21) is the normal form of the Möbius transformation. The constant K is called the **multiplier** of the Möbius transformation. Observe that it depends solely on the sum $(a+d)$. The importance of this fact will become clear subsequently.

(ii) ***Finite, coincident fixed points*:** Recall that this case corresponds to $c \neq 0$ and $(a+d) = \pm 2$. The fixed point is given by $\xi_1 = \xi_2 = \xi = (a-d)/(2c)$. The multiplier reduces to $K = 1$. Starting with $w = (az+b)/(cz+d)$, a little algebra leads to the normal form in this case, which is

$$\boxed{\frac{1}{w-\xi} = \frac{1}{z-\xi} \pm c, \quad \text{depending on whether } a+d = \pm 2.} \tag{27.22}$$

(iii) ***One fixed point at*** ∞**:** This is the case $c = 0,\ a \neq d$. As you know, the Möbius transformation now reduces to the linear transformation $z \mapsto w = (az+b)/d$, with fixed points at $\xi_1 = b/(d-a)$ and $\xi_2 = \infty$. It is easy to see that we now have the normal form

$$\boxed{w - \xi_1 = K(z-\xi_1) \quad \text{where} \quad K = a/d.} \tag{27.23}$$

(iv) ***Both fixed points at*** ∞**:** Now $c = 0$ and $a = d$. The transformation reduces to a shift of the origin,

$$\boxed{w = z \pm b, \quad \text{depending on whether } a = d = \pm 1.} \tag{27.24}$$

Again, this is the normal form in the present case. The multiplier $K = 1$ in this case, of course.

★ **4.** Verify that the normal forms in the four cases listed above are as given in Eqs. (27.21)–(27.24).

27.3.2 Iterates of a Möbius Transformation

It is easy to check that the result of making two Möbius transformations in succession are again a Möbius transformation. Hence, the effect of an arbitrary number on them performed in succession is also a Möbius transformation. In particular, we may ask for the result of a given transformation, $z \mapsto w = (az+b)/(cz+d)$, iterated an arbitrary number of times. Let $z \mapsto z^{(n)}$ after n repeated applications of a Möbius transformation, i.e.,

$$z^{(n)} = \underbrace{f(f(\cdots f(z))\cdots)}_{n-\text{fold iteration}} \quad \text{where} \quad f(z) = \frac{az+b}{cz+d}. \tag{27.25}$$

In its original form, the final expression for $z^{(n)}$ will obviously be quite complicated. But if you use the normal form, you can write down the answer by inspection, in each of the four cases listed above.

(i) ***Finite, distinct fixed points*:** Equation (27.21) leads to

$$\boxed{\frac{z^{(n)} - \xi_1}{z^{(n)} - \xi_2} = K^n \left(\frac{z - \xi_1}{z - \xi_2}\right).} \tag{27.26}$$

(ii) ***Finite, coincident fixed points*:** Equation (27.22) yields

$$\boxed{\frac{1}{z^{(n)} - \xi} = \frac{1}{z - \xi} \pm n\,c, \quad \text{for } a + d = \pm 2.} \tag{27.27}$$

(iii) ***One fixed point at*** ∞**:** It follows from Eq. (27.23) that

$$\boxed{z^{(n)} - \xi_1 = K^n\,(z - \xi_1), \quad \text{where} \quad K = a/d.} \tag{27.28}$$

(iv) ***Both fixed points at*** ∞**:** In this case, Eq. (27.24) gives

$$\boxed{z^{(n)} = z \pm n\,b, \quad \text{for } a = d = \pm 1.} \tag{27.29}$$

The reason why K is called the multiplier should be clear now. The relations (27.26)–(27.29) are trivially valid for $n = 0$ as well, with $z^{(0)} \equiv z$. Interestingly, they are also valid for *negative* integer values of n, with $z^{(-1)}$ interpreted as the inverse map of z, and $z^{(-n)}$ as the n-fold iterate of the inverse map of z:

$$z^{(-1)} = f^{-1}(z) = \frac{dz - b}{-cz + a}, \quad z^{(-n)} = f^{-1}\big(z^{-(n-1)}\big). \tag{27.30}$$

It is often useful to re-write the formula (27.26) for $z^{(n)}$ in the form

$$z^{(n)} = \frac{(\xi_1 - K^n\,\xi_2)\,z + \xi_1\xi_2\,(K^n - 1)}{(1 - K^n)\,z + (K^n\,\xi_1 - \xi_2)}. \tag{27.31}$$

This form shows explicitly that the result of an n-fold iteration of a Möbius transformation is again a Möbius transformation. Equation (27.31) is easily inverted to express $z \equiv z^{(0)}$ in terms of $z^{(n)}$. All you have to do is to replace K^n by K^{-n}. Inspection shows that this merely entails the interchange $\xi_1 \longleftrightarrow \xi_2$ in Eq. (27.31). Thus,

$$z^{(0)} = \frac{(\xi_2 - K^n\,\xi_1)\,z^{(n)} + \xi_1\xi_2\,(K^n - 1)}{(1 - K^n)\,z^{(n)} + (K^n\,\xi_2 - \xi_1)}\,. \tag{27.32}$$

27.3.3 Classification of Möbius Transformations

The results in Eqs. (27.26)–(27.29) lead to the systematic classification of Möbius transformations. Owing to limitations of space, I restrict the discussion to some introductory remarks on the topic.

For any value of z, the set of points $\{z^{(n)}\}$, where n runs over the integers, is an **orbit** under the map represented by a Möbius transformation. Such an orbit is analogous to the orbit of a point in the phase space of a dynamical system in discrete time, n playing the role of time. The collection of orbits corresponding to different values of $z = z^{(0)}$ (or the initial conditions, in the case of a dynamical system) gives us a *flow* in the complex plane $\hat{\mathbb{C}}$ (analogous to a phase portrait in phase space). The nature of this flow is essentially determined by the fixed points, which are the analogs of the **critical points** or equilibrium points in phase space. This is the starting point of the classification of Möbius transformations.

Let us consider the general case in which $c \neq 0$ (so that the transformation does not reduce to a mere linear transformation), and there are two distinct, finite fixed points. It is convenient to write Eq. (27.26) for $z^{(n)}$ in the form

$$z^{(n)} = \left\{\xi_1 - \xi_2\,K^n\left(\frac{z^{(0)} - \xi_1}{z^{(0)} - \xi_2}\right)\right\}\Big/\left\{1 - K^n\left(\frac{z^{(0)} - \xi_1}{z^{(0)} - \xi_2}\right)\right\}. \tag{27.33}$$

It is now easy to see, at least in most cases, what happens as $n \to \infty$.

(a) If $|K| > 1$, the factor K^n will grow in magnitude with increasing n. Therefore, $z^{(n)} \to \xi_2$ as $n \to \infty$, for all initial points $z^{(0)}$ (other than ξ_1, of course). In the spirit of dynamical systems, we may regard ξ_2 as an *asymptotically stable* fixed point, or **attractor**; the flow is generally *toward* this point. On the other hand, ξ_1 acts like a *unstable* fixed point, or **repellor**; the flow is generally *away* from this point.
(b) If $|K| < 1$, the factor K^n tends to zero with increasing n. Hence, $z^{(n)} \to \xi_1$ as $n \to \infty$, for all initial points $z^{(0)}$ (other than ξ_2). The attractor is now ξ_1, and the flow is toward this point. The repellor is ξ_2, and the flow leads away from this point.
(c) When $|K| = 1$, we have the analog of **marginal fixed points**, and the flow is neither toward nor away from the fixed points.
(d) The case when the two fixed points coincide at $\xi = (a - d)/(2c)$ occurs when $(a + d)^2 = 4$, and corresponds to $K = 1$. This is a special case of $|K| = 1$.

These statements can be corroborated as follows: Under the transformation from z to $w = f(z)$, we have $dw = f\,'(z)dz$. The Jacobian of the transformation is, therefore, $f\,'(z)$. Its magnitude $|f\,'(z)|$ is the local **stretch factor** (or contraction factor). It is this quantity that essentially characterizes the local flow in the neighborhood of any point. Now,

$$f(z) = \frac{az+b}{cz+d} \quad \Rightarrow \quad |f'(z)| = \frac{1}{|cz+d|^2}, \tag{27.34}$$

on using the fact that $ad - bc = 1$. (Remember that we are considering the nontrivial case in which $c \neq 0$.) At the fixed points ξ_1 and ξ_2, the respective stretch factors become

$$\left.\begin{aligned} |f'(\xi_1)| &= \frac{4}{\left|a+d+\sqrt{(a+d)^2-4}\,\right|^2} = |K|^2, \\ |f'(\xi_2)| &= \frac{4}{\left|a+d-\sqrt{(a+d)^2-4}\,\right|^2} = \frac{1}{|K|^2}. \end{aligned}\right\} \tag{27.35}$$

It can be shown that a fixed point is

- unstable if the stretch factor at that point is greater than unity;
- asymptotically stable if it is less than unity; and
- marginal when it is equal to unity.

The statements made above then follow.

★ **5.** Verify Eqs. (27.35).

The crucial point is this: in all cases, K depends only on the trace

$$T = (a+d) \tag{27.36}$$

of the matrix $\left(\begin{smallmatrix} a & b \\ c & d \end{smallmatrix}\right)$ made up of the coefficients of the transformation. From Eq. (27.20),

$$\boxed{K = \frac{T - \sqrt{T^2-4}}{T + \sqrt{T^2-4}}.} \tag{27.37}$$

This property enables the classification of Möbius transformations into different types. Four types of transformations are possible, *depending solely on the value of* T. I merely list them here, without going into further detail. The first three types correspond to real values of the trace T.

Type 1: Elliptic transformation, when T is real and $-2 < T < 2$. Then $|K| = 1$, but $K \neq 1$.

Type 2: Parabolic transformation, when $T = \pm 2$. The two fixed points now coincide, and we have $K = 1$.

Type 3: Hyperbolic transformation, when T is real and $|T| > 2$. In this case K is a positive number other than unity (i.e., $K \neq 1$).

Type 4: Loxodromic transformation, when T is any complex number such that $T \notin [-2, 2]$. A loxodrome is a spiral-shaped curve drawn on the surface of a sphere such that it intersects every line of longitude at some constant angle ψ. (The planar analog of such a curve is an equiangular spiral, discussed in Chap. 1, Sect. 1.1.4.) The name "loxodromic transformation" arises from the shape of the flow on the Riemann sphere under the transformation.

The case $T = 0$ (which implies that $K = -1$) is sometimes referred to as a **circular transformation**. It is a special case of an elliptic transformation. Similarly, hyperbolic transformations comprise a sub-class of loxodromic transformations, corresponding to real values of T such that $T^2 > 4$.

27.3.4 The Isometric Circle

Equation (27.34), which says that $|f'(z)| = 1/|cz + d|^2$, has a direct geometric interpretation. As stated earlier, $|f'(z)|$ is the local stretch or contraction factor associated with the mapping of an infinitesimal neighborhood of any point z by the Möbius transformation $z \mapsto (az + b)/(cz + d)$. It follows that there is *no* distortion at all in the map of the curve given by $|f'(z)| = 1$. This is the circle

$$|cz + d| = 1, \quad \text{or} \quad |z + (d/c)| = 1/|c|. \tag{27.38}$$

(Remember that we are considering the general transformation with $c \neq 0$.) This circle, with center at $-d/c$ and radius equal to $1/|c|$, is called the **isometric circle** corresponding to the transformation. It plays an important role in the theory of Möbius transformations.

Points inside the isometric circle satisfy $|cz + d| < 1$. Hence infinitesimal area elements inside the circle are *magnified* by the mapping, by a factor $1/|cz + d|^4$. Similarly, points outside the isometric circle satisfy $|cz + d| > 1$. Hence, infinitesimal area elements in this region are *shrunk* by the transformation, by a factor $1/|cz + d|^4$.

★ **6.** Consider the Möbius transformation $z \mapsto w = (az + b)/(cz + d)$ (where $c \neq 0$).

(a) Show that the transformation maps its isometric circle to the isometric circle of the inverse transformation $w \mapsto z = (dw - b)/(-cw + a)$. Further, the interior (respectively, exterior) of the isometric circle in the z-plane is mapped to the exterior (respectively, interior) of its image in the w-plane.

(b) Consider the case of finite, distinct fixed points ξ_1 and ξ_2, and (i) $|K| > 1$ and (ii) $|K| < 1$, respectively. What is the isometric circle of the nth iterate of the transformation? What happens as $n \to \infty$?

27.4 Group Properties

An important feature of Möbius transformations or maps $f : z \mapsto w$ is the fact that they form a *group*, the so-called **Möbius group** Möb $(2, \mathbb{C})$. In this section, I give a brief account of some salient properties of this group.

27.4.1 The Möbius Group

It is easy to check that Möbius transformations satisfy the axioms that define a group:

(a) The successive application of two Möbius transformations is another Möbius transformation. More formally, if f and g are $\hat{\mathbb{C}} \to \hat{\mathbb{C}}$ maps representing Möbius transformations, so is their composition $f \circ g$.
(b) The composition of transformations is associative. If f, g and h are maps, then $f \circ (g \circ h) = (f \circ g) \circ h$.
(c) There is an identity transformation under which every z is mapped to itself.
(d) Each transformation $f : z \mapsto w$ has an inverse $f^{-1} : w \mapsto z$. That is, if $w = f(z) = (az + b)/(cz + d)$, then $z = f^{-1}(w) = (dw - b)/(-cw + a)$.

Here is how the transformation obtained by composing two successive transformations can be read off easily:

- The transformation $z \mapsto w = (az + b)/(cz + d)$ can be associated with the matrix of coefficients, $\left(\begin{smallmatrix} a & b \\ c & d \end{smallmatrix}\right)$.
- The composition of transformations then corresponds to the product of the associated matrices (in the right order).

Thus, if

$$z \mapsto z' = \frac{a_1 z + b_1}{c_1 z + d_1} \quad \text{is followed by} \quad z' \mapsto z'' = \frac{a_2 z' + b_2}{c_2 z' + d_2}, \tag{27.39}$$

then

$$z'' = \frac{(a_2 a_1 + b_2 c_1) z + a_2 b_1 + b_2 d_1}{(c_2 a_1 + d_2 c_1) z + c_2 b_1 + d_2 d_1}. \tag{27.40}$$

But the set of all nonsingular (2×2) matrices with complex elements constitutes a group. This group is called the general linear group over the complex numbers in 2 dimensions, and is denoted by $GL(2, \mathbb{C})$. The group composition law is just matrix multiplication. Does this mean that there is a one-to-one correspondence between Möbius transformations and the elements of $GL(2, \mathbb{C})$?

The answer is no. As you know, the coefficients of a Möbius transformation can always be chosen such that $ad - bc = 1$. What is relevant here, therefore, is the set of (2×2) matrices with complex elements and determinant equal to 1. These matrices form a *subgroup* of $GL(2, \mathbb{C})$, namely, the unimodular or **special linear group** over the complex numbers, denoted by $SL(2, \mathbb{C})$. Can we conclude that there is a one-to-one correspondence between Möbius transformations and the elements of $SL(2, \mathbb{C})$?

Once again, the answer is no. There is one more point to be taken into account. Changing the sign of all the four coefficients in a Möbius transformation does not alter it, because $(az + b)/(cz + d) \equiv (-az - b)/(-cz - d)$. Hence, to each Möbius transformation

$$z \mapsto w = \frac{az + b}{cz + d} \quad \text{with} \quad ad - bc = 1, \tag{27.41}$$

there correspond *two* matrices belonging to $SL(2, \mathbb{C})$, namely,

$$\begin{pmatrix} a & b \\ c & d \end{pmatrix} \quad \text{and} \quad \begin{pmatrix} -a & -b \\ -c & -d \end{pmatrix}. \tag{27.42}$$

The identity transformation $z \mapsto z$ thus corresponds to the (2×2) identity matrix I, as well as its negative, $-I$. Recall, now, the connection between the special unitary group $SU(2)$ and the special orthogonal group $SO(3)$ that was discussed in Chap. 15, Sect. 15.3.3. What we have now is an analogous situation.

- There is a 2-to-1 correspondence, or homomorphism, from the group $SL(2, \mathbb{C})$ to the group Möb $(2, \mathbb{C})$.
- The matrices I and $-I$ form the kernel of the homomorphism. That is, $\{I, -I\}$ is the set of matrices in $SL(2, \mathbb{C})$ whose image in Möb $(2, \mathbb{C})$ is the identity transformation (under which every point $z \mapsto z$ itself).

But the two matrices I and $-I$ themselves form a group, under matrix multiplication. As you know, this group is just the cyclic group of order 2, which is isomorphic to $\mathbb{Z}_2$, the additive group of integers modulo 2. The the quotient group $SL(2, \mathbb{C})/\mathbb{Z}_2$ is essentially the group of unimodular 2×2 matrices up to the overall sign (or "modulo the sign") of each matrix. It is this group, also known as the **projective special linear group** $PSL(2, \mathbb{C})$, with which the group of Möbius transformations is in one-to-one correspondence.

- Möb $(2, \mathbb{C})$ is *isomorphic* to $SL(2, \mathbb{C})/\mathbb{Z}_2$. This group isomorphism is written as

$$\boxed{\text{Möb}\,(2, \mathbb{C}) \cong SL(2, \mathbb{C})/\mathbb{Z}_2 \equiv PSL(2, \mathbb{C}).} \tag{27.43}$$

- The group $SL(2, \mathbb{C})$ is the universal covering group of Möb $(2, \mathbb{C})$. The latter is a subgroup of $SL(2, \mathbb{C})$.

Finally, I merely mention the following remarkable fact that connects Möbius transformations to special relativity:

- The group of Möbius transformations is isomorphic to $SO(3, 1)$, the group of proper orthochronous Lorentz transformations, in the usual spacetime comprising three spatial dimensions and one time dimension!

27.4.2 The Möbius Group Over the Reals

As mentioned above, the Möbius group Möb $(2, \mathbb{C})$ is isomorphic to certain other important groups such as the projective linear group $PSL(2, \mathbb{C})$ and the homogeneous Lorentz group $SO(3, 1)$. Further, it is a subgroup of the special linear group $SL(2, \mathbb{C})$, and hence of the general linear group $GL(2, \mathbb{C})$. In turn, the Möbius group itself has some important and interesting subgroups.

Möbius transformations with real parameters, in which the coefficients a, b, c and d are restricted to *real* numbers, comprise a group on their own, Möb $(2, \mathbb{R})$. . This group is isomorphic to the projective linear group $PSL(2, \mathbb{R})$ over the reals, a subgroup of $PSL(2, \mathbb{C})$. It is also a quotient group, being the special linear group $SL(2, \mathbb{R})$ modulo $\mathbb{Z}_2$: analogous to (27.43), we have

$$\boxed{\text{Möb}\,(2, \mathbb{R}) \cong SL(2, \mathbb{R})/\mathbb{Z}_2 \equiv PSL(2, \mathbb{R}).} \tag{27.44}$$

Möbius transformations with real parameters have some additional properties that are of importance. It is easy to show that

- a real Möbius transformation maps the upper (respectively, lower) half of the complex plane to the upper (respectively, lower) half-plane.

★ **7.** Verify the foregoing statement.

***The modular group*:** Among Möbius transformations with real parameters (a, b, c, d) where $ad - bc = 1$, those with *integer* values of the parameters form a group by themselves! This is the group $PSL(2, \mathbb{Z})$, called the **modular group**. Similarly, the set of Möbius transformations in which (a, b, c, d) are **Gaussian integers**, i.e., each parameter is of the form $m + ni$ where m and n are integers, also forms a group. The modular group has numerous remarkable properties, and it plays a role in many areas of mathematics including hyperbolic geometry, number theory, elliptic functions, etc. It also appears in certain topics in theoretical physics such as conformal field theory.

27.4.3 *The Invariance Group of the Unit Circle*

Here is another important example of a continuous subgroup of the Möbius group. Consider all Möbius transformations that leave the *unit circle* unchanged, i.e., those that map the circle $|z| = 1$ in the z-plane to the circle $|w| = 1$ in the w-plane. These transformations are either of the form

$$z \mapsto w = \frac{az + b}{b^* z + a^*}, \quad \text{where} \quad |a|^2 - |b|^2 = 1, \tag{27.45}$$

or of the form

$$z \mapsto w = \frac{az + b}{-b^* z - a^*}, \quad \text{where} \quad -|a|^2 + |b|^2 = 1. \tag{27.46}$$

Let us call these Type A and Type B transformations, respectively, for convenience. They form a subgroup of the group Möb $(2, \mathbb{C})$ of Möbius transformations. Transformations of Types A and B are distinguished by the following property:

- A Type A transformation maps the interior of the unit circle in the z-plane to its interior in the w-plane, and similarly for the exterior of the unit circle.
- A Type B transformation maps the interior of the unit circle in the z-plane to its exterior in the w-plane, and *vice versa*.

★ **8.** Show that

(a) the general form of a Möbius transformation $z \mapsto w$ that maps the unit circle to the unit circle is given by Eq. (27.45) or Eq. (27.46);
(b) all such transformations form a group;
(c) $|z| \lessgtr 1 \Rightarrow |w| \lessgtr 1$ for Type A transformations, while $|z| \lessgtr 1 \Rightarrow |w| \gtrless 1$ for Type B transformations.

Connection with the pseudo-unitary group $SU(1, 1)$: The matrices corresponding to Möbius transformations of Types A and B are, respectively, of the general forms

$$U_+ = \begin{pmatrix} a & b \\ b^* & a^* \end{pmatrix} \quad \text{where} \quad \det U_+ = |a|^2 - |b|^2 = 1, \tag{27.47}$$

and

$$U_- = \begin{pmatrix} a & b \\ -b^* & -a^* \end{pmatrix} \quad \text{where} \quad \det U_- = -|a|^2 + |b|^2 = 1. \tag{27.48}$$

But these are precisely the general forms of the matrices belonging to the indefinite or pseudo-unitary group $SU(1, 1)$, as you have seen already in Eqs. (15.101) and (15.102) of Chap. 15, Sect. 15.4.4. Further, $SU(1, 1)$ turns out to be isomorphic to the special linear group $SL(2, \mathbb{R})$. The **symplectic group** $Sp(2, \mathbb{R})$ of canonical transformations of a 1-freedom Hamiltonian system is *also* isomorphic to $SL(2, \mathbb{R})$!

$$\boxed{SU(1,1) \cong SL(2,\mathbb{R}) \cong Sp(2,\mathbb{R}).} \tag{27.49}$$

These relationships have important consequences. The group of Möbius transformations that leaves the unit circle invariant is, therefore, isomorphic to $SL(2, \mathbb{R})/\mathbb{Z}_2$. In other words, it is Möb (2, $\mathbb{R}$) once again.

$$\boxed{\text{Invariance group of the unit circle} = SL(2,\mathbb{R})/\mathbb{Z}_2 \cong \text{Möb}\,(2,\mathbb{R}).} \tag{27.50}$$

27.4.4 The Group of Cross-Ratios

The group Möb (2, $\mathbb{C}$) of Möbius transformations of a complex variable z has an interesting *discrete* subgroup of transformations called the **group of cross-ratios**. It comprises the following set of six special Möbius transformations that we may denote by e_i, where i runs over the set 1, 2, ..., 6:

$$\left.\begin{array}{lll} e_1 & : z \mapsto z & \text{(the identity transformation)} \\ e_2 & : z \mapsto 1/z & \text{(inversion)} \\ e_3 & : z \mapsto 1 - z & \text{(rotation about the point } \tfrac{1}{2} \text{ through an angle } \pi) \\ e_4 & : z \mapsto 1 - (1/z) & \text{(inversion followed by rotation)} \\ e_5 & : z \mapsto 1/(1 - z) & \text{(rotation followed by inversion)} \\ e_6 & : z \mapsto z/(z - 1) & \text{(inversion-rotation-inversion).} \end{array}\right\} \tag{27.51}$$

The name "group of cross-ratios" arises from the following fact: These transformations are precisely the ones leading to the identities (27.14) and (27.15) satisfied by the cross-ratio of any four distinct points in $\hat{\mathbb{C}}$.

★ **9.** Write down (a) the values of the parameters a, b, c, d corresponding to the foregoing transformations (such that $ad - bc = 1$ in each case); (b) the "multiplication table" for the group; and (c) the inverse of each element other than the identity element e_1.

The transformation e_3 can also be regarded as successive reflections about the lines $\text{Re}\, z = \frac{1}{2}$ and $\text{Im}\, z = 0$ (the real axis), performed in either order. (Recall, for instance, that the parity transformation $x \mapsto -x,\; y \mapsto -y$ in the Euclidean plane is just a rotation about the origin through an angle π.) The last element $e_6 = e_2\, e_3\, e_2$ (inversion-rotation-inversion) or, alternatively, $e_6 = e_3\, e_2\, e_3$ (rotation-inversion-rotation).

The group of cross-ratios is *not* an abelian or commutative group, since $e_i\, e_j \neq e_j\, e_i$ in general. Several identities follow at once from the multiplication table: for instance,

$$(e_4)^3,\ (e_5)^3,\ (e_6)^2,\ (e_3\,e_4)^2,\ (e_4\,e_3)^2,\ (e_3\,e_5)^2,\ (e_5\,e_3)^2,\ \ldots$$

are all equal to the identity transformation. The group is also called the **anharmonic group**. It is isomorphic to the **symmetric group** S_3, which is the group of permutations of 3 objects or symbols. It is also isomorphic to the group of symmetries of an equilateral triangle under rotations and reflections, which is called the **dihedral group** of order 6.

27.5 Solutions

1. (a) Using the identities in (27.14), the 24 cross-ratios can be arranged in 3 sets of 8 each:

$$\begin{aligned}
[12;34] &= 1/[12;43] = 1/[21;34] = [21;43] \\
&= [34;12] = 1/[34;21] = 1/[43;12] = [43;21]. \\
[13;24] &= 1/[13;42] = 1/[31;24] = [31;42] \\
&= [24;13] = 1/[24;31] = 1/[42;13] = [42;31]. \\
[14;23] &= 1/[41;23] = 1/[14;32] = [41;32] \\
&= [23;14] = 1/[23;41] = 1/[32;14] = [32;41].
\end{aligned}$$

Further, from (27.15) it follows that

$$[13;24] = 1 - [12;34] \quad \text{and} \quad [14;23] = 1 - 1/[12;34].$$

Hence, all the 24 cross-ratios are written, for instance, in terms of $[12;34] = [z_1,\ z_2;\ z_3,\ z_4]$.

(b) You have only to note that

$$(w_i - w_j) = \frac{(z_i - z_j)}{(cz_i + d)(cz_j + d)},$$

and Eq. (27.17) follows readily. ▶

2. Consider the cross-ratio $[z, z_1; z_2, z_3]$. If $z \mapsto w$ under the transformation, then this cross-ratio must be equal to $[w, w_1; w_2, w_3]$. That is,

$$\frac{(z - z_2)(z_1 - z_3)}{(z - z_3)(z_1 - z_2)} = \frac{(w - w_2)(w_1 - w_3)}{(w - w_3)(w_1 - w_2)}.$$

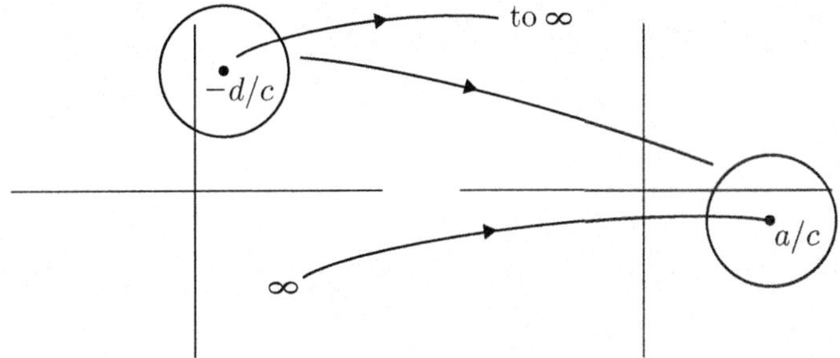

Fig. 27.3 The isometric circle of the Möbius transformation $z \mapsto w = (az + b) / (cz + d)$, and its image under the transformation

Solve this simple equation for w, to show that it is in the form of a Möbius transformation of z. ▶

3. Consider the cross-ratio $[z, z_1; z_2, z_3]$. The invariance of the cross-ratio implies that this quantity must be equal to $[w, 0; 1, \infty]$. That is,

$$\frac{(z - z_2)(z_1 - z_3)}{(z - z_3)(z_1 - z_2)} = [w, 0; 1, \infty] = 1 - w,$$

from which the desired expression for w follows at once. ▶

6. (a) The inverse transformation is $z = (dw - b)/(-cw + a)$. Substituting this expression for z in the equation $|cz + d| = 1$ gives $|cw - a| = 1$. This is precisely the isometric circle of the inverse transformation. Its center is at a/c and its radius is again $1/|c|$. Similarly, it is trivially seen that

$$|cz + d| \lessgtr 1 \Rightarrow |cw - a| \gtrless 1.$$

Figure 27.3 illustrates these facts. As we know already from Eq. (27.5), the center $-d/c$ of the isometric circle is mapped to ∞, while ∞ is mapped to the center a/c of the image of the isometric circle under the mapping.

(b) (i) Suppose $|K| > 1$. It follows at once from Eq. (27.31) that the isometric circle sought is given by

$$\left| z - \frac{(K^n \xi_1 - \xi_2)}{(K^n - 1)} \right| = \frac{1}{|K^n - 1|}.$$

As $n \to \infty$, the center of the isometric circle above tends to ξ_1, while its radius tends to zero. This is completely consistent with what we know already about this case: ξ_2 acts as an attractor for the flow; all points z flow toward this attractor under the iteration of the map, except for the point ξ_1 itself.

(ii) When $|K| < 1$, K^{-n} becomes unbounded as $n \to \infty$. You must therefore re-write Eq. (27.31) in the form

$$z^{(n)} = \frac{(K^{-n}\,\xi_1 - \xi_2)\,z + \xi_1\xi_2\,(1 - K^{-n})}{(K^{-n} - 1)\,z + (\xi_1 - K^{-n}\,\xi_2)}\,.$$

The isometric circle is given by

$$\left| z - \frac{(K^{-n}\,\xi_2 - \xi_1)}{(K^{-n} - 1)} \right| = \frac{1}{|K^{-n} - 1|}\,.$$

It is now clear that, as $n \to \infty$, the center of the isometric circle tends to ξ_2, while its radius tends to zero. Once again, this is consistent with what you know already: when $|K| < 1$, ξ_1 acts as an attractor for the flow; all points z flow toward this point under the iteration of the map, except for the point ξ_2 itself. ▶

7. Consider the Möbius transformation $z \mapsto w = (az + b)/(cz + d)$ where a, b, c and d are real numbers, and $ad - bc = 1$. Let $z = x + iy$ and $w = u + iv$. Simple algebra then gives $v = y/|cz + d|^2$, so that $\text{Im}\, w \gtrless 0$ according as $\text{Im}\, z \gtrless 0$. ▶

8. (a) In the z-plane, the unit circle is given by $z^*z - 1 = 0$. In the w-plane this becomes, on setting $z = (dw - b)/(-cw + a)$,

$$(|d|^2 - |c|^2)\, w^*w + (a^*c - b^*d)\, w + (ac^* - bd^*)\, w^* - (|a|^2 - |b|^2) = 0.$$

In order for this to be the equation of the unit circle, given by $w^*w - 1 = 0$, the following conditions must be satisfied:

$$a^*c - b^*d = 0, \;\; ac^* - bd^* = 0,$$

as well as,

$$|d|^2 - |c|^2 = |a|^2 - |b|^2 \neq 0.$$

Hence, $a/d^* = b/c^* = k$, say. The last equation above then gives $|k|^2 = 1$. But the condition $ad - bc = 1$ implies that $k(|d|^2 - |c|^2) = 1$. Therefore, k must be real, so that $k = \pm 1$. This leads to two types of transformations:

Type A: If $k = +1$, then $d = a^*$ and $c = b^*$. The general form of the Möbius transformation is

$$w = \frac{az + b}{b^*z + a^*}\,, \quad \text{where} \;\; |a|^2 - |b|^2 = 1.$$

Type B: If $k = -1$, then $d = -a^*$ and $c = -b^*$. The general form of the Möbius transformation is now

$$w = \frac{az + b}{-b^* z - a^*}, \quad \text{where} \quad -|a|^2 + |b|^2 = 1.$$

(b) The inverse transformations corresponding to the Möbius transformations of Types A and B above are given, respectively, by

$$z = \frac{a^* w - b}{-b^* w + a}, \quad \text{where} \; |a|^2 - |b|^2 = 1,$$

and

$$z = \frac{-a^* w - b}{b^* w + a}, \quad \text{where} \quad -|a|^2 + |b|^2 = 1.$$

Further, you can verify that two successive transformations of Type A , or of Type B, are equivalent to another transformation of Type A. Similarly, two successive transformations of Types A and B, or *vice versa*, are equivalent to another transformation of Type B. These Möbius transformations, therefore, comprise a group that is a subgroup of Möb (2, $\mathbb{C}$).
(c) Again, using the respective inverse transformations that express z in terms of w in the two cases, the inequalities sought are easily obtained. ▶

9. (a), (b) The set (a, b, c, d) is only determined up to an overall sign in each case. The multiplication table is given on the right. Here the (ij)th element of the array stands for the transformation $e_i \, e_j$ (i.e., the transformation e_i following the transformation e_j), in accord with the usual convention.

	a	b	c	d
e_1	1	0	0	1
e_2	0	$-i$	$-i$	0
e_3	i	$-i$	0	$-i$
e_4	1	-1	1	0
e_5	0	1	-1	1
e_6	i	0	i	$-i$

	e_1	e_2	e_3	e_4	e_5	e_6
e_1	e_1	e_2	e_3	e_4	e_5	e_6
e_2	e_2	e_1	e_5	e_6	e_2	e_4
e_3	e_3	e_4	e_1	e_2	e_6	e_5
e_4	e_4	e_3	e_6	e_5	e_1	e_2
e_5	e_5	e_6	e_2	e_1	e_4	e_3
e_6	e_6	e_5	e_4	e_3	e_2	e_1

(c) It follows from the multiplication table by inspection that e_2, e_3 and e_6 are their own inverses, while e_4 and e_5 are each other's inverses. ▶

Chapter 28
Laplace Transforms

28.1 Definition and Properties

28.1.1 *Definition of the Laplace Transform*

Recall that the Fourier integral expansion of a function $f(t)$ (where $t \in \mathbb{R}$) is

$$f(t) = \frac{1}{2\pi} \int_{-\infty}^{\infty} d\omega \, e^{-i\omega t} \, \widetilde{f}(\omega). \tag{28.1}$$

As you know, the Fourier transform is an example of an integral operator $\mathcal{F}$ acting on the elements of a function space: given a function $f(t)$, we have

$$(\mathcal{F}f)(\omega) = \widetilde{f}(\omega) = \int_{-\infty}^{\infty} dt \, e^{i\omega t} \, f(t). \tag{28.2}$$

The kernel of the integral operator $\mathcal{F}$ is $e^{i\omega t}$. The function space to which $f(t)$ belongs is $\mathcal{L}_1(-\infty\,,\,\infty)$, the space of all integrable functions. A necessary (but not sufficient) condition for integrability over an infinite range is that $f(t) \to 0$ as $t \to \pm\infty$.

This condition is quite restrictive, whereas we are often concerned with functions $f(t)$ that tend to a constant or even diverge as $t \to \infty$. Moreover, when t represents the time, we frequently encounter functions that are only defined on a half-line, say $0 \leq t < \infty$. An example is a causal response function. More generally, this is the natural domain for an initial value problem. In such cases, it is helpful to define another integral transform, the **Laplace transform**, according to

$$\boxed{(\mathcal{L}f)(s) \equiv \widetilde{f}(s) \overset{\text{def.}}{=} \int_0^{\infty} dt \, e^{-st} \, f(t).} \tag{28.3}$$

V. Balakrishnan, *Mathematical Physics*,
https://doi.org/10.1007/978-3-030-39680-0_28

It is clear that the factor e^{-st} provides a convergence factor as long as $\text{Re}\, s > 0$: the integral in Eq. (28.3) converges even if $f(t)$ increases like any arbitrary power of t for large values of t. In fact, even if $f(t)$ increases exponentially with t, so that $f(t) \sim e^{\lambda t}$ (where λ is a positive number) as $t \to \infty$, its Laplace transform as given by the integral (28.3) is well-defined, as long as s is held in the region $\text{Re}\, s > \lambda$. Thus:

- The Laplace transform $\widetilde{f}(s)$ of a function $f(t)$, as given by its defining integral representation, is an analytic function of s for a sufficiently large positive value of $\text{Re}\, s$. That is, the region of analyticity of a Laplace transform $\widetilde{f}(s)$ is typically *some right half-plane* $\text{Re}\, s > \lambda$ of the complex variable s, where λ is a real number.

We may then expect to be able to continue $\widetilde{f}(s)$ analytically to the left of this region in the complex s-plane. It is evident that $\widetilde{f}(s)$ would have one or more singularities in the left half-plane (as otherwise, it can only be a trivial constant). It should also be evident that there do exist functions that do not possess a Laplace transform. For example, if $f(t) \sim \exp{(t^{1+\alpha})}$ (where $\alpha > 0$) as $t \to \infty$, then $\widetilde{f}(s)$ does not exist for any value of s.

***The iterate of a Laplace transform*:** Recall Eq. (18.16) of Chap. 18, Sect. 18.2.1: the square of the Fourier transform operator is essentially the parity operator, apart from a multiplicative constant. In the same way, we may ask what the operator $\mathcal{L}^2$ amounts to. It is easy to see that, provided the integrals concerned exist,

$$(\mathcal{L}^2 f)(u) = \int_0^\infty ds\, e^{-us} \int_0^\infty dt\, e^{-st}\, f(t) = \int_0^\infty dt\, \frac{f(t)}{(t+u)}\,. \tag{28.4}$$

The integral on the right-hand side is the so-called **Stieltjes transform** of $f(t)$. I will not discuss this transform further here.

28.1.2 Transforms of Some Simple Functions

The Laplace transforms of simple functions are easily written down. Consider, for instance, the function $f(t) = t^\alpha\, e^{-at}$. The convergence of the integral defining its transform $\widetilde{f}(s)$ places some conditions on the parameters α and a. The lower limit of integration, $t = 0$, does not pose any problem provided $\text{Re}\, \alpha > -1$. Similarly, the upper limit ($t = \infty$) is taken care of as long as $\text{Re}\, s > -\text{Re}\, a$. The integral is easily read off—it is a gamma function. We thus have

$$\begin{aligned} \mathcal{L}[t^\alpha\, e^{-at}] &= \int_0^\infty dt\, t^\alpha\, e^{-(s+a)t} \quad (\text{Re}\, \alpha > -1,\ \text{Re}\, s > -a) \\ &= \frac{\Gamma(\alpha+1)}{(s+a)^{\alpha+1}}\,. \end{aligned} \tag{28.5}$$

The first line on the right-hand side is an analytic function in the right half-plane $\text{Re}\, s > -a$. The second line is its analytic continuation to the whole of the complex s-plane. Observe that this representation shows that $\widetilde{f}(s)$ has a singularity at $s = -a$, which lies on the *boundary* of the region of validity of the original representation. This is a particular example of a very general feature that is the analog of a similar result in the case of power series. Recall that a power series in a complex variable z converges absolutely in the interior of its circle of convergence, and that its analytic continuation (if it exists) has at least one singularity on the boundary of this circle, i.e., *on* the circle itself.

A number of simpler cases may be read off from Eq. (28.5). For instance,

$$\mathcal{L}[t^\alpha] = \frac{\Gamma(\alpha+1)}{s^{\alpha+1}}, \quad \text{and hence} \quad \mathcal{L}[t^n] = \frac{n!}{s^{n+1}} \; (n = 0,\, 1,\, \ldots). \tag{28.6}$$

Similarly,

$$\mathcal{L}[e^{-at}] = \frac{1}{(s+a)}, \tag{28.7}$$

the right-hand side being the analytic continuation, to arbitrary complex values of a, of the integral defined by the left-hand side. It follows from Eq. (28.7) that

$$\left.\begin{aligned} \mathcal{L}[\cos(at)] &= \frac{s}{s^2+a^2}, \quad & \mathcal{L}[\cosh(at)] &= \frac{s}{s^2-a^2} \\ \mathcal{L}[\sin(at)] &= \frac{a}{s^2+a^2}, \quad & \mathcal{L}[\sinh(at)] &= \frac{a}{s^2-a^2}\,. \end{aligned}\right\} \tag{28.8}$$

In general,

$$\boxed{\mathcal{L}[f(t)] = \widetilde{f}(s) \quad \Longrightarrow \quad \mathcal{L}[f(t)\,e^{-at}] = \widetilde{f}(s+a).} \tag{28.9}$$

The following is a very useful identity: integrate both sides of the defining equation (28.3) with respect to s, from 0 to ∞. Then, *provided the corresponding integrals exist,*

$$\boxed{\int_0^\infty \frac{dt\, f(t)}{t} = \int_0^\infty ds\, \widetilde{f}(s).} \tag{28.10}$$

Numerous definite integrals may be evaluated with the help of this identity.

★ **1.** Use the identity (28.10) to show that

$$\int_0^\infty dt\, \frac{\sin(bt)}{t} = \tfrac{1}{2}\,\pi\, \varepsilon(b) \quad (b \text{ real}).$$

The Dirichlet integral (Eq. (2.18) of Chap. 2, Sect. 2.3) has thus been recovered, once again.

***The canonical partition function as a Laplace transform*:** Consider a classical system with a continuous energy spectrum from some finite ground state energy E_0 to infinity. Without loss of generality, we may shift the energy scale so that $E_0 = 0$. Let $\rho(E)$ denote the density of states of the system: that is, the number of accessible microstates of the system in the energy range $(E,\ E + dE)$ is equal to $\rho(E)\,dE$. Now suppose the system is in thermal equilibrium with a heat bath at absolute temperature. T Then, according to equilibrium statistical mechanics, the canonical partition function of the system when it is in thermal is given by

$$Z(\beta) = \int_0^\infty dE\, \rho(E)\, e^{-\beta E}, \tag{28.11}$$

where $\beta = 1/(k_B T)$ and $e^{-\beta E}$ is the Boltzmann factor.

- The partition function of a system in the canonical ensemble is essentially the Laplace transform of its density of states. The reciprocal temperature β plays the role of the transform variable s.

A monatomic classical ideal gas of N atoms in a container of volume V at a temperature T provides a simple physical example. The energy of each particle is $\varepsilon = p^2/(2m)$, so that $4\pi p^2\, dp = 2\pi(2m)^{3/2}\, \varepsilon^{1/2}\, d\varepsilon$. The density of states of a particle is, therefore, proportional to $\varepsilon^{1/2}$. (Remember the conditions under which this is so: free particles of nonzero rest mass, moving non-relativistically in three-dimensional space.) The partition function "per atom" is given by

$$\begin{aligned} z(V,T) &= \frac{1}{h^3}\int_V d^3r \int d^3p\, e^{-\beta\varepsilon} = \frac{2\pi V}{h^3}(2m)^{3/2}\int_0^\infty d\varepsilon\, \varepsilon^{1/2}\, e^{-\beta\varepsilon} \\ &= \frac{2\pi V}{h^3}(2m)^{3/2}\,\Gamma(\tfrac{3}{2})\beta^{-3/2} = \frac{V}{h^3}\,(2\pi m k_B T)^{3/2}. \end{aligned} \tag{28.12}$$

The canonical partition function of the gas, with the correct Boltzmann counting, is therefore, given by

$$Z(V,T,N) = \frac{1}{N!}\Big[\frac{V}{h^3}(2\pi m k_B T)^{3/2}\Big]^N, \tag{28.13}$$

and all the thermodynamic properties of the classical ideal gas follow.

28.1.3 The Convolution Theorem

As in the case of the Fourier transform (Eq. (18.9) of Chap. 18, Sect. 18.1.6), there exists a convolution theorem for Laplace transforms. It is of considerable use in the numerous applications of these transforms. If $f(t)$ and $g(t)$ have Laplace transforms $\widetilde{f}(s)$ and $\widetilde{g}(s)$, respectively, their convolution has the Laplace transform $\widetilde{f}(s)\,\widetilde{g}(s)$.

That is,

$$\boxed{\mathcal{L}\left[\int_0^t dt'\, f(t')\, g(t-t')\right] = \mathcal{L}\left[\int_0^t dt'\, f(t-t')\, g(t')\right] = \widetilde{f}(s)\,\widetilde{g}(s).} \tag{28.14}$$

Note the limits of integration in the definition of the convolution of $f(t)$ and $g(t)$ in the case of Laplace transforms. The special case $g(t) = 1$ gives

$$\mathcal{L}\left[\int_0^t dt'\, f(t')\right] = \frac{\widetilde{f}(s)}{s}. \tag{28.15}$$

Repeated application of the convolution theorem gives

$$\mathcal{L}\left[\int_0^t dt_n \int_0^{t_n} dt_{n-1} \cdots \int_0^{t_2} dt_1\, f(t-t_n)\, f(t_n - t_{n-1}) \cdots f(t_2 - t_1)\right] = \frac{\left[\widetilde{f}(s)\right]^n}{s}. \tag{28.16}$$

★ **2.** Establish Eqs. (28.14) and (28.16).

28.1.4 Laplace Transforms of Derivatives

Using integration by parts, it is easy to see that

$$\boxed{\mathcal{L}\left[df(t)/dt\right] = s\,\widetilde{f}(s) - f(0).} \tag{28.17}$$

Observe the occurrence of the "initial" value $f(0)$ in the formula above. Now let $f^{(n)}(t)$ denote $d^n f(t)/dt^n$, with $f^{(0)}(t) \equiv f(t)$. Then

$$\mathcal{L}\left[f^{(n)}(t)\right] = s\,\mathcal{L}\left[f^{(n-1)}(t)\right] - f^{(n-1)}(0), \quad n \geq 1. \tag{28.18}$$

It follows that

$$\mathcal{L}\left[f^{(n)}(t)\right] = s^n\,\widetilde{f}(s) - s^{n-1} f(0) - s^{n-2} f^{(1)}(0) - \cdots - s\, f^{(n-2)}(0) - f^{(n-1)}(0). \tag{28.19}$$

Hence, the generalization of Eq. (28.17) is the formula

$$\boxed{\mathcal{L}\left[d^n f(t)/dt^n\right] = s^n\,\widetilde{f}(s) - \sum_{j=1}^{n} s^{n-j}\left[d^{j-1} f(t)/dt^{j-1}\right]_{t=0}.} \tag{28.20}$$

Note that the initial values of all the lower order derivatives of $f(t)$ appear in the formula, multiplied by powers of s.

★ **3.** Verify the identities (28.17) and (28.20).

Thus,

- the Laplace transform essentially converts differentiation (with respect to t) to multiplication (by s).

This fact is of great use in the applications of Laplace transforms. In particular, the solution of linear differential equations with constant coefficients and specified initial conditions is reduced to simple algebra, as you will see.

28.2 The Inverse Laplace Transform

28.2.1 The Mellin Formula

We know that the Fourier transform operator $\mathcal{F}$ and its inverse operator $\mathcal{F}^{-1}$ are integral operators, with kernels proportional to $e^{i\omega t}$ and $e^{-i\omega t}$, respectively. The range of integration in both cases is the whole of the real number line (in t and in ω, respectively). Similarly, both the Laplace transform operator $\mathcal{L}$ and its inverse $\mathcal{L}^{-1}$ are integral operators, with kernels proportional to e^{-st} and e^{st}, respectively. The range of integration over t is, of course, 0 to ∞ on the real line. But the range of integration over s in the inverse Laplace transform is a little more complicated.

The inverse Laplace transform that expresses $f(t)$ in terms of $\widetilde{f}(s)$ is given by the **Mellin formula**,

$$\boxed{(\mathcal{L}^{-1}\widetilde{f})(t) = f(t) = \frac{1}{2\pi i}\int_C ds\, e^{st}\, \widetilde{f}(s) = \frac{1}{2\pi i}\int_{c-i\infty}^{c+i\infty} ds\, e^{st}\, \widetilde{f}(s).} \qquad (28.21)$$

Here, the contour of integration C is an infinite line in the complex s-plane that runs parallel to the imaginary axis (see Fig. 28.1). C stays entirely in the original region of analyticity of $\widetilde{f}(s)$ as determined by its defining integral, Eq. (28.3). As you know, this region is always some right half-plane in s. Thus, C cuts the real axis at a point c such that the contour stays to the right of *all* the singularities of $\widetilde{f}(s)$. In the finite part of the plane, C can be deformed or moved about such that it does not leave the region of analyticity $\widetilde{f}(s)$ (i.e., such that it does not encounter any singularity of this function), without changing the value of the integral. In particular, C can be moved parallel to itself (thus changing c), as long as it does not leave the region of analyticity.

But we can do more. Using the Mellin formula, inverse Laplace transforms can be explicitly found on many cases, by actually evaluating the line integral in Eq. (28.21) using contour integration. For this purpose, it is necessary to convert C to a *closed* contour, so that we may apply the residue theorem, for instance. Closing it with an infinite semicircle to the *right* of C is ruled out, because the factor e^{st} in the integrand diverges as $\text{Re}\, s \to +\infty$ (remember that $f(t)$ is only defined for $t \geq 0$). Therefore, C can only be closed, if at all, by adding the zero contribution from an infinite semicircle to the *left* of C (that is, the semicircle on which $\text{Re}\, s \to -\infty$).

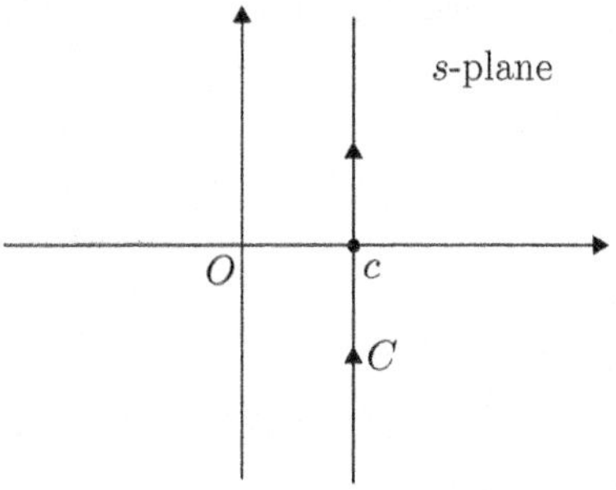

Fig. 28.1 The integration contour C for the inverse Laplace transform is the line $\text{Re}\, s = c$ in the s-plane. Here c is any real number such that all the singularities of $\widetilde{f}(s)$ lie to the left of C

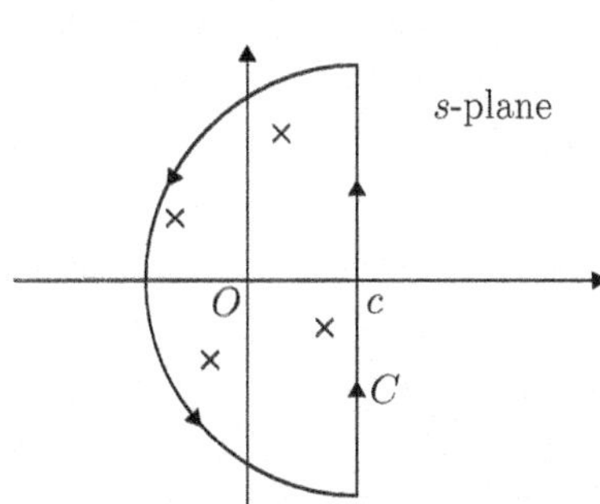

Fig. 28.2 Closing the contour C for the inverse Laplace transform, to evaluate the integral. The crosses denote all the poles of the integrand

It is obvious that C can be closed in this manner only if $\widetilde{f}(s)$ has isolated singularities in the left half-plane, and not any branch cuts running all the way to infinity. In particular, it is easy to invert the Laplace transform whenever $\widetilde{f}(s)$ is a rational function of s. Adding a large semicircle in the left half-plane to C gives a closed contour that encircles *all* the poles of $\widetilde{f}(s)$, as shown in Fig. 28.2. The residue theorem then gives the value of the contour integral, and hence $f(t)$ is determined.

★ **4.** Here are some simple cases in which $f(t)$ can be evaluated as just described. Let a be a real constant. Show that

(a) $\widetilde{f}(s) = \dfrac{1}{s(s^2 + a^2)} \implies f(t) = \dfrac{1 - \cos at}{a^2}.$

(b) $\widetilde{f}(s) = \dfrac{1}{(s^2 + a^2)^2} \implies f(t) = \dfrac{\sin at - at \cos at}{2a^3}.$

(c) $\widetilde{f}(s) = \dfrac{s}{(s^2 + a^2)^2} \implies f(t) = \dfrac{t \sin at}{2a}.$

Is the restriction on a (to real values) necessary?

28.2.2 *LCR Circuit Under a Sinusoidal Applied Voltage*

As another standard example from elementary physics, consider the response of an LCR series circuit when a sinusoidal voltage of amplitude V_0 and (angular) frequency ω is applied to it. The LCR series circuit has already been considered in Chap. 24, Sect. 24.2.2, where we determined its complex admittance.

The differential equation satisfied by the charge $q(t)$ on the capacitor is

$$L\,\ddot{q} + R\,\dot{q} + (1/C)\,q = V_0 \sin \omega t. \tag{28.22}$$

As you know, this equation is precisely the equation of motion of a sinusoidally forced damped simple harmonic oscillator. The damping constant is $R/L = \gamma$, which is just the reciprocal of the time constant of an LR circuit. The natural frequency of the circuit in the absence of the resistor is $1/\sqrt{LC} = \omega_0$. The condition $\omega_0 > \frac{1}{2}\gamma$ corresponds to the underdamped case. Let us consider this case, for definiteness. It should be clear, by now, that it suffices to solve the problem for this case. The solutions in the critically damped ($\omega_0 = \frac{1}{2}\gamma$) and overdamped ($\omega_0 < \frac{1}{2}\gamma$) cases may then be written down by simple analytic continuation. Typically, trigonometric functions will become hyperbolic functions as a consequence.

Let

$$\omega_{\rm u} = \big(\omega_0^2 - \tfrac{1}{4}\gamma^2\big)^{1/2}, \tag{28.23}$$

as already defined in Eq. (24.23), Chap. 24, Sect. 24.2.2. Suppose the initial conditions are such that both the initial charge and the initial current are equal to zero, i.e.,

$$q(0) = 0 \quad \text{and} \quad \dot{q}(0) = 0. \tag{28.24}$$

The solution of Eq. (28.22) can be written as the sum of a *transient part* and a *steady state part*,

$$q(t) = q^{\rm tr}(t) + q^{\rm st}(t). \tag{28.25}$$

These are given, respectively, by

$$q^{\rm tr}(t) = \Big(\frac{V_0\,\omega}{L\,\omega_{\rm u}}\Big)\left\{\frac{(\omega^2 - \omega_0^2 + \frac{1}{2}\gamma^2)\,\sin(\omega_{\rm u}t) + \omega_{\rm u}\,\gamma\,\cos(\omega_{\rm u}t)}{(\omega^2 - \omega_0^2)^2 + \omega^2\gamma^2}\right\}e^{-\gamma t/2} \tag{28.26}$$

and

$$q^{\rm st}(t) = -\Big(\frac{V_0}{L}\Big)\left\{\frac{(\omega^2 - \omega_0^2)\,\sin(\omega t) + \omega\,\gamma\,\cos(\omega t)}{(\omega^2 - \omega_0^2)^2 + \omega^2\gamma^2}\right\}. \tag{28.27}$$

★ **5.** Obtain the solution quoted in Eqs. (28.25)–(28.27).

Observe that the transient component of the solution, $q^{\rm tr}(t)$, is characterized by the frequency $\omega_{\rm u}$ that depends on the circuit parameters L, C and R. This part decays to zero exponentially in time, owing to the dissipation in the system. The steady state component $q^{\rm st}(t)$, on the other hand, oscillates with the same frequency ω as the applied voltage, and does not damp out with time. These statements also apply to the current in the circuit, given by $I(t) = \dot{q}(t)$.

The statements just made are actually independent of the initial conditions. In order to check this out, consider general values $q(0)$ and $\dot{q}(0) = I(0)$, respectively, of the initial charge on the capacitor and the initial current in the circuit. The complete solution for $q(t)$ is now given by Eqs. (28.25)–(28.27) with an extra term $q^{\rm tr}_{\rm hom}(t)$

added to $q^{\text{tr}}(t)$, where

$$q^{\text{tr}}_{\text{hom}}(t) = \left\{ \left(\frac{I(0) + \frac{1}{2}\gamma\, q(0)}{\omega_{\text{u}}} \right) \sin(\omega_{\text{u}} t) + q(0)\, \cos(\omega_{\text{u}} t) \right\} e^{-\gamma t/2}. \qquad (28.28)$$

That is, the complete solution of Eq. (28.22) for general initial values $q(0)$ and $I(0)$ is

$$q(t) = \underbrace{q^{\text{tr}}_{\text{hom}}(t)}_{\text{CF}} + \underbrace{q^{\text{tr}}(t) + q^{\text{st}}(t)}_{\text{PI}}. \qquad (28.29)$$

★ **6.** Given the initial values $q(0)$ and $I(0)$, verify that the extra term in the transient part of the solution is given by Eq. (28.28).

Complementary function and particular integral: I have used the notation $q^{\text{tr}}_{\text{hom}}(t)$ for the expression in Eq. (28.28) because it is precisely the solution to the *homogeneous* differential equation

$$L\ddot{q} + R\dot{q} + (1/C)\, q = 0 \qquad (28.30)$$

that is satisfied by the charge on the capacitor in the *absence* of any applied voltage. With reference to the inhomogeneous differential equation (28.22), the solution given by Eqs. (28.25)–(28.27) represents the **particular integral** (PI), while the expression in (28.28) represents the **complementary function** (CF). The purpose of adding "the right amount" of the complementary function to the particular integral (in the solution of an inhomogeneous differential equation) is to ensure that the boundary conditions (in this case, the initial conditions) are satisfied by the solution. The example just considered ought to help you see this quite clearly.

28.3 Bessel Functions and Laplace Transforms

28.3.1 Differential Equations and Power Series Representations

Bessel functions (of the first and second kinds, and combinations of these) comprise another very important family of special functions that appear in an enormous variety of problems in physics and applied mathematics. For instance, just as the spherical harmonics appear naturally in problems involving spherical symmetry, Bessel and modified Bessel functions appear in those involving axial or cylindrical symmetry. These functions satisfy Bessel's differential equation

$$z^2 \frac{d^2\phi}{dz^2} + z\, \frac{d\phi}{dz} + (z^2 - \nu^2)\, \phi = 0, \qquad (28.31)$$

where ν is, in general, a complex constant; or the modified Bessel equation

$$z^2 \frac{d^2\phi}{dz^2} + z \frac{d\phi}{dz} - (z^2 + \nu^2)\,\phi = 0. \tag{28.32}$$

The Bessel function of the first kind, $J_\nu(z)$, and the modified Bessel function of the first kind, $I_\nu(z)$, are the regular solutions of Eqs. (28.31) and (28.32), respectively. However, since these solutions do not reduce to polynomials even when the order ν is an integer, they were not included in Chap. 16 on orthogonal polynomials. On the other hand, a mention of at least a few properties of some Bessel functions is necessary at some point, in view of their frequent appearance in physical problems. This is a good place for this purpose. Owing to limitations of space, we shall restrict our attention to functions of the first kind. For ease of identification, I will use ν to denote an arbitrary complex value of the order, and l to denote a nonnegative integer value of the order.

A power series representation of $I_l(z)$ has already been written down in Eq. (19.44) of Chap. 19, Sect. 19.2.5. For all complex values of the order ν as well, as long as it is not a negative integer, we have

$$\boxed{I_\nu(z) = \sum_{n=0}^{\infty} \frac{1}{\Gamma(\nu + n + 1)\, n!} \left(\tfrac{1}{2}z\right)^{2n+\nu} \quad (\nu \neq -1,\ -2,\ \ldots).} \tag{28.33}$$

Similarly, $J_\nu(z)$ may be defined by the power series

$$\boxed{J_\nu(z) = \sum_{n=0}^{\infty} \frac{(-1)^n}{\Gamma(\nu + n + 1)\, n!} \left(\tfrac{1}{2}z\right)^{2n+\nu} \quad (\nu \neq -1,\ -2,\ \ldots)} \tag{28.34}$$

For negative integer values of the order, these functions are defined by the relations

$$I_{-l}(z) \stackrel{\text{def.}}{=} I_l(z), \quad l = \text{integer} \tag{28.35}$$

which has already been written down (see Eq. (19.46) of Chap. 19, Sect. 19.2.5); and

$$\boxed{J_{-l}(z) \stackrel{\text{def.}}{=} (-1)^l\, J_l(z), \quad l = \text{integer}.} \tag{28.36}$$

Observe that the only difference between Eqs. (28.33) and (28.34) is a factor of $(-1)^n$ in the summand in the latter case. It should, therefore, come as no surprise that I_ν and J_ν are related according to

$$\boxed{J_\nu(z) = e^{-i\pi\nu/2}\, I_\nu(iz), \quad \text{or} \quad I_\nu(z) = e^{i\pi\nu/2}\, J_\nu(-iz).} \tag{28.37}$$

When $\nu = l$, a nonnegative integer, the series for the Bessel function of the first kind reduces to

$$J_l(z) = \sum_{n=0}^{\infty} \frac{(-1)^n}{(n+l)!\,n!} \left(\tfrac{1}{2}z\right)^{2n+l}. \qquad (28.38)$$

The modified Bessel function of the first kind, $I_l(z)$, has the same power series representation as in Eq. (28.38), but without the factor $(-1)^n$ in the summand. Once again, this has already been written down in Eq. (19.45) of Chap. 19, Sect. 19.2.5. Note the presence of the two factorials in the denominator of the summand. It is, therefore, evident that $J_l(z)$ is an entire function of z. Recall that the modified Bessel function $I_l(z)$ has already been shown to be an entire function (of order 1), in Chap. 22, Sect. 22.6.2. That result can essentially be taken over to the case of $J_l(z)$. When the order ν is not an integer, of course, the sums in Eqs. (28.33) and (28.34), *apart from the overall factor* z^ν, continue to be entire functions of z. The presence of the factor z^ν implies that

- both $I_\nu(z)$ and $J_\nu(z)$ have algebraic branch points at $z = 0$ and $z = \infty$ whenever ν is not an integer.

28.3.2 Generating Functions and Integral Representations

The generating function for the modified Bessel function $I_l(z)$ has already been written down, in Eq. (19.48) of Chap. 19, Sect. 19.2.5. I repeat it here (with some minor changes of variables convenient for our present purposes), alongside the generating function for the Bessel function $J_l(z)$:

$$\boxed{\sum_{l=-\infty}^{\infty} I_l(z)\,\zeta^l = e^{z(\zeta+\zeta^{-1})/2}} \qquad (28.39)$$

and

$$\boxed{\sum_{l=-\infty}^{\infty} J_l(z)\,\zeta^l = e^{z(\zeta-\zeta^{-1})/2}} \qquad (28.40)$$

Using the transformation $\zeta \to 1/\zeta$, it is easy to see that these expansions are obviously consistent with the definitions in Eqs. (28.36) and (28.35) of the Bessel and modified Bessel functions of negative integer order.

Bessel functions and modified Bessel functions satisfy a number of recursion relations. A very useful pair follows immediately from Eqs. (28.39) and (28.40), on differentiating with respect to z:

$$\left.\begin{aligned} 2\,I_l'(z) &= I_{l-1}(z) + I_{l+1}(z) \\ 2\,J_l'(z) &= J_{l-1}(z) - J_{l+1}(z), \end{aligned}\right\} \qquad (28.41)$$

where a prime denotes the derivative with respect to the argument. In particular, in the case $l = 0$ we have the useful relations

$$I_1(z) = I_0'(z) \quad \text{and} \quad J_1(z) = -J_0'(z). \tag{28.42}$$

Similarly, differentiation of Eqs. (28.39) and (28.40) with respect to ζ leads to the pair of recursion relations

$$\left.\begin{aligned} 2l\, I_l(z) &= z\left[I_{l-1}(z) - I_{l+1}(z)\right] \\ 2l\, J_l(z) &= z\left[J_{l-1}(z) + J_{l+1}(z)\right]. \end{aligned}\right\} \tag{28.43}$$

★ **7.** Derive Eqs. (28.41) and (28.43) from Eqs. (28.39) and (28.40).

It is useful to regard the series in Eqs. (28.39) and (28.40) as Laurent expansions, in powers of the complex variable ζ, of the exponential functions on the right-hand sides of these equations. The expansions are convergent in the region $0 < |\zeta| < \infty$. The corresponding inversion formulas provide integral representations for $I_l(z)$ and $J_l(z)$:

$$I_l(z) = \frac{1}{2\pi i}\oint_C \frac{d\zeta}{\zeta^{l+1}}\, e^{z(\zeta+\zeta^{-1})/2} \tag{28.44}$$

and

$$J_l(z) = \frac{1}{2\pi i}\oint_C \frac{d\zeta}{\zeta^{l+1}}\, e^{z(\zeta-\zeta^{-1})/2}, \tag{28.45}$$

where the contour C encircles the origin $\zeta = 0$ once, in the positive sense. In turn, these representations lead to numerous useful integral formulas for $I_l(z)$ and $J_l(z)$. For example,

$$\boxed{I_l(z) = \frac{1}{\pi}\int_0^\pi d\theta\, e^{z\cos\theta}\cos l\theta} \tag{28.46}$$

and

$$\boxed{J_l(z) = \frac{1}{\pi}\int_0^\pi d\theta\, \cos(l\theta - z\sin\theta),} \tag{28.47}$$

where $l = 0, 1, 2, \ldots$.

★ **8.** Derive Eqs. (28.46) and (28.47) from Eqs. (28.44) and (28.45).

Other notable contour integral representations exist for the Bessel, modified Bessel, and related functions, that are valid for noninteger values of the order ν. I shall not discuss these owing to limitations of space.

***Asymptotic behavior*:** Although it might appear that $I_\nu(z)$ and $J_\nu(z)$ are not very different from each other, they differ drastically in several respects. This is clearly illustrated, for instance, by the behavior of the two functions when their argument is real positive ($z = x > 0$). $I_l(x)$ increases monotonically with x, while $J_l(x)$ oscillates with an amplitude that falls off like $1/\sqrt{x}$. The leading asymptotic behaviors of these functions are, respectively, as follows:

$$I_l(x) \sim \frac{e^x}{\sqrt{2\pi x}}, \quad \text{independent of } l. \tag{28.48}$$

On the other hand,

$$J_l(x) \sim \begin{cases} \dfrac{(-1)^{l/2}}{\sqrt{\pi x}}(\sin x + \cos x), & l \text{ even} \\ \\ \dfrac{(-1)^{(l-1)/2}}{\sqrt{\pi x}}(\sin x - \cos x), & l \text{ odd.} \end{cases} \tag{28.49}$$

28.3.3 Spherical Bessel Functions

When the order ν is a half-odd-integer, the Bessel function $J_{l+\frac{1}{2}}(z)$ simplifies to a trigonometric function modulated by a factor $1/\sqrt{z}$. It turns out to be very convenient to multiply this function by another factor of $1/\sqrt{z}$, and to define the **spherical Bessel function** of the first kind and of order l as

$$\boxed{j_l(z) \stackrel{\text{def.}}{=} \sqrt{\frac{\pi}{2z}}\, J_{l+\frac{1}{2}}(z).} \tag{28.50}$$

These functions occur frequently in various physical problems—for example, in the quantum theory of scattering, as you will see in Chap. 32, Sect. 32.3. It is, therefore, helpful to write down a few of their properties.

The first two spherical Bessel functions of the first kind are

$$j_0(z) = \frac{\sin z}{z} \quad \text{and} \quad j_1(z) = \frac{\sin z - z\cos z}{z^2}. \tag{28.51}$$

A useful representation of $j_l(z)$ is given by

$$j_l(z) = (-z)^l \left(\frac{1}{z}\frac{d}{dz}\right)^l \left(\frac{\sin z}{z}\right). \tag{28.52}$$

$j_l(z)$ is the regular solution of the differential equation

$$\frac{d}{dz}\left(z^2 \frac{d\phi}{dz}\right) + \left[z^2 - l(l+1)\right]\phi = 0. \tag{28.53}$$

Compare this with the differential equation satisfied by $J_l(z)$ (Eq. (28.31) with $\nu = l$). From the power series for $J_\nu(z)$ in Eq. (28.34), it follows that the power series representation of $j_l(z)$ is

$$j_l(z) = 2^l \sum_{n=0}^{\infty} \frac{(-1)^n (l+n)!}{(2l+2n+1)!\, n!}\, z^{2n+l}. \tag{28.54}$$

This series converges absolutely for all finite z, so that $j_l(z)$ is an entire function of z.

When $z = x$, the leading asymptotic behavior of $j_l(x)$ as $x \to \infty$ is given by

$$\boxed{j_l(x) \sim \frac{\sin\left(x - \frac{1}{2} l\pi\right)}{x} = \begin{cases} (-1)^{l/2}\,(\sin x)/x, & l \text{ even} \\ (-1)^{(l+1)/2}\,(\cos x)/x, & l \text{ odd}. \end{cases}} \tag{28.55}$$

Once again, compare this with the asymptotic behavior of $J_l(x)$, as given by (28.49).

We will have an occasion to recall Eqs. (28.54) and (28.55) in Chap. 32, Sect. 32.3.2, when we discuss the expansion of a plane wave in terms of Legendre polynomials, in the context of the *partial wave analysis* of the scattering of a particle in a potential field.

28.3.4 Laplace Transforms of Bessel Functions

It is easy to write down formal power series representations for the Laplace transforms of the functions $I_\nu(at)$ and $J_\nu(at)$, where $\mathrm{Re}\,\nu > -1$, $0 \le t < \infty$, and a is (for definiteness) a positive constant. Keep $\mathrm{Re}\, s > 0$, and take the transform of the series in Eq. (28.33) term by term using Eq. (28.6). This is permissible because the series converges for all finite values of t. Thus,

$$\mathcal{L}\left[I_\nu(at)\right] = \frac{1}{s} \sum_{n=0}^{\infty} \frac{\Gamma(2n+\nu+1)}{\Gamma(n+\nu+1)!\, n!} \left(\frac{a}{2s}\right)^{2n+\nu}. \tag{28.56}$$

The transform of $J_\nu(at)$ is given by the same series, with an extra factor $(-1)^n$ in the summand. But what is the region of convergence in the s-plane of this formal series in powers of (a^2/s^2), and is it a useful representation?

★ **9.** Find the region of convergence, in the s-plane, of the infinite series on the right-hand side of Eq. (28.56).

Consider the function $\widetilde{f}(s) = (s^2 - a^2)^{-1/2}$, where a is a positive constant. Keeping $\mathrm{Re}\, s > a$, expand it in a binomial series in inverse powers of s, and invert the transform term by term using the fact that the inverse transform of $1/s^{2n+1}$ is $t^{2n}/(2n)!$. The result is

$$\mathcal{L}^{-1}\left[\frac{1}{\sqrt{s^2 - a^2}}\right] = \sum_{n=0}^{\infty} \frac{1}{(n!)^2} \left(\tfrac{1}{2} at\right)^{2n} = I_0(at). \tag{28.57}$$

Hence,

$$\mathcal{L}\left[I_0(at)\right] = \int_0^\infty dt\, e^{-st}\, I_0(at) = \frac{1}{\sqrt{s^2 - a^2}}\,. \tag{28.58}$$

Similarly, the inverse transform of the function $(s^2 + a^2)^{-1/2}$ is found to be $J_0(at)$, so that

$$\mathcal{L}\left[J_0(at)\right] = \int_0^\infty dt\, e^{-st}\, J_0(at) = \frac{1}{\sqrt{s^2 + a^2}}\,. \tag{28.59}$$

From the leading large-t-behavior of $J_0(at)$ (read off from Eq. (28.49)), it follows that the integral on the left-hand side of Eq. (28.59) converges for $\text{Re}\, s > 0$. Interestingly enough, however, the integral continues to be convergent even if we set $s = 0$ in the integrand. Although $J_0(at)$ only falls off as slowly as $t^{-1/2}$ as $t \to \infty$, the oscillatory (sign-changing) nature of the function enables the integral to converge. It does not converge *absolutely*, of course (This should remind you at once of what happens in the case of the Dirichlet integral $\int_0^\infty dx\, (\sin\, x)/x$.) Hence,

$$\int_0^\infty dt\, J_0(at) = \frac{1}{a} \quad (a > 0). \tag{28.60}$$

Can we similarly set $s = 0$ in Eq. (28.58), and proceed to conclude that $\int_0^\infty dt\, I_0(at) = i/a$? Certainly not! Since $I_0(at) \sim e^{at}/t^{1/2}$ for large t, the integral on the left-hand side in Eq. (28.58) converges only if $\text{Re}\, s > a\ (> 0)$. This is reflected in the fact that the "master function" in this case, $(s^2 - a^2)^{-1/2}$, has a *singularity* at $s = a$.

A number of related results can be deduced from Eqs. (28.58) and (28.59). Using the relations (28.42) and the formula (28.17) for the Laplace transform of the derivative of a function, we get

$$\mathcal{L}\left[I_1(at)\right] = \frac{\left(s - \sqrt{s^2 - a^2}\right)}{a\,\sqrt{s^2 - a^2}}\,, \quad \mathcal{L}\left[J_1(at)\right] = \frac{\left(\sqrt{s^2 + a^2} - s\right)}{a\,\sqrt{s^2 + a^2}}\,. \tag{28.61}$$

Further, using the method of induction, or the recursion relations (28.41), we find (for $l = 0,\ 1,\ 2,\ \ldots$)

$$\mathcal{L}\left[I_l(at)\right] = \frac{\left(s - \sqrt{s^2 - a^2}\right)^l}{a^l\,\sqrt{s^2 - a^2}}\,, \quad \mathcal{L}\left[J_l(at)\right] = \frac{\left(\sqrt{s^2 + a^2} - s\right)^l}{a^l\,\sqrt{s^2 + a^2}}\,. \tag{28.62}$$

★ **10.** Derive Eqs. (28.61) and (28.62) following the steps indicated above.

The Laplace transforms of $I_\nu(at)$ and $J_\nu(at)$ (for any value of ν satisfying $\text{Re}\, \nu > -1$) are given by the expressions in Eqs. (28.62), with l replaced by ν. Once again, we can set $s = 0$ in the expression for $\mathcal{L}\left[J_\nu(at)\right]$, to obtain the formula $\int_0^\infty dt\, J_\nu(at) = 1/a$, where $\text{Re}\, \nu > -1$ and $a > 0$.

28.4 Laplace Transforms and Random Walks

Laplace transforms find several applications in the analysis of stochastic processes. For instance, the master equations for many Markov processes are solved most easily using Laplace transforms. Some examples of such equations were considered in Chap. 21, Sect. 21.5. In particular, you have seen in Sect. 21.5.2 how the modified Bessel function occurs in the solution of the random walk problem on a linear lattice. Let us now consider two other random walk problems, related to the foregoing, in which the modified Bessel function appears naturally.

28.4.1 Random Walk in d Dimensions

Consider an unbiased simple random walk on an infinite hypercubic lattice in d dimensions. Each site is labeled by a set of d integers, $(j_1, \ldots, j_d) \equiv \mathbf{j}$. The walker jumps from any given site to any one of the $2d$ nearest-neighbor sites with a probability $1/(2d)$, with a mean transition rate λ. Let $P(\mathbf{j}, t)$ be the probability that the walker is at $\mathbf{j}$ at time t, given that the walk started at the origin $\mathbf{0}$ at $t = 0$. Then

$$P(\mathbf{j}, t) = e^{-\lambda t}\, I_{j_1}(\lambda t/d)\, I_{j_2}(\lambda t/d) \cdots I_{j_d}(\lambda t/d). \tag{28.63}$$

★ **11.** Derive Eq. (28.63).

As always, the leading long-time behavior of $P(\mathbf{j}, t)$ is of interest. This follows directly from the asymptotic behavior of the modified Bessel function, Eq. (28.48). We find

$$P(\mathbf{j}, t) \sim \left(\frac{d}{2\pi\lambda t}\right)^{d/2}. \tag{28.64}$$

The asymptotic power-law decay in time ($\sim t^{-d/2}$) of the probability is characteristic of diffusion in the absence of drift, as you will see in Chap. 30.

The generalization of the solution above to include any directional bias is straightforward. Let α_i and β_i be the probabilities of a jump in which the coordinate j_i changes to $j_i + 1$ and $j_i - 1$, respectively. Then $\sum_{i=1}^{d}(\alpha_i + \beta_i) = 1$. The solution for $P(\mathbf{j}, t)$ is given by

$$\boxed{P(\mathbf{j}, t) = e^{-\lambda t} \prod_{i=1}^{d} (\alpha_i/\beta_i)^{j_i/2}\, I_{j_i}\big(2\lambda t\sqrt{\alpha_i\,\beta_i}\,\big).} \tag{28.65}$$

This is the d-dimensional generalization of the solution on a linear lattice, given by Eq. (21.61) of Chap. 21, Sect. 21.5.2. The asymptotic behavior of $P(\mathbf{j}, t)$ is no longer a pure power-law fall-off in time, but also involves a decaying exponential. Up to a multiplicative constant, we have

$$P(\mathbf{j},t) \sim t^{-d/2} \exp\Big[-\lambda t\Big\{1 - 2\textstyle\sum_1^d (\alpha_i\,\beta_i)^{1/2}\Big\}\Big]. \tag{28.66}$$

★ **12.** Write down the master equation satisfied by $P(\mathbf{j},t)$ for the foregoing biased random walk in d dimensions (the generalization of Eq. (21.56)), and derive the solution given in Eq. (28.65).

28.4.2 The First-Passage-Time Distribution

Here is a random walk problem that is a neatly-tailored application of Laplace transforms: the determination of a **first-passage-time distribution**.

Consider, once again, a biased random walk on an infinite linear lattice whose sites are labeled by $j \in \mathbb{Z}$. The respective probabilities of a jump from j to $j+1$ and $j-1$ are α and β, with $\alpha+\beta = 1$. The mean jump rate is λ. It is convenient, for our present purposes, to allow for an arbitrary initial position, and to indicate it explicitly in the probability distribution P. Let $P(k,t\,|\,j)$ denote the probability that the walker is at site k at time t, given that she started from the site j at $t=0$. As we have already found in Eq. (21.61) of Chap. 21, Sect. 21.5.2, this probability is

$$P(k,t\,|\,j) = e^{-\lambda t}\,(\alpha/\beta)^{(k-j)/2}\,I_{k-j}\big(2\lambda t\sqrt{\alpha\,\beta}\,\big). \tag{28.67}$$

The question we now ask is: what is the probability density function (PDF) in *time*, that the walker starts at j at time 0, and reaches the site k *for the first time*? Denote this PDF (very often loosely termed a probability *distribution*, rather than a *density*) by $Q(k,t\,|\,j)$. The probability that the walker reaches k for the *first* time in the time interval $(t, t+dt)$ is then given by $Q(k,t\,|\,j)\,dt$. I reiterate that $Q(k,t\,|\,j)$ is the probability density function of the random variable representing the *time* of first passage to the site k from the starting point j. First-passage-time distributions of this sort are of great importance in applications of random walks. In the present instance, there is more than one way to determine $Q(k,t\,|\,j)$. The method to be described below shows how useful Laplace transforms are in the solution of this problem.

The renewal equation: For definiteness, let $j < k$, and let l be any arbitrary site $\geq k$, as shown in Fig. 28.3. The random walk is a stationary Markov process. This fact implies that there is a certain relationship connecting P and Q. In order to start at j at $t=0$ and reach l at time t, the random walker must definitely hit the site k (where $j < k \leq l$) for the first time at some instant t' in the interval $[0,t]$, and then spend the rest of the time in a random walk from k to l. The crucial point is the following: Hitting the site k for the *first* time at *different* instants are *mutually exclusive* events. Hence the probabilities of all such events must be summed up, by integrating over t'. We, therefore, have

Fig. 28.3 Sites connected by the renewal equation in a random walk on a linear lattice

$$\boxed{P(l,t\,|\,j) = \int_0^t dt'\,P(l,t-t'\,|\,k)\,Q(k,t'\,|\,j).} \tag{28.68}$$

This is called a **renewal equation**. Taking Laplace transforms and using the convolution theorem immediately yields

$$\widetilde{Q}(k,s\,|\,j) = \frac{\widetilde{P}(l,s\,|\,j)}{\widetilde{P}(l,s\,|\,k)}. \tag{28.69}$$

But the left-hand side of this equation does not involve l at all. This relation implies, therefore, that the ratio on the right-hand side must actually be independent of the site index l. This is indeed borne out, as you will see shortly. Using Eqs. (28.9) and (28.62), the Laplace transform of the solution for $P(k,t\,|\,j)$ in Eq. (28.67) is given by

$$\widetilde{P}(k,s\,|\,j) = \frac{\left[s+\lambda-\left\{(s+\lambda)^2-4\lambda^2\alpha\beta\right\}^{1/2}\right]^{k-j}}{(2\lambda\beta)^{k-j}\left\{(s+\lambda)^2-4\lambda^2\alpha\beta\right\}^{1/2}}. \tag{28.70}$$

Observe how the dependence of $\widetilde{P}(k,s\,|\,j)$ on the initial site j and the final site k occurs in a factorized form. This is a general property of such probabilities for stationary Markov processes. It ensures that the l-dependence cancels out in the ratio on the right-hand side of Eq. (28.69). Using the analogs of Eq. (28.70) for $\widetilde{P}(l,s\,|\,j)$ and $\widetilde{P}(l,s\,|\,k)$ in (28.69), we obtain

$$\widetilde{Q}(k,s\,|\,j) = \frac{\left[s+\lambda-\left\{(s+\lambda)^2-4\lambda^2\alpha\beta\right\}^{1/2}\right]^{k-j}}{(2\lambda\beta)^{k-j}}. \tag{28.71}$$

It only remains to invert this Laplace transform. That, too, can be done with the information we already have.

First, if $\widetilde{f}(s) = \mathcal{L}\big[f(t)\big]$, then $-d\widetilde{f}(s)/ds = \mathcal{L}\big[tf(t)\big]$, provided the integral exists. Next, note the simple identity

$$-\frac{d}{ds}\left(s-\sqrt{s^2-a^2}\right)^l = \frac{l\left(s-\sqrt{s^2-a^2}\right)^l}{\sqrt{s^2-a^2}}. \tag{28.72}$$

Hence,

$$\mathcal{L}^{-1}\left[\left(s-\sqrt{s^2-a^2}\right)^l\right] = (l/t)\,I_l(at). \tag{28.73}$$

This suffices to deduce that the inverse Laplace transform of $\widetilde{Q}(k, s \mid j)$ is

$$\boxed{Q(k,t \mid j) = \frac{(k-j)}{t}\left(\frac{\alpha}{\beta}\right)^{(k-j)/2} e^{-\lambda t}\, I_{k-j}\big(2\lambda t\sqrt{\alpha\beta}\big).} \tag{28.74}$$

Observe that

$$Q(k,t \mid j) = \frac{(k-j)}{t}\, P(k,t \mid j). \tag{28.75}$$

But bear in mind that $P(k, t \mid j)$ is the probability distribution of the *location* k of the random walker at any given time t, whereas $Q(k, t \mid j)$ is the PDF of the *time* t of first passage to any given k.

★ **13.** Start with Eq. (28.68) and work through the steps above, to arrive at the result in Eq. (28.74).

A number of rather remarkable conclusions follow readily from the closed-form solution found above.

The total probability of a first passage from j to k is obtained by integrating the PDF $Q(k, t \mid j)$ over t from zero to infinity. But this is just the Laplace transform $\widetilde{Q}(k, s \mid j)$ at $s = 0$. Thus,

$$\text{Prob (first passage from } j \text{ to } k) = \int_0^\infty dt\, Q(k,t \mid j) \equiv \widetilde{Q}(k,0 \mid j). \tag{28.76}$$

Setting $s = 0$ in Eq. (28.71), we find

$$\begin{aligned}\int_0^\infty dt\, Q(k,t \mid j) &= \left(\frac{1-|2\beta-1|}{2\beta}\right)^{k-j} \\ &= \begin{cases} 1, & \text{for } \beta \le \frac{1}{2} \\ (\alpha/\beta)^{k-j} < 1, & \text{for } \beta > \frac{1}{2}. \end{cases}\end{aligned} \tag{28.77}$$

Remember that these results have been derived for $k > j$. Interchanging α and β in Eq. (28.77) gives the corresponding results in the case $k < j$. We are, therefore, led to the following conclusions:

- The first passage from any site j to any other site k on an infinite linear lattice occurs *almost surely*, i.e., the probability of its occurrence is equal to 1, if the random walk is unbiased.
- Likewise, the first passage from j to k occurs almost surely if the random walk is biased in the direction pointing from j to k.
- The first passage from j to k is *not* a sure event, i.e., the total probability of its occurrence is less than 1, if the bias is in the direction opposite to that from j to k.

In the last case the time of first-passage from j to k is not a "proper" random variable, in the sense that its probability distribution is not normalized to unity.

***The mean first passage time*:** The expectation value of the time of first passage from j to k, also called the mean first-passage time, is given by

$$\langle t(j \to k)\rangle \stackrel{\text{def.}}{=} \int_0^\infty dt\, t\, Q(k,t\,|\,j) \Big/ \int_0^\infty dt\, Q(k,t\,|\,j). \tag{28.78}$$

In terms of the Laplace transform $\widetilde{Q}(k, s\,|\,j)$, this reduces to

$$\langle t(j \to k)\rangle = \frac{\big[-d\widetilde{Q}(k,s\,|\,j)/ds\big]_{s=0}}{\big[\widetilde{Q}(k,s\,|\,j)\big]_{s=0}}. \tag{28.79}$$

(When the first passage concerned is a sure event, the denominator on the right-hand side reduces to unity, of course.) Using Eq. (28.71) for $\widetilde{Q}(k, s\,|\,j)$ and simplifying, the final result is

$$\boxed{\langle t(j \to k)\rangle = \frac{(k-j)}{\lambda\,|\alpha-\beta|}} \tag{28.80}$$

for any $k > j$. Thus, on an infinite linear lattice, we have the following results:

- If the random walk is biased, the mean first-passage time from one site to another located in the direction of the bias is finite.
- In an unbiased random walk, the mean first-passage time from one site to another is *infinite*, although such first passage occurs with a probability 1.
- The first passage from one site to another located in the direction opposed to the bias is not a sure event. But you can take the average of the first-passage time over all those realizations of the random walk in which such first passage does occur (hence the denominators in Eqs. (28.78) and (28.79)). The result is again a finite quantity.

Similar results can be established for random walks on two-dimensional lattices. On a three-dimensional lattice, first-passage from one site to another is not a sure event even for an unbiased random walk.

★ **14.** Starting from Eq. (28.79), carry out the algebra to arrive at the result in Eq. (28.80).

A quick way to see that the mean first-passage time (from any site j to any other site k) is infinite for an unbiased random walk on a linear lattice which is as follows: When $\alpha = \beta$, the PDF of the first-passage time is

$$Q(k,t\,|\,j) = \frac{|k-j|}{t}\, e^{-\lambda t}\, I_{k-j}(\lambda t).$$

The asymptotic ($\lambda t \gg 1$) behavior of the modified Bessel function is $\sim t^{-1/2}\, e^{\lambda t}$. Hence, the leading long-time behavior of $Q(k, t \mid j)$ is $\sim t^{-3/2}$. The first moment of this distribution will, therefore, diverge like $\int^{\infty} dt\, t^{-1/2}$.

★ **15.** Show that, when $\alpha > \beta$, the relative fluctuation in the time of first passage from a site j to a site $k > j$ is equal to $1/\sqrt{(k-j)(\alpha-\beta)}$.

28.5 Solutions

4. The singularities of $\widetilde{f}(s)$ are (a) simple poles at $0, \pm ia$, (b) double poles at $\pm ia$, and (c) double poles at $\pm ia$. Close the contour C with a large semicircle in the left half-plane, and pick up the residues at the poles of $\widetilde{f}(s)$ enclosed by the contour. It is clear that this does not require a to be restricted to real values. Whatever be the locations of the poles of $\widetilde{f}(s)$, the Mellin contour C lies, by definition, to the right of all these poles. ▶

5. Take the Laplace transform of both sides of the differential equation (28.22). With the initial conditions $q(0) = 0$ and $\dot{q}(0) = 0$, the transform of $q(t)$ is

$$\widetilde{q}(s) = \frac{(V_0\, \omega/L)}{(s^2+\omega^2)(s^2+\gamma s+\omega_0^2)}\,.$$

Resolve the right-hand side into partial fractions, and invert the Laplace transform. All you need to use is the fact that the inverse Laplace transform of $(s+a)^{-1}$ is e^{-at}. ▶

8. Take the contour C to be the unit circle. Set $\zeta = e^{i\theta}$, where θ runs from $-\pi$ to π. ▶

9. Use the doubling formula for the gamma function, Eq. (25.53) of Chap. 25, Sect. 25.2.8, for the factor $\Gamma(2n+\nu+1)$. Then use the fact that the leading asymptotic ($n \to \infty$) behavior of the ratio of two gamma functions is given by

$$\frac{\Gamma(n+\alpha)}{\Gamma(n+\beta)} \sim n^{\alpha-\beta}.$$

This shows that the asymptotic behavior of the magnitude of the general term in the series in Eq. (28.56) is $\sim n^{-1/2}\,(a^2/s^2)^n$. By the ratio test, we conclude that the series converges absolutely for $|s| > a$.

Remark What this result actually suggests is that a closed-form expression for $\mathcal{L}\left[I_\nu(at)\right]$ may be expected to have one or more singularities on the circle $|s| = a$ in the s-plane. This is borne out, as we see in the case of the Laplace transform of $I_l(z)$ where l is a nonnegative integer. ▶

11. The master equation for $P(\mathbf{j}, t)$ is

$$\frac{dP(\mathbf{j}, t)}{dt} = \frac{\lambda}{2d} \sum_{\delta} P(\mathbf{j} + \delta, t) - \lambda P(\mathbf{j}, t),$$

where δ stands for a nearest-neighbor vector of the site $\mathbf{j}$. (That is, each δ has one of its d components equal to ± 1, and all other components equal to 0.) Define the generating function

$$f(z_1, \ldots, z_d, t) = \sum_{\mathbf{j}} P(\mathbf{j}, t)\, z_1^{j_1} z_2^{j_2} \cdots z_d^{j_d},$$

where each component of $\mathbf{j}$ is summed over all the integers. From the master equation for $P(\mathbf{j}, t)$, it follows that the equation satisfied by this generating function is

$$\frac{\partial f}{\partial t} = \lambda \left\{ \frac{1}{2d} \left(z_1 + z_1^{-1} + \cdots + z_d + z_d^{-1} \right) - 1 \right\} f,$$

with the initial condition $f = 1$ at $t = 0$. The solution is $e^{-\lambda t}$ times an exponential that factors into a product of generating functions of modified Bessel functions,

$$f(z_1, \ldots, z_d, t) = e^{-\lambda t} \prod_{i=1}^{d} e^{\lambda t (z_i + z_i^{-1})/(2d)}.$$

The result quoted above for $P(\mathbf{j}, t)$ follows at once. ▶

13. The algebra can be simplified a bit! The renewal equation (28.68) is valid for any $l \geq k$, and the l-dependence in Eq. (28.69) cancels out. Therefore, you may as well set $l = k$ to obtain

$$\widetilde{Q}(k, s \mid j) = \widetilde{P}(k, s \mid j) / \widetilde{P}(k, s \mid k),$$

and proceed as before. ▶

15. Note that the mean squared value of the first passage time is given by the second derivative of the Laplace transform of the first passage time distribution at $s = 0$, i.e.,

$$\langle t^2(j \to k) \rangle = \left[d^2 \widetilde{Q}(k, s \mid j) / ds^2 \right]_{s=0}.$$

Recall that the relative fluctuation is the ratio of the standard deviation to the mean value. ▶

Chapter 29
Green Function for the Laplacian Operator

29.1 The Partial Differential Equations of Physics

In Chap. 6, Sect. 6.2.6, and Chap. 8, Sect. 8.2.3, I have explained why the Laplacian operator is so fundamental, and why it occurs so commonly in physical problems. The Laplacian appears in all the standard and important partial differential equations of elementary mathematical physics: Laplace's equation, Poisson's equation, Helmholtz's equation, the diffusion or heat equation, and the wave equation. All of these are linear equations. Second-order partial differential equations are classified as elliptic equations (e.g., the Laplace, Poisson, and Helmholtz equations), parabolic equations (e.g., the diffusion equation), and hyperbolic equations (e.g., the wave equation). There are fundamental mathematical differences between these three classes of equations. A vast literature exists on these equations, the methods of solving them, and their solutions. My objective here is much more restricted in scope: it is to derive the fundamental solutions to these equations, by determining the corresponding Green functions. Now that we have discussed both Fourier transforms and Laplace transforms, the machinery required for this purpose is in place. In this chapter, we shall consider Poisson's equation, which requires the determination of the Green function of the Laplacian operator. In Chaps. 30, 31 and 32, we will derive the fundamental Green functions for the diffusion equation, the wave equation and the Helmholtz equation, respectively.

29.2 Green Functions

It is helpful to start with a brief account of the essentials of the Green function method.

V. Balakrishnan, *Mathematical Physics*,
https://doi.org/10.1007/978-3-030-39680-0_29

29.2.1 Green Function for an Ordinary Differential Operator

Consider an inhomogeneous, linear, ordinary differential equation of the form

$$\mathcal{D}_x\, f(x) = g(x) \tag{29.1}$$

in some interval (a, b) in the real variable x. Here $\mathcal{D}_x$ is a differential operator involving derivatives of various orders with respect to x, and $g(x)$ is a given function. It is required to find $f(x)$.

Let us write Eq. (29.1) in abstract form in terms of the elements $|\, f\,\rangle$ and $|\, g\,\rangle$ of a linear space that are represented by $f(x)$ and $g(x)$ in a suitable function space. As you know by now, this means that $\langle x\,|\, f\,\rangle = f(x)$ and $\langle x\,|\, g\,\rangle = g(x)$. Let D denote the abstract operator that is represented by $\mathcal{D}_x$ in function space. Then Eq. (29.1) is just

$$\mathsf{D}\,|\, f\,\rangle = |\, g\,\rangle. \tag{29.2}$$

The formal, general solution of this equation is given by

$$|\, f\,\rangle = \mathsf{D}^{-1}\,|\, g\,\rangle + \sum_i c_i\,|h_i\rangle, \tag{29.3}$$

where the c_i are constants, and the kets $|h_i\rangle$ are the linearly independent solutions of the *homogeneous* equation

$$\mathsf{D}\,|h_i\rangle = 0 \quad \text{or} \quad \mathcal{D}_x\, h_i(x) = 0. \tag{29.4}$$

Equation (29.3) involves the inverse of the operator D. In function space, this means that we need the inverse of the differential operator $\mathcal{D}_x$. In general, this inverse is an integral operator. (We will consider integral operators at greater length in Chap. 32.) To see this, take the scalar product of both sides of the formal solution for $|\, f\,\rangle$ above with $\langle\, x\,|$. We have, inserting the identity operator $I = \int_a^b dx'\,|\,x'\,\rangle\langle x'|$ appropriately,

$$\begin{aligned}
\langle x\,|\, f\,\rangle \equiv f(x) &= \langle\, x\,|\mathsf{D}^{-1}\,|\, g\,\rangle + \sum_i c_i\,\langle\, x\,|h_i\rangle \\
&= \int_a^b dx'\,\langle\, x\,|\mathsf{D}^{-1}|\,x'\,\rangle\langle x'|\, g\,\rangle + \sum_i c_i\, h_i(x) \\
&= \underbrace{\int_a^b dx'\, G(x, x')\, g(x')}_{\text{PI}} + \underbrace{\sum_i c_i\, h_i(x)}_{\text{CF}},
\end{aligned} \tag{29.5}$$

where

$$\boxed{G(x,x') \stackrel{\text{def.}}{=} \langle x\,|\mathsf{D}^{-1}|x'\rangle.} \tag{29.6}$$

$G(x, x')$ is the **Green function** of the differential operator $\mathcal{D}_x$. It is just the "matrix element" of the inverse operator D^{-1} between the states $\langle\, x \mid$ and $\mid x'\,\rangle$, and is called the **kernel** of the integral operator D^{-1}. The final expression in (29.5) is the formal solution to the inhomogeneous differential equation (29.1). As I have mentioned earlier (in Chap. 28, Sect. 28.2.2), we learn in elementary treatments of differential equations that the general solution of an inhomogeneous differential equation of this sort is made up of two parts: a particular integral (PI) that depends on $g(x)$, and a complementary function (CF) that does not. The first term in the final expression on the right-hand side in Eq. (29.5) is the PI, while the second term is the CF. The right combination of the two is determined by fixing the values of the constant c_i using the boundary conditions.

In order to write it down the solution explicitly, we need to find the Green function $G(x, x')$. This quantity satisfies the same differential equation as $f(x)$, but with a δ-function as the inhomogeneous term. To establish this, we start with the identity $\mathsf{D}\,\mathsf{D}^{-1} = I$. Therefore,

$$\langle\, x\,|\mathsf{D}\,\mathsf{D}^{-1}|\,x'\,\rangle = \langle x|I|\,x'\,\rangle = \delta(x - x'). \tag{29.7}$$

Inserting the identity operator $\int_a^b dy \mid y\,\rangle\langle\, y\mid$, the left-hand side becomes

$$\begin{aligned}\int_a^b dy\,\langle\, x\,|\mathsf{D}|\,y\,\rangle\langle\, y\,|\mathsf{D}^{-1}|\,x'\,\rangle &= \int_a^b dy\,\mathcal{D}_x\,\langle x|\,y\,\rangle G(y,x')\\ &= \mathcal{D}_x \int_a^b dy\,\delta(x-y)G(y,x') = \mathcal{D}_x\,G(x,x').\end{aligned} \tag{29.8}$$

Therefore,

$$\boxed{\mathcal{D}_x\,G(x,x') = \delta(x - x').} \tag{29.9}$$

This equation must be solved, and the values of the constants c_i must be adjusted, using the appropriate boundary conditions. In this manner, we arrive at the unique solution of (29.1) that satisfies the given boundary conditions on $f(x)$.

29.2.2 An Illustrative Example

A simple example helps illustrate how the Green function method works. Consider the ordinary differential equation

$$\frac{d^2}{dx^2} f(x) = g(x), \tag{29.10}$$

where $x \in [0, 1]$, and $g(x)$ is a given function of x. Suppose it is required to find the solution $f(x)$ that satisfies the general linear boundary conditions

$$f(0) + c_1 f'(0) = c_2 \quad \text{and} \quad f(1) + c_3 f'(1) = c_4, \tag{29.11}$$

where $c_1,\ c_2,\ c_3$ and c_4 are given constants. The differential equation satisfied by the Green function $G(x, x')$ is

$$\frac{d^2 G}{dx^2} = \delta(x - x'). \tag{29.12}$$

G (regarded as a function of x) must satisfy the same boundary conditions as $f(x)$. It is intuitively clear that, since the second derivative of G has a δ-function singularity at $x = x'$, its first derivative must have a *finite* discontinuity at that point, i.e., it must involve $\theta(x - x')$. In turn, this implies that G itself must be continuous at $x = x'$, but it must have a "cusp" (or abrupt change of slope) at that point.

Since, $d^2G/dx^2 = 0$ at all points except the point $x = x'$, it follows that

$$G(x, x') = \begin{cases} A_1 x + A_2 & \text{for} \quad 0 \le x < x' \\ A_3 x + A_4 & \text{for} \quad x' < x \le 1. \end{cases} \tag{29.13}$$

It remains to determine $A_1,\ A_2,\ A_3$ and A_4. (These quantities will be functions of x', of course.) We have precisely four conditions for this purpose: the boundary conditions on G at $x = 0$ and $x = 1$, the continuity of G at $x = x'$, and the fact that

$$\left[dG/dx\right]_{x \downarrow x'} - \left[dG/dx\right]_{x \uparrow x'} = 1. \tag{29.14}$$

Equation (29.14) follows upon integrating the differential equation (29.12) over x from $x' - \epsilon$ to $x' + \epsilon$, and passing to the limit $\epsilon \to 0$. Carrying out the algebra, we find

$$\left.\begin{aligned} A_1 &= \frac{x' - 1 - c_2 - c_3 + c_4}{1 - c_1 + c_3}, \\ A_2 &= \frac{-c_1 x' + (c_1 + c_2)(1 + c_3) - c_1 c_4}{1 - c_1 + c_3}, \\ A_3 &= \frac{x' - c_1 - c_2 + c_4}{1 - c_1 + c_3}, \\ A_4 &= \frac{-(1 + c_3)x' + (c_1 + c_2)(1 + c_3) - c_1 c_4}{1 - c_1 + c_3}. \end{aligned}\right\} \tag{29.15}$$

★ **1.** Verification of the foregoing is a helpful exercise.

(a) Work out the steps to check that the Green function is as given in Eqs. (29.13) and (29.15).

(b) Find the explicit solution to Eq. (29.10) in the simple case in which the boundary conditions are $f(0) = 0,\ f(1) = 0$.

(c) As you know, Green functions are frequently of the form $G(x - x')$, a function of the *difference* two arguments x and x'. In more than one dimension, Green functions are often of the form $G(\mathbf{r} - \mathbf{r}')$. This must follow from translational invariance. Consider the equation (29.12) satisfied by $G(x, x')$ in the example at hand. The operator d^2/dx^2 is translationally invariant: if we set $\xi = x - x'$, then for any given x' we have $d^2/dx^2 = d^2/d\xi^2$. Moreover, the function on the right-hand side, $\delta(x - x')$, is also a function of the difference $(x - x')$. Why is it, then, that the solution for $G(x, x')$ found above is a function of both x and x', and not of the difference $(x - x')$ alone?

In essence, this is the Green function method. All that has been said above can be generalized to linear ordinary differential equations in a complex variable z, and further, to the case of linear partial differential equations in several variables—e.g., when x is replaced by $\mathbf{r}$ in any number d of spatial dimensions, or by $(\mathbf{r}, t)$ in $(d + 1)$ dimensions.

29.3 The Fundamental Green Function for ∇^2

We turn now to a problem of great physical importance: the determination of the Green function of the Laplacian operator, required for the solution of Poisson's equation. In effect, we seek the inverse of the ∇^2 operator. To start with, we consider the standard three-dimensional case. Subsequently, we shall go on to the cases $d = 2$ and $d \geq 4$.

29.3.1 Poisson's Equation in Three Dimensions

A particular integral of Poisson's equation has already been written down in the context of electrostatics: Equation (5.57) of Chap. 5, Sect. 5.3.5, for the electrostatic potential $\phi(\mathbf{r})$ in the presence of a given static charge density $\rho(\mathbf{r})$. Poisson's equation in this case is $\nabla^2 \phi(\mathbf{r}) = -\rho(\mathbf{r})/\epsilon_0$. The solution satisfying natural boundary conditions, namely, $\phi(\mathbf{r}) = 0$ as $r \to \infty$ along any direction in space, is $\phi(\mathbf{r}) = (4\pi\epsilon_0)^{-1} \int d^3r'\, \rho(\mathbf{r}')/|\mathbf{r} - \mathbf{r}'|$. This expression was written down as a known result, on the basis of Coulomb's Law together with the superposition principle. But this solution can be derived systematically, as follows:

Consider the general form of Poisson's equation together with natural boundary conditions, namely,

$$\boxed{\nabla^2 f(\mathbf{r}) = g(\mathbf{r}), \quad \text{with} \quad f(\mathbf{r}) \to 0 \quad \text{as} \quad r \to \infty.} \tag{29.16}$$

Here $g(\mathbf{r})$ is a given function that acts as the "source" term for the scalar field $f(\mathbf{r})$. We are interested here in the particular integral of Eq. (29.16). This is given by

$$f(\mathbf{r}) = \int d^3r\,'\, G^{(3)}(\mathbf{r}, \mathbf{r}\,')\, g(\mathbf{r}\,'), \tag{29.17}$$

where the integration runs over all space. The superscript on the Green function $G^{(3)}$ serves as a reminder of the spatial dimensionality. This function satisfies the differential equation

$$\nabla_{\mathbf{r}}^2\, G^{(3)}(\mathbf{r}, \mathbf{r}\,') = \delta^{(3)}(\mathbf{r} - \mathbf{r}\,'). \tag{29.18}$$

The subscript $\mathbf{r}$ on the gradient operator indicates the variable with respect to which the differentiation is to be performed. We require the solution of Eq. (29.18) that vanishes as $r \to \infty$. Note that

- $G^{(3)}$ may be interpreted, apart from a constant of proportionality, as the Coulomb potential due to a point charge at the source point $\mathbf{r}\,'$.

The significance of this fact will become clear shortly.

29.3.2 *The Solution for* $G^{(3)}(\mathbf{r}, \mathbf{r}\,')$

Observe that

(i) the operator $\nabla_{\mathbf{r}}^2$ is translationally invariant: shifting $\mathbf{r}$ by $\mathbf{r}\,'$ does not change the operator;
(ii) the δ-function on the right-hand side of Eq. (29.18) is a function of the difference $\mathbf{r} - \mathbf{r}\,'$; and finally,
(iii) the boundary condition imposed involves $r \to \infty$, which is the same as $|\mathbf{r} - \mathbf{r}\,'| \to \infty$.

Taken *together*, these facts ensure that the solution $G^{(3)}(\mathbf{r}, \mathbf{r}\,')$ is a function of the difference $\mathbf{r} - \mathbf{r}\,'$ alone. Let us, therefore, set $\mathbf{R} = \mathbf{r} - \mathbf{r}\,'$, and write the Green function as $G^{(3)}(\mathbf{R})$.

Based on what we have learned in the preceding chapters, solving Eq. (29.18) is quite easy. Note that $\nabla_{\mathbf{r}}^2 = \nabla_{\mathbf{R}}^2$, because $\mathbf{R}$ is just a shift of the variable $\mathbf{r}$. Equation (29.18) becomes

$$\nabla_{\mathbf{R}}^2\, G^{(3)}(\mathbf{R}) = \delta^{(3)}(\mathbf{R}). \tag{29.19}$$

Now define the Fourier transform pair

$$G^{(3)}(\mathbf{R}) = \frac{1}{(2\pi)^3}\int d^3k\, e^{i\mathbf{k}\cdot\mathbf{R}}\,\widetilde{G}^{(3)}(\mathbf{k}) \quad\Longleftrightarrow\quad \widetilde{G}^{(3)}(\mathbf{k}) = \int d^3R\, e^{-i\mathbf{k}\cdot\mathbf{R}}\,G^{(3)}(\mathbf{R}). \tag{29.20}$$

The Fourier representation of the δ-function in Eq. (29.19) is of course

$$\delta^{(3)}(\mathbf{R}) = \frac{1}{(2\pi)^3}\int d^3k\; e^{i\mathbf{k}\cdot\mathbf{R}}. \tag{29.21}$$

Moreover, recalling an elementary result from vector calculus (Eq. (6.64) of Chap. 6, Sect. 6.2.6), we have

$$\nabla^2_{\mathbf{R}}\,(e^{i\mathbf{k}\cdot\mathbf{R}}) = -k^2\,(e^{i\mathbf{k}\cdot\mathbf{R}}). \tag{29.22}$$

Therefore, Eq. (29.19) reduces to a trivial algebraic equation for $\widetilde{G}^{(3)}(\mathbf{k})$, and we get

$$\widetilde{G}^{(3)}(\mathbf{k}) = -1/k^2. \tag{29.23}$$

Inverting the Fourier transform,

$$G^{(3)}(\mathbf{R}) = -\frac{1}{(2\pi)^3}\int d^3k\,\frac{e^{i\mathbf{k}\cdot\mathbf{R}}}{k^2}. \tag{29.24}$$

This multiple integral can be evaluated explicitly. Note that the integrand $e^{i\mathbf{k}\cdot\mathbf{R}}/k^2$ is a scalar, invariant under rotations. So are the volume element d^3k and the region of integration (all of $\mathbf{k}$-space). The orientation of axes in $\mathbf{k}$-space can, therefore, be chosen as we please. In particular, we can use spherical polar coordinates (k, θ, φ), and choose the polar axis to lie in the direction of $\mathbf{R}$. Then

$$G^{(3)}(\mathbf{R}) = -\frac{1}{(2\pi)^3}\int_0^\infty dk\,k^2\int_{-1}^{1} d(\cos\theta)\int_0^{2\pi} d\varphi\,\frac{e^{ikR\cos\theta}}{k^2}. \tag{29.25}$$

Integrating over φ and θ, we get

$$G^{(3)}(\mathbf{R}) = -\frac{1}{2\pi^2 R}\int_0^\infty dk\,\frac{\sin kR}{k}. \tag{29.26}$$

But the integral over k is precisely the familiar Dirichlet integral, Eq. (2.18) of Chap. 2, Sect. 2.3. (This fact is of significance, as you will see shortly.) Substituting its value $(=\frac{1}{2}\pi)$ in Eq. (29.26), we arrive at the fundamental Green function of the ∇^2 operator in three dimensions,

$$\boxed{G^{(3)}(\mathbf{r}-\mathbf{r}') = -\frac{1}{4\pi R} = -\frac{1}{4\pi|\mathbf{r}-\mathbf{r}'|}}\,. \tag{29.27}$$

This is the solution that was written down in Eq. (16.134) in Chap. 16, Sect. 16.4.8. Note that it is actually a function of R, the magnitude of $\mathbf{R}$.

★ **2.** Verify the steps leading from Eq. (29.24) to Eq. (29.27).

In the mathematics literature, the "free" Green function corresponding to a differential operator, i.e., the Green function obtained without any boundary conditions (or often, natural boundary conditions such as the one we have used above for the Laplacian), is called the **fundamental solution** corresponding to the differential operator. I will use the slightly redundant term **fundamental Green function** in order to avoid any confusion. The expression in Eq. (29.27) is the fundamental Green function for the Laplacian operator in three-dimensional Euclidean space.

29.3.3 Solution of Poisson's Equation

Substituting for $G^{(3)}$ from Eq. (29.27) in Eq. (29.17), the particular integral of Poisson's equation (29.16) is given by

$$\boxed{f(\mathbf{r}) = -\frac{1}{4\pi}\int d^3r'\,\frac{g(\mathbf{r}')}{|\mathbf{r}-\mathbf{r}'|}\,.} \tag{29.28}$$

Simplification for a spherically symmetric source: When the source function $g(\mathbf{r}) = g(r)$, i.e., when it is spherically symmetric, the solution (29.28) can be simplified further. For this purpose we need the expansion of the Coulomb kernel in spherical harmonics, written out in Eq. (16.139) of Chap. 16, Sect. 16.4.8. Repeating it for ready reference,

$$\frac{1}{|\mathbf{r}-\mathbf{r}'|} = \frac{1}{r_g}\sum_{n=0}^{\infty}\sum_{m=-n}^{n}\frac{4\pi}{2n+1}\left(\frac{r_s}{r_g}\right)^n Y^*_{nm}(\theta',\varphi')\,Y_{nm}(\theta,\varphi), \tag{29.29}$$

where $r_s = \min(r, r')$ and $r_g = \max(r, r')$. Inserting this in Eq. (29.28) and interchanging the order of summation and integration,

$$f(\mathbf{r}) = -\sum_{n=0}^{\infty}\sum_{m=-n}^{n}\frac{Y_{nm}(\theta,\varphi)}{(2n+1)}\int d^3r'\,\frac{g(r')}{r_g}\left(\frac{r_s}{r_g}\right)^n Y^*_{nm}(\theta',\varphi'). \tag{29.30}$$

The integrations over the angular variables θ' and φ' can then be carried out immediately by using Eq. (16.128) of Chap. 16, Sect. 16.4.7. Repeating it for ready reference,

$$\int d\Omega'\,Y^*_{nm}(\theta',\varphi') = \sqrt{4\pi}\,\delta_{n,0}\,\delta_{m,0}\,. \tag{29.31}$$

Because of the Kronecker deltas, the dependence of $f(\mathbf{r})$ on the angles θ and φ also disappears, so that $f(\mathbf{r}) = f(r)$. The solution simplifies, finally, to

$$f(r) = -\int_0^\infty dr'\,\frac{r'^2\,g(r')}{r_>}$$
$$= -\frac{1}{r}\int_0^r dr'\,r'^2\,g(r') - \int_r^\infty dr'\,r'\,g(r'). \qquad (29.32)$$

★ **3.** Verify that Eq. (29.30) simplifies to Eq. (29.32) when $g(\mathbf{r}) = g(r)$. Now start with Eq. (29.32) and verify that $\nabla^2 f(r) = g(r)$.

29.3.4 Connection with the Coulomb Potential

The fact that the fundamental Green function of the Laplacian operator in three-dimensional Euclidean space is $-1/(4\pi|\mathbf{r} - \mathbf{r}'|)$ is an important result. To repeat a point made earlier: this Green function is (apart from a constant factor) the Coulomb potential due to a point charge. This connection lends a deeper significance to the Coulomb potential, and hence to the inverse-square central force. I have already mentioned, in Chap. 8, Sect. 8.1.5, that the inverse-square law leads directly to Gauss's Law in electrostatics: The flux of the electrostatic field over a closed surface is equal to the total charge enclosed by the surface, apart from a multiplicative constant ($= 1/\epsilon_0$, in SI units). A counterpart obviously exists in gravitation as well. The flux of the gravitational field across a closed surface is equal to the total mass enclosed by the surface, apart from a multiplicative constant.

- The inverse-square central force is the only force law (in three-dimensional space) for which such a property holds good.

Interestingly enough, there exist force laws with an analogous property in spaces of other dimensionalities as well. These laws arise from the counterparts of the Coulomb potential in those spaces. The case $d = 2$ is somewhat exceptional, as you will see. It will be dealt with after we discuss the case $d > 3$.

29.4 The Coulomb Potential in $d > 3$ Dimensions

29.4.1 Simplification of the Fundamental Green Function

What is the analog of the Coulomb potential (or the inverse-square central force) in a space of an arbitrary number of dimensions? The connection just pointed out leads to a consistent way to *define* such a potential, via Poisson's equation for the potential due to a point source (or charge). In other words

- The Coulomb potential in d-dimensional space is, apart from a constant of proportionality, the fundamental Green function of the Laplacian operator in d dimensions.

We, therefore, look for the fundamental solution of the equation

$$\nabla_{\mathbf{R}}^2 G^{(d)}(\mathbf{R}) = \delta^{(d)}(\mathbf{R}), \quad \text{with} \quad G^{(d)}(\mathbf{R}) \underset{R\to\infty}{\longrightarrow} 0. \tag{29.33}$$

The superscript in $G^{(d)}$ is to remind us that we are concerned with the Green function in d-dimensional space. Proceeding exactly as before, the Fourier transform of $G^{(d)}(\mathbf{R})$ is once again given by

$$\widetilde{G}^{(d)}(\mathbf{k}) = -1/k^2\,. \tag{29.34}$$

Therefore,

$$\boxed{G^{(d)}(\mathbf{R}) = -\frac{1}{(2\pi)^d}\int \frac{d^d k}{k^2}\, e^{i\mathbf{k}\cdot\mathbf{R}}.} \tag{29.35}$$

- The Coulomb potential in d dimensions is essentially the inverse Fourier transform of $-1/k^2$.

The computation of the d-dimensional multiple integral in (29.35) is very instructive. Clearly, the integral is evaluated most conveniently in terms of (hyper)spherical polar coordinates in d dimensions. These coordinates have been defined and discussed in Chap. 6, Sect. 6.1.3. Following Eq. (6.15), the volume element $d^d k$ is given by

$$d^d k = k^{d-1}\,(\sin^{d-2}\theta_1)\,(\sin^{d-3}\theta_2)\,\ldots\,(\sin\theta_{d-2})\,dk\,d\theta_1\,\ldots\,d\theta_{d-2}\,d\varphi. \tag{29.36}$$

Recall that the ranges of the variables are given by $0 \le k < \infty,\ 0 \le \theta_j \le \pi$, and $0 \le \varphi < 2\pi$. We can choose the orientation of the axes such that $\mathbf{R}$ is along the direction of the first coordinate k_1, so that $\mathbf{k}\cdot\mathbf{R} = kR\cos\theta_1$. The multiple integral in Eq. (29.35) then simplifies as follows:

(i) The integration over φ is trivial, and yields a factor of 2π.
(ii) The integration over each of the "polar" angles $\theta_2\,,\ \ldots\,,\ \theta_{d-2}$ can be carried out using Eq. (3.21) of Chap. 3, Sect. 3.1.5 for the definite integral $\int_0^\pi d\theta\,\sin^l\theta$, where l is a nonnegative integer. Integrating over $\theta_2\,,\ \ldots\,,\ \theta_{d-2}$, therefore, yields a factor

$$\prod_{l=1}^{d-3}\left\{\frac{\sqrt{\pi}\,\Gamma\left(\frac{1}{2}(l+1)\right)}{\Gamma\left(1+\frac{1}{2}l\right)}\right\} = \frac{\pi^{(d-3)/2}}{\Gamma\left(\frac{1}{2}(d-1)\right)}, \tag{29.37}$$

after simplification.
(iii) It remains to integrate over k and θ_1. Interestingly, the full R-dependence of the Green function can be extracted by a simple scaling! Change variables of integra-

tion from k to the dimensionless variable $\kappa = kR$. The remaining integrals are then given by

$$\int_0^\infty dk \, \frac{k^{d-1}}{k^2} \int_0^\pi d\theta_1 \, e^{ikR \cos\theta_1} \sin^{d-2}\theta_1 = \frac{1}{R^{d-2}} \int_0^\infty d\kappa \, \kappa^{d-3} \times$$
$$\times \int_0^\pi d\theta_1 \, e^{i\kappa \cos\theta_1} \sin^{d-2}\theta_1. \quad (29.38)$$

Hence, $G^{(d)}(\mathbf{R})$ is only a function of the magnitude R of $\mathbf{R}$. I will, therefore, denote it by $G^{(d)}(R)$ from now on. It is evident that $G^{(d)}(R) \propto 1/R^{d-2}$, with a d-dependent constant of proportionality.

(iv) The integral over θ_1 is not an "elementary" one. It turns out to be essentially a representation of the Bessel function of the first kind. The formula required is[1]

$$J_\nu(z) = \frac{\left(\frac{1}{2}z\right)^\nu}{\sqrt{\pi}\,\Gamma\left(\nu + \frac{1}{2}\right)} \int_0^\pi d\theta \, e^{\pm iz\cos\theta} \sin^{2\nu}\theta, \quad \mathrm{Re}\,\nu > -\tfrac{1}{2}. \quad (29.39)$$

Applying this formula, we get

$$\int_0^\pi d\theta_1 \, e^{i\kappa\cos\theta_1} \sin^{d-2}\theta_1 = \sqrt{\pi}\,\Gamma\left(\tfrac{1}{2}(d-1)\right) (2/\kappa)^{(d-2)/2} J_{\frac{d}{2}-1}(\kappa). \quad (29.40)$$

Putting together the results of **(i)**–**(iv)**, Eq. (29.35) reduces to

$$G^{(d)}(R) = -\frac{1}{(2\pi)^{d/2} R^{d-2}} \int_0^\infty d\kappa \, \kappa^{(d-4)/2} J_{\frac{d}{2}-1}(\kappa). \quad (29.41)$$

★ **4.** Start with Eq. (29.35) and work through all the subsequent steps to arrive at Eq. (29.41).

29.4.2 Power Counting and a Divergence Problem

The integral in (29.41), namely,

$$\int_0^\infty d\kappa \, \kappa^{(d-4)/2} J_{\frac{d}{2}-1}(\kappa), \quad (29.42)$$

is a function of d alone. Evaluating it will complete the computation of $G^{(d)}(R)$ in explicit form. But we need to consider, first, the conditions under which the integral in (29.42) converges absolutely. The argument is called **power counting**, and is quite

[1]Gradshteyn and Ryzhik (see Bibliography), p. 953, formula **8.411** (7).

simple in this instance. We must look at the behavior of the integrand at both the end-points of integration, and make sure that it is not too singular to be integrable at these points. Here is where problems arise.

***An ultraviolet divergence*:** Consider the upper limit first. As you know,

$$\int^{\infty} d\kappa\, \kappa^{r} < \infty \text{ provided } r < -1, \text{ or, more generally, } \operatorname{Re} r < -1. \qquad (29.43)$$

Now, for large values ($\gg 1$) of κ, the leading asymptotic behavior of the Bessel function is as follows: $J_\nu(\kappa) \sim \left(\frac{1}{2}\pi\kappa\right)^{-1/2}$ *independent* of the order ν of the Bessel function, multiplied by a cosine function (whose magnitude remains ≤ 1, of course). Hence, the integral (29.42) converges at the upper limit of integration if

$$\operatorname{Re}\left\{\tfrac{1}{2}(d-4) - \tfrac{1}{2}\right\} < -1, \quad \text{or} \quad \operatorname{Re} d < 3. \qquad (29.44)$$

This seems to pose a problem even for $d = 3$ (which is certainly a very physical case), although we have found a finite expression for $G^{(3)}(R)$! In general, the unboundedness or "divergence" of integrals of this kind, arising from the behavior of the integrand as $k \to \infty$, is an example of what is known as an **ultraviolet divergence** in physics, especially in the context of quantum field theory. As d increases beyond 3, the divergent behavior of the integral presumably gets worse, because the volume element $d^d k$ involves a factor k^{d-1}. This factor increases rapidly with k for larger values of d. We, therefore, face an ultraviolet divergence when we try to compute the Green function in a space of dimension $d > 3$. At the same time, the finiteness of $G^{(3)}(R)$ is also puzzling, and needs to be explained.

***An infrared divergence*:** At the lower limit of integration in (29.42), a different problem is encountered. We know that

$$\int_0 d\kappa\, \kappa^{r} < \infty \text{ provided } \operatorname{Re} r > -1. \qquad (29.45)$$

Now, the leading behavior of the Bessel function as $\kappa \to 0$ is given by $J_\nu(\kappa) \sim \kappa^{\nu}$. Hence the integral converges at the lower limit of integration if

$$\operatorname{Re}\left\{\tfrac{1}{2}(d-4) + \tfrac{1}{2}d - 1\right\} > -1, \quad \text{or} \quad \operatorname{Re} d > 2. \qquad (29.46)$$

For $\operatorname{Re} d \leq 2$, the integral exhibits an **infrared divergence**. Although this does not affect the Green function in any space of dimension ≥ 3, it warns us that there is a problem in the case $d = 2$, in particular.

The terms *ultraviolet divergence* and *infrared divergence* originate in quantum field theory. There, the variable k is associated with the linear momentum of some particle (recall that it is the Fourier conjugate of a position variable). Large k implies a small de Broglie wavelength ("ultraviolet"), while small k corresponds to a large de Broglie wavelength ('infrared').

To sum up:

- The derivation of a fully explicit formula for the fundamental Green function $G^{(d)}(R)$ of the Laplacian operator has been reduced to the evaluation of the integral in (29.42).
- Power counting shows that this integral is only convergent when the parameter d lies in the range $2 < \text{Re } d < 3$. It is formally divergent when d lies outside this range.
- But d is the dimensionality of space. We expect finite expressions for $G^{(d)}(R)$ to exist for positive integer values ≥ 2 of d. In particular, the cases $d = 2$ and $d = 3$ are of direct physical interest.
- A way has to be found to get around the ultraviolet and infrared divergences. The solution must be consistent with the result $G^{(3)}(R) = -1/(4\pi R)$ already derived in Sect. 29.3.2. The reason why $G^{(3)}(R)$ is finite, in spite of the ultraviolet divergence of (29.42) if Re d is not *less* than 3, must also be understood.

29.4.3 Dimensional Regularization

The way to deal with such divergences in physical problems, and to extract meaningful results for physical quantities, is called **regularization**. A valid regularization procedure must meet several nontrivial requirements. In relativistic quantum field theory, for instance, such requirements include gauge invariance and Lorentz invariance. There are no such complications in the simple problem at hand. But it is an instructive exercise. I shall, therefore, spell out the details.

There are many possible regularization methods. Here, I have chosen **dimensional regularization**, which relies on the idea of analytic continuation, for two reasons: first, you are already familiar with analytic continuation from the preceding chapters; second, the method is a powerful one, and is used quite commonly in modern quantum field theory. The basic procedure (or prescription) is as follows:

(i) The dimension d itself is treated as a *complex variable*. The integral (29.42) converges in the region $2 < \text{Re}\, d < 3$ of the complex d-plane. In this region, the integral defines an analytic function of d.

(ii) Another representation of this function is found, that can be analytically continued to values of d that are of physical relevance, but which lie outside the region of convergence of the original integral.

(iii) As in the case of an analytic function that is defined by a power series, there are likely to be singularities (specifically, poles) present at the boundaries of the original region of convergence. If such a singularity occurs at a physical value of d, the *regular part* of the function (i.e., the function with the singular part subtracted out) is supposed to represent the value of the function at the point concerned.

How this works will become clear as we proceed.

The first step is to evaluate the definite integral (29.42), *keeping d in the region* $2 < \text{Re}\, d < 3$ in which the integral is well-defined. How this is done is a separate and peripheral issue, and I will not digress into it. For our present purposes, it suffices to note that it is a "known" integral, given by the general formula[2]

$$\int_0^\infty d\kappa\, \kappa^\mu\, J_\nu(\kappa) = 2^\mu\, \frac{\Gamma\left(\frac{1}{2}(\nu+1+\mu)\right)}{\Gamma\left(\frac{1}{2}(\nu+1-\mu)\right)}\quad \left(\text{Re}\,\mu < -\tfrac{1}{2}\,,\ \text{Re}\,(\mu+\nu) > -1\right). \tag{29.47}$$

We have merely to apply this formula to the integral (29.42), setting $\mu = \frac{1}{2}d - 2$ and $\nu = \frac{1}{2}d - 1$. The condition Re $\mu < -\frac{1}{2}$ translates to Re $d < 3$, while the condition Re $(\mu+\nu) > -1$ becomes Re $d > 2$. Not surprisingly, these are precisely the conditions on d that specify the region of convergence of the integral (29.42), as we have already deduced by power counting. Thus,

$$\int_0^\infty d\kappa\, \kappa^{(d-4)/2}\, J_{\frac{d}{2}-1}(\kappa) = 2^{(d-4)/2}\,\Gamma\left(\tfrac{1}{2}d - 1\right),\quad 2 < \text{Re}\ d < 3. \tag{29.48}$$

But the right-hand side of this equation is an analytic function of d that is defined in the whole of the d-plane. It has a simple pole (arising from the gamma function) at each of the points $d = 2, 0, -2, -4, \ldots$, and is finite elsewhere. It is the analytic continuation sought. Substituting this result into Eq. (29.41), we obtain the relatively simple expression

$$G^{(d)}(R) = -\,\frac{\Gamma\left(\frac{1}{2}d - 1\right)}{4\pi^{d/2}\, R^{d-2}}\,. \tag{29.49}$$

Using the identity $\Gamma(z-1) = \Gamma(z)/(z-1)$, this may be re-written in the form

$$\boxed{G^{(d)}(R) = -\,\frac{\Gamma\left(\frac{1}{2}d\right)}{2\pi^{d/2}\,(d-2)\, R^{d-2}}\,.} \tag{29.50}$$

I reiterate that this expression for $G^{(d)}(R)$ is now in an explicit analytic function is defined in the whole of the complex d-plane. In particular, it has a simple pole at $d = 2$ in the complex d-plane (more on this below), and no singularities at all to the right of that pole. Setting $d = 3$, it trivially checked that $G^{(3)}(R) = -1/(4\pi R)$, as we have found already. You can now proceed to set $d = 4, 5, \ldots$ in Eq. (29.49) to write down the fundamental Green function of the Laplacian in these dimensions. In 4, 5 and 6 dimensions, for instance, the Coulomb potential is given by

$$G^{(4)}(R) = -\frac{1}{4\pi^2 R^2}\,,\quad G^{(5)}(R) = -\frac{1}{8\pi^2 R^3}\ \text{ and }\ G^{(6)}(R) = -\frac{1}{4\pi^3 R^4}\,, \tag{29.51}$$

respectively. We have established an important result:

---

[2]Gradshteyn and Ryzhik, p. 684, formula **6.561** (14).

- The fundamental Green function of the Laplacian, i.e., the Coulomb potential in $d \geq 3$ spatial dimensions, is proportional to $1/R^{d-2}$. The exact expression is given by Eq. (29.50).

I have mentioned that we may expect the analytic function representing $G^{(d)}(R)$ to have singularities (poles) in the complex d-plane on the boundaries of the region of convergence of the original integral. In the present instance, these boundaries are the lines Re $d = 3$ and Re $d = 2$, respectively. In the light of Eq. (29.50), two questions remain:

(i) Why is there *no* pole on the right boundary? In particular, $G^{(3)}(R)$ is finite. Does it mean that there is no ultraviolet divergence in this case?
(ii) There *is* is a pole at $d = 2$ on the left boundary, the signature of an infrared divergence that makes $G^{(2)}(R)$ formally infinite. How do we extract the correct Green function in two-dimensional space from Eq. (29.50)?

The first of these questions are answered below. The second requires actual implementation of dimensional regularization. This will be done in Sect. 29.5.1.

The reason why $G^{(d)}(R)$ ***is not singular at*** $d = 3$ is as follows. The kind of simple power counting we have carried out to determine the range of parameters in which an integral converges is termed **naive power counting**, and rightly so! The integral (29.42) does diverge when Re d is *greater* than 3. Right *on* the boundary, however, naive power counting is not reliable, because the factor $J_{\frac{d}{2}-1}(\kappa)$ in the integrand changes sign in an oscillatory manner, while it decays to zero like $\kappa^{-1/2}$ with increasing κ. The cancellations between the positive and negative contributions to the integral improve the convergence of the integral to such an extent that it converges when Re d is *equal* to 3. As you have seen in Eq. (29.26) in Sect. 29.3.2, $G^{(3)}(R)$ involves the Dirichlet integral $\int_0^\infty dk\,(\sin kR)/k$, which is finite but not *absolutely* convergent. The changes in sign of the sine function in the integrand help overcome the *logarithmic* divergence that would ensue if the integrand had been $|\sin kR|/k$. This is why $G^{(d)}(R)$ has no pole at $d = 3$. When Re $d > 3$, we have a power-law divergence. The oscillatory nature of the Bessel function no longer suffices to make the integral finite. But this does not matter anymore, since the integral itself is not a valid representation of $G^{(d)}(R)$ in that region. Instead, we have the analytic continuation in Eq. (29.50), which is well-defined and which yields finite expressions for $G^{(d)}(R)$ upon setting $d = 4, 5, \ldots$.

An example analogous to the Dirichlet integral is provided by the integral in Eq. (28.60) of Chap. 28, Sect. 28.3.4, namely, $\int_0^\infty dt\, J_0(at) = 1/a$, where $a > 0$. By naive power-counting, this integral should diverge, because $J_0(t) \sim t^{-1/2}$ for large t, and $\int^\infty dt\, t^{-1/2}$ diverges. But the sign changes of the Bessel function lead to a finite value for the integral.

The *infrared* divergence in $G^{(d)}(R)$, however, shows up explicitly. The representation in (29.50) has a simple pole at $d = 2$. We must, therefore, use the prescription of dimensional regularization to extract $G^{(2)}(R)$ from this expression. This will be done in Sect. 29.5.1. Before we do that, however, let us see how a far simpler procedure than the one given above leads us directly to the answer in Eq. (29.50) for $G^{(d)}(R)$.

29.4.4 A Direct Derivation

Let us go back to Eq. (29.33), and regard it now as Poisson's equation for the electrostatic potential due to a unit charge (in suitable units) located at $R = 0$. Let $\mathbf{F} \equiv \nabla G^{(d)}$ denote the "field" due to this charge. Integrate both sides of Eq. (29.33) over a hypersphere of radius R. The right-hand side of the equation becomes equal to unity, being the integral of a δ-function. On the left-hand side, use Gauss's Theorem in vector calculus to write the volume integral $\int dV\,(\nabla\cdot\mathbf{F})$ as a "surface" integral $\int \mathbf{F}\cdot d\mathbf{S}$. (This theorem is not restricted to three-dimensional space!) Thus,

$$\int dV\,\nabla^2 G^{(d)} = \int dV\,\nabla\cdot\nabla\,G^{(d)} = \int dV\,(\nabla\cdot\mathbf{F}) = \int \mathbf{F}\cdot d\mathbf{S}\,. \tag{29.52}$$

But the field required is spherically symmetric. Therefore, it has only a radial component F_R which, moreover, depends only on the magnitude R. Hence,

$$\int \mathbf{F}\cdot d\mathbf{S} == F_R \int dS = F_R\,S_d(R) = 1, \tag{29.53}$$

where $S_d(R)$ is the surface "area" of a hypersphere of radius R in d-dimensional space. But this quantity has already been calculated (in Eq. (6.21) of Chap. 6, Sect. 6.1.3) to be

$$S_d(R) = \frac{2\pi^{d/2} R^{d-1}}{\Gamma\left(\frac{1}{2}d\right)}\,. \tag{29.54}$$

Therefore,

$$F_R = (\nabla G^{(d)})_R = \frac{dG^{(d)}}{dR} = \frac{1}{S_d(R)} = \frac{\Gamma\left(\frac{1}{2}d\right)}{2\pi^{d/2}\,R^{d-1}}\,. \tag{29.55}$$

Integrating from R to ∞ and imposing the boundary condition $G^{(d)}(R) \to 0$ as $R \to \infty$, we get

$$\begin{aligned}\int_R^\infty dR'\,\frac{dG^{(d)}(R')}{dR'} = -\,G^{(d)}(R) &= \frac{\Gamma\left(\frac{1}{2}d\right)}{2\pi^{d/2}}\int_R^\infty \frac{dR'}{R'^{\,d-1}}\\ &= \frac{\Gamma\left(\frac{1}{2}d\right)}{2\pi^{d/2}\,(d-2)\,R^{d-2}}\,.\end{aligned} \tag{29.56}$$

This yields for $G^{(d)}(R)$ precisely the expression in Eq. (29.50). Note how the boundary condition has been incorporated.

The underlying reason why an inverse-square central force (in three dimensions) leads to an integral theorem like Gauss's Law is also obvious now. The surface area of a sphere increases with its radius like R^2. If the field drops off like $1/R^2$, there is obviously an exact compensation in the flux of the field across a sphere centered at the origin, and the total flux becomes independent of the radius R. Precisely the same thing happens in d (> 3) dimensions, provided the field drops off like $1/R^{d-1}$ because the "surface" of a hypersphere increases like R^{d-1}. The potential must then decrease like $1/R^{d-2}$, exactly as we have deduced.

29.5 The Coulomb Potential in $d = 2$ Dimensions

29.5.1 *Dimensional Regularization*

The case $d = 2$ requires a separate treatment. The reason can be traced, ultimately, to the simple fact that $\int dR/R = \ln R$, rather than a power of R. As shown above, the analytic formula (29.50) for $G^{(d)}(R)$ has a simple pole at $d = 2$. This is a reflection of the fact that the original integral representation (29.41) for $G^{(d)}(R)$ has an infrared divergence when $\text{Re}\, d \leq 2$. The prescription of dimensional regularization, as applicable to the problem at hand, is as follows:
(i) Start with Eq. (29.50), namely,

$$G^{(d)}(R) = -\frac{\Gamma\left(\frac{1}{2}d\right)}{2\pi^{d/2}\,(d-2)\,R^{d-2}}\,. \tag{29.57}$$

Treating this as an analytic function of d, write it in the form of a Laurent series about the point $d = 2$, i.e.,

$$G^{(d)}(R) = \underbrace{\frac{\text{residue}}{(d-2)}}_{\text{singular part}} + \underbrace{\sum_{n=0}^{\infty} c_n\,(d-2)^n}_{\text{regular part}}\,. \tag{29.58}$$

(ii) Subtract out the singular part, and set $d = 2$ in the regular part. This leaves behind just the coefficient c_0, which is guaranteed to be the Green function $G^{(2)}(R)$ that we seek.
Expand each of the d-dependent factors in Eq. (29.57), except the pole factor $(d-2)^{-1}$, in a Taylor series about $d = 2$, and retain only terms up to the first order in $(d-2)$. We then get

$$G^{(d)}(R) = -\frac{1}{2\pi(d-2)} + \frac{1}{2\pi}\ln R + \frac{1}{4\pi}(\ln\pi + \gamma) + \mathcal{O}\big((d-2)^1\big), \tag{29.59}$$

where γ is the Euler-Mascheroni constant (defined in Eq. (25.20) of Chap. 25, Sect. 25.2.4). The first term on the right-hand side must be discarded, according to the prescription described above. The true fundamental Green function of ∇^2 in two dimensions is then

$$G^{(2)}(R) = \frac{1}{2\pi} \ln R + \text{constant}. \tag{29.60}$$

The constant is actually arbitrary, and is fixed by specifying a boundary condition. Note that

- the boundary condition $G^{(2)}(R) \to 0$ as $R \to \infty$ is not possible in $d = 2$, owing to the logarithmic R-dependence of the potential.

You will recognize that this logarithmic potential is essentially the same as the electrostatic potential ϕ due to an uniformly charged, infinitely long straight line in three-dimensional space, with R replaced by ϱ, the *axial* distance from the line. Recall that, in this problem too, the potential does not vanish, but instead diverges, as $\varrho \to \infty$. What is done then is to specify that the potential at some axial distance a has the value ϕ_a. The potential difference $\phi(\varrho) - \phi_a$ is then $(\lambda/2\pi\epsilon_0) \ln(\varrho/a)$, where λ is the line charge density.

★ **5.** Start with Eq. (29.57) and show that the Laurent series about $d = 2$ is as given in Eq. (29.59), leading to the result in Eq. (29.60).

29.5.2 Direct Derivation

Should we believe the result in Eq. (29.60) , as it seems to have been derived using *prescription* that appears to be arbitrary? The answer is "yes". Corroboration comes from the same direct physical argument as was given in Sect. 29.4.4 for the case $d \geq 3$. As in that case, start with the differential equation

$$\nabla_{\mathbf{R}}^2 \, G^{(2)}(\mathbf{R}) = \delta^{(2)}(\mathbf{R}). \tag{29.61}$$

Once again, regard this as Poisson's equation for the electrostatic potential due to a unit charge at $R = 0$. Let $\mathbf{F} \equiv \nabla G^{(2)}$ denote the planar vector field due to this charge. Integrate both sides of Eq. (29.61) over a circle of radius R. The right-hand side of the equation becomes unity, of course. Gauss Theorem, applied to the left-hand side, gives (using the circular symmetry of the field)

$$2\pi R \, F_R = 1, \quad \text{so that} \quad F_R = 1/(2\pi R). \tag{29.62}$$

Since, $F_R = dG^{(2)}/dR$, integrating with respect to R immediately gives $G^{(2)}(R) = (1/2\pi) \ln R + \text{constant}$, exactly as in Eq. (29.60).

- The fundamental Green function $G^{(2)}(R)$ of the Laplacian, i.e., the Coulomb potential in 2-dimensional Euclidean space, is proportional to the *logarithm* of R.

This seemingly simple fact has remarkably profound consequences in diverse areas of physics, such as condensed matter physics and quantum field theory, among others. It even seems to have a bearing on the phenomenon of quark confinement!

29.5.3 An Alternative Regularization

A crucial point to be kept in mind regarding regularization is the following. Different regularization procedures must lead to the same physically acceptable result—or else the procedure would be arbitrary, and its result unreliable. Let us, therefore, re-work the case at hand using an alternative procedure, in order to satisfy ourselves that the same final result, $G^{(2)}(R) = (1/2\pi) \ln R + \text{constant}$, is obtained once again.

Going back to Eq. (29.35) and setting $d = 2$ we have, formally,

$$G^{(2)}(R) = -\frac{1}{(2\pi)^2} \int \frac{d^2k}{k^2}\, e^{i\mathbf{k}\cdot\mathbf{R}}. \tag{29.63}$$

Since $d^2k = k\, dk\, d\varphi$ is in plane polar coordinates, we are now faced with $\int_0 dk/k$. But such an integral is logarithmically divergent. The problem arises from the behavior of the integrand near $k = 0$—an infrared divergence, as you already know. A regularization procedure that almost suggests itself is the replacement of the factor k^2 in the denominator of the integrand in Eq. (29.63) by $(k^2 + \mu^2)$, where μ is a positive constant. The idea is to carry out the integration with this modified integrand, and then pass to the limit $\mu \to 0$ to extract the Green function. Let us, therefore, define the function

$$G^{(2)}(R\,,\,\mu) \stackrel{\text{def.}}{=} -\frac{1}{(2\pi)^2} \int \frac{d^2k}{(k^2 + \mu^2)}\, e^{i\mathbf{k}\cdot\mathbf{R}}. \tag{29.64}$$

As before, it makes sense to work in plane polar coordinates (k, φ) in $\mathbf{k}$-space. Choosing the k_1-axis to lie along $\mathbf{R}$,

$$G^{(2)}(R\,,\,\mu) = -\frac{1}{(2\pi)^2} \int_0^\infty \frac{k\, dk}{(k^2 + \mu^2)} \int_0^{2\pi} d\varphi\, e^{ikR\cos\varphi}. \tag{29.65}$$

As you might expect (e.g., from Eq. (29.39), in the case $\nu = 0$), the angular integral is just 2π times $J_0(kR)$, the Bessel function of order zero. Thus,

$$G^{(2)}(R\,,\,\mu) = -\frac{1}{2\pi} \int_0^\infty \frac{k\, dk}{(k^2 + \mu^2)}\, J_0(kR). \tag{29.66}$$

The integral is again a standard one,[3] and yields

$$G^{(2)}(R\,,\,\mu) = -\,(1/2\pi)\, K_0(\mu R), \tag{29.67}$$

[3]Gradshteyn and Ryzhik, p. 717, formula **6.532** (4).

Here K_0 is the modified Bessel function of the *second* kind, of order zero. It is the singular solution of Bessel's differential equation, Eq. (28.32) of Chap. 28, Sect. 28.3.1, in the case $\nu = 0$. A representation of this function that is convenient for our purposes is[4]

$$K_0(z) = -I_0(z)\,\ln\left(\tfrac{1}{2}z\right) + \sum_{m=0}^{\infty} \frac{\psi(m+1)}{(m!)^2}\left(\tfrac{1}{2}z\right)^{2m}. \qquad (29.68)$$

Here $I_0(z)$ is the familiar modified Bessel function of the first kind, of order zero, and $\psi(m+1)$ is the digamma function (the logarithmic derivative of the gamma function). Recall that $I_0(z)$ is an entire function of z. So, too, is the second term (the infinite sum) on the right-hand side of Eq. (29.68). We may, therefore, conclude that $K_0(z)$ has a logarithmic singularity at $z = 0$. Hence, we have, using the fact that $I_0(0) = 1$,

$$G^{(2)}(R\,,\,\mu) \xrightarrow[\mu\to 0]{} (1/2\pi)\,\ln\,R + (1/2\pi)\,\ln\left(\tfrac{1}{2}\mu\right) + \mathcal{O}(\mu^2). \qquad (29.69)$$

We must again discard a formally infinite constant (namely, $\ln\,\mu$ as $\mu \to 0$), just as the pole term at $d = 2$ was discarded when we used dimensional regularization. (This teaches us a general lesson about regularization: No matter what the regularization procedure is, a genuine infinity cannot be avoided merely by changing the method of regularization.) The R-dependent part of the asymptotic form in (29.69), $(1/2\pi)\ln\,R$, is the Green function we seek (apart from an additive constant, of course). But this is precisely the expression found earlier using dimensional regularization. This result supports the assertion that $G^{(2)}(R)$ is indeed independent of the method of regularization, as required.

29.6 Solutions

1. (a) The derivation of the expressions in (29.15) is straightforward, along the lines already spelled out in the text.
(b) The boundary conditions $f(0) = 0,\ f(1) = 0$ correspond to specifying that $c_1 = c_2 = c_3 = c_4 = 0$ in Eq. (29.11). Then

$$G(x, x') = \begin{cases} x\,(x'-1) & \text{for}\quad 0 \le x < x' \\ (x-1)\,x' & \text{for}\quad x' < x \le 1. \end{cases}$$

It is also obvious that the CF is identically equal to zero for the initial conditions given. The solution of the differential equation is, therefore, given by

---

[4]Gradshteyn and Ryzhik, p. 961, formula **8.447** (3).

$$\begin{aligned} f(x) &= (x-1)\int_0^x dx'\,x'\,g(x') + x\int_x^1 dx'\,(x'-1)\,g(x') \\ &= x\int_0^1 dx'\,x'\,g(x') - \int_0^x dx'\,x'\,g(x') - x\int_x^1 dx'\,g(x'). \end{aligned}$$

It is easily checked that the solution satisfies the differential equation $f'' = g(x)$, as well as the given boundary conditions.

(c) Even though the differential equation satisfied by the Green function is invariant under a shift from x to $x - x'$, the *boundary conditions* in the case at hand pertain to the finite points $x = 0$ and $x = 1$, and these are not translationally invariant. As a result, $G(x, x')$ is not a function of $x - x'$ alone. ▶

3. You will need the expression for ∇^2 in spherical polars in Eq. (6.63) of Chap. 6, Sect. 6.2.6, and the formula for differentiation under the integral sign in Eq. (3.3) of Chap. 3, Sect. 3.1.1. ▶

5. Use the fact that $\Gamma'(1) = -\gamma$ (Eq. (25.23) of Chap. 25, Sect. 25.2.4). ▶

Chapter 30
The Diffusion Equation

30.1 The Fundamental Gaussian Solution

30.1.1 Fick's Laws of Diffusion

Diffusion is the process by which an uneven concentration of a substance gets gradually smoothed out spontaneously—e.g., a concentration of a chemical species (such as a drop of ink or dye) in a beaker of water spreads out "by itself", even in the absence of stirring. The microscopic mechanism of diffusion involves a very large number of collisions of the dye molecules with those of the fluid, which cause the dye molecules to move essentially randomly and disperse throughout the medium, even without any stirring of the fluid. The observation of pure diffusion in a fluid is not as simple as it appears to be at first sight. Other processes tend to dominate the mixing of a dye, in practice. For instance, **convection** currents set up by small thermal gradients, **advection** due to the initially present velocity gradients, and so on. Hence, the fluid should not only be stirred, but also not shaken!

A macroscopic description of the process of diffusion can, however, be given on simple physical grounds. This description is based on a fundamental partial differential equation, the diffusion equation. It serves as a basic model of phenomena that exhibit **dissipation**, a consequence of the **irreversibility** of macroscopic systems in time.

The diffusion equation arises as follows: The local, instantaneous concentration $\rho(\mathbf{r}, t)$ of dye molecules satisfies the equation of continuity, which is called Fick's First Law in this context:

$$\frac{\partial \rho}{\partial t} + \nabla \cdot \mathbf{j} = 0, \quad \text{(Fick's I Law)} \tag{30.1}$$

where $\mathbf{j}$ is the "diffusion current density". Equation (30.1) is simply the statement of the conservation of matter at the local level. The crucial physical input is the specification of $\mathbf{j}$. We assume that $\mathbf{j}$ is proportional to the local difference in concentrations,

V. Balakrishnan, *Mathematical Physics*,
https://doi.org/10.1007/978-3-030-39680-0_30

i.e., to the gradient of the concentration itself (Fick's Second Law). Thus

$$\mathbf{j}(\mathbf{r}, t) = -D\,\nabla\,\rho(\mathbf{r}, t) \quad \text{(Fick's II Law)}. \tag{30.2}$$

The positive constant D called the **diffusion coefficient**. It has the physical dimensions of $(\text{length})^2/\text{time}$. The minus sign on the right-hand side of Eq. (30.2) signifies the fact that the diffusion occurs from a region of *higher* concentration to a region of *lower* concentration: That is, the diffusion current tends to make the concentration uniform.

Eliminating $\mathbf{j}$ between Eqs. (30.1) and (30.2), we get the diffusion equation for the concentration $\rho(\mathbf{r}, t)$:

$$\boxed{\frac{\partial}{\partial t}\,\rho(\mathbf{r}, t) = D\,\nabla^2 \rho(\mathbf{r}, t).} \tag{30.3}$$

Equation (30.3) is a first-order differential equation in the time variable, and a second-order one in the spatial variables. It is a *parabolic equation* in the standard classification of second-order partial differential equations. In order to find a unique solution to it, you need an initial condition that specifies the initial concentration profile $\rho(\mathbf{r}, 0)$, as well as boundary conditions that specify $\rho(\mathbf{r}, t)$, for all $t \geq 0$, at the boundaries of the region in which the diffusion is taking place. The presence of the first-order time derivative in the diffusion equation implies that the equation is not invariant under the time reversal transformation $t \mapsto -t$. Irreversibility is thus built into the description of the phenomenon.

30.1.2 *Further Remarks on Linear Response*

Fick's II Law is an example of a very general feature of diverse physical systems called a linear response, aspects of which I have discussed in Chap. 24, in particular in Sect. 24.1.1. A few more comments are in order here.

In broad terms, linear response means that, under suitable conditions, the average response of a system to an applied stimulus or "force" is directly proportional to the stimulus. The "average" here refers to *averaging over fluctuations at the microscopic or molecular level* obtain a response at the macroscopic level. I shall not go any further into this important point here. But it is necessary to mention it for correctness. In the present case, linear response implies that the diffusion current that is set up in the medium as a result of the unequal concentrations at different points is proportional to the gradient of ρ, rather than the gradient of some nonlinear function of ρ (such as ρ^α where $\alpha \neq 1$, for instance). Other common examples of linear response are: the current in a conductor is proportional to the applied voltage or potential difference (Ohm's law); the strain in an elastic medium is proportional to the applied stress (Hooke's law); the magnetization of a paramagnetic material is proportional to the applied magnetic field; the polarization of a dielectric medium is proportional to the applied electric field; and so on.

Another example of linear response, leading to an exact analog of the diffusion equation, is provided by the phenomenological description of heat conduction. Given the initial temperature distribution $T(\mathbf{r}, 0)$ of a body, the problem is to find the temperature distribution $T(\mathbf{r}, t)$ at any later time $t > 0$. Analogous to Fick's second law, it is assumed that the heat flux is proportional to the negative of the temperature gradient (a linear response). The constant of proportionality in this case is the thermal conductivity of the body, κ. The equation for $T(\mathbf{r}, t)$ reads

$$\frac{\partial}{\partial t} T(\mathbf{r}, t) = \kappa \nabla^2 T(\mathbf{r}, t). \tag{30.4}$$

It should be clear, now, why the diffusion equation is also known as the **heat equation**.

The response of a system need not remain linear for an arbitrarily strong applied stimulus. It is plausible on physical grounds that the response is linear only if the strength of the applied stimulus is "not too large". For instance, sufficiently strong stress on a solid will take it beyond the elastic regime, in which the stress–strain relationship is no longer a linear one. Similarly, as the applied magnetic field increases, the magnetization of a paramagnetic material tends to a saturation value rather than increasing proportionately with the field—the response is no longer a linear one. The electric polarizability of a medium may become a nonlinear function of the applied electric field at sufficiently large field strengths. The entire field of nonlinear optics, involving many novel phenomena, is based on such higher-order effects.

30.1.3 The Fundamental Solution in d Dimensions

In Chap. 20, Sect. 20.4.1, we saw how a random walk goes over into diffusion in the limit of vanishing step sizes in space and time.

- At the level of individual particles, it turns out that the positional probability density function $p(\mathbf{r}, t)$ of a particle satisfies exactly the same diffusion equation as the concentration $\rho(\mathbf{r}, t)$ does in the macroscopic description of the diffusion process.

In Sect. 30.2.1, we shall see how this comes about in the case of diffusion in one spatial dimension. The extension of this result to an arbitrary number of dimensions is straightforward.

Our concern here is to find the basic solution of the diffusion equation. Accordingly, consider this equation in a (Euclidean) space of an arbitrary number of spatial dimensions, d. We can subsequently set $d = 1, 2, 3, \ldots$ in the solution, as required. We, therefore, begin with

$$\boxed{\frac{\partial}{\partial t} p(\mathbf{r}, t) = D \nabla^2 p(\mathbf{r}, t).} \tag{30.5}$$

The simplest instance corresponds to natural boundary conditions, i.e., the PDF $p(\mathbf{r}, t) \to 0$ as $r \to \infty$ along any direction. We may start with the initial condition

$$p(\mathbf{r}, 0) = \delta^{(d)}(\mathbf{r}), \tag{30.6}$$

where $\delta^{(d)}(\mathbf{r})$ is the d-dimensional δ-function. This means that the diffusing particle starts at the origin at $t = 0$. In the context of the diffusion equation for the concentration $\rho(\mathbf{r}, t)$, such an initial condition represents a point source of unit concentration at the origin. In a space of infinite extent, we may take the starting point to be the origin of coordinates without any loss of generality. (In any case, the generalization to an arbitrary initial source point $\mathbf{r}_0$ is trivial.) The solution thus obtained is the fundamental solution (or Green function) of the diffusion equation. It can, therefore, be used to write down the solution corresponding to an arbitrary initial PDF $p(\mathbf{r}, 0)$ (or an initial concentration profile $\rho_{\text{init}}(\mathbf{r})$). I will consider the diffusion equation for the probability density $p(\mathbf{r}, t)$, but all the results that follow are applicable, as they stand, to the case of $\rho(\mathbf{r}, t)$.

The diffusion equation presents an *initial-value problem*. Moreover, it is a linear equation in the unknown function $p(\mathbf{r}, t)$. It is, therefore, well-suited for the application of a Laplace transform with respect to the time variable t, and a Fourier transform with respect to the spatial variable(s) $\mathbf{r}$. Let $\widetilde{p}(\mathbf{r}, s)$ and $\phi(\mathbf{k}, t)$ denote, respectively, the Laplace transform and Fourier transform of $p(\mathbf{r}, t)$. Further, let $\widetilde{\phi}(\mathbf{k}, s)$ denote the Laplace transform of $\phi(\mathbf{k}, t)$ (or the Fourier transform of $\widetilde{p}(\mathbf{r}, s)$), in an obvious notation. Taking the Laplace transform of both sides of the diffusion equation (30.5), we get

$$s\, \widetilde{p}(\mathbf{r}, s) - p\,(\mathbf{r}, 0) = D\nabla^2\, \widetilde{p}(\mathbf{r}, s). \tag{30.7}$$

Using Eq. (30.6) for $p(\mathbf{r}, 0)$,

$$(s - D\nabla^2)\, \widetilde{p}(\mathbf{r}, s) = \delta^{(d)}(\mathbf{r}). \tag{30.8}$$

Expand $\widetilde{p}(\mathbf{r}, s)$ in a Fourier integral with respect to the spatial variable $\mathbf{r}$, according to

$$\widetilde{p}(\mathbf{r}, s) = \frac{1}{(2\pi)^d} \int d^d k\; e^{i\mathbf{k}\cdot\mathbf{r}}\, \widetilde{\phi}(\mathbf{k}, s). \tag{30.9}$$

The δ-function, of course, has the familiar Fourier representation

$$\delta^{(d)}(\mathbf{r}) = \frac{1}{(2\pi)^d} \int d^d k\; e^{i\mathbf{k}\cdot\mathbf{r}}. \tag{30.10}$$

Use these expressions in Eq. (30.8), and equate the coefficients of the basis vector $e^{i\mathbf{k}\cdot\mathbf{r}}$ (in the space of functions of $\mathbf{r}$). We must then have, for each $\mathbf{k}$,

$$(s + Dk^2)\, \widetilde{\phi}(\mathbf{k}, s) = 1, \quad \text{or} \quad \widetilde{\phi}(\mathbf{k}, s) = 1/(s + Dk^2). \tag{30.11}$$

We thus obtain a very simple expression for the double transform $\widetilde{\phi}(\mathbf{k}, s)$. The transforms must be inverted to find the PDF $p(\mathbf{r}, t)$. It is easier to invert the Laplace transform first, and then the Fourier transform. The Laplace transform is trivially inverted: recall that $\mathcal{L}^{-1}[(s+a)^{-1}] = e^{-at}$. Therefore,

$$\phi(\mathbf{k}, t) = e^{-Dk^2 t}, \quad \text{so that} \quad p\,(\mathbf{r}, t) = \frac{1}{(2\pi)^d} \int d^d k\, e^{i\mathbf{k}\cdot\mathbf{r}}\, e^{-Dk^2 t}. \tag{30.12}$$

This d-dimensional integral factors into a product of d integrals upon writing $\mathbf{r}$ and $\mathbf{k}$ in Cartesian coordinates. Each of the factors is the familiar shifted Gaussian integral (Eq. (2.4) of Chap. 2, Sect. 2.1) that we have encountered many times. The final result is

$$\boxed{p\,(\mathbf{r}, t) = \frac{1}{(4\pi D t)^{d/2}}\, e^{-r^2/(4Dt)}.} \tag{30.13}$$

This is the fundamental Gaussian solution to the diffusion equation in d spatial dimensions. Its three-dimensional version is already familiar to you, as it has already been written down in Eq. (20.80) of Chap. 20, Sect. 20.4.1.

★ **1.** Work out the steps starting from Eq. (30.7) to arrive at Eq. (30.13) for $p(\mathbf{r}, t)$.

Observe that the fundamental solution (30.13) for $p(\mathbf{r}, t)$ is a function of r alone, as far as its position-dependence is concerned. It is important to understand why this comes about:

- The basic reason, of course, is that the diffusion equation involves the scalar operator ∇^2, which is rotationally invariant.
- But it is *also* necessary for the boundary conditions and the initial condition to be spherically symmetric. These requirements are satisfied in the present instance: recall that the boundary condition is $p(\mathbf{r}, t) \to 0$ as $r \to \infty$, while the initial condition is $p(\mathbf{r}, 0) = \delta^{(d)}(\mathbf{r})$.

30.1.4 Solution for an Arbitrary Initial Distribution

With the fundamental solution at hand, it is easy to write down the particular integral that solves the diffusion equation (30.5) for a general, arbitrarily specified PDF at any initial instant t'. We have

$$\left.\begin{aligned} p(\mathbf{r}, t) &= \int d^d r\; G(\mathbf{r}, t\,;\, \mathbf{r}', t')\, p(\mathbf{r}', t') \quad (t > t'), \\ G(\mathbf{r}, t\,;\, \mathbf{r}', t') &= \frac{1}{[4\pi D(t-t')]^{d/2}} \exp\left\{-\frac{(\mathbf{r}-\mathbf{r}')^2}{4D(t-t')}\right\}. \end{aligned}\right\} \tag{30.14}$$

Thus, the PDF at any time $t > t'$ is an integral transform of the PDF at time t'. The kernel of the transform is, apart from the t-dependent normalization factor, just the fundamental Green function for the diffusion operator $(\partial/\partial t - D\nabla^2)$. We have seen that the heat conduction equation (30.4) has exactly the same form as the diffusion equation. The Green function in (30.14) is called the **heat kernel** in the mathematical literature. Recall the remarks made in Chap. 8, Sect. 8.2.3, on the extension of the Laplacian operator to curved spaces and to differential forms of arbitrary order. The corresponding generalizations of the heat kernel play an important role in these mathematical studies.

30.1.5 Moments of the Distance Travelled in Time t

Let us return to the solution in Eq. (30.13) for the PDF of a diffusing particle that starts from the origin of coordinates at $t = 0$. Since, $p(\mathbf{r}, t)$ does not depend on the angular coordinates in $\mathbf{r}$, the mean *displacement* of the diffusing particle vanishes for all t, as you would expect:

$$\langle \mathbf{r}(t)\rangle = \int d^d r\, \mathbf{r}\, p(\mathbf{r}, t) = 0. \tag{30.15}$$

(Integration over the angular coordinates of the vector $\mathbf{r}$ yields zero.) The mean *distance* travelled by the particle does not vanish, of course, for any $t > 0$. All the moments of the distance travelled in a given time interval t can be written down easily. We find

$$\langle r^l(t)\rangle = \int d^d r\, r^l\, p(\mathbf{r}, t) = \int_0^\infty dr \int d\Omega_d\, r^{d+l-1}\, p(\mathbf{r}, t), \tag{30.16}$$

on going over to polar coordinates in d dimensions ($d\Omega_d$ denotes the "solid angle" element in d-dimensional space). Since the PDF $p(\mathbf{r}, t)$ does not involve any angular coordinates, the integration over these variables can be carried out immediately. The formula in Eq. (6.21) of Chap. 6, Sect. 6.1.3 gives

$$\int d\Omega_d = S_d(1) = \frac{2\pi^{d/2}}{\Gamma\left(\frac{1}{2}d\right)}, \tag{30.17}$$

the surface "area" of a sphere of unit radius in d-dimensional space. Next, we use the Gaussian integral (3.16) of Chap. 3, Sect. 3.1.4 to evaluate the integral over r. This yields

$$\langle r^l(t)\rangle = \frac{\Gamma\left(\frac{1}{2}(d+l)\right)}{\Gamma\left(\frac{1}{2}d\right)}\,(4Dt)^{l/2}. \tag{30.18}$$

An important point: the moments grow with time according to $\langle r^l(t)\rangle \sim t^{l/2}$, *independent* of the dimensionality d of the space in which the diffusion takes place. In particular, the variance of the displacement is obtained by setting $l = 2$ in Eq. (30.18). We get the important formula

$$\boxed{\text{Var}\,(\mathbf{r}) = \langle \mathbf{r}^2\rangle - \langle \mathbf{r}\rangle^2 = \langle r^2\rangle = 2d\,Dt.} \tag{30.19}$$

This result is the leading signature of normal diffusion in the absence of any drift.

★ **2.** Starting from (30.16), establish Eq. (30.18).

Note, in passing, that the variance of the displacement (found above) is not the same as the variance of r, the *distance* from the origin. The latter is Var $(r) = \langle r^2\rangle - \langle r\rangle^2$, which is also proportional to t, but with a different coefficient.

30.2 Diffusion in One Dimension

I turn now to the case $d = 1$, i.e., diffusion in one spatial dimension. The calculations are simpler in this instance, and at the same time, we obtain a number of useful insights into the nature of the diffusion problem. In particular, the effects of boundary conditions can be examined in some detail.

30.2.1 Continuum Limit of a Biased Random Walk

In order to see how diffusion arises as to the continuum limit of a random walk, let us go back to the biased random walk on a linear lattice considered in Chap. 19, Sect. 19.4.1. This walk is a little more general than an unbiased one. It will lead to a diffusion equation that is appropriate an important physical situation: namely, diffusion on a line in the presence of a constant external force field. The force-free case is a special case of this equation.

The random walker starts at the site $j = 0$ at time 0. The probability of a step to the right is α, while that of a step to the left is $\beta = 1 - \alpha$. Let a and τ denote, respectively, the lattice constant and the size of a time step. Repeating Eq. (19.58) for the probability that the walker is at the site j (or the point ja on the line) at time step n (or at time $n\tau$),

$$P(ja,\, n\tau) = \alpha\, P(ja - a,\, n\tau - \tau) + \beta\, P(ja + a,\, n\tau - \tau). \tag{30.20}$$

a and τ have been introduced explicitly, in order to see how the continuum limit arises. In Chap. 21, Sect. 21.5.2, and again in Chap. 28, Sect. 28.4.2, we considered a random walk on a discrete lattice, but in continuous time. Here, we are going to pass to the continuum limit in both space and time. This happens when the quantities $a,\ \tau$ and $(\alpha - \beta)$ tend to zero simultaneously and in a specific manner, as follows:

Subtract $P(ja,\ n\tau - \tau)$ from each side of Eq. (30.20). On the right-hand side, rewrite the coefficient of $P(ja + a,\ n\tau - \tau)$ as $\beta = \alpha - (\alpha - \beta)$, and the coefficient of $P(ja,\ n\tau - \tau)$ as $-1 = -2\alpha + (\alpha - \beta)$. Collecting terms suitably, this yields

$$\begin{aligned} P(ja,\ n\tau) - P(ja,\ n\tau - \tau) \\ &= \alpha\big\{P(ja - a,\ n\tau - \tau) - 2P(ja,\ n\tau - \tau) + P(ja + a,\ n\tau - \tau)\big\} \\ &\quad - (\alpha - \beta)\big\{P(ja + a,\ n\tau - \tau) - P(ja,\ n\tau - \tau)\big\}. \end{aligned} \tag{30.21}$$

Divide both sides by τ. Multiply and divide the first term on the right-hand side by a^2, and the second term by a. Now let $a \to 0,\ \tau \to 0$ and $\alpha - \beta \to 0$ (that is, let $\alpha \to \frac{1}{2},\ \beta \to \frac{1}{2}$), such that

$$\lim \frac{a^2\alpha}{\tau} = D \quad \text{and} \quad \lim \frac{a(\alpha - \beta)}{\tau} = c, \tag{30.22}$$

where D and c are finite, nonzero constants. Since $\alpha \to \frac{1}{2}$, the constant D is essentially $\lim a^2/(2\tau)$. D has the physical dimensions of $(\text{length})^2/(\text{time})$, while c has the physical dimensions of (length)/(time), i.e., a velocity. Further letting $j \to \infty$ and $n \to \infty$ such that ja and $n\tau$ tend to the continuous variables x and t respectively, Eq. (30.21) leads to the following partial differential equation for the probability *density* function (PDF) $p(x, t)$ of the position:

$$\boxed{\frac{\partial p(x,t)}{\partial t} = -c\,\frac{\partial p\,(x,t)}{\partial x} + D\,\frac{\partial^2 p\,(x,t)}{\partial x^2}.} \tag{30.23}$$

★ **3.** Work through the steps detailed above to obtain Eq. (30.23) starting from the biased random walk Eq. (30.20).

Equation (30.23) is called the **Smoluchowski equation**. It describes the diffusion (e.g., of colloidal particles) in one dimension under the influence of a constant field of force (e.g., gravity). The first term on the right-hand side is the **drift term**, and the second is the **diffusion term**. The parameter c represents the **mean drift velocity**, while D is the diffusion coefficient, as usual. I will refer to diffusion in the absence of a drift as *free diffusion*. Note that Eq. (30.23) can be written in the form of a continuity equation, according to

$$\frac{\partial p}{\partial t} + \frac{\partial j}{\partial x} = 0, \quad \text{where} \quad j(x,t) = c\,p(x,t) - D\,\frac{\partial p\,(x,t)}{\partial x}. \tag{30.24}$$

Observe that the ratio D/c is a natural *length scale* in the problem. It is a measure of the relative importance of the drift and diffusion contributions to the motion of the colloidal particles. When multiplied by a characteristic length scale L, such as that which occurs in the definition of the **Reynold's number** in fluid flow, the quantity Lc/D yields a dimensionless number called the **Péclet number**. This is essentially the ratio of the **advective transport rate** to the **diffusive transport rate** in fluid flow under a force field.

An important physical application of the Smoluchowski equation is provided by the phenomenon of **sedimentation** under gravity. I will return to this phenomenon in Sect. 30.3.

★ **4.** ***Random walk with a sojourn probability at each site***: Here is a minor extension of the foregoing. Consider a biased random walk on a linear lattice, with the following specifications: At each time step, the random walker jumps a step to the right or left with respective probabilities α and β, or else *stays* at the same site with a sojourn probability $\gamma = 1 - \alpha - \beta$. (Recall that we have already considered this possibility in the case of a random walk in continuous time, in Chap. 21, Sect. 21.5.2.)

(a) Write down the recursion relation satisfied by the probability $P(ja,\, n\tau)$ that the walker is at the site j at time $n\tau$ in this case.
(b) Show that the Smoluchowski equation is once again obtained in the continuum limit, with no change other than a reduction of the diffusion coefficient by the factor $(1 - \gamma)$.

30.2.2 Free Diffusion on an Infinite Line

Let us consider first the diffusion on a line in the absence of a drift. This is the continuum limit of an unbiased random walk on a linear lattice. The diffusion equation satisfied by the PDF $p(x, t)$ is, of course, Eq. (30.23) with c set equal to zero, i.e.,

$$\frac{\partial p(x,t)}{\partial t} = D\,\frac{\partial^2 p\,(x,t)}{\partial x^2}. \tag{30.25}$$

We already know the normalized solution on the infinite line $(-\infty < x < \infty)$, with natural boundary conditions $p(x,t) = 0$ for $x \to \pm\infty$ and the initial condition $p(x, 0) = \delta(x)$: set $d = 1$ in the general Gaussian solution in Eq. (30.13). We have

$$p\,(x,t) = \frac{1}{(4\pi Dt)^{1/2}}\, e^{-x^2/(4Dt)}. \tag{30.26}$$

More generally, the solution for a prescribed initial PDF $p(x, 0)$ is the $d = 1$ counterpart of (30.14), namely,

$$p(x,t) = \frac{1}{(4\pi Dt)^{1/2}} \int_{-\infty}^{\infty} dx\,'\, e^{-(x-x\,')^2/(4Dt)}\; p(x\,', 0). \tag{30.27}$$

★ **5.** Consider an initial PDF of the position of the diffusing particle (or the initial concentration of a diffusing substance) that is given by the rectangular distribution

$$p(x,0) = \begin{cases} 1/(2b) & \text{for } -b \le x \le b \\ 0 & \text{for } |x| > b, \end{cases}$$

where b is a positive constant.

(a) Show that

$$p(x\,,\,t) = (4b)^{-1}\left\{\operatorname{erf}\left[(x+b)/\sqrt{4Dt}\,\right] - \operatorname{erf}\left[(x-b)/\sqrt{4Dt}\,\right]\right\},$$

where erf (x) is the error function (defined in Chap. 3, Sect. 3.1.2).

(b) Sketch $p(x,t)$ schematically as a function of x for a given value of t.

30.2.3 Absorbing and Reflecting Boundary Conditions

In physical applications, the region in which diffusion occurs is finite rather than infinite. We then need to specify appropriate boundary conditions at the end points of the region. It is important to recognize that the solutions of a given partial differential equation with the same initial conditions, but different boundary conditions, may be very different functions.

Suppose the diffusion occurs in a one-dimensional "box" given by $b_1 \le x \le b_2$. If the diffusing particle gets absorbed at the ends of the box, or can leak out through the ends and thus leave the region of physical interest, we must impose **absorbing boundary conditions**, for which we shall use the abbreviation ABC. The corresponding PDF will be denoted by $p_{\text{abs}}(x,t)$, for the sake of clarity:

$$p_{\text{abs}}(x\,,\,t) = 0 \;\text{ at }\; x = b_1 \;\text{ and }\; x = b_2 \;\;(\text{ABC}). \tag{30.28}$$

On the other hand, if the substance cannot leak out of the ends and stays confined in the region $b_1 \le x \le b_2$ at all times, the diffusion *current* must vanish at the end points. Hence, we must impose **reflecting boundary conditions**, for which we shall use the abbreviation RBC. Again, the corresponding PDF will be denoted by $p_{\text{ref}}(x,t)$, for the sake of clarity. In the case of free diffusion, i.e., when $c = 0$, the boundary conditions are

$$\frac{\partial}{\partial x}\, p_{\text{ref}}(x,t) = 0 \;\text{ at }\; x = b_1 \;\text{ and }\; x = b_2 \;\;(\text{RBC}). \tag{30.29}$$

In the case of the Smoluchowski equation (30.23) or (30.24), RBCs are a little more involved. They now read

$$c\, p_{\text{ref}} - D\,(\partial\, p_{\text{ref}}/\partial x) = 0 \;\text{ at }\; x = b_1 \;\text{ and }\; x = b_2 \;\;(\text{RBC}). \tag{30.30}$$

Other boundary conditions, or combinations of these, may be applicable, depending on the physical problem.

There is a drastic difference in the long-time behavior of the respective solutions for RBC and ABC. In the case of reflecting boundaries, the particle never leaves the region between the boundaries. (Alternatively, the total amount of the diffusing substance remains conserved.) Therefore, the normalization condition, $\int_{b_1}^{b_2} dx\; p_{\text{ref}}(x,t) = 1$, remains valid for all $t \geq 0$. In the case of absorbing boundaries, the particle is absorbed when it happens to hit either boundary, and the diffusion process comes to an end (or the diffusing substance leaks out through the end points). In this case it is clear that $p_{\text{abs}}(x,t) \to 0$ as $t \to \infty$, at any point x inside the box. At any time t, the total probability $\int_{b_1}^{b_2} dx\; p_{\text{abs}}(x,t)$ then represents the **survival probability** inside the box. It starts with an initial value equal to unity, and decreases to zero with increasing time. The question of precisely how the survival probability tends to zero as $t \to \infty$ is of interest in its own right. I will consider it in Sect. 30.2.5, in the case of free diffusion.

Free diffusion on a semi-infinite line: As a concrete example, consider first the case of free diffusion in the semi-infinite region $-\infty < x < b$, where b is a positive constant. On the left, we have the natural boundary condition $p(x,t) \to 0$ as $x \to -\infty$. On the right, we have either (i) $p_{\text{abs}}(b,t) = 0$ if the barrier at $x = b$ is an absorber, or (ii) $(\partial p_{\text{ref}}/\partial x)_{x=b} = 0$ if the barrier is a reflector. For the same initial condition $p(x,0) = \delta(x)$, the respective solutions in the two cases are given by

$$\boxed{\begin{aligned} p_{\text{ref}}(x,t) &= (4\pi D t)^{-1/2}\left\{e^{-x^2/(4Dt)} + e^{-(2b-x)^2/(4Dt)}\right\}, \\ p_{\text{abs}}(x,t) &= (4\pi D t)^{-1/2}\left\{e^{-x^2/(4Dt)} - e^{-(2b-x)^2/(4Dt)}\right\}. \end{aligned}} \qquad (30.31)$$

★ **6.** Verify that the solutions given in Eqs. (30.31) satisfy the free diffusion equation (30.25) with the boundary conditions specified above.

30.2.4 *Finite Boundaries: Solution by the Method of Images*

The solutions in Eqs. (30.31) have a striking interpretation. They are superpositions (the sum and the difference, respectively) of (Gaussian) solutions of the diffusion equation with *natural* boundary conditions, but with x and $(2b - x)$, respectively, as the coordinate argument. Now, $(2b - x)$ is precisely the coordinate of the *image* of the point x reflected in a mirror located at the boundary b. (See Fig. 30.1.) This is no coincidence! The solutions above can indeed be obtained by a general technique called **the method of images**. You have come across this method in the context of electrostatics, where it yields the solution to Poisson's equation in the presence of specific boundary conditions provided the problem has a convenient symmetry.

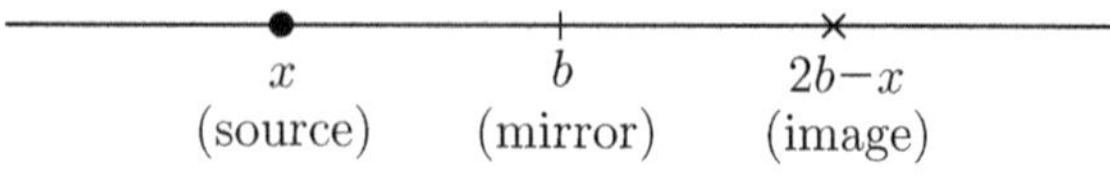

Fig. 30.1 The image of a source point x reflected in a mirror at the point b is at the point $2b - x$

However, the method is quite general, and is applicable in other instances as well. In broad terms:

- The method of images is a technique to find the Green function of a linear differential operator with specific boundary conditions, provided the problem has some symmetry that can be exploited.
- It is based on the fact that the differential equation concerned, taken together with appropriate initial and boundary conditions, has a *unique* solution.

My purpose in using the method of images in the present context of the diffusion equation is two-fold. The first is to demonstrate how powerful the method is, when the conditions are right. The second is to show that the method is also applicable to time-*dependent* problems, and not just time-independent problems as in electrostatics.

Consider free diffusion in the finite line segment $-b \leq x \leq b$, where $b > 0$. For simplicity, we assume once again that the diffusing particle starts from $x = 0$ at $t = 0$, so that $p(x, 0) = \delta(x)$. We want the solutions to the diffusion equation (30.25) subject to this initial condition and the boundary conditions

$$\text{(i)}\ \big[\partial\, p_{\text{ref}}(x,t)/\partial x\big]_{x=\pm b} = 0\ \text{(RBC)};\ \text{ or }\ \text{(ii)}\ p_{\text{abs}}(\pm b, t) = 0\ \text{(ABC)}.$$

Imagine placing mirrors at both boundaries, facing each other. A source point with coordinate x lying between the two mirrors, therefore, has an infinite number of images. The successive images in the mirror at b are located at

$$2b - x,\ 4b + x,\ 6b - x,\ 8b + x, \ldots ,$$

while the successive images in the mirror at $-b$ are located at

$$-2b - x,\ -4b + x,\ -6b - x,\ -8b + x,\ \ldots ,$$

as we move away from the source point on both sides of the x-axis. (Draw a figure and mark the successive image points.) The solution is a superposition of the fundamental Gaussian solution with each of these points as the coordinate argument. The boundary conditions above are incorporated by the following simple prescriptions:

(i) For RBC, the coefficient of the contribution from each image is just 1.
(ii) For ABC, the coefficient of the contribution from an image arising from n reflections is $(-1)^n$.

The complete solutions in the two cases may, therefore, be written down. After a bit of simplification, they are given, respectively, by

$$p_{\text{ref}}(x,t) = \frac{1}{\sqrt{4\pi Dt}} \sum_{n=-\infty}^{\infty} e^{-(x+2nb)^2/(4Dt)} \quad \text{(RBC at } x = \pm b) \tag{30.32}$$

and

$$p_{\text{abs}}(x,t) = \frac{1}{\sqrt{4\pi Dt}} \sum_{n=-\infty}^{\infty} (-1)^n \, e^{-(x+2nb)^2/(4Dt)} \quad \text{(ABC at } x = \pm b). \tag{30.33}$$

★ **7.** Verify that the solutions given in Eqs. (30.32) and (30.33) satisfy the free diffusion equation (30.25) with RBC and ABC, respectively.

It is natural, now, to ask: can each of these solutions be written as the fundamental Gaussian solution on the infinite line, plus an "extra" piece arising from the presence of the finite boundaries? Indeed they can. It is easily shown that Eqs. (30.32) and (30.33) can be written, respectively, as

$$p_{\text{ref}}(x,t) = \frac{e^{-x^2/(4Dt)}}{\sqrt{4\pi Dt}} \left\{ 1 + 2\sum_{n=1}^{\infty} e^{-n^2b^2/(Dt)} \cosh\left(\frac{nbx}{Dt}\right) \right\} \quad \text{(RBC)} \tag{30.34}$$

and

$$p_{\text{abs}}(x,t) = \frac{e^{-x^2/(4Dt)}}{\sqrt{4\pi Dt}} \left\{ 1 + 2\sum_{n=1}^{\infty} (-1)^n \, e^{-n^2b^2/(Dt)} \cosh\left(\frac{nbx}{Dt}\right) \right\} \quad \text{(ABC)}. \tag{30.35}$$

It is important to remember, however, that the physical region in which these solutions are valid is restricted to the line interval $[-b\,,\, b]$.

★ **8.** Show that Eqs. (30.32) and (30.33) can be re-expressed as in Eqs. (30.34) and (30.35), respectively.

It is remarkable that the *only* difference between the solutions for reflecting and absorbing boundary conditions is the extra factor $(-1)^n$ in the summand in the latter case. But this is sufficient to alter completely the long-time behavior of the PDF $p(x,t)$, as you will see.

30.2.5 Finite Boundaries: Solution by Separation of Variables

You are undoubtedly familiar with an elementary method for the solution of partial differential equations, namely, the method of separation of variables. The diffusion equation (30.25), with an initial condition $p(x,t) = \delta(x)$ and either reflecting or absorbing boundary conditions at $x = \pm b$, is tailor-made for solution by this method. The details are straightforward, and are left to you as an exercise (see below). The solutions one obtains are

$$p_{\text{ref}}(x,t) = \frac{1}{2b} + \frac{1}{b}\sum_{n=1}^{\infty} e^{-n^2\pi^2 Dt/b^2} \cos\left(\frac{n\pi x}{b}\right) \quad \text{(RBC)} \tag{30.36}$$

and

$$p_{\text{abs}}(x,t) = \frac{1}{b}\sum_{n=0}^{\infty} e^{-(2n+1)^2\pi^2 Dt/(4b^2)} \cos\left(\frac{(2n+1)\pi x}{2b}\right) \quad \text{(ABC).} \tag{30.37}$$

★ **9.** Using the method of separation of variables, derive the solutions (30.36) and (30.37) of the one-dimensional diffusion equation (30.25), with the initial condition $p(x,0) = \delta(x)$ and RBC and ABC, respectively, at both the points $x = b$ and $x = -b$.

The infinite series in Eqs. (30.36) and (30.37) involve the exponentials of *quadratic* functions of the summation index n. Such series appear quite frequently in solutions to the diffusion problem in finite regions. Sums of this kind cannot be expressed in closed form in terms of elementary algebraic or transcendental functions, such as trigonometric, hyperbolic or logarithmic functions. The Gaussian sums mentioned in Chap. 18, Sect. 18.4.2 are special cases of these sums. They are related to a class of special functions of considerable importance in mathematical physics, namely, **Jacobi theta functions** and their derivatives. Owing to limitations of space, I will not consider these functions in this book.

30.2.6 Survival Probability and Escape-Time Distribution

The representations of $p_{\text{ref}}(x,t)$ and $p_{\text{abs}}(x,t)$ given by Eqs. (30.36) and (30.37) are useful in reading off the *long-time* behavior of the PDFs, because they are superpositions of decaying exponential functions of time. It follows by inspection from Eq. (30.36) that $p_{\text{ref}}(x,t) \to 1/(2b)$ as $t \to \infty$ in the case of reflecting boundaries. This uniform distribution in the interval $[-b, b]$ is precisely what we would expect on physical grounds as the asymptotic PDF. On the other hand, in the case of absorbing boundaries, Eq. (30.37) shows that $p_{\text{abs}}(x,t) \to 0$ as $t \to \infty$ for every value of x. This is an indication of the fact that absorption at one boundary or the other is a *sure event* (i.e., it will occur with probability 1) for the random process concerned.

The **survival probability** $S(t, \pm b\,|\,0)$ is the probability that, starting from the origin at $t = 0$, the diffusing particle survives in the open interval $(-b, b)$ till time t, without hitting either of the boundary points. Its definition is obvious:

$$S(t, \pm b\,|\,0) \stackrel{\text{def.}}{=} \int_{-b}^{b} dx\; p_{\text{abs}}(x,t). \tag{30.38}$$

More generally, the survival probability in any region should be computed by averaging overall possible starting points as well, but I shall not go into this detail here. Using Eq. (30.37) in (30.38), we get (check this out)

$$S(t, \pm b \,|\, 0) = \frac{4}{\pi} \sum_{n=0}^{\infty} \frac{(-1)^n}{(2n+1)} e^{-(2n+1)^2 \pi^2 D t/(4b^2)}. \qquad (30.39)$$

As a check, note that $S(0, \pm b \,|\, 0) = 1$, as required. You need to use the well-known Madhava–Leibniz formula $\sum_0^\infty (-1)^n/(2n+1) = \frac{1}{4}\pi$.

The survival probability $S(t, \pm b \,|\, 0)$ is a superposition of an infinite number of decaying exponentials that decay monotonically to zero, with a leading asymptotic behavior $\sim e^{-\pi^2 D t/(4b^2)}$. The fact that $S(t, \pm b \,|\, 0) \to 0$ as $t \to \infty$ implies that absorption at one or the other of the two boundary points is a *sure* event, i.e., it will occur with probability 1.

***Escape rate and mean escape time*:** A diffusing particle starting from $x = 0$ will hit one of the end points $\pm b$ for the *first* time at some random instant of time, which is usually called the first passage time. Other names are the *escape time*, *exit time*, and *hitting time*, depending on the specific phenomenon being modeled by the diffusion process. What is the distribution of this time? This sort of question occurs in numerous applications—for instance, in reaction-diffusion problems in chemical physics.

What we seek is the PDF $Q(t, \pm b \,|\, 0)$ of the time when the diffusing particle, starting at $x = 0$ at $t = 0$, reaches either b or $-b$ for the first time, without ever having hit either of those two points at any earlier time. (We have already solved the analogous problem for a random walk in continuous time, but on a *discrete* linear lattice, in Chap. 28, Sect. 28.4.2.) It is obvious that the difference

$$[S(t, \pm b \,|\, 0) - S(t + \delta t, \pm b \,|\, 0)]$$

is the probability that the particle reaches either b or $-b$ for the first time in the time interval between t and $t + \delta t$. Hence, the PDF $Q(t, \pm b \,|\, 0)$ is just the rate of decrease of the probability $S(t, \pm b \,|\, 0)$ of survival in the open interval $(-b, b)$. That is,

$$\begin{aligned} Q(t, \pm b \,|\, 0) &\equiv -\frac{d}{dt} S(t, \pm b \,|\, 0) \\ &= \frac{\pi D}{b^2} \sum_{n=0}^{\infty} (-1)^n (2n+1)\, e^{-(2n+1)^2 \pi^2 D t/(4b^2)}. \end{aligned} \qquad (30.40)$$

It is easily checked that $\int_0^\infty dt\; Q(t, \pm b \,|\, 0) = 1$, i.e., the PDF is normalized to unity. In other words, the first passage to one or the other of the end points is a sure event, as we have already verified.

The mean first passage time

$$\langle t(0 \to \pm b)\rangle \stackrel{\text{def.}}{=} \int_0^\infty dt\, t\, Q(t, \pm b \,|\, 0) \tag{30.41}$$

is also easily determined. By dimensional considerations, it must be proportional to b^2/D, which is the only time scale in the problem. The actual value turns out to be

$$\boxed{\langle t(0 \to \pm b)\rangle = b^2/(2D),} \tag{30.42}$$

justifying the term "diffusion time" for this mean value. $b^2/(2D)$ is the mean time taken to diffuse through a distance b.

★ **10.** Verify that the PDF $Q(t, \pm b \,|\, 0)$ is normalized to unity, and derive the result in Eq. (30.42). For the latter, you will need the sum

$$\sum_{n=0}^{\infty} \frac{(-1)^n}{(2n+1)^3} = \frac{1}{32}\pi^3.$$

30.2.7 *Equivalence of the Solutions*

Compare the solutions obtained for the PDFs $p_{\text{ref}}(x, t)$ and $p_{\text{abs}}(x, t)$ by the method of images, Eqs. (30.32)–(30.33), with those obtained by the separation of variables, Eqs. (30.36)–(30.37). A striking feature emerges. In the former set, the time variable t appears in the *denominator* of the argument of the exponential, whereas in the latter set, t appears in the *numerator* in the exponent. This feature immediately suggests that useful approximations to the PDFs are provided by the former set at short times ($t \ll b^2/D$), and by the latter set at long times ($t \gg b^2/D$). In any case, we have two distinct representations for $p(x, t)$ in each instance, apparently with very different kinds of t-dependence. How can these be reconciled with each other?

As you may have guessed by now, the answer lies in the *Poisson summation formula*, Eq. (18.43) of Chap. 18, Sect. 18.4.1. For ready reference, let us write down the relevant formulas once again. Let q and k be real variables in $(-\infty\,, \infty)$, and let $f(q)$ and $\widetilde{f}(k)$ be a Fourier transform pair,[1] i.e.,

$$f(q) = \frac{1}{2\pi}\int_{-\infty}^{\infty} dk\, e^{ikq}\, \widetilde{f}(k) \iff \widetilde{f}(k) = \int_{-\infty}^{\infty} dq\, e^{-ikq}\, f(q). \tag{30.43}$$

[1] I have used the symbol q rather than x, since x has already been used to denote the position variable in the diffusion problem at hand.

Then the Poisson summation formula asserts that

$$\sum_{n=-\infty}^{\infty} f(nL) = (1/L) \sum_{n=-\infty}^{\infty} \widetilde{f}(2\pi n/L), \tag{30.44}$$

where L is any positive number. Set $L = b/(Dt)^{1/2}$. Consider, first, the case of reflecting boundary conditions. Identifying the right-hand side of Eq. (30.32) with $\sum_{-\infty}^{\infty} f(nL)$ implies that $f(q)$ must be the Gaussian

$$f(q) = \frac{1}{\sqrt{4\pi Dt}} \exp\left\{-\left(q + \frac{x}{\sqrt{4Dt}}\right)^2\right\}. \tag{30.45}$$

The Fourier transform of this function with respect to q is

$$\widetilde{f}(k) = \frac{1}{\sqrt{4Dt}} \exp\left(-\frac{k^2}{4} + \frac{ikx}{\sqrt{4Dt}}\right). \tag{30.46}$$

Applying the Poisson summation formula now yields Eq. (30.36), after a bit of simplification.

★ **11.** Work through the steps above to show the equality of the expressions in Eqs. (30.32) and (30.36). I have written down Eqs. (30.45) and (30.46) in order to enable you to check the algebra.

The case of absorbing boundaries proceeds similarly. The only different feature is the presence of the factor $(-1)^n$ in the summand in Eq. (30.33). It should be obvious to you that this factor must now be written as $(-1)^n = e^{i\pi n}$ $(= e^{i\pi(nL)/L})$, so that it can be extended to non-integer values of n. The function $f(q)$ can then be identified. It is given by

$$f(q) = \frac{1}{\sqrt{4\pi Dt}} \exp\left\{-\left(q + \frac{x}{\sqrt{4Dt}}\right)^2 + \frac{i\pi q\sqrt{Dt}}{b}\right\}. \tag{30.47}$$

Its Fourier transform is

$$\widetilde{f}(k) = \frac{1}{\sqrt{4Dt}} \exp\left\{-\frac{1}{4}\left(k - \frac{\pi\sqrt{Dt}}{a}\right)^2 + \frac{ix}{2\sqrt{Dt}}\left(k - \frac{\pi\sqrt{Dt}}{a}\right)\right\}. \tag{30.48}$$

As before, application of the Poisson summation formula followed by some simplification leads to Eq. (30.37).

★ **12.** Work through the steps above to show the equality of the expressions in Eqs. (30.33) and (30.37). As before, Eqs. (30.47) and (30.48) have been written down so as to help you to check the algebra. Would it make any difference to the final result if we write $(-1)^n$ as $e^{-i\pi n}$ rather than $e^{i\pi n}$ in identifying the function $f(q)$?

30.3 Diffusion with Drift: Sedimentation

30.3.1 The Smoluchowski Equation

As the last example of diffusion in one dimension, we consider the phenomenon of sedimentation: the diffusion of colloidal particles in a vertical column of liquid, under the influence of gravity. For simplicity, we assume the column to be semi-infinite in extent. The vertical position coordinate z lies in the range $[0, \infty)$. The Smoluchowski equation for the PDF $p(z, t)$ of a diffusing particle is

$$\boxed{\frac{\partial p\,(z,t)}{\partial t} = c\,\frac{\partial p\,(z,t)}{\partial z} + D\,\frac{\partial^2 p\,(z,t)}{\partial z^2},} \tag{30.49}$$

where D is the diffusion constant and c is the *downward* drift velocity (terminal velocity) under gravity. This is why the sign of the first term on the right-hand side in Eq. (30.49) is *opposite* to that of the corresponding term in Eq. (30.23). It is clear that the only relevant spatial variable is z so that the problem is effectively one-dimensional in this respect. The same equation holds good for the particle concentration $\rho(z, t)$, in a macroscopic description.

As I have pointed out earlier, the Smoluchowski equation (30.49) can be written in the form of a continuity equation involving the current $j(z, t)$:

$$\frac{\partial p}{\partial t} + \frac{\partial j}{\partial z} = 0, \quad \text{where} \quad j(z,t) = -c\,p(z,t) - D\,\frac{\partial p\,(z,t)}{\partial z}. \tag{30.50}$$

Since there is no leakage through the floor of the column, $z = 0$ is a reflecting boundary. Hence, the appropriate boundary condition at $z = 0$ is the vanishing of the current, i.e.,

$$\left[D\,\frac{\partial p}{\partial z} + c\,p\right]_{z=0} = 0 \quad (\text{RBC at } z = 0). \tag{30.51}$$

For simplicity of notation, the PDF has been written as p, without the subscript "ref". The other boundary condition is evidently the natural boundary condition $p(z, t) \to 0$ as $z \to \infty$. The task is to solve for $p(z, t)$, given an arbitrary initial PDF $p_{\text{init}}(z)$. We will do this by finding the Green function for the problem, which means that we first need the solution corresponding to the "sharp" initial condition

$$p(z, 0) = \delta(z - z_0), \tag{30.52}$$

where z_0 is any positive number. The PDF is normalized to unity according to $\int_0^\infty dz\, p(z, t) = 1$.

30.3.2 Equilibrium Barometric Distribution

Before solving Eq. (30.49), it is instructive to show that the concentration $p(z, t)$ settles down to a steady-state or equilibrium PDF $p_{\text{eq}}(z)$ as $t \to \infty$, regardless of the initial condition. It is easy to find $p_{\text{eq}}(z)$. Since it is (by definition) independent of t, it satisfies the *ordinary* differential equation

$$D\,\frac{d^2 p_{\text{eq}}(z)}{dz^2} + c\,\frac{d\,p_{\text{eq}}(z)}{dz} \equiv -\frac{d\,j_{\text{eq}}(z)}{dz} = 0. \tag{30.53}$$

Therefore, the equilibrium current $j_{\text{eq}}(z)$ must be a constant, independent of z. On the other hand, the current must vanish at $z = 0$, by the boundary condition. Hence, $j_{\text{eq}}(z)$ must vanish identically for all z. That is,

$$D\,\frac{d\,p_{\text{eq}}(z)}{dz} + c\,p_{\text{eq}}(z) = 0. \tag{30.54}$$

The unique *normalized* solution to this simple first-order equation is just

$$p_{\text{eq}}(z) = (c/D)\,e^{-cz/D}. \tag{30.55}$$

The significance of the natural length scale D/c is now evident. Note, also, that the boundary condition $p_{\text{eq}}(z) \to 0$ as $z \to \infty$ is automatically satisfied.

In a physical context, the constants c and D have specific connotations. Consider spherical particles of mass m and radius R moving under gravity in a fluid of viscosity η. Then the terminal drift velocity c is determined, as you know, by **Stokes' formula**

$$mg = 6\pi R\,\eta c. \tag{30.56}$$

On the other hand, the diffusion constant D is related to both η and the (absolute) temperature T of the fluid via an important relationship that follows from the **fluctuation-dissipation theorem**. In the case at hand, this relationship is deduced easily, as follows: In thermal equilibrium at a given temperature T, the probability density $p_{\text{eq}}(z)$ must be proportional to the Boltzmann factor $e^{-mgz/(k_B T)}$, since mgz is the potential energy of a particle at a height z above ground level. Comparison with Eq. (30.55) shows that c/D must be equal to $mg/(k_B T)$. Using Eq. (30.56) for mg, we get

$$D = \frac{k_B T}{6\pi R\,\eta}. \tag{30.57}$$

This is the celebrated **Stokes–Einstein formula** for D.

***A digressionary remark*:** This is a good place to mention that the familiar factor 6π in Stokes' formula (30.56) is only valid for **stick boundary conditions** (or no-slip boundary conditions) at the surface of the particles—that is, if the velocity of the fluid flowing past a (spherical) particle drops to zero right at its surface. Other boundary conditions would change this numerical factor. For instance, **slip boundary conditions**, which hold good if the fluid does not wet the particles at all, lead to a factor 4π instead of 6π. Similarly, the overall factor multiplying $\eta\, c$ in the expression for the viscous drag force changes if the shape of the particles is not spherical.

Since $c/D = mg/(k_B T)$, the normalized PDF (30.55) can also be written as

$$\boxed{p_{\text{eq}}(z) = (mg/k_B T)\, e^{-mgz/(k_B T)}.} \tag{30.58}$$

This is, of course, the well-known **barometric distribution** that gives (in a zeroth-order approximation) the variation of the density of the atmosphere with the altitude. It follows immediately from Eq. (30.58) that the mean, variance, and relative fluctuation of z are given by

$$\langle z \rangle_{\text{eq}} = \frac{k_B T}{mg}, \quad \big[\text{Var}\,(z)\big]_{\text{eq}} = \left(\frac{k_B T}{mg}\right)^2, \quad \frac{(\Delta z)_{\text{eq}}}{\langle z \rangle_{\text{eq}}} = 1. \tag{30.59}$$

An important conclusion can be drawn from the fact that $[\text{Var}\,(z)]_{\text{eq}}$ has a constant, finite value: there is actually no long-range diffusion of particles in this case! As you know, pure diffusion implies a linear increase of the variance of the position with time. The force of gravity that causes the downward drift, therefore, has a very important effect: it suppresses the diffusion and makes the PDF settle down to a nontrivial equilibrium distribution as $t \to \infty$.

30.3.3 *The Time-Dependent Solution*

We turn now to the solution of the Smoluchowski equation (30.49) with the initial condition $p(z, 0) = \delta(z - z_0)$ and the boundary conditions already stated: namely, $[D\partial p/\partial z + cp]_{z=0} = 0$ and $p \to 0$ as $z \to \infty$.

Let $\widetilde{p}(z, s)$ denote the Laplace transform (with respect to t) of $p(z, t)$. Taking the Laplace transforms of both sides of Eq. (30.49), we get

$$\frac{d^2\widetilde{p}}{dz^2} + \frac{c}{D}\frac{d\widetilde{p}}{dz} - \frac{s}{D}\,\widetilde{p} = -\frac{1}{D}\,\delta(z - z_0). \tag{30.60}$$

But this has precisely the form of a differential equation for a Green function! The solution is as follows: It is convenient to define the quantity

$$r(s) = (c^2 + 4sD)^{1/2}. \tag{30.61}$$

Then

$$\widetilde{p}(z,s) = \frac{2}{r(r-c)}\left(r\cosh\frac{rz_<}{2D} - c\sinh\frac{rz_<}{2D}\right)\exp\left\{-\frac{c(z-z_0)+rz_>}{2D}\right\}, \tag{30.62}$$

where $z_< = \min(z_0\,,\, z)$ and $z_> = \max(z_0\,,\, z)$.

★ **13.** Start with the Smoluchowski equation (30.49), take Laplace transforms to obtain Eq. (30.60), and solve it to arrive at the result quoted in Eq. (30.62).

★ **14.** The normalization condition on the PDF is $\int_0^\infty dz\, p(z,t) = 1$. Taking Laplace transforms, we get $\int_0^\infty dz\, \widetilde{p}(z,s) = 1/s$. Verify that this relation is satisfied by the expression for $\widetilde{p}(z,s)$ given in Eq. (30.62).

What is remarkable is that the complicated expression for the Laplace transform $\widetilde{p}(z,s)$ in Eq. (30.62) can be inverted to arrive at a rather involved, but explicit, expression for the PDF $p(z,t)$. I shall merely quote the result[2] here. In order to remind ourselves that this is the solution corresponding to the specific starting point z_0, let us write the PDF in proper notation as the *conditional* PDF $p(z,t\,|\,z_0)$. Then

$$\begin{aligned} p(z,t\,|\,z_0) = {} & \frac{e^{-[2c(z-z_0)+c^2t]/(4D)}}{\sqrt{4\pi Dt}}\left\{e^{-(z-z_0)^2/(4Dt)} + e^{-(z+z_0)^2/(4Dt)}\right\} \\ & + \frac{c}{2D}\, e^{-cz/D}\,\mathrm{erfc}\left\{(z+z_0-ct)/\sqrt{4Dt}\right\}, \end{aligned} \tag{30.63}$$

where erfc (x) is the complementary error function $1 - \mathrm{erf}\,(x)$, defined in Eq. (3.8) of Chap. 3, Sect. 3.1.2. Consider the $t \to \infty$ limit of the expression in Eq. (30.63). Since erfc $(-\infty) = 2$, we find

$$\lim_{t\to\infty} p(z,t\,|\,z_0) = (c/D)\, e^{-cz/D} = p_{\mathrm{eq}}(z), \tag{30.64}$$

no matter what z_0 is. This establishes explicitly the assertion made earlier. Note also that the ratio D/c^2 is the natural time scale in the sedimentation problem. You can see from Eq. (30.63) that it is essentially this time scale that controls the asymptotic approach of $p(z,t)$ to the equilibrium distribution $p_{\mathrm{eq}}(z)$.

Finally, it only remains to recall that the PDF $p(z,t)$ corresponding to a general initial PDF $p_{\mathrm{init}}(z)$ is given by

$$p(z,t) = \int_0^\infty dz_0\; p(z,t\,|z_0)\; p_{\mathrm{init}}(z_0), \tag{30.65}$$

where $p(z,t\,|\,z_0)$ is given by Eq. (30.63). You should have no difficulty, now, in recognizing that $p(z,t\,|\,z_0)$ is precisely the Green function of the diffusion operator $(\partial/\partial t - D\,\partial^2/\partial z^2)$ in the region $0 \le z < \infty$, with the reflecting boundary condition (30.51) at $z = 0$.

---

[2] S. Chandrasekhar, Rev. Mod. Phys. **15**, 1 (1943).

Question: Why is $p(z, t\,|z_0)$ a function of both z and z_0, rather than a function of the difference $(z - z_0)$ alone, as you may have come to expect a Green function?[3]

30.4 The Schrödinger Equation for a Free Particle

30.4.1 Connection with the Free-Particle Propagator

The general solution (30.14) of the diffusion equation for an arbitrary initial PDF $p(\mathbf{r}, 0)$ enables us to write down a similar solution for another very important equation. This is the Schrödinger equation for the position-space wave function $\psi(\mathbf{r}, t)$ of a nonrelativistic free particle of mass m in d-dimensional space, which reads

$$\boxed{\frac{\partial\psi(\mathbf{r}, t)}{\partial t} = \frac{i\hbar}{2m}\,\nabla^2\psi(\mathbf{r}, t).} \tag{30.66}$$

It is obvious that Eq. (30.66) has exactly the same form as the diffusion equation (30.5), with the difference that the *real* diffusion constant D in the latter is replaced by the *pure imaginary* constant $i\hbar/(2m)$. This is a very important difference indeed, and I will comment on it shortly. Ignoring it for the moment, we can now write down the formal solution to the Schrödinger equation (30.66) by analogy with the solution (30.14). Given that the wave function is $\psi(\mathbf{r}, t\,')$ at some instant of time $t\,'$, the solution at any time $t > t\,'$ is given by

$$\psi(\mathbf{r}, t) = \left[\frac{m}{2\pi i\hbar(t - t\,')}\right]^{d/2} \int d^d r\,' \exp\left\{\frac{im(\mathbf{r} - \mathbf{r}\,')^2}{2\hbar(t - t\,')}\right\} \psi(\mathbf{r}\,', t\,'). \tag{30.67}$$

But we know from quantum mechanics that the solution must have the form

$$\psi(\mathbf{r}, t) = \int d^d r\,'\, K(\mathbf{r}, t\,;\, \mathbf{r}\,', t\,')\,\psi(\mathbf{r}\,', t\,') \quad (t > t\,'), \tag{30.68}$$

where $K(\mathbf{r}, t\,;\, \mathbf{r}\,', t\,')$ is the free particle **Feynman propagator**. This quantity is defined as

$$\boxed{K(\mathbf{r}, t\,;\, \mathbf{r}\,', t\,') \stackrel{\text{def.}}{=} \left\langle\mathbf{r}\right| e^{-iH(t-t\,')/\hbar} \left|\mathbf{r}\,'\right\rangle,} \tag{30.69}$$

where H is the free particle Hamiltonian $\mathbf{p}^2/(2m)$. The ket $|\mathbf{r}\rangle$ is the position eigenstate of the particle corresponding to the position eigenvalue $\mathbf{r}$, as already introduced in Chap. 13, Sect. 13.2.5. We conclude that the explicit form of the free particle propagator for a nonrelativistic particle moving in d-dimensional Euclidean space is

---

[3] Because the existence of the boundary at the *finite* point $z = 0$ breaks the translational invariance possessed by the whole of the real axis.

$$\boxed{K(\mathbf{r},t\,;\,\mathbf{r}',t') = \left[\frac{m}{2\pi i\hbar(t-t')}\right]^{d/2} \exp\left\{\frac{im(\mathbf{r}-\mathbf{r}')^2}{2\hbar(t-t')}\right\} \quad (t>t').} \tag{30.70}$$

The replacement of the real constant D by the pure imaginary constant $i\hbar/(2m)$ actually represents a drastic change. The kernel

$$e^{-(\mathbf{r}-\mathbf{r}')^2/4D(t-t')}$$

decays to zero as $|\mathbf{r}-\mathbf{r}'| \to \infty$. On the other hand, the kernel

$$e^{im(\mathbf{r}-\mathbf{r}')^2/2\hbar(t-t')}$$

is an oscillatory function with a modulus equal to unity. As a result, questions of convergence arise, and these require careful handling. I do not go into these aspects here, but merely mention that the formal similarity between the diffusion equation and the Schrödinger equation has an important consequence: it serves as the starting point for the **path integral formulation** of quantum mechanics.

30.4.2 *Spreading of a Quantum Mechanical Wave Packet*

As you know from elementary quantum mechanics, the wave packet representing a free particle *disperses*, i.e., broadens or spreads in time, even though the particle moves in a vacuum. The physical reason for this fact is easy to see. Let ε and p denote the energy and the magnitude of the momentum of the particle, respectively. Wave-particle duality is expressed by the Einstein–de Broglie relations $\varepsilon = \hbar\omega$ and $p = \hbar k$, where ω is the angular frequency and k is the wave number. Hence, the relation $\varepsilon = p^2/(2m)$ becomes $\omega = \hbar k^2/(2m)$. This *nonlinear* dispersion relation immediately implies that the phase velocity ω/k is not equal to the group velocity $d\omega/dk$, i.e., there is **dispersion**. In other words, an initial wave packet will change shape and spread with time, even though the particle is free, and is not under the influence of any force.

***A one-dimensional example*:** In order to see this quantitatively, consider the one-dimensional counterpart of Eq. (30.67). Setting $t' = 0$ for simplicity, we have

$$\psi(x,t) = \left(\frac{m}{2\pi i\hbar t}\right)^{1/2} \int_{-\infty}^{\infty} dx\, e^{im(x-x')^2/(2\hbar t)}\, \psi(x',0). \tag{30.71}$$

We need an initial state that represents the quantum mechanical counterpart of a classical free particle moving with some constant momentum p_0. The wave function corresponding to the momentum *eigenstate* $|p_0\rangle$ is a plane wave, proportional to $\exp(ip_0\,x/\hbar)$. But this wave function is not normalizable in $(-\infty, \infty)$, whereas we would like to work with square-integrable functions throughout. This is achieved by *modulating* the plane wave with a Gaussian, centered at the origin, say. The initial wave function is then given by a **Gaussian wave packet** of the form

$$\psi(x,0) = (2\pi\sigma^2)^{-1/4}\, e^{ip_0x/\hbar}\, e^{-x^2/4\sigma^2}, \tag{30.72}$$

where σ is a positive constant with the physical dimensions of a length. This wave function is normalized to unity: $\int_{-\infty}^{\infty} dx\, |\psi(x,0)|^2 = 1$. It is easily verified that

$$\left.\begin{aligned} \langle x(0)\rangle &= \int_{-\infty}^{\infty} dx\, x\, |\psi(x,0)|^2 = 0, \\ \langle x^2(0)\rangle &= \int_{-\infty}^{\infty} dx\, x^2\, |\psi(x,0)|^2 = \sigma^2. \end{aligned}\right\} \tag{30.73}$$

Therefore, the initial uncertainty (i.e., the standard deviation) in the position of the particle is $\Delta x(0) = \sigma$. Further,

$$\left.\begin{aligned} \langle p(0)\rangle &= \int_{-\infty}^{\infty} dx\, \psi^*(x,0)\, (-i\hbar)\, \frac{d}{dx}\, \psi(x,0) = p_0\,, \\ \langle p^2(0)\rangle &= \int_{-\infty}^{\infty} dx\, \psi^*(x,0)\, (-i\hbar)^2\, \frac{d^2}{dx^2}\, \psi(x,0) = p_0^2 + \frac{\hbar^2}{4\sigma^2}\,. \end{aligned}\right\} \tag{30.74}$$

Hence, the initial uncertainty (or standard deviation) in the momentum is $\Delta p(0) = \hbar/(2\sigma)$. It follows that $\Delta x(0)\,\Delta p(0) = \frac{1}{2}\hbar$. In other words, the initial state is a minimum uncertainty state.

The Hamiltonian of a free particle is given by just the kinetic energy term, i.e., $H = p^2/(2m)$. As H is a quadratic function of the dynamical variables, **Ehrenfest's Theorem** implies that the expectation value $\langle p(t)\rangle$ remains equal to p_0 for all t, while $\langle x(t)\rangle$ is given by p_0t/m: these are precisely the expressions that we would obtain for the momentum and position of a classical free particle with an initial momentum p_0 and an initial position $x_0 = 0$. These statements can be checked out directly, once we find the time-dependent wave function $\psi(x,t)$. When Eq. (30.72) is substituted in Eq. (30.71), the resulting integral is a shifted Gaussian integral. Evaluating the integral, we find that the wave function at any time t is given by the following expression:

$$\psi(x,t) = \frac{e^{\phi(x,t)}}{(2\pi\sigma^2)^{1/4}[1 + (i\hbar t/2m\sigma^2)]^{1/2}}, \tag{30.75}$$

where the exponent $\phi(x,t)$ is given by

$$\phi(x.t) = \frac{imx^2}{2\hbar t} - \frac{im(x - p_0t/m)^2}{2\hbar t[1 + (\hbar^2t^2/4m^2\sigma^4)]} - \frac{(x - p_0t/m)^2}{4\sigma^2[1 + (\hbar^2t^2/4m^2\sigma^4)]}\,. \tag{30.76}$$

Note that the first two terms on the right-hand side of Eq. (30.76) are purely imaginary. Hence, they only contribute pure phase factors to $\psi(x,t)$ upon exponentiation. These do not affect the probability density function $|\psi(x,t)|^2$. This PDF is given by

$$|\psi(x,t)|^2 = \left\{\frac{1}{2\pi[\sigma^2 + (\hbar^2 t^2/4m^2\sigma^2)]}\right\}^{1/2} \exp\left\{-\frac{(x - p_0 t/m)^2}{2[\sigma^2 + (\hbar^2 t^2/4m^2\sigma^2)]}\right\}. \tag{30.77}$$

Thus, the PDF of x at any time $t > 0$ remains a normalized Gaussian, with its peak at $x = p_0 t/m$, and a variance that increases with time. The mean and variance of the position are now found to be

$$\left.\begin{aligned} \langle x(t)\rangle &= \frac{p_0 t}{m}, \\ \langle x^2(t)\rangle - \langle x(t)\rangle^2 \equiv \big(\Delta x(t)\big)^2 &= \sigma^2 + \frac{\hbar^2 t^2}{4m^2\sigma^2}. \end{aligned}\right\} \tag{30.78}$$

We also find that

$$\left.\begin{aligned} \langle p(t)\rangle &= \int_{-\infty}^{\infty} dx\, \psi^*(x,t)\,(-i\hbar)\frac{d}{dx}\psi(x,t) = p_0, \\ \langle p^2(t)\rangle &= \int_{-\infty}^{\infty} dx\, \psi^*(x,t)\,(-i\hbar)^2\frac{d^2}{dx^2}\psi(x,t) = p_0^2 + \frac{\hbar^2}{4\sigma^2}. \end{aligned}\right\} \tag{30.79}$$

The expectation value of the Hamiltonian (the energy of the particle) is, of course, constant in time. For the state under consideration, it is given by

$$E \equiv \langle H\rangle = \frac{1}{2m}\langle p^2\rangle = \frac{p_0^2}{2m} + \frac{\hbar^2}{8m\sigma^2}. \tag{30.80}$$

★ **15.** The explicit expression for $\psi(x.t)$ given by Eqs. (30.75) and (30.76) enables us to find all quantities of interest. You will find it instructive to work out all the steps required to derive the results quoted above.

(a) Carry out the integration in Eq. (30.71) using (30.72) for the initial wave function $\psi(x',0)$, to obtain the solution quoted in Eqs. (30.75) and (30.76).
(b) Establish the results in Eqs. (30.78) and (30.79).
(c) Why do the expectations values of p and p^2 remain unchanged from their initial values, while those of x and x^2 are functions of time?
(d) The expressions derived above for $\Delta x(t)$, $\Delta p(t)$ and E become infinite in the limit $\sigma^2 \to 0$. Why does this happen?

30.4.3 The Wave Packet in Momentum Space

Consider, now, the behavior of the free particle wave packet in momentum space. The initial state we have chosen is not a position eigenstate, because the wave function $\psi(x, 0)$ in Eq. (30.72) is an extended wave packet. But it is not a momentum eigenstate, either. The momentum-space wave function corresponding to the initial state is essentially the Fourier transform of $\psi(x, 0)$. It is easily determined to be

$$\widetilde{\psi}(p, 0) = \left(\frac{2\sigma^2}{\pi\hbar^2}\right)^{1/4} e^{-\sigma^2(p-p_0)^2/\hbar^2}. \tag{30.81}$$

This wave function is also a Gaussian, centered at the value p_0 of the momentum. It is normalized to unity, i.e., $\int_{-\infty}^{\infty} dp\, |\widetilde{\psi}(p, 0)|^2 = 1$.

It is much easier, of course, to find expectation values of functions of p in the momentum representation. The action of the momentum operator on any state of the particle is just multiplication of the corresponding momentum-space wave function by p. The PDF in momentum space is $|\widetilde{\psi}|^2$. Equations (30.74) can be written down essentially by inspection, given Eq. (30.81). What is more, the wave function at any time $t > 0$ is itself determined very easily from the Schrödinger equation in momentum space, namely,

$$i\hbar\frac{\partial\widetilde{\psi}(p, t)}{\partial t} = \frac{p^2}{2m}\widetilde{\psi}(p, t). \tag{30.82}$$

The solution is simply

$$\widetilde{\psi}(p, t) = e^{ip^2t/(2m\hbar)}\,\widetilde{\psi}(p, 0). \tag{30.83}$$

The momentum-space wave function at any time t is, therefore, just a phase factor times the initial wave function. It is immediately evident that $|\widetilde{\psi}(p, t)|^2 = |\widetilde{\psi}(p, 0)|^2$ for all t. This implies, once again, that the expectation values of all functions of p remain unchanged from their initial values.

To sum up: consider a free nonrelativistic particle moving in one dimension, with an initial position-space wave function that is a Gaussian wave packet. Let the variance of the position be σ^2. Then:

(i) The initial momentum-space wave function is also a Gaussian wave packet. The variance of the momentum is $\hbar^2/(4\sigma^2)$.
(ii) Hence, the initial state of the particle is a minimum uncertainty state.
(iii) The position-space wave function remains a Gaussian wave packet for all time, but its width broadens as t increases.
(iv) The uncertainty in the position of the particle increases with time according to

$$\boxed{\Delta x(t) = \sigma\left(1 + \frac{\hbar^2 t^2}{4m^2\sigma^4}\right)^{1/2}}. \tag{30.84}$$

(v) The momentum-space wave function remains a Gaussian wave packet that does not spread with time. The uncertainty in the momentum at any time remains equal to its initial value, $\hbar/(2\sigma)$.
(vi) Hence, an initial minimum uncertainty state loses this property for $t > 0$.

We started in the foregoing with an initial state represented by Gaussian wave packets for the PDFs in both x and p. The wave packet in position space broadens, while that in momentum space does not. Why does an "asymmetry" develop between the position and momentum variables as t increases? The reason is that the Hamiltonian $H = p^2/(2m)$, which governs the time evolution of the particle, is not a symmetric function of the operators x and p. It is obvious that, more generally, the same sort of asymmetry (between position-space and momentum-space wave packets) will develop even when a potential $V(x)$ is present in the Hamiltonian, with one notable exception. Remember that we are only considering standard, nonrelativistic, one-particle Hamiltonians of the form $H = p^2/(2m) + V(x)$.

The exception occurs in the case of the linear harmonic oscillator, for which $H = \frac{1}{2}(p^2 + x^2)$ (in suitable units). This Hamiltonian is a symmetric function of x and p, i.e., it is unchanged under the interchange of x and p. As you know, it is then possible to have *coherent states*, with probability densities that are Gaussians both in position space and in momentum space, as discussed in Chap. 14, Sect. 14.4.2.

- The PDFs in position space and in momentum space corresponding to these oscillator coherent states do not broaden under time evolution. The states remain minimum uncertainty states for all t.
- In fact, it is precisely this property that is responsible for the very name, *coherent* state.
- The x^2 potential provides just the right degree of "confinement" to maintain coherence and the minimum uncertainty property of the states concerned, as the system evolves in time.

30.5 Solutions

4. (a) The recursion relation is obviously

$$P(ja,\, n\tau) = \alpha\, P(ja - a,\, n\tau - \tau) + \beta\, P(ja + a,\, n\tau - \tau) + \gamma\, P(ja,\, n\tau - \tau).$$

(b) The recursion relation can again be re-written as in Eq. (30.21), on using the fact that $\alpha + \beta + \gamma = 1$. The limits to be taken are once again $a \to 0,\ \tau \to 0$ and $\alpha - \beta \to 0$. The same Smoluchowski equation (30.23), is obtained in the continuum limit. However, in the present case $\alpha \to \frac{1}{2}(1 - \gamma)$, rather than $\frac{1}{2}$. Therefore,

$$D = \lim_{a,\tau \to 0} \frac{a^2 \alpha}{\tau} = \lim_{a,\tau \to 0} \frac{a^2(1 - \gamma)}{2\tau}$$

in this instance. ▶

9. Set $p(x,t) = T(t)\,X(x)$, as usual, in the diffusion equation. It follows that

$$\frac{1}{DT}\frac{dT}{dt} = \frac{1}{X}\frac{dX}{dx} = -C^2,$$

a constant. The boundary conditions yield the allowed values of C. The general solution is a superposition of solutions for the various allowed values of C, because the diffusion equation is a linear equation. The calculations are simplified by noting that the diffusion equation, the set of boundary conditions, as well as the initial condition, are *all* unchanged under the transformation $x \mapsto -x$. As a consequence, the solution $p(x,t)$ remains an even function of x at all times. This fact will help you select the correct solution in each case. You will also need the following representation of the initial PDF,

$$\delta(x) = \frac{1}{2b} + \frac{1}{b}\sum_{n=1}^{\infty} \cos\frac{n\pi x}{b},$$

which follows from the relation $\delta\big(x/(2b)\big) = \sum_{n=-\infty}^{\infty} e^{\pi n i x/b}$. ▶

10. It is convenient to work with the Laplace transform of the first passage time density $Q(t, \pm b \,|\, 0)$. Equation (30.40) yields the expression

$$\widetilde{Q}(s, \pm b \,|\, 0) = 4\pi D \sum_{n=0}^{\infty} \frac{(-1)^n (2n+1)}{4b^2 s + (2n+1)^2 \pi^2 D}.$$

It is easy to see that

$$\int_0^\infty dt\; Q(t, \pm b \,|\, 0) \equiv \widetilde{Q}(0, \pm b \,|\, 0) = 1,$$

establishing the normalization of the PDF. The mean first passage time, defined in Eq. (30.41), can also be obtained from the formula

$$\langle t(0 \to \pm b)\rangle = -\big[d\widetilde{Q}/ds\big]_{s=0}.$$

The result $\langle t(0 \to \pm b)\rangle = b^2/(2D)$ follows readily. ▶

13. To refresh your memory, here is a step-by-step recapitulation of the usual procedure. The δ-function on the right-hand side of Eq. (30.60) vanishes for $z > z_0$ and also for $z < z_0$. In each of these two regions, therefore, Eq. (30.60) becomes a *homogeneous* differential equation with constant coefficients, whose solution is a linear combination of two exponentials. Thus, there are four constants of integration to be determined. The solution for $z > z_0$ cannot involve an increasing exponential in z because p must vanish as $z \to \infty$. That makes one of the constants zero. To find the remaining three, match the two solutions at $z = z_0$, and use the fact that the discontinuity in the first derivative $d\widetilde{p}/dz$ at $z = z_0$ is equal to $-1/D$. Finally, impose the reflecting boundary condition (30.51) at $z = 0$ (this condition obviously holds good for the Laplace transform $\widetilde{p}$ as well). The solution is then determined completely. ▶

15. (a), (b) The calculations required are straightforward.
(c) The Hamiltonian of a free particle, $H = p^2/(2m)$, commutes with the momentum operator p. Therefore, the expectation values of p and all functions of p must remain unchanged in time: if $f(p)$ is such a function, then, in any normalized state $|\Psi(t)\rangle$ of the particle,

$$\begin{aligned}\langle f(p)\rangle(t) &= \langle\Psi(t)|\, f(p)\, |\Psi(t)\rangle = \langle\Psi(0)|\, e^{iHt/\hbar}\, f(p)\, e^{-iHt/\hbar}\, |\Psi(0)\rangle \\ &= \langle\Psi(0)|\, f(p)\, |\Psi(0)\rangle = \langle f(p)\rangle(0).\end{aligned}$$

On the other hand, since $[x, H] \neq 0$, the expectation values of x and functions of x will be time-dependent.
(d) In the limit $\sigma^2 \to 0$, the initial wave function $\psi(x, 0)$ becomes a Dirac δ-function at the origin. Such a wave function represents a position eigenstate, and is not square-integrable. On the other hand, the foregoing discussion is applicable only to normalizable states of the particle. The limit $\sigma^2 = 0$ is, therefore, a *singular* one. This fact shows up as divergences in the expressions for $\Delta x(t)$, $\Delta p(t)$ and E when we let $\sigma^2 \to 0$. ▶

Chapter 31
The Wave Equation

31.1 Causal Green Function of the Wave Operator

31.1.1 Formal Solution as a Fourier Transform

In Chaps. 29 and 30, you have seen how Fourier and Laplace transforms help us solve Poisson's equation and the diffusion equation. We turn our attention now to another very important equation of mathematical physics, the wave equation. As you might expect, this is a well-studied topic with a vast literature. But my objective here is quite restricted. To repeat what has been stated in Chap. 29, Sect. 29.1, we want to find the fundamental Green function for the wave operator (or the d'Alembertian operator) that leads to the so-called causal, retarded solution of the wave equation. This is the fundamental solution that is of direct physical significance for the propagation of signals.

Before going on, it must be mentioned that there is also the Helmholtz equation, which stands in between Poisson's equation and the wave equation in order of complexity. The fundamental Green function for the Helmholtz operator will be derived in Chap. 32, Sect. 32.2.3, in the context of nonrelativistic quantum mechanical scattering theory.

The wave equation for a scalar function (or signal) $f(\mathbf{r}, t)$ is given by

$$\boxed{\left(\frac{1}{c^2}\frac{\partial^2}{\partial t^2} - \nabla^2\right) f(\mathbf{r}, t) = g(\mathbf{r}, t),} \tag{31.1}$$

where $g(\mathbf{r}, t)$ is a specified function of space and time that represents the *source* of the signal, and c is the speed of the signal. Although I have used c to denote this speed, what follows is not restricted to the propagation of light in free space. (It will be obvious from the context when the signal concerned is indeed light propagating in free space.) We shall first consider, formally, wave propagation in a general d-dimensional Euclidean space of infinite extent. ∇^2 in Eq. (31.1) then stands for the

V. Balakrishnan, *Mathematical Physics*,
https://doi.org/10.1007/978-3-030-39680-0_31

Laplacian operator in d dimensions. Subsequently, we shall work out the explicit solutions for the cases $d = 1, 2$ and 3 that are of direct physical interest. Finally, some comments will be made on what happens when $d > 3$.

The crucial difference between the wave equation and Poisson's equation is the relative minus sign in Eq. (31.1) between the time derivative $\partial^2/\partial t^2$ and the spatial derivatives in ∇^2. This makes all the difference in the world (literally!), and is a reflection of the fact that the spacetime we consider has d space-like dimensions and 1 time-like dimension. From the mathematical point of view, the wave equation is a *hyperbolic* partial differential equation, while Poisson's equation is an elliptic equation and the diffusion equation is a parabolic equation, as I have mentioned earlier.

The particular integral of the inhomogeneous wave equation (31.1) is given by the integral representation

$$f(\mathbf{r}, t) = \int d^d r' \int_{-\infty}^{\infty} dt'\, G(\mathbf{r}, t\,;\, \mathbf{r}', t')\, g(\mathbf{r}', t'), \tag{31.2}$$

where $d^d r'$ is the volume element in d dimensions. The Green function G satisfies the equation

$$\Big(\frac{1}{c^2}\frac{\partial^2}{\partial t^2} - \nabla^2\Big) G(\mathbf{r}, t\,;\, \mathbf{r}', t') = \delta^{(d)}(\mathbf{r} - \mathbf{r}')\, \delta(t - t'), \tag{31.3}$$

where $\delta^{(d)}$ denotes the d-dimensional δ-function. In physical terms, $G(\mathbf{r}, t\,;\, \mathbf{r}', t')$ represents the signal at time t at the point $\mathbf{r}$, arising from a sharply-pulsed point source of unit strength that is switched on at the point $\mathbf{r}'$ at the instant of time t'.

We are interested in the Green function that satisfies natural boundary conditions, i.e., $G \to 0$ as $|\mathbf{r} - \mathbf{r}'| \to \infty$. Further, we seek the *causal* Green function that vanishes identically for all $t < t'$, so that there is no signal anywhere *before* the source is switched on: this is the **principle of causality**. Hence G must satisfy the conditions

$$G = 0 \text{ and } \partial G/\partial t = 0 \text{ for all } t < t', \text{ at all points.} \tag{31.4}$$

The differential equation (31.3) is invariant under a shift of the origin of the spatial coordinates and of the time. The boundary conditions and the initial conditions are also translation-invariant in space and time. We, therefore, expect to find that the Green function has the form[1]

$$G(\mathbf{r}, t\,;\, \mathbf{r}', t') \equiv G(\mathbf{r} - \mathbf{r}', t - t') = \theta(t - t')\, K(\mathbf{r} - \mathbf{r}', t - t'). \tag{31.5}$$

[1]In a region of *finite* extent, in the presence of boundary conditions at finite values of r, the dependence of G on $\mathbf{r}$ and $\mathbf{r}'$ cannot be reduced in general to a dependence on the difference $\mathbf{r} - \mathbf{r}'$ alone. I have already made this comment in the context of Poisson's equation, in Chap. 29, Sect. 29.3.2.

The quantity $K(\mathbf{r} - \mathbf{r}', t - t')$ is sometimes referred to as the **propagator**. As we shall see, causality imposes an even stronger constraint. Since the propagation of the signal occurs with a finite speed c, a signal emanating from the point $\mathbf{r}'$ at time t' cannot reach the point $\mathbf{r}$ till time $t' + |\mathbf{r} - \mathbf{r}'|/c$. In other words, it is *retarded* in time. This feature will also emerge automatically in the solution for the Green function.

It is obviously convenient to shift variables from $\mathbf{r}$ and t to

$$\mathbf{R} = \mathbf{r} - \mathbf{r}' \quad \text{and} \quad \tau = t - t', \tag{31.6}$$

respectively. Then

$$\boxed{\left(\frac{1}{c^2}\frac{\partial^2}{\partial \tau^2} - \nabla^2\right) G(\mathbf{R}, \tau) = \delta^{(d)}(\mathbf{R})\, \delta(\tau),} \tag{31.7}$$

where ∇^2 now stands for the Laplacian operator with respect to $\mathbf{R}$. We must impose the conditions $G = 0$ and $\partial G/\partial \tau = 0$ for all $\tau < 0$, and the boundary condition $G \to 0$ as $R \to \infty$. Now express G as a Fourier integral, according to

$$G(\mathbf{R}, \tau) = \int \frac{d^d k}{(2\pi)^d} \int_{-\infty}^{\infty} \frac{d\omega}{2\pi}\, e^{i(\mathbf{k}\cdot\mathbf{R} - \omega\tau)}\, \widetilde{G}(\mathbf{k}, \omega). \tag{31.8}$$

Observe that we use a Fourier transform in both space *and* time, rather than a Fourier transform with respect to $\mathbf{r}$ and a Laplace transform with respect to t. This is helpful in imposing the conditions (31.4), as you will see. The inverse relation is of course

$$\widetilde{G}(\mathbf{k}, \omega) = \int d^d\mathbf{R} \int_{-\infty}^{\infty} d\tau\, e^{-i(\mathbf{k}\cdot\mathbf{R} - \omega\tau)}\, G(\mathbf{R}, \tau). \tag{31.9}$$

Further,

$$\delta^{(d)}(\mathbf{R})\, \delta(\tau) = \int \frac{d^d k}{(2\pi)^d} \int_{-\infty}^{\infty} \frac{d\omega}{2\pi}\, e^{i(\mathbf{k}\cdot\mathbf{R} - \omega\tau)}. \tag{31.10}$$

Equation (31.7) then gives

$$\int \frac{d^d k}{(2\pi)^d} \int_{-\infty}^{\infty} \frac{d\omega}{2\pi}\, e^{i(\mathbf{k}\cdot\mathbf{R} - \omega\tau)} \left\{(\omega^2 - c^2k^2)\, \widetilde{G}(\mathbf{k}, \omega) + c^2\right\} = 0. \tag{31.11}$$

It follows that the expression in curly brackets in the equation above must be equal to zero.[2] Hence

[2]The reason should be obvious by now, but let us spell out the argument once again. The set of functions $\{e^{i(\mathbf{k}\cdot\mathbf{R} - \omega\tau)}\}$, where ω and the Cartesian components of $\mathbf{k}$ run over all real values, forms a complete orthonormal basis in the space of integrable functions of τ and $\mathbf{R}$. Therefore, the left-hand

$$\widetilde{G}(\mathbf{k},\omega) = -\frac{c^2}{(\omega^2 - c^2k^2)}, \quad \text{where } k^2 = |\mathbf{k}|^2 = k_1^2 + \cdots + k_d^2. \qquad (31.12)$$

As expected, the introduction of the Fourier transform has converted the partial differential equation for G into a trivially-solved algebraic equation for its Fourier transform $\widetilde{G}(\mathbf{k},\omega)$. Inverting the Fourier transforms, we obtain the formal solution for $G(\mathbf{R}\,,\,\tau)$, namely,

$$G(\mathbf{R}\,,\,\tau) = -c^2 \int \frac{d^d k}{(2\pi)^d}\, e^{i\mathbf{k}\cdot\mathbf{R}} \int_{-\infty}^{\infty} \frac{d\omega}{2\pi}\, \frac{e^{-i\omega\tau}}{(\omega^2 - c^2k^2)}. \qquad (31.13)$$

31.1.2 *Simplification of the Formal Solution*

We now encounter a difficulty: the formal solution in (31.13) does not make sense as it stands! The integral over ω diverges because the integrand has poles at $\omega = -ck$ and $\omega = ck$ that lie *on* the path of integration. The difficulty is overcome by an application of the "trick" that has already been described in Chap. 23, Sect. 23.3.4, namely, an $i\epsilon$-prescription:

- Displace the poles away from the real axis into the complex ω-plane by infinitesimal amounts $\pm i\epsilon$, evaluate the integrals, and then pass to the limit $\epsilon \to 0$.

But this procedure will now yield more than one Green function: each pole can be displaced into the upper or lower half-plane in ω, giving four possible answers. The one we want here is the Green function satisfying the physical requirement of causality. This condition enables us to select the correct $i\epsilon$-prescription, and hence the correct Green function, unambiguously.

The prescription might appear to be an arbitrary one, but it is not. The following points are noteworthy:

- $i\epsilon$-prescriptions are ways of incorporating initial conditions or boundary conditions in the evaluation of Green functions.
- The method is based on the fact that the transforms (Fourier, Laplace, etc.) of Green functions can often be represented as *boundary values* of analytic functions of the transform variables.

Displacing a pole of the integrand on the real axis by an infinitesimal amount $i\epsilon$ [respectively, $-i\epsilon$] is equivalent to *indenting* the contour into an infinitesimal semicircle in the lower [respectively, upper] half of the complex plane. Once the poles are displaced off the real axis in the ω-plane, the integration over ω can be carried out using contour integration. Let Ω be a large positive constant. Consider a closed contour comprising a straight line from $-\Omega$ to $+\Omega$ along the real axis in the ω-plane,

---

side in Eq. (31.11) can vanish only if the coefficient of $e^{i(\mathbf{k}\cdot\mathbf{R}-\omega\tau)}$ vanishes, for every set of values of ω and the components of $\mathbf{k}$.

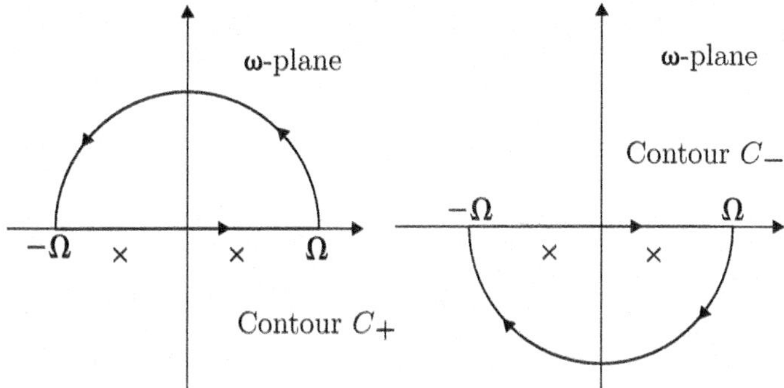

Fig. 31.1 Contours C_+ (used when $\tau < 0$) and C_-(used when $\tau > 0$). The poles of the integrand, located at $-ck - i\epsilon$ and $ck - i\epsilon$, are indicated by crosses

and a semicircle of radius Ω that takes us back from $+\Omega$ to $-\Omega$ in either the upper or lower half-plane. See Fig. 31.1. The limit $\Omega \to \infty$ is to be taken after the contour integral is evaluated. *Provided* the contribution from the semicircle vanishes in the limit $\Omega \to \infty$, the original line integral from $-\infty$ to $+\infty$ over ω is guaranteed to be precisely equal to the integral over the closed contour. You will recall that this sort of indentation of the contour to avoid a pole of the integrand, and completion of the contour to obtain a closed contour, was precisely the technique used in the derivation of dispersion relations for the generalized susceptibility, in Chap. 24, Sect. 24.2.1.

Now, for $\tau < 0$, this semicircle *must* lie in the *upper* half-plane in ω, because it is only in this region that the factor $e^{-i\omega\tau}$ in the integrand vanishes exponentially as $\Omega \to \infty$. The addition of the semicircle to the contour would then simply add a vanishing contribution to the original line integral that we want to evaluate. Therefore, provided *no* singularities of the integrand lie on the real axis or in the *upper* half-plane in ω, the integral over the contour C_+ is guaranteed to vanish identically for $\tau < 0$. But this is precisely what is required by causality: namely, that $G(\mathbf{R}, \tau)$ be equal to 0 for all $\tau < 0$.

On the other hand, for $\tau > 0$, we do expect to have a signal that does not vanish identically. But now the semicircle closing the contour *must* lie in the *lower* half-plane, because it is only then that the factor $e^{-i\omega\tau}$ in the integrand vanishes exponentially as $\Omega \to \infty$; and hence so does the contribution from the semicircle to the integral over the contour C_-. Therefore, provided all the singularities of the integrand are in the *lower* half-plane, all our requirements are satisfied. This is ensured by displacing each of the poles of the integrand at $\omega = -ck$ and $\omega = +ck$ by an infinitesimal *negative* imaginary quantity $-i\epsilon$ where $\epsilon > 0$, and then passing to the limit $\epsilon \to 0$ after the integral is evaluated. Equivalently, we may replace ω by $\omega + i\epsilon$ in the denominator of the integrand, and take the limit $\epsilon \to 0$ after carrying out the integration. The causal Green function we seek is, therefore, given by

$$G(\mathbf{R}, \tau) = -c^2 \lim_{\epsilon\to 0} \int \frac{d^d k}{(2\pi)^d}\, e^{i\mathbf{k}\cdot\mathbf{R}} \int_{-\infty}^{\infty} \frac{d\omega}{2\pi}\, \frac{e^{-i\omega\tau}}{(\omega + i\epsilon)^2 - c^2k^2}, \tag{31.14}$$

where ϵ is a positive infinitesimal. But, as discussed above,

$$\int_{-\infty}^{\infty}\frac{d\omega}{2\pi}\,\frac{e^{-i\omega\tau}}{(\omega+i\epsilon)^2-c^2k^2}=\lim_{\Omega\to\infty}\int_{-\Omega}^{\Omega}\frac{d\omega}{2\pi}\,\frac{e^{-i\omega\tau}}{(\omega+i\epsilon)^2-c^2k^2}$$
$$=\lim_{\Omega\to\infty}\int_{C_\pm}\frac{d\omega}{2\pi}\,\frac{e^{-i\omega\tau}}{(\omega+i\epsilon)^2-c^2k^2}. \tag{31.15}$$

Recall that we must use the contour C_+ for $\tau<0$, and C_- for $\tau>0$. But C_+ does not enclose any singularity of the integrand, and so the corresponding integral vanishes, just as we want it to. In the case of C_-, the integral is $(-2\pi i)$ times the sum of the residues of the integrand at the two poles enclosed. (The extra minus sign arises because C_- is traversed in the clockwise or negative sense.) We thus obtain, after simplification,

$$\boxed{G(\mathbf{R}\,,\,\tau)=c\,\theta(\tau)\int\frac{d^dk}{(2\pi)^d}\,\frac{\sin c\tau k}{k}\,e^{i\mathbf{k}\cdot\mathbf{R}}.} \tag{31.16}$$

Note how the step function $\theta(\tau)$ required by causality has emerged automatically in the solution for $G(\mathbf{R}\,,\,\tau)$.

★ **1.** Go through the steps to arrive at Eq. (31.16), starting from Eq. (31.14).

★ **2.** Each of the two poles of the integrand in the divergent integral

$$\int_{-\infty}^{\infty}\frac{d\omega}{2\pi}\,\frac{e^{-i\omega\tau}}{(\omega^2-c^2k^2)}$$

can be displaced by an infinitesimal imaginary part $\pm i\epsilon$ so as to lie either in the upper or lower half-plane. This leads to four possible ways of making the divergent integral finite: the limit $\epsilon\to 0$, taken after the integral is evaluated, gives finite boundary values. Find these values.

31.2 Explicit Solutions for $d = 1$, 2 and 3

We are now in a position to consider the solutions for different values of d, the number of spatial dimensions. In order to make this explicit, let us denote the Green function by $G^{(d)}(\mathbf{R}\,,\,\tau)$ instead of $G(\mathbf{R}\,,\,\tau)$, from now on.

31.2.1 The Green Function in (1 + 1) *Dimensions*

The case of one spatial dimension ($d=1$) is somewhat distinct from the others, and simpler too. Note that the symbol k in the factor $(\sin c\tau k)/k$ in Eq. (31.16) stands for $|\mathbf{k}|$. When $d = 1$, therefore, we should remember to write $|k|$ instead of just k in this factor. Further, $\mathbf{k} \cdot \mathbf{R}$ is just kX in this case, where $X = x - x'$. Therefore,

$$G^{(1)}(X,\tau) = c\ \theta(\tau) \int_{-\infty}^{\infty} \frac{dk}{2\pi} \frac{\sin\ c\tau|k|}{|k|}\ e^{ikX} = c\ \theta(\tau) \int_{-\infty}^{\infty} \frac{dk}{2\pi} \frac{\sin\ c\tau k}{k}\ e^{ikX}. \tag{31.17}$$

It is obvious from this expression that $G^{(1)}(X,\tau)$ is a symmetric function of X, i.e.,

$$G^{(1)}(-X,\tau) = G^{(1)}(X,\tau). \tag{31.18}$$

(Change the variable of integration in Eq. (31.17) from k to $-k$, and the result follows.) As we shall see shortly, $G^{(1)}(X,\tau)$ is actually a function of $|X|$. Setting $e^{ikX} = \cos\ kX + i\ \sin\ kX$, the contribution from the $\sin\ kX$ term vanishes because the integrand is an odd function of k. We are left with the integral

$$G^{(1)}(X,\tau) = c\ \theta(\tau) \int_{-\infty}^{\infty} \frac{dk}{2\pi k}\ \sin\,(c\tau k)\ \cos\,(kX). \tag{31.19}$$

Carrying out the integration, we get

$$G^{(1)}(X,\tau) = \tfrac{1}{4} c\,\theta(\tau)\ [\varepsilon(c\tau + X) + \varepsilon(c\tau - X)]\,, \tag{31.20}$$

where $\varepsilon(x) = \pm 1$ for $x \gtrless 0$. (Recall the signum function, defined in Eq. (2.19) of Chap. 2, Sect. 2.3). Equation (31.20) can be further simplified to yield

$$\boxed{G^{(1)}(X,\tau) = \tfrac{1}{2} c\,\theta(\tau)\,\theta(c\tau - |X|).} \tag{31.21}$$

★ **3.** Starting with Eq. (31.19), derive Eq. (31.21).

The factor $\theta(c\tau - |X|)$ in Eq. (31.21) ensures that the signal arising from the source point x' at time t' does not reach any point x until time $t' + |x - x'|/c$, as required by causality and the finite speed of signal propagation. Another aspect of the solution is also noteworthy. An observer at x does not receive a *pulsed* signal, even though the sender had sent out such a signal (Recall that the initial disturbance was a Dirac δ-function). In fact, the signal received at any x *persists* thereafter for all time. And it does so without diminishing in strength, a feature that is unique to the case $d = 1$.

31.2.2 The Green Function in (2 + 1) *Dimensions*

We return to the general formula (31.16) in more than one spatial dimension, i.e., $d \geq 2$. Writing it down once again for ready reference,

$$G^{(d)}(\mathbf{R}\,,\,\tau) = c\,\theta(\tau)\int \frac{d^d k}{(2\pi)^d}\,\frac{\sin c\tau k}{k}\,e^{i\mathbf{k}\cdot\mathbf{R}}. \tag{31.22}$$

The integrand on the right-hand side is a scalar, and so is the volume element $d^d k$. Further, the region of integration (namely, all of **k**-space) is invariant under rotations of the coordinate axes. Hence $G^{(d)}(\mathbf{R}\,,\,\tau)$ is a scalar, i.e., it is unchanged under rotations of the spatial coordinate axes about the origin. This remains true for all integer values of $d \geq 2$. Moreover, as a result of this rotational invariance, $G^{(d)}(\mathbf{R}\,,\,\tau)$ is actually a function of R and τ (where R stands for $|\mathbf{R}|$, as usual). This means that we can choose the orientation of the axes in **k**-space according to our convenience, without affecting the result. This fact is of help in evaluating the integral. We write the Green function as $G^{(d)}(R, \tau)$, henceforth.

Let us now consider the case $d = 2$. Clearly, it is most convenient to work in plane polar coordinates in **k**-space, choosing one of the axes (say, the k_1-axis) along the vector **R**. Then

$$G^{(2)}(R, \tau) = c\;\theta(\tau)\int_0^\infty \frac{k\,dk}{(2\pi)^2}\,\frac{\sin c\tau k}{k}\int_0^{2\pi} d\varphi\; e^{ikR\,\cos\varphi}. \tag{31.23}$$

But the angular integral is precisely one of the integral representations of the Bessel function of order zero: we have[3]

$$\int_0^{2\pi} d\varphi\, e^{iz\cos\varphi} = 2\pi J_0(z). \tag{31.24}$$

Using this formula,

$$G^{(2)}(R, \tau) = \frac{c\,\theta(\tau)}{2\pi}\int_0^\infty dk\;\sin{(c\tau k)}\;J_0(kR). \tag{31.25}$$

The integral on the right-hand side is a little tricky to evaluate. As we know from Eq. (28.49) of Chap. 28, Sect. 28.3.1, $J_0(kR)$ decays quite slowly, like $1/\sqrt{k}$, as $k \to \infty$. However, it is an oscillatory function that changes sign. So is the factor $\sin{(c\tau k)}$. As a result of the partial cancelation of positive and negative contributions, it turn out that the definite integral in Eq. (31.25) has a finite value. (This should remind you once again of what happens in the case of the Dirichlet integral.) Here is one way to find it.[4] Since $\sin{(c\tau k)}$ is the imaginary part of $e^{ic\tau k}$, and $J_0(kR)$ is real for real values of the argument kR, we have

$$G^{(2)}(R, \tau) = \frac{c\,\theta(\tau)}{2\pi}\mathrm{Im}\,\Big\{\int_0^\infty dk\,e^{ic\tau k}\;J_0(kR)\Big\}. \tag{31.26}$$

---

[3]GR, ...

[4]You could try to look up the definite integral in a table of integrals, but computing it on the basis of the information we already have is certainly more instructive!

But we may regard the integral above as the *analytic continuation* to $s = -ic\tau$ of the Laplace transform of the Bessel function (see Eq. (28.59) of Chap. 28, Sect. 28.3.4),

$$\int_0^\infty dk\, e^{-sk}\, J_0(kR) = \frac{1}{\sqrt{s^2 + R^2}}\,. \tag{31.27}$$

The analytic continuation can be justified properly, but I shall not digress to do so here. (It is equivalent to introduce a decaying exponential factor or "regulator" like $e^{-\lambda k}$ into the integrand, and then passing to the limit $\lambda \to 0$ after the integral is evaluated.) It follows from Eq. (31.27) that

$$\text{Im}\left\{\int_0^\infty dk\, e^{ic\tau k}\, J_0(kR)\right\} = \begin{cases} 0 & \text{for } c^2\tau^2 < R^2 \\ 1/\sqrt{c^2\tau^2 - R^2} & \text{for } c^2\tau^2 > R^2. \end{cases} \tag{31.28}$$

★ **4.** Derive Eq. (31.28), given the integral in Eq. (31.27).

Since we are only interested in non-negative values of τ and R, the condition $c^2\tau^2 > R^2$ yields a factor $\theta(c\tau - R)$. We get, finally,

$$\boxed{G^{(2)}(R, \tau) = \frac{c\ \theta(\tau)}{2\pi}\ \frac{\theta(c\tau - R)}{\sqrt{c^2\tau^2 - R^2}}\,.} \tag{31.29}$$

The sharply-pulsed signal emanating from $\mathbf{r}'$ at time t' thus reaches any point $\mathbf{r}$ only at time $t' + |\mathbf{r} - \mathbf{r}'|/c$, in accordance with causality and the finite velocity of propagation of the disturbance. But once again, the signal received at any field point $\mathbf{r}$ is no longer a sharply-pulsed one: it persists for all $t > t' + |\mathbf{r} - \mathbf{r}'|/c$, although its strength slowly decays as t increases, like t^{-1} at very long times.

- Thus, in both one and two spatial dimensions, there is an *after-effect* (or "after-glow") at any observation point, even for a sharply-pulsed initial signal emanating from the source.

31.2.3 The Green Function in (3 + 1) *Dimensions*

Something entirely different happens in the most important case of three-dimensional space. We now have

$$G^{(3)}(R, \tau) = c\,\theta(\tau) \int \frac{d^3\mathbf{k}}{(2\pi)^3}\, \frac{\sin c\tau k}{k}\, e^{i\mathbf{k}\cdot\mathbf{R}}\,. \tag{31.30}$$

Evaluating the integral over $\mathbf{k}$ yields the expression

$$\boxed{G^{(3)}(R, \tau) = \frac{\theta(\tau)\,\delta(\tau - R/c)}{4\pi R}\,.} \tag{31.31}$$

★ **5.** Establish this result.

The crucial point about the solution (31.31) in three-dimensional space is this:

- If the source pulse is a δ-function impulse emanating from $\mathbf{r}'$ at time t', the signal at any field point $\mathbf{r}$ is *also* a δ-function pulse that reaches (and passes) this point at precisely the instant $t' + |\mathbf{r} - \mathbf{r}'|/c$.

Hence there is no after-effect that lingers on at $\mathbf{r}$, in stark contrast to the situation in $d = 1$ and $d = 2$.

- Moreover, the amplitude of the pulse at $\mathbf{r}$ drops with the distance from the source like $1/R$, in exactly the way the Coulomb potential falls off.

These features are unique to three-dimensional space.

Another interesting reduction takes place in the case $d = 3$. Formally, if the limit $c \to \infty$ is taken in the wave equation, the wave operator reduces to the negative of the Laplacian operator. We might, therefore, expect the solution for $G^{(3)}(R, \tau)$ to reduce to the corresponding Green function for $-\nabla^2$. And indeed it does so: recall that the fundamental Green function for ∇^2 is precisely $-1/(4\pi R)$ in three spatial dimensions (Eq. (29.27) of Chap. 29, Sect. 29.3.2).

31.2.4 Retarded Solution of the Wave Equation

The fact that the causal Green function corresponding to the wave operator in $(3 + 1)$ dimensions is a δ-function has an immediate application. It enables us to write down the retarded solution to the inhomogeneous wave equation (31.1). Using Eq. (31.31) for $G^{(3)}(R, \tau)$ in Eq. (31.2), we obtain the following expression for the particular integral in the case of natural boundary conditions in infinite space:

$$f(\mathbf{r}, t) = \frac{1}{4\pi} \int \frac{d^3r'}{|\mathbf{r} - \mathbf{r}'|} \left[g(\mathbf{r}', t')\right]_{\text{ret}}, \tag{31.32}$$

where the notation $[g(\mathbf{r}', t')]_{\text{ret}}$ denotes the fact that the time argument is to be set equal to the *retarded time* according to

$$t' = t - \frac{|\mathbf{r} - \mathbf{r}'|}{c}. \tag{31.33}$$

Equation (31.32) gives the *causal, retarded* solution of the wave equation.

Electromagnetism in the Lorenz gauge: The foregoing result is immediately applicable to electromagnetism. You have seen in Chap. 9, Sect. 9.1.7, that the equations of electromagnetism take on a particularly simple and symmetric form in the Lorenz gauge, specified by the condition

$$\frac{1}{c^2}\frac{\partial \phi}{\partial t} + \nabla \cdot \mathbf{A} = 0 \tag{31.34}$$

on the scalar potential ϕ and vector potential $\mathbf{A}$. In this gauge, each of these potentials satisfies (in free space) an inhomogeneous wave equation. Repeating Eqs. (9.25) and (9.24) for ready reference, we have

$$\frac{1}{c^2}\frac{\partial^2\phi}{\partial t^2} - \nabla^2\phi = \frac{\rho(\mathbf{r},t)}{\epsilon_0} \quad \text{and} \quad \frac{1}{c^2}\frac{\partial^2\mathbf{A}}{\partial t^2} - \nabla^2\mathbf{A} = \mu_0\,\mathbf{j}(\mathbf{r},t). \tag{31.35}$$

The solutions to these equations with natural boundary conditions are

$$\left.\begin{aligned} \phi(\mathbf{r},t) &= \frac{1}{4\pi\epsilon_0}\int\frac{d^3r\,'}{|\mathbf{r}-\mathbf{r}\,'|}\,\big[\rho(\mathbf{r}\,',t\,')\big]_{\text{ret}}, \\ \mathbf{A}(\mathbf{r},t) &= \frac{\mu_0}{4\pi}\int\frac{d^3r\,'}{|\mathbf{r}-\mathbf{r}\,'|}\,\big[\mathbf{j}(\mathbf{r}\,',t\,')\big]_{\text{ret}}. \end{aligned}\right\} \tag{31.36}$$

These are called **retarded potentials**. The formal solutions for the electric field $\mathbf{E}(\mathbf{r},t)$ and magnetic field $\mathbf{B}(\mathbf{r},t)$ arising from arbitrary sources $\rho(\mathbf{r},t)$ and $\mathbf{j}(\mathbf{r},t)$ can now be written down from the relations $\mathbf{E} = -\partial\mathbf{A}/\partial t - \nabla\phi$ and $\mathbf{B} = \nabla\times\mathbf{A}$. You must bear in mind that the solutions for ϕ and $\mathbf{A}$ in (31.36) are only valid in the Lorenz gauge. But the expressions obtained for the physical fields $\mathbf{E}$ and $\mathbf{B}$ are of course gauge-invariant, and hold good in any gauge.

31.3 Remarks on Propagation in Dimensions $d > 3$

The results derived above essentially imply that the basic, linear wave equation permits the propagation of sharp pulses in three-dimensional space, but not in one- or two-dimensional space. I end this chapter with some comments on what happens in a general number of spatial dimensions $d > 3$.

Interestingly enough, it turns out that the propagation of sharp signals is possible in all *odd*-dimensional spaces with $d \geq 3$, while it fails for all *even* values of d. This is yet another manifestation of the fundamental differences that exist between Euclidean spaces of even and odd dimensionalities, respectively. Consider a sharp δ-function pulse emitted at the origin of coordinates at the instant t_0. If the subsequent propagation of this signal is controlled by the wave equation (31.1), then:
(i) In spaces of dimension $d = 3, 5, \ldots$, the signal received at any point $\mathbf{r}$ is sharply-pulsed. It arrives at that point at time $t_0 + |\mathbf{r}-\mathbf{r}\,'|/c$ and passes on instantaneously, with no after-effect.
(ii) In spaces of dimension $d = 2, 4, \ldots$, the signal reaches $\mathbf{r}$ at time $t_0 + |\mathbf{r}-\mathbf{r}\,'|/c$, but lingers on thereafter, slowly decaying in strength (like $1/t^{d-1}$ at long times). We have already seen an example of such an after-effect in the case $d = 2$.
(iii) There is, however, one feature that is quite unique to $d = 3$: this is the only case in which the original δ-function pulse is transmitted without any distortion, namely, as a pure δ-function pulse.

There is an elegant and powerful way to solve the problem of finding the Green function for a general value of d. It is based on the **relativistic invariance** of the wave operator and of the solution sought, i.e., on the fact that these remain invariant under rotations of the coordinate axes and under boosts to other inertial frames. (Here I have identified c with the speed of light in a vacuum). I shall merely write down the final answer here. Let

$$\xi^2 = c^2\tau^2 - R^2. \tag{31.37}$$

The causal Green function vanishes identically for $\xi^2 < 0$ (or *space-like* intervals). For *time-like* ($\xi^2 > 0$) and *light-like* ($\xi^2 = 0$) intervals, it turns out to be proportional to a derivative of a δ-function, according to

$$G^{(d)}(R, \tau) \propto \frac{d^{(d-3)/2}}{d\xi^{(d-3)/2}}\,\delta(\xi^2). \tag{31.38}$$

The order of the derivative, $\frac{1}{2}(d-3)$, is an integer when d is an odd number. The Green function remains a sharply-pulsed quantity in this case, although it only for $d = 3$ that you get just a δ-function. For larger odd values of d $(= 5, 7, \ldots)$, the fundamental solution is given by higher and higher derivatives of the δ-function. These are increasingly singular quantities. When d is an even integer, on the other hand, the solution is a **fractional derivative** of a δ-function. Fractional derivatives are *nonlocal* objects, defined in terms of suitable integral transforms (such as the Fourier transform). This is how the extended nature of the Green function arises in the case of wave propagation in even-dimensional spaces, leading to the after-effects mentioned earlier.

There is another interesting connection between the Green functions in spaces of different dimensions. The solution in $(d+2)$ spatial dimensions is related to that in d spatial dimensions by

$$G^{(d+2)} = -\frac{1}{2\pi d}\,\frac{\partial^2 G^{(d)}}{\partial \xi^2}\,. \tag{31.39}$$

This relationship shows how the solutions in $d = 5, 7, \ldots$ can be generated from the solution in $d = 3$, while those in $d = 4, 6, \ldots$ can be generated from that in $d = 2$.

Dispersion and nonlinearity: Finally, I mention that there are two important additional aspects of wave or signal propagation that can be adjusted so as to modify the fundamental solution considered here. The first is **dispersion**, which occurs because waves of different wavelengths propagate with different speeds in a medium. The corresponding dispersion relation (or frequency-wave number relationship) can be quite complicated. The second aspect is **nonlinearity**. The simple wave equation we have considered here is linear in the signal $f(\mathbf{r}, t)$. Physical situations, however, often need to be described by nonlinear wave equations of various kinds. The interplay between dispersion and nonlinearity can be extremely intricate and interesting, and a vast variety of new phenomena can arise as a result. Among these are the so-called **solitary waves** and **propagating solitons**, which represent very robust pulsed disturbances. They comprise a whole subject in their own right. Limitations of space preclude a discussion of these important and fascinating topics in this book.

31.4 Solutions

3. Use the identity

$$\sin(c\tau k)\,\cos(kX) = \tfrac{1}{2}[\sin k(c\tau + X) + \sin k(c\tau - X)].$$

The integral in (31.19) becomes a sum of two Dirichlet integrals (Eq. (2.18) of Chap. 2, Sect. 2.3). ▶

4. A little care is needed in getting the overall sign right in the result. Note that the Laplace transform (31.27), regarded as an analytic function of s, has branch points at $s = \pm iR$, and a branch cut running between these points. Identify the phases of the analytic function $(s^2 + R^2)^{-1/2}$ at different points in the complex s-plane, and use this information to arrive at the result quoted in Eq. (31.28). ▶

5. Use spherical polar coordinates $(k\,,\,\theta\,,\,\varphi)$ in $\mathbf{k}$-space. Exploit the rotational invariance of the integral to choose the polar axis along the direction of the vector $\mathbf{R}$. The integration over φ is then trivial. Integrate over θ from 0 to π. Use the symmetry of the integrand to extend the range of integration over k to $(-\infty, \infty)$. This gives

$$G^{(3)}(R,\tau) = \frac{c\,\theta(\tau)}{2(2\pi)^2 R}\int_{-\infty}^{\infty} dk\,\big\{\cos[(c\tau - R)k] - \cos[(c\tau + R)k]\big\}.$$

Next, replace $\cos[(c\tau \pm R)k]$ in the integrand by $\exp[i(c\tau \pm R)k]$, since the contributions from the sine functions vanishes by symmetry. Use the Fourier representation of the Dirac δ-function to obtain

$$G^{(3)}(R,\tau) = \frac{\theta(\tau)}{4\pi R}\big[\delta(\tau - R/c) - \delta(\tau + R/c)\big].$$

The term involving $\delta(\tau + R/c)$ can be dropped, because we are only concerned with non-negative values of τ and R. ▶

Chapter 32
Integral Equations

Integral equations appear in many contexts in physical applications. Integral and differential equations are closely related to each other, because (as already pointed out in Chap. 29, Sect. 29.2) the inverse of a differential operator is an integral operator, in general. As you know, the Green function method of solving differential equations is based precisely on this observation. Broadly speaking, a differential equation together with specified boundary conditions can be reduced to an integral equation or an integral representation for the solution, with an appropriate Green function (or kernel of an integral operator).

32.1 Fredholm Integral Equations

32.1.1 Equation of the First Kind

A **Fredholm integral equation** of the first kind is an equation of the form

$$\boxed{\int_a^b dy\, K(x, y)\, f(y) = g(x),} \tag{32.1}$$

where the quantities given are (i) the range $[a, b]$, (ii) the **kernel** $K(x, y)$, and (iii) the function $g(x)$. It is required to solve for the function $f(x)$. In order to pose the problem more sharply, one specifies $g(x)$ as an element of some function space (for example, the space $L_2[a, b]$), and seeks a solution that belongs to that space. It is then convenient to regard the integral equation (32.1) as an equation between vectors $|\, f\,\rangle$ and $|\, g\,\rangle$ in the linear space, written out in the "position basis". Recall from our discussion of linear vector spaces that $f(x)$ and $g(x)$ are just inner products, i.e.,

$$f(x) \equiv \langle x|\, f\,\rangle, \quad g(x) \equiv \langle x|\, g\,\rangle. \tag{32.2}$$

V. Balakrishnan, *Mathematical Physics*,
https://doi.org/10.1007/978-3-030-39680-0_32

The kernel $K(x, y)$ is the "matrix element" in the position basis of some abstract operator K, i.e.,

$$\boxed{K(x, y) \equiv \langle x | \mathsf{K} | y \rangle.} \tag{32.3}$$

The completeness relation in the position basis is, of course, $\int_a^b dy \,|\, y \rangle\langle y| = I$, the unit operator in the linear space. It is then obvious that the integral equation (32.1) is simply the equation

$$\mathsf{K} \,|\, f \rangle = |\, g \rangle \tag{32.4}$$

written out in the position basis. The formal solution to this equation is

$$|\, f \rangle = \mathsf{K}^{-1} \,|\, g \rangle, \tag{32.5}$$

provided the inverse operator K^{-1} exists. It is, of course, quite another problem to find this operator explicitly.

32.1.2 Equation of the Second Kind

***The inhomogeneous equation*:** An *inhomogeneous* Fredholm integral equation of the second kind is an equation of the form

$$\boxed{f(x) - \lambda \int_a^b dy\, K(x, y)\, f(y) = g(x),} \tag{32.6}$$

where the range $[a, b]$, the kernel $K(x, y)$, and the function $g(x)$ are given. Here λ is a constant (more about its significance shortly). Observe that the unknown function f now occurs both outside the integral and inside it. Once again, writing the equation in its abstract form as an equation between elements of a function space, we have

$$|\, f \rangle - \lambda \mathsf{K} \,|\, f \rangle = (I - \lambda \mathsf{K}) \,|\, f \rangle = |\, g \rangle, \tag{32.7}$$

where I is the unit operator. The formal solution to this equation is given by

$$|\, f \rangle = (I - \lambda \mathsf{K})^{-1} \,|\, g \rangle, \tag{32.8}$$

provided the operator inverse $(I - \lambda \mathsf{K})^{-1}$ exists. You will recognize that the operator $(I - \lambda \mathsf{K})^{-1}$ is very closely related to the *resolvent* of the operator K. Recall Eq. (11.54) of Chap. 11, Sect. 11.4.4 for the definition of the resolvent of a matrix. Exactly as in the case of a matrix, the resolvent of an operator K is defined as

$$R_{\mathsf{K}}(z) \stackrel{\text{def.}}{=} (zI - \mathsf{K})^{-1}, \tag{32.9}$$

where z is a complex variable. Hence, the inverse operator $(I - \lambda \mathsf{K})^{-1}$ is just $(1/\lambda)\, R_{\mathsf{K}}(1/\lambda)$.

★ **1.** In writing down (32.8) as the general solution of Eq. (32.7), we seem to have forgotten to add the possible solutions to the corresponding *homogeneous* equation. Suppose there exist one or more non-null vectors $|h_i\rangle$, such that

$$(I - \lambda \mathsf{K})\,|h_i\rangle = 0.$$

It would appear that the general solution of Eq. (32.7) must be of the form

$$|\, f\,\rangle = (I - \lambda \mathsf{K})^{-1}\,|\, g\,\rangle + \sum_i c_i\,|h_i\rangle,$$

because applying the operator $(I - \lambda \mathsf{K})$ to both sides of the last equation correctly yields the original equation (32.7). Why is there no additional term of the form $\sum_i c_i\,|h_i\rangle$ present on the right-hand side of Eq. (32.8)?

***The homogeneous equation*:** If $g(x)$ is absent (i.e., identically equal to zero), we have a *homogeneous* Fredholm integral equation of the second kind, namely,

$$\boxed{f(x) = \lambda \int_a^b dy\, K(x, y)\, f(y).} \tag{32.10}$$

In abstract form,

$$|\, f\,\rangle = \lambda \mathsf{K}\,|\, f\,\rangle, \quad \text{or} \quad \mathsf{K}\,|\, f\,\rangle = \lambda^{-1}\,|\, f\,\rangle. \tag{32.11}$$

The significance of the constant λ is now clear. The homogeneous equation is an *eigenvalue equation*. It has a nonvanishing solution only when $1/\lambda$ is an eigenvalue of the operator K, and $|\, f\,\rangle$ is the corresponding eigenfunction. Moreover, when $1/\lambda$ equals one of the eigenvalues of K, the inverse $(I - \lambda \mathsf{K})^{-1}$ does not exist, exactly as in the case of a matrix. In that case the *inhomogeneous* equation $|\, f\,\rangle - \lambda \mathsf{K}\,|\, f\,\rangle = |\, g\,\rangle$ does not have a solution.

The different points made in the foregoing are brought together in the so-called **Fredholm alternative**, which may be summarized as follows:

(i) If $1/\lambda \neq$ an eigenvalue of K, then:

(a) The *inhomogeneous* integral equation (32.6) has a *unique* solution for every $g(x) \not\equiv 0$.
(b) The *homogeneous* equation (32.10) has only the *trivial* solution $f(x) = 0$.

(ii) If $1/\lambda$ belongs to the eigenvalue spectrum of K, then:

(a) The *inhomogeneous* integral equation (32.6) has *no* solution.
(b) The *homogeneous* equation (32.10) has *nontrivial* solutions, namely, the eigenfunctions corresponding to the eigenvalues concerned.

Obviously, the solution of the homogeneous equation can only be specified up to a multiplicative constant. The latter must be determined (if required) by imposing an additional requirement such as some normalization condition.

The Fredholm alternative should remind you of the simple **Cramer's rule** that applies to linear simultaneous equations in elementary algebra. The Fredholm alternative is actually a very general statement, and applies to operator equations in even more general function spaces than the ones we are concerned with here.

A minor point: We could, of course, have started by writing the original integral equation as $\lambda \,|\, f\,\rangle = \mathsf{K} \,|\, f\,\rangle$, so that λ (rather than $1/\lambda$) denotes an eigenvalue of K. But the notation used above is the conventional one in the context of integral equations, and I shall retain it.

32.1.3 Degenerate Kernels

$K(x, y)$ is a **degenerate kernel** or **separable kernel** if it can be expressed as a *finite* sum of products of functions of x and functions of y, according to

$$K(x, y) = \sum_{j=1}^{r} \phi_j(x)\, \psi_j^*(y), \tag{32.12}$$

for some *finite* positive integer value of r. Although we shall restrict ourselves to real-valued functions and kernels for most part, it is convenient to write down the formalism in the more general case when these are complex-valued. The positive integer r is the **rank** of the kernel. It is immediately clear that the operator K must be of the form

$$\mathsf{K} = \sum_{j=1}^{r} |\phi_j\rangle\langle\psi_j|, \tag{32.13}$$

so that $K(x, y) \equiv \langle\, x\,|\mathsf{K}|\, y\,\rangle$ has the separable form written down in Eq. (32.12).

A kernel that is not degenerate (or separable) is said to be **nondegenerate**. This is the general case. Note that *every* kernel (i.e., every function of two variables x and y) can always be written in the form

$$K(x, y) = \sum_{j=1}^{\infty} \phi_j(x)\, \psi_j^*(y). \tag{32.14}$$

When the sum terminates after a *finite* number of terms, we have a degenerate kernel.

The general solution of Eq. (32.6) in the case of a *non*degenerate kernel is quite intricate, as we shall see a little later. (This solution was Fredholm's great achievement.) But the solution in the case of a degenerate kernel is straightforward. It is simpler to work with the abstract form of the equation, notation-wise. Substituting the expression in Eq. (32.13) in Eq. (32.7), we have

$$|\,f\,\rangle - \lambda \sum_{j=1}^{r} \big(\langle \psi_j |\, f\rangle\big)\, |\phi_j\rangle = |\, g\,\rangle, \tag{32.15}$$

or

$$|\,f\,\rangle - \lambda \sum_{j=1}^{r} C_j\, |\phi_j\rangle = |\, g\,\rangle, \quad \text{where} \quad C_j \equiv \langle \psi_j |\, f\rangle. \tag{32.16}$$

The inner products C_j $(1 \le j \le r)$ are as yet unknown constants, since they involve $|\,f\,\rangle$. In order to determine them, all you have to do is to take the inner product of the integral equation with each $\langle \psi_k|$ in turn. This yields the set of r *inhomogeneous* linear simultaneous equations

$$C_k - \lambda \sum_{j=1}^{r} \langle \psi_k|\phi_j\rangle\, C_j = \langle \psi_k|g\rangle \quad (1 \le k \le r) \tag{32.17}$$

for the coefficients $C_j,\ 1 \le j \le r$. Solve this set of equations and substitute for the constants C_j in Eq. (32.16) to obtain $|\,f\,\rangle$. Going back to the position basis, the solution for $f(x) \equiv \langle x|\, f\,\rangle$ is

$$f(x) = g(x) + \lambda \sum_{j=1}^{r} C_j\, \phi_j(x). \tag{32.18}$$

Remember that the constants C_j will also involve λ. Note also that the completeness relation $\int_a^b dx\, |\,x\,\rangle\langle\, x\,| = I$ leads to the expressions

$$C_j = \int_a^b dx\, \psi_j^*(x)\, f(x), \quad \langle \psi_k|\phi_j\rangle = \int_a^b dx\, \psi_k^*(x)\, \phi_j(x). \tag{32.19}$$

★ **2.** Solve the integral equation $f(x) - \lambda \int_0^1 dy\, K(x, y)\, f(y) = 1$ when the kernel $K(x, y)$ is

(i) xy (ii) $\sin(\pi x)\,\sin(\pi y)$ (iii) e^{x+y} (iv) $x + y$ (v) $\sin \pi(x + y)$.

In each case, identify the value(s) of λ at which the solution breaks down.

32.1.4 The Eigenvalues of a Degenerate Kernel

Now consider the homogeneous equation with a degenerate kernel of rank r,

$$f(x) = \lambda \int_a^b dy\, K(x, y)\, f(y) \quad \text{where} \quad K(x, y) = \sum_{j=1}^{r} \phi_j(x)\, \psi_j^*(y). \tag{32.20}$$

This just the eigenvalue equation for the finite-rank operator K, namely,

$$|\, f\,\rangle = \lambda \mathsf{K}\,|\, f\,\rangle = \lambda \sum_{j=1}^{r} \big(|\phi_j\rangle\langle\psi_j|\big)\,|\, f\,\rangle. \tag{32.21}$$

(Remember that λ^{-1} stands for the eigenvalue.) Once again, take the inner product of both sides of this equation with each of the bra vectors $\langle\psi_k|$. The constants $C_j = \langle\psi_j|\, f\rangle$ now satisfy the set of r *homogeneous* linear simultaneous equations

$$C_k - \lambda \sum_{j=1}^{r} \langle\psi_k|\phi_j\rangle\, C_j = 0, \quad 1 \le k \le r. \tag{32.22}$$

Equation (32.22) is of the form $M_{kj}\, C_j = 0$, where M is the $(r \times r)$ matrix with elements

$$M_{kj} = \delta_{kj} - \lambda\langle\psi_k|\phi_j\rangle = \delta_{kj} - \lambda \int_a^b dx\, \psi_k^*(x)\, \phi_j(x). \tag{32.23}$$

A nontrivial solution for the set $\{C_j\}$ exists if and only if the determinant of M vanishes. Hence, λ must be a root of the secular equation

$$\det M = 0. \tag{32.24}$$

At each of these special values of λ, the homogeneous integral equation has a nontrivial solution that is essentially a linear combination of the functions $\phi_j(x)$. (An overall multiplicative constant may be fixed by some normalization condition.) As already stated, when λ takes on any of these values, the *in*homogeneous equation does *not* have a solution.

★ **3.** Find the value(s) of λ for which the following homogeneous integral equations have nontrivial solutions, as well as the corresponding solutions:

(a) $f(x) = \lambda \int_{-\infty}^{\infty} dx\, e^{-(x^2+y^2)}\, f(y)$.

(b) $f(x) = \lambda \int_{-\infty}^{\infty} dx\, e^{-(x^2+y^2)}\, (x + y)\, f(y)$.

32.1.5 Iterative Solution: Neumann Series

When a kernel is nondegenerate, it has an infinite number of eigenvalues (including the multiplicity of possible repeated eigenvalues). As I have already stated, we can always write a nondegenerate kernel in the form $K(x, y) = \sum_{j=1}^{\infty} \phi_j(x) \psi_j^*(y)$. Doing so and following the procedure given above for separable kernels, we are faced with an infinite-dimensional matrix problem. In general, this is not very helpful.

On the other hand, the inhomogeneous equation

$$f(x) = g(x) + \lambda \int_a^b dy\, K(x, y)\, f(y) \tag{32.25}$$

can be solved, in principle, by an *iterative method* that is valid for sufficiently small values of λ. Once again, it is easier to see how this works in abstract notation. Recall the formal solution in Eq. (32.8), $|\, f\,\rangle = (I - \lambda\, \mathsf{K})^{-1}\, |\, g\,\rangle$. It turns out that, as long as $|\lambda|$ is smaller than a certain positive number Λ, the operator $(I - \lambda\, \mathsf{K})^{-1}$ can be expanded in a convergent "binomial series", i.e.,

$$(I - \lambda\, \mathsf{K})^{-1} = I + \lambda\, \mathsf{K} + \lambda^2\, \mathsf{K}^2 + \cdots . \tag{32.26}$$

The convergence of the right-hand side of Eq. (32.26) to the operator on the left-hand side is in the sense of the operator norms of the two sides. The value of Λ will be specified shortly. The solution for $|\, f\,\rangle$ is then given by the convergent infinite series

$$|\, f\,\rangle = |\, g\,\rangle + \lambda\, \mathsf{K}\, |\, g\,\rangle + \lambda^2\, \mathsf{K}^2\, |\, g\,\rangle + \cdots . \tag{32.27}$$

Like K, the powers K^n are also integral operators with corresponding kernels that we may denote by $K_n(x\,,\,y)$. That is,

$$K_n(x\,,\,y) \stackrel{\text{def.}}{=} \langle\, x\,|\mathsf{K}^n|\, y\,\rangle. \tag{32.28}$$

Using the completeness relation in the position basis, it is easy to see that the *once-iterated kernel* $K_2(x, y)$ is given by

$$\boxed{K_2(x\,,\,y) = \int_a^b dy_1\, K(x\,,\,y_1)\, K(y_1\,,\,y).} \tag{32.29}$$

Similarly, for all $n \geq 2$, the successive iterates of the kernel are given by

$$K_n(x\,,\,y) = \int_a^b dy_1 \int_a^b dy_2 \ldots \int_a^b dy_{n-1}\, K(x\,,\,y_1)\, K(y_1\,,\,y_2)\, \ldots\, K(y_{n-1}\,,\,y). \tag{32.30}$$

In a more compact form, and with $K_1(x\,,\,y) \stackrel{\text{def.}}{=} K(x\,,\,y)$, we have

$$\boxed{K_n(x\,,\,y) = \int_a^b dy_1\, K(x\,,\,y_1)\, K_{n-1}(y_1\,,\,y), \quad n \geq 2.} \tag{32.31}$$

The series solution (32.27) for $|\, f\,\rangle$ is then given, in the position basis, by

$$\boxed{f(x) = g(x) + \sum_{n=1}^{\infty} \lambda^n \int_a^b dy\, K_n(x\,,\, y)\, g(y).} \tag{32.32}$$

The power series solution in Eq. (32.32) is called a **Neumann series**. It is convenient to write the solution in the form

$$f(x) = g(x) + \lambda \int_a^b dy\, H(x\,,\, y\,;\, \lambda)\, g(y), \tag{32.33}$$

where

$$\begin{aligned} H(x\,,\, y\,;\, \lambda) &= K(x\,,\, y) + \lambda\, K_2(x\,,\, y) + \lambda^2\, K_3(x, y) + \cdots \\ &= \sum_{n=1}^{\infty} \lambda^{n-1}\, K_n(x\,,\, y). \end{aligned} \tag{32.34}$$

The quantity $H(x\,,\, y\,;\, \lambda)$ is called the **resolvent kernel** corresponding to the kernel $K(x\,,\, y)$. As I have already mentioned, the series on the right-hand side of Eq. (32.34) is convergent in the operator norm when the parameter λ lies within a circle centered at the origin and of a certain radius Λ in the complex plane.

- The Neumann series solution of the inhomogeneous Fredholm equation is valid for $|\lambda| < \Lambda$.
- Λ is the magnitude of the smallest value of λ for which the *homogeneous* integral equation has a nontrivial solution. In other words, $\Lambda = 1/|\mu_{\text{max}}|$, where μ_{max} is the eigenvalue of the operator K with the largest magnitude.

Note that the series solution and the statements made above remain valid in the simpler case of a degenerate kernel of any finite rank r.

★ **4.** Find the Neumann series solution of the inhomogeneous integral equation

$$f(x) = 1 + \lambda \sqrt{\frac{2}{\pi}} \int_{-\infty}^{\infty} dy\, e^{-(x^2+y^2)}\, f(y).$$

What is the region of convergence of this solution in the λ-plane?

32.2 Nonrelativistic Potential Scattering

32.2.1 The Scattering Amplitude

The quantum theory of scattering provides a physical application of Fredholm integral equations in general, and of Neumann series in particular. What follows is a brief discussion of time-independent scattering theory in the simplest case: the elastic scattering of a nonrelativistic particle of mass m from a static central potential $V(r)$ that vanishes as $r \to \infty$. I assume that the reader has some familiarity with the physics underlying this problem, and focus mainly on the aspects related to integral equations. In the process, I will digress to derive the fundamental "outgoing wave" Green function for the Helmholtz operator $\nabla^2 + k^2$, where k is a positive constant. This will round off the set of Green functions already derived for the Poisson equation, the diffusion equation and, the wave equation in Chaps. 29, 30 and 31, respectively.

In order to have a tunable parameter measuring the "strength" of the potential, it is customary to write the potential as $\lambda\, V(r)$, where λ is a constant. This is merely for convenience. In what follows, the product $\lambda\, V(r)$ has the physical dimensions of energy. Various physical quantities can then be expressed as power series in λ, and approximations to different orders in λ can be made. The time-independent Schrödinger equation for the wave function corresponding to a stationary state with energy eigenvalue E is given by

$$-\frac{\hbar^2}{2m}\nabla^2\psi(\mathbf{r}) + \lambda V(r)\,\psi(\mathbf{r}) = E\,\psi(\mathbf{r}). \tag{32.35}$$

We are interested here in the **scattering states** of the particle, with eigenvalues belonging to the *continuous* part $E \geq 0$ of the spectrum of the Hamiltonian.

As you know, the corresponding wave functions are not normalizable. There is more than one way to deal with this drawback, and to work with mathematically more tractable normalizable wave functions. For instance, one could use the so-called *box normalization*; or else we could work with normalizable *wave packets* instead of the plane waves representing the momentum eigenstates. I shall not digress into this complication here.

Accordingly, the initial state of the particle is taken to be a momentum eigenstate, with eigenvalue $\mathbf{p} = \hbar\mathbf{k}$. It is, therefore, represented by the position-space wave function (or "incident" plane wave)

$$\psi_{\text{inc}}(\mathbf{r}) \equiv \langle \mathbf{r}|\mathbf{p}\rangle = e^{i\mathbf{p}\cdot\mathbf{r}/\hbar} = e^{i\mathbf{k}\cdot\mathbf{r}}, \tag{32.36}$$

as discussed in Chap. 13, Sect. 13.2.5. Thus, $\mathbf{k}$ is the direction of the momentum of the incident particle. After scattering, the wave vector changes direction without changing its magnitude, because the scattering is elastic: that is, the energy remains

$$E = \frac{\hbar^2 k^2}{2m} \tag{32.37}$$

throughout. The scattered wave vector can be directed along any direction in space. The quantity of interest is the *probability* of scattering in any particular direction. This is measured by the **differential cross-section** $d\sigma/d\Omega$, defined as

$$\frac{d\sigma}{d\Omega} \stackrel{\text{def.}}{=} \frac{\text{flux of particles per unit solid angle in the direction concerned}}{\text{incident flux}}. \tag{32.38}$$

Sufficiently far away from the scattering center (i.e., as $r \to \infty$, or, more precisely, for $kr \gg 1$), the scattered wave has the form of an outgoing spherical wave. Figure 32.1 depicts the scattering process schematically. The total wave function is a superposition of the incident and scattered wave functions. Its asymptotic form is as follows: We use spherical polar coordinates, taking the incident wave vector **k** to be the polar axis. Then

$$\boxed{\psi(\mathbf{r}) = \psi_{\text{inc}}(\mathbf{r}) + \psi_{\text{sc}}(\mathbf{r}) \xrightarrow[kr \gg 1]{} e^{i\mathbf{k}\cdot\mathbf{r}} + f(k, \theta)\,\frac{e^{ikr}}{r}.} \tag{32.39}$$

This is a fundamental relation in scattering theory. It serves to define the energy and angle-dependent factor $f(k, \theta)$ called the **scattering amplitude**. This quantity has no dependence on the azimuthal angle φ because the scattering potential is spherically symmetrical. But it does have θ-dependence, because the *incident* wave vector **k** singles out a special direction. In other words:

- The potential is spherically symmetric, but the symmetry of the scattered wave function is reduced to an axial or cylindrical symmetry about the direction of the initial momentum of the particle.

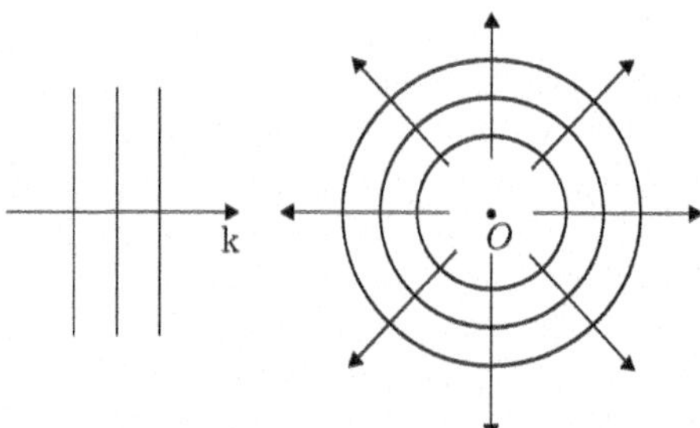

Fig. 32.1 Scattering of an incident plane wave (whose wavefronts are represented by parallel lines) from a central potential. The dot indicates the scattering center $r = 0$. The scattered wave is an outgoing spherical wave (whose wavefronts are represented by concentric circles)

The differential cross-section is easily shown to be given by

$$\boxed{\frac{d\sigma}{d\Omega} = |f(k, \theta)|^2.} \tag{32.40}$$

The total cross-section for scattering is then

$$\sigma = \int d\Omega \, \frac{d\sigma}{d\Omega} = \int d\Omega \, |f(k, \theta)|^2 = 2\pi \int_{-1}^{1} d\,(\cos\,\theta)\,|f(k, \theta)|^2. \tag{32.41}$$

★ **5.** Derive the formula (32.40) for the differential cross-section.

32.2.2 Integral Equation for Scattering

In order to find the scattering amplitude, we need the asymptotic form of the solution for the wave function. For this purpose, it is convenient to convert the time-independent Schrödinger equation (32.35) from a differential equation to an integral equation. We first write the differential equation in the form

$$(\nabla^2 + k^2)\,\psi(\mathbf{r}) = \lambda\, U(r)\,\psi(\mathbf{r}), \quad \text{where} \quad U(r) = (2m/\hbar^2)\,V(r). \tag{32.42}$$

Suppose, for a moment, that we treat the right-hand side $\lambda U(r)\psi(\mathbf{r})$ as a "source" term in an inhomogeneous differential equation. The general solution will then be the sum of the complementary function (CF) and the particular integral (PI). The CF is a solution of the *homogeneous* equation $(\nabla^2 + k^2)\,\psi(\mathbf{r}) = 0$, chosen so as to satisfy the boundary conditions. In the present case, the complementary function is just the incident wave $e^{i\mathbf{k}\cdot\mathbf{r}}$. The PI involves $G(\mathbf{r}, \mathbf{r}\,')$, the Green function for the **Helmholtz operator** $(\nabla^2 + k^2)$. Thus,

$$\psi(\mathbf{r}) = \underbrace{e^{i\mathbf{k}\cdot\mathbf{r}}}_{\text{CF}} + \underbrace{\lambda \int d^3 r\,'\, G(\mathbf{r}, \mathbf{r}\,')\, U(r\,')\,\psi(\mathbf{r}\,')}_{\text{PI}}. \tag{32.43}$$

The presence of the unknown function ψ in the integrand on the right-hand side makes (32.43) an integral equation, rather than a formula, for ψ.

The Green function $G(\mathbf{r}, \mathbf{r}\,')$ we require is the one that corresponds asymptotically to an outgoing wave $\sim e^{ikr}/r$. As you will see in Sect. 32.2.3 below, it is given by

$$G(\mathbf{r}, \mathbf{r}\,') \equiv G(R) = -\frac{e^{ikR}}{4\pi R}, \quad \text{where} \quad R = |\mathbf{r} - \mathbf{r}\,'|. \tag{32.44}$$

Inserting this expression into Eq. (32.43), we get the integral equation for scattering:

$$\boxed{\psi(\mathbf{r}) = e^{i\mathbf{k}\cdot\mathbf{r}} - \frac{\lambda}{4\pi}\int d^3r\,'\,\frac{e^{ik|\mathbf{r}-\mathbf{r}\,'|}}{|\mathbf{r}-\mathbf{r}\,'|}\,U(r\,')\,\psi(\mathbf{r}\,').} \tag{32.45}$$

This is an inhomogeneous Fredholm equation of the second kind, with a kernel given by

$$K(\mathbf{r},\mathbf{r}\,') = -\frac{e^{ik|\mathbf{r}-\mathbf{r}\,'|}}{4\pi|\mathbf{r}-\mathbf{r}\,'|}\,U(r\,'). \tag{32.46}$$

32.2.3 Green Function for the Helmholtz Operator

The fundamental Green function for the Helmholtz operator $(\nabla^2 + k^2)$, quoted in Eq. (32.44), is derived as follows: $G(\mathbf{r},\mathbf{r}\,')$ satisfies the equation

$$(\nabla_{\mathbf{r}}^2 + k^2)G(\mathbf{r},\mathbf{r}\,') = \delta^{(3)}(\mathbf{r}-\mathbf{r}\,'). \tag{32.47}$$

As in the case of the Laplacian operator discussed in Chap. 29, Sect. 29.3.2, all three of the following are translation invariant:

(i) the differential operator $(\nabla_{\mathbf{r}}^2 + k^2)$,
(ii) the δ-function $\delta^{(3)}(\mathbf{r}-\mathbf{r}\,')$, and
(iii) the free boundary condition (the vanishing of G as $r \to \infty$ along all directions).

Hence, G will once again turn out to be a function of the difference $\mathbf{R} = \mathbf{r} - \mathbf{r}\,'$. Introducing the Fourier transform of $G(\mathbf{R})$ and following the same steps as in Sect. 29.3.2, we now get

$$G(\mathbf{R}) = -\frac{1}{(2\pi)^3}\int d^3q\,\frac{e^{i\mathbf{q}\cdot\mathbf{R}}}{q^2 - k^2}\,. \tag{32.48}$$

I have used $\mathbf{q}$ to denote the Fourier transform variable conjugate to $\mathbf{R}$, since $\mathbf{k}$ has already been used for the incident wave vector. As usual, we work in spherical polar coordinates in $\mathbf{q}$-space, and choose the polar axis along the direction of the vector $\mathbf{R}$. Carrying out the angular integrations, we get (noting that the answer is a function of then magnitude R alone)

$$G(R) = -\frac{1}{2\pi^2 R}\int_0^\infty dq\,\frac{q\,\sin qR}{q^2 - k^2} = -\frac{1}{8\pi^2 Ri}\int_{-\infty}^\infty dq\,\frac{q\,(e^{iqR} - e^{-iqR})}{q^2 - k^2}\,. \tag{32.49}$$

The integrand has nonintegrable singularities on the path of integration, at the points $q = \pm k$. We must, therefore, use an appropriate $i\epsilon$-prescription (recall Chap. 23, Sect. 23.3.4) to make the integral well-defined.

The next step, of course, is to close the contour of integration by adding a semicircle to the line integral. The contribution of this semicircle to the integral must vanish as its radius tends to infinity. It must, therefore, lie in the *upper* half-plane for the term

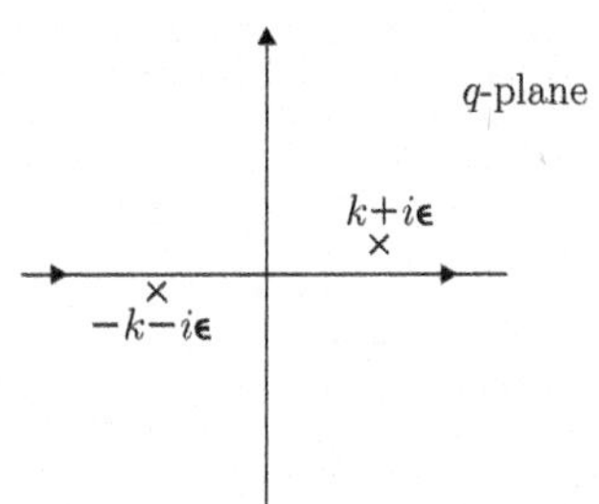

Fig. 32.2 The correct $i\epsilon$ prescription to determine the scattering Green function of the Helmholtz operator. The poles of the integrand are displaced so as to lie at $q = -k - i\epsilon$ and $q = k + i\epsilon$, respectively,

involving e^{iqR}, and in the *lower* half-plane for the term involving e^{-iqR}. Moreover, we want $G(R)$ to have an asymptotic behavior $\sim e^{ikr}/r$ as $r \to \infty$. This means that only the pole at $q = k$ must contribute to the term proportional to e^{iqR}. Similarly, only the pole at $q = -k$ must contribute to the term proportional to e^{-iqR}. Therefore, the pole at $q = k$ must be displaced into the upper half-plane; and the pole at $q = -k$ must be displaced into the lower half-plane. Hence, the correct $i\epsilon$-prescription is implemented by setting

$$G(R) = -\frac{1}{8\pi^2 Ri} \lim_{\epsilon \to 0} \int_{-\infty}^{\infty} dq \, \frac{q\,(e^{iqR} - e^{-iqR})}{q^2 - (k + i\epsilon)^2}\,, \tag{32.50}$$

where ϵ is an infinitesimal positive number. Figure 32.2 shows the displaced poles.

Closing the contour as described above, and using the residue theorem, we get the result quoted in Eq. (32.44), namely,

$$\boxed{G(R) = -\frac{e^{ikR}}{4\pi R} = -\frac{e^{ik|\mathbf{r}-\mathbf{r}'|}}{4\pi|\mathbf{r}-\mathbf{r}'|}}\,. \tag{32.51}$$

★ **6.** Starting from Eq. (32.47), work out the steps to arrive at Eq. (32.51).

The $i\epsilon$-prescription in Eq. (32.50) replaces k by $k + i\epsilon$ in the factor $(q^2 - k^2)^{-1}$ in the integral over q. In contrast, the $i\epsilon$-prescription in the case of the retarded Green function for the wave operator, Eq. (31.14) of Chap. 31, Sect. 31.1.2, involved replacing the integration variable ω by $\omega + i\epsilon$ in the factor $(\omega^2 - c^2k^2)^{-1}$ in the integral over ω. This difference offers a simple way of remembering the correct prescription in each case.

★ **7.** You will find it a simple but instructive exercise to work out the Green functions obtained from Eq. (32.48) by adopting the three other ways in which the poles at $q = \pm k$ can be given infinitesimal imaginary displacements. Let ϵ be a positive infinitesimal, as usual. Show that

$$-\frac{1}{(2\pi)^3} \lim_{\epsilon \to 0} \int d^3q \, \frac{e^{i\mathbf{q}\cdot\mathbf{R}}}{q^2 - (k - i\epsilon)^2} = 0,$$

while

$$-\frac{1}{(2\pi)^3}\lim_{\epsilon\to 0}\int d^3q\,\frac{e^{i\mathbf{q}\cdot\mathbf{R}}}{(q\pm i\epsilon)^2-k^2}=-\frac{\cos kR}{4\pi R}.$$

Remark The first of the foregoing corresponds to displacing the pole at k into the lower half-plane in q, and the pole at $-k$ into the upper half-plane. The Green function then vanishes identically. The other two cases correspond to displacing both poles into the upper or the lower half-plane. These results show you how drastically different the solutions that follow from different $i\epsilon$-prescriptions can be from each other.

32.2.4 Formula for the Scattering Amplitude

Let us return to the scattering problem. Our immediate objective here is to extract the asymptotic behavior of the wave function from the integral equation (32.45), in order to identify the scattering amplitude.

As $r\to\infty$ (more precisely, for $kr\gg 1$), we may replace the factor $1/|\mathbf{r}-\mathbf{r}'|$ in the integrand by just $1/r$. The implicit assumption here is that the potential $V(r')$ decays rapidly for very large values of r', so that the contribution to the integral from the region in which *both* r and r' are large is quite negligible. In the pure *phase* factor $\exp(ik|\mathbf{r}-\mathbf{r}'|)$, however, we need to be more careful. The quantity $|\mathbf{r}-\mathbf{r}'|$ cannot be approximated by r alone. The next term in the expansion must be retained, since it is finite as $r\to\infty$. We have

$$\begin{aligned} ik\,|\mathbf{r}-\mathbf{r}'| &= ik\left(r^2-2\mathbf{r}\cdot\mathbf{r}'+r'^2\right)^{1/2}\simeq ikr\left(1-\frac{2\mathbf{r}\cdot\mathbf{r}'}{r^2}\right)^{1/2}\\ &\simeq ikr\left(1-\frac{\mathbf{r}\cdot\mathbf{r}'}{r^2}\right)=ikr-ik\,\mathbf{e}_r\cdot\mathbf{r}', \end{aligned} \tag{32.52}$$

using the fact that $\mathbf{r}/r=\mathbf{e}_r$. Now, the general field point $\mathbf{r}$ at which the wave function is being calculated is the point at which the particle detector is located, in order to measure the flux per unit solid angle in that direction. Thus, the wave vector of the scattered particle $\mathbf{k}'$, is along $\mathbf{r}$. That is, $k\,\mathbf{e}_r=\mathbf{k}'$ (see Fig. 32.3). Therefore,

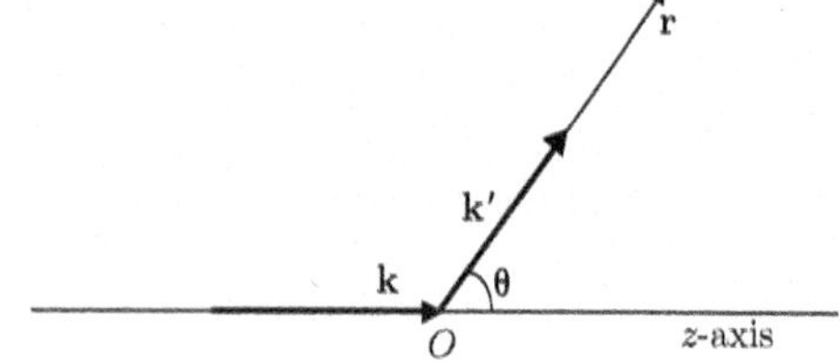

Fig. 32.3 An arbitrary "field point" **r** at which the scattered flux is measured; the direction $\mathbf{k}'$ of the scattered wave is along **r**

$$ik\,|\mathbf{r} - \mathbf{r}'| \simeq ikr - i\mathbf{k}' \cdot \mathbf{r}'. \tag{32.53}$$

Inserting these approximations in the integral equation (32.45) for $\psi(\mathbf{r})$, we get

$$\psi(\mathbf{r}) \to e^{i\mathbf{k}\cdot\mathbf{r}} - \frac{e^{ikr}}{r}\,\frac{\lambda}{4\pi}\int d^3r'\, e^{-i\mathbf{k}'\cdot\mathbf{r}'}\, U(r')\,\psi(\mathbf{r}'). \tag{32.54}$$

Compare this result with the asymptotic form of the wave function in Eq. (32.39). It follows at once that the scattering amplitude is given by the formula

$$\boxed{f(k,\theta) = -\frac{\lambda}{4\pi}\int d^3r'\, e^{-i\mathbf{k}'\cdot\mathbf{r}'}\, U(r')\,\psi(\mathbf{r}').} \tag{32.55}$$

This is an *exact* formula for the scattering amplitude. But it is a formal expression that is not of much use as it stands, because it involves the unknown total wave function $\psi(\mathbf{r})$ at *all* points in space.

***Scattering geometry*:** For scattering in any given direction, the vector

$$\hbar\mathbf{Q} = \hbar\,(\mathbf{k}' - \mathbf{k}) \tag{32.56}$$

is the final momentum of the particle minus its initial momentum. That is, $\mathbf{Q}$ is the wave vector corresponding to the **momentum transfer** associated with the scattering process. Recall that the energy E is related to $k = |\mathbf{k}| = |\mathbf{k}'|$ according to $E = \hbar^2k^2/(2m)$. The relation between k, Q, and θ is also easy to derive. Figure 32.4 shows the relation between the incident and scattered wave vectors for a given scattering angle θ. It is easily seen that

$$Q^2 = 2k^2\,(1 - \cos\,\theta) = 4k^2\,\sin^2\left(\tfrac{1}{2}\theta\right), \quad \text{or} \quad \boxed{Q = 2k\,\sin\left(\tfrac{1}{2}\theta\right).} \tag{32.57}$$

$\sin\frac{1}{2}\theta$ is nonnegative since $0 \le \theta \le \pi$. The limiting values $\theta = 0$ and $\theta = \pi$ correspond to **forward scattering** and **backward scattering**, respectively.

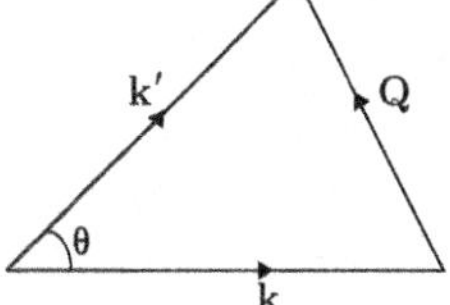

Fig. 32.4 The relation between the incident and scattered wave vectors, the momentum transfer wave vector, and the scattering angle for elastic scattering

32.2.5 The Born Approximation

Going back to the inhomogeneous integral equation (32.45) for the wave function, we can solve it iteratively for sufficiently small values of the "strength" $|\lambda|$ of the potential—that is, we can write down the Neumann series solution. This is given by

$$\begin{aligned}\psi(\mathbf{r}) = e^{i\mathbf{k}\cdot\mathbf{r}} &- \frac{\lambda}{4\pi}\int d^3r\,'\,\frac{e^{ik|\mathbf{r}-\mathbf{r}\,'|}}{|\mathbf{r}-\mathbf{r}\,'|}\,U(r\,')\,e^{i\mathbf{k}\cdot\mathbf{r}\,'}\\ &+\Big(\frac{\lambda}{4\pi}\Big)^2\int d^3r\,'\int d^3r\,''\,\frac{e^{ik|\mathbf{r}-\mathbf{r}\,'|}}{|\mathbf{r}-\mathbf{r}\,'|}\,U(r\,')\,\frac{e^{ik|\mathbf{r}\,'-\mathbf{r}\,''|}}{|\mathbf{r}\,'-\mathbf{r}\,''|}\,U(r\,'')\,e^{i\mathbf{k}\cdot\mathbf{r}\,''}+\cdots \quad (32.58)\end{aligned}$$

The precise conditions on the potential under which this is a convergent series are discussed in textbooks on quantum mechanics, and I shall not go into this aspect here. Broadly speaking, the series solution is valid *when the effect of the potential is weak as compared to the kinetic energy of the incident particle*. Substitution of the solution (32.58) in the formula (32.55) for $f(k,\theta)$ gives an expression for the scattering amplitude as a power series in λ, called the **Born series**. The utility of the parameter λ is now obvious: it helps us keep track of the order of the approximation when the iterative solution for the wave function or the scattering amplitude is truncated in practice.

Since, the constant factor outside the integral in Eq. (32.55) already has a factor of λ, it is trivial to write down the scattering amplitude to first order in λ. All we have to do is to approximate $\psi(\mathbf{r}\,')$ in the integrand by the incident wave $e^{i\mathbf{k}\cdot\mathbf{r}\,'}$ itself. This is called the (first) **Born approximation** or the Born formula. Let us denote the scattering amplitude in this approximation by $f_{\mathrm{B}}(k,\theta)$. Dropping the prime on the variable of integration $\mathbf{r}\,'$ for simplicity of notation, we have

$$\boxed{f_{\mathrm{B}}(k,\theta) = -\frac{\lambda}{4\pi}\int d^3r\, e^{-i\,\mathbf{Q}\cdot\mathbf{r}}\,U(r), \quad \text{where} \quad \mathbf{Q}=\mathbf{k}'-\mathbf{k}.} \quad (32.59)$$

The Born formula is a good approximation when the incident speed is sufficiently large: specifically, when

$$\hbar^2 k/m \gg r\,|\lambda V(r)| \quad (32.60)$$

for all r. It remains valid even when this condition is not satisfied, provided the potential drops off more rapidly than $1/r^2$ as $r\to\infty$: more precisely, when

$$|\lambda V(r)| \ll \hbar^2/(mr^2). \quad (32.61)$$

The formula (32.59) has an easily-remembered physical interpretation:

- The scattering amplitude in the Born approximation is, up to a constant factor, the (three-dimensional) Fourier transform of the potential, with the momentum transfer wave vector $\mathbf{Q}$ playing the role of the variable conjugate to $\mathbf{r}$.

Now, $e^{i\mathbf{k}\cdot\mathbf{r}} = e^{i\mathbf{p}\cdot\mathbf{r}/\hbar} = \langle \mathbf{r} \,|\, \mathbf{p} \rangle$ is the position space wave function corresponding to the initial momentum eigenstate $|\mathbf{p}\rangle$ of the particle. Similarly, $e^{-i\mathbf{k}'\cdot\mathbf{r}} = e^{-i\mathbf{p}'\cdot\mathbf{r}\hbar} = \langle \mathbf{r} \,|\, \mathbf{p}' \rangle^* = \langle \mathbf{p}' \,|\, \mathbf{r} \rangle$. Using the completeness relation $\int d^3r \, |\mathbf{r}\rangle \, \langle \mathbf{r}| = I$, we see that

$$\boxed{f_{\mathrm{B}}(k, \theta) = -\frac{\lambda}{4\pi} \langle \mathbf{p}' | U | \mathbf{p} \rangle = -\frac{m\lambda}{2\pi\hbar^2} \langle \mathbf{p}' | V | \mathbf{p} \rangle.} \tag{32.62}$$

In other words:

- The scattering amplitude in the Born approximation is essentially the matrix element of the potential energy operator between the initial and final free-particle momentum eigenstates.

The angular integration in Eq. (32.59) for $f_{\mathrm{B}}(k, \theta)$ is easily carried out. The outcome is

$$\boxed{f_{\mathrm{B}}(k, \theta) = -\frac{2m\lambda}{\hbar^2 \, Q} \int_0^\infty dr \, r \, \sin(Qr) \, V(r).} \tag{32.63}$$

★ **8.** Derive Eq. (32.63) from Eq. (32.59).

★ **9.** Consider the finite potential barrier given by

$$V(r) = \begin{cases} V_0 & \text{for } r \le a \\ 0 & \text{for } r > a. \end{cases}$$

(a) Show that $f_{\mathrm{B}}(k, \theta) = 2m\lambda V_0 \, (\sin \, Qa - Qa \, \cos \, Qa)/(\hbar^2 Q^3)$.
(b) Hence show that the forward scattering amplitude in the Born approximation is $f_{\mathrm{B}}(k, 0) = (2m\lambda V_0 \, a^3)/(3\hbar^2)$.

32.2.6 Yukawa and Coulomb Potentials; Rutherford's Formula

According to quantum field theory, the forces between elementary particles arise from the exchange of other particles which are the quanta of **gauge fields**. In the nonrelativistic limit of scattering from a static potential, such a force generically reduces to a form called the **Yukawa potential**, given by

$$V(r) = \frac{e^{-r/\xi}}{r}. \tag{32.64}$$

Here $\xi = h/(\mu c)$ is the Compton wavelength of the exchanged particle, μ being its rest mass. In physical terms, ξ represents the "range" of the potential. The functional form of the Yukawa potential arises in other contexts as well. For instance, the **screened Coulomb potential** in a dielectric medium has the Yukawa form.

In the Born approximation, the scattering amplitude for the Yukawa potential is

$$f_{\rm B}(k,\theta) = -\frac{2m\lambda}{\hbar^2(Q^2+\xi^{-2})} = -\frac{2m\lambda}{\hbar^2[2k^2(1-\cos\theta)+\xi^{-2}]}. \tag{32.65}$$

Correspondingly, the total cross-section in the Born approximation is

$$\sigma_{\rm B}(E) = \frac{4\pi^2 m\lambda^2\xi^2}{E\hbar^2}\left(\frac{4mE\xi^2+\hbar^2}{8mE\xi^2+\hbar^2}\right). \tag{32.66}$$

★ **10.** The results above are established easily.

(a) Insert the Yukawa potential (32.64) in Eq. (32.63) to obtain Eq. (32.65).
(b) Now evaluate the angular integral $\int d\Omega\,|f_{\rm B}(k,\theta)|^2$ to obtain Eq. (32.66).

For very large values of the incident energy E, the total cross-section falls off like $1/E$ (provided ξ is finite). This is a general feature of scattering from a wide class of potentials. Remember, however, that the formalism developed above is only applicable to particles moving at nonrelativistic velocities.

The Coulomb potential: The Yukawa potential reduces to the Coulomb potential in the limit $\xi \to \infty$ or $\mu \to 0$, i.e., when the mass of the exchanged particle is zero. (Electromagnetic interactions are mediated by the exchange of photons, which are massless quanta.) From Eq. (32.65), it follows that the scattering amplitude in the Born approximation is now $f_{\rm B}(k,\theta) = -(2m\lambda)/(\hbar^2\,Q^2)$. The differential cross-section is, therefore,

$$\boxed{\left[\frac{d\sigma}{d\Omega}\right]_{\rm B} = \frac{\lambda^2}{16E^2}\,\text{cosec}^4\left(\tfrac{1}{2}\theta\right).} \tag{32.67}$$

This is precisely the famous **Rutherford scattering formula**. There are several remarkable features about this result:

- Equation (32.67) is also the *exact* expression for the differential cross-section for scattering in the Coulomb potential λ/r, because all the higher order corrections to the first Born approximation happen to vanish in this instance.
- Moreover, the expression is actually independent of Planck's constant, and is exactly the same as the differential scattering cross-section of *classical* particles in a Coulomb potential.
- The differential cross-section for *forward* scattering ($\theta = 0$) diverges, i.e., becomes unbounded. So does the *total* cross-section, as the singularity of $d\sigma/d\Omega$ at $\theta = 0$ in Eq. (32.67) is not integrable.

All these "minor miracles" and drawbacks are related to the long-range nature of the Coulomb potential. It turns out that the assumption of an initial *plane* wave state ($\psi_{\rm inc} = e^{i\mathbf{k}\cdot\mathbf{r}}$) is incompatible with a potential that decays as slowly as $1/r$. The scattering problem has to be re-done in this case. The Schrödinger equation in the presence of a Coulomb potential can be solved exactly, using parabolic cylindrical coordinates. The solution involves confluent hypergeometric functions. The initial state is found to be a plane wave that is amplitude-modulated by a factor of $1/r$, and phase-modulated by a term involving $\ln r$. I will not go any further into these issues here. The details can be found in any good textbook dealing with nonrelativistic scattering theory.

32.3 Partial Wave Analysis

I digress a little further into nonrelativistic scattering theory, to consider partial wave analysis in brief. This is an important physical application of the expansion of a function in terms of a family of orthogonal polynomials—in this case, Legendre polynomials. The basic features of such an expansion have already been dealt with in Chap. 16, Sect. 16.4.3.

32.3.1 The Physical Idea Behind Partial Wave Analysis

Consider, for a moment, the elastic scattering of a *classical* particle in a central (or spherically symmetric) potential. The particle is incident with an initial momentum $\mathbf{p}$, and is scattered to some final momentum $\mathbf{p}'$, such that $|\mathbf{p}| = |\mathbf{p}'|$. As you know, during this process the orbital angular momentum of the particle about the origin, $\mathbf{L} = \mathbf{r} \times \mathbf{p}$, remains constant in time. This helps reduce the dynamical problem to the solution of an ordinary second-order differential equation for the particle's radial coordinate r, for each given value of $|\mathbf{L}|$. In this sense, the motion for any value of the angular momentum is automatically decoupled from that for any other value.

In quantum mechanics, too, the spherical symmetry of the particle's Hamiltonian implies the conservation of angular momentum. We should, therefore, be able to analyze the scattering for each allowed value of the orbital angular momentum of the particle, independent of the process for the other values. But before this separation can be done, it is necessary to deal with an important fact. The initial state of the incident particle is an eigenstate of its *linear* momentum—and this observable does not commute with the *angular* momentum of the particle. The initial momentum eigenstate $|\mathbf{p}\rangle$, represented by the plane wave $e^{i\mathbf{k}\cdot\mathbf{r}}$, is *not* an eigenstate of the orbital angular momentum. Rather, it is an infinite *superposition* of angular momentum eigenstates. The latter are labeled by the quantum number ℓ, since the orbital angular momentum is quantized. We must, therefore, write down this superposition, and then analyze what happens to each angular momentum component under the action of the potential. This is called partial wave analysis. The special feature of scattering by a central potential is the following:

- The scattering process does not "mix up" different angular momentum components, precisely because the potential is spherically symmetric.

This is the idea underlying partial wave analysis, and the reason why it is useful. Recall from Chap. 16, Sect. 16.4.3, that a general function of θ, where $0 \le \theta \le \pi$, can be expanded in a series of Legendre polynomials. When the scattering amplitude $f(k, \theta)$ is expanded in this manner, all the information about the scattering is carried by the set of coefficients $\{f_\ell(k)\}$ in the expansion, as you will see in what follows:

32.3.2 Expansion of a Plane Wave in Spherical Harmonics

The first step is to expand the incident wave function $\psi_{\text{inc}}(\mathbf{r}) = e^{i\mathbf{k}\cdot\mathbf{r}} = e^{ikr\cos\theta}$ in terms of Legendre polynomials $P_\ell(\cos\theta)$. This is a standard formula that is written down in texts on scattering theory. But let us derive it, as we now have the machinery to do so. We have

$$\begin{aligned} e^{ikr\cos\theta} &= \sum_{n=0}^{\infty} \frac{(ikr\cos\theta)^n}{n!} \\ &= \sum_{n=0}^{\infty} \frac{(ikr)^n}{n!} \sum_{\substack{\ell=0 \\ (n-\ell)\text{ even}}}^{n} (2\ell+1) \frac{2^\ell\, n!\,\left(\frac{1}{2}(n+\ell)\right)!}{(n+\ell+1)!\,\left(\frac{1}{2}(n-\ell)\right)!} P_\ell(\cos\theta), \end{aligned} \tag{32.68}$$

where I have used the expansion of $(\cos\theta)^n$ in Legendre polynomials given by Eqs. (16.111) and (16.115) of Chap. 16, Sect. 16.4.4. After some algebra (see below), this expression can be brought to the form

$$e^{ikr\cos\theta} = \sum_{\ell=0}^{\infty} (2\ell+1)\, i^\ell\, P_\ell(\cos\theta) \left\{ 2^\ell \sum_{s=0}^{\infty} \frac{(-1)^s\,(\ell+s)!}{(2\ell+2s+1)!\,s!} (kr)^{2s+\ell} \right\}. \tag{32.69}$$

But the expression inside the curly brackets is precisely the spherical Bessel function $j_\ell(kr)$, as you can see by comparing it with Eq. (28.54) of Chap. 28, Sect. 28.3.1. We therefore get, finally,

$$\boxed{e^{ikr\cos\theta} = \sum_{\ell=0}^{\infty} (2\ell+1)\, i^\ell\, j_\ell(kr)\, P_\ell(\cos\theta).} \tag{32.70}$$

This is an important formula.

★ **11.** Work this derivation out explicitly.

★ **12.** Let us take this opportunity to write down the values of a couple of definite integrals that follow from the expansion (32.70). Let ξ be a real variable, and n a non-negative integer. Show that

$$\int_0^1 dx\, P_{2n}(x)\, \cos(\xi x) = (-1)^n\, j_{2n}(\xi)$$

and

$$\int_0^1 dx\, P_{2n+1}(x)\, \sin(\xi x) = (-1)^n\, j_{2n+1}(\xi).$$

To return to the matter at hand: What we need here is the asymptotic (large r) behavior of the expression in Eq. (32.70). Recall the leading asymptotic behavior of $j_\ell(kr)$ from Eq. 28.55 of Chap. 28, Sect. 28.3.1. Using this result, we get

$$\begin{aligned}\psi_{\text{inc}}(\mathbf{r}) = e^{ikr\cos\theta} &\xrightarrow[kr\gg 1]{} \frac{1}{kr}\sum_{\ell=0}^{\infty}(2\ell+1)\, i^\ell\, \sin\left(kr - \tfrac{1}{2}\ell\pi\right) P_\ell(\cos\theta) \\ &= \frac{1}{2ikr}\sum_{\ell=0}^{\infty}(2\ell+1)\left\{e^{ikr} - (-1)^\ell\, e^{-ikr}\right\} P_\ell(\cos\theta).\end{aligned} \tag{32.71}$$

The incident plane wave is, therefore, a superposition of a set of incoming spherical waves (proportional to e^{-ikr}/r) and a set of outgoing spherical waves (proportional to e^{ikr}/r). What the scattering by the potential does is to affect the *phase* of each member of the set of outgoing waves by adding a certain **phase shift** δ_ℓ to it, as we shall see.

32.3.3 Partial Wave Scattering Amplitude and Phase Shift

Next, consider the scattered wave function, whose leading asymptotic behavior is given by $\psi_{\text{sc}}(\mathbf{r}) \sim f(k,\theta)\, e^{ikr}/r$. The obvious step is to expand the scattering amplitude $f(k,\theta)$ in terms of Legendre polynomials. Let this expansion have the form

$$\boxed{f(k,\theta) = \frac{i}{2k}\sum_{\ell=0}^{\infty}(2\ell+1)\, f_\ell(k)\, P_\ell(\cos\theta).} \tag{32.72}$$

The factor $i/(2k)$ has been separated out merely for convenience: $[1 - f_\ell(k)]$ then turns out to be a pure phase factor, as you will see in Eq. (32.81). The coefficient $f_\ell(k)$ is called the **partial wave scattering amplitude**. Inverting the expansion (32.72) we have, formally,

$$f_\ell(k) = -ik\int_{-1}^{1} d(\cos\theta)\, f(k,\theta)\, P_\ell(\cos\theta). \tag{32.73}$$

★ **13.** Show that, for the Yukawa potential $V(r) = e^{-(r/\xi)}/r$ (Eq. (32.64)), the partial wave scattering amplitude in the Born approximation is given by

$$\left[f_\ell(k)\right]_{\text{B}} = \frac{2im\lambda}{\hbar^2 k}\, Q_\ell\left(1 + \frac{1}{2k^2\xi^2}\right),$$

where Q_ℓ is the Legendre function of the second kind.

Using Eqs. (32.71) and (32.72) in Eq. (32.39), the asymptotic expression for the full wave function $\psi = \psi_{\text{inc}} + \psi_{\text{sc}}$ becomes

$$\begin{aligned}\psi(\mathbf{r}) &\xrightarrow[kr\gg 1]{} e^{i\mathbf{k}\cdot\mathbf{r}} + f(k,\theta)\,\frac{e^{ikr}}{r} \\ &\sim \frac{1}{2ikr}\sum_{\ell=0}^{\infty}(2\ell+1)\left\{[1 - f_\ell(k)]\,e^{ikr} - (-1)^\ell\, e^{-ikr}\right\} P_\ell(\cos\theta).\end{aligned} \tag{32.74}$$

Let us set this aside for a moment, and go back to the Schrödinger equation (32.42) satisfied by the wave function $\psi(\mathbf{r})$, namely,

$$[\nabla^2 + k^2 - \lambda\, U(r)]\,\psi(\mathbf{r}) = 0. \tag{32.75}$$

As you know, this equation becomes separable in spherical polar coordinates because $U(r)$ is a central potential. The angular dependence of the wave function can be expressed as a sum over Legendre polynomials. In order to match the expansion of $\psi_{\text{inc}}(\mathbf{r})$ in the first line of Eq. (32.71), we write this expansion as

$$\psi(\mathbf{r}) = \frac{1}{kr}\sum_{\ell=0}^{\infty}(2\ell+1)\,i^\ell\, u_\ell(r)\, P_\ell(\cos\theta). \tag{32.76}$$

The radial wave function has been written in the form $u_\ell(r)/r$. This is a standard trick: it ensures that there is no first derivative term, i.e., a term proportional to du_ℓ/dr, in the differential equation satisfied by $u_\ell(r)$. This is the well-known radial Schrödinger equation in a central potential, namely,

$$\frac{d^2u_\ell}{dr^2} + \left\{k^2 - \lambda U(r) - \frac{\ell(\ell+1)}{r^2}\right\} u_\ell = 0. \tag{32.77}$$

Incidentally, this trick (to eliminate the first derivative term) only works in three dimensions. The term $\ell(\ell+1)/r^2$ in Eq. (32.77) is, of course, the well-known **centrifugal barrier**. It acts like an additional repulsive potential for non-zero values of ℓ.

★ **14.** Derive Eq. (32.77) from Eq. (32.75).

We are interested in the *regular* solution to Eq. (32.77), i.e., the solution that satisfies $u_\ell(0) = 0$. (Recall that the radial wave function is actually $u_\ell(r)$ *divided* by r.) At the other end, as $r \to \infty$, both the potential $U(r)$ and the term $\ell(\ell+1)/r^2$ tend to zero. Hence, the leading asymptotic behavior of the solution must be that of the solutions of $u_\ell'' + k^2 u_\ell = 0$, namely, a linear combination of e^{ikr} and e^{-ikr}. This linear combination can be written in the general form

$$u_\ell(r) \sim a_\ell \sin\left(kr - \tfrac{1}{2}\ell\pi + \delta_\ell\right), \tag{32.78}$$

where both the coefficient $a_\ell(k)$ and the phase shift $\delta_\ell(k)$ are, in general, functions of k. Substituting for $u_\ell(r)$ from (32.78) in Eq. (32.76), the asymptotic behavior of the wave function is given by

$$\psi(\mathbf{r}) \sim \frac{1}{2ikr}\sum_{\ell=0}^{\infty}(2\ell+1)\left\{a_\ell\, e^{ikr+i\delta_\ell} - (-1)^\ell\, a_\ell\, e^{-ikr-i\delta_\ell}\right\} P_\ell(\cos\theta). \tag{32.79}$$

Comparing the respective coefficients of e^{ikr} and e^{-ikr} in Eqs. (32.74) and (32.79), we have

$$a_\ell\, e^{i\delta_\ell} = 1 - f_\ell \quad \text{and} \quad a_\ell\, e^{-i\delta_\ell} = 1. \tag{32.80}$$

Eliminating a_ℓ, we get

$$\boxed{f_\ell(k) = 1 - e^{2i\delta_\ell(k)}.} \tag{32.81}$$

This important relation connects the partial wave scattering amplitude and the corresponding phase shift. It can be shown that δ_ℓ is real, but I will not do so here. Inserting Eq. (32.81) in Eq. (32.72), we obtain the relation between the scattering amplitude and the set of phase shifts:

$$\boxed{f(k,\theta) = \frac{i}{2k}\sum_{\ell=0}^{\infty}(2\ell+1)\left(1 - e^{2i\delta_\ell(k)}\right) P_\ell(\cos\theta).} \tag{32.82}$$

The physical significance of the phase shift δ_ℓ is as follows. Consider the radial wave equation (32.77) in the *absence* of the potential $U(r)$. The regular solution of the resulting equation is then proportional to $r\, j_\ell(r)$. This function has an asymptotic behavior $\sim \sin\left(kr - \frac{1}{2}\ell\pi\right)$. Comparing this expression with that in Eq. (32.78), we see that δ_ℓ is precisely the extra phase shift arising from the influence of the potential.

32.3.4 The Optical Theorem

There exists an important consistency condition on the scattering amplitude, called the **Optical Theorem**. This relation connects the imaginary part of the *forward* scattering amplitude $f(k, 0)$ to the total cross-section, according to

$$\boxed{\mathrm{Im}\, f(k,0) = \frac{k}{4\pi}\,\sigma_{\text{tot}}\,.} \tag{32.83}$$

The expansion of the scattering amplitude in terms of Legendre polynomials makes the derivation of this relation quite trivial.

★ **15.** Use the formula (32.82) for the scattering amplitude to derive Eq. (32.83).

It is evident from Eq. (32.82) that all the information about the scattering process is contained in the set of phase shifts $\{\delta_\ell(k) \mid \ell = 0, 1, \ldots\}$. The focus now shifts to the determination of these quantities. But I end this digression into scattering theory here, to return to the topic of integral equations.

32.4 The Fredholm Solution

32.4.1 The Fredholm Formulas

Let us go back to the inhomogeneous Fredholm equation of the second kind, Eq. (32.25). Repeating it for ready reference,

$$f(x) = g(x) + \lambda \int_a^b dy\, K(x, y)\, f(y). \tag{32.84}$$

Recall, from Eqs. (32.33) and (32.34) of Sect. 32.1.5, that this equation has the Neumann series solution

$$f(x) = g(x) + \lambda \int_a^b dy\, H(x, y\,;\, \lambda)\, g(y), \tag{32.85}$$

where the resolvent kernel $H(x, y\,;\, \lambda) = \sum_1^\infty \lambda^{n-1}\, K_n(x, y)$. The power series in λ converges absolutely inside a circle of radius Λ centered at the origin in the complex λ-plane. The question that arises naturally is: does there exist an *analytic continuation* of the solution to the region *outside* this circle?

Fredholm derived such a solution (which I shall merely quote here). The case of greatest interest in physical applications corresponds to **square-integrable kernels**,, i.e., kernels satisfying the condition

$$\| \mathsf{K} \|^2 = \int_a^b dx \int_a^b dy\, |K(x, y)|^2 < \infty. \tag{32.86}$$

The main results are as follows:

(i) The solution for $f(x)$ has the general form given by Eq. (32.85).

(ii) The resolvent kernel $H(x, y\,;\, \lambda)$ is actually *the ratio of two entire functions* of λ, and is of the form

$$\boxed{H(x, y\,;\, \lambda) = \frac{D(x, y\,;\, \lambda)}{D(\lambda)}}\,. \tag{32.87}$$

$H(x, y; \lambda)$ is, therefore, a *meromorphic* function of λ, with no singularities other than poles in the finite part of the complex λ-plane.

$D(x, y; \lambda)$ and $D(\lambda)$ are called the Fredholm numerator and **Fredholm denominator**, respectively. They are defined as follows: In the functions below, all the variables are assumed to lie in the range $[a, b]$. Define, first, the determinant of kernels

$$K\begin{pmatrix} y_1 & y_2 & \cdots & y_n \\ z_1 & z_2 & \cdots & z_n \end{pmatrix} = \begin{vmatrix} K(y_1, z_1) & K(y_1, z_2) & \cdots & K(y_1, z_n) \\ K(y_2, z_1) & K(y_2, z_2) & \cdots & K(y_2, z_n) \\ \cdots & \cdots & \cdots & \cdots \\ K(y_n, z_1) & K(y_n, z_2) & \cdots & K(y_n, z_n) \end{vmatrix}. \quad (32.88)$$

For $n = 1$, $K\begin{pmatrix} y \\ z \end{pmatrix} \equiv K(y, z)$. Then the Fredholm numerator is given by

$$D(x, y; \lambda) = K(x, y) + \sum_{n=1}^{\infty} \frac{(-\lambda)^n}{n!} \int_a^b dy_1 \int_a^b dy_2 \cdots \cdots \int_a^b dy_n\, K\begin{pmatrix} x & y_1 & \cdots & y_n \\ y & y_1 & \cdots & y_n \end{pmatrix}. \quad (32.89)$$

The Fredholm denominator is given by

$$D(\lambda) = 1 + \sum_{n=1}^{\infty} \frac{(-\lambda)^n}{n!} \int_a^b dy_1 \int_a^b dy_2 \cdots \int_a^b dy_n\, K\begin{pmatrix} y_1 & y_2 & \cdots & y_n \\ y_1 & y_2 & \cdots & y_n \end{pmatrix}. \quad (32.90)$$

The remarkable fact is that the infinite series in both (32.89) and (32.90) converge absolutely for all finite values of λ. Hence they are entire functions of λ, as stated above.

(iii) The zeroes of $D(\lambda)$ are *poles* of the resolvent kernel. They are located at the reciprocals of the eigenvalues of the kernel.

(iv) If $K(x, y)$ a degenerate kernel of rank r, $D(\lambda)$ is just a polynomial of degree r in λ.

(v) If $K(x, y)$ a nondegenerate kernel, $D(\lambda)$ has an infinite number of zeroes, with an accumulation point at $\lambda = \infty$.

32.4.2 Remark on the Application to the Scattering Problem

The formalism described above is applicable to Fredholm integral equations in a single variable. It is, however, extended to functions of several independent variables. In particular, we are concerned here with the integral equation for the wave function of a nonrelativistic particle undergoing scattering in a central potential. Repeating

Eqs. (32.45) and (32.46) for ready reference, we have

$$\psi(\mathbf{r}) = e^{i\mathbf{k}\cdot\mathbf{r}} + \lambda \int d^3r'\, K(\mathbf{r}, \mathbf{r}')\, \psi(\mathbf{r}'), \tag{32.91}$$

where the kernel is given by

$$K(\mathbf{r}, \mathbf{r}') = -\frac{e^{ik|\mathbf{r}-\mathbf{r}'|}}{4\pi|\mathbf{r}-\mathbf{r}'|}\, U(r'). \tag{32.92}$$

I stated in Sect. 32.2.2 that this was a Fredholm equation, implying that the Fredholm solution and its associated properties were applicable to it. There are, however, some technical difficulties to be overcome before this can be done.

It has already been mentioned that the square-integrability of the kernel is a sufficient condition for the theory to be applicable. This requires a small modification in the integral equation above. The kernel $K(\mathbf{r}, \mathbf{r}')$ is not a *symmetric* kernel as it stands, i.e., it is not a symmetric function of $\mathbf{r}$ and $\mathbf{r}'$, owing to the presence of the factor $U(r')$. This flaw is easily remedied for a potential $U(r)$ that remains either positive or negative throughout the range of r. Consider the integral equation for the function $\phi(\mathbf{r}) = \psi(\mathbf{r})/\sqrt{|U(r)|}$, rather than that for $\psi(\mathbf{r})$ itself. We have

$$\phi(\mathbf{r}) = \sqrt{|U(r)|}\, e^{i\mathbf{k}\cdot\mathbf{r}} + \lambda \int d^3r'\, K_{\mathrm{s}}(\mathbf{r}, \mathbf{r}')\, \phi(\mathbf{r}'), \tag{32.93}$$

with the symmetric kernel

$$K_{\mathrm{s}}(\mathbf{r}, \mathbf{r}') = \mp \frac{e^{ik|\mathbf{r}-\mathbf{r}'|}}{4\pi|\mathbf{r}-\mathbf{r}'|}\, \sqrt{U(r)\, U(r')}. \tag{32.94}$$

The $-$ and $+$ signs correspond, respectively, to the cases $U(r) > 0$ and $U(r) < 0$. For the class of potentials for which $K_{\mathrm{s}}(\mathbf{r}, \mathbf{r}')$ is square-integrable, i.e., when

$$\int d^3r \int d^3r'\, \frac{U(r)\, U(r')}{|\mathbf{r}-\mathbf{r}'|^2} < \infty, \tag{32.95}$$

it is guaranteed that the Fredholm solution of the integral equation (32.93) exists, and that the results of the associated theory are applicable. Note that the kernel $K_{\mathrm{s}}(\mathbf{r}, \mathbf{r}')$ is always *non*degenerate.

The second problem that arises is the following. Consider the determinant of kernels that occurs in the formula (32.90) for the Fredholm denominator. From the definition in Eq. (32.88) for this determinant, it is evident that the "diagonal matrix element" $K(y_i\,,\, y_i)$ appears in it. On the other hand, it is obvious from Eq. (32.94) that $K_{\mathrm{s}}(\mathbf{r}, \mathbf{r})$ is infinite. Therefore, the Fredholm formulas as they stand cannot be applied to the integral equation (32.93). The way out is an extension of the Fredholm theory due to Poincaré and Hilbert. In essence, this modification says that the divergent quantity $K_s(\mathbf{r}, \mathbf{r})$ can simply be replaced by 0 in all the Fredholm formulas without affecting the applicability of the formalism! The Fredholm denominator corresponding to the integral equation (32.93) is, therefore, given by

$$D(\lambda) = 1 + \sum_{n=1}^{\infty} \frac{(-\lambda)^n}{n!} \int d^3r_1 \int d^3r_2 \cdots \int d^3r_n \, K\begin{pmatrix} \mathbf{r}_1 \; \mathbf{r}_2 \ldots \mathbf{r}_n \\ \mathbf{r}_1 \; \mathbf{r}_2 \ldots \mathbf{r}_n \end{pmatrix}, \tag{32.96}$$

where

$$K\begin{pmatrix} \mathbf{r}_1 \; \mathbf{r}_2 \cdots \mathbf{r}_n \\ \mathbf{r}_1 \; \mathbf{r}_2 \cdots \mathbf{r}_n \end{pmatrix} = \begin{vmatrix} 0 & K_s(\mathbf{r}_1, \mathbf{r}_2) & \cdots & K_s(\mathbf{r}_1, \mathbf{r}_n) \\ K_s(\mathbf{r}_1, \mathbf{r}_2) & 0 & \cdots & K_s(\mathbf{r}_2, \mathbf{r}_n) \\ \cdots & \cdots & \cdots & \cdots \\ K_s(\mathbf{r}_1, \mathbf{r}_n) & K_s(\mathbf{r}_2, \mathbf{r}_n) & \cdots & 0 \end{vmatrix}. \tag{32.97}$$

Here $K_s(\mathbf{r}_i, \mathbf{r}_j)$ is the symmetric, square-integrable kernel defined in Eq. (32.94). A similar modification occurs, of course, in the formula for the Fredholm numerator.

32.5 Volterra Integral Equations

While Fredholm integral equations involve integrals between constant limits (such as a and b), Volterra integral equations involve a variable limit of integration. In what follows, all the functions concerned are assumed to be defined in some interval $[0, b]$, where b is a positive number. (The lower limit of the range has been chosen to be zero, for convenience.) The Volterra integral equation of the second kind is given by

$$\boxed{f(x) - \lambda \int_0^x dy \, K(x, y) \, f(y) = g(x),} \tag{32.98}$$

where $g(x)$ is a given function of x. This is the standard form of an integral equation of this type. Note that the upper limit of integration in the second term is x itself. We can keep track of the variable upper limit of integration by taking the kernel $K(x, y)$ to be identically equal to zero whenever its second argument exceeds the first, i.e., by *defining*

$$K(x, y) \equiv 0 \quad \text{for} \quad y > x. \tag{32.99}$$

As in the case of the inhomogeneous Fredholm equation, we can write down an iterative solution to this equation:

$$\begin{aligned} f(x) &= g(x) + \lambda \int_0^x dy_1 \, K(x, y_1) \, g(y_1) + \lambda^2 \int_0^x dy_1 \int_0^{y_1} dy_2 \, K(x, y_1) K(y_1, y_2) \, g(y_2) \\ &\quad + \lambda^3 \int_0^x dy_1 \int_0^{y_1} dy_2 \int_0^{y_2} dy_3 \, K(x, y_1) K(y_1, y_2) K(y_2, y_3) \, g(y_3) + \cdots . \end{aligned} \tag{32.100}$$

Note the manner in which the range of integration "shrinks" as you go from y_1 to y_2 to y_3 to .... Once again, this solution can be expressed in the form

$$f(x) = g(x) + \lambda \int_0^x dy\, H(x, y\,;\, \lambda)\, g(y), \tag{32.101}$$

where $H(x, y\,;\, \lambda)$ is the resolvent kernel. It is obviously of the form

$$H(x, y\,;\, \lambda) = K(x, y) + \lambda K_2(x, y) + \lambda^2 K_3(x, y) + \cdots = \sum_{n=1}^{\infty} \lambda^{n-1} K_n(x, y). \tag{32.102}$$

Here $K_1(x, y) \equiv K(x, y)$, as usual. The question is: what is the n-fold iterated kernel $K_n(x, y)$ for $n \geq 2$, keeping in mind the fact that the kernel $K(x, y)$ vanishes (by definition) when $y > x$?

Observe that it is $g(y_n)$ that occurs in the integrand of the general term in the iterated solution, whereas we want to write this as $g(y)$ and integrate y from 0 to x. We must, therefore, *invert* the order of integration in each term of the series. This is easy, because all the variables concerned run over nonnegative values. Recall the simple formula given in Eq. (3.24) of Chap. 3, Sect. 3.2 for interchanging the order of integration in a double integral. Using it, we have

$$\int_0^x dy_1 \int_0^{y_1} dy_2\, (\cdots) = \int_0^x dy_2 \int_{y_2}^x dy_1\, (\cdots)\,. \tag{32.103}$$

Applying the formula repeatedly, it is easy to verify that

$$\int_0^x dy_1 \int_0^{y_1} dy_2 \cdot\cdot \int_0^{y_{n-2}} dy_{n-1} \int_0^{y_{n-1}} dy_n\, (\cdot\cdot) = \int_0^x dy_n \int_{y_n}^x dy_{n-1} \cdot\cdot \int_{y_3}^x dy_2 \int_{y_2}^x dy_1\, (\cdot\cdot). \tag{32.104}$$

On the right-hand side of this identity, all the upper limits of integration have become x, while the lower limits have become variable. The n-fold iterated kernel is, therefore, given by

$$K_n(x, y) = \int_y^x dy_{n-1} \int_{y_{n-1}}^x dy_{n-2} \cdot\cdot \int_{y_2}^x dy_1\, K(x, y_1)\, K(y_1, y_2) \cdot\cdot K(y_{n-1}, y). \tag{32.105}$$

In this form it is obvious that, like $K(x, y)$ itself, the iterates $K_n(x, y)$ also vanish for $y > x$.

Alternatively, we can bring the integration over y_n alone all the way to the left, by successively interchanging the order of integration between y_{n-1} and y_n, followed by that between y_{n-2} and y_n,, and so on. The result is the identity

$$\int_0^x dy_1 \int_0^{y_1} dy_2 \cdot\cdot \int_0^{y_{n-2}} dy_{n-1} \int_0^{y_{n-1}} dy_n\, (\cdot\cdot) = \int_0^x dy_n \int_{y_n}^x dy_1 \int_{y_n}^{y_1} dy_2 \cdot\cdot \int_{y_n}^{y_{n-2}} dy_{n-1}\, (\cdot\cdot). \tag{32.106}$$

The n-fold iterated kernel is now identified to be

$$K_n(x, y) = \int_y^x dy_1 \int_y^{y_1} dy_2 \cdots \int_y^{y_{n-2}} dy_{n-1}\, K(x, y_1)\, K(y_1, y_2) \cdots K(y_{n-1}, y). \tag{32.107}$$

This is entirely equivalent to the formula in Eq. (32.105).

Once again, we consider only square-integrable kernels, i.e., kernels such that

$$\int_0^b dx \int_0^b dy\, |K(x, y)|^2 < \infty. \tag{32.108}$$

The important difference between the Volterra and Fredholm cases can be summarized as follows:

(i) The resolvent kernel $H(x\,,\, y\,;\, \lambda)$ in the case of the Volterra equation is an *entire* function of λ. (In contrast, it is a meromorphic function of λ in the case of a Fredholm equation.)
(ii) Hence, the Neumann series solution of the Volterra integral equation is convergent for all finite λ, and is a unique solution in the space $L_2[0\,,\, b]$ for all $g(x) \in L_2[0\,,\, b]$.
(iii) As a consequence, the *homogeneous* Volterra equation

$$f(x) = \lambda \int_0^x dy\, K(x\,,\, y)\, f(y) \tag{32.109}$$

has no solutions other than the trivial solution $f(x) = 0$ in the space $L_2[0\,,\, b]$.

The proofs of these statements are not too difficult, but I shall not go into them here.

★ **16.** Consider the inhomogeneous Volterra equation

$$f(x) - \lambda \int_0^x dy\, K(x, y)\, f(y) = g(x)$$

where (apart from the step function) the kernel has the separable form

$$K(x, y) = \phi(x)\, \psi(y)\, \theta(x - y).$$

Here ϕ, ψ and g are given functions.

(a) Show that the iterated kernels are given by

$$K_{n+1}(x\,,\, y) = K(x\,,\, y)\, T_n(y\,,\, x) \quad \text{for} \quad n \geq 1,$$

where

$$T_n(y\,,\, x) = \int_y^x dz\, K(z\,,\, z)\, T_{n-1}(z\,,\, x) \quad \text{and} \quad T_0(y, x) \equiv 1.$$

(b) Show that $f(x)$ satisfies the inhomogeneous second-order differential equation

$$\phi'\,\frac{d^2 f}{dx^2} - (\lambda\phi\phi'\psi + \phi'')\,\frac{df}{dx} + \lambda(\phi\phi'' - 2\phi'^2\psi - \phi\phi'\psi')\,f = (g'\phi')',$$

where a prime denotes the derivative with respect to x. (Assume that all the derivatives involved exist.) What are the boundary conditions satisfied by $f(x)$?

★ **17.** Consider the Volterra integral equation

$$f(x) = 1 + \lambda \int_0^x dy\, x\, y\, f(y).$$

(a) Show that the solution is given by

$$f(x) = 1 + \sum_{n=1}^{\infty} \frac{(\lambda x^3)^n}{2\cdot 5\cdot 8\cdots(3n-1)}.$$

(b) Show that, for any given value of x, the series above converges for all finite values of λ.

32.6 Solutions

1. $(I - \lambda\mathsf{K})\,|h_i\rangle = 0 \Rightarrow \mathsf{K}\,|h_i\rangle = (1/\lambda)\,|h_i\rangle$. Nontrivial vectors $|h_i\rangle$ satisfying this eigenvalue equation can only exist for special values of λ, namely, whenever $1/\lambda$ is an *eigenvalue* of the operator K. It is precisely for these values of λ that the inverse operator $(I - \lambda\mathsf{K})^{-1}$ does *not* exist (see below). But it has been stated already that (32.8) is the general solution of Eq. (32.7) *provided the inverse operator* $(I - \lambda\mathsf{K})^{-1}$ *exists.* ▶

2. $K(x, y)$ is of rank 1 in cases (i)–(iii), and of rank 2 in cases (iv) and (v).
(i) $f(x) = 1 - (3\lambda x)/[2(\lambda - 3)]$, provided $\lambda \neq 3$.
(ii) $f(x) = 1 - [4\lambda \sin(\pi x)]/[\pi(\lambda - 2)]$, provided $\lambda \neq 2$.
(iii) $f(x) = 1 - [2\lambda(e-1)e^x]/[\lambda(e^2-1) - 2]$, provided $\lambda \neq 2/(e^2-1)$.
(iv) $f(x) = 1 - [\lambda(\lambda + 6 + 12x)]/[\lambda^2 + 12\lambda - 12]$, provided $\lambda \neq -6 \pm 4\sqrt{3}$.
(v) $f(x) = 1 - [4\lambda(\lambda \sin \pi x + 2\cos \pi x)]/[\pi(\lambda^2 - 4)]$, provided $\lambda \neq \pm 2$. ▶

3. Note that the rank of the kernel is (a) $r = 1$, and (b) $r = 2$.
(a) It is clear that $f(x) = \lambda C\, e^{-x^2}$, where the constant $C = \int_{-\infty}^{\infty} dy\, e^{-y^2} f(y)$. Multiply both sides of the equation by e^{-x^2} and integrate over x from $-\infty$ to ∞:

$$\int_{-\infty}^{\infty} dx\, e^{-x^2} f(x) = C = \lambda C \int_{-\infty}^{\infty} dx\, e^{-2x^2}.$$

Therefore, either $C = 0$, in which case $f(x) = 0$ identically, and the integral equation has only a trivial solution. Or else, $C \neq 0$, in which case the value of λ is fixed by the condition

$$\lambda \int_{-\infty}^{\infty} dx\, e^{-2x^2} = 1, \quad \text{or} \quad \lambda = (2/\pi)^{1/2}.$$

When λ has this value (and no other), the equation has a nontrivial solution $f(x) = a\, e^{-x^2}$, where a is an arbitrary constant.
(b) We now have $f(x) = \lambda\,(C_1\, x + C_2)\, e^{-x^2}$, where

$$C_1 = \int_{-\infty}^{\infty} dy\, e^{-y^2}\, f(y) \quad \text{and} \quad C_2 = \int_{-\infty}^{\infty} dy\, e^{-y^2}\, y\, f(y).$$

Multiply both sides of the expression for $f(x)$ by e^{-x^2} and $x\, e^{-x^2}$, in turn, and integrate over all x. This gives

$$C_1 = \lambda(\pi/2)^{1/2} C_2 \quad \text{and} \quad C_2 = (\lambda/4)(\pi/2)^{1/2} C_1 \quad \Longrightarrow \quad \lambda = \pm(8/\pi)^{1/2}.$$

The corresponding solutions are

$$f(x) = \begin{cases} a_1\,(2x+1)\, e^{-x^2} & \text{for} \quad \lambda = (8/\pi)^{1/2} \\ a_2\,(2x-1)\, e^{-x^2} & \text{for} \quad \lambda = -(8/\pi)^{1/2}, \end{cases}$$

where a_1 and a_2 are arbitrary constants. $f(x) \equiv 0$ for all other values of λ. ▶

4. $g(x) = 1$ in this instance. The kernel $K(x, y) = (2/\pi)^{1/2}\, e^{-x^2}\, e^{-y^2}$ is degenerate and of rank 1. Moreover, it has the remarkable property of "reproducing" itself upon iteration: we find $K_2(x\,,\, y) = K(x, y)$, which implies that $K_n(x, y) = K(x, y)$ for all positive integer values of n. Therefore, the series (32.32) becomes, since $g(x) = 1$ in this case,

$$f(x) = 1 + \sum_{n=1}^{\infty} \lambda^n \int_{-\infty}^{\infty} dy\, K(x, y) = 1 + \sqrt{2}\, e^{-x^2} \sum_{n=1}^{\infty} \lambda^n.$$

It is obvious that the Neumann series converges in the region $|\lambda| < 1$ in this instance. The series is trivially summed to obtain the closed-form solution

$$f(x) = 1 + \frac{\lambda\sqrt{2}}{(1-\lambda)}\, e^{-x^2}.$$

Remark This expression has a pole in the λ-plane at $\lambda = 1$, as expected: when $\lambda = 1$, the resolvent $(I - \lambda \mathsf{K})^{-1}$ does not exist. For this value of λ, and for this value alone, the *homogeneous* integral equation

$$f(x) = \sqrt{(2/\pi)} \int_{-\infty}^{\infty} dy \, e^{-(x^2+y^2)} \, f(y)$$

has a nontrivial solution, given by $f(x) = (\text{const.})\, e^{-x^2}$. ▶

5. The incident current density is

$$\mathbf{j}_{\text{inc}} = \frac{\hbar}{2mi} \left(\psi_{\text{inc}}^* \, \nabla \psi_{\text{inc}} - \psi_{\text{inc}} \, \nabla \psi_{\text{inc}}^* \right) = \frac{\hbar \mathbf{k}}{m}.$$

Its magnitude, $\hbar k/m$, is the incident flux. The scattered current density is given by

$$\mathbf{j}_{\text{sc}} = \frac{\hbar}{2mi} \left(\psi_{\text{sc}}^* \, \nabla \psi_{\text{sc}} - \psi_{\text{sc}} \, \nabla \psi_{\text{sc}}^* \right).$$

What is needed here is the radially outward scattered flux through a cone of solid angle $d\Omega$ located at any angular position (θ, φ). This is given by

$$\mathbf{j}_{\text{sc}} \cdot d\mathbf{S} = \mathbf{j}_{\text{sc}} \cdot \mathbf{e}_r \, r^2 d\Omega = \frac{\hbar}{2mi} \left(\psi_{\text{sc}}^* \, \frac{\partial \psi_{\text{sc}}}{\partial r} - \psi_{\text{sc}} \, \frac{\partial \psi_{\text{sc}}^*}{\partial r} \right) r^2 d\Omega.$$

Using the asymptotic form in Eq. (32.39) for ψ_{sc}, this quantity simplifies to $(\hbar k/m) \, |f(k, \theta)|^2 \, d\Omega$. Dividing out the incident flux, it follows that $d\sigma/d\Omega = |f(k, \theta)|^2$. ▶

8. Obviously, it is most convenient to carry out the angular integrals using spherical polar coordinates for $\mathbf{r}$, with the polar axis chosen to lie along the vector $\mathbf{Q}$. ▶

11. Invert the order of the summation in the second line of Eq. (32.68), noting that

$$\sum_{n=0}^{\infty} \sum_{\ell=0}^{n} (\cdots) \rightarrow \sum_{\ell=0}^{\infty} \sum_{n=\ell}^{\infty} (\cdots).$$

Now set $n - \ell = 2s$, and convert the sum over n to a sum over s. ▶

12. Equation (32.70) $\Longrightarrow i^\ell \, j_\ell(\xi) = \frac{1}{2} \int_{-1}^{1} dx \, e^{i\xi x} \, P_\ell(x)$. Equate the respective real and imaginary parts. ▶

13. The scattering amplitude in the Born approximation, $f_{\text{B}}(k, \theta)$, has already been found for the Yukawa potential. It is given by Eq. (32.65) of Sect. 32.2.6. Substitute this expression in Eq. (32.73). Now use the integral representation for Q_ℓ given in Eq. (26.30) of Chap. 26, Sect. 26.2.5.

Remark As you know, the Yukawa potential goes over into the Coulomb potential in the limit $\xi \to \infty$. If this limit is used in the expression for $[f_\ell(k)]_{\text{B}}$, the argument of Q_ℓ becomes unity. But we know that $Q_\ell(z)$ has a logarithmic branch point at $z = 1$, and diverges as $z \to 1$. In the present instance, this divergence is a reflection of the divergence of the differential cross-section for the Coulomb potential at $\theta = 0$ (forward scattering), already discussed at the end of Sect. 32.2.6. As I have stated there, a more careful treatment is needed in the case of the Coulomb potential. ▶

14. The standard "separation of variables" procedure, of course, is to assume the wave function to be the product of functions of r, θ and φ, respectively, and to work through the steps. Here is a quicker way. Write the kinetic energy operator in the form

$$\frac{\mathbf{p}^2}{2m} = \frac{p_r^2}{2m} + \frac{\mathbf{L}^2}{2mr^2},$$

where p_r is the radial momentum operator.[1] But we know that the orbital angular momentum is quantized. The operator $\mathbf{L}^2$ can, therefore, be replaced by its eigenvalue $\ell(\ell+1)\hbar^2$. The operator p_r^2 is, of course, represented by $(-i\hbar)^2$ times the *radial* part of the Laplacian operator, i.e.,

$$p_r^2 \to -\frac{\hbar^2}{r^2}\frac{\partial}{\partial r}\left(r^2 \frac{\partial}{\partial r}\right),$$

as in Eq. (6.63) of Chap. 6, Sect. 6.2.6. Write the radial wave function in the form $u_\ell(r)/r$, and Eq. (32.77) follows. ▶

15. Set $\theta = 0$ in Eq. (32.82), and recall that $P_\ell(1) = 1$ for every integer ℓ. It follows that

$$\text{Im}\, f(k, 0) = (1/k) \sum_\ell (2\ell + 1)\, \sin^2 \delta_\ell.$$

Now write down $|f(k, \theta)|^2$ from Eq. (32.82), and integrate over the angular coordinates, using the orthonormality relation for the Legendre polynomials. The result is

$$\int d\Omega\, |f(k, \theta)|^2 = (4\pi/k^2) \sum_\ell (2\ell + 1)\, \sin^2 \delta_\ell.$$

Equation (32.83) follows at once.

Remark The Optical Theorem is, in fact, a direct consequence of the conservation of probability, even though the particular derivation given here does not shed much light on how this comes about. ▶

16. (a) The result sought follows in a straightforward manner on using Eq. (32.107) for the successive iterated kernels.
(b) Differentiate both sides of the integral equation

$$f(x) - \lambda\, \phi(x) \int_0^x dy\, \psi(y)\, f(y) = g(x)$$

[1] This operator has already been discussed in Chap. 29, Sect. 14.3.6.

twice with respect to x, using the formula for differentiation under the integral sign (Eq. (3.3) of Chap. 3, Sect. 3.1.1). Eliminate the quantity $\int_0^x dy\, \psi(y)\, f(y)$ using the expression for $f'(x)$.

The boundary conditions on $f(x)$ follow upon setting $x = 0$ in the integral equation and in the equation obtained by differentiating it once with respect to x. This gives the conditions $f(0) = g(0)$ and $f'(0) = g'(0) + \lambda\, \phi(0)\psi(0)g(0)$. ▶

17. (a) The solution is easily derived by iteration. (b) The ratio test shows that the radius of convergence in the λ-plane of the iterative solution is infinite. Hence, the solution is an entire function of λ, as required of the solution of a Volterra integral equation.

Remark It follows from the integral equation for $f(x)$ that it is the solution of the differential equation

$$\frac{d^2 f}{dx^2} - \lambda x^2 \frac{df}{dx} - 3\lambda x f = 0$$

satisfying the boundary conditions $f(0) = 1,\ f'(0) = 0.$ ▶

Bibliography and Further Reading

The list that follows includes many classic texts and monographs on the topics in mathematical methods that are discussed in this book. In general, I have tried to cite relatively recent editions/reprints of these classics. The list is meant to be representative, and is by no means exhaustive.

Ablowitz, M. J., & Fokas, A. S. (2003). *Complex variables*. Cambridge: Cambridge University Press.

Ahlfors, L. (1979). *Complex analysis* (3rd ed.). New York: McGraw-Hill.

Akhiezer, N. I., & Glazman, I. M. (1993). *Theory of linear operators in Hilbert space* (Vols. 1 and 2). New York: Dover.

Bamberg, P., & Sternberg, S. (1996). *A course in mathematics for students of physics* (Vols. 1 and 2) Cambridge: Cambridge Univesity Press.

Barton, G. (1989). *Elements of Green's functions and propagation*. Oxford: Oxford University Press.

Conway, J. B. (1978). *Functions of one complex variable I* (2nd ed.). Berlin: Springer.

Cornwell, J. F. (1986). *Group theory in physics* (Vols. 1 and 2). Cambridge: Academic.

Courant, R., & Hilbert, D. (1989). *Methods of mathematical physics* (Vols. 1 and 2). Weinheim: Wiley-VCH.

Cox, D. R., & Miller, H. D. (1977). *The theory of stochastic processes*. United Kingdom: Chapman & Hall.

de Lange, O. L., & Raab, R. E. (1991). *Operator methods in quantum mechanics*. Oxford: Clarendon Press.

Dennery, P., & Krzywicki, A. (1996). *Mathematics for physicists*. New York: Dover.

Dunford, N., & Schwartz, J. T. (1958). *Linear operators. Part I: General theory*. Woburn: Interscience.

Feller, W. (2008). *An introduction to probability theory and its applications* (Vols. 1 and 2). Wiley Student Edn.

Ford, L. R. (1951). *Automorphic functions*. White River Junction: Chelsea Publishing.

Friedman, B. (1990). *Principles and techniques of applied mathematics*. New York: Dover.

Gardiner, C. W. (1997). *Handbook of stochastic methods*. Berlin: Springer.

V. Balakrishnan, *Mathematical Physics*,
https://doi.org/10.1007/978-3-030-39680-0

Gilmore, R. (2008). *Lie groups, physics, and geometry: An introduction for physicists, engineers and chemists*. Cambridge: Cambridge University Press.

Gradshteyn, I. S., & Ryzhik, I. M. (1980). *Table of integrals, series and products, corrected and enlarged* (4th ed.). Cambridge: Academic.

Grimmett, G., & Stirzaker, D. (2001). *Probability and random processes*. Oxford: Oxford University Press.

Halmos, P. R. (2015) *Finite-dimensional vector spaces*. Benediction Classics.

Hille, E. (2012 & 2002). *Analytic function theory* (Vols. I and II). American Mathematical Society.

Iranpour, R., & Chacon, P. (1988). *Basic stochastic processes: The Mark Kac lectures*. New York: MacMillan Publishing.

Jordan, T. F. (2006). *Linear operators for quantum mechanics*. New York: Dover.

Kato, T. (1995). *Perturbation theory for linear operators*. Berlin: Springer.

Körner, T. W. (1993). *Fourier analysis*. Cambridge: Cambridge University Press.

Lewis, D. W. (1995). *Matrix theory*. New Delhi: Allied Publishers.

Lighthill, M. J. (1975). *Introduction to fourier analysis and generalised functions*. Cambridge: Cambridge University Press.

Markushevich, A. I., & Silverman, R. A. (2005). *Theory of functions of a complex variable* (2nd ed.). American Mathematical Society.

Morse, P. M., & Feshbach, H. (1986). *Methods of theoretical physics* (Parts 1 and 2). Feshbach Publishing.

Nikiforov, A. F., & Uvarov, V. B. (1988). *Special functions of mathematical physics*. Birkhäuser.

Papoulis, A. (1991). *Probability, random variables, and stochastic processes* (International ed.). New York: McGraw-Hill.

Perelomov, A. (1986). *Generalized coherent states and their applications*. Berlin: Springer.

Pogorzelski, W. (1966). *Integral equations and their applications* (Vol. 1). Bergama: Pergamon.

Reed, M., & Simon, B. (1972–77). *Methods of modern mathematical physics* (Vols. 1–4). Cambridge: Academic.

Sansone, G., & Gerretsen, J. (1969). *Lectures on the theory of functions of a complex variable* (Vols. 1 and 2). Groningen: Wolters-Noordhoff.

Sattinger, D. H., & Weaver, O. L. (1986). *Lie groups and algebras with applications to physics, geometry, and mechanics*. Berlin: Springer.

Stratonovich, R. L. (1963). *Topics in the theory of random noise* (Vol. 1). Gordon & Breach.

Szegö, G. (1975). *Orthogonal polynomials*. American Mathematical Society.

Titchmarsh, E. C. (1976). *The theory of functions*. Oxford: Oxford University Press.

Tricomi, F. G. (1957). *Integral equations*. Woburn: Interscience.

Van Kampen, N. G. (2007). *Stochastic processes in physics and chemistry* (3rd ed.). Amsterdam: Elsevier.

Watson, G. N. (1995). *Theory of bessel functions*. Cambridge: Cambridge University Press.

Whittaker, E. T., & Watson, G. N. (1996). *A course of modern analysis*. Cambridge: Cambridge University Press.

Wybourne, B. G. (1974). *Classical groups for physicists*. New Jersey: Wiley.

Index

V. Balakrishnan, *Mathematical Physics*,
https://doi.org/10.1007/978-3-030-39680-0

S

T

The manufacturer's authorised representative in the EU is Springer Nature Customer Service Centre GmbH, Europaplatz 3, 69115 Heidelberg, Germany. If you have any concerns regarding our products, please contact ProductSafety@springernature.com

Printed and bound by CPI Group (UK) Ltd, Croydon, CR0 4YY
15/07/2026
02167630-0001